High-Productivity Drilling Tools

This completely updated volume covers the design, manufacturing, and inspection of high-productivity drilling tools (HPDT) and addresses common issues with drilling system components. It discards old notions and beliefs as it introduces scientifically and technically sound concepts and rules with detailed explanations and multiple practical examples.

High-Productivity Drilling Tools: Design and Geometry introduces the development of the concept of high-productivity (HP) drill design and its manufacturing and application features. This book continues to develop the concept of a drilling system in the new edition and includes new practical examples. It explains how to properly design and manufacture drilling tools for a specific application and includes a detailed explanation of the design features, tool manufacturing and implementation practices, metrology of drilling and drilling tools, and the tool failure analysis. Using the coherency law as the guidelines introduced in the first edition, the new edition shows how to formulate the requirements for the components of the drilling system, pointing out that the drilling tool is the key component to be improved.

This practical book should be on the shelves of all industrial engineers, those working in production and manufacturing, process designers, tool material designers, cutting tool designers, and quality specialists. Researchers, senior undergraduate students, and graduate students will also find this book full of very helpful reference information.

High-Productivity Drilling Tools
Design and Geometry

Second Edition

Viktor P. Astakhov

CRC Press
Taylor & Francis Group
Boca Raton London New York

CRC Press is an imprint of the
Taylor & Francis Group, an **informa** business

Designed cover image: iStock

Second edition published 2024
by CRC Press
2385 NW Executive Center Drive, Suite 320, Boca Raton FL 33431

and by CRC Press
4 Park Square, Milton Park, Abingdon, Oxon, OX14 4RN

CRC Press is an imprint of Taylor & Francis Group, LLC

© 2024 Viktor P. Astakhov

First edition published by CRC Press 2014

ISBN: 978-1-032-20353-9 (hbk)
ISBN: 978-1-032-20354-6 (pbk)
ISBN: 978-1-003-26329-6 (ebk)

DOI: 10.1201/9781003263296

Typeset in Times
by codeMantra

Contents

Preface

Why don't you write books people can read?

Nora Joyce to her husband James (1882–1941)

MODERN OBJECTIVE OF MACHINING

Reduction of direct manufacturing costs associated with machining operations is a never-ending challenge for manufacturing plants that have this problem more pressing in recent years because of two prime reasons. The first one is the increased use of special alloys with advanced properties and significant tightening of quality requirements for machined parts. The second one is increasing global competition, which is changing the environment facing most companies today. Previously sheltered from the global market, many manufacturing companies now concern about how to hold and even increase profitability against international competitors.

To meet these challenges, many metal-machining manufacturing companies strive to reduce cycle times and costs-per-parts/units through investing heavily in the increased use of high-speed, highly-efficient machining operations and thus changing the whole metal-machining culture which was around for more than a hundred years. These changes include the utilization of machines with powerful, digitally controlled, truly high-speed motor spindles; the application of high-pressure, through-tool, metal-working fluid (hereafter MWF) supply; the implementation of high-precision hydraulic, shrink fit, and steerable tool holders; the integration of advanced cutting process monitoring; wider use of advanced cutting tool materials; and so on.

These changes can be called the fourth "silent" industrial revolution as they happened in a rather short period of time currently becoming well known as Industry 4.0 initiative. The implementation of the listed developments led to a stunning result: For the first time in manufacturing history, the machining operating time became a bottleneck in the part machining cycle time. As discussed in Chapter 1, in shops with stand-alone computer numerical controlled (CNC) machines, the machining time is 20% of the operating time, whereas in automotive shops, this time reaches 60%. The latter is due to aggressive tool use strategy, elimination of any tool/part inspection in the machine, adjustable pre-setting of the MWF pressure for each individual tool, automated (robotic) workpiece/part loading/unloading, using tools with RFID chips, and so on. In both cases, however, the actual machining time is the largest (commonly called as a bottleneck). Knowing these data, one should realize that the implementation of high-penetration rate tools and well-designed machining operations to reduce machining time has become a vital necessity.

Therefore, the best, most reliable cutting tools of advanced designs, with the best tool materials and the highest possible tool manufacturing quality capable of highly-efficient machining, should be used in the modern metal-working industry regardless of their cost as this cost is still virtually insignificant compared to the gain due to increased productivity as the major objective of modern cutting operations. These tools are referred to in this book as high-productivity (HP) tools.

HIGH-PRODUCTIVITY DRILLING TOOLS

Various studies and surveys indicate that hole making (drilling) is one of the most time-consuming metal cutting operations in the typical shop. It is estimated that 36% of all machine hours (40% of CNC machines) is spent performing hole-making operations, as opposed to 25% for turning and 26% for milling, producing 60% of chips. Therefore, the use of HP drilling tools could significantly reduce the time required for machining operations and thus reduce costs.

Over the past decade, the tool materials and coatings used for drills have improved dramatically. New, powerful, high-speed spindles, rigid machines, proper tooling including precision workholding, and high-pressure, high-concentration metal working fluid (MWF) have enabled a significant improvement in the quality of drilled holes and an increase in the cutting speed and penetration rate in drilling operations. In modern machine shops, as, for example, in the automotive industry, the quality requirements for drilled holes today are the same as they used to be for reamed holes just a decade ago. The cutting speed over the same time period has tripled and the penetration rate has doubled.

Despite all these new developments, many drilling operations even in the most advanced manufacturing facilities remain the weakest link among other machining operations. Moreover, there is still a significant gap in the efficiency, quality, and reliability of drilling operations between advanced and common machine shops. This is due to a lack of understanding of not only the process and its challenges, but primarily of the design, manufacturing, and application methods of HP drilling tools. It is totally forgotten that process capability, quality, and efficiency are primarily decided on the cutting edges of the drill as this tool does the actual machining, while all other components of the drilling system play supporting roles presumably assuring the drill best working conditions. Therefore, properly designed and manufactured drilling tools for a given application is the key to achieving high efficiency in drilling operations. However, successful implementation of such tools to achieve high-productivity drilling operation is achieved if and only if the system concept of drilling operation (discussed in Chapter 1) is clearly understood and thus properly implemented.

AIM OF THIS BOOK

This book aims to provide in-depth explanations of the most important aspects in the design and manufacturing of HP drilling tools (with multiple practical examples) for various applications as well as to address the common issues with various important components of the drilling systems to enable one to design/develop effective drilling operations based on the coherency law for drilling systems introduced and exemplified in this book. It sharply discards many old notions and beliefs while introducing scientifically and technically sound notions/concept/rules with detailed explanations.

UNIQUENESS OF THIS VOLUME

UNDERSTANDING

One of the most distinguished features of the proposed book is providing explanation to the recommendations made. It provides a detailed description of the physical/machinal/technological rationale behind these suggested recommendations/designs/processes. A real question, however, is "Do we really need understanding of the theory behind HP drilling, and thus the whole essence of metal cutting?" The answer is not as simple as one might think. As long as things proceed well, we do not need to explain or to understand the essence. We drive our cars, watch TV, or use computers without fully understanding of the processes involved. Moreover, explanations and quests for understanding would retard and disturb "normal" course of manufacturing activities including buying and using drilling tools, and receiving awards for excellence in manufacturing. In other words, the explanations to encourage people to ask inconvenient questions and thus challenge comfortable status quo are discouraged. However, when something goes wrong or breaks down in a production or research activity, or when a competing industrial or scientific group, company, country, etc. achieve better results, so that one is confronted with losing the market or the funding, or even a sector of national economy, then the understanding of the involved processes becomes urgent. It is common, unfortunately, that such urgency becomes evident too late. In fully applicable to metal cutting and tool design research: when machining time was insignificant in the whole time of part manufacturing, any research on its improvement was not funded; when machining time became a bottleneck, no research results and specialists are available for the design of efficient machining operations.

Another vital but routinely ignored aspect of the importance of proper explanation belongs to the plane of human psychology. It can be explained with an example from the world of science. In 1847, decades before the germ theory of infection, physician Ignaz Semmelweis suggested that, by washing their hands before examinations, doctors could save the lives of many maternity ward patients. He published the results of his many-year study showing that hand-washing produced a reduction in maternal mortality to less than 1%, compared to 10%–35% in general practice. Surprisingly, his results were totally rejected by the medical community as many doctors feel offended by his pure evidence-based finding, and thus a clear suggestion. Needless to say that this suggestion was not followed. The prime reason for that is that Dr. Semmelweis did not offer any scientific rationale why his suggestion works regardless of a great body of meticulously carried experimental study data. Only decades later, when the germ theory of infection was widely adopted in medicine, hand-washing became a routine mandatory procedure as doctors understood why they needed to do so. This example shows the role of the widely accepted theory (or simply, explanations) for even seemingly obvious experimental data or even the result of observation that everyone can see.

FEATURES

The major feature of this book is the introduction and development of the concept of high-productivity (HP) drilling tool design and its manufacturing and application features. The uniqueness of this book can be summarized as follows:

1. For the first time, a clear quantitative detailed explanation of the major objective of modern manufacturing as the underlying vital necessity of the design/selection/use of most advanced HP drilling tools is provided. It argues that *in the direct opposite of prevailing common notion* in the world-wide manufacturing and scientific (metal cutting and cutting tools) worlds, the highest quality, and thus the most expensive and reliable, cutting tools must be used in modern manufacturing as a way to reduce the manufacturing cost due to high productivity of such tools.
2. A unique concept of the drilling tool quality is introduced and used as the framework of the book content development. It is to say that throughout this book, the scientific and technological synergy that has been developed among the basic components of drill quality such as the tool material, tool design, and tool manufacturing quality is in the focus of attention. This synergy stems from the wide, and thus stable knowledge base and experience. The concept and its development throughout this book instill the comprehension of conceptualizing the whole process of the design of HP drilling tools in potential readers.
3. The concept of a drilling system, introduced in the first edition of this book, is further developed with practical examples. The interinfluence of the components of this system such as a drill, drilling machine, tool holder, fixture, and coolant, is analyzed. The coherency law is introduced.
4. Comprehensive analyses of the drill force balance, drill geometry, and metal-working fluid (MWF) flow allowed to introduce pioneering HP tool designs named VPA1©, VPA2©, VPA3©, and VPA4©. In the author's opinion, the developed high-stability drills (VPA1©, VPA2©) should be a new standard for the whole industry replacing up to 80% of the existing drill designs.

SOURCE

This book summarized many-year experience of the author in industry including research, development, manufacturing, implementation, and failure analysis of cutting tools. The author serves as the cutting tool corporate specialist in Production Service Management, Inc. (PSMi) company, which is a part of the EWIE Group of companies that provides tool commodity management services (including cutting tool development, procurement, application optimization and cost reduction, and failure analysis) for many industries/manufacturing facilities, such as the automotive, aerospace, power, medical, and in many countries that give him truly unique opportunities to visit various manufacturing facilities including cutting tool manufacturing facilities, learn their machining process and cutting tool manufacturing and management practices, investigate cutting tool failures, and understand their needs and goal in the implementation of advanced cutting tools. The author learned that advanced manufacturing runs on tight timelines and high demand, so to gain a competitive edge, optimization of various machining operations including the use of advanced cutting tools is absolutely paramount. The author has also led a unique assessment program of cutting tool manufacturers for the automotive industry.

I hope that reading this book is worth your time as it covers a number of important pieces of knowledge and information in a systematic manner never covered before. In other words, I hope that the material covered in this book can help you achieve your goals such as improved productivity and quality and reduced unscheduled downtime in performing hole-making operations.

HOW THIS BOOK IS ORGANIZED

The structure of this book is unusual for the literature in the field because its logic is governed by HP drill design, manufacturing, and implementation theory and practice. A summary of the contents of the chapters is listed in the following.

Chapter 1: Basics of Drilling and Drilling System

This chapter consists of two logically connected parts. The first introductory part presents a short classification of drilling operations. It discusses the components of the drilling regime: cutting speed, cutting feed, feed rate, and material removal rate with practical examples; that is, it sets the scene for the other chapters by introducing the correct terminology and precise definitions of the parameters of the drilling regime. The second part first introduces the basics of the system approach. The structure of the drilling system is defined, and the coherency law is formulated. Then, the chapter introduces a unique concept of cutting tool quality as a synergy of the cutting tool material, tool design, and tool manufacturing quality. The prime objective of the drilling system is established, and major constraints to achieve this objective are considered. The case for HP drills is considered and discussed. The design procedure for drilling systems is considered with practical examples.

Chapter 2: Geometry and Design Components of Drilling Tools

This chapter begins with a section "Drills." It includes a detailed classification of drills and defines the basic terms involved. Then, this chapter is split into two major parts. The first part presents a comprehensive description with detailed explanations and examples of the most essential features of drills that an advanced user of HP should know. As such, a set of practical application rules to be used in practical drilling to help end users to select the proper drill for the application and to optimize the drilling process are formulated and explained. The second part of this section considers drill from the tool design/development/research perspective. A detailed explanation of the major constraints on the penetration rate imposed by the drilling tool itself and the correlation of these constraints with drill design and geometry parameters are provided. The force balance is defined as the major prerequisite feature in HP drill design/manufacturing. Pioneering HP tool designs named VPA1©, VPA2©, and VPA3© were developed and explained. It is discussed that the proposed designs should be implemented in most of the "standard" straight and twist drills to assure their entrance stability and to improve the quality, i.e., meeting tight tolerances on diametric accuracy, hole shape and straightness, position, and so on.

The basic designs and proper geometry of reamers and taps are considered correcting many old notions prevailing in the industry today. Detailed classifications of the features of these tools as well as the proper definitions of the terms involved are given. The advantages and limitations of form taps are also discussed.

Chapter 3: Deep-Hole Drilling Tools

This chapter discusses classification, geometry, and design of deep-hole drilling tools as so-called self-piloting tools (SPTs). Two concepts, namely prevailing and original, of self-piloting are explained showing that depending upon the geometry of a given SPT, the common or original concept is prevailing, whereas the second one is supplemental. The major emphasis is placed on gundrills. The design and geometry of these drills are explained at basic and tool design/development/research levels. Based upon the analysis of the force balance, the unbalanced moment is revealed so that the application of the VPA2d© design (introduced in Chapter 2) in gundrilling is explained showing its practical applications. Based upon the MWF flow distribution in the machining zone, the VPA4© design is introduced and explained in great detail and its practical realization is discussed with examples. Basic design and geometry of single tube system (STS) drills are considered. The cutting force balance, unbalanced moments, and power distribution in STS drilling are discussed. Two major problems in the design/geometry of STS drills are revealed, namely MWF pressure distribution in the machining zone and cutting of the core. A new STS drill, utilizing the VPA3© and VPA2© designs introduced in Chapter 2, was developed to solve the revealed problems. The underlying principles of the ejector drills are discussed showing the advantages and limitations of this type of drills. Three basic myths about ejector drilling circulated in the trade (and unfortunately sometimes in scientific) literature and textbooks are debunked. An ejector drill head geometry with improved chip-breaking ability is introduced.

Chapter 4: PCD Drilling Tools

This chapter discusses PCD drills according to the concept of cutting tool quality introduced in Chapter 1. It is to say that the synergy of fundamentals and particularities of the polycrystalline (PCD) tool material, drill manufacturing quality, and drilling tool design is considered. The chapter covers PCD tool material in unparalleled depth from its origin and technology to all facets of its implementation in various drilling tool designs. Thermal stability of PCD was revealed as its weakest feature causing the vast majority of premature failures of PCD tools. It is conclusively proven that overheating on brazing and improper finishing of PCD tools are prime causes of such failures. The practical recommendations on the best practices in brazing and finishing are discussed. The chapter considers various PCD drill designs including the PCD-tipped drill, full-face (cross) PCD drills, and full-head PCD drills. The advantages and disadvantages of these designs/constructions as well as their application particularities including tool drawings are discussed.

Appendix: Basics of the Tool Geometry

The major objective of this appendix is to familiarize potential readers with the basic notions and definitions used in the analysis of tool geometry. It provides the fundamentals and definitions of the involved terminology for better comprehension of Chapters 2–4. Three systems of consideration of the cutting tool geometry are considered, namely tool-in-hand (T-hand-S) and tool-in-use (T-use-S), in the proposed, namely the tool-in-holder (T-hold-S), systems. The relevancy of these systems is explained, and the corresponding angles are clearly defined with graphical supports. The geometrical relationships among the introduced angles are established. The chapter explains with examples that among many angles of the cutting tools, the clearance angle is the major distinguishing feature. The influence of other angles on the cutting process and its outcome are also discussed.

Acknowledgments

I express my gratitude to all the people who provided support, helped with the testing and implementation of advanced tools, shared their viewpoints and discussed technical issues, allowed me to use their lab, inspection, and production equipment, and assisted in the editing of this book.

I thank all my former and present teachers, colleagues, and students who have contributed to my knowledge of the subject.

My special thanks go to Mr. Scott Burke, CEO of EWIE Group of companies, and Mr. Todd Markel, CEO of PSMi company, for their continuous support over many years, toleration of my unorthodox ideas, and providing me with multiple opportunities to research, test, and implement advanced cutting tools, and thus to develop my knowledge, skills, and abilities within the industry.

Above all, I thank my wife, Sharon, and the rest of my family, who supported and encouraged me in spite of the numerous days, evenings, and weekends devoted to writing this book, they provided a loving family environment that afforded me the tranquility and peace of mind that made writing this book possible. This book is dedicated to them.

Author

Viktor P. Astakhov earned his PhD in mechanical engineering from Tula State Polytechnic University, Tula-Moscow, Russia, in 1983. He was awarded a DSc designation (Dr. Habil., Docteur d'État) and the title "State Professor of Ukraine" in 1991 for the outstanding service rendered during his teaching career and for the profound impact his work had on science and technology. An internationally recognized educator, researcher, and mechanical engineer, he has won a number of national and international awards for his teaching and research. In 2011, he was elected to the SME College of Fellows.

Besides his teaching engagements, Dr. Astakhov currently serves as the tool research and application manager of Production Service Management Co., which is a part of EWIE Group of companies, a large international tool management company that provides tooling services to many industries such as the automotive, aerospace, power, medical, and others. He has published monographs and textbooks, book chapters, and many papers in professional journals as well as in trade periodicals. He has authored the following books: *Drills: Science and Technology of Advanced Operations*, *Metal Working Fluids: Fundamentals and Recent Advances*, *Geometry of Single-Point Turning Tools and Drills: Fundamentals and Practical Applications*, *Tribology of Metal Cutting*, *Physics of Strength and Fracture Control*, and *Mechanics of Metal Cutting*. He also serves as the editor in chief, associate editor, board member, reviewer, and advisor for many international journals and professional societies.

1 Basics of Drilling and Drilling System

Everything is designed. Few things are designed well.

Brian Reed, American graphic designer

1.1 BASIC DRILLING OPERATIONS

Drilling is a hole-making machining operation accomplished using a drilling tool. Figure 1.1a shows a common drilling arrangement in a drilling machine. The workpiece is clamped on the machine table with a vice equipped with jaws that clamp against the workpiece, holding it secure. The drill is clamped in the machine spindle that provides the rotation and the feed motions. Figure 1.1b shows a common drilling arrangement on a lathe. The workpiece is clamped in a self-centering three-jaw lathe chuck installed on the machine spindle that provides rotation and the tool is installed on the tailstock engaged with the lathe carriage that provides the feed motion.

A drilling tool is defined as an end-cutting tool indented for one of the hole-making operations. Such a tool has the terminal (working) end and the rear end for its location in a tool holder. In all drilling operations, the primary motion is rotation of the workpiece or the tool or both (counter-rotation drilling) and secondary motion, which is translational feed motion (Figure 1.2), which can be applied either to the tool or to the workpiece depending on the particular design of the machine tool used.

(a)   (b)

FIGURE 1.1 Generic drilling: (a) on a vertical drilling machine and (b) on a lathe.

DOI 10.1201/9781003263296-1

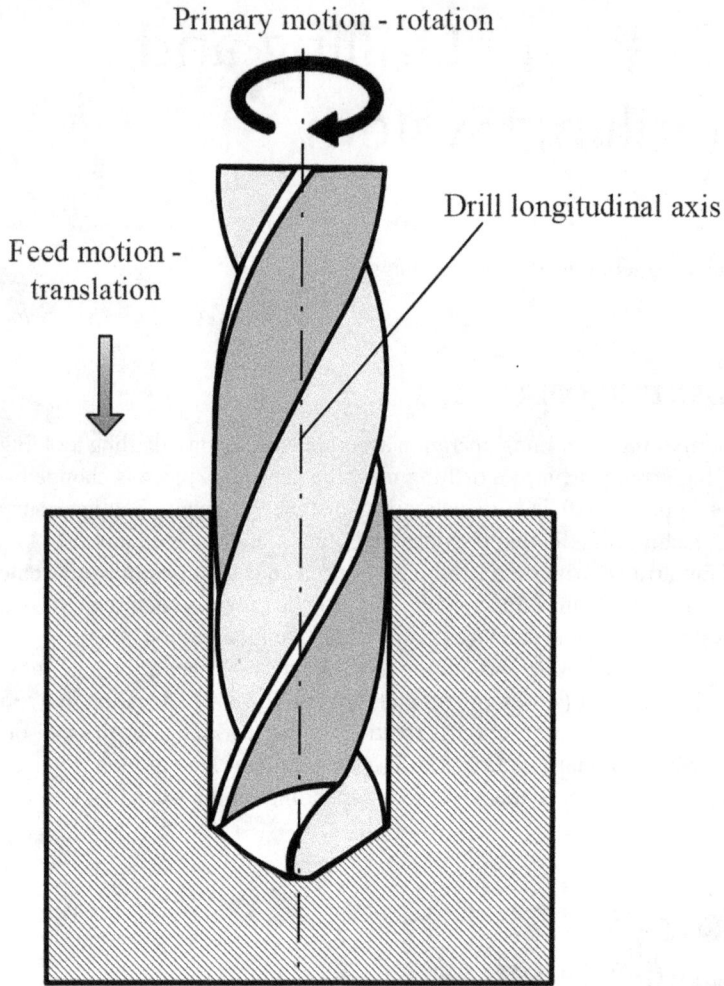

FIGURE 1.2 Motions in drilling.

There are a great number of drilling operations used in modern industry. Figure 1.3 shows some of the most frequently used. Although all these operations use the same kinematic motions and generic drilling tool definition, the particular tool designs, machining regimes, and many other features of the drilling tools involved are operation specific. These basic operations are defined as follows:

1. Drilling is the making of a hole in a workpiece where none previously existed. In this case, the operation is referred to as solid drilling. If an existing hole (e.g., a cored hole in die casting) is drilled, then the operation is referred to as core drilling. A cutting tool called the drill enters the workpiece axially through the end and cuts a hole with a diameter equal to that of the tool. Drilling may be performed on a wide variety of machines such as a lathe/turning center and drilling/milling/boring machine known as a machining center.

2. Boring (Figure 1.4) is the enlarging of an existing hole. A boring tool enters the workpiece axially and cuts along an internal surface to form different features, such as steps, tapers, chamfers, and contours. Boring is commonly performed after drilling a hole in order to enlarge the diameter, making steps and special features, or to improve hole geometrical quality (i.e., to obtain high-precision diameter and shapes in the transverse [e.g., roundness] and longitudinal [e.g., position deviation] directions). Nowadays, however, many modern boring tools are multi-edge tools allowing a significant increase in boring productivity and accuracy.

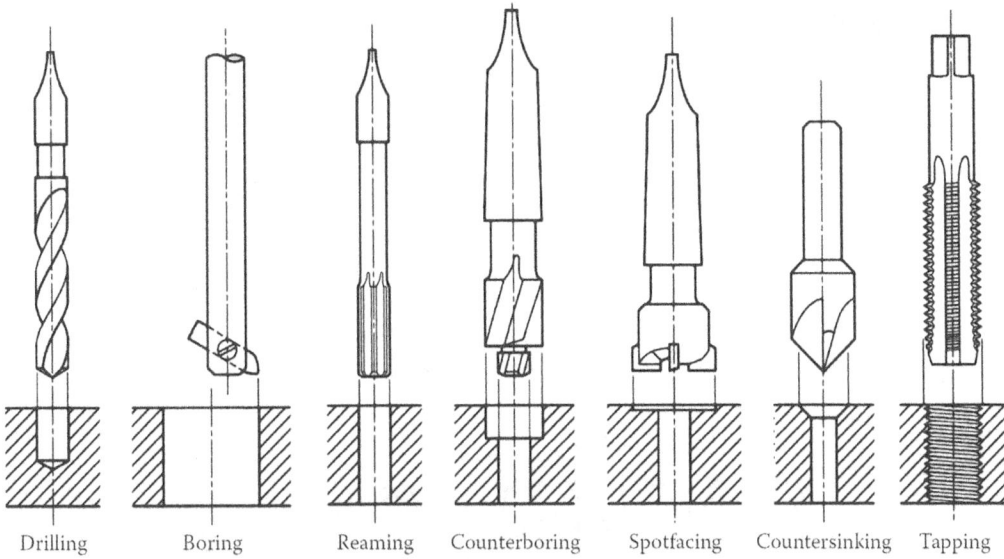

FIGURE 1.3 Basic drilling operations.

FIGURE 1.4 Boring.

3. Reaming (Figure 1.5) is the enlarging of an existing hole to accurate size and shape. An end-cutting tool called the reamer enters the workpiece axially through the end and enlarges an existing hole to the diameter of the tool. Reaming is often performed after drilling or boring to obtain a more accurate diameter, better surface roughness, and shape in the transverse direction.

4. Counterboring is a flat-bottomed cylindrical enlargement of the mouth of a hole, usually of slight depth, as for receiving a cylindrical screw head. An end-cutting tool referred to as the counterbore enters the workpiece axially and enlarges the top portion of an existing hole to the diameter of the tool. Counterboring is often performed after drilling to provide space for the head of a fastener, such as a bolt, to sit flush with the workpiece surface.

5. Countersinking is the process of making a cone-shaped enlargement at the entrance of a hole. An end-cutting tool called the countersink enters the workpiece axially and enlarges

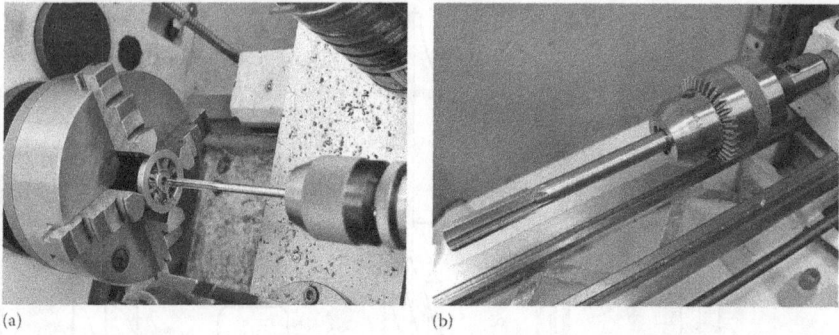

(a)   (b)

FIGURE 1.5   Reaming: (a) location of the part in the machine spindle and (b) location of the reamer in the tailstock.

FIGURE 1.6   Tapping.

the top portion of an existing hole to a cone-shaped opening. Countersinking is often performed after drilling to provide space for the head of a fastener, such as a screw, to sit flush with the workpiece surface. Common included angles for a countersink include 60°, 82°, 90°, 100°, 118°, and 120°.

6. Spotfacing is a drilling operation performed where it is assumed that there will be a highly irregular face surface around a hole. This is common with castings. The spotface may be either below the surface of the surrounding metal or placed on the top of a boss, as is typical with castings. The purpose of spotfacing can be either to provide a flat surface to accommodate a screw head, nut, or washer or to make true face to start other drilling operations. The spotface tool resembles an end mill cutter. A pilot in the center of the cutting surface is often added if the alignment of the existing hole and the spotface is important.

7. Tapping (Figure 1.6) is a drilling operation of cutting internal threads with an end-form tool referred to as the threading tap. A tap enters the workpiece axially through the end and cuts internal threads into an existing hole. The existing hole is typically drilled by the required tap drill size that will accommodate the desired tap.

1.2 MACHINING REGIME IN DRILLING OPERATIONS

The cutting speed and cutting feed are prime or basic parameters that constitute the machining regime in drilling operations.

1.2.1 CUTTING SPEED

In metric units of measure (the SI system), the cutting speed is calculated as

$$v = \frac{\pi d_{dr} n}{1,000} \quad \text{(m/min)} \tag{1.1}$$

where $\pi = 3.141$, d_{dr} is the drill diameter in millimeters, and n is the rotational speed in rpm or rev/min no matter which rotates, the drill or the workpiece. If both the drill and the workpiece rotate in opposite directions (the so-called counterrotation), then n is the sum of the rotational speeds of the drill, n_{dr}, and the workpiece, n_w, that is, $n = n_{dr} + n_w$.

For example, if $d_{dr} = 10\,\text{mm}$ and drill rotates with $n = 2,170\,\text{rpm}$ while the workpiece is stationary, then $v = \pi d_{dr} n/1,000 = 3.141 \times 10 \times 2,170/1,000 = 68.15$ m/min.

In the imperial units of measure, the cutting speed is calculated as

$$v = \frac{\pi d_{dr} n}{12} \quad \text{(sfm or ft/min)} \tag{1.2}$$

where $\pi = 3.141$, d_{dr} is the drill diameter in inches, and n is the rotational speed in rpm or rev/min.

For example, if $d_{dr} = \frac{3}{4}$ in. (19.05 mm) and the drill rotates with $n = 1,220\,\text{rpm}$ while the workpiece is stationary, then $v = \pi d_{dr} n/12 = 3.141 \times 3/4 \times 1,220/12 = 239.5$ sfm.

Although Eqs. (1.1) and (1.2) are exemplified for drills, they are perfectly valid for all drilling tools shown in Figure 1.3 having the basic motions shown in Figure 1.2. To calculate the cutting speed properly, the relevant diameter should be used in Eqs. (1.1) and (1.2) instead of d_{dr}. For example, for reaming, counterboring, spotfacing, and tapping, this diameter is equal to the outside tool diameter. For boring, diameter d_{br} equal to the finish diameter of the hole being bored should be used. When one deals with a multi-stage boring tool, this diameter is equal to the largest finish diameter of the hole being bored. In countersinking, this diameter is equal to the largest diameter of the cone-shaped enlargement at the entrance of a hole.

Normally in the practice of machining, the cutting speed v is selected for a given tool design, tool material, work material, and particularities of a given drilling operation. Then, the spindle rotational speed should be calculated using Eq. (1.1) and the given diameter as

$$n = \frac{1,000v}{\pi d_{dr}} \tag{1.3}$$

1.2.2 FEED, FEED PER TOOTH, AND FEED RATE

The feed motion is provided to the tool or the workpiece, and when added to the primary motion leads to a repeated or continuous chip removal and the formation of the desired machined surface. In all drilling tools, the feed is provided along the rotational axis as shown in Figure 1.2.

Figure 1.7 provides visualization of the basic components of the drilling regime such as the cutting feed, depth of cut, and uncut chip thickness commonly referred to as the chip load in professional literature. Designations of the components of the drilling regime are shown according to the International Organization for Standardization (ISO) Standard 3002-3.

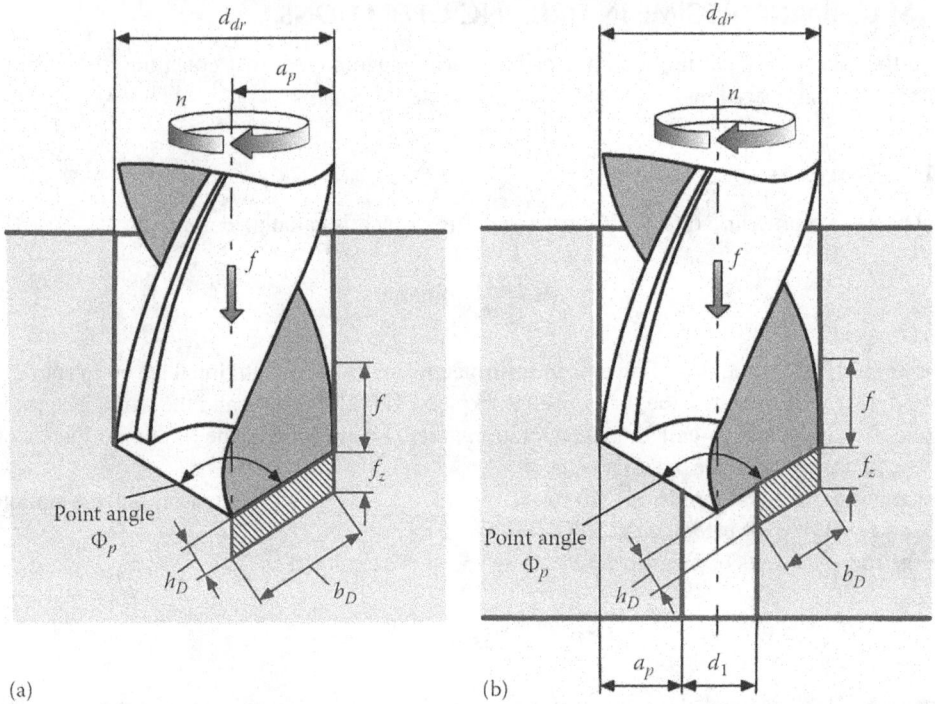

FIGURE 1.7   Visualization of the components of the drilling regime: (a) solid drilling and (b) core drilling.

The cutting feed, f, is the distance in the direction of feed motion at which the drilling tool advances into the workpiece per one revolution, and thus, the feed is measured in millimeters per revolution (inches per revolution). The feed per tooth, f_z (the subscript z came from German *zahn*, i.e., a tooth), is determined as

$$f_z = f/z \tag{1.4}$$

where z is the number of cutting teeth. For example, if a two-flute drill is fed with the cutting feed $f=0.360$ mm/rev, then the feed per tooth (the chip load) is $f_z=0.360/2=0.180$ mm/z; if a four-flute reamer is fed with the cutting feed $f=0.240$ mm/rev, then the feed per tooth (the chip load) is $f_z=0.240/4=0.06$ mm/z.

The feed speed (ISO Standard 3002-3) commonly referred to in the literature as the feed rate, v_f, is the velocity of the tool in the feed direction. It measures in millimeters per minute (mm/min) or inches per minute (ipm) and is calculated as

$$v_f = f \cdot n \tag{1.5}$$

where f is the feed (mm/rev or ipm) and n is the rotational speed (rpm).

The feed speed (feed rate) is often referred to as the penetration rate in the professional literature on drilling. It is used as a measure of drilling productivity. Substituting Eq. (1.3) into Eq. (1.5) and arranging the terms, one can obtain

$$v_f = k_{dt} f v \tag{1.6}$$

where $k_{dt} = 1,000/(\pi d_{dr})$ is a constant for a given drill.

It directly follows from Eq. (1.6) that the penetration rate depends equally on the cutting speed and feed. This fact should be kept in mind when designing a drilling operation/drill and selecting the tool material and components of a drilling system, for example, the metal-working fluid (MWF) (coolant) supply system.

Although Eqs. (1.4) and (1.5) are exemplified for drills, they are perfectly valid for all drilling tools shown in Figure 1.3 having the basic motions shown in Figure 1.2.

1.2.3 DEPTH OF CUT AND MATERIAL REMOVAL RATE

The depth of cut in solid drilling is calculated as $a_p = d_{dr}/2$. In the case of core or *pilot* hole drilling shown in Figure 1.7b, the depth of cut is calculated as $a_p = (d_{dr} - d_1)/2$, where d_1 is the diameter of the pilot (core) hole.

The material removal rate is known as *MRR*, which is the volume of work material removed by the tool per unit time. Figure 1.8 presents visualization of the volume of the work materials removed in solid and core drilling. It directly follows from this figure that *MMR* (measured in mm³/min) in solid drilling is calculated as

$$MMR = \frac{\pi d_{dr}^2}{4} v_f \tag{1.7}$$

Substituting Eq. (1.6) into Eq. (1.7), one can obtain

$$MMR = \frac{\pi d_{dr}^2}{4} v_f = \frac{\pi d_{dr}^2}{4} \frac{1,000}{\pi d_{dr}} fv = 250 \, fv d_{dr} \tag{1.8}$$

Referring to Figure 1.8, one can calculate *MMR* (measured in mm³/min) in core drilling as

$$MMR = \frac{\pi \left(d_{dr}^2 - d_1^2 \right)}{4} v_f = \frac{\pi \left(d_{dr}^2 - d_1^2 \right)}{4} \frac{1,000}{\pi d_{dr}} fv = 250 \, fv \frac{d_{dr}^2 - d_1^2}{d_{dr}} \tag{1.9}$$

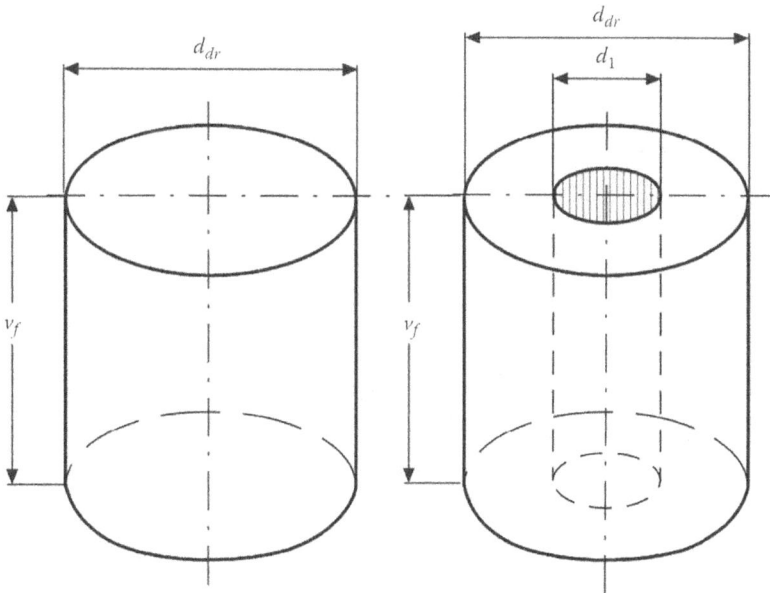

FIGURE 1.8 Visualization of the volume of the work materials removed in solid and core drilling.

1.2.4   Cut and Its Dimensions

Standard ISO 3002-3 defines the cut as a layer of the workpiece material to be removed by a single action of a cutting part. It was pointed out that for a given cutting tooth, geometrical parameters of the cut are

- Nominal cross-sectional area, A_D;
- Nominal thickness, h_D;
- Nominal width, b_D.

For solid drilling, these parameters are calculated referring to Figure 1.7a as follows.

The nominal thickness of cut known in the literature as the uncut (undeformed) chip thickness or chip load is calculated as

$$h_D = f_z \sin\left(\Phi_p/2\right) = \left(f/z\right)\sin\left(\Phi_p/2\right) \tag{1.10}$$

where Φ_p is the drilling tool point angle (discussed later in this book in Chapter 2).

The nominal width of the cut known in the literature as the uncut (undeformed) chip width is calculated as

$$b_D = a_p \big/ \sin\left(\Phi_p/2\right) = \left(d_{dr}/2\right)\big/\sin\left(\Phi_p/2\right) \tag{1.11}$$

The nominal cross-sectional area known in the literature as the uncut (undeformed) chip cross-sectional area is calculated as

$$A_D = h_D b_D \tag{1.12}$$

Substituting Eqs. (1.10) and (1.11) into Eq. (1.12), one can obtain

$$A_D = \left(f_z d_{dr}\right)/2 = \left(f d_{dr}\right)/(2z) \tag{1.13}$$

In the case of core or *pilot* hole drilling shown in Figure 1.7b, these parameters are calculated as

$$h_D = f_z \sin\left(\Phi_p/2\right) = \left(f/z\right)\sin\left(\Phi_p/2\right) \tag{1.14}$$

$$b_D = \left(d_{dr} - d_1\right)\big/ 2\sin\left(\Phi_p/2\right) \tag{1.15}$$

$$A_D = h_D b_D = f_z\left(d_{dr} - d_1\right)/2 = f\left(d_{dr} - d_1\right)/(2z) \tag{1.16}$$

The foregoing considerations reveal that the MRR and undeformed chip cross-sectional area do not depend on the drill point angle, while the uncut chip thickness and its width do.

Example 1.1

Problem

Determine the drill rotational speed, feed speed (the feed rate), depth of cut, MRR, nominal thickness of cut (uncut [undeformed] chip thickness or chip load) and width, and nominal cross-sectional area (the uncut [undeformed] chip cross-sectional area) for a drilling operation with a two-flute drill ($z=2$) having $\Phi_p=120°$ if the selected cutting speed is $v=80$ m/min, drill diameter is $d_{dr}=8$ mm, and feed is $f=0.15$ mm/rev.

Solution

The spindle rotational speed is calculated using Eq. (1.3) as

$$n = \frac{1,000v}{\pi d_{dr}} = \frac{1,000 \cdot 80}{3.141 \cdot 8} = 3184.7 \, \text{rpm}.$$

For practical purposes, $n = 3,185$ rpm is adopted.

The feed speed (the feed rate) is calculated using Eq. (1.5) as

$$v_f = fn = 0.15 \cdot 3,185 = 477.75 \text{ mm/min}.$$

The depth of cut is $d_w = d_{dr}/2 = 8/2 = 4$ mm.

The MRR is calculated using Eq. (1.8) as

$$MRR = 250 \, fvd_{dr} = 250 \cdot 0.15 \cdot 80 \cdot 8 = 24,000 \text{ mm}^3/\text{min}.$$

The nominal thickness of cut (uncut [undeformed] chip thickness or chip load) is calculated using Eq. (1.10) as

$$h_D = (f/z)\sin(\Phi_p/2) = \frac{0.15}{2}\sin 60° = 0.065 \, \text{mm}.$$

The nominal width of the cut (the uncut [undeformed] chip width) is calculated using Eq. (1.11) as

$$b_D = (d_{dr}/2)/\sin(\Phi_p/2) = \frac{8}{2}\Big/\sin 60° = 4.619 \, \text{mm}.$$

The nominal cross-sectional area (the uncut [undeformed] chip cross-sectional area) is calculated using Eq. (1.12) as

$$A_D = h_D b_D = 0.065 \cdot 4.619 = 0.300 \, \text{mm}^2.$$

Except for some special and form tools (e.g., the tap), the discussed formulae for calculating geometrical parameters of the cut are valid for all drilling tools. For counterboring and spotfacing, the point angle $\Phi_p=180°$ should be used. For countersinking, these parameters vary over the cutting cycle so that the maximum nominal thickness and width of cut should be determined.

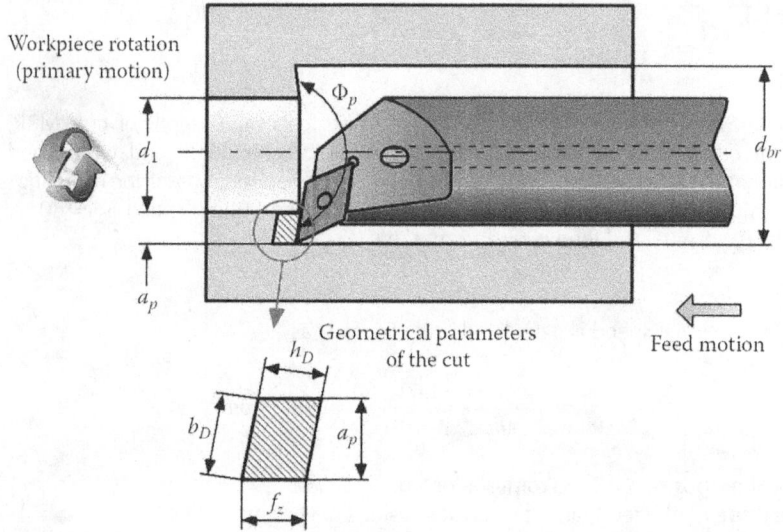

FIGURE 1.9   Boring operation.

Example 1.2

Problem

Determine the drill rotational speed, feed speed (the feed rate), depth of cut, nominal thickness of cut (uncut [undeformed] chip thickness or chip load) and width, and nominal cross-sectional area (the uncut [undeformed] chip cross-sectional area) for a single-cutter boring tool shown in Figure 1.9. The tool has $\Phi_p = 190°$, the selected cutting speed $v = 240$ m/min, bored hole diameter $d_{br} = 40$ mm, diameter of the hole to be bored $d_1 = 35$ mm, and cutting feed $f = 0.10$ mm/rev.

Solution

The spindle rotational speed is calculated using Eq. (1.3) as

$$n = \frac{1,000v}{\pi d_{dr}} = \frac{1,000 \cdot 240}{3.141 \cdot 40} = 1,910.8 \text{ rpm.}$$

For practical purpose, $n = 1,911$ rpm is adopted.
   The feed speed (the feed rate) is calculated using Eq. (1.5) as

$$v_f = fn = 0.10 \cdot 1,911 = 191.1 \text{ mm/min.}$$

The depth of cut is $a_p = (d_{br} - d_1)/2 = (40 - 35)/2 = 2.5$ mm.
   The nominal thickness of cut (uncut [undeformed] chip thickness or chip load) is calculated using Eq. (1.10) as

$$h_D = \left(f/z\right)\sin\left(\Phi_p/2\right) = \frac{0.10}{1}\sin\left(190°/2\right) = 0.10 \text{ mm.}$$

The nominal width of the cut (the uncut [undeformed] chip width) is calculated using Eq. (1.15) as

$$b_D = b_D = \left(d_{br} - d_1\right)\big/2\sin\left(\Phi_p/2\right) = \frac{40 - 35}{2\sin 190°} = 2.49 \text{ mm.}$$

The nominal cross-sectional area (the uncut [undeformed] chip cross-sectional area) is calculated using Eq. (1.12) as

$$A_D = h_D b_D = 0.10 \cdot 2.49 = 0.249 \text{ mm}^2.$$

1.2.5 SELECTING MACHINING REGIME: GENERAL IDEA

Selecting the proper speed and feed for a particular drilling application is critical to reduce drill wear and breakage as well as to achieve high drilling efficiency in terms of cost per machined hole. In this author's opinion, the latter is the most proper measure of a drilling tool performance as well as the efficiency of the drilling operation. Therefore, the cutting speed and feed selection is not just technical as it used to be but rather a process economy-driven issue to achieve the system objective. However, such a selection is not as straightforward as it used to be a few decades ago.

It used to be that speed and feed recommendations were selected as provided by the literature on the field. For example, one of the most popular resources is *Machinery's Handbook*, which celebrated with its 31st edition, nearly 100 years as *The Bible of the Mechanical Industries*. The values selected in this way are always subject to specific job conditions so they were always considered as estimates to give the process designer/manufacturing specialist/operator an approximate starting point.

What makes selecting the right parameters so difficult is that there is little margin for error. Speeds and feeds that are too high, as well as speeds and feeds that are too low, can result in low efficiency of the whole operation and can cause drilling tool breakage. Moreover, the rapid change of tool material properties and tool coatings as well as drill design specifics, including the MWF application technique, *overrun* the recommendation provided in the reference literature so that, in many cases, the data provided can no longer be considered a good starting point.

Tables 1.1 and 1.2 give some recommendations for the selection of drilling speeds for the purpose of tool layout design (discussed later in this chapter). Once the design of the whole drilling operation is complete, and thus the parameters of the drill and drilling operation (tool holder, method of MWF supply, machine capabilities in terms of achievable speeds and feeds, etc.) have been selected, a standard or a special drill from at least two drill manufacturers should be quoted, asking them (besides the tool cost and lead time) to suggest speed and feed for the given designed operation as well as the estimation of tool life and tool reliability. If a special high-volume operation is to be designed, and thus a special drilling tool is to be used, a tool manufacturer should be involved in the design of the tool layout and then the drilling tool to achieve the maximum efficiency of this operation.

Although speeds and feed rates are determined by the type of material being drilled and the depth of the hole, there are two other important system considerations to keep in mind: tool holding and work holding. A frequent cause of drill breakage is a loose or poorly designed tool holder that imparts wobble to the drill. Even at slow speeds and feeds, a wobble will quickly break a drill. There are many types of tool-holding devices to choose from, but hydraulic and shrink-fit tool holders provide the most secure method of tool holding, because their use results generally in the least amount of runout. A precision collet tool holder is the next best option.

Workpiece clamping is also important. If the workpiece is not clamped properly, chatter or workpiece shifting due to cutting forces can be the case during drilling, which results in lower tool life, poor quality of the machined surface, and even breakage of the drill. If a drilling operation has been proceeding normally, and drills suddenly begin breaking, the first areas to check are the tool holder and the workpiece clamping.

TABLE 1.1
Speed and Feed Recommendations: HSS Drills

Work Material	Hardness HB	HSS Grade	Cutting Speed (m/min)			Feed (mm/rev) for Drill Diameter (mm)			
			TiN	TiAlN	TiCN	9–12	13–17	18–24	25–35
Free machining steel (1112, 12L14, etc.)	100–150	M4	60	85	72	0.17	0.22	0.30	0.35
	151–200	M4	55	80	70	0.17	0.22	0.30	0.35
	201–250	M4	50	72	65	0.15	0.15	0.30	0.35
Low-carbon steel (1010, 1020, 1025, etc.)	85–125	M4	52	75	67	0.15	0.22	0.30	0.38
	126–175	M4	50	73	65	0.15	0.22	0.30	0.38
	176–225	M4	45	70	60	0.13	0.20	0.25	0.35
	226–275	M4	42	65	55	0.13	0.20	0.25	0.35
Medium-carbon steel (1045, 1140, 1151, etc.)	125–175	M4	50	73	65	0.15	0.15	0.30	0.38
	176–225	M4	45	70	60	0.13	0.13	0.25	0.35
	226–275	M4	42	65	55	0.13	0.13	0.25	0.35
	276–325	T15	40	60	52	0.10	0.10	0.22	0.33
Alloy steel (4130, 4140, 4150, 5140, 8640, etc.)	125–175	M4	45	65	60	0.15	0.20	0.25	0.35
	176–225	M4	42	60	55	0.13	0.20	0.25	0.35
	226–275	M4	40	55	52	0.13	0.17	0.25	0.35
	276–325	T15	35	52	47	0.10	0.15	0.22	0.30
	326–375	T15	32	48	43	0.08	0.15	0.22	0.30
High-strength alloy (4340, 4330V, 300 M, etc.)	225–300	M4	42	58	55	0.15	0.25	0.30	0.35
	301–350	M4	35	52	48	0.13	0.22	0.25	0.30
	351–400	T15	30	42	40	0.10	0.20	0.22	0.25
Structural steel (A36, A285, A516, etc.)	100–150	M4	42	58	55	0.15	0.25	0.30	0.35
	151–250	M4	35	52	48	0.13	0.22	0.25	0.30
	251–350	T15	30	42	40	0.10	0.20	0.22	0.25
High-temp. alloy	140–220	T15	9	12	10	0.07	0.15	0.20	0.25
Hastelloy, Inconel	221–310	M48	7	10	9	0.07	0.13	0.17	0.20
Stainless steel	135–185	M4	23	32	28	0.15	0.20	0.22	0.27
303,416,420,17–4 PH, etc.	186–275	M4	18	28	25	0.13	0.17	0.20	0.25
Tool steel	150–200	T15	25	32	32	0.10	0.15	0.20	0.25
H-13, H-21, A-4, 0–2, 5–3, etc.	201–250	M48	18	28	26	0.10	0.15	0.20	0.25
Aluminum <8% Si		M4		180–250		0.20	0.33	0.40	0.50
>8% Si		M4		120–170		0.20	0.33	0.40	0.45

TABLE 1.2

Speed and Feed Recommendations: Carbide Drills

Work Material	Hardness HB	Grade	Cutting Speed (m/min)			Feed (mm/rev) for Drill Diameter (mm)			
			TiN	TiAlN	TiCN	9–12	13–17	18–24	25–35
Free machining steel (1112, 12L14, etc.)	100–150	P30	98	128	115	0.20	0.30	0.38	0.38
	151–200	P30	85	110	100	0.17	0.28	0.35	0.35
	201–250	P30	80	104	90	0.15	0.25	0.25	0.33
Low-carbon steel (1010, 1020, 1025, etc.)	85–125	P20	92	120	110	0.20	0.25	0.33	0.42
	126–175	P20	104	104	90	0.17	0.25	0.33	0.40
	176–225	P20	95	95	82	0.15	0.22	0.30	0.38
	226–275	P20	82	82	75	0.13	0.22	0.30	0.35
Medium-carbon steel (1045, 1140, 1151, etc.)	125–175	P20	80	104	90	0.17	0.25	0.30	0.40
	176–225	P20	73	95	84	0.15	0.22	0.28	0.38
	226–275	P20	64	82	72	0.15	0.22	0.28	0.38
	276–325	P20	55	70	63	0.13	0.22	0.25	0.35
Alloy steel (4130, 4140, 4150, 5140, 8640, etc.)	125–175	P20	76	100	87	0.17	0.25	0.33	0.40
	176–225	P20	70	92	80	0.15	0.22	0.30	0.38
	226–275	P20	64	82	72	0.15	0.22	0.30	0.38
	276–325	P20	60	76	69	0.13	0.20	0.27	0.35
	326–375	P20	52	67	60	0.10	0.17	0.25	0.33
High-strength alloy (4340, 4330V, 300 M, etc.)	225–300	P20	48	60	55	0.15	0.22	0.25	0.30
	301–350	P20	43	55	48	0.13	0.20	0.22	0.27
	351–400	P20	37	48	43	0.10	0.17	0.20	0.25
						0.20			
Structural steel A36, A285, A516, etc.	100–150	P20	73	95	84	0.15	0.27	0.35	0.40
	151–250	P20	60	76	69	0.15	0.25	0.30	0.35
	251–350	P20	55	70	63	0.13	0.22	0.27	0.30
High-temp. alloy Hastelloy, Inconel	140–220	M10	80	105	90	0.10	0.17	0.22	0.27
	221–310	M10	60	85	70	0.10	0.15	0.20	0.25
Stainless steel 303, 416, 420, 17–4 PH, etc.	135–185	M30	48	64	56	0.17	0.22	0.30	0.35
	186–275	M30	37	48	43	0.15	0.20	0.27	0.20
Tool steel (H-13, H-21, A-4, O-2, 5-3, etc.)	150–200	M10	48	67	58	0.10	0.17	0.22	0.27

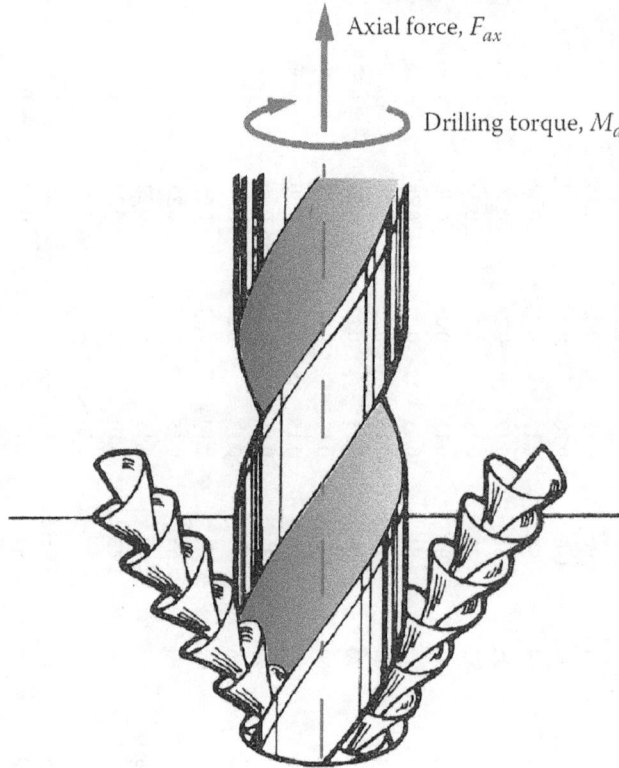

FIGURE 1.10 Total axial force and drilling torque.

1.2.6 CUTTING FORCE AND POWER IN DRILLING OPERATIONS

The complete force balance in drilling tools with the definitions of its components is discussed in Chapter 2. It is shown that the resultant force system in drilling can be represented by the axial force F_{ax} and drilling torque M_{dr} as shown in Figure 1.10. As can be seen, the drilling torque acts about the rotation axis, whereas the axial force acts along this axis. These parameters are used in the selection/design of the machine tool- and work-holding fixture design.

The drilling torque and axial force vary in rather wide ranges depending upon the type of workpiece material, its metallurgical state, and its mechanical properties; tool design, geometry, and material; machining regime; MWF type and supply parameters; and many other particularities of the machining system. Out of these two components, only the drilling torque contributed to the drilling power. The power required in drilling operations $P_{c\text{-}dr}$ is calculated using the drilling torque as follows:

In metric units of measure (the SI system),

$$P_{c-dr} = \frac{M_{dr}n}{9,950} \tag{1.17}$$

where $P_{c\text{-}dr}$ is in kW, n is in rpm, and M_{dr} is in Nm.

In the imperial units of measure,

$$P_{c-dr} = \frac{M_{dr}n}{5,252} \tag{1.18}$$

where $P_{c\text{-}dr}$ is in horsepower, n is in rpm, and M_{dr} is in foot-pounds.

1.3   DRILLING SYSTEM

1.3.1   SYSTEM CONCEPT

Modern technological concepts make it possible to define the present stage of development as the system era. Management makes use of *system concept, system philosophy,* and *system approach.* Engineers and physical scientists speak of *system analysis, system engineering,* and *system theory.* Even in medicine or biology, the specialists speak of the *nervous system,* the *homeostatic system,* the *gene system,* etc. However, the picture is not as bright as it seemed to be in the 1960s when the system approach began to boom. Only in certain fields, for example, computer science, has the system concept been developing rapidly with great practical significance. As a result, only this field has system specialists (system analysts, system programmers, and system managers).

With the emergence of the concept of system engineering, the traditional role of specialization in engineering has been broadened or even completely changed. Traditionally, engineers specialized in a certain branch of engineering. At the system level, however, an engineer is not as much concerned with mechanics or even physics as he or she is with organization, information, and communication, with the mathematical, logical, or even phenomenological relationships among system components, whether they are physical or not. At this level, his or her principal enemy is always the complexity of a system under consideration so wide knowledge not only in a certain engineering field but rather *broad-brush* education and experience is very useful in dealing with system problems.

System problems are often aptly described as a *can of worms,* because it is difficult to discriminate between the different elements of the problem such as the system's boundaries, the system's components and their levels, the system organization, and the interrelationships between the levels. The whole problem seems to be constantly in motion; the components are hopelessly intertwined, so much so that there may be only one indivisible component. It is difficult to grasp any one of the slippery components, and the problem is partly immersed in obscuring debris overshadowed by old beliefs, improper notions, and *experience-based* rules from the past often developed for considerably different systems or for an older (sometimes really *ancient*) set of system components.

The process is somewhat as follows: the system engineer, faced with a problem derived from some system phenomenon, attempts to describe the structure of the system as a set of components. He or she assigns various relationships among the components and attempts to build the model of the system. Then, he or she experiments with the model both mathematically and deductively, all the while checking the results of such comparisons and experiments with the requirements of the problem and experimental or heuristic evidence concerning the phenomenon itself. He or she modifies the model and experiments some more. Finally, he or she arrives at a satisfactory model and proceeds to analyze using various mathematical and computational techniques in order to arrive at an engineering decision. Though the previously discussed procedure looks relatively simple and logical, the chief problem here is to distinguish the system to be analyzed, its boundaries, and its components. Intuition and experience at this stage are essential.

The modern general system theory is intended to reveal similarities in structures and functions for different systems, independent of the particular domain in which the system exists (Skyttner 2006). It is based on the assumption that there are universal principles of organization, which hold for all systems, be they physical, chemical, biological, mental, or social. The mechanistic worldview seeks universality by reducing everything to its material constituents. The systemic worldview, on the contrary, seeks universality by ignoring the concrete material out of which systems are made, so that their abstract organization comes into focus. When it comes to manufacturing systems, a suitable fragile balance between these two approaches should be found depending on a particular objective of system analysis.

This chapter provides a brief consideration of the drilling system and its components in terms of their suitability for HP drilling. It introduces basic system objectives and rules to help practical tool and manufacturing engineers in the intelligent selection of system components. The main attention is paid to the systemic selection of the cutting tool (drills) to achieve the system objective.

Although the following chapter mainly discusses the system approach to drilling, its main ideas are fully applicable to any cutting tool and tooling in a modern high-efficiency machining environment. In this author's opinion, such an approach should be used in machining system design, retrofitting, and components and tool selection, that is, in any aspect of manufacturing.

1.3.2 STRUCTURE OF THE DRILLING SYSTEM

Significant progress in drilling has been achieved that resulted in the introduction of high-penetration-rate drilling. It has emerged during the past 10 years as the process that allows a penetration rate of more than 5 m/min for aluminum alloys, more than 2 m/min for cast irons, and more than 1 m/min for alloy steels. It became possible due to significant improvements in the manufacturing quality of drills, including the quality of their components, the implementation of better drilling machines equipped with advanced controllers as well as their proper maintenance, the application of better MWFs, better training of engineers and operators, and many other factors. However, the actual penetration rate and drilling process efficiency (the cost per unit length of drilled holes) vary significantly from one application to another, from one manufacturing plant to the next, depending on an overwhelming number of variables. Optimum drill performance in drilling is achieved when the combination of the cutting speed (rpm), feed, tool geometry, tool material grade, and MFW parameters is selected properly depending upon the work material (its hardness, composition, and structure), drilling machine conditions, and the quality requirements of the drilled holes (Astakhov 2001). To get the most out of a drilling job, one must consider the complete drilling system, which includes everything related to the operation (Figure 1.11). Such a consideration is known as the system engineering approach according to which the drilling system should be distinguished and analyzed for coherency of its components.

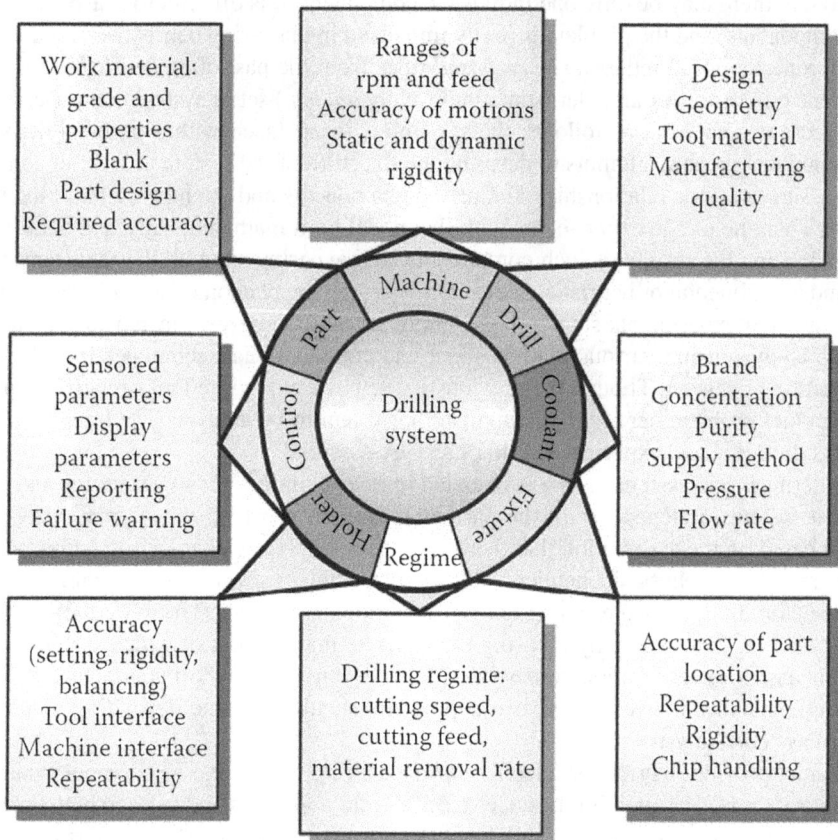

FIGURE 1.11   Drilling system structure.

According to system engineering theory, it is improper to consider any component of a drilling system separately, thereby ignoring the system's properties. The so-called component approach is a common manufacturing practice in today's environment, where different manufacturers produce the various components of the machining system but no one seems to be responsible for system coherency. Low efficiency, subpar quality, and tool failure are direct results of such an approach. Any lack of coherence in the machining system normally leads to visible tool failure. Such failures can easily turn a drilling operation into the bottleneck operation in the automotive industry as a complete production line or a manufacturing cell can be down for a long time due to the failure of a single tool. The direct consequences are significant downtime, low efficiency of an operation, and poor quality of machined parts.

To understand the performance of the drilling system and thus the root cause for many drilling-related problems, one should always consider the components of the drilling system shown in Figure 1.11 in a systemic way. One can appreciate the system properties of the drilling system if one realizes that the same drill used in different drilling machines shows a wide range of results from breakage to excellent performance; the same drill used on the same machine exhibits different results for different work materials; the same drill used on the same machine for drilling the same work material performs differently depending upon a particular brand of MWF used for the operation, the MWF flow rate, filtration, and temperature; and the performance of the same drill used on the same machine for drilling the same work material using the same MWF parameters would depend largely on the type and conditions of the tool holder. The same drill used on the same machine for drilling the same work material using the same MWF parameters and the same tool holder depends significantly on the machining regime. Moreover, the quality parameters of machined parts and drill performance are also affected by the part-holding fixture, namely, its accuracy, repeatability, and rigidity. The drill performance also depends on the extent of the operator's experience and training. The latter is particularly true if the control system provides relevant information to the operator and/or when this operator uses manual gages to inspect the quality of the machined hole. As can be seen, each individual system component can affect the system performance dramatically. The key here is to assure system coherency, that is, the condition when all the system components work as a *team* to achieve the ultimate system's objective.

Unfortunately, the drill manufacturer is often unfairly blamed for failures of the drilling system as the sole culprit because the drilling tool, as the weakest link in the drilling system, fails as a result of improper performance of various system components. For example, one manufacturer of gundrills for the automotive industry was blamed for gundrill breakage occurring at the tip-shank brazed joint. For over 5 years, this manufacturer tried to improve the strength of this joint. When this strength became sufficiently great, breakage of gundrill's carbide tips began to occur. An analysis of the root cause of this problem showed that the lack of the MWF flow rate supplied to the drilling zone caused the drill breakage. Because the root cause was not properly determined, the increased strength of the discussed brazed joint shifted the breakage to the carbide tip as a new weakest link.

1.3.3 Coherency Law

Although there are a number of important system laws, for example, the deformation and optimal temperature laws, the coherency law is always of prime importance in practice design and analysis of high-productivity (HP) drilling systems. For example, if one buys and installs on an old car advanced sidewall design tires specially designed for use at the Indianapolis 500 car race, the performance of the car will be worse than with its old *native* tires although new tires come with the speedway's distinctive *Wing and Wheel* official logo in full color. Although these tires are probably the best that the tire manufacturers can offer (not to mention their cost), they are not suitable for this old car. The same analogy can be made for cutting tools – the best and most expensive drill will not perform well if the machining system does not support its performance. In other words, a drill may have the best geometry and tool material and can be perfectly designed and manufactured, but it may not perform well for a given application. If, for example, the quality of the machined hole allows a drill with 20 µm runout and the drill is selected with 10 µm runout, then the total runout of the tool holder and spindle system should not exceed 10 µm, that is, a high-precision tool holder and spindle should be the case. If, for example, an HP drilling tool is made of an advanced tool material, that is, submicrograin carbide with tantalum-hafnium additives (Ta_4HfC_5), and has internal channels for MWF supply, then the drilling system should be able to run the drill at optimal cutting speed for this tool material, to assure drill installation runout <10 µm, to supply MWF with the pressure sufficient to achieve the flow rate needed for optimal drill performance, etc.

The coherency law states that all components of the drilling (machining) system should be coherent. It implies the following:

1. All components should be logically connected to achieve the prime system objective.
2. All components should be of the same quality.

The latter requires explanation as it is routinely ignored. If one uses a component of higher quality/capability than other components of the drilling system, then he or she loses money as this more expensive component cannot perform to its full capability due to restriction of other components of the drilling system. For example, if one uses an expensive high-quality hydraulic holder on an old machine have a subpart quality spindle, this holder does not improve drilling performance. On the other hand, if one uses a drill with no through coolant capability on a modern drilling machine equipped with a high-pressure MWF delivery system with pressure/flow rate and temperature control, then he or she cannot achieve HP drilling. The MWF unit is wasted. As saying goes: "The speed of the caravan is the speed of its slowest camel."

1.3.4 SYSTEM OBJECTIVE

1.3.4.1 Background

In many books and research papers, drill tool life is of prime concern. In practical tool management, the cost of the drill, tool life, and other miscellaneous articles are of prime concern because these parameters are easy to measure and report. In manufacturing reality, this is not nearly the case, although it is not clearly understood by many as the interpretation of the structure of manufacturing cost is incorrect.

Figure 1.12 shows the manufacturing time structure in modern metal-working shops. As can be seen, in shops with stand-alone computer numerical control (CNC) machines, the machining time is 20% of the operating time, whereas in automotive shops, this time reaches 60%. The latter is due to aggressive tool use strategy, elimination of any tool/part inspection in the machine, adjustable presetting of the MWF pressure for each individual tool, automated (robotic) workpiece/part loading/unloading, using tools with RFID chips, and so on. In both cases, however, the actual machining time is the largest (commonly called as bottleneck).

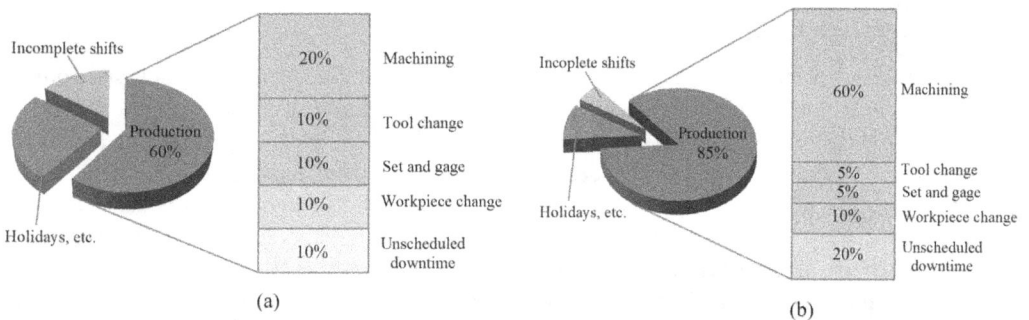

(a)   (b)

FIGURE 1.12 Time structure in the modern metal-working industry: (a) modern machine shop with stand-alone CNC machines and (b) automotive manufacturing shop.

Knowing these data, one should realize that the implementation of HP tools and well-designed machining operations to reduce machining time is a necessity. Unfortunately, some management-level authorities of metal-machining companies misinterpreted the essence of the above-presented data, and thus their multiple implications. In the author's opinion, this is because management deal with the data presented in the cost format as shown in Figure 1.13. This picture/number blinds "effective managers" as they look at the data presented as the driving force to send manufacturing operations to overseas countries where the labor cost is 5%. Moreover, it implies not to invest in the development of new cutting tools/advanced cutting processes – what to bother with this pathetic 3% of costs.

Such a management perception stems from pre-4th industrial revolution experience when the machining time in a cycle time of manufacturing part was so insignificant due to manual part loading–unloading, part and tool setting on the machine, part gaging in the machine, etc. took most of this time. As a result, a reduction of the machining operating time due to the use of advanced (and thus more expensive) cutting tools and optimization of machining processes was not requested, and, therefore, discouraged as manufacturing professionals did not see any benefits of such activities. As a result, a number of old perceptions are still in full swing even in the most advanced industries. The most profound are strict requirements in the automotive industry for a 10%–15% yearly reduction of tool cost and strong drive for increasing tool life. It is not realized that even a 30% decrease in the tool cost results or a 50% increase in tool life in 1% of decrease in the manufacturing cost per part.

In the author's opinion, the only proper way for the modern metal-machining industry to progress is to increase productivity of machining operations via the implementation of high-penetration-rate, high-quality tools and well-designed machining operations. A 20% increased productivity reduces the manufacturing cost per part by 15% as the other costs shown in Figure 1.13 are fixed costs so their relative impact per part reduces.

The structure of manufacturing cost shown in Figure 1.13 should not surprise anyone if he or she analyses the data. An example is presented here to facilitate understanding. Even 20 years ago, an operating cost of $0.8/min was the commonly used benchmark for machine tools. Today with the introduction of modern, multi-axis, high-speed CNC machines with linear drives and excessively ridged spindle units, this cost is almost $3/min. For 2,200 operating hours per year, $3/min means that the operating cost is 396,000 per year for just one machine. Even factoring in 80% efficiency for loading/unloading, changing tools, and setup, an increase in productivity by 50% amounts to a potential yearly saving of $158,400 per CNC machining center per year. As the machining operating time became a bottleneck in the part machining cycle time, this can be achieved only through increasing the productivity of the cutting tool. In other words, an item that, according to Figure 1.13, accounts for only 3% of manufacturing cost can affect this cost significantly as productivity is

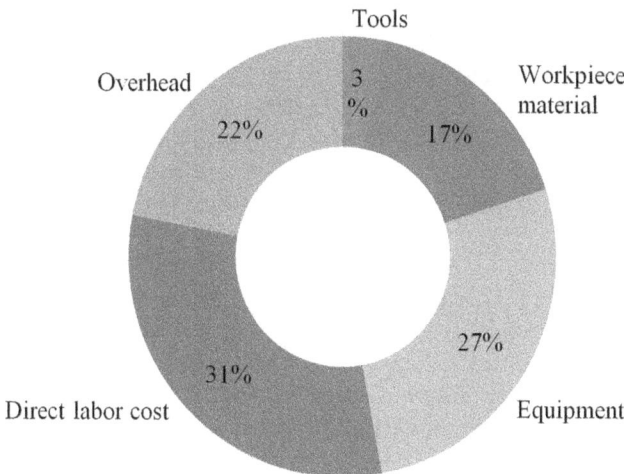

FIGURE 1.13 Manufacturing cost structure in a typical automotive plant.

literarily determined on *the tool cutting edge*. Therefore, the best, most reliable cutting tools of advanced designs with the best tool materials and the highest possible tool manufacturing quality capable of highly-efficient machining should be used in the modern metal-working industry regardless of their cost as this cost is still virtually insignificant compared to the gain due to increased productivity.

Another important but routinely ignored aspect is unscheduled downtime. Figure 1.12 shows that it is up to 10% for stand-alone CNC machines, whereas it reaches 20% in highly-automated industries (Astakhov 2014). Nevertheless, it is considered a kind of unavoidable nuisance. As a result, the unscheduled/uncontrolled downtime due to cutting tool failures has not been in the center of interest of researchers and engineers for years because (1) it was insignificant compared to the scheduled downtime; (2) speed and cutting feed were relatively low so that cutting tool failed by wear, which was periodically controlled; and (3) the costs of tool failure were low as cutting tools were relatively inexpensive and their failures did not bring much damage to the machining system (machine, part fixture, and workpiece). However, times have changed so that the uncontrolled downtime has gradually become significant. Tool failure today is a major cause of unscheduled stoppage in a machining environment and is costly not only in terms of time lost but also in terms of capital destroyed (Rehorn et al. 2005). Some estimates state that the amount of downtime due to tool breakage on an average machine tool is on the order of 6.8%, while when tool failures are considered, the figure is closer to 12% (Astakhov 2014). Even if the tool does not fail during machining, the use of excessively worn or damaged tools can put extra strain on the machine tool system and cause a loss of quality of the finished part.

1.3.4.2 System Objective Formulation

The prime system objective is an increase in the drill penetration rate, that is, drilling productivity while maintaining quality requirements to the machined holes. As such, the drill penetration rate is entirely a combined system rather than a component characteristic.

1.3.4.3 Major Constraints

There are two major kinds of constraints imposed to achieve system objective:

1. *Technical constraints*: The major technical constraints are quality requirements (i.e., surface finish, straightness, and shape) to drilled holes and force factors (the drilling torque and axial force as discussed in Chapter 2). Therefore, all system parameters should be selected so that they fully support this system objective, pushing the boundary of the major constraint as far as physically possible. Unless such an understanding settles in the minds of researchers, tool manufacturers, and process designers/manufacturing professionals, no significant progress in drilling efficiency can be achieved.
2. *Constraints caused by cognitive bias* based on multiple non-system perceptions of the performance of drilling tools that, unfortunately, dominate in the industry. According to Cherry (2020), a cognitive bias is a systematic error in thinking that occurs when people are processing and interpreting information in the world around them and affects the decisions and judgments that they make.

According to the author's experience, the listed technical constraints are relatively easy to deal with as they can be explained and thus understood with clear directions to overcome or reduce severity of these constraints can be developed for each and every particular case found in various drilling operations, whereas constraints caused by cognitive bias due to non-system thinking are difficult, sometimes almost impossible, to deal with at various management levels at many manufacturing companies.

The cognitive bias-related problems start with an operator of a simple manual machine who asks to give him or her a drilling tool, which works much better compared to what he or she currently has. Moreover, this tool should be preferably cheaper. In this context, the "much better" stands for longer tool life and easy in-house (sometimes manual) re-sharpening. Unfortunately, this is not considered properly in many big companies, for example, in an automotive company, where all decisions on buying machines, MWF systems and MWF, tooling (part fixture and tool holders, for example) and cutting tools are made by the purchasing. The purchasing has an objective to save money (compared to the last year, for example) so often seemingly "good deals" on the components of the drilling system are taken in this respect.

To buy a cheaper tool, and thus to save money, and expect its better performance compared to the existing tool constitutes the most common cognitive bias in industry. Everyone seems to understand that you cannot buy a luxury/high-performance car at the price of a compact car – if someone disagrees, please consult your local car dealer. However, this understanding disappears when it comes to HP drilling systems and the corresponding drilling tools. If one wants better productivity/quality, then an HP drilling system should include better (and thus more expensive) components so that such a system and thus HP drilling tools are unavoidably more expensive. However, higher productivity of such a drilling system results in cost savings that significantly overlap the initial high cost.

Moreover, many automotive companies outsource the machine maintenance, MWF business (often called the chemical management), and cutting tool management. Normally, three different outside companies having different objectives are selected to carry these businesses, which, in the author's opinion, presents the major problem. For example, the chemical management objective is to save money on the MWF and its maintenance – it is often done by buying cheaper MWF, keeping the lowest allowable MWF concentration, and not having a control plan with schedules on the laboratory checking of MWF vital parameters. Under these conditions, drilling tools suffer due to improper parameters of MWF as these parameters are often most essential for the performance of these tools. The latter, however, is the responsibility of the tool management company. Another example is machine maintenance, particularly maintenance of high-speed spindles and their replacement. Such maintenance and replacement are expensive and, moreover, require to take machines out of production. As a result, there is always a temptation to run spindles longer, thus with higher runouts and lower rigidity. Once again, under these conditions, drilling tools suffer (lower tool life, poorer machining quality, etc.), but the tool management company is blamed for subpar tool performance. It is always the tool to blame.

When it comes to the tool management company, its objective called assessment matrix does not include an increase in productivity and a decrease in downtime. Rather, the cost saving on the cutting tool via buying cheaper tools and an increase in tool life are set as the performance goals. Reading this, one may ask a question: "Do they really need high-productivity and low-downtime machining systems?"

1.4 CONCEPT OF CUTTING TOOL QUALITY

This section presents a unique concept of tool quality based on the author's many years of experience in the field. The concept is unusual based on a unique analogy to facilitate its understanding by many specialists of various qualification levels. The author hopes that this analogy will be a guideline in any activity on ordering/tool drawing approvals, quality assessments, and failure analyses of various cutting tools.

According to old Hindu belief, the earth is supported by four world elephants resting on a world turtle. Although it is an arguable concept not-sufficiently supported by further geographic discoveries, it holds nicely when one explains the concept of quality of the cutting tool. In this sense, the first front elephant is the tool material (30%), the second is the tool design (30%), and the third one is tool manufacturing quality (30%) as shown in Figure 1.14. The fourth elephant (poorly visible) represents miscellaneous (about 10%), which may include handling, packing particularities, etc. Note that these four elephants are supported by an old wise world turtle.

Considering only three front elephants, one can see that they are equally important in supporting cutting tool quality. Moreover, for the tool quality to be stable, they should be of the same height. It is to say that if one uses the best tool material, but the tool design and its manufacturing quality are

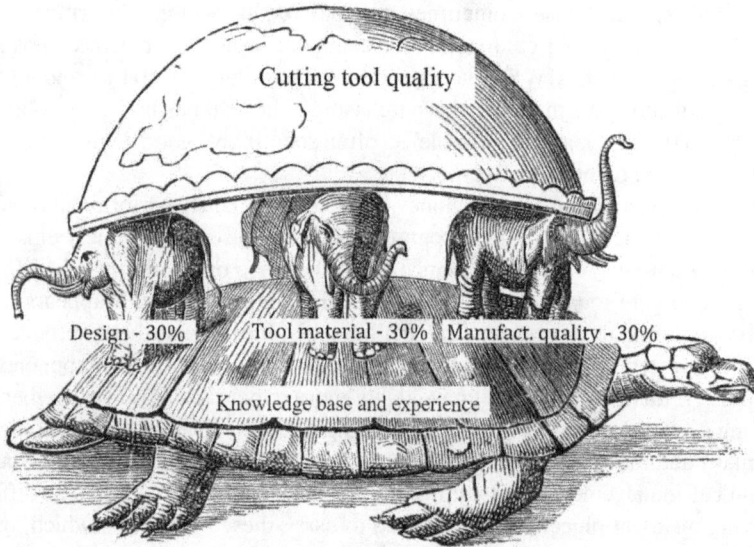

FIGURE 1.14 Graphical representation of the concept of cutting/drilling tool quality.

poor, the tool quality is crooked, i.e., the tool would not perform as intended. The best tool design and perfect manufacturing quality cannot make up for an improper tool material; the best tool material and design cannot make up for poor manufacturing quality; and so on. Only by binding together the front "elephants," as a single effort, one can achieve high tools quality.

The stability of all four elephants, and thus tool quality, is also defined by the area of the world turtle, which represents knowledge and experience. The greater the turtle, the more stable the elephants. Moreover, a great area of the turtle allows the elephants to move around to achieve the required stability of the earth, i.e., the required tool quality. In other words, it allows us to find the best/optimal way to achieve this quality. No matter how good/modern the design tool and manufacturing/metrology equipment a company has, if its "turtle" is small, the quality of manufacturing tools cannot be achieved. On the other hand, if a tool manufacturing company has a great "turtle" then it can manage to manufacture quality tool with less advanced equipment. The author is aware of many practical examples to support this point. Therefore, a good tool manufacturing company should have qualified personnel at different levels and only then good manufacturing/inspection equipment.

1.5   CASE FOR HP DRILLS

Two things can be said about HP drills:

1. They drill better and faster, and last longer, than traditional high-speed, carbide, and poly-crystalline diamond (PCD) drills.
2. They are more expensive than traditional on-shelf drills due to special grades of tool materials, special grinds, closer tolerances on practically all design and geometry features, much higher requirements for surface roughness on the ground surface, etc. In other words, their design, manufacturing, and inspection require higher qualifications of the engineers, many extra operations, advanced equipment, and well-trained operators.

Do the benefits justify the increased cost of HP drills? Not always. And not always in a way that can be seen just by analyzing short-term costs with no system considerations. Evaluating the economics of an HP drill involves a range of different factors. The bulk evaluation of drilling costs presented in the previous section does not account for many particularities of the drilling economy so a more detailed analysis (at least, a methodology for such an analysis) is needed to understand the principal sources of HP drill efficiency and thus to carry out this analysis when a case for deployment of HP drills is developed/discussed.

To make the further analysis more detailed, a simple but realistic example of drilling work is considered as a model. The numbers used to analyze this job may not apply directly to a particular shop or application, but they are realistic as based upon the current industrial practice. Moreover, the logic of the analysis can be applied to a wide variety of drilling operations. What this analysis shows is that the cost-effectiveness of a given drill may be determined by factors that are far removed from the tool's initial price. The price often blinds the purchasing department/personnel as their objective is the purchasing cost reduction and not the total gains obtained by a company in the implementation of HP drills.

Table 1.3 presents the results of economy analysis. The analysis is based on a simple drilling job. Figure 1.15 shows a part where six holes are to be machined. Figure 1.16 shows particularities of the tool layout. The specifics of the job are listed as follows:

- The workpiece is made of high-silicon aluminum alloy, and it is run on a single manufacturing cell. Six cells are used in the shop.
- The production rate for all of the work on this part is 20 pieces/h/cell.
- The piece needs six drilled step holes having diameters as shown in Figure 1.16.
- The job runs three shifts for a total of 20 h/day, 5 days/week.
- The shop runs at a flat rate of $60/h including machine costs, labor, and overhead.

TABLE 1.3
Result of Economy Analysis/Comparison of the Standard and HP Drills

Item	Standard Drill	HP Drill
Feed rate (mm/min)	3,600	9,600
Total stroke length (mm)	26	26
Machining time (min)	0.072	0.027
Machining time cost per minute ($)	1	1
Machining time cost per hole ($)	0.072	0.027
Cost per part ($)	0.432	0.162
Drill life (holes)	5,000	7,000
Drill purchasing cost ($)	95	145
Drill cost per hole ($)	0.02	0.02
Combined cost per hole ($)	0.09	0.05
Combined cost per part ($)	0.55	0.29
Holes to drill per year	662,640.00	662,640.00
Number of re-sharpenings	8	8
Total drill life	40,000.00	56,000.00
Total drills needed	17	12
Average drill breakage per year due to casting and machining system defects	6	3
Actual total drills needed	23	15
Total drill purchase cost ($)	2,185.00	2,175.00
Re-sharpening cost per drill (S)	20.00	60.00
Total re-sharpening cost ($)	3,680.00	7,200.00
Total drills cost	5,865.00	9,375.00
Cutting cost per year ($)	**60,300.24**	**31,617.39**
Total cost ($)	**66,165.24**	**40,992.39**
Net cost per part ($)	0.60	0.37
Net cost per hole ($)	0.10	0.06

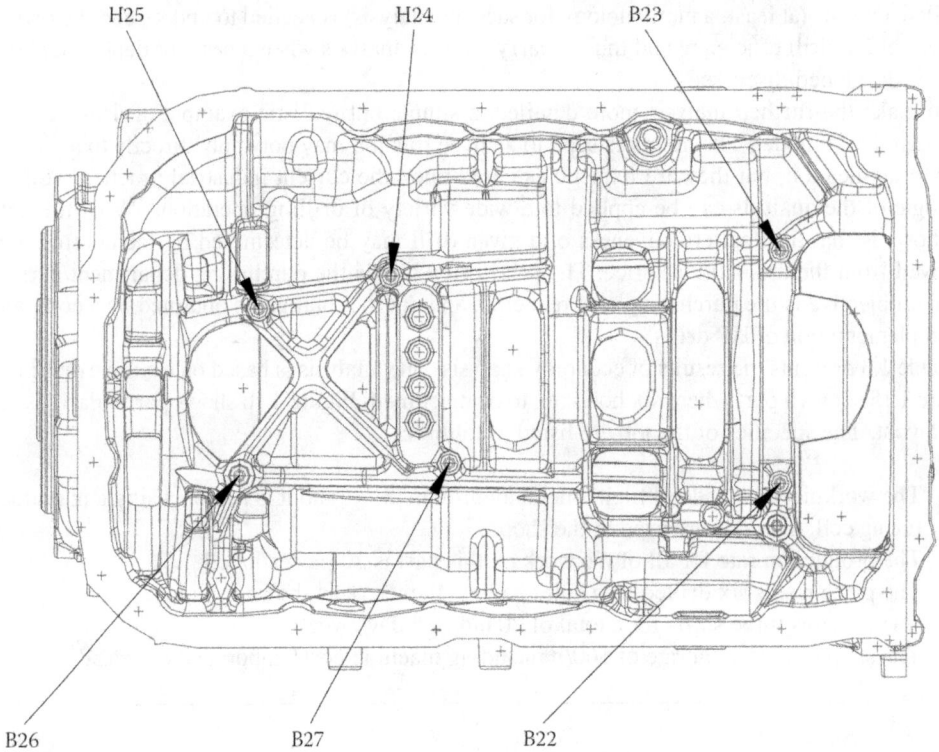

FIGURE 1.15   Six holes to be machined in each part.

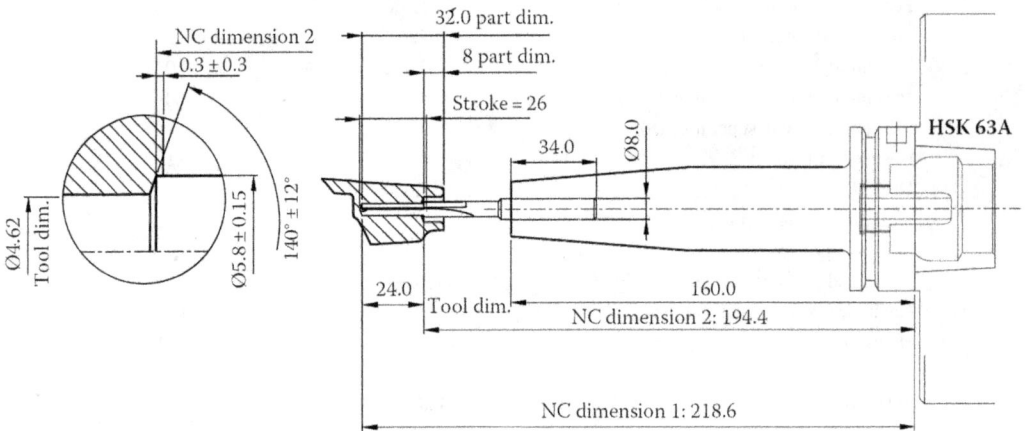

FIGURE 1.16   Particularities of the tool layout.

The analysis begins with looking at a basic measure of drill performance: the feed rate. The two tools compared are a typical, good-quality standard carbide drill and an HP carbide drill designed as the application-specific tool. The cutting regimes used for these drills were determined as follows:

1. The standard drill allows the spindle rotational speed $n = 12,000$ rpm and cutting feed $f = 0.3$ mm/rev. The speed is selected according to the drill manufacturer's specification. It is limited by the grade of carbide and coating used. The cutting feed is limited by two major factors: (1) the maximum allowed axial force and (2) the ability of drill to reliably

remove the formed chips with no clogging of the chip flutes. Naturally, this drill was tried for high speed with no success as tool life reduced significantly due to accelerated drill corner wear. Any attempt to increase the feed per revolution resulted in clogging of the chip flute and drill breakage.

2. The special drill is designed to the requirements of VPA©'s designs (explained in Chapter 2 of this book and in Chapter 3 in Astakhov (2024)). It allows the spindle rotational speed $n=24,000$ rpm and cutting feed $f=0.4$ mm/rev.

The purchasing cost of the standard drill is $95 and its re-sharpening cost is $20, while those of the HP drill are $145 and $60, respectively. The difference in the initial cost absorbs the cost per development and testing, the higher cost of application-specific carbide grade (in the considered case, the generic HP carbide grade is changed to newly developed CERATIZIT grade with custom coolant hole diameters and location), and additional operations of finish grinding get a mirror surface finish. The higher re-sharpening cost is explained by additional grinding operations with specially dressed diamond wheels and detailed inspection after each re-sharpening.

The analysis presented in Table 1.3 does not account for the inventory cost and cost of the drill in the flow. These two will reduce the saving margin. However, this analysis has not accounted for increased productivity, that is:

- The yearly program, while using HP drills, can be completed in one-third of the time needed if the standard drills are used. As a result, the CNC machine can be used for manufacturing other parts. For example, implementation of a few HP drills for the part shown in Figure 1.15 allowed sufficient cycle time so it was found that there is no need to purchase another manufacturing cell to handle the increased program. The resulting savings were much greater than that shown in Table 1.3.
- The reliability of the HP drill is at least twice greater than that of the standard drills. Moreover, these drills can handle casting porosity and small inclusions with no excessive wear and drill breakage. Saved downtime and cost of scrapped parts can be significant when the HP drills are used.

It is worth to mention that the automotive plant in the considered example runs six parallel manufacturing cells so the saving due to development and implementation of the single small HP drill shown in Table 1.3 is approx. $150,000/year. Note that the presented numbers are taken from a real-life case.

The real advantage of HP drills developed and manufactured using the above-introduced concept of tool quality and VPA©'s designs is that it can be designed and made to address a particular need, for example, to increase productivity, to increase tool life, and to improve the quality of the drilled holes. Moreover, three- or two-pass operations used to achieve high quality of drilled holes can be accomplished in one-pass saving machining time and tool costs.

1.6 DESIGN OF DRILLING SYSTEMS

Although the design of each drilling system should follow its unique path depending on given practical conditions, the basic common features of the drilling system design are the same. As an example, Figure 1.17 shows a simplified flow chart for the HP drilling system design for the existing drilling machine (Astakhov 2014), that is, for the most common practical case.

In this flow chart, the machine selection, fixture design, controller selection and programming, and verification stages are well covered in the literature (Campbell 1994, Parkesh 2003, Smid 2003, Nee et al. 2004, Koening 2007). The selection of MWF parameters for drilling operations, design of the internal coolant channels, and chip removal parameters are discussed in Chapter 3 in Astakhov (2024). The following sections are concerned with the drilling tool-related items as they not properly covered in the literature on the matter.

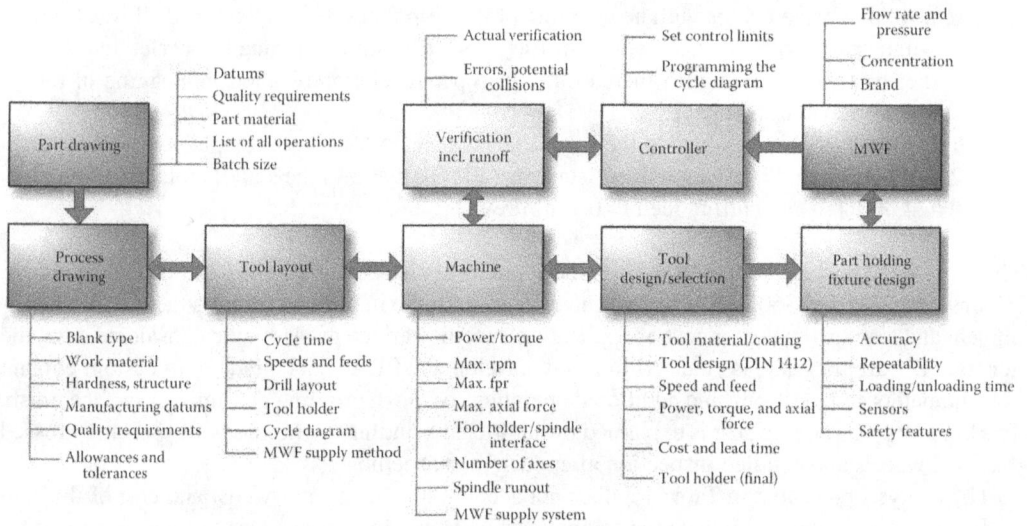

FIGURE 1.17 Flow chart of the HP drilling system design.

As mentioned above, in modern HP drilling systems, all the considerations start with the tool as the productivity of the whole drilling/machining system is literally determined on its cutting edge. The tool layout and drill selection particularities are discussed in this section.

1.6.1 PART DRAWING ANALYSIS AND DESIGN OF THE TOOL LAYOUT

The design of a drilling operation should always begin with a part drawing analysis. In the considered example, it is assumed the batch size and the list of all manufacturing operations (preceding and subsequent to the considered drilling operation) are known. Moreover, it is assumed that a specific type of blank (casting, forging, bar stock, etc.) has been selected. The analysis of the part drawing includes the following major steps:

1. Analysis of the part material: chemical composition, mechanical properties, and metallurgical state (including allowable inclusions, porosity, and cavities). Special attention should be paid to allowable variation of these characteristics.
2. Datums and datum features are related to the hole to be machined.
3. Quality requirements for the considered hole that include size, shape, location tolerances, and surface roughness (or surface integrity if specified as discussed in Chapter 1 in Astakhov (2024)).

The results of this analysis are used in the construction of the process drawing.

The next step is the design of the process drawing. This drawing is *derived* from the part drawing and contains essential information relevant to the hole-making operation. This information includes all the requirements for the drilling operation. Moreover, it contains suggestions on the proper part location to assure such requirements. Figure 1.18 shows an example of the process drawing. As can be seen, it includes the diameter of holes to be drilled and information on the location of this hole with respect to the part design datums. Technological datums as, for example, location of the hole with respect to the locating pins, are also added to assure the required hole location with respect to the designed datum. The sense and meaning of GD&T designations are explained in Chapter 1 and Appendix A in Astakhov (2024) in detail.

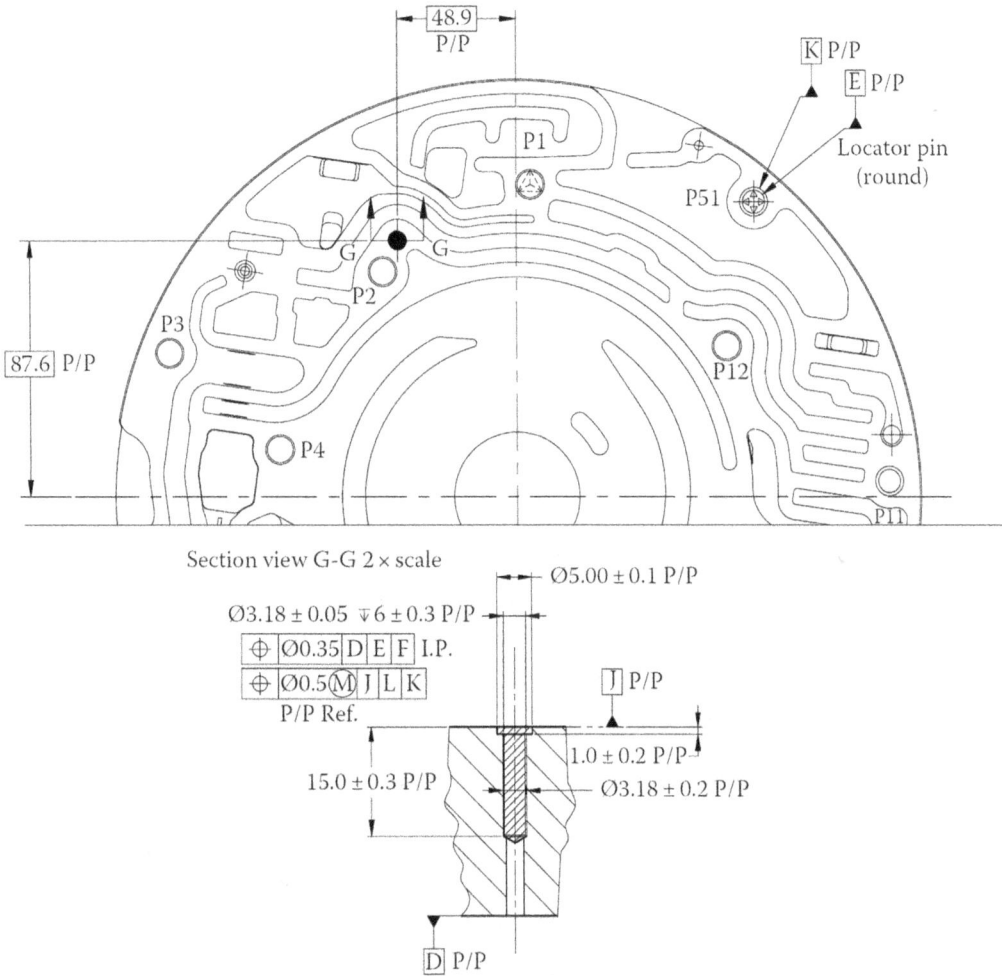

FIGURE 1.18 An example of process drawing.

The next step is to design the tool layout. Usually, the tool layout is the handover document transferred from engineering to the shop floor. A tool layout captures the tool information in the language of engineering, consisting of drawings, bills of material, and parameter lists. A single-tool layout refers to a single-tool assembly for a certain operation performed with a specific spindle on a specific machine tool. The layout documents the components of the tool assembly, including spare parts.

An example of the tool layout in the automotive industry is shown in Figure 1.19, and the corresponding parameters are shown in Table 1.4. As can be seen, the following information is derived from the tool layout:

- Drill starting and end positions that define (1) the drilling length and (2) the length of the drill working part.
- The diameter of the drill (as derived from the diameter of the hole to be drilled) and drill point angle as equal to that requested by the part drawing.
- The drilling regime in terms of the cutting speed and feed. As the development of a tool layout is an iterative process, the initial assignment of these two regime parameters is

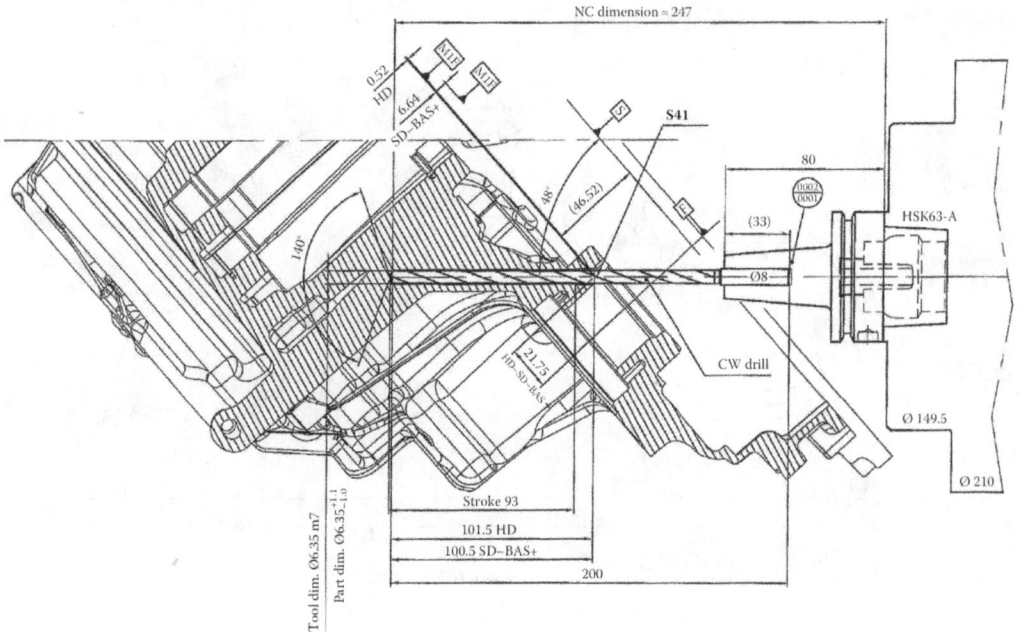

FIGURE 1.19 An example of the tool layout.

normally based upon (1) the cycle time available for the operation, (2) the data available in the company's tooling database (for similar applications), and (3) recommendation of the leading tool suppliers available in their online catalogs. The so selected cutting speed and feed then can be changed several times in the process of the final revision of the drilling operation.

- The general tool holder that depends on the machine available for the operation. At this stage, the machine–holder interface is fixed (by HSK63-A in Figure 1.19) while the holder/ tool interface is still open. In the considered case, it can be shrink-fit, hydraulic, collet, etc., interfaces depending upon the accuracy required, availability of a particular tool setting machine, and many other technical and logistic factors.
- The cycle diagram of the drilling operation is shown in Figure 1.20. This diagram allows one to calculate the machining cycle time (the time needed for the machining part of the drilling operations). When one adds this time to the time needed for loading–unloading of a part, the drilling cycle time is obtained.
- The drill sketch is shown in Figure 1.21. Note that this is not a drill drawing. Rather, it represents a general idea of what kind of drill is needed for the drilling operation. Later on, the particular drill parameters (drill geometry, design, particular grade of tool material, coating, etc.) will be selected.

In the considered procedure, the machine is assumed to be known so that the cutting regime including the cutting power, cycle programming, tool holder, and stroke length is selected to be feasible for the chosen machine.

Tool layouts are a great way to communicate and brainstorm ideas back and forth between tool manufacturers (suppliers) and tool users. They are also used in programming the machine and tool presetting equipment. In continuous improvement efforts, on-site cost reduction teams improve cycle times – changing speed and feed rate or calling for alternate tooling – and thereby changing the tool specification based upon information available in the tool layout.

TABLE 1.4
Tool Layout Particularities

Work Material	Tool Material
Aluminum ANSI A380Mod	Sintered coated carbide
Si 7.5%–10%, Cu 2%–4%,	Tool type
Mg 0.3% max	Twist drill
Optional material	Tool holder
Aluminum ANSI A383Mod	Per layout
Si 12% max, Mg 0.3% max	
Blanks: die castings	

Machining Regime as Recommended by a Leading Cutting Tool Supplier			
Cutting diameter	d	mm	6.35
Cutting speed	v	m/min	250
Spindle rotational speed	n	rpm	12,532
Cutting feed	f	mm/rev	0.22
Feed per tooth	f_z	mm/rev/tooth	0.11
Uncut chip thickness (chip load)	h_D	mm	$h_D = t_z \sin(\Phi_p/2) = 0.11 \sin(140/2) = 0.103$
Feed rate	v_f	mm/min	2,757
Coolant			
Type			Water soluble
Concentration		%	8 min
Supply method			Internal through tool
Flow rate		1/min	18
Pressure		MPa	5.5

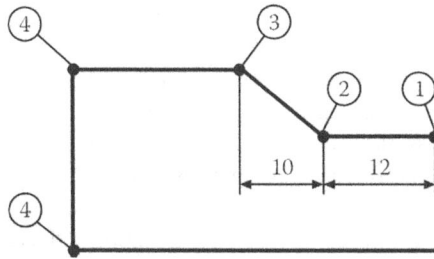

① $n = 500$ rpm, $f = 1$ mm/rev	③ $n = 12532$ rpm, $f = 0.15$ mm/rev
② $n = 6500$ rpm, $f = 0.08$ mm/rev	④ G0 (rapid retract) to $n = 500$ rpm

FIGURE 1.20 Drilling cycle diagram.

1.6.2 DRILL SELECTION

1.6.2.1 Design

According to the tool layout, the feed rate $v_f = 2,757$ mm/min and the cutting speed $v = 250$ m/min are selected based upon the required (economy-justified) cycle time. At this stage, an important decision is to be made, namely, what tool to use: standard or custom. Although more than 70% of cutting tools in the automotive industry are custom, the use of relatively inexpensive standard drills seems to be an attractive option.

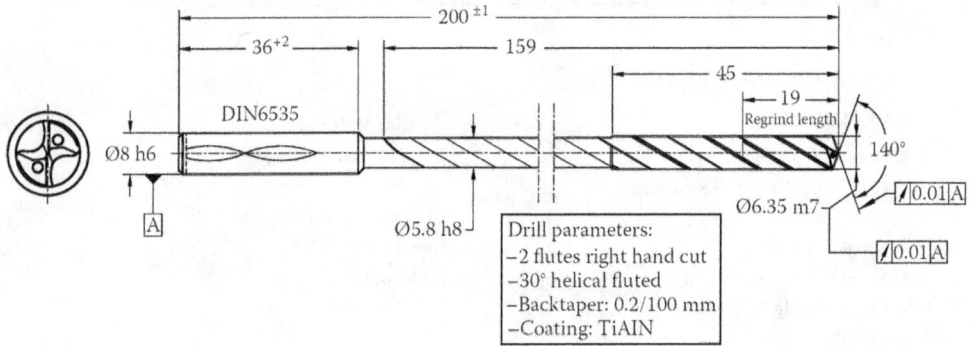

FIGURE 1.21 Drill sketch.

What exactly determines if a tool is special is a matter for discussion. The basic definition used in practice defines a custom drill as to be anything other than a jobber length drill. The custom tool is developed to address a particular need, for example, particular feature configuration in an engine or transmission part. However, when such a part goes into production, and the car manufacturer starts buying the tools in mass quantities, then this tool technically is no longer a custom tool. It can be referred to as a *standard custom tool* (Kennedy 2010).

Even when produced in large quantities, custom tools lack the economy of scale characteristic of standard tools. A part maker's willingness to pay a premium and take advantage of specials depends largely on the shop culture. Standard tooling is cheaper, but there is value in the more expensive custom tools; they offer more capability. If a shop uses the custom tools properly, their productivity can go way up and their profits can go way up too. If a shop is reluctant to make large capital purchases, a custom tool may enable this shop to increase productivity without purchasing new equipment.

Large aerospace and automotive manufacturers regularly use custom drills because they generally provide longer tool life and better hole quality, and the cost of reworking out-of-spec holes can easily exceed the custom tools' cost. Those manufacturers plan months ahead in most cases. Other shops may not be as proactive with inventory. A shop may deplete its supply of a certain diameter custom drill and then rush to find a standard drill in that diameter. This drill is simply issued to the shop floor, and the machine operator fights the consequences of such a decision because it is convenient.

Therefore, there are two possible outcomes of the considered step: (1) a standard drill is selected, and (2) a custom-made drill is selected based upon the economy analysis similar to that presented in Section 1.5.

If a standard drill is selected, then the further consideration is the choice of a tool supplier as the selected standard drill can be manufactured by a number of companies. Selection of a particular supplier of the standard drill should be based on the comparison of the following items:

- Suitability of a standard drill for the considered operation in terms of work material and selected machining regime;
- Track record of suppliers (if available) for supplying/supporting standard drills for other operations in the considered manufacturing facility or to the other manufacturing facilities of the company;
- Quality rating of the suppliers;
- Customer technical support provided by suppliers;
- Lead time or on-shelf supply basis;
- Cost of the drill and cost of re-sharpening;
- Lead time for a re-sharpening including the location of the nearest re-sharpening facility.

Often, however, the initial cost of the drill is the major driving rationale in the selection of the supplier as such a selection is carried out by the purchasing department. Fortunately, such a short-sighted strategy has been changing over the last decades so more and more direct manufacturing personnel have the decisive say in the selection of the tool supplier.

Once the supplier is chosen, the drill design, tool material, and coating are fixed by the chosen tool manufacturer. Technically, no further actions on the drill are needed. In the considered case, however, one problem, namely, the reliability of chip removal from the long hole, remains. This is because the drill is long so standard long drills have the thick web (discussed in Chapter 2) at the expense of reduced cross-sectional area of the chip flute. It may present a problem as the feed rate is high; thus, a great amount of chip is to be removed over the flutes.

Often, to deal with the problem of chip removal from long holes, the so-called peck drilling – a drilling operation that periodically retracts the tool to clear chips or flood the hole with MWF – is recommended by many literature sources on drilling so that the peck drilling cycle is a standard part of many drilling machine CNC controllers. Experienced practitioners, however, are of the opinion that peck drilling is the second most efficient way to destroy carbide drills. Putting them on a nice hard surface and then smashing them with a big hammer is the first way. There are two major reasons:

1. Many work materials harden almost instantly if rubbed with tools, instead of cutting continuously. Even a couple of revolutions without continuous penetration can leave the bottom of the hole so hard, so shiny, and so perfectly armored and burnished that the drill has a hard time to start cutting again.
2. The loose chip left in the chip flute can drop back into the partially drilled hole as the drill retracts. Note that the chip is much harder than the work material due to severe plastic deformation during chip formation. When the drill returns, severe recutting takes place that is not beneficial to the tool wear.

When a custom-made drill is selected based upon the economy analysis similar to that presented in Section 1.5, a tool supplier capable of manufacturing the special drill has to be chosen. Often, the special tool is considered as having some additional steps or other non-standard features, not HP drills capable of high-efficiency drilling. When, however, the latter is the case, a small development project may be needed.

To carry out this project, a certain qualification/ability of the chosen supplier is required. On the one hand, a local drilling tool manufacturer may not have sufficient design, manufacturing, and metrological capabilities to carry out this project. On the other hand, big tool manufacturers may not be interested in the development of one special tool. Their local sales and application specialists may not understand the essence of the project whereas their real engineering and R&D are located far away, often overseas. Moreover, language, terminology, real understanding of tool design, etc. are more and more become issues in online meetings with their specialists.

1.6.2.2 Drill Material Selection

To design a tool layout for an HP drilling operation, a drill material should be selected. It is an important decision to make at this stage because drilling accounts for a large percentage of the total cost of manufacturing, and one of the primary ways to cut drilling costs is to reduce drill breakage and premature wear.

Drill breakage is the most common problem associated with drilling operations, and it is a major cause of machine downtime. When a drill breaks, it must be removed from the workpiece using another drilling operation. This wastes valuable production time and can produce out-of-tolerance holes, scarred surface finishes, and other quality problems in the finished component.

Premature drill wear also reduces the efficiency of drilling operations. Typically, a drill wears too fast when for a given drilling operation (drilling system and properties of the work material), the tool material, cutting speed, and feed rate are not selected properly for the application. The effect of an excessive cutting speed is normally easy to recognize as it causes the excessive wear of the drill periphery corners as discussed later in Chapter 2 of this book and Chapter 4 in Astakhov (2024). In some cases, too slow speeds and feeds have the effect of burnishing the hole, creating unwanted bright spots on the hole walls, and causing tool wear. Premature wear also can lead to drill breakage as the drilling torque and axial force increase with increasing tool wear.

Chapter 4 in this book provides information on PCD tool material, whereas Chapter 2 in Astakhov (2024) provides help in the proper selection of other common tool materials used to make drilling tools. In the considered case of drilling due to high cutting speed, sintered carbide or PCD is the material of choice. As the work material is high-silicon aluminum alloy then a special low cobalt carbide grade of high quality (as discussed in Chapter 2 in Astakhov (2024) in great detail) should be considered for the application. Alternatively, a PCD drill (see Chapter 4) can be used. A PCD drill is approximately five times more expensive and the lead time for such drills normally doubles that for special carbide drills. However, a PCD drill can offer at least twice greater cutting speed, and thus, the penetration rate combined with up to five times tool life increases. Although the listed advantages of a PCD drill are attractive, one needs to know the following before considering an implementation of such drills:

- Many leading drill manufacturers include PCD drills in their catalogs. The problem is that only a few of them are capable of manufacturing PCD drills of high quality and reliability.
- The drilling system for implementation of PCD drills should be of much greater quality than that for cemented carbide drills in terms of structural rigidity, available speeds and feeds, purity of MWF as well as its available flow rate at high pressure, low casting inclusions and cavities count (Astakhov and Patel 2018), etc.
- The tool re-sharpening should always be done by the tool supplier.

1.6.3   HP DRILL DESIGN/GEOMETRY SELECTION

As pointed out in the Preface, the most frequently asked question is about the difference between HP drills and usual drills. In other words, what makes these drills more productive and efficient particularly in demanding applications. The answer is simple and straightforward: It is a synergy of the system approach with the quality of HP drills. Both are explained in this book. It is to say that the components of the drilling system and its design as the combination of these components are aimed to support drill performance to achieve the system objective. The quality of the HP drill is understood as discussed in Section 1.4 and graphically represented in Figure 1.14. This can be explained as follows:

1. The design and point geometry are selected for a particular work material so that it assures
   a. Minimum axial force and drilling torque.
   b. Balanced design of the major cutting and chisel edges to assure chip flows from the different cutting edges with no crossing/interference and providing sufficient room to assure that there are no restrictions on these flows (Chapter 2).
   c. The use of the newly developed pioneering design/geometry concepts such as VPA1©, VPA2©, VPA3© (Chapter 2), VPA4© (Chapter 3), VPA5© design for drills and reamers, and VPA5© reverse for gundrills (discussed in Chapter 3 in Astakhov (2024)) developed and tested for HP drills.
   d. Perfect force balance of HP drills due to proper assigning of their design and manufacturing tolerances (Chapter 1 in Astakhov (2024)).
   e. Efficient use of the MWF flow supplied to the drill through its proper distribution in the machining zone (Chapter 3 in Astakhov (2024)).
2. The drill material is selected to be
   a. Of high quality and consistency (Chapter 2 in Astakhov (2024)).

b. Application specific. Chapter 2 in Astakhov (2024) provides the guidelines for making the proper choice of the tool material based upon its manufacturing practices and relevant properties.

3. High manufacturing quality (Chapter 2) including the tool final inspection (discussed in Chapter 1 in Astakhov (2024)).

Although an HP drill designed following the above guidelines is fully capable to assure the highest possible productivity of a given drilling operation, this capability is realized if and only if it is fully supported by the drilling system and its components.

Are HP drills universal? No, they are not. They can be compared to high-performance cars that require certain road conditions, high qualification of drivers, intelligent maintenance, etc. High-performance cars may not be suitable for country roads, low-cost maintenance, and for carrying heavy loads. Moreover, they are not meant for just everyday commuting *home-workplace and back* at an average speed of 40 km/h. However, when properly used, they deliver high comfort, joy of driving, and incredible reliability. As more and more people appreciate these benefits, the market segment of these cars increases fast. Almost the same can be said about HP drills. They are suitable for high-quality, well-maintained drilling systems. They require high-quality tool materials and special, application-specific coatings, intelligent tool manufacturing including complicated inspection equipment and procedures, computerized presetting in high-quality drill holders, etc. The benefits of their use include high penetration rate (high productivity), long tool life, great tool reliability, and much better quality of the machined hole – often two- or even three-tool hole operations can be reduced to single-tool operation. HP drilling is the major trend for the future where unattended fully computerized manufacturing lines with greatly reduced labor cost, HP, and high consistency of quality of the manufactured parts are the ultimate goals.

Reading the previous paragraph, a casual reader may think that this is a dream or at least something from the distant future. In the author's opinion, it is today's reality, for example, in the automotive industry, where manufacturing cells and production lines deliver HP, with minimum attendance, and great quality.

To finish the introduction of HP drills, the author wants to remind potential readers that only 30 years ago, carbide drills were debatable tools as they cost much more than high-speed steel (HSS) drills; they require much more intelligent handling and re-sharpening, better tool holders and machine spindles, etc.; and many specialists at that time thought that carbide drills would have very limited use in practical manufacturing. Well, today these drills are common standard tools used universally while HSS drill usage has reduced. The author is deeply convinced that the same will happen with HP drills as the economy of modem manufacturing dictates so. That is why this book is written to accelerate wider acceptance of such drills.

REFERENCES

Astakhov, V. P. (2001). "Gundrilling know how." *Cutting Tool Engineering* **52**: 34–38.

Astakhov, V. P. (2014). *Drills: Science and Technology of Advanced Operations*. Boca Raton, FL, CRC Press.

Astakhov, V. P. (2024). *High-Productivity Drilling Tools: Materials, Metrologyn and Failure Analysis*. Boca Raton, FL, CRC Press.

Astakhov, V. P., Patel, S. (2018). Efficient drilling of high-silicon aluminum alloys: Fundamentals and recent advances. *Drilling Technology: Fundamentals and Recent Advances*, J. P. Davim. Berlin, Germany, De Gruyter Oldenbourg.

Campbell, P. D. Q. (1994). *Basic Fixture Design*. New York, Industrial Press.

Cherry, K. (2020). "What Is Cognitive Bias?", from https://www.verywellmind.com/what-is-a-cognitive-bias-2794963.

Kennedy, B. (2010). "Custom connection." *Cutting Tool Engineering* **62**(3): 37–42.

Koening, D. T. (2007). Machine tool and equipment selection and implementation. *Manufacturing Engineering: Principles for Optimization*. D. T. Koening. New York, ASME.

Nee, A. Y. C., Tao, Z.J., Kumar, A.S. (2004). *An Advanced Treatise on Fixture Design and Planning*. Singapore, World Scientific Publications.

Parkesh, J. (2003). *Jigs and Fixtures Design Manual*. New York, McGraw-Hill.

Rehorn, A. G., Jin, J., Orban, P.E., (2005). "State-of-the-art methods and results in tool condition monitoring: A review." *The International Journal of Advanced Manufacturing Technology* **26**: 693–710.

Skyttner, L. (2006). *General Systems Theory: Problems, Perspectives, Practice*. Singapore, World Scientific Publishing Company.

Smid, P. (2003). *CNC Programming Handbook*. New York, Industrial Press.

2 Geometry and Design Components of Drilling Tools

Good design goes to heaven; bad design goes everywhere.

Mieke Gerritzen (NL), a designer and co-founder of All Media Foundation

2.1 DRILLS

2.1.1 CLASSIFICATION

A drill is an end-cutting tool for machining holes having one or more major cutting edges (often referred to as lips) and having one or more helical or straight chip removal flutes. A great variety of drills are used in industry. They can be classified as follows:

Classification based on construction:

1. *Homogeneous drills*: those made of one piece of tool material such as sintered carbide or high-speed steel (HSS). Most of the HSS and carbide drills used in practice today are homogeneous drills.
2. *Tipped drills*: those having a body of one material with cutting edges (or their parts as the periphery corners) made of other materials brazed, mechanically fixed, or otherwise bonded in place. Cemented-carbide-tipped drill was used as an alternative to HSS for drilling difficult-to-machine materials, for woodworking drilling operations, and for drilling specific work materials such as concrete. Nowadays, however, the use of carbide-tipped drills is significantly reduced with the wide implementation of homogeneous (often referred to as solid-carbide) drills. On the contrary, the use of polycrystalline diamond (PCD)-tipped drills has been growing. The PCD tips in such tools are placed at the periphery regions of the major cutting edges (see Chapter 4).
3. *Indexable-insert drills*: those having cutting portions or indexable cutting inserts (cartridges) held in place. An example of this type of drill is shown in Figure 2.1. The indexable inserts can be installed in the drill body or placed in the cartridges that are installed into the drill body. The former designs allow smaller drill diameters but when severe breakage occurs, the whole drill body should be replaced. The latter design is more expensive and suitable for drills of larger diameters. However, when severe breakage occurs, only the cartridge is normally damaged. Moreover, a much greater adjustment range in terms of setting the exact drill diameter is possible with this design. This type of drill can drill rapidly and produce holes up to five times the diameter of the drill length with close tolerance.
4. *Interchangeable (replaceable)-tip drills*: those made with a replaceable drill head. An example of an indexable (replaceable) drill head is shown in Figure 2.2. Clamping methods for interchangeable-tip drilling systems range from internal or external screws to turn-and-click systems. Compared to the typical ±0.12 mm accuracy of an indexable-insert drill, an interchangeable-tip drill can hold ±0.05 mm in most cases, compared to the ±0.025 mm typical with solid-carbide tools and ±0.015 mm typical with HP drills. The use of interchangeable-tip drills eliminates more than re-sharpening costs. During re-sharpening, the cost of having drills in float can be considerable. For example, one may have two or three drills in operation and two or three drills as backups, but then he or

DOI 10.1201/9781003263296-2

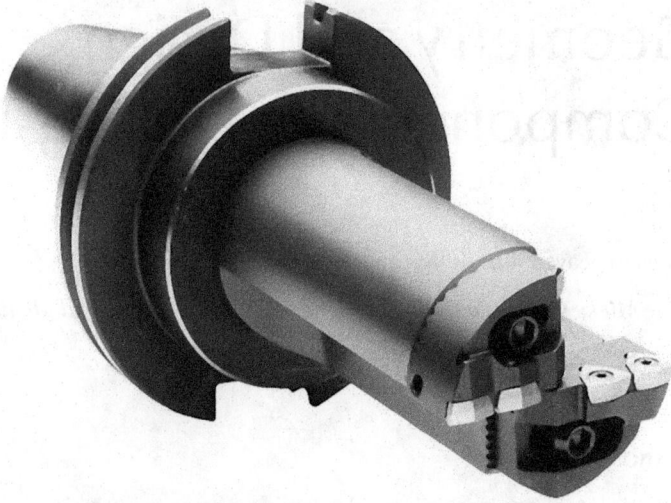

FIGURE 2.1   Drill with indexable carbide inserts.

FIGURE 2.2   Drill with an interchangeable (replaceable) tip.

she might have to have four more drills being sharpened, and four drills on the way back from the re-sharpener. Moreover, interchangeable-tip drills provide a further degree of flexibility in that the bodies can be fitted with tip geometries that maximize productivity for a specific workpiece material. For example, among the specialized geometries offered by Sumitomo for its SMD tools are MTL-style tips for steels and general-purpose applications and MEL-geometry tools designated for work materials like stainless steel and high-temperature alloys. It is also possible to combine the advantages of indexable-insert and interchangeable-head drill designs.

Classification based on shank configuration:

1. *Straight shank drills*: those having cylindrical shanks, which may have the same (Figure 2.3) or different diameter than the body of the drill. The shank can be made with or without driving flats, tang, neck, grooves, or threads.
2. *Taper shank drills*: those having conical shanks suitable for direct fitting into tapered holes in machine spindles, driving sleeves, or sockets (Figure 2.4). Tapered drills with Morse taper shanks are most common for normal-precision applications and generally have a tang meant exclusively to facilitate drill removal from the machine with a drift.

FIGURE 2.3 Drill with a cylindrical shank.

FIGURE 2.4 Drill with a conical (taper) shank.

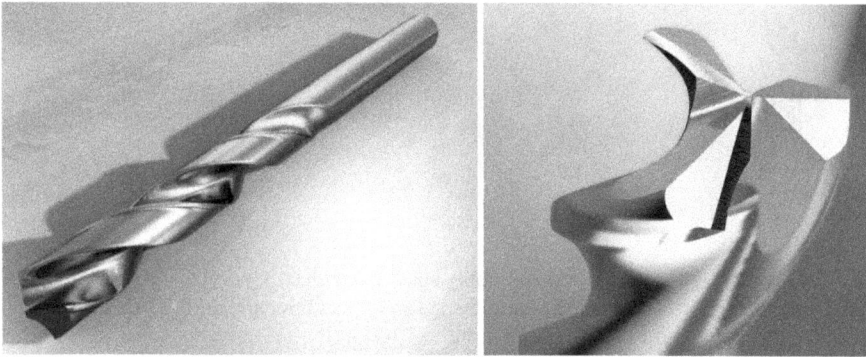

| (a) | (b) |

FIGURE 2.5 Drills: (a) two-flute and (b) three-flute.

Classification based on the number of flutes:

1. *Single-flute drills*: those having only one chip removal flute, for example, gundrills (see Chapter 3).
2. *Two-flute drills*: those having two chip removal flutes, for example, the conventional type of straight-flute and twist drills (Figure 2.5a).
3. *Multiple-flute drills*: those having more than two flutes. This drill type is commonly used for HP drilling of relatively brittle work materials, enlarging and finishing, cored, cast, or punched holes (Figure 2.5b).

Classification based on MWF (coolant) supply:

1. *Drills with an external coolant supply*: those having no special means for coolant supply.
2. *Drills with an internal coolant supply*: those having internal coolant supply holes (Figure 2.6) or passages and those having coolant supply passages separated from the chip removal passages.

Classification based on assumed force balance:

1. *Transiently balanced drills*: those having only side margins adjacent to the corners of the major cutting edges as supporting means in the radial direction. The radial stability of such drills completely relies on the force balance in drilling.
2. *Transiently balanced drills with additional supports*: those relying on the complete force balance in drilling while having additional supporting margins normally located on the

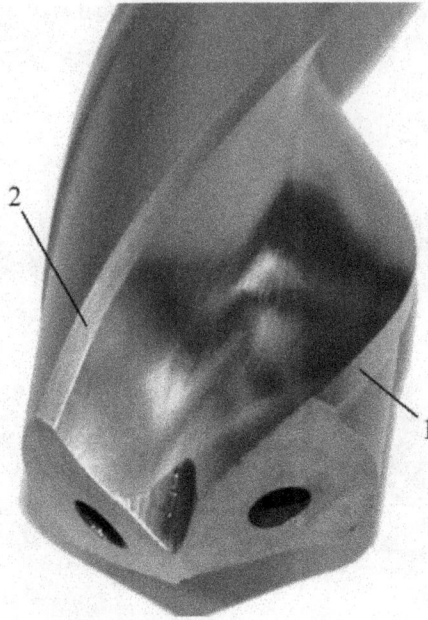

FIGURE 2.6   Twist drill with internal coolant supply holes and with (auxiliary) margins located on the heels:
(1) side margin adjacent to the corner of the cutting edge and (2) additional margin made on the heel.

heels to improve drilling stability (Figure 2.6). Although the advantages of such a design
in terms of improving drilling stability were known since the end of the nineteenth century, its wide use became feasible with improving drill manufacturing technology and
drill setting accuracy. In other words, when computer numerical control (CNC) drill OD
grinding became common, both runouts of spindles and drill holders were decreased
dramatically.

3. *Self-piloting drills*: those drills designed so that the intentionally unbalanced radial force
   generated by the cutting edge in drilling acts on the supporting elements (often referred
   to as guide pads). As a result, a self-piloting drill guides or steers itself during a drilling
   operation using the walls of the hole being drilled as the pilot surface. Chapter 3 discusses
   the self-piloting principle and various designs of self-piloting tools.

Classification based on functions and applications:

1. *Solid drills*: those making holes in a solid workpiece with no previously made holes
   (Figures 2.5 and 2.6).
2. *Spot drills*: short and rigid solid drills used to drill a starting hole (an indent) for the secondary (larger size) drill to enter, acting as a guide. An example of spot drills is shown in
   Figure 2.7.
3. *Center drills*: those for making a conical indentation in the end of a workpiece to mount it
   between centers for subsequent machining operations. An example is shown in Figure 2.8.
   Often, they are used as spot drills although it is not advisable.
4. *Trepanning drills*: drill for cutting circular holes around a center. In trepanning, also
   known as trephination, the drill cuts only a small groove instead of solid core. Trepanning
   is a great alternative to solid drilling as it requires less cutting power, and in the majority of
   cases, the cores that remain can be used to produce other parts. Figure 2.9 shows examples
   of trepanning drills.

FIGURE 2.7   Spot drill.

FIGURE 2.8   Center drill.

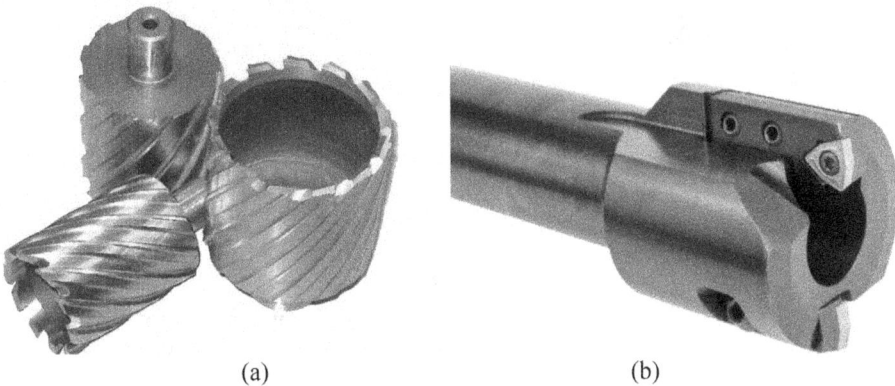

(a)   (b)

FIGURE 2.9   Trepanning drills: (a) common and (b) modern high productivity.

5. *Micro drills*: drills used for small holes mainly to drill circuit boards for electronic equipment. Often, micro drills are *pivot drills* as shown in Figure 2.10. *Pivot drills* were designed for watch and clock repair a long time ago. The pivot is the small diameter cylindrical end on the arbor that carries a train wheel (gear) in a watch or clock. The pivot was formed by turning down the two ends of the hardened steel arbor or staff. Sometimes this small end breaks off or becomes badly worn. Complete replacement arbors have never been easily available or easy to make. As a result, one way that repair people have coped is to drill a shallow hole in the end of the arbor and insert a piece of steel pivot wire. That process is called re-pivoting. The hole is drilled with a pivot drill, which is usually a short bit with a larger, long shank. The end of the drill is sharpened for optimum use in relatively hard steel. The tool design may be a spade type or, if modern, it may be a twist type. It may be made of carbon steel, or it may be HSS or carbide. Watchmakers could make their own pivot drills and replacement pivot inserts out of sewing needles. Many designs of pivot drills were developed, and most are probably not used in watch or clockwork nowadays.

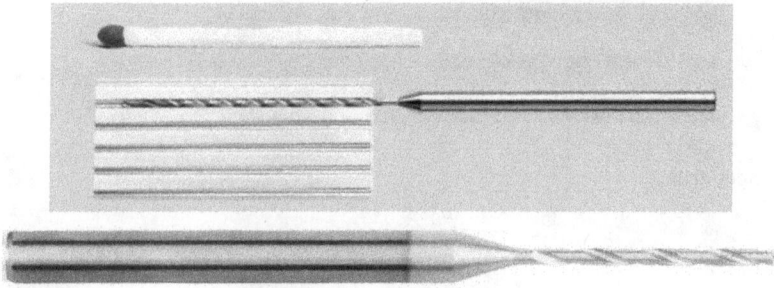

FIGURE 2.10   Microdrill.

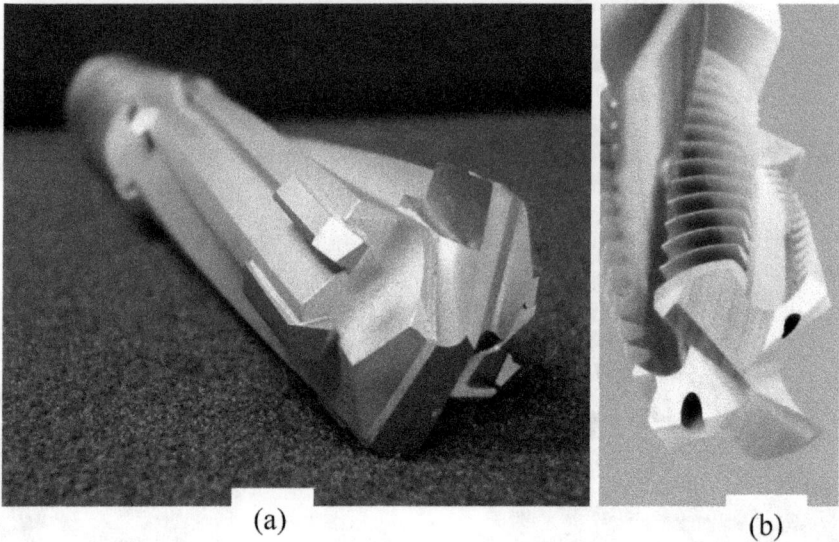

(a)   (b)

FIGURE 2.11   Combined drills: (a) drill reamer and (b) thriller (drill-thread mill).

6. *Combined drills*: although probably the simplest example of combined drills is the center drill shown in Figure 2.8, which combines the drill portion and an adjacent countersink portion, modern combined drills include the drill portion that can be combined with the reamer, cold rolling, thread, and other portions. Figure 2.11 shows two common examples. A drill reamer (Figure 2.11a) and, popular in the automotive industry, a thriller (Figure 2.11b) are used to reduce the cycle time. Thrillers enable drilling, chamfering, and thread milling to be performed with one tool, reducing the machining time.

Classification based on a specific work material a drill is made for:

A number of specific work materials such as wood, rubber, ceramics, and glass have been used for years. Nowadays, a great number of manmade materials, e.g., fiber-reinforced composites, are used in the airspace, automotive, energy, and other industries. Efficient drilling of these materials requires some specific drill materials, designs, and geometries. Although many of the drills meant for special work materials might have a very distinctive appearance, they use *the same principles* as those for metals. Figure 2.12 shows an example of woodworking drills. Figure 2.13 shows a modern hammer drill for machining holes in concrete. Figure 2.14 shows a drill for glass and ceramics.

2.1.2 NOMENCLATURE/TERMINOLOGY

Understanding the proper terminology as related to drills is a key to understanding the drill design, manufacturing, and proper applications as all specialists involved in this chain should speak the same language. This is particularly true in drilling as many lay-language and shop-convenience terms developed over many years are used so that the same drill components may have multiple names and, what is more undesirable, understood differently by different specialists. Therefore, this section provides visualization and definitions of the basic terms related to drills.

The basic terms used for straight-flute and for twist drills are the same so the basic terms related to the twist drill are considered in this section. A twist drill is defined as an end cutting tool having

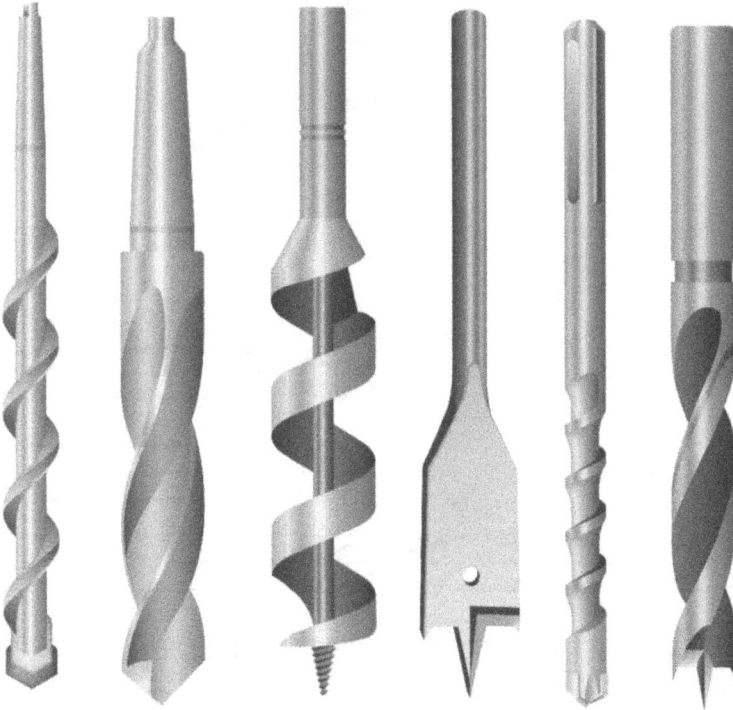

FIGURE 2.12 Wood working drills.

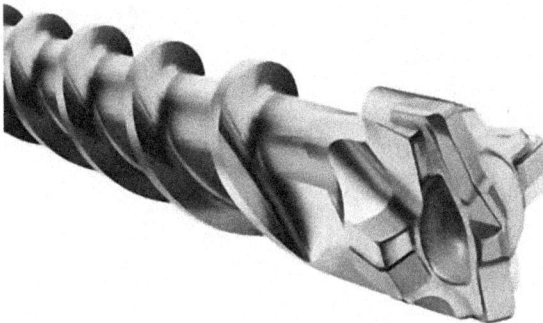

FIGURE 2.13 Hammer carbide drill for concrete.

one or more cutting teeth (often referred to as cutting lips in the trade literature) formed by the corresponding number of helical chip removal (transportation) flutes. A common twist drill is shown in Figure 2.15. It consists of the body, neck (optional), and shank.

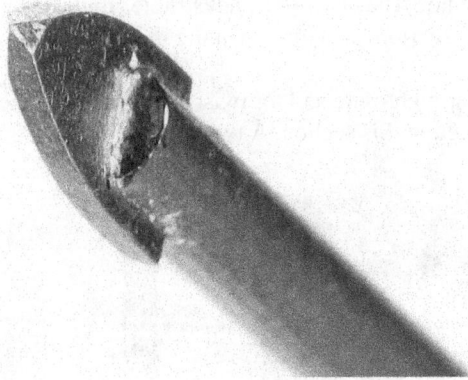

FIGURE 2.14   Glass and tile drill.

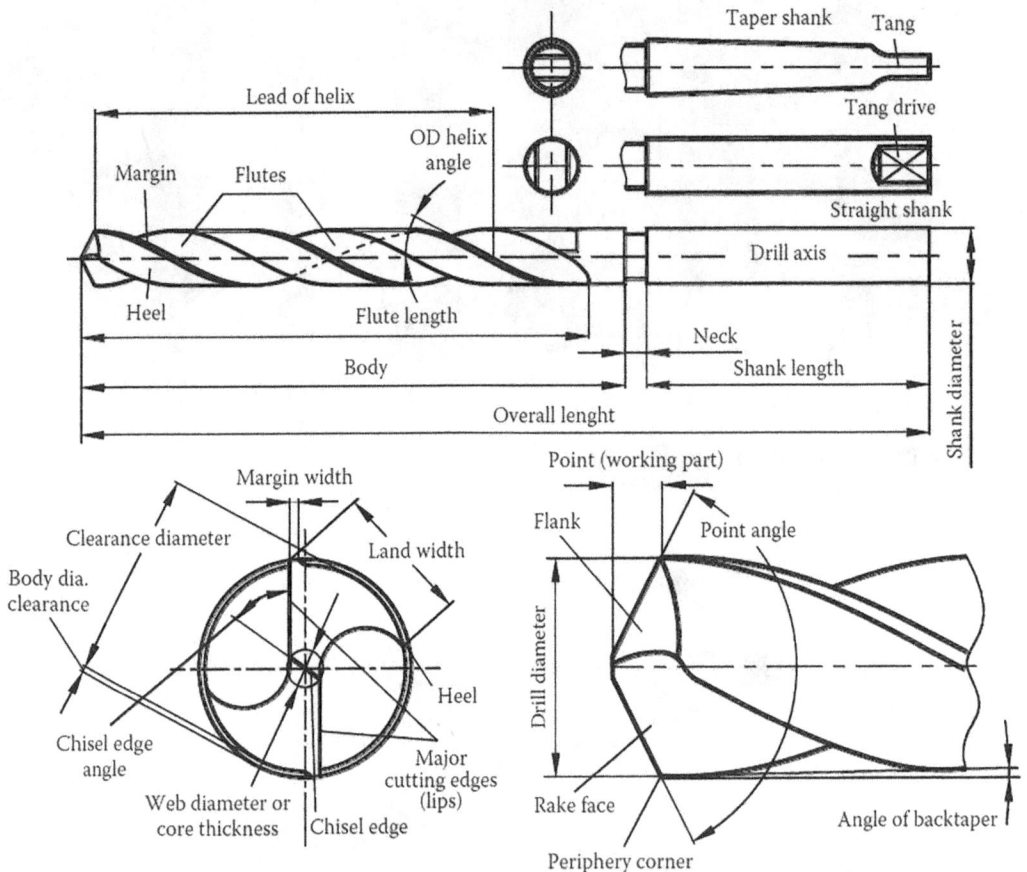

FIGURE 2.15   Illustration of terms applying to twist drills.

The working part has at least two helical flutes called the chip removal flutes. The lead of the helix of the flute depends on many factors including the properties of the work material so it varies from 10° up to 45° for high-helix twist drills (the reason for this will be explained later). The flute profile and its location with respect to the drill longitudinal axis determine many facets of twist drill performance because (1) it determines the geometry of the drill rake face (the shape of the cutting edge (lip), the rake angle and its variation along this edge, the cutting edge inclination angle and its variation along this edge); (2) it determines the reliability of chip removal, that is, chip breakage into pieces (sections) suitable for transportation and the ease of such transportation; and (3) together with the web diameter, it directly affects the buckling stability of the drill; together with the flute helix angle, it determines the torsional stability of the drill. As a result, a great number of various flute profiles have been developed, and many of them are available as applied to twist drills produced by various drill manufacturers (Astakhov 2010). Among them, the three shown in Figure 2.16 are basic ones.

Some important terms related to the twist drill design and geometry are defined using general guidelines provided by United States Cutting Tool Institute (1989) as follows:

Back taper: a slight decrease in diameter from front to back in the body of the drill.

Body: the portion of the drill extending from the shank or neck to the periphery corners of the cutting lips.

Body diameter clearance: that portion of the land that has been cut away to prevent its rubbing against the walls of the hole being drilled.

Chip packing: the failure of chips to pass through the flute during the cutting action.

Chisel edge: the edge at the end of the web that connects the major cutting edges, normally their inner ends. Note that this definition is purely geometrical as in reality, and there is more than one chisel edge as explained later in this chapter.

Chisel edge angle: the angle included between the chisel edge and the cutting drill transverse axis (a.k.a. the centerline), as viewed from the end of the drill.

Clearance: the space provided to eliminate undesirable contact (interference) between the drill and the workpiece.

Cutter sweep: the section of the flute formed when the tool that is used to generate the flute exits the cut. Often, it is called the flute washout.

Cutting tooth: a part of the body bounded by the rake and flank surfaces and by the land.

Double margin drill: a drill whose body diameter clearance is produced to leave two margins on each land and is normally made with margins on the leading edge and on the heel of the land (Figure 2.6), although any other angular location of two or more auxiliary margins is possible.

Drill axis: the imaginary straight line that forms the longitudinal center line of the drill. Often, it is referred to as the drill rotational axis. Note that each cylindrical/tapered part of a drill has its own geometrical axis. Among them, the axis of the shank is often considered

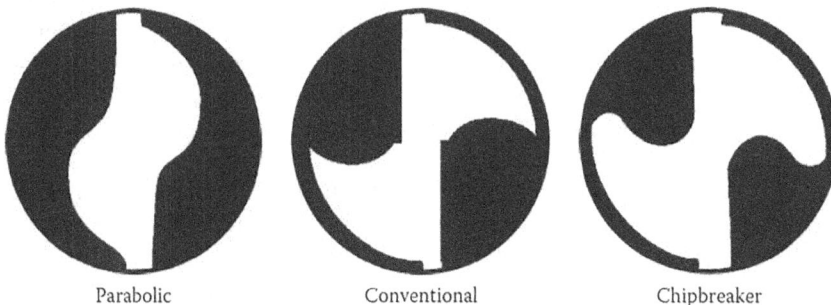

Parabolic   Conventional   Chipbreaker

FIGURE 2.16 Common flute profiles.

as the drill axis so that drill shape tolerances and symmetry of its features are considered with respect to this axis.

Drill diameter: the diameter over the margins of the drill measured at the periphery corners.

Flute length: the length from the drill point to the extreme back end of the flutes. It includes the sweep of the tool that is used to generate the flutes and, therefore, does not indicate the usable length of flutes.

Flutes: helical or straight grooves cut or otherwise formed in the body of the drill to provide the major cutting edges (often also called the lips), to permit removal of chips, and to allow metal cutting fluid (MWF) to reach the cutting zone if no other means of MWF delivery is provided.

Galling: an adhering deposit of nascent work material on the margin adjacent to the periphery corner of the cutting edge.

Helix angle: the angle made by the leading edge of the land with the axis of the drill shown as OD helix angle in Figure 2.15.

Heel: the trailing edge of the land.

Land: the peripheral portion of the cutting tooth and drill body between adjacent flutes.

Land clearance: see the preferred term, body diameter clearance.

Land width: the distance between the leading edge and the heel of the land measured at right angles to the leading edge.

Lead: the axial advance of a helix for one complete turn or the distance between two consecutive points at which the helix is tangent to a line parallel to the drill axis.

Lip (major cutting edge): a cutting edge that extends from the drill periphery corner to the vicinity of the drill center. The cutting edges of a two-flute drill extend from the chisel edge to the periphery.

Lip clearance (often referred to as relief): the clearance made from the cutting edge to form the flank surface. There can be several consecutive flank surfaces with different clearance angles as the primary clearance, secondary clearance, etc., made to clear the lip as well as to prevent interference between the flank surface and the bottom of the hole being drilled.

Lip relief angle: obsolete term for the lip clearance angle.

Margin: the cylindrical portion of the land that is not cut away to provide clearance.

Neck: the section of reduced diameter between the body and the shank of a drill.

Overall length: the length from the extreme end of the shank to the outer corners of the cutting lips.

Periphery: the outside circumference of a drill.

Periphery corner: the point of intersection of the major cutting edge (the lip) and the margin. In a two-flute drill, the drill diameter is measured as the radial distance between two periphery corners. Note that this diameter should be indicated in drill drawings, which is not always the case in reality.

Relative lip height: the difference in indicator reading between the cutting lips. Lips axial runout is another commonly used term (Chapter 7 in Astakhov (2014)).

Relief: the result of the removal of tool material behind or adjacent to the cutting edge and leading edge of the land to provide clearance and prevent interference (commonly called rubbing or heel drag) between the cutting tooth and the bottom of the hole being drilled.

Shank: the part of the drill by which it is held and driven.

Web: the central portion of the body that joins the lands. The extreme end of the web forms the chisel edge on a two-flute drill.

Web modification: modification of the web from its ordinary thickness, shape, and/or location to reduce drilling thrust, enhance chip splitting, and change chip flow direction. The simplest modification is web thinning.

Web diameter/Core thickness: the diameter of the web at the point, unless another specific location is indicated.

2.1.3 DESIGN AND GEOMETRY: ADVANCED USERS' PERSPECTIVE

This section aims to correct a methodological mistake made by the author in the first edition of the book where the development of drills geometry, design, and application item is blended together with deep detailed explanations needed for developing HP drills. As such, advanced users found it difficult to follow the text extracting portions relevant to the proper application of HP drills.

Although the classical paper on drills (e.g., Galloway 1957) contains many pieces of relevant information and thus can be of great help to many users, they include a lot of outdated information, which does not account for the abilities of the modern CNC point grinding machines (point grind variety and accuracy to reproduce important features on drills in their manufacturing) and inspection equipment including digital tool microscopes and CNC inspection machines. Moreover, old papers are not readily available for reading online even for users with access to scientific databases. Modern papers on the subject present only the final results in the vectorial/matrix/tensor forms that make these results incomprehensible to many specialists, e.g., Ren and Ni (1999).

Section 2.1.3 presents the minimal set of knowledge needed for advanced users as the number of them is much-much greater than that of the HP designers, developers, and researchers so that the success of wide implementations of HP drill rests mainly with advanced users.

2.1.3.1 Appeal

This section is written to explain what an advanced drill user, who wants to understand the application side of drills/drilling, and thus to acquire the full control of his or her drilling operations, should know about drills. The idea of this section stems from multiple feedbacks on the previous book (Astakhov 2014), where the research, design, development, and application aspects were logically blended. This, however, presents some challenges for multiple drill users who want to know only the essentials of the application side of drilling in order to select proper tools for particular applications, make reasonable justification to the selection made, inspect drills, and find and avoid drill failures at least at the level "what not to do." The authors hope that this will be of essential help to specialists and professionals in manufacturing.

The presentation of the material in this section is unique as the drill business application side has never been presented in such a practical, simple yet correct manner. In presenting material in this section, the author used a saying by Albert Einstein: "Everything should be made as simple as possible, but no simpler." The mathematics used includes only some basic trigonometrical functions such as sin, cos, and tan and only basics of geometry are used to introduce/explain in a practical manner with pictural examples the fundamentals of drill geometry and design. The most important suggestions/results to follow in application practice are formulated as *Application Rules*. The author hopes that he passed his half-way and thus would like to encourage advanced users to pass their part to achieve the astonishing results.

2.1.3.2 Straight-Flute and Twist Drills Particularities

Two basic types of drills in terms of the chip removal flute orientation about the drill rotational axis are used: straight-flute and twist drills. Straight-flute drills find wide application in industry (Figure 2.17). The effectiveness of today's straight-flute drills changed traditional assumptions about the indispensability of spiral-flute drills known as twist drills. Traditionally, the so-called die drills worked well because the work materials they were engineered to machine typically produced short chips. Chip control was a key issue in the application of straight-flute drills, which work best in materials that do not generate long, stringy chips. These include various grades of

FIGURE 2.17   Modern straight-flute drill used in the automotive industry.

FIGURE 2.18   Twist drills.

cast iron, powder metals, and medium- to high-silicon aluminum alloys widely used in the automotive industry. However, a wide implementation of new machine tools with high rpm spindles and high-pressure MWF (coolant) delivery technologies, as well as enhanced tool geometries, have expanded the application range of straight-flute drills to 10 and even more length-to-diameter ratios. These drills are cheaper than twist drills, their manufacturing and metrology are simpler, and multi-step constructions in one-pass tools for precession hole drilling are easy to implement. As a result, the use of straight-flute drills increased dramatically over recent years in the automotive and aerospace industries.

Twist drills (Figure 2.18) are the most widely used drills for general applications. When properly designed and made, the helical flute facilitates chip removal from the machining zone. The coupling of the axial force (thrust) and drilling torque enhances drilling stability and increases the allowable penetration rate compared to straight-flute drills that are particularly important in drilling difficult-to-machine work materials (explained in Section 2.1.5.5). That is why twist drills dominate in such applications.

The twist drill bit was invented by Steven A. Morse who received US Patent No. 38,119 for his invention "Improvements of Drill-Bits" in 1863. The proposed original method of manufacture was to cut two grooves in opposite sides of a round bar, then to *twist* the bar to produce the helical flutes. This gave the tool its name, which stays although the proper name for this tool should be a drill with helical chip removal flutes or simply helical flute drill. Nowadays, a flute is usually made by rotating the bar while moving it past a grinding wheel with its axis inclined at the helix angle to the axis of the bar and the profile of which corresponds to the flute profile in the normal cross section. For larger-diameter drills, other manufacturing methods of forming helical flutes in the drill body are also used.

Morse's aim in transforming a straight-flute drill into a twist drill was to enhance the chip transportation from the machining zone. The idea came from the Archimedes screw pump shown in Figure 2.19 developed around 250 BC (named after the Greek mathematician and inventor, Archimedes, 287–212 BC) and used today not only for pumping water but also in screw conveyors for transporting solids, including the chip in manufacturing plants. The use of the Archimedes screw principle in drilling brought more advantages than Morse's initial thought, that is, the net result was much greater than he was bargained for.

It is discussed later in this chapter that high periphery rake angles and chip shaping ability distinguish twist drills from straight-flute drills, allowing the machining of difficult-to-machine materials that produce the so-called long chips. Moreover, due to preferable chip shape, high feeds can be used. Reduction of the amount of plastic deformation due to a high rake angle reduces the temperature at the drill periphery so that higher cutting speeds can be used for the same tool life. A higher penetration rate achieved with twist drills is due to the torque–thrust coupling effect (discussed later in this chapter) because its existence allows greater critical axial force and torques. One should understand, however, that all these advantages can be achieved if the following parameters are selected properly: (1) the web diameter, (2) flute shape, (3) clearance angle distribution over the major cutting edge, (4) chisel edge geometry, and (5) surface roughness of the chip flutes and flank faces.

The major parameter affecting the previously discussed advantages of twist drills is the helix angle (Figure 2.15). In terms specific to twist drills, the helix angle can be found by unraveling the helix from the drill diameter, representing the section as a right triangle, and calculating the angle that is formed as shown in Figure 2.20. As follows from this figure that

$$\omega_d = \arctan \frac{\pi d_{dr}}{p_{hl}} \tag{2.1}$$

FIGURE 2.19 An Archimedes screw pump.

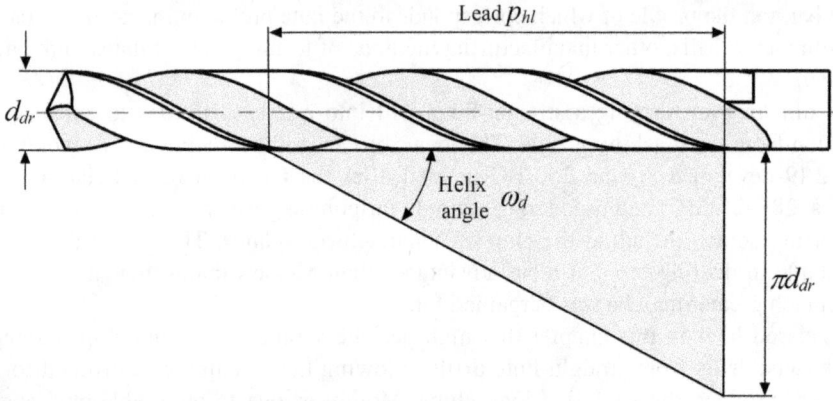

FIGURE 2.20 Unraveled helix corresponding drill diameter.

TABLE 2.1
Variation of ω_i with Radius $r_i\omega_i$

r_i/r_{dr}	$\omega_d=15°$	$\omega_d=30°$	$\omega_d=45°$	$\omega_d=60°$
1	15°	30°	45°	60°
0.8	12.1°	24.8°	38.7°	54.2°
0.6	9.1°	19.1°	31.0°	46.1°
0.4	6.1°	13.0°	21.9°	34.7°
0.2	3.1°	6.6°	11.3°	19.1°

where p_{hl} is the lead of the helix.

Normally in tool drawings, the helix angle, ω_d, corresponding to the drill outside diameter ($d_{dr}=2r_{dr}$), is indicated. Knowing this angle, one can calculate the helix angle corresponding to any point i of the cutting edge located at radius r_i as

$$\omega_i = \arctan\left(\frac{r_i}{r_d}\tan\omega_d\right) \tag{2.2}$$

It follows from this equation that the angle of helix reduces as a point of consideration is moved closer to the drill center as shown in Table 2.1.

The flutes of most twist drills have a standard helix angle (28°–32°), and this drill can be used to drill almost all work materials. For general-purpose work, twist drills having non-standard helix angles are not required. However, in cases where certain specific work materials most frequently are drilled, non-standard-helix-angle drills may have significant advantages. Also called fast-spiral drills, high-helix-angle (34°–40°) drills have an excellent chip evacuation ability that is essential in drilling difficult-to-machine work materials and in deep-hole drilling. There are also slow-helix angle (12°–22°) drills that are used to drill brass, soft bronze, and sheet materials. These materials are also drilled using straight-flute drills, having a zero-degree helix angle, because they will not tend to pull or run ahead of the feed and have less tendency to grab when opening up a hole. Reamers are normally provided with a slow helix angle.

2.1.3.3   Number of the Cutting Edges of a Common Drill

The design of a drill corresponds to its dual purpose: (1) cut cylindrical holes in a solid workpiece and (2) transport the chip formed in such cutting outside the hole being cut. Therefore, the optimality of the drill design for a given drilling conditions should always be considered keeping in mind how well this drill fulfills the stated purposes.

The cutting portion of a drill is known as the drill point. Figure 2.21 shows a common drill point for a two-flute twist drill. Such a point has six cutting edges, i.e., three cutting edges per flute, namely

1. Two major cutting edges, *1–2* and *5–6*.
2. Two chisel edges, *2–3* and *3–5*.
3. Two minor cutting edges, *1–4* and *6–7*.

As discussed in the Appendix, a cutting edge is actually a line of intersection of the rake and flank faces so that its shape is entirely defined by the shape of these faces. In metal cutting, the geometry of the cutting edge is understood as a set of certain angles, namely the rake, clearance, tool cutting edge, and inclination angles as discussed in the Appendix. Therefore, these angles in terms of their meaning and significance to drill performance should be considered, and thus unambiguously understood by drill advanced users for the generic drill point shown in Figure 2.21. Besides geometry parameters, the parameters of the uncut chip geometry, namely the uncut chip thickness (a.k.a. the chip load in industry) and its width, should also be determined as functions of the drilling feed (discussed in Chapter 1). The latter is of prime importance in the consideration of drill microgeometry discussed in Section 2.1.4. The next sections define the listed parameters for the drill cutting edges shown in Figure 2.21.

2.1.3.4   Major Cutting Edge – Point Angle

2.1.3.4.1   *Definition and Recommended Values*

As discussed in the Appendix, the clearance angle α distinguishes the cutting tool from other separating tools, while the tool cutting edge angle κ_r is the most important geometrical parameter of this tool. In drilling tools, the point angle Φ_p (or the half-point angle φ) is considered instead

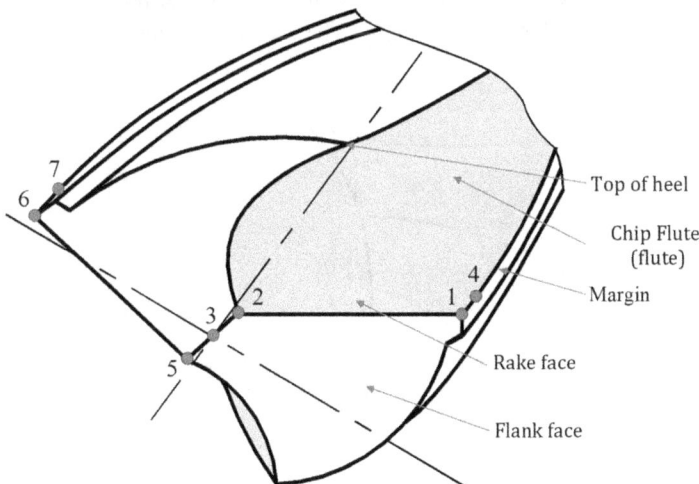

FIGURE 2.21   Components and cutting edges of the drill point of a two-flute drill.

of κ_r for convenience of measurements. The point angle Φ_p is defined as the angle between projections of the major cutting edges *1–2* and *5–6* into the reference plane P_r (explained in the Appendix). The half-point angle φ is defined as the angle between the projection of the major cutting edge into the reference plane P_r and the direction of the cutting feed/drill rotational excise. Figure 2.22 shows the representation/sense of the point Φ_p and the half-point angle φ. These angles are related as

$$\frac{\Phi_p}{2} = \varphi = \kappa_r \tag{2.3}$$

Therefore, the point angle Φ_p should be regarded at the same level as κ_r significance of which is discussed in the Appendix. Unfortunately, this is not nearly the case in the practice of drill design, selection, and applications where the real influence of this angle is underestimated.

The 118° point angle is the standard angle for most drills used in industry although the 140° point angle recently became standard for cemented carbide drills. Since the beginning of the twentieth century, 118° point angle is considered to be a good compromise for general-purpose drills for a variety of different work metals. Although carbon-steel drills were eventually replaced by HSS drills and now cemented carbide drills have taken over, and although the grinding technique has changed from hand grinding to simple fixture grinding and then to specialized drill fixture grinding, and eventually to CNC drill grinding, the standard 118° point angle has not been changed. It mysteriously suits many users who do not want to deal with the application-specific drill geometry thinking that it is up to the drill manufacturers to suggest the optimum drill geometry, drill material, and its coating for a given application.

The author's many-year experience, however, shows otherwise. Drill manufacturers always try to sell the so-called on-the-shelf products or stock items, which they can be produced in mass quantity at low manufacturing cost and with decent quality. With such drills, the lead time (the time between a purchase order and actual tool delivery) is minimal and the tool is relatively inexpensive. These real-world conveniences often overshadow the potential gains in efficiency (tool life, productivity, etc.) that can be achieved with application-specific drills. The logic of many production managers, engineers, and practitioners is rather simple: "We just buy more drills."

Experienced practitioners in the field who care about the system efficiency and the quality of the machined parts normally pay more attention to application-specific point angles. Table 2.2 shows the application-specific values for the point angle. The use of these application-specific point angles

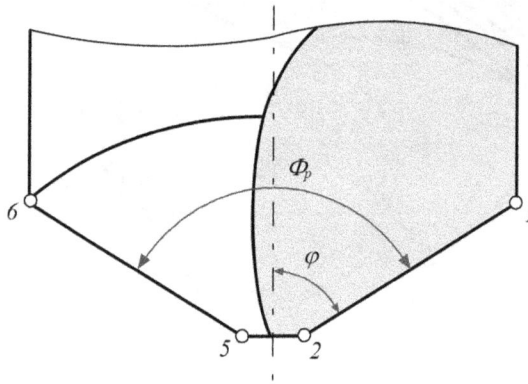

FIGURE 2.22   Representation/sense of the point Φ_p and the half-point angle φ.

TABLE 2.2
Recommended Point Angles for Drills

Work Material	Tensile Strength (MPa)	Hardness (HB)	Point Angle (±3°)
Soft steels	<400	<120	110–120
Structural steels. Ordinary carbon steels with low to medium carbon content (<0, 5% C)	<550	<200	118–125
Carbon steels with high carbon content (>0, 5% C)			
Ordinary low alloy steels. Ferritic and martensitic stainless steels	500–700	180–260	130–135
Tool steels and difficult-to-machine alloys	>700	240–320	140
Cast irons		140–200	90–100
		200–240	110–115
		>240	118–125
Brass			110–129
Copper			120–130
Aluminum alloys			
Low or no silicon (long-chip type)			130–140
High silicon (short-chip type)			115–120
Magnesium alloys			130–140
Nickel			118–125
Zink alloys			100–115
Molded plastics			118–125
Laminated plastics			125–135
Carbon			80–90

increases productivity and the quality of the drilling operation that, considered together, results in higher efficiency of drilling operations. One may argue that the use of application-specific point angles increases the cost of tools as some adjustment should be made in dill manufacturing. In reality of modern manufacturing, it is not so as many drills are application-specific, e.g., having a specific customer-defined step(s) length. To manufacture these tools and, especially, tailored HP drills, a number of adjustments should be made anyway in drill manufacturing so that the use of the application-specific point angles does not increase the cost of such tools.

Although the point angle affects almost every facet of drill performance, the further brief analysis of the drill point optimality concerns with the following: the axial/radial force ratio, uncut chip thickness, exit burr formation, and cycle time as these are of concern in applications of any drills including HP drills.

2.1.3.4.2 Axial/Radial Force Ratio and Force Balance

When the drill works, the cutting force is generated at each point of the major cutting edge. It is customary to represent the cutting force by its resultant applied at a certain point *m* of this edge. Generally, the cutting force is a 3D vector, which is commonly resolved in three orthogonal components as shown in Figure 2.23. They are as follows:

1. Power component F_p, often referred to as the cutting force in the literature. It is directed along the tangent to the rotation motion of point m. That is why is called sometimes as the tangential force. Normally, being the largest component, it is used to calculate drilling torque/power so that it directly affects the allowable penetration rate, and thus drilling productivity (Chapter 1). As a result, prime attention is paid to this component in the literature on drilling.

2. Axial component F_a, often referred to as the axial force. It is directed parallel to the drill rotation axis in the sense shown in Figure 2.23. As discussed in Section 2.1.5.1, it directly affects the allowable penetration rate, and thus drilling productivity. As a result, a lot of attention is paid to this component in the literature on drilling.

3. Radial component, F_r, often referred to as the radial force. It is directed perpendicular to the drill rotation axis. Because this component does not affect neither drilling torque nor drilling productivity, it is almost neglected in the literature on drilling as not many specialists know about its significance in the design and application of drill, and especially, HP drills.

The next critical issue to be clarified is about the ratio of the axial and radial forces. For a given cutting condition including work and tool materials, cutting speed and feed and so on, the magnitude and direction of the power component of the cutting force F_p as well as the point of its application m on the cutting edge *1–2* does not depend on the point angle. Moreover, under the same conditions, the magnitude of the so-called horizontal force, F_n, perpendicular to the cutting edge at point m is also independent of point angle. However, because the direction of the major cutting edge *1–2* varies with the half-point angle, φ, the direction of F_n also varies to keep it perpendicular to the major cutting edge. Figure 2.24 visualizes these considerations. As the direction of F_n varies with φ, the ratio of the axial and radial forces also varies as

$$\frac{F_a}{F_r} = \tan \varphi \qquad (2.4)$$

As follows from this equation, the radial component increases and the axial component decreases when the point angle decreases. This can be a valuable means at tool/process designer's disposal

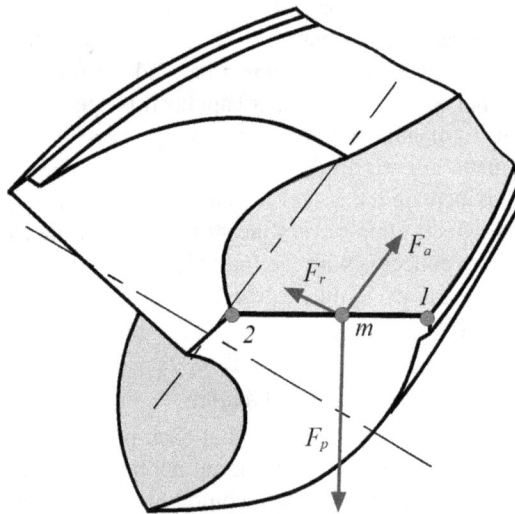

FIGURE 2.23   Force components acting on the major cutting edge *1–2*.

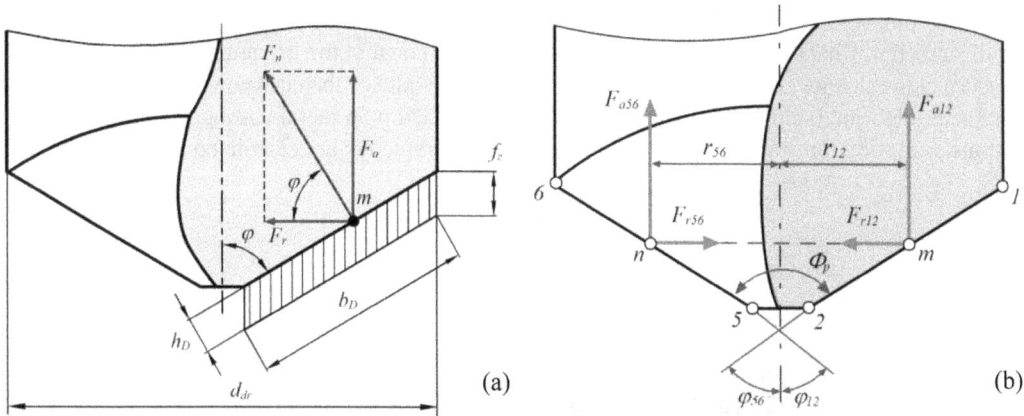

FIGURE 2.24 Axial and radial forces: (a) ratio and (b) force balance.

to increase the allowable drill penetration rate normally restricted by the axial force particularly for long drills. According to Figure 2.24a and Eq. (2.4), when point angle $\Phi_p=90°$, then $\varphi=45°$, $F_r=F_a=0.7F_n$; when $\Phi_p=118°$, then $\varphi=59°$, $F_r=0.6F_a=0.52F_n$; when point angle $\Phi_p=140°$, then $\varphi=70°$, $F_r=0.36F_a=0.34F_n$.

There are two major restrictions on increasing the point angle:

1. The smaller the point angle, the more symmetrical (with respect to its rotation axis) the drill should be. The word "symmetrical" implies that the force balance shown in Figure 2.24b is almost perfectly maintained. To maintain this balance, the following conditions should be met:
   - The power and radial forces on the major cutting edges should be equal, i.e., $F_{p12}=F_{p56}$ and $F_{r12}=F_{r56}$.
   - Radial forces F_{r12} and F_{r56} should share the same line of action as shown in Figure 2.24b.
   - Radii r_{12} and r_{56} should be equal.
      These conditions suggest that the half-point angles, φ_{12} and φ_{56}, of the major cutting edges 1–2 and 5–6 should be equal. A small violation of drill symmetry in a drill with small point angle may result in a significant unbalanced radial force that lowers tool life and ruins the quality of drilled holes. Although the wide use of modern CNC drill grinders significantly improved drill symmetry, it is still not perfect depending on the drill head holding conditions on point grinding. Among others, the most important parameter to be maintained is the lip-height variation (Astakhov 2014).
2. The chip formed by the major cutting edge moves/flows in the direction approximately that of the normal force F_n (Figure 2.24a). The smaller the point angle, the greater the chance that this chip flow interferes with the side wall of the flute creating an additional force acting on the drill, and thus can potentially violate the discussed force balance. Such an interference causes poor conformance of the chip and flute shapes, thus creating problems with chip transportation over the flute. The problem with chip flow occurs in drilling of ductile and tough work materials known as long-chip materials. To improve chip flow conditions, the point angle is increased so that the formed chip is directed more to the hypothetic center of the flute. For short-chip work materials, e.g., cast iron, zinc alloys, the chip breaks into fragments before reaching the wall of the flute so that the discussed interference is not a problem so smaller point angles can be used.

2.1.3.4.3 Uncut Chip Thickness

As discussed in Chapter 1, the nominal thickness of cut known in the literature as the uncut (undeformed) chip thickness or chip load, h_D, and the nominal width of the cut known in the literature as the uncut (deformed) chip width, b_D, are two important factors in metal cutting affecting its major outputs as cutting force and tool life. In drilling, these parameters are calculated as (Figure 2.24a)

$$h_D = f_z \sin \varphi \qquad (2.5)$$

$$b_D = \frac{d_{dr}/2}{\sin \varphi} \qquad (2.6)$$

where f_z is the feed per tooth (mm/rev).

It follows from Eqs. (2.5) and (2.6) that when the point angle decreases, the uncut (undeformed) chip thickness decreases that sets less pressure on the unit length of the cutting edge (actually, the drill rake face) and thus may result in greater tool life or, alternatively, may allow a higher feed per tooth and thus a higher drill penetration rate. On the other hand, the proportionally wider chip is formed by the cutting edge that requires special means of its handling (both breaking and transportation over the chip flute). Therefore, there is an optimal point angle for a given work material and drill parameters (tool material and geometry). The latter can be adjusted in a rather wide range to achieve higher drilling efficiency.

As discussed above, small point angles not only cause wider chips but also significantly change the direction of chip flow. Special shapes of the chip flute as discussed in Section 2.1.5.10 should be used to handle the formed chip.

2.1.3.4.4 Exit Burr and Delamination

The ISO-13715 standard defines a burr as the deviation outside the ideal geometrical shape of an edge. Ko et al. (2003) and Ko and Chang (2003) distinguished entrance burr and exit burr in drilling operations. According to them, the entrance burr is formed around the edges of the hole being drilled when the drill enters the workpiece whereas the exit burr forms when the drill exits the hole. These two places of burr occurrence are discussed by many other researchers. Their results are summarized by Sreenivasulu and Srinivasa Rao (2019).

In the author's opinion, the entrance burr should not form at all with quality components of the drilling system. It is true that in some cases of drilling, the entrance burr forms due to drill wandering, commonly referred to as drill walking. It happens, however, if and only if the following is the case:

- Severe problems with drill point design, e.g., due to lack of its self-centering ability, and/or manufacturing quality, e.g., the drill excessive runout and lip-height variation.
- Problem with non-rigid machining system and/or workpiece clamping.

The exit burr forms due to the mutual action of the portions of the major cutting edges adjacent to the drill corner and the axial force. The burr starts to form as the drill approaches the drilled hole exit. The section of the bottom of the hole being drilled becomes thinner so it deflects under the action of the axial force and then plastically deforms. When the drill exits the workpiece, the remaining material deflects creating an exit burr. Its appearance is shown in Figure 2.25.

Burr formation in drilling is one of the serious problems in precision engineering and mass production where the productivity of advanced manufacturing systems is often reduced due to additional deburring operations. Therefore, understanding the drilling burr formation and its dominant parameters is essential for controlling the burr size.

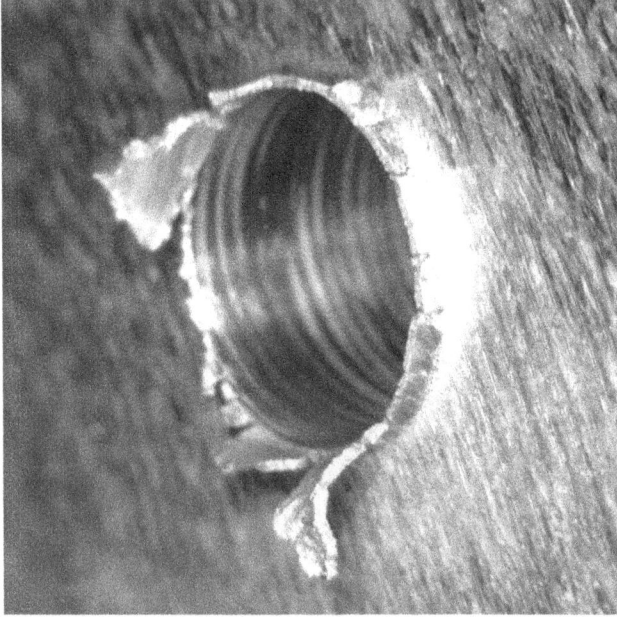

FIGURE 2.25 Example exit burr.

Standard ISO13715 "Technical product documentation – Edges of undefined shape – Indication and dimensioning" states that in technical drawings, the ideal geometric shape is represented without any deviation and, in general, without considerations of the conditions of the edges. Nevertheless, for many purposes (the functioning of a part or out of safety considerations, for example), particular conditions of the edges need to be indicated. Such conditions include those of external edge free from burr or those with a burr of limited size, and internal edges with passing.

For internal edges, found in drilled holes, the passing deviation is introduced by this standard. It is defined as the deviation outside the ideal geometrical shape of an edge defined by two tangents (meeting at point *1*) outside straight lines of the adjacent feature of the zone of the undefined edge. This deviation is shown by *2* in Figure 2.26. It is pointed out that a burr or flash can be considered to be a special case of external passing (Figure 2.27).

As explained by Pres et al. (2014), due to the diversity of forms, the burr shape classification is difficult. The appearance of the burr is influenced by many drill design and drilling operation parameters/factors. Standard ISO 13715 defines only the burr height H_p as shown in Figure 2.28. As can be seen, the measurement of the burr height is carried out in a direction perpendicular to the ideal geometrical shape of an external edge of the workpiece.

For holes where the burr is critical, H_p is shown using the special graphical symbol and a number as shown in Figure 2.29. In this figure, H_p=+0.3 mm on the left face of the hole and H_p=+0.3 mm on the right.

Experience shows that in drilling, particularly when deburring operations are applied, not only the height, H_b, but also the burr thickness, T_b (see Figure 2.30) is of importance. Studies have shown that both H_b and T_b increase when the point angle increases. As an example, in drilling of 304L steel, when the drill exits the workpiece, a cup, as a part of the bottom of the hole being drilled, is pushed/fractured forward by the axial force applied by the chisel edge. The small amount of the remaining work material is deformed forward by the drill corners so that an exit burr is formed as

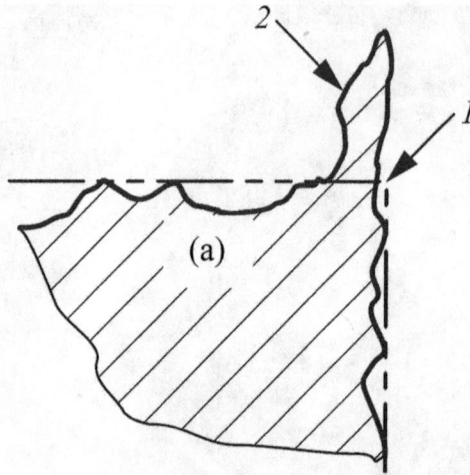

FIGURE 2.26 Passing on an external edge (flash or burr).

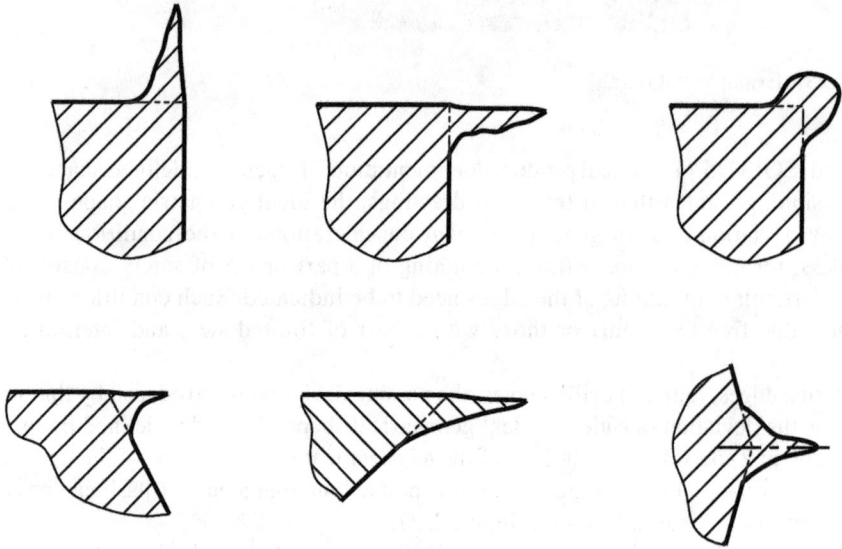

FIGURE 2.27 Examples of passing on external edge (burr or flash).

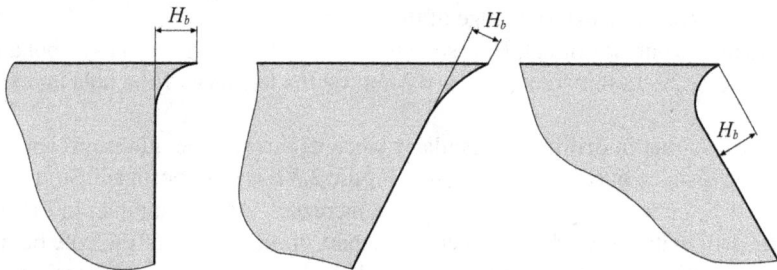

FIGURE 2.28 Sense of H_p measurements.

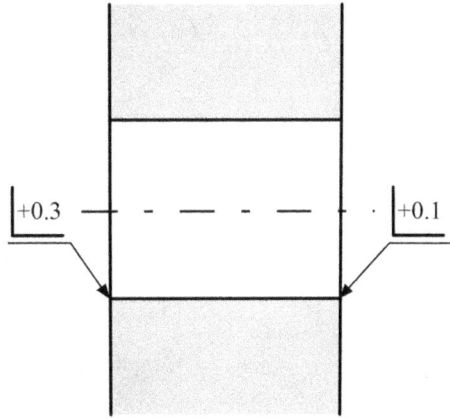

FIGURE 2.29 Example of the representation of H_p in part drawings.

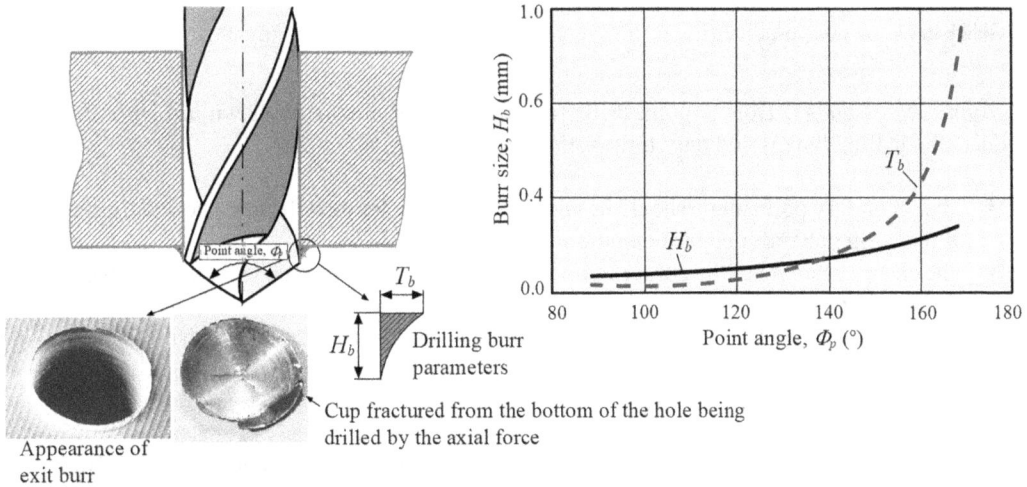

FIGURE 2.30 Showing the appearance of the exit burr, fractured cup, and the influence of the point angles on the drilling burr parameters.

shown. Figure 2.30 also shows the influence of the drill point angle on the burr parameters in drilling ANSI 304L steel (Kim and Dornfeld 2002).

As follows from Figure 2.30, smaller point angles are beneficial in terms of reduction of the exit burr parameters. Often, it is not feasible to use smaller point angles in many real-life applications as drills with small point angles have above-discussed severe performance disadvantages. To solve this problem, and thus keeping the exit burr within the allowable limits, the well-known modification of the standard point grind of a carbide drill can be used. It is shown in Figure 2.31. This drill has a primary point angle of 140° and the secondary point angle of 70°–75° to minimize the existing burr parameters as the value of the point angles in the region of the drill corner is of significance in terms of burr formation. Section 2.1.3.8.5 provides an explanation of its cutting kinematics. In the scientific literature, this modification is known as the double-point drill and in trade literature it is known as corner breaks. Its implementation increases the tool life and reduces the drilling exit burr when drilling a wide variety of engineering materials (Pilny et al. 2012).

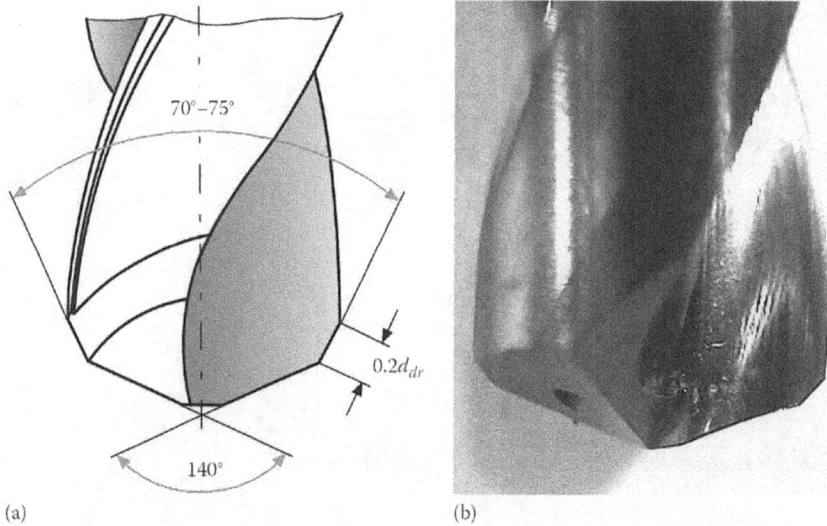

FIGURE 2.31 A double-point drill: (a) the point geometry and (b) a real drill (right).

Application rule #1: Double-point drills (i.e., with corner breaks as shown in Figure 2.31) can reduce the drilling burr if and only if the following conditions are met:

- The secondary point angle should be approx. twice smaller than the first one and the length of the corner chamfer edge should be at least $0.2d_{dr}$;
- The clearance angle of the corner chamfer edge should be at least equal or greater than that of the major cutting edge;
- The drill lip-height variation and runout should be as recommended for HP drills (Astakhov 2014).

The selection of the optimal point angle tends to be the real challenge of machining composite parts. A drill cutting through a metal part simply has to remove the material and clear the hole. By contrast, a drill cutting through a layered composite structure is likely to push the layers ahead of it, producing unacceptable delamination on the exit side (Khashaba 2012). The problem is that there are a great number of composite and plastic materials with considerably irreverent drillability that require a special drill geometry, primarily point and rake angles for a given application (Davim and Reis 2003, Davim et al. 2004a, b).

The quality of the drilled holes such as roughness/waviness of its wall surface, roundness, and axial straightness of the hole section causes a number of problems in service life of composite parts. As pointed out by Khashaba (2012), bolting and riveting are extensively used as a primary method of forming structural joints of composite parts in the aerospace and automobile industries. High stress on the rivet, for example, is often found as the root cause of its failure. Microcracking and delamination left after drilling significantly reduce the composite strength. Therefore, the quality of the drilled holes is found to be critical to the life of the riveted joints. Reduced mechanical properties of fiber-reinforced plastics (known as FRP), due to the stress concentration caused by the softening and re-solidification of the matrix material that has different thermal properties than the fiber, is another problem that occurs in drilling of FRP. The low thermal conductivity and sharp temperature gradients in FRP lead to thermal damages and burning of the matrix.

Damages associated with drilling FRP composites were observed at the entrance as well as at the exit of the drilled hole, in the form of peel-up and push-out delaminations, respectively (Khashaba 2012).

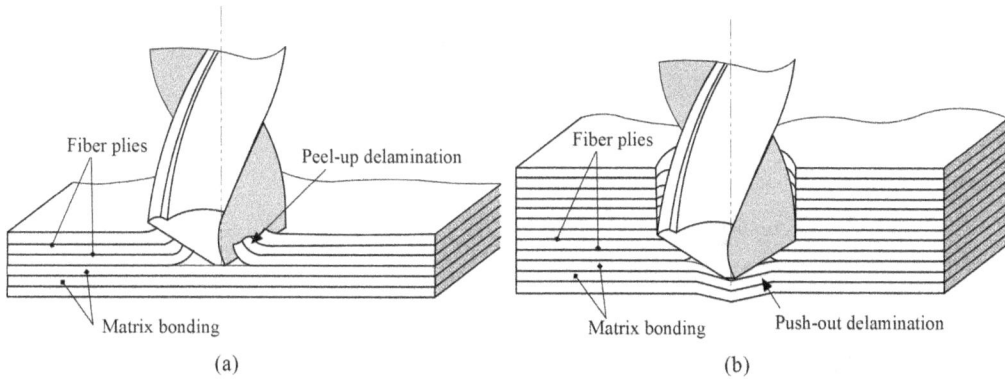

FIGURE 2.32 Delamination in FRP: (a) peel-up at the drill entrance and (b) push-out at the exit.

Peel-up delamination occurs as the drill enters the laminate and is shown schematically in Figure 2.32a. After the cutting edge of the drill makes contact with the laminate, the cutting force at the drill corners is the driving force for peel-up delamination. It generates a peeling force in the axial direction through the slope of the drill flute. The flute tends to pull away the upper laminas and the material spirals up before it is machined completely. This action results in separating the upper laminas from the uncut portion held by the downward acting thrust force and forming a peel-up delamination zone at the top surface of the laminate. The peeling force is a function of tool geometry (the drill point angle) and friction conditions at the chip flute defined by both the surface conditions of the drill rake face and flute profile.

Push-out delamination occurs as the drill reaches the exit side of the material as is shown schematically in Figure 2.32b. As the drill approaches the end, the wall thickness of the material becomes smaller so that its resistance to deformation due to the axial (thrust) force imposed by the drill reduces. At some point, the thrust force exceeds the interlaminar bond strength causing an exit delamination zone as the tool pierces through the exit side. This happens before the laminate is completely penetrated by the drill. In practice, it has been found that the push-out delamination is more severe than that of peel-up (Khashaba 2012). Large point angles are beneficial at the drill entrance, while small ones cause little damage at the drill exit (Heisela and Pfeifroth 2012).

An intelligent selection of the proper parameters of drill geometry and suitable cutting conditions can significantly reduce the push-out delamination by lowering the axial force (Abrao and Aspinwall 1996, Ahmad 2012). Unfortunately, the optimum geometrical and machining conditions found for one FRP may not be suitable for a seemingly similar FRP. For example, the standard point angle of 118° was found to be optimal in drilling AS4/PEEK CFRP, while the effect of the point angle was found to be marginal in drilling T300/5208 CFRP (Ahmad 2012). In the author's opinion, however, the influence of the chisel edge on the axial force was not properly accounted for that led to the obtained result. In any case of drilling, the high rake angle (and thus fast helix angle) combined with a high clearance angle of the major cutting edge seem to be beneficial in drilling of carbon FRPs. A small point angle tends to produce better exit quality. However, too small a point angle can give the tool poor strength. For CFRP, the optimal compromise seems to be a point angle of 90°.

Although a small point angle combined with an aggressive rake angle of the major cutting edge is beneficial for FRPs, it may not be suitable for other plastics. For example, according to Basic Car Audio Electronics Co., the best results in drilling acryl (transparent hole wall with no feed marks; no chipping on the drill entrance and exit) as shown in Figure 2.33 are achieved when a small point drill having a zero T-hand-S rake angle over the entire major cutting edge and standard helix angle (30°) is used.

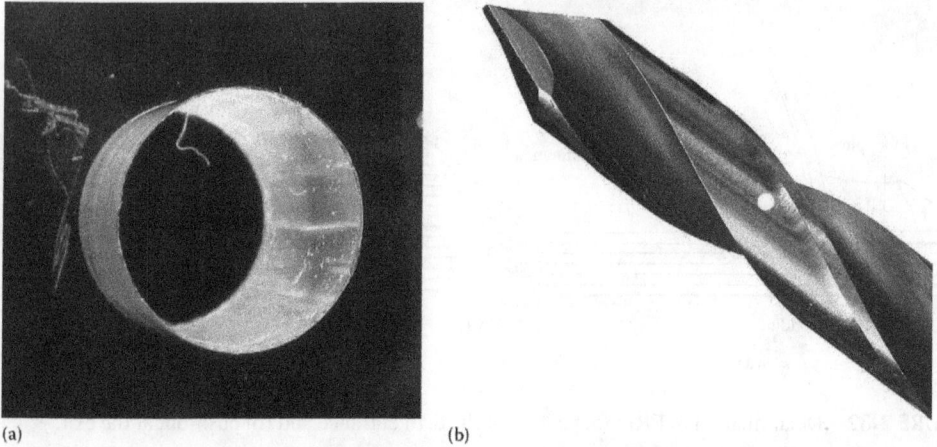

FIGURE 2.33 Drilling of acryl: (a) achievable quality and (b) the optimal geometry drill.

2.1.3.4.5 Other Important Considerations

2.1.3.4.5.1 Starter Drills   To prevent the formation of a significant bell mouth (a tapered enlargement of the hole diameter with rough surface) at the hole entrance, drill wandering, commonly referred to as drill walking, should be minimized. This usually involves starting a hole with a starter drill, known as a spot drill, followed by a successively longer secondary drill until the desired hole depth is attained. Spot drilling creates a location for the secondary drill to enter, acting as a guide. A spot drill (see Figure 2.7) is short and rigid, with a very short flute. It has just enough point geometry to drill an indent in the workpiece.

There are just two choices to spot the workpiece: the center drill or the spot drill. Despite the popularity in the job-shop environment, center drills (see Figure 2.8) are not designed for starting holes. Their purpose is to drill holes with 60° angles in the ends of parts. These holes are used to secure parts in the tailstock of a lathe or to mount a workpiece between centers for a grinding operation. The center drill is inefficient as its thick web puts a great deal of pressure on the tool. Pressure leads to heat, and heat leads to premature drill failure. The tip of the center drill is also prone to breaking.

The best tool for starting a hole is a spot drill, which is also referred to as a spotting/centering/starting drill. Spot drills come in several styles, but they all share similar characteristics. All are short and rigid and can handle much higher feed rates than comparably sized center drills. Spot drills made of HSS or carbide and normally have zero back taper.

According to a common notion, which appears in many trade/technical literature sources and tool manufacturers' catalogs, the point angle of a spot drill should be greater or at least equal to that of the drill to follow. If this condition is not met, a damage to the drill due to the "shock loading" occurs over some regions adjacent to drill corners as shown in Figure 2.34a. According to this notion, a drill should enter a previously drilled spot with a slightly larger angle than its point. As a result, its chisel edge comes first to contact with the conical hole made by the spot drill (Figure 2.34b).

Experience, however, shows that a condition when the chisel edge makes the initial drill contact as shown in Figure 2.34b is not the greatest as a mismatch of the shape left by the chisel edge of the spot drill and that of the drill to follow can create some problems including drill vibration. Moreover, standard spot drills have a point angle of 90° (Figure 2.35). A more detailed analysis of the recommendation shown in Figure 2.34 reveals that it stems from old times when drills were ground by hands or by hand-operated simple grinding accessories/fixtures. A common outcome of such a grinding was a great lip-height variation. When this is the case, only one major cutting edge

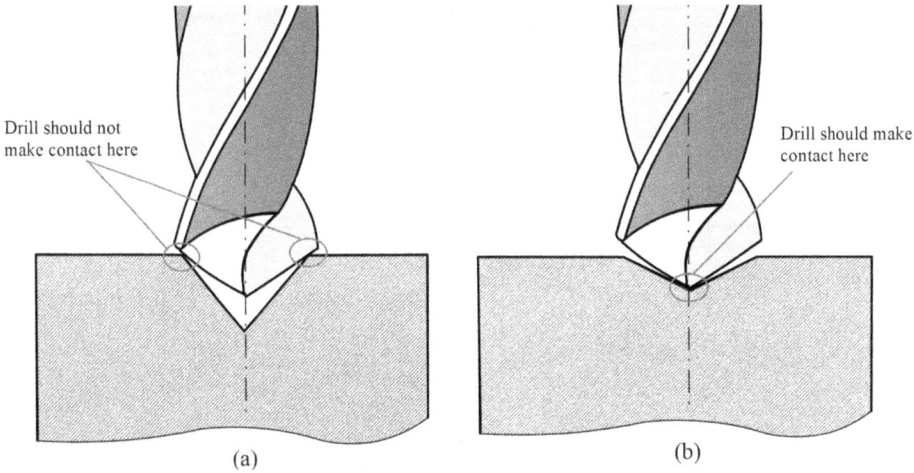

FIGURE 2.34 Common recommendations for selection of the point angle of a spot drill: (a) improper is when the point angle is greater than that of the secondary dill and (b) proper (by the prevailing notion).

FIGURE 2.35 NC spot drill (V3090) 2-flute, point angle 90° by UniCut Co.

comes into contact with the hole made by a spot drill having a smaller point angle (see Figure 2.34a) that can damage this edge.

Nowadays, however, most of drills and all HP drills are ground and re-sharpened using modern CNC drill point grinding machines so that the lip-height variation is not an issue. Therefore, the proper spot drilling application should be as shown in Figure 2.36. It is to say that a standard 90° spot drill having the diameter $d_{con} = 0.5d_{dr}$ should be used. As such, both major cutting edges of the drill to follow are engaged in drilling by their middle points that assure smooth entrance of the second drills with no shocks.

2.1.3.4.5.2 Restriction on the Point Angle Special cases can occur in practice where there are some restrictions on the values of the point angle. It often happens in the automotive industry. An example is shown in Figure 2.37. As can be seen, the distance between the hole exit and the wall (the next vain) is small so that the choice of the point angle is dictated by this distance h_g.

To deal with the problem, special Multifaced Drills (known as MFDs) can be used as discussed by the author earlier (Astakhov 2014).

2.1.3.4.5.3 Cycle Time The drilling operation cycle time includes the length/depth of the hole to be drilled and depends on the length of the drill point. Moreover, the cutting feed is

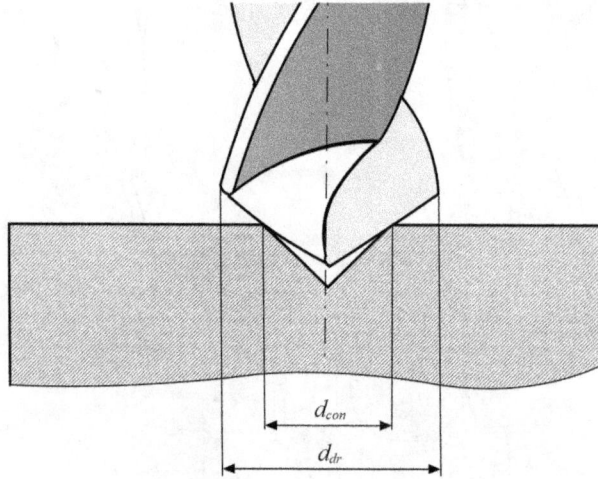

FIGURE 2.36 Proper engagement of a drill in the spotted hole.

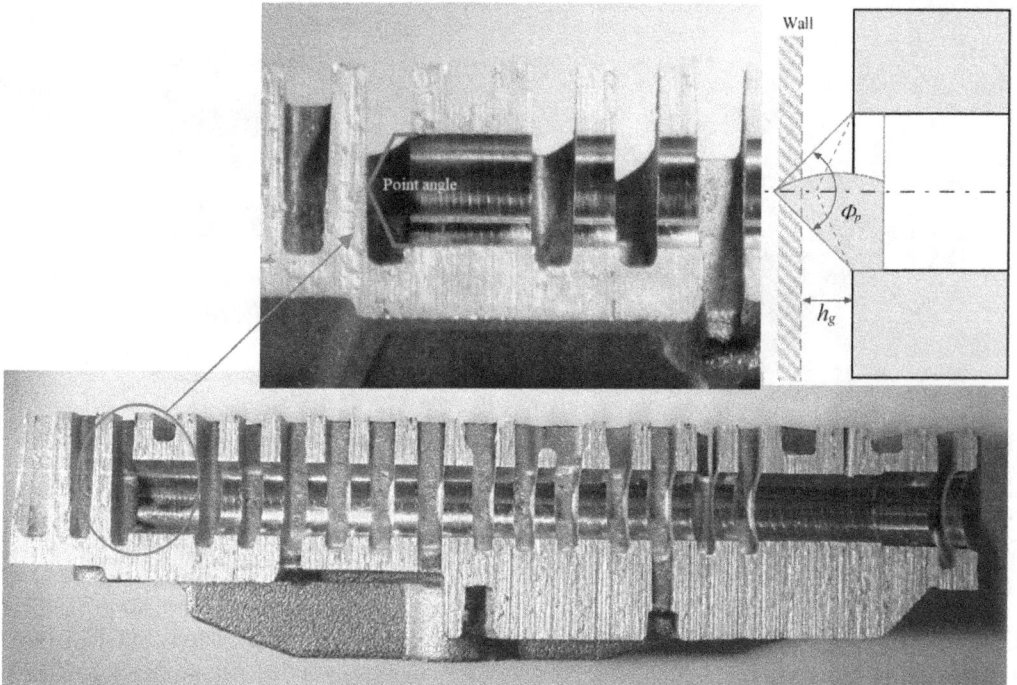

FIGURE 2.37 Typical hole made in an automotive aluminum part (the valve body of an automatic transmission).

significantly reduced from the beginning of drilling till the whole point is emerged in the hole and reduced again on the drill exit till the whole drill point is out of the drilled hole. Figure 2.38 shows clearly that the drilling cycle time is greater for a drill of 90° point angle than that for a drill of 140° because point length l_{p-90} is greater than point length l_{p-140}. Although one may think that this difference is neglectable small, it makes a huge difference in mass production, e.g., in the automotive industry.

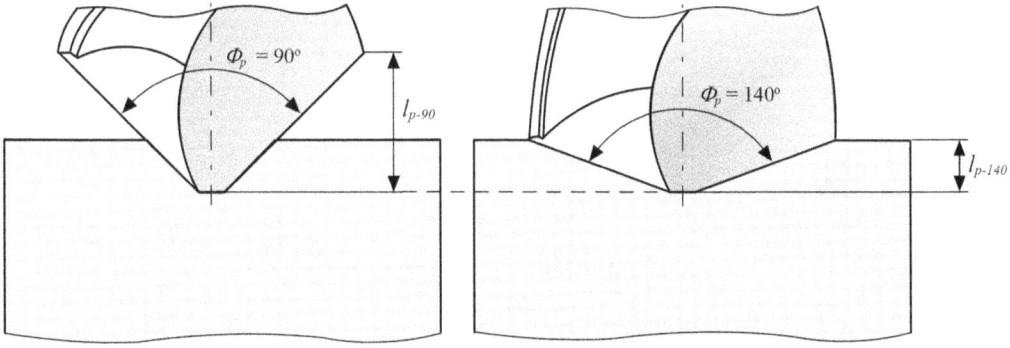

FIGURE 2.38 Drill point length for 90° and 140° drills.

2.1.3.5 Major Cutting Edge - Clearance Angle

It is discussed in the Appendix that the clearance angle is the only distinguished feature of *the cutting edge* of any type of cutting tool. Its sole purpose is to clear the motion of the cutting edge, i.e., to prevent any unwanted contact between the tool and machined surface. Consequently, whereas the other angles of this edge can be positive, neutral, or negative, the clearance angle *MUST* always be positive to clear the motion of the cutting edge. According to the author's experience, when it comes to drilling tools, in general, and to drills, in particular, the clearance angle is probably the most important yet least understood, and thus not properly assigned parameters of the tool geometry. Therefore, this section aims to resolve these issues, i.e., to explain in a unique manner the true meaning of the clearance angle in drills, its assigning in tool drawings, inspection/measurement, and recommended values for optimal drill performance.

2.1.3.5.1 Clearing Drill Motion

The first step is to understand what does it mean "to clear the motion of the cutting edge." As explained in Chapter 1, in operation, the drill rotates about its axis and translates along this axis. To understand the result of this combined motion, let us consider a point a having the radius of rotation r_a as shown in Figure 2.39. As this point rotates and translates, its trajectory is a helical line having constant parameters: radius r_a, velocity v_a, and pitch equal to the cutting feed $p = f$ as shown in Figure 2.40. The latter is true as the drill is a sold body so that each of its point advances in the axial direction by the distance equal to the cutting feed f over one revolution of the drill.

When the combined motion of the whole major cutting edge *1–2* is considered, the path traced by this edge has the shape of helicoid as shown in Figure 2.41. Its pitch is equal to the cutting feed (the axial advance per one revolution). The clearance angle is meant to prevent the contact of any point of the flank face with the depicted path. Obviously, if a drill has two major cutting edges, two helicoid surfaces are the case, three cutting edges – three and so on. Regardless of how many helicoid surfaces are traced, the outside diameter of helicoids is equal to the drill diameter, its inside diameter is equal to the web diameter, and its pinch – to the cutting feed. In further considerations, the helicoid formed by the major cutting edge *1–2* will be referred to as the surface of cut.

To understand the concept of "clearing path," one needs to consider a cylindrical cross section of the drill that is the cross section by a cylinder having the axis coincident with the drill rotation axis. Figure 2.39a shows such a cross section drawn through point a on the major cutting edge. Let us consider only a part of this cylindrical cross section, namely cross section *A-A*, which covers the drill tooth from the entrance point a to exit point b on the edge of the heel. This is shown in Figure 2.39b to facilitate understanding. Unfolding this cross section into a plane shown in Figure 2.39a, one can obtain the various unfoldings of the shape of the drill flank face.

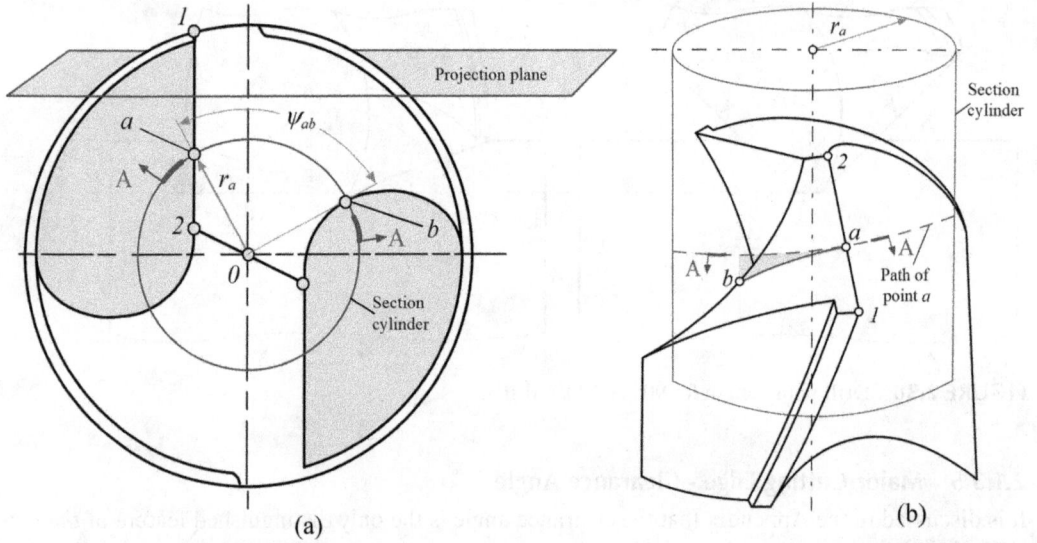

FIGURE 2.39 Cylindrical cross section A-A through point *a* on the major cutting edge *1-2*: (a) face view and (b) 3D representation.

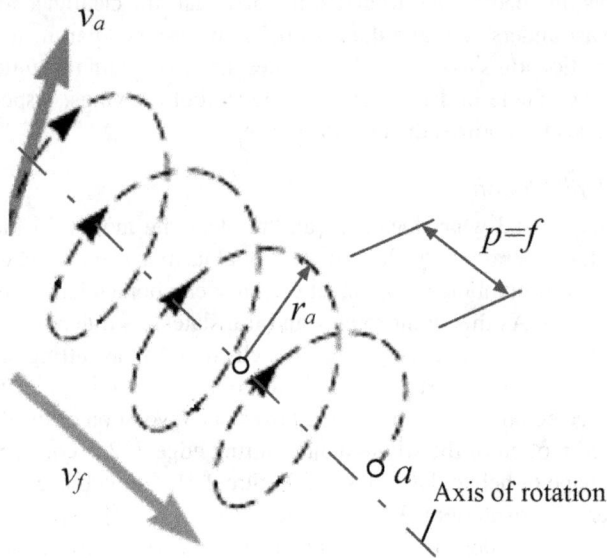

FIGURE 2.40 Trajectory of point *a*.

The common shapes are shown in Figure 2.42. In these figures,

- Surface of cut in this figure is the trace of the surface of cut, i.e., the line of its intersection with the cylindrical cross section *A-A*, is a straight line as it follows from the basic properties of the helical surface (Figure 2.20). This line is inclined at angle ξ_a calculated as

$$\xi_a = \arctan \frac{f}{2\pi r_a} \qquad (2.7)$$

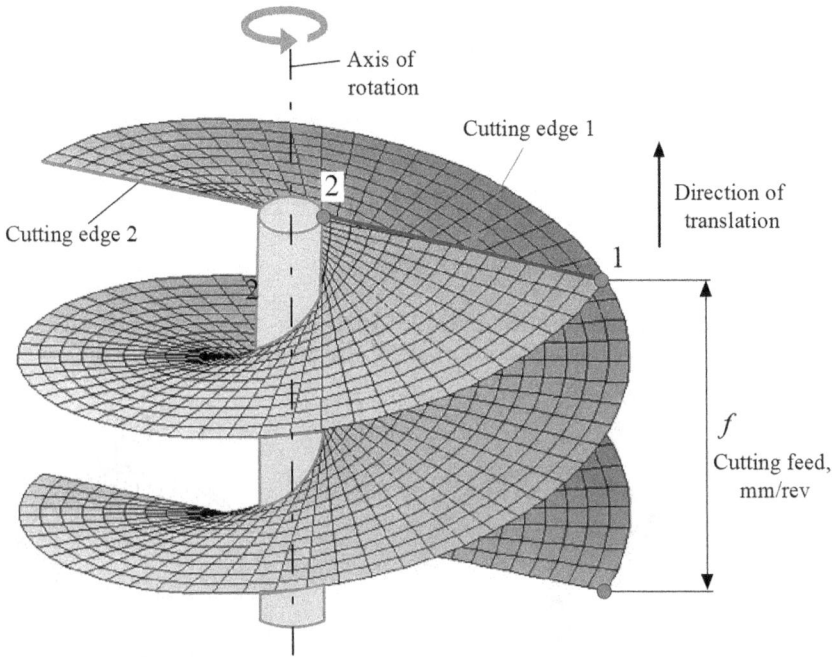

FIGURE 2.41 Helicoids as the trajectories of the drill major cutting edges.

- Δf is a part of f for length ab, i.e.,

$$\Delta f_a = f \frac{\psi_{ab}}{360°} \qquad (2.8)$$

where ψ_{ab} is the central angle between points a and b as shown in Figure 2.39.
- q_a is the decline of the flank face from point a to point b. Its meaning is clearly indicated in Figure 2.43, where the point of a dial indicator is set at 0 at point a and then the drill is rotated by angle ψ_{ab} so that the point of the indicator is reached point b. The reading of the indicators is equal to q_a.
- α_α is the drill clearance angle at point a in the cylindrical cross section $A\text{-}A$.

As shown in Figure 2.42, the flank face in the unfolded section $A\text{-}A$ can be a straight line when the flank face is helical, convex surface when the flank face is a conical surface, or concave when the flank face is planar. In any case, the contact of the drill and the surface of cut must occur only at point a, i.e., at the drill major cutting edge. A clear gap between other point/areas of the flank face should be the case regarding any particular type/method of flank face grind. Otherwise, drilling becomes impossible.

Any contact of the flank face points/areas with the surface of cut (besides the cutting edge) is known as *interference*. It follows from Figure 2.42 that to avoid interference, the following condition for any point i on the major cutting edge 1–2

$$q_i > \Delta f_i = f \frac{\psi_{ii'}}{360°} \qquad (2.9)$$

where $\psi_{ii'}$ is the central angle in the sense shown in Figure 2.39 for point a.

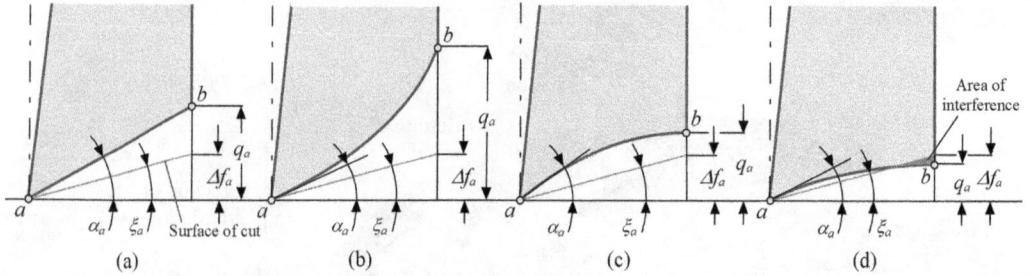

FIGURE 2.42 Unfoldings of the cylindrical cross section A-A into a plane having different shaper of segment a-b: (a) straight, (b) convex, (c) concave when $q_a > \Delta f_a$, and (d) concave when $q_a > \Delta f_a$.

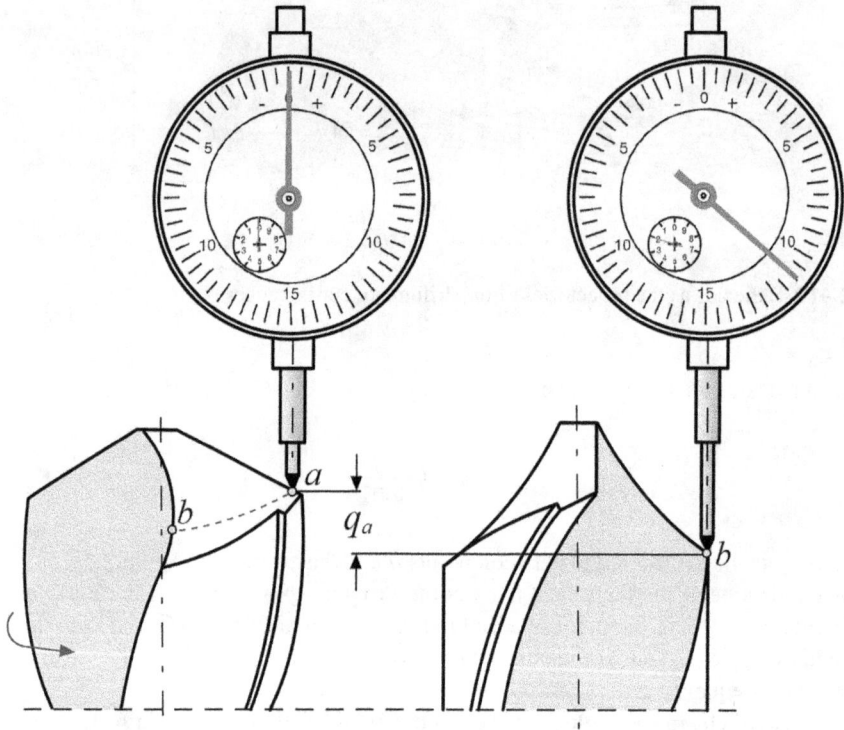

FIGURE 2.43 Sense of the decline distance q_a.

Experience and calculations showed that point *1* of the major cutting edge is the most vulnerable for interference (Figure 2.44). Therefore, for common shapes of the flank face, it is sufficient to check the condition set by Eq. (2.9) only for this point, i.e.,

$$q_1 > \Delta f_1 = f \frac{\psi_{11'}}{360°} \qquad (2.10)$$

The decline of the flank face at point *1* should be sufficient to assure the absence of interference of this face with the surface of cut and penetration of the MWF (the coolant) to the drill flank in the areas close to the cutting edge. On the other hand, it should not be excessive to keep a sufficient section of the cutting wedge for: (1) heat transfer into the tool, and thus avoid excessive temperatures over the tool–chip and tool–workpiece contact areas and (2) its rigidity and thus vibration in cutting. Experience shows that the optimal decline should be $q_1 = (0.04 - 0.15) d_{dr}$.

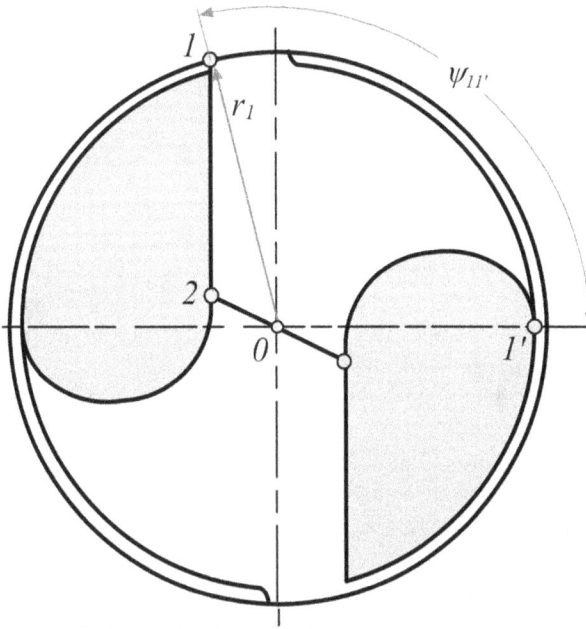

FIGURE 2.44 Point 1 as the most vulnerable to interference.

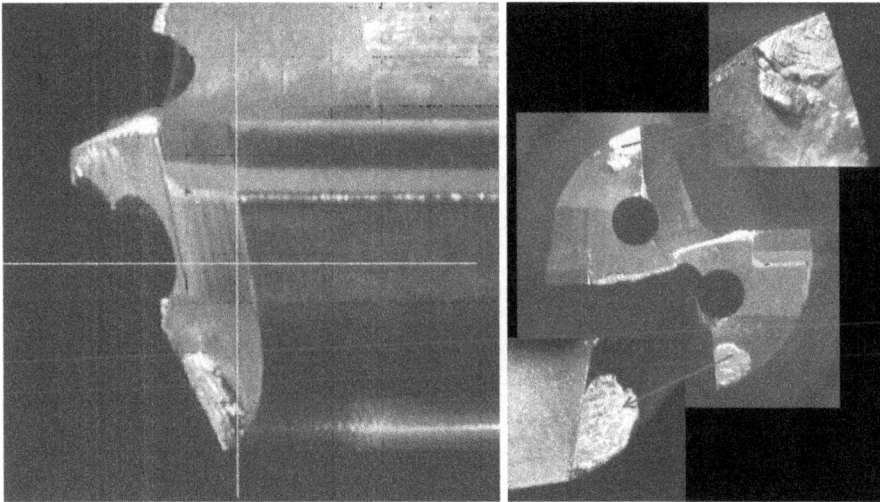

FIGURE 2.45 Appearance of the drill's flank face when the interference of point *1* takes place.

To exemplify the discussed problem, Figure 2.45 shows the appearance of the drill's flank face when interference of point *1* takes place. As can be seen, the drill has planar flank faces with no secondary flanks place applied. An important fact is that this figure is obtained in the machining of aluminum alloy A308, a relatively soft work material. As an increase in the axial force was caught by the advanced machine controller and the operation was stopped, the drill did not break. Normally, when even "light" interference takes place, the drilling tool unavoidably breaks. Because broken tools are not normally analyzed for the root cause of the breakage, many specialists and practitioners in the field did not see the discussed interference as a problem. This problem occurs rather often in many drilling tools (drills, reamers, taps).

Application Rule #2: Drill points of broken/failed drilling tools should first undergo visual inspection for interference marks. The same inspection should be applied for: (1) test tools after some number of machined parts and (2) tools removed from the machine due to high axial force and/or drilling torques as spotted by the controller of the machine.

2.1.3.5.2 Clearance Angles Shown in Drawings

The clearance angle of any point i of the major cutting edge in the cylindrical cross section A-A, α_i is the only clearance angle that makes the kinematic and physics of cutting sense so its consideration is totally sufficient from these two standpoints. The problem, however, is that it is almost impossible to show this angle in a drill drawing, and thus to reproduce it in drill point grinding. Therefore, this section introduces and explains the relevance of two other clearance angles used in practice, namely the normal clearance angle, α_n, and working clearance angle, α_f.

A few basic cases are possible in the defining of the flank face geometry in tool drawings:

1. The primary flank is planar (flat). When the primary flank face is flat, i.e., consists of one or more planes as shown in Figure 2.46a, then regardless of a particular drill type, the primary clearance angle is the same for all points of the major cutting edge in the so-called tool-in-hand system (T-hand-S), which is the tool geometry measuring/ inspection system as discussed in the Appendix. It is defined as the normal (i.e., perpendicular to the cutting edge, i.e., in the normal section N-N) clearance angle as shown in Figure 2.47.
2. The primary flank is not planar as shown in Figure 2.46b. Commonly, conical, cylindrical, and helical surfaces are used as the tool flank surface (Astakhov 2010). In this case, the so-called working clearance angle, α_f (discussed in the Appendix), is indicated at the drill corner in the plane parallel to the drill axis in the manner shown in Figure 2.47.

When the flank face is planar, the meaning, assigning in a drawing, and measuring the normal clearance angle do not present any problem. That is why this type of point grind is the most used today. Its particularities and proper application are discussed in Section 2.1.5.8.

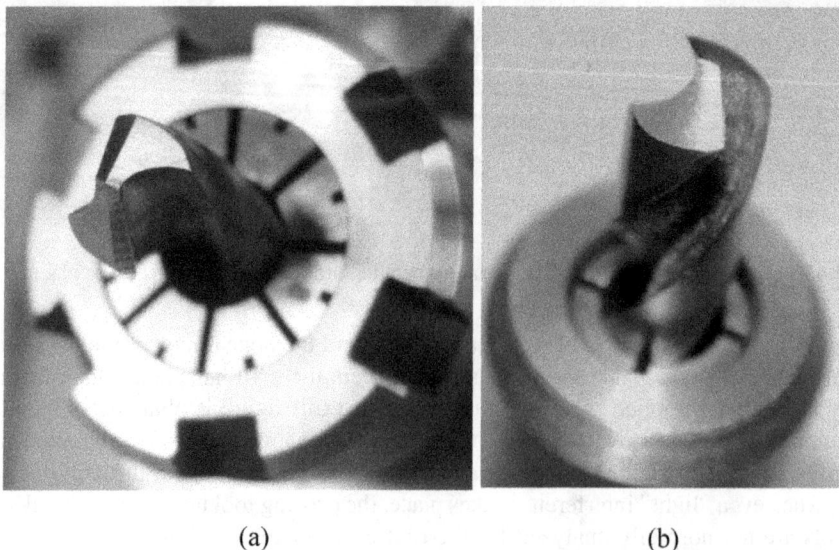

(a)   (b)

FIGURE 2.46 Flank faces: (a) planar and (b) non-planar.

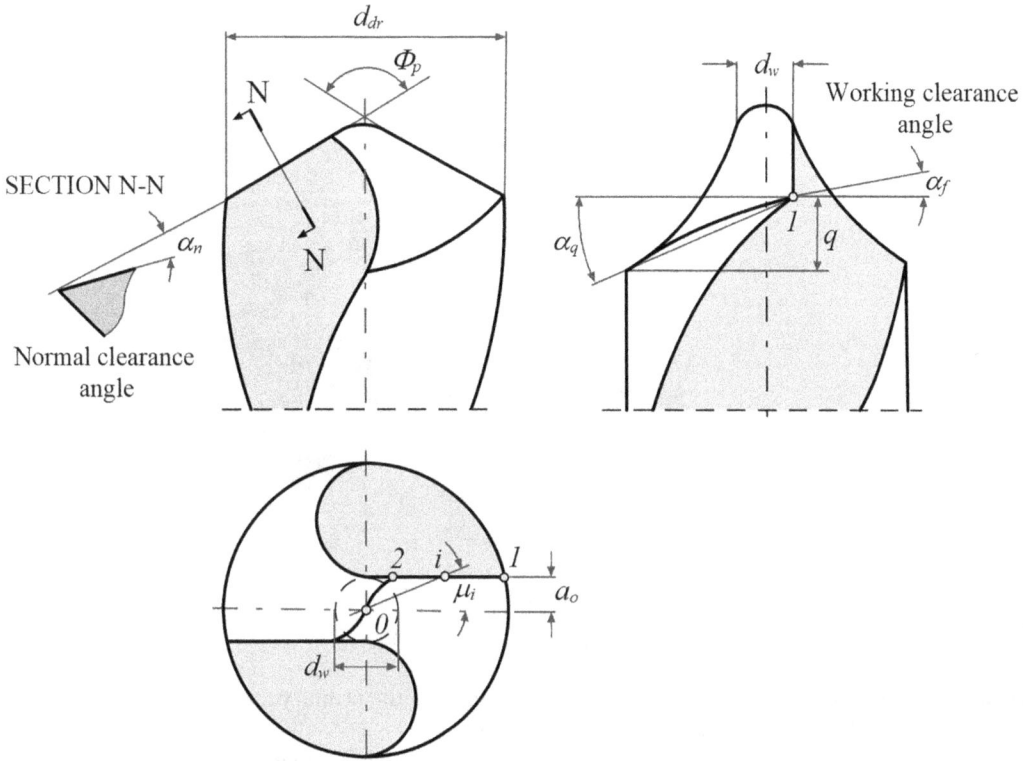

FIGURE 2.47 The sense of the normal α_n and working α_f clearance angles.

The clearance angle of any point i of the major cutting edge (see Figure 2.47) in the cylindrical cross section A-A, α_i correlates with the normal clearance angle, α_n as (Astakhov 2010)

$$\tan \alpha_{n-i} = \frac{\tan \alpha_i \sin \Phi_p - \sin \mu_i \cos \Phi_p}{\cos \mu_i} \qquad (2.11)$$

where Φ_p is the drill point angle and μ_i is the angle of the radius of point i. The sense of this angle is clearly shown in Figure 2.47. When the major cutting edge is straight, it can be calculated as

$$\sin \mu_i = \frac{d_w}{2r_i} \quad \text{so} \quad \sin \mu_1 = \frac{d_{dr}}{2r_i} \qquad (2.12)$$

where d_w is the web diameter, and $\mu_1 = \mu_i$ when a point i coincides with point 1. It follows from Figure 2.47 that

$$\mu_1 = \arctan \frac{a_o}{d_{dr}/2} \qquad (2.13)$$

where a_o is the distance of point 1 from the so-called central line, often referred to as its location ahead of the centerline. It follows from Figure 2.47 that $a_o = d_w/2$. This is the case for many common drills.

Indication of the clearance angles (primary and secondary) in drill drawing and appearance of the drill face for a straight-flute drill having planar flanks are shown in Figure 2.48a and b. Figure 2.48a shows an example of its designation in tool drawings. The primary and secondary

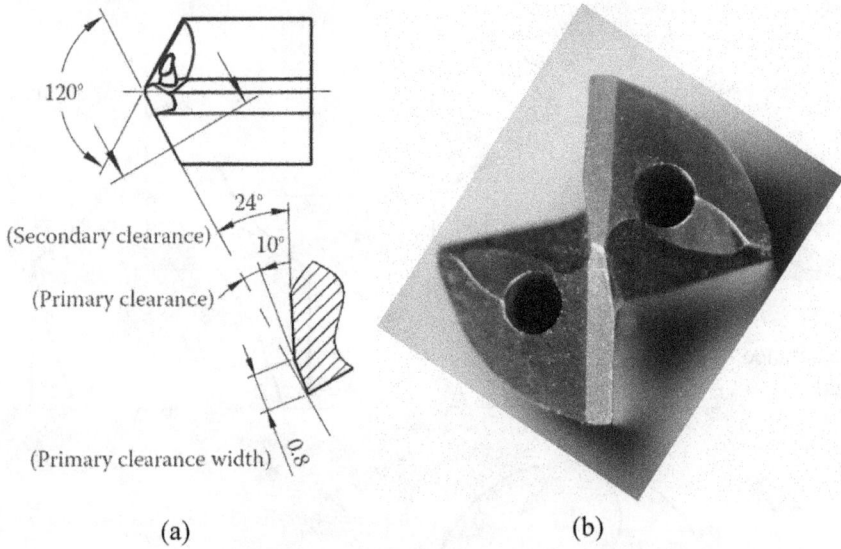

FIGURE 2.48 Planar flank face: (a) clearance angle indication in tool drawings and (b) appearance of the drill face.

clearance angles are shown as the normal clearance angle of the primary and secondary planar drill faces. Figure 2.48b shows the appearance of the drill face.

In this case, the normal primary clearance angle can be easily inspected/measured using a common tool benchtop microscope (i.e., PG 1000) in the manner shown in Figure 2.49 where the major cutting edge is set to be perpendicular to the display. As can be seen, the measured primary clearance angle is 13.5°.

When it comes to non-planar flank faces, the meaning, assigning in a drawing, and measuring the normal clearance angle become blurry even for seasoned specialists. Figures 2.50a and b show clearance angle indication in tool drawings and appearance of the drill face. As can be seen, being variable along the major cutting edge, the working clearance angle is indicated only at the drill corner (8°). The type of flank surface (conical, cylindrical, helical) and its setting parameters, which define the variation of the clearance angle along the cutting edge and geometry of the chisel edge, are never indicated in tool drawings.

When the flank face is not planar, the normal clearance angle, α_n, in the sense shown in Figure 2.47 is not constant over the major cutting edge 1–2. For any point i of this edge, it correlates with the working clearance angle in this point as

$$\alpha_{n-i} = \arctan\left(\tan\alpha_{f-i}\sin\left(\Phi_p/2\right)\right) \tag{2.14}$$

and thus can be correlated with the clearance angle of any point i of the major cutting edge in the cylindrical cross section A-A, α_i through Eq. (2.11).

For point 1 for which the working clearance angle is shown in tool drawing, this correlation is

$$\alpha_{n-1} = \arctan\left(\tan\alpha_{f-1}\sin\left(\Phi_p/2\right)\right) \tag{2.15}$$

Equation (2.15) clearly shows that α_{n-1} is smaller than α_{f-1}. It does not present problems for larger point angles ($\geq 130°$) whereas it can be a problem for drill with small point angles ($\leq 120°$).

The author found that there is a major problem with measurements of the working clearance angle on a drill with non-planar flanks. It is to say that the angle α_q between lines connecting the

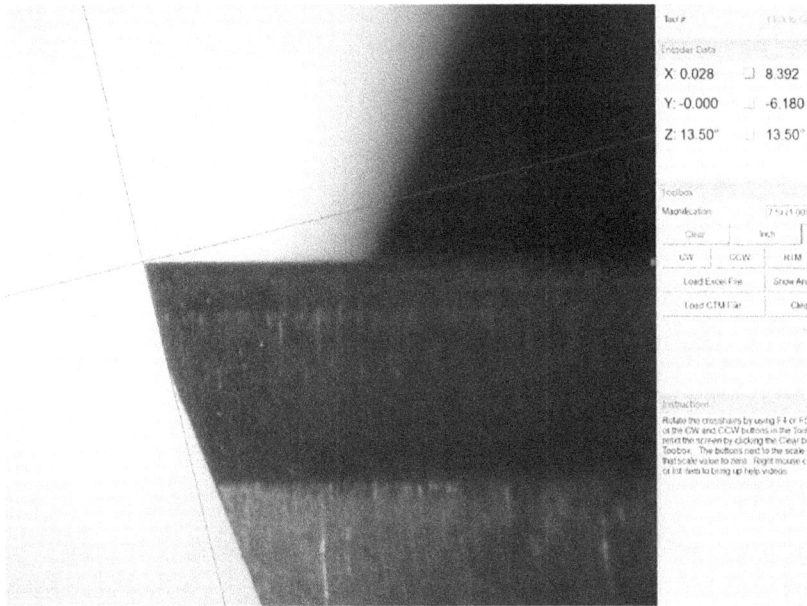

FIGURE 2.49 Measuring the primary clearance angle of a planar flank face.

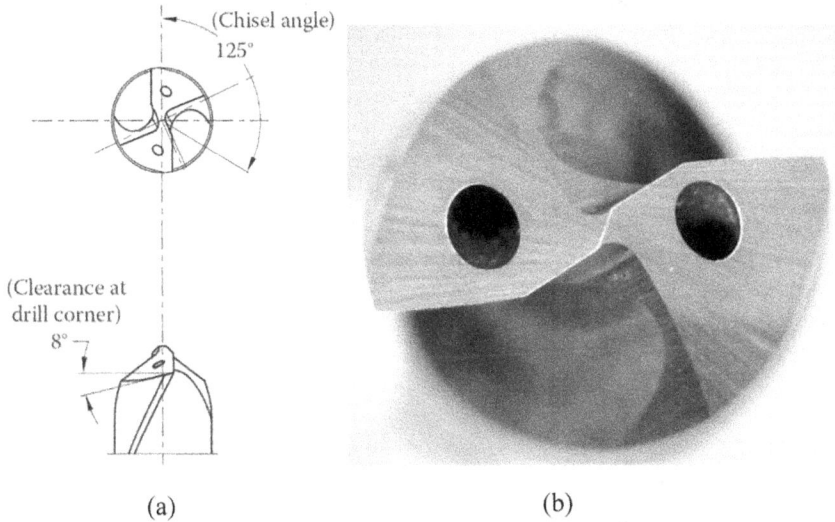

(Chisel angle)
125°

(Clearance at
drill corner)
8°

(a)

(b)

FIGURE 2.50 Non-planar (helical) flank face: (a) clearance angle indication in tool drawings and (b) appearance of the drill face.

point *1* and the top of the heel (shown in Figure 2.47) is actually measured as recommended by *Metal Cutting Tool Handbook* (page 106, Figure 3 in (1989)). One needs to realize that α_q is not α_f shown in Figure 2.47, although the former is routinely measured/inspected instead of the latter in industry. As such, excessively small clearance angles are often applied to the major cutting edge as shown in Figure 2.51a and next to zero clearance angles are applied to the second and other stages of many drills as shown in Figure 2.51b.

The important information for advanced uses on how to deal with the problem is given here because the discussed problem is a common problem for many drills in the automotive industry with

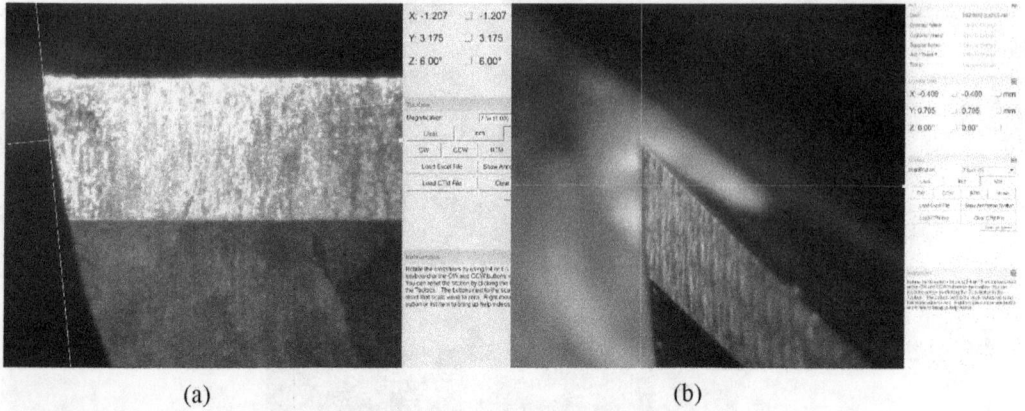

(a)   (b)

FIGURE 2.51 Showing (a) a small clearance angle applied to the major cutting edge and (b) next to zero clearance angles are applied to the second stage of the drill.

two- or three-step drills where the second/third stage of many drills is the chamfering stage. As a result, the cutting edge of this stage forms unwanted burr around the drilled hole that often limits tool life as the height of the burr becomes the tool life criterion. When this is the case, the intended tool life is reduced by 50% or even more.

Figure 2.52 shows the drill position with respect to a microscope for proper measurement of angle α_f. It should be measured as the angle between a vertical (or horizontal depending on a particular tool microscope setting) line and a tangent to the curved flank face at the drill corner (point 1). This angle should be measured only for a part of the curved flank face having width smaller or equal to the margin width. Such a measurement/inspection requires significant experience of an inspector so that extensive training to anyone involved in the discussed measurement/inspection should be provided. According to the author's experience, when modern CNC tool inspection machines are used, the matter gets worse as these machines include an incorrect built-in procedure for the measurement of angle α_f (discussed in Chapter 1 in Astakhov (2024)).

The ultimate solution of the problem is to make planar flank faces of all stages of a drill. It is discussed in Astakhov (2014) that non-planar flanks including helical grind have no advantages in the performance of modern drill design. Moreover, they show inferior performance for HP drills with high-pressure MWF (the coolant) supply as well as minimum quantity lubricant (MQL) drills. Therefore, planar flank faces of various designs should be used in modern industry. Fortunately, modern CNC drill grinding machines have this option. Figure 2.53(a) shows the appearance of a planar flank face of the second stage on the monitor of the controlled ANCA drill grinding machine. Figure 2.53(b) shows a planar flank face applied to the second stage of a drill.

The recommended clearance angles for HP drills and reamers are given in Table 2.3. Application experience shows that these values should be considered good starting values in the optimization of the drill geometry parameters. Unfortunately, many drill manufacturers and users do not follow these conclusions/recommendations, resulting in lower tool life, inferior quality of drilled holes, and drill breakage. According to the author's experience, optimization (making them close to the values shown in Table 2.3) of the clearance angles of HP drills, that is, meant for high-speed, high-penetration drilling of high-silicon automotive aluminum in the setting of the largest transmission plant, resulting in a two-time increase in tool life, three-time reduction of the drill breakage and their premature failures, and a 30% increase in the allowable penetration rate. Such an optimization first met great resistance from the largest drill manufacturing companies. When they were forced to manufacture such drills, the Oskar worth performance (the best dramatic acting) "No it is impossible because it is impossible ever" of their technical/sales personnel became a daily routine. The test and then implementation results, however, were rather stunning so that the optimized drills

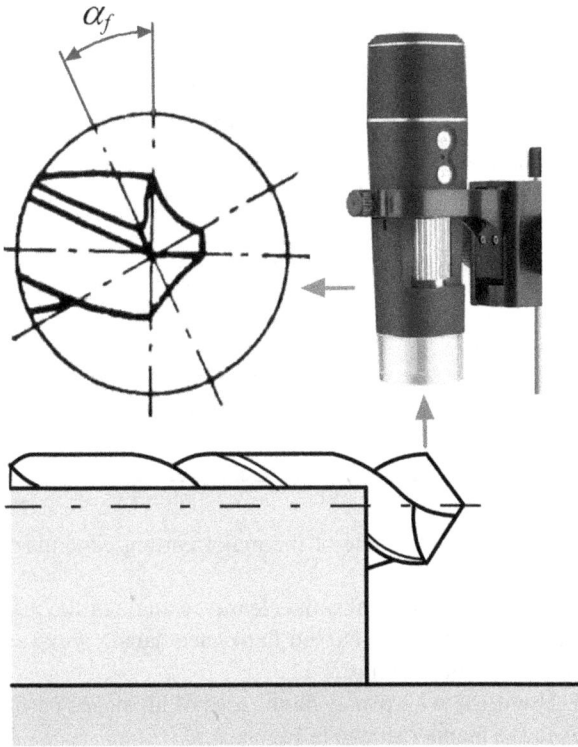

FIGURE 2.52 Proper measurement of angle α_f.

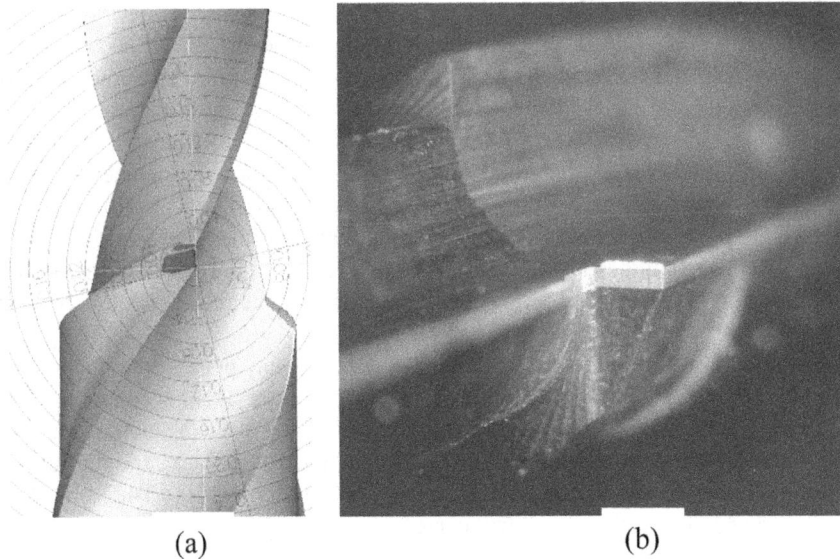

FIGURE 2.53 Planar flank face: (a) appearance in the controller of an ANCA drill grinding machine and (b) applied to the second stage of a drill.

became standard for the application. After the first addition of this book was published, a gradual but steady change from old to new standard values of the clearance angle has been taking place due to irrefutable performance results.

TABLE 2.3

Recommended Clearance Angles for Drills (for Twist Drill: The Periphery Clearance Angle)

Drill Diameter Range (mm)	Suggested Clearance Angle (°)		
	General-Purpose Drills	Drills for Tough and Hard Materials	Drills for Soft and Free Machining Materials
0.35–1.00	24	20	26
1.05–2.50	21	18	24
2.55–3.00	18	16	22
3.05–6.50	16	14	20
6.55–8.95	14	12	18
9.00–13.00	12	10	16
13.10–20	10	8	14
>20	8	7	12

Application rule #3: The clearance angle of the major cutting edge must be included in ANY drill tool drawing.

Application rule #4: For non-planar flanks, the clearance angle of the major cutting edge angle should be measured only for a part of the curved flank face having width smaller or equal to the margin width in the manner shown in Figure 2.52.

Application rule #5: Use drills with planar flank faces of all stages particularly for the second and other (if any) stages in the manner shown in Figure 2.53.

Application rule #6: For HP drills, the values of clearance angle given in Table 2.3 should be used.

2.1.3.6 Major Cutting Edge - Rake Angle

The two most important systems of the consideration of the drilling tool geometry are the tool-in-hand system (T-hand-S) and the tool-in-holder system (T-hold-S) defined and explained in detail in the Appendix are to be considered in any meaningful consideration of a drilling tool. The drilling tool geometry parameters in T-hand-S are used in the tool design, and thus assigned in the tool drawing, used in tool manufacturing and its inspection. The geometry parameters of a drilling tool in T-hold-S define working conditions of this tool, i.e., its performance including tool life, allowable penetration rate/productivity, and quality of the machined holes. Obviously, the geometry parameters in these systems are uniquely related through the particularities of the drilling tool design. Therefore, the sections to follow consider the relevant geometry parameters in both systems.

The rake angle of a straight-flute drill in T-hand-S is zero at any point of the major cutting edge so it is normally not indicated in tool drawings. The T-hold-S rake angle accounts for the position of the cutting edge defined by the distance a_o (Figure 2.47), i.e., by the location of this edge ahead of the centerline and the point angle. Rodin (1971) calculated that for a straight-flute drill with $\Phi_p = 120°$, web diameter $d_w = 0.15 d_{dr}$ and the T-hand-S rake angle $\gamma_n = 0°$, the T-hold-S rake angle γ_{nh} is always negative and decreases toward drill center as

r_i/r_{dr}	1	0.8	0.6	0.4	0.2
γ_{nh}	−4°20′	−5°30′	−7°20′	−11°30′	−29°20′

As can be seen, angle γ_{nm} decreases most noticeably for $r_i < 0.5 r_{dr}$. To deal with the problem, gashes are commonly applied to the rake face. Figure 2.54 shows a fragment of the drill drawing where the gash, applied on the rake face, is properly dimensioned. The T-hand-S rake angles γ_{f-g} of the gashed part are normally but not always indicated on the drill drawing.

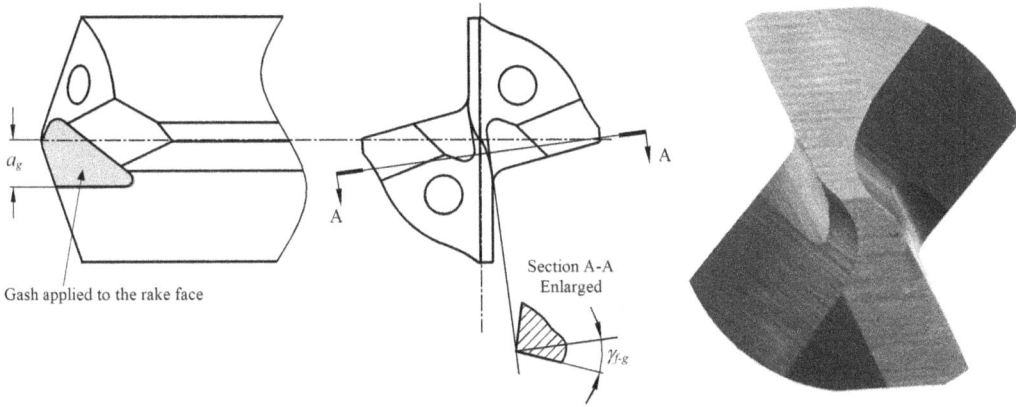

FIGURE 2.54 A gash is applied to the rake face to modify its geometry (a straight-flute drill).

Application rule #7: Gashes should be applied to the rake faces of the straight-flute and twist drills. Distance a_g Figure 2.54 should be approximately equal to $1/4d_{dr}$.

For straight-flute and twist drills, the rake angle of gashes, $\gamma_{f\text{-}g}$ should be as follows:

- For difficult-to-machine materials: $-20°$.
- For general-purpose drills: $-10°$.
- For high-penetration drills used for aluminum alloys: $-5°$.

The determination of the rake angle in twist drills is a bit more complicated as it depends on the actual shape of the rake face. When the rake face is not modified, this face is a part of the helical chip flute made on the tool. In this case, the rake angle of any kind is not indicated in the tool drawing. Obviously, the rake angle of the cutting edge, γ_f in T-hand-S at drill corner is equal to the helix angle ω_d as shown in Figure 2.55. According to Eq. (2.2), the helix angle decreases from the drill corner toward its center so that the T-hand-S rake angle decreases and at any point of the major cutting edge it is $\gamma_{fi}=\omega_i$.

In T-hold-S, the rake angle also depends on the position of the cutting edge defined by the distance a_o (Figure 2.47), i.e., by the location of this edge ahead of the centerline and the point angle. Rodin (1971) calculated that for a twist drill with $\Phi_p=120°$, web diameter $d_w=0.15d_{dr}$, helical angle $\omega_d=30°$, the real rake angle γ_{nm} varies along the major cutting edge as follows:

r_i/r_{dr}	1	0.8	0.6	0.4	0.2
γ_{nm}	30°25′	10°50′	14°30′	3°15′	−24°05′

As can be seen, the variation of the T-hold-S rake angle along the major cutting edge is much greater for twist drills compared to that in straight-flute drills. To correct this variation, particularly at the inner end of the major cutting edge, gashes must be applied. The application of a gash to the rake face of twist drills is known as web thinning as it reduces the length of the chisel edge (Astakhov 2010) whereas its true meaning is to correct the rake face geometry/rake angle variation of the major cutting edge. Such an application is absolutely needed in machining of difficult-to-machine materials and for HP drills. Standard DIN 1412 defines the extent of gashes as shown in Figure 2.56.

Form A point was initially intended for use on drills of over 20 mm to reduce the pressure on the web. Normally, the chisel edge is thinned up to 8% of diameter. Nowadays, with CNC grinding machines, this type becomes the most popular point for general applications. Form B point allows cutting edge runout correction and improving rake angle in the regions adjacent to the chisel edge. It was initially developed for brittle and difficult-to-machine work materials. Form C, or as it is widely known, Split Point was intended to use on drills with a heavy web to give better starting and

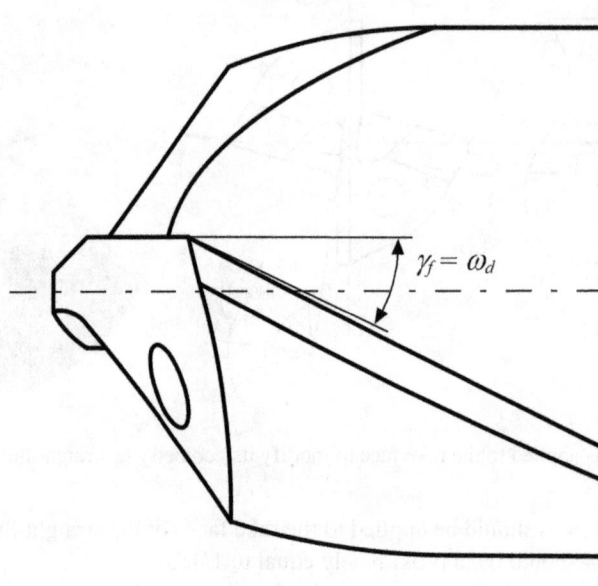

FIGURE 2.55 T-hand-S rake angle of the major cutting edge at the drill corner of a helical drill.

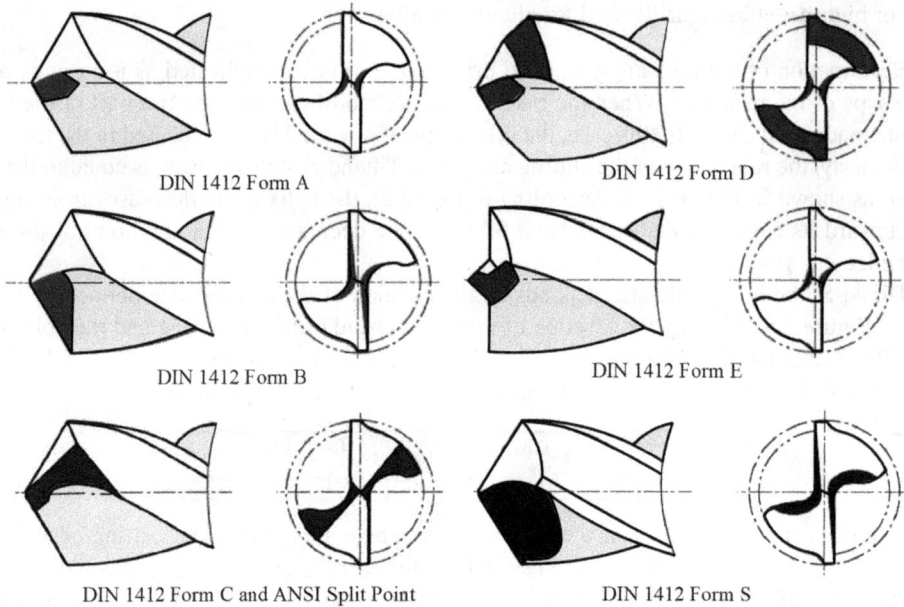

DIN 1412 Form A

DIN 1412 Form D

DIN 1412 Form B

DIN 1412 Form E

DIN 1412 Form C and ANSI Split Point

DIN 1412 Form S

FIGURE 2.56 Gash extends according to the DIN 1412 standard.

thus produce more accurate holes. Form D is known as Cast Iron Point as its outer corners prevent frittering of the iron on breakthrough. Soon, it was found that this point grind is very useful for a wide variety of work materials, particularly when the exit burr is of concern. Form E was developed for use on sheet metal. It was soon found that various modifications of this grind are also useful for many applications. Form S is normally used for drills with parabolic flutes.

Although this DIN classification gives initial ideas for available point grinds and it is used for many modern CNC multi-axis grinders, it does not compare drills in terms of the axial force that determines the penetration rate. Moreover, the considerations of the applications of various gashes made according to this DIN standard do not account for one important and potentially significant issue with twist drills.

As discussed in Section 2.1.3.2, fast-spiral drills are used for modern difficult-to-machine materials to facilitate chip removal over helical flutes. As such, the helix angle can be as high as 40° or even greater. The problem is that the working rake angle in T-hand-S and T-hold-S angles is equal to the helix angle as shown in Figure 2.55. This creates two problems. The first one is that the drill corners become weak (see Figure 2.51b), and thus susceptible to chipping and high temperatures due to a small wedge angle, i.e., low mechanical strength and small section for heat removal. The second problem is that the discussed drills have a tendency to *dig* or *grab* or run ahead of the feed particularly when breaking through the hole (Astakhov 2011a). It happens, for example, in drilling long holes (e.g., oil holes in crankshafts) and/or when drilling system is not rigid, for example, when hand-held drilling machines are used. It also happens when drilling very thin stock, owing to the tendency of the drill to hook into the work when breaking through a hole. The reason for that is discussed in the Appendix, Section 8.2.2.2. In drilling a soft work material, a chatter of the drill can also occur. Distinctive chatter marks on the hole bottom similar to that shown in Figure 2.57 often occur. The reason for that is explained by the author earlier (Astakhov 2010, 2014).

To prevent this from happening, the high rake angle in the vicinity of the drill corners is reduced by grinding a flat facet on the rake face in the region near the drill corner where the rake angle is the highest. An example of such a modified twist drill is shown in Figure 2.58. The extent of the facet and its rake angle in this case are application-specific and thus should be determined experimentally for a given application by maintaining the balance between the occurrence of chatter and increased cutting force factors, primarily the drilling torque. Commonly, the rake face in such a case is ground with a zero-rake angle for simplicity. The drill ground in this manner is also very effective for drilling unannealed steel or hard spots in cast iron.

Application rule #8: The application of flat facets on the rake face of the twist drill corners for a long twist drill (length-to-diameter ratio >10) with a helical angle of more than 30° is beneficial as it improves drill performance. Note that the facets applied to the rake faces of two or more major cutting edges should be the same and symmetrical with respect to the drill axis.

The last item to be clarified in this section is the curved shape of the major cutting edge of twist drills made by leading tool manufacturers. Figure 2.59 shows CoroDrill®860 by Sandvik Coromant as an example. The explanation of a curved major cutting edge is simple – it improves chip breakability. To understand why it happens, one needs to recall that most of car hoods are built with a 0.7 mm steel sheet (called gauge). When this sheet is flat, it does not have much bending rigidity.

FIGURE 2.57   Examples of chatter marks on the bottom of the hole being drilled.

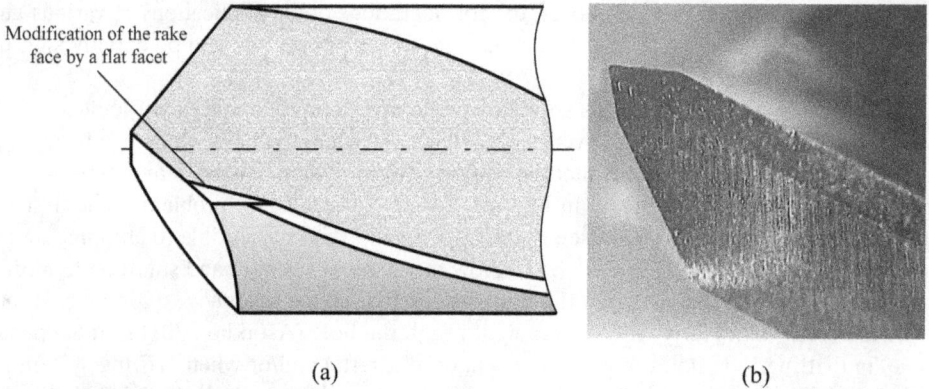

Modification of the rake
face by a flat facet

(a)   (b)

FIGURE 2.58   Modification of the rake face adjacent to the drill corner: (a) the model and (b) a real drill.

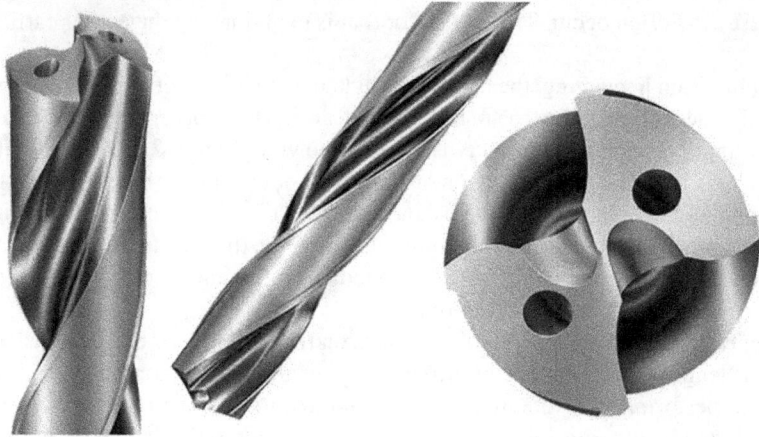

FIGURE 2.59   CoroDrill®860 by Sandvik Coromant with curved major cutting edges.

These sheets are reinforced by forming them into curved patterns that include hood geometry characteristics such as the hood angle and curvature that significantly increase structural stiffness. The same is with chip, i.e., when curved to the shape of the major cutting edge, it becomes much rigid so that, when bended by the wall of the flute, it breaks easier than a "flat-shaped" chip. Moreover, due to distances a_o (see Figure 2.47), the T-hold-S rake angle varies over the major cutting edge as discussed above. As a result, chip deformation also varies along this edge. This non-uniform chip deformation adds to improved chip breakability.

Important particularities of modified/improved rake face of twist drills often used for specific application conditions and/or requirements are discussed in Section 2.1.5.10.

2.1.3.7   Chisel Edge

One of the most important yet least understood design components of a drill is the chisel edge. Its design and geometry parameters play an important role in the performance of any drill whereas its role in the performance of HP drill is vital. Commonly, the chisel edge is responsible for about 60% of the drilling axial force and 10% of drilling torques. Experience shows that the proper design/optimization of the chisel edge geometry can significantly reduce the drilling axial force and thus increase the allowable rate of penetration (known as ROP in oil and gas drilling). The reduction of the drilling torque can also be achieved although its contribution to the total drilling torque is small.

This section aims to explain in a simple manner what is that "proper design of the chisel edge" in various applications.

2.1.3.7.1 Common Perceptions/Notions of the Chisel Edge

The standard drill nomenclature defined by standard ISO 5419:2015 "Twist drills – Terms, definitions, and types presents the chisel edge" depictures the chisel edge as shown in Figure 2.60a and then defines the following terms as related to the chisel edge:

- 3.26 – *Chisel edge.* The edge formed by the intersection of the flanks.
- 3.27 – *Chisel edge corner.* The corner formed by the intersection of a major cutting edge and the chisel edge.
- 3.28 – *Chisel edge length.* The distance between the chisel edge corners.

Standard ISO 5419 and thus practically all literature sources on drills present the chisel edge as a single design component of a drill as shown in Figure 2.60b. Because literature sources do not discuss the geometry of the chisel edge (e.g., its rake and clearance angles), a great number of practical engineers experience difficulties in understanding that the chisel edge is the cutting edge, not an indenter penetrating the workpiece as it is often presented in the professional literature. Moreover, a picture of the chisel edge that acts as an indenter penetrating into the workpiece (Figure 2.61) taken

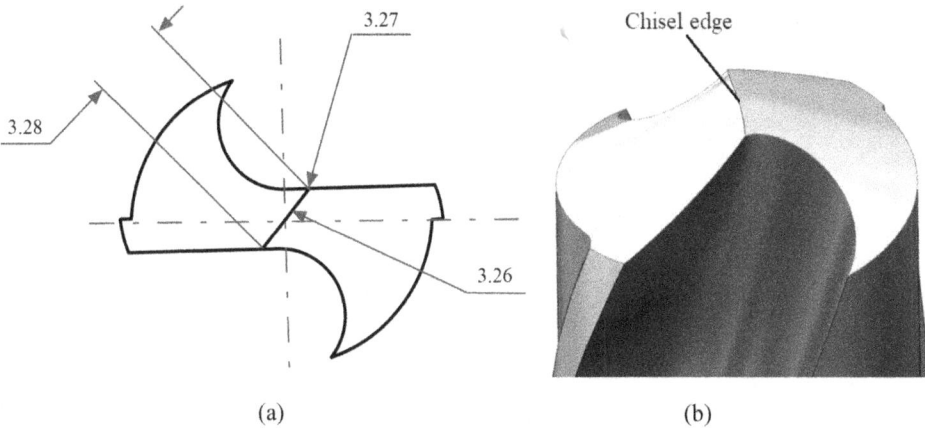

(a)   (b)

FIGURE 2.60 Chisel edge: (a) visualization of the clauses related to the chisel edge according to Standard IS0 5419 and (b) as commonly depicted in the literature.

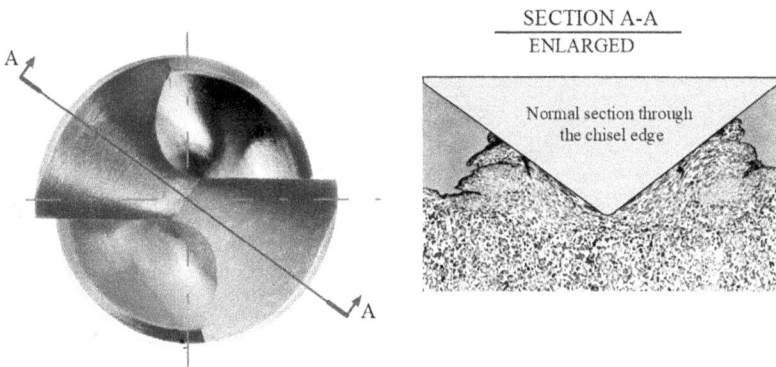

FIGURE 2.61 Chisel edge penetration into the workpiece commonly presented in manufacturing books and articles.

out of the context from a classical paper by Galloway (1957) is presented in many manufacturing books and there is a prevailing notion circulating in professional literature that this edge smashes or extrudes the work material.

As early as in 1956, Ernst and Haggerty (1958) studied chip formation by various points of the major and chisel edges using a ½″ (12.7 mm) drill. Figure 2.62 represents a reconstruction of their experimental results. According to this figure, the chisel edge cuts with a constant highly negative rake angle so the partially formed chips are heavily deformed as shown by micrographs Ch1, Ch2, and Ch3. Although this picture is important, it does not explain directly a great axial force generated at the chisel edge. The clearance angle of the chisel edge is rather great so there is no contact of the flank face of the chisel edge with the chip and/or with the machined surface (the bottom of the hole being drilled).

As discussed in Section 2.1.3.3, a common two-flute drill point has two major cutting edges, *1–2* and *5–6*, and two chisel edges, *2–3* and *3–5* (Figure 2.21). Figure 2.63a shows the model of chip formation on the major and chisel edges. As discussed by Astakhov (2014), the chip flow direction is in the direction of the cutting speed. For major cutting edges *1–2* and *5–6*, the cutting speeds v_{12} and v_{56} are shown in the middle of these edges. For chisel edges *2–3* and *3–5*, *the corresponding cutting speeds v_{23} and v_{35} are perpendicular to these edges but have the opposite directions.* An important conclusion follows from Figure 2.63a is that the directions of the cutting speed on chisel edges *2–3* and *3–5* are opposite as they are located on the opposite sides of the drill rotation axis (point *3*). It is to say that if the chisel edge passes through the axis of rotation, there are two chisel edges – each one starts from the inner end of the corresponding major cutting edge and extends to the center of rotation (edges *2–3* and *3–5*). Figure 2.63b provides an experimental proof for the model shown in

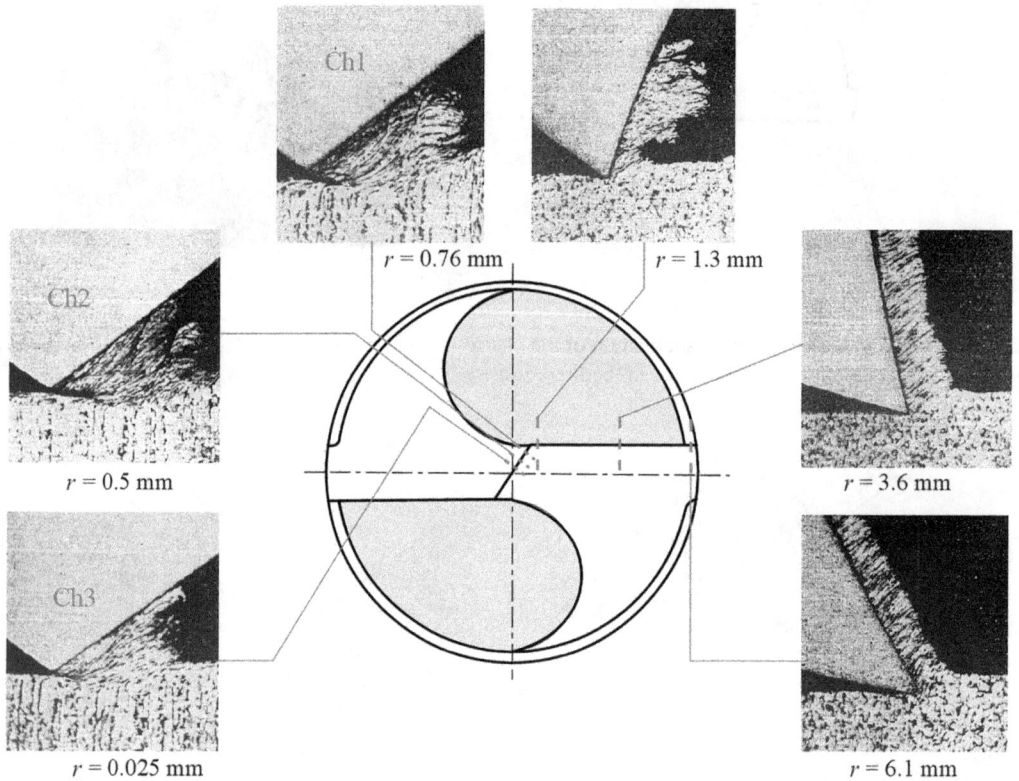

FIGURE 2.62 Showing images of the partially formed chip at successive points of the major and chisel edges.

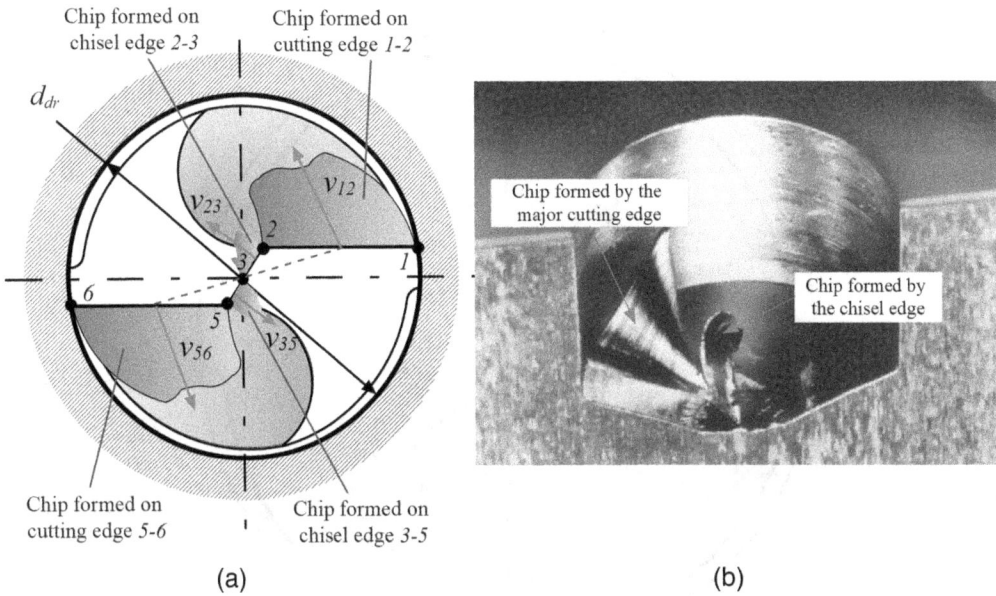

FIGURE 2.63 Chips formed by the major and chisel edges: (a) model and (b) partially formed chips obtained using a quick-stop device.

Figure 2.63a as it shows partially formed chips obtained using a quick-stop device. As can be seen, distinctive chips are formed by the major and chisel edges. Therefore, the above-mentioned prevailing notion of "indentation" of the work material by the chisel edge is false when a drill is designed and made properly.

As the chisel edges *2–3* and *3–5* are legitimate cutting edges, each of these two chisel edges has its rake and flank faces with the corresponding rake and clearance angles (see the Appendix). By the standard definitions, the rake face of the chisel edge *2–3* is the face over which the chip formed in drilling slides and the flank face, which faces the machined surface. The chisel edge *2–3* is formed by the intersection of the flank face of the major cutting edge *5–6*, which serves as the rake face for this edge, and the flank face of the major cutting edge *1–2*, which is the flank face for this chisel edge. Similarly, the chisel edge *3–5* is formed by the intersection of the flank face of the major cutting edge *1–2*, which serves as the rake face for this chisel edge, and the flank face of the major cutting edge *5–6*, which is the flank face for this chisel edge. Figure 2.64 provides visualization of the thus introduced rake and flank faces.

2.1.3.7.2 Length of the Chisel Edge

The notions of the length of the chisel edge are depicted in Figure 2.60. This length depends first on the web diameter/core thickness of the drill as no other means to reduce this length (e.g., the above-discussed gashes) are applied. As such, the greater the web diameter, the greater the length of the chisel edge, and thus the axial force.

There are two kinds of drills in terms of the web diameter-to-the drill diameter ratio (Figure 2.65). The web diameter comprises 10%–20% of a standard drill's diameter for conventional drills. On a wide-web drill, web diameter is increased to about 25%–39% of the drill diameter. Wide-web drills are more rigid and stronger than conventional drills; thus, higher penetration rates (up to 20%) and length-to-diameter ratio and better accuracy of machined holes are achieved with these drills, particularly for long drills. Wide-web drills are normally used to produce holes in difficult-to-machine and heat-treated work materials. An apparent drawback to wide-web drills is the reduced cross-sectional area of the chip flutes. However, a wide use of drills

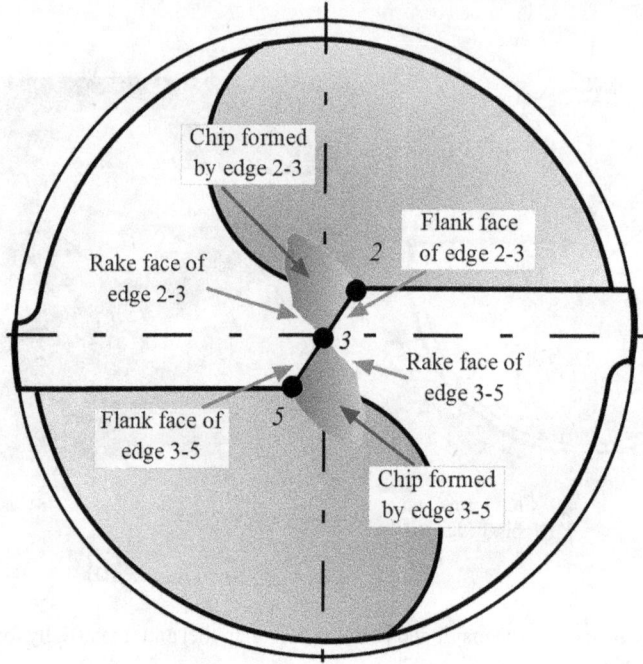

FIGURE 2.64 Rake and flank faces of two chisel edges *2–3* and *3–5*.

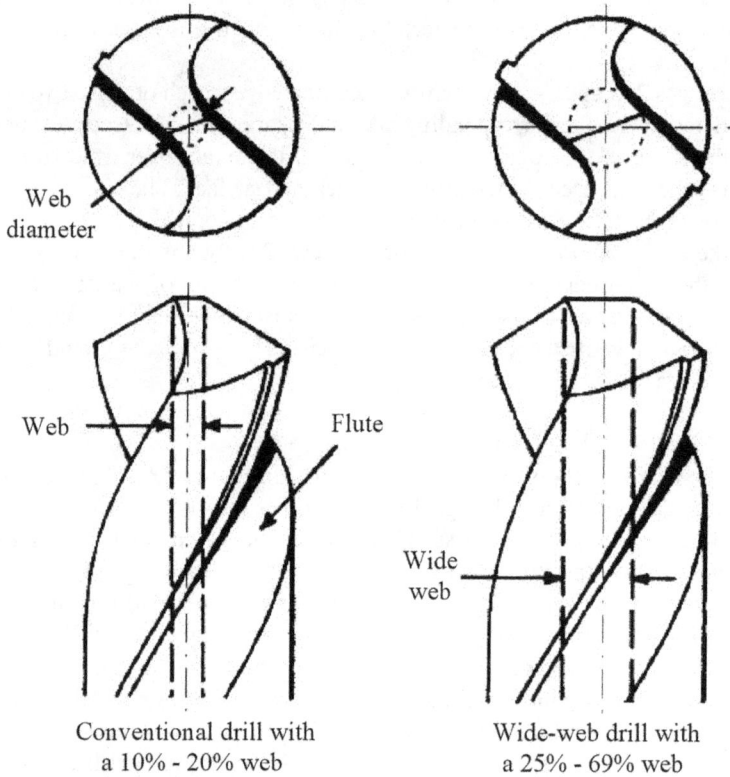

FIGURE 2.65 Conventional and wide-web drills.

with high-pressure internal MWF supply solved this problem. Moreover, the penetration rate in drilling of difficult-to-machine and heat-treated work materials is rather low, which results in a much smaller amount of chip produced and thus needed to be transported from the machining zone through these flutes.

When the web diameter increases, the length of the chisel edge increased as can be seen in Figure 2.47. According to Fiesselmann (1993), it is seldom understood that a drill with 30% web requires almost twice the axial force of the 20% web drill. Further, the 40% web drill found in drills recommended for harder, tougher alloys, or which result from using parabolic flute drills for holes deeper than 10 times the drill diameter in depth, has an axial force almost four times greater than that of the 20% web drill. This is shown in Figure 2.66.

Normally, gashes on the rake face (see Section 2.1.3.6) are applied to reduce the length of the chisel edge, and thus the drilling axial force. As mentioned above, such an application is commonly known as web thinning. The application of web thinning is mandatory for wide-web drills – its proper application for this case is shown in Figure 2.50b. It is also beneficial for conventional (15%–20% web) drills (Figure 2.67).

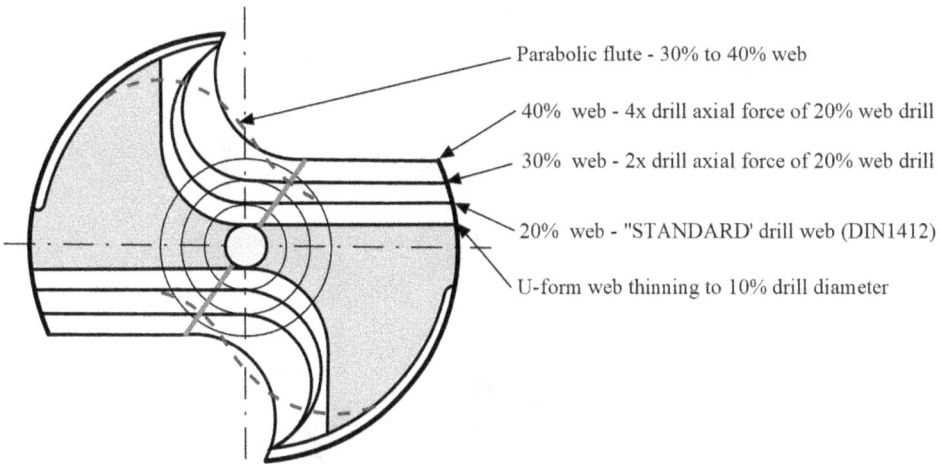

Parabolic flute - 30% to 40% web

40% web - 4x drill axial force of 20% web drill

30% web - 2x drill axial force of 20% web drill

20% web - "STANDARD' drill web (DIN1412)

U-form web thinning to 10% drill diameter

FIGURE 2.66   Effect of the web diameter on the axial force.

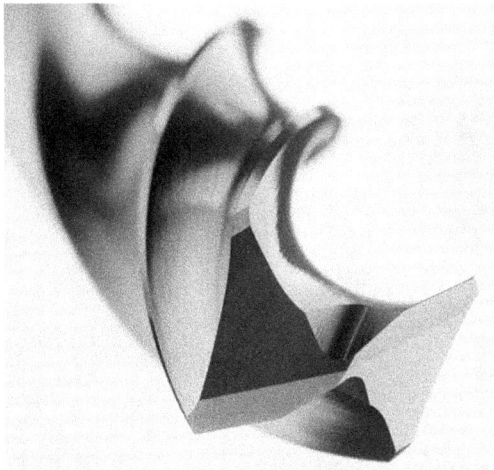

FIGURE 2.67   Example of the proper application of web thinning for a conventional drill.

Application rule #9: Gashes must be applied to the rake faces of wide-web drills to reduce the length of the chisel edge. Their application is also highly beneficial for conventional HP drills working with high feeds.

The next item to be clarified is the resultant length of the chisel edge after web thinning. If this thinning does not go far enough so that the chisel edge is still too long, the drilling axial force is not reduced enough to allow a higher drill penetration rate. If web thinning goes too far, the region around the chisel edge can be subjected to bulk fracture in the manner shown in Figure 2.68. The extent to which the length of the chisel edge can be reduced depends on the work material, particularly its strength and hardness and the quality of drill (drill material and its manufacturing quality). The rigidity of the drilling system plays also a significant role.

Experience shows that the most common "sin" in applications of web thinning is rough rake face of the applied gashes that often cause the built-up edge on this rake face. Examples are shown in Figure 2.69. As a result, the drilling axial force increases and tool life decreases.

Application rule #10: Surface roughness (often called finish) on the rake face of web-thinning gashes should be the same or even better than that on the drill flute.

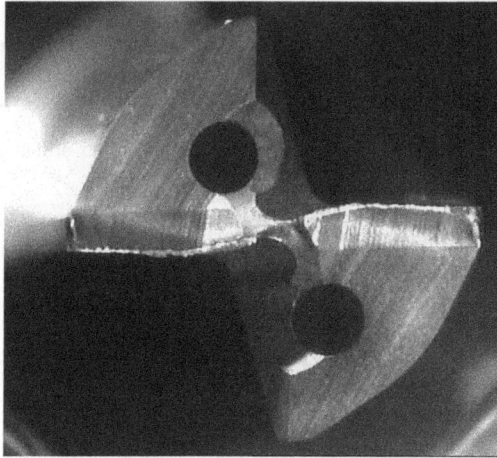

FIGURE 2.68   Bulk fracture of the chisel edge region.

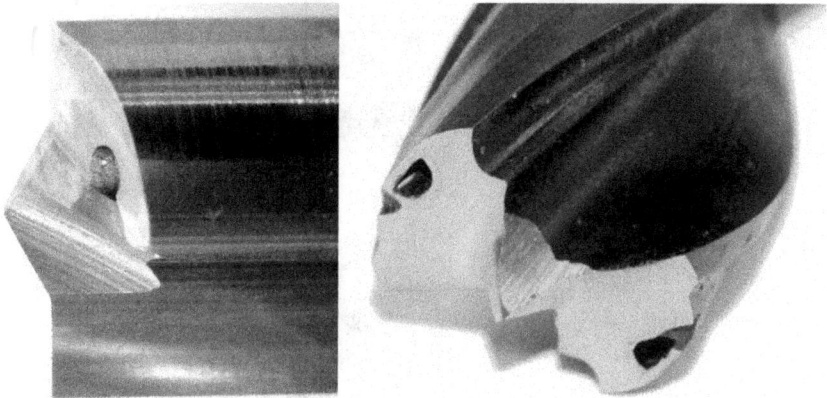

FIGURE 2.69   Rough surface of the rake face of gashes.

2.1.3.7.3 Simple Geometry of the Chisel Edge

The angle of the chisel edge, ψ_{cl} is the starting angle in considerations of the chisel edge geometry. This angle is often shown in drill drawing in different manners, but its proper indication is as shown in Figure 2.70 (Astakhov 2010). Although the chisel edge angle is commonly shown in drill drawing, it is not an independent angle achieved in point grinding. Rather, it depends on the type of the point grind and on the parameters of this grind. In other words, it is achieved "automatically" by other geometry parameters of the drill, not independently as discussed later.

If no modification to the rake face of the chisel edge is made, the rake angle of the chisel edge, γ_{cl}, is always negative as shown in Section A-A in Figure 2.70 although its "negativity" varies from one type of the point grind to the next. This is also shown by micrographs Ch1–Ch3 in Figure 2.62. The clearance angle of the chisel edge, α_{cl}, is normally high, i.e., it is much greater than the recommended values (see Table 2.3). This is because the rake and flank faces of the chisel edge are part of flank face(s) of the major cutting edge. If one tries to improve the rake angle of the chisel edge, i.e., to make it less negative, then the clearance angle of the chisel edge increases correspondingly. This makes the chisel wedge weaker, and thus more susceptible to breakage.

It is a very important but routinely ignored feature of the chisel edge is its so-called self-centering ability. In simple terms, it means that point 3 of the chisel edge should make the first contact with the surface of the workpiece to drill a small starter hole to guide the drill. The shape of the chisel edge and its desired location of point 3 is shown by View B in Figure 2.70. Self-steering ability is judged by a half-point angle ν_{cl} of the chisel edge and its extent s_{cl}. The smaller the half-point angle, the greater the self-centering ability of the drill. Unfortunately, the half-point angle ν_{cl} has never been considered in comparison of various point grind and in analyses of drill performance. Often, the lack of self-centering ability is compensated by the introduction of an additional drilling operation using the starter drills as discussed in Section 2.1.3.4.5. It is to say that the lack of knowledge always comes at addition often avoidable costs of tools, tooling, and production time.

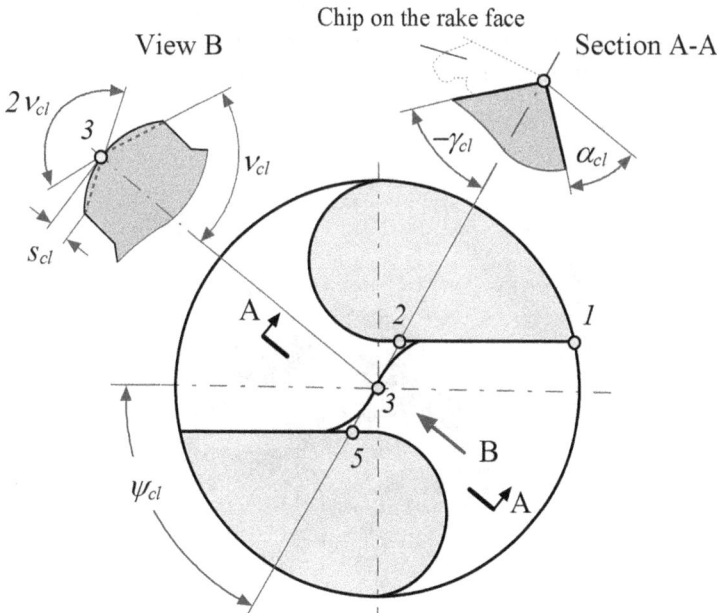

FIGURE 2.70 Generic chisel geometry.

Figure 2.71 shows the chisel edge shape for a conical drill point. As can be seen, this drill possesses self-centering ability as point *3* is ahead of points *2* and *5* in the axial direction. The proper selection of the setting parameters of this type of point grind can improve both the rake angle of the chisel edge and its self-centering as discussed later. This figure shows that one can observe the particularities of the chisel edge geometry using a simple tool benchtop microscope (e.g., PG 1000, ZOLLER »pomBasic«, etc.).

2.1.3.7.4 Modern Point Grinds and the Chisel Edge Geometry

There are a number of various types of drill point grinds as discussed by Astakhov (2010). Today, however, with wide use of CNC point grinding machines, as, for example, ANCA, Star NXT, and Walter HELITRONIC, four-facet and helical types of point grinds are mostly used in manufacturing. The sections to follow aim to provide advanced users with the information essential for the selection of one or another type of point grinds.

2.1.3.7.5 Standard Four-Facet Point Grind

According to the standard four-facet point grind depicted in Figure 2.72a and b, each major cutting edge is provided with the primary planar flank called primary facet having the width, a_{p-fl} equal to or greater than distance $2a_o$ (Figure 2.47a). This first facet is ground with the primary clearance angle, α_{n2}. The second facet is ground behind the primary facet. As such, the clearance angle of the second facet α_{n1} is always greater than that of the primary facet. The clearance angle of the first facet is to assure the proper cutting conditions at the major cutting edge whereas the clearance angle of the second facet is selected large enough to prevent interference of the drill flank faces with the bottom of the hole being drilled.

According to Figure 2.72a and as can be seen in Figure 2.72b, the secondary flank face (facet) does not have any influence on the geometry of the chisel edge formed only by the intersection primary flank facets. As these facets are planes, their line of intersection is a straight line. As a result, two parts of the chisel edge, namely *2–3* and *3–5* are aligned as shown in view B in Figure 2.72a. As such, a half-point angle v_{cl} (Figure 2.70) becomes equal to 90° so that the point angle of the chisel edge $2v_{cl}=180°$. Therefore, a drill with such a chisel edge does not possess any self-centering ability. No matter what are the values of the primary clearance angle and the point angle of a standard four-facet grind, the chisel angle is straight and the rake angle of the chisel edge is highly negative.

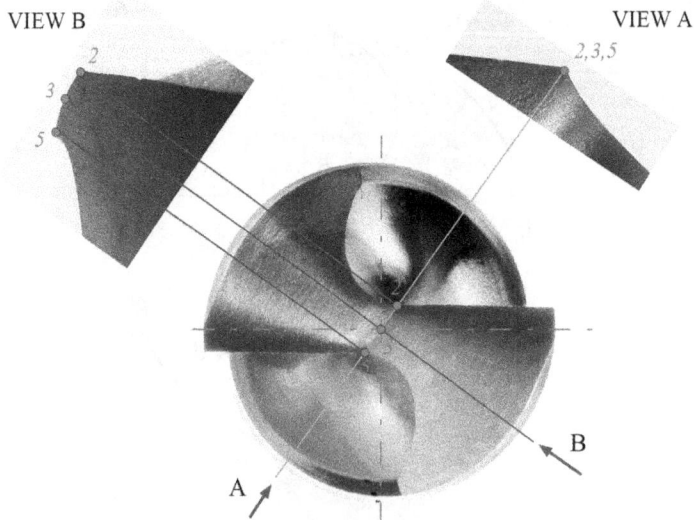

FIGURE 2.71 Chisel edge shape for a conical drill point.

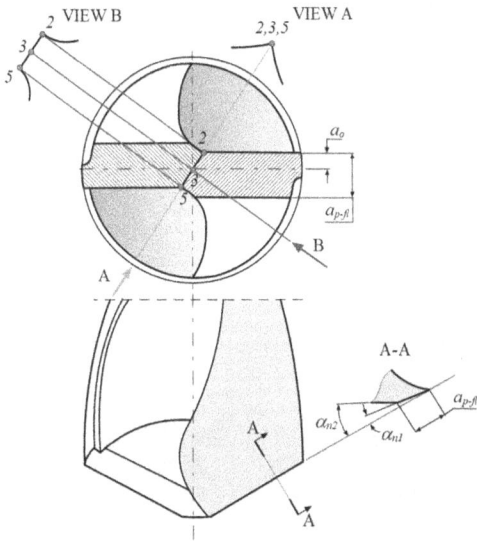

FIGURE 2.72 Particularities of the flank facets and chisel edge of the standard four-facet point grind: (a) model and (b) a real drill.

Figure 2.73 shows the shape of the chisel edge shape for a real standard four-facet point grind. As can be seen, two parts of the chisel edge *2–3* and *3–5* are aligned so that the point angle of the chisel edge $2v_{cl}=180°$. One can observe the discussed particularities of the chisel edge geometry using a simple tool benchtop microscope, i.e., to see the "flat" chisel edge and its highly negative rake angle.

Although drills with the standard four-facet point grind can be used in machining of steel with relatively low feeds, their applications are accompanied by high axial force and low tool life. When this point grind is used with HP drills, particularly those used in the automotive industry for various aluminum alloys, they are often failing prematurely due to high axial force, chipping, and subpart quality of drilled hole (both shape and roughness). As an example of such an application, Figure 2.74a shows the face view of a new PCD-tipped drill having the standard four-facet point. Figure 2.74b shows the appearance of the chisel edge after 8,115 drilling cycles. As can be seen, the whole vicinity of the chisel edge is covered by a thick layer of extruded aluminum (the soft matrix material for 380 alloy). This aluminum deposit caused a low-frequency radial vibration of the drill, and thus oversized condition of drilled holes – the most common cause for the premature failure of this tool.

Application rule #11: Avoid using drills, especially HP drills, with the standard four-facet point as they do not possess any self-centering ability and have the worst (compared to other common point grinds) cutting conditions at the chisel edge.

2.1.3.7.6 True Four-Facet Drill Point Grind
Significant improvements in drill self-centering and in the geometry of the chisel edge can be achieved when

- The width of the primary flank facet is equal to a_o, i.e., when $a_{p-fl}=a_o$.
- The secondary flank facet is applied with the normal clearance angle greater than that of the primary clearance angle, i.e., $\alpha_{n2}>\alpha_{n1}$.

This type of drill point grind is known as the true four-facet grind. A model that visualizes the essential geometry parameters of such a drill is shown in Figure 2.75a. Real drills having the true four-facet point are shown in Figures 2.75b and 2.67.

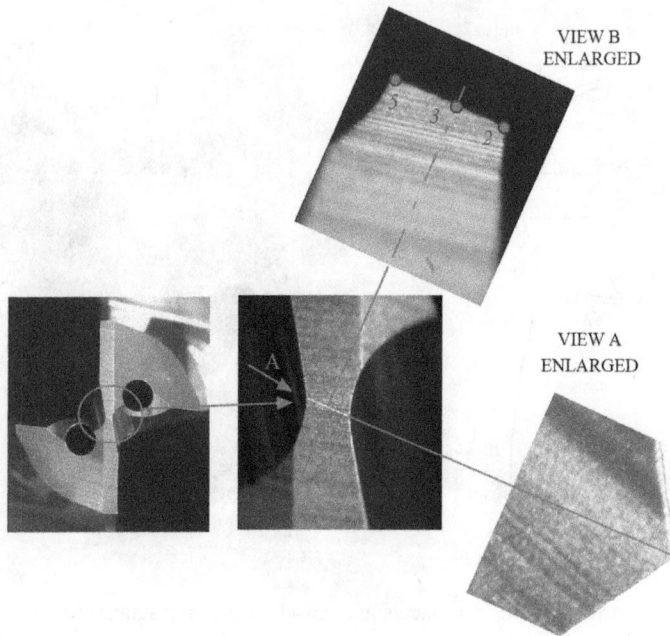

FIGURE 2.73 Chisel edge shape for a standard four-facet point grind.

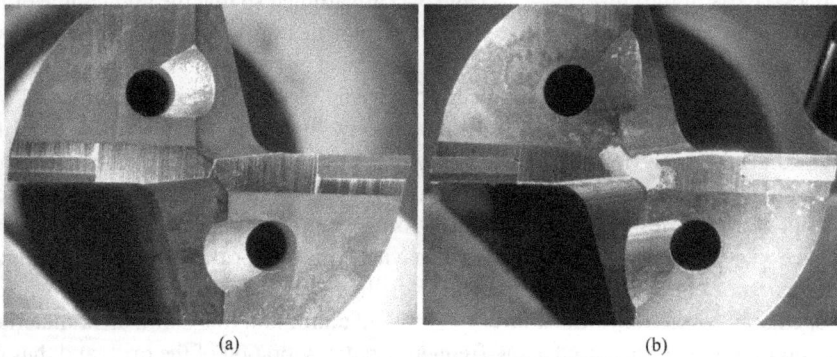

FIGURE 2.74 Showing (a) a new drill and (b) the appearance of the chisel edge after 8,115 drilling cycles.

Referring to Figure 2.75a, one can see the same as in the above-discussed four-facet point grind, and the flank face of each major cutting edge (lip) (*1–2* and *5–6*) consists of two planes. The major cutting edge *1–2* is provided with the primary flank plane *1* having the clearance angle α_{n1} and secondary flank plane *2* having the clearance angle α_{n2}. Respectively, the major cutting edge *5–6* is provided with the primary flank plane *2* and secondary flank plane *3* having the same clearance angles as primary *1* and secondary *2* flank planes. Therefore, the difference between the standard and true four-facet point drills is in the width of the primary clearance plane: it is equal to or greater than $2a_o$ in the former whereas it is exactly equal to a_o in the latter.

As a result of the discussed arrangement of the primary and secondary flank facet/planes, the rake face of the chisel edge *2–3* is formed by the secondary flank plane *2* and the rake face of the chisel edge *3–5* is formed by the secondary flank plane *1*. Because the clearance angles of the secondary flank planes are significantly larger than those of the primary flank planes, and, moreover it can be chosen as great as needed as it does not affect any cutting action, the rake angles of both

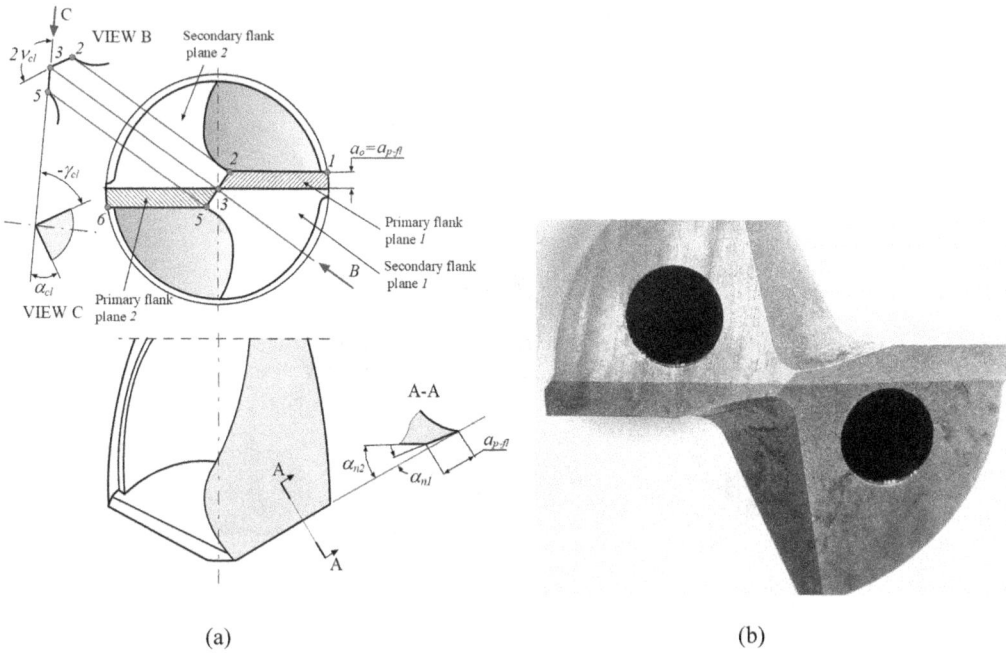

FIGURE 2.75 Particularities of the flank facets and chisel edge of the true four-facet point grind: (a) model and (b) a real drill.

chisel edges (2–3 and 3–5) can be made with much less negative rake angles γ_{cl} compared to those achieved with the standard four-facet point grind. Therefore, the axial drilling force is reduced that allows greater drill penetration rates, and thus higher productivity.

Another significant difference in the considered point grind is that the chisel edges 2–3 and 3–5 are not aligned as in the previously discussed standard four-facet point grind. Rather, as can be seen in Figure 2.75a, point 3 appears to be the drill apex with the point angle $2v_{cl}$ known as the point angle of the chisel edge. This apex first touches the workpiece at the beginning of drilling (provided that this surface is perpendicular to the drill rotating axis) so that it helps to reduce drill wandering and thus reduces drill transverse vibrations at the hole entrance. Therefore, a drill with such a point has gain some self-centering ability. It was also found that this shape of the chisel edge makes the chisel wedge stronger and less susceptible to chipping. Figure 2.76 shows the discussed feature of the chisel edge. It is discussed in Section 2.1.5.11.8 that angle $2v_{cl}$ (and thus the drill's self-centering ability) depends on the primary α_{n1} and secondary α_{n2} clearance angles as well as on drill point angle Φ_p. The smaller the chisel edge point angle $2v_{cl}$, the better the self-centering ability of a drill.

Figure 2.77 shows a drill with a true four-facet point after drilling soft aluminum alloy to obtain the chip adhered to the rake faces of the chisel edge. As can be seen, both chisel edges worked properly so that this figure resembles the modeling of chip formation shown in Figure 2.64. It is to say that this point grind works well even for soft work materials prone to significant plastic deformation in cutting, and thus to the formation of a significant built-up edge. Figure 2.74b shows that this is not nearly the case with the standard four-facet point grind.

To gain the self-centering advantage with less negative rake angle and thus a greater allowable penetration rate, the point grinding machine setup should be very accurate, particularly for small drill diameters (less than 8 mm). This is needed to make the lines of intersection of the primary and secondary facets aligned at the drill rotation axis (point 3 in Figure 2.75a) as shown in Figure 2.78a. Unfortunately, this is often not the case in reality. Figure 2.78b shows a common point grinding flaw – the second flank planes do not form the intended chisel edge as the lines of their

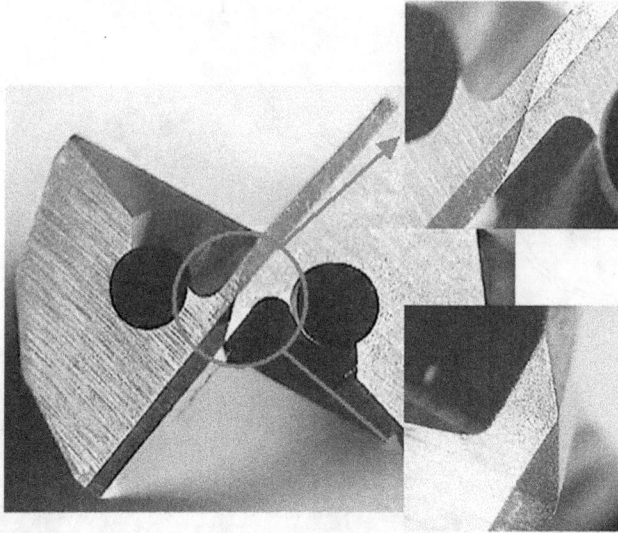

FIGURE 2.76   Chisel edge shape achieved in the true four-facet point grind..

FIGURE 2.77   Showing the built-up edge on the rake faces of the chisel edge.

(a)   (b)

FIGURE 2.78   Face view of the true four-facet point: (a) properly ground and (b) the common point grinding flaw.

intersection with the primary flank faces do not pass through the axis of rotation. The net result is a non-straight chisel edge consisting of three segments. This ruins all the discussed advantages of the true four-facet point grind.

Application rule #12: Drills with the true four-facet point work well in many applications, including HP applications for steels and cast irons as the true four-facet geometry assures the sufficient strength of the region around the chisel edge. However, it happens when the drill point design and geometry parameters are determined/selected properly as discussed in Section 2.1.5.11.8.

2.1.3.7.7 Split-Point Grind

It can be seen in Figure 2.77 that the built-up edge (aluminum deposit) on the rake faces of the chisel edge is much greater than that on the rake faces of the major cutting edges. This is because the rake faces of the chisel edges are still with negative rake angles so that the deformation of the chip in its formation on the rake face of the chisel edge is much greater. Obviously, it causes greater axial force in drilling restricting the allowable penetration rate. To solve such a problem, some modifications of the rake face geometry of the chisel edge can be used. Normally, the rake face of the chisel edge is modified to improve its rake angle. Such a modification is known as the split-point geometry.

As early as in 1923, Oliver patented (US Patent No. 1,467,491) (1923) a very distinctive drill with split point shown in Figure 2.79a. The drill has two major cutting edges 1 and 2 and two chisel edges 3 and 4. The rake faces of these edges are provided with two depressions 5 and 6. As a result, the rake face 7 obtains the rake angle, which can be selected according to particular work material. As claimed in the patent description, such *geometry allows at least a 50% higher penetration rate compared to a drill with the standard geometry.* Although it is a great design, it was, in the author's opinion, much ahead of the time as grinding fixtures available at that time did not allow for reproducing such a flank geometry with any reasonable accuracy that is crucial for its proper implementation. Moreover, the available drilling systems were not capable of a significant increase in the drill penetration rate due to the lack of power/strength of the feeding mechanisms and feed accuracy. As a result, this great flank geometry was not requested for more than 50 years.

The full realization of the potential of the split-point geometry is offered by US Patent Nos. 4,556,347 (1985) and 4,898,503 (1990). Such geometry is shown in Figure 2.79b. According to these patents, the chisel edge 1 is provided with the rake face (notch) 2 having a rake angle between 5° and 10°, while the angle of the notch to the drill axis is selected to be between 32° and 38°. As claimed by US Patent No. 4,556,347 (1985), the comparison of the performance of this drill with a

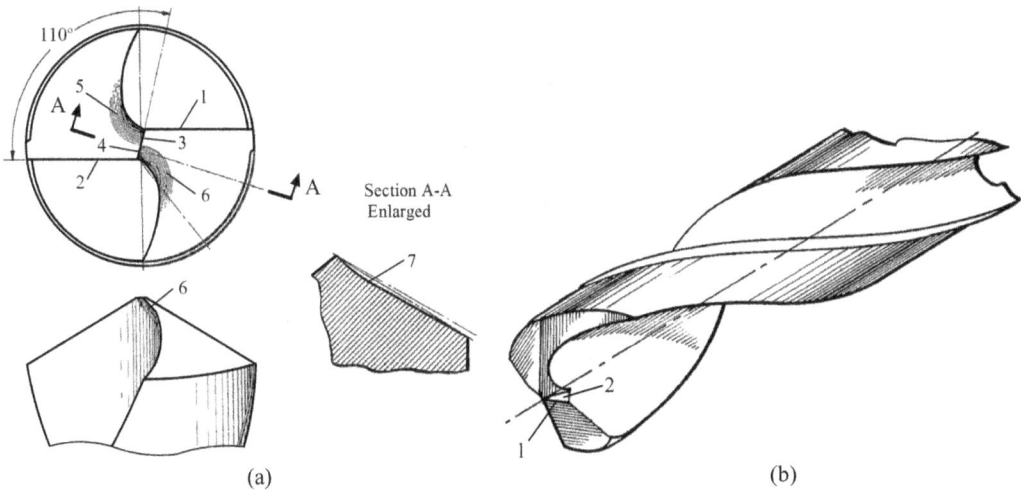

FIGURE 2.79 Split-point geometry: (a) Original geometry according to US Patent No. 1,467,491 (1923) and (b) geometry according to US Patent Nos. 4,556,347 (1985) and 4,898,503 (1990).

commercially available precision twist drill conforming to NAS 907 standard showed a significant improvement in tool life when drilling difficult-to-machine materials. For example, a 5.8 times increase in tool life was achieved in drilling Inconel 718 of 44 HRC.

American Cutting Tool Institute (ACTI) recommends (1989) the following split-point geometries shown in Figure 2.80. Figure 2.80a shows a split point with notches ground right along and coinciding with the chisel edge. Figure 2.80b shows the split point similar to Figure 2.80a, except that the notches are at an angle to the original chisel edge. This type of the split point enables the drill to center itself as the apex of the drill point contacts the surface of the workpiece first. Note that the half-point angle of the apex, φ_2, is greater than that of the drill (φ_1). Figure 2.81 shows WSTAR solid-carbide drills for machining of aluminum developed and manufactured by Mitsubishi Materials Corporation according to Figure 2.64b. Note also that the true four-facet point grind is used to achieve drill self-centering. Figure 2.80c shows a split point with the original chisel edge intact. The notches are parallel but separated. While the highly negative rake at the chisel edge remains, it is negative only on a small part of the contact length of the rake face. Moreover, such geometry provides a means of egress for the chip formed by the chisel edge.

Further improvements in split-point drills are achieved when the sharp corner between the chisel edge and the inner end of the major cutting edge is rounded as shown in Figure 2.82a. As can be seen

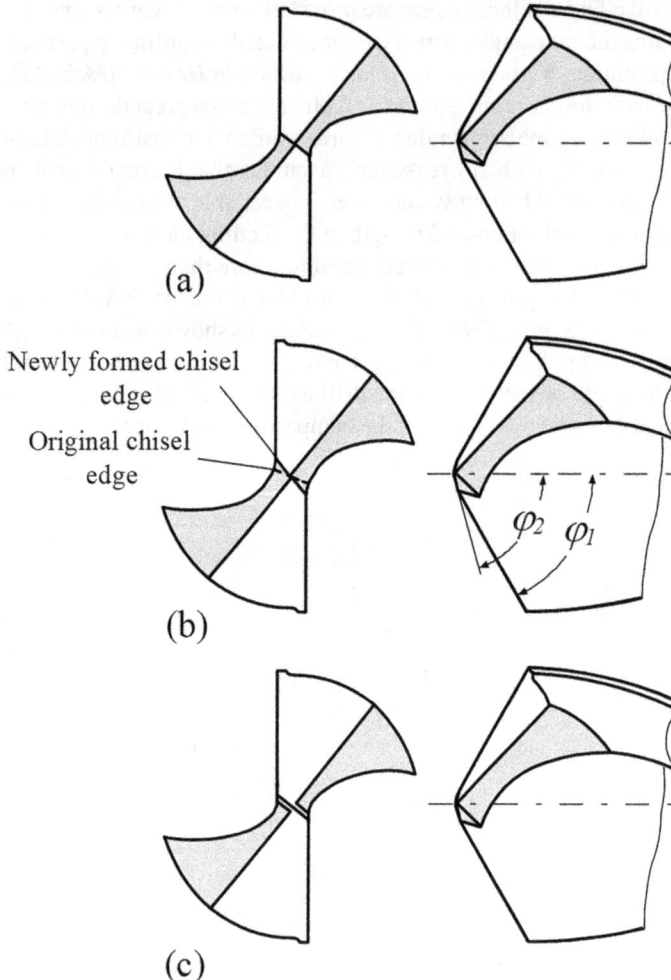

(a)

Newly formed chisel edge

Original chisel edge

φ_2 φ_1

(b)

(c)

FIGURE 2.80   Split-point geometries recommended by American Cutting Tool Institute (1988).

FIGURE 2.81 WSTAR drill by Mitsubishi Materials Corporation.

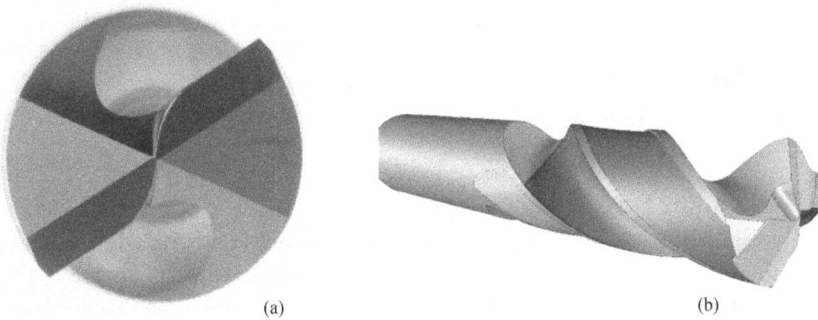

(a)   (b)

FIGURE 2.82 A split-point drill with S-shaped chisel edge: (a) a real drill and (b) appearance in the controller of ANCA drill grinding machine.

in this figure, the chisel edge becomes S shaped due to a special grind of the tertiary flank surface that is obviously not a plane. A reduction of the axial force by 20% and smooth chip flow from the chisel edge is achieved with such a design of the chisel edge and its rake face. This S-shaped split point is included in the library of modern CNC grinders, e.g., ANCA MX7 machine (Figure 2.82b). The use of modern drill point grinding machines assures symmetry of both parts of the chisel edge that is a crucial requirement to achieve the intended performance of such drills.

Although the split-point geometry is a modern trend in drill design, manufacturing, and application, one should understand that such a geometry weakens the chisel edge region of the drill. To avoid chipping of this region due to unwanted forces in drilling, the drill should be made perfectly symmetrical, used with the minimum possible runout (tool holder and machine spindle) in the rigid drilling system and so on. Because there are a number of various split-point designs, the final assessment of how good is a particular/chosen design can be made by observing the built-up edge on the rake faces of the major and chisel edges in drilling of a work material tend to form the built-up edge easily. Figure 2.83a shows the aluminum A309 built-up edge on rake faces of the major and chisel edges in drilling with the standard four-facet point grind drill. As can be seen, the rake and flank faces of the chisel edges are completely covered by aluminum deposit, which grew much beyond the normal built-up edge. Figure 2.83b shows the aluminum A309 built-up edge on rake faces of the major and chisel edges in drilling with a split-point grind. Note that for a fair comparison, the drill

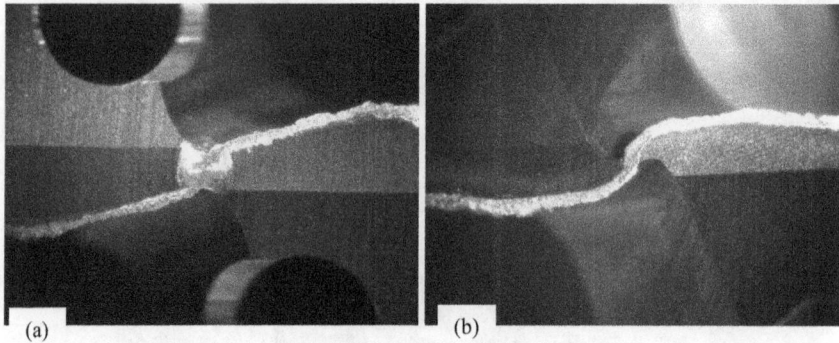

FIGURE 2.83 Showing the aluminum built-up edge on rake faces of the major and chisel edges: (a) standard four-facet point grind and (b) split-point grind.

with a standard four-facet point grind was used. As can be seen, the height and deformation pattern of the built-up edge on these edges are the same. This is the best indicator that the design and geometrical parameters of the split-point geometry are selected properly.

Application rule #13: Applications of drills with split-point grind can significantly increase the allowable penetration rate and tool life and improve the quality of matched holes in terms of their shape, location accuracy, and surface roughness (finish). One can obtain these benefits if and only if the conditions of the drilling system, including tool holder, spindle and part fixture rigidity, MWF (coolant) supply methods, and its quality support application of such drills (Astakhov and Joksch 2012).

2.1.3.8 Minor Cutting Edge and Back Taper

Two other areas of drill design that impact its performance are two closely related items, namely the minor cutting edge (see Figure 2.21) and back taper. As both these items are not self-evident and, moreover, very poorly presented in the literature, they require some explanations because they are of critical importance in drilling, and thus should be clearly understood by application specialists.

As discussed by the author earlier (Astakhov 2006), when the work material is being cut, the material is deliberately overstressed beyond the elastic limit in order to induce a permanent deformation and then separation of the stock to be removed. As such, the fracture of the chip from the wall of the hole being drilled occurs so the strain and stress at fracture are achieved at any point of separation of the work material. After such a separation occurs, the work material recovers elastically or springs back (Astakhov 2018).

2.1.3.8.1 Springback: Affecting Factors

Figure 2.84 shows a simple model where no drill runout is considered for the sake of simplifying further considerations. In Figure 2.84a, the drill corners *1* and *5* separate the chip from the rest of the workpiece forming the wall of the hole being drilled. The diameter of this wall is always equal to that of the drill, that is, d_{dr}. After corners *1* and *5* of the drill pass a certain part of the machined hole, the load due to cutting is removed so that the applied stress returns to zero. As a result, the machined hole shrinks due to springback (elastic recovery) because elastic deformation is recoverable deformation (Astakhov 2018). Although known as elastic recovery, it is actually not only due to the mechanical action of corners *1* and *5* but also due to heating and subsequent cooling of the work material adjacent to the hole being drilled. In other words, the thermal energy due to plastic deformation of the work material and friction between the tool and the workpiece in their relative motion causes thermal expansion of the work material around the drill terminal end. When corners *1* and *5* advance further, the work material contracts due to cooling by MWF (coolant) so that thermal

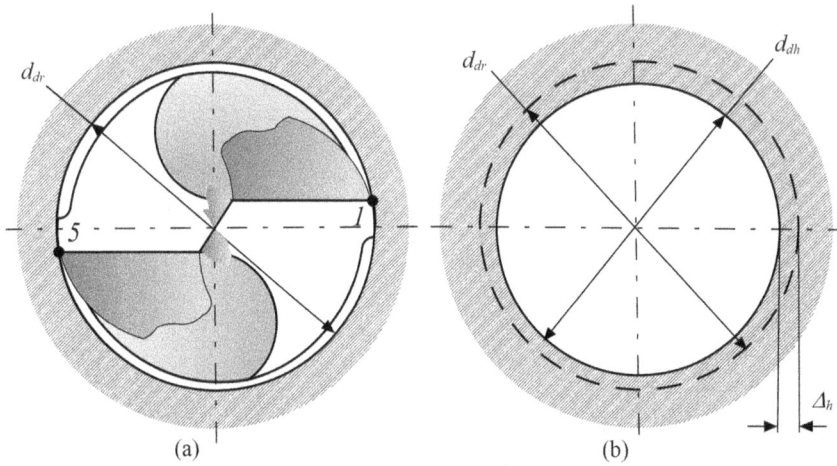

FIGURE 2.84 Model of springback in drilling operations: (a) the diameter of the hole being drilled at the point of cut and (b) elastic and thermal recovery behind this point.

shrinkage takes place. As a result of the mentioned mechanical and thermal factors, the diameter of the hole being machined d_{dh} becomes smaller than that of the diameter of the drill, d_{dr} by $2\Delta_h$ as shown in Figure 2.84b.

2.1.3.8.2 Springback: Causes

Because the two mentioned factors affecting springback, namely mechanical factor, known as elastic recovery, and thermal factor, known as thermal shrinkage, are of great importance in the design and application of HP drilling tools, their underlining causes should be well understood.

As defined by Astakhov (1998/1999) and then supported by many others, e.g., by Atkins (2003, 2009), the process of metal cutting is the purposeful fracture of the work material as physical separation of the workpiece into two portions, namely the machined part and the chip, takes place. To understand the concept of elastic recovery, one should consider the stress–strain diagram of the work material up to fracture (Astakhov 2018). A generalized and thus simplified stress–strain diagram is shown in Figure 2.85 (Astakhov and Outeiro 2019). The most important aspects relevant to the foregoing analysis are segments between the numbered points. When a solid material is subjected to small stresses, the bonds between the atoms are stretched. When the stress is removed, the bonds relax and the material returns to its original shape. This reversible deformation is called elastic deformation represented by segment *1–2* in the diagram. For most materials, segment *1–2* is linear. Segment *1–3* is known as elastic strain, i.e., the maximum elastic deformation the material can be subjected to. The stress at point 2 is known as the yield strength of the material.

The slope of this linear segment (tan θ) is called the modulus of elasticity or Young's modulus, E, determined by selecting two points on segment *1–2* (*a* and *b*) in the manner shown in Figure 2.85 so that

$$E = \tan \theta = \frac{S_b - S_a}{e_b - e_a} = \frac{S_{ab}}{e_{ab}} \tag{2.16}$$

where S_a, S_b, e_a, and e_b stresses and strains at points a and b, respectively.

Segment *2–4* in the diagram (Figure 2.85) is called the strain hardening region. If the applied stress exceeds the yield strength, the material exhibits a combination of the elastic (reversible) and plastic (irreversible) deformations. If applied stress can grow further up to point *4* on the diagram, then fracture occurs. The strain corresponding to point *4* is known as the strain at fracture.

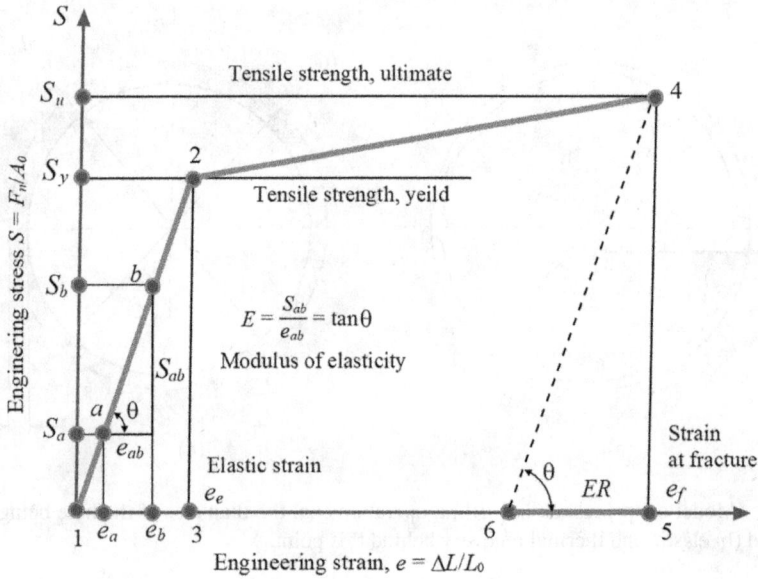

FIGURE 2.85 A simplified stress–strain diagram.

In Figure 2.85, it is represented by segment *1–5* on the strain axis. After fracture, however, the applied stress is released and the permanent strain found in the work material (represented by segment *1–6* in Figure 2.85) is less than that at fracture by the elastic strain represented by distance *6-5* in the stress–strain diagram. As such, the location of point *6* is readily found by drawing a line from point *4* parallel to line *1–2*, i.e., at angle θ. Segment *5–6* on the strain axis is known as elastic recovery, *ER*, which is also known as springback in materials processing although the latter also includes thermal shrinkage.

Several variables influence the amount of elastic recovery. Among others, the stress at fracture (defines the height of the starting point of the unloading line represented by point *4* in Figure 2.85), the modulus of elasticity (defines the slope of the unloading line represented by line *4–6* in Figure 2.85), and thermal expansion are of importance. It is obvious from the diagram shown in Figure 2.85 that the higher the strength of the work material, the greater the elastic recovery; the lower the elasticity modulus, the higher the elastic recovery. Consider a few practical examples. As follows from the diagram shown in Figure 2.85, the said elastic recovery of a unit volume of the work material, termed as the unit elastic recovery, *UER*, can be determined using triangle *456* as the ratio of the ultimate strength of the work material, S_u, and its modulus of elasticity, E, that is, $UER = S_u/E$.

Consider practical examples:

1. *Elastic recovery*: As the modulus of elasticity is almost the same for a wide group of steels ($E = 200\,\text{GPa}$), *UER* is determined by the strength of the steel. For cold-drawn steel AISI 1012, having $S_u = 270\,\text{MPa}$, $UER = 0.00135$, while for annealed steel AISI 1095, having $S_u = 650\,\text{MPa}$, $UER = 0.00325$. The matter gets worse when titanium or aluminum alloys are drilled. For commonly used annealed titanium alloy Ti-Al6-4V (Grade 5), $S_u = 880\,\text{MPa}$, $E = 113.8\,\text{GPa}$, and $UER = 0.00772$, that is, *four times greater than that of medium-carbon steels*. This explains known difficulties with drill binding in machining of titanium alloys. For common in the automotive industry, aluminum alloy A380 $S_u = 320\,\text{MPa}$, $E = 71\,\text{GPa}$, and $UER = 0.00450$, which is greater than that of steels.
2. *Thermal contraction*: The coefficient of thermal expansion for steel AISI 1012 is $12.0\,\mu\text{m/m°K}$, whereas for A380 is $21.8\,\mu\text{m/m°K}$, i.e., it is almost twice greater for steel AISI 1012.

As a result, the combined (elastic recovery and thermal shrinkage) hole springback is much (more than threefold) greater for A380 than for steel AISI 1012.

2.1.3.8.3 Back Taper: Notion

As defined in Section 2.1.2, back taper is a slight decrease in diameter from front to back in the body of the drilling tool. The purpose of back taper is to reduce the heat due to friction while the tool is engaged in the workpiece, thus to prevent binding of a drill in the hole being drilled. If no back taper is applied to the drill's margins, then drill will be gripped by the contracted wall of the machined hole. Although such binding is one of the most common failure modes of twist drills, its real cause is not well explained, and thus measures for its prevention are not well understood.

In drilling tools such as drills and reamers, the back taper Δ_{bt} (included) is assigned on the drawing as a diameter decrease $(d_{dr} - d_{drl})$ per certain length l_{bt} (Figure 2.86), i.e.,

$$\Delta_{bt} = \frac{d_{dr} - d_{dr1}}{l_{bt}} \tag{2.17}$$

It was customary for many years to use $l_{bt} = 100\,\text{mm}$ in the assignees of the backtaper value. Conventional drill designs have backtaper values that correspond to standards established within the industry (0.03–0.12 mm/100 mm). Normally, back taper is not applied for the whole length of the drill rather for a certain length starting from the drill corners. This length (20–50 mm depending on the drill length) sometimes referred to as the effective length should always be a part of any drill drawing. For many years, this standard value of back taper was not of any concern as the runout of drills, drill holders, and spindles of machines (the so-called system runout Astakhov (2014)) was enormous from the modern standpoint resulting in the diameter of drilled hole much greater than that of drills. Moreover, the cutting speed used in drilling was low so that thermal expansion/contraction of the hole being drilled was neglectable.

Times have changed and so have the components of drilling system as discussed in Chapter 1. It is to say that better drills with small runouts, better tool holders, and machines were introduced in the industry over relatively recent times. Moreover, the use of high cutting speeds in drilling, particularly with carbide and PCD drills and reamers, became commonplace. What was not changed is the amount of back taper applied to standard drilling tools.

When the back taper is insufficient, the result depends on how small actual back taper is and many other particularities of the hole-making operations. A common indication of insufficient back taper is a poor quality of drilled holes. An example of hole excessive roughness is shown in Figure 2.87a. Figure 2.87b shows feed and retraction marks on the hole surface. Figure 2.87c shows chatter marks (known as "tiger strips" in the industry) and retraction marks on the hole surface.

Yet another clear indication of insufficient back taper is the appearance of the major and auxiliary margins of the drill/reamer. It is shown in Figure 2.88. As can be seen in Figure 2.88a, the major margin shows rubbing marks over a significant distance from the drill corner. Besides rubbing marks, the leading edge of the auxiliary margin shows evidence of "undocumented" cutting activities, which should never happen when the drill back taper is sufficient. It is sometimes referred

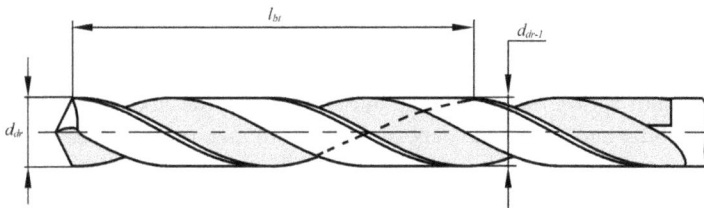

FIGURE 2.86 Showing the parameters for assigning back taper.

FIGURE 2.87 Poor quality of the drilled hole due to lack of back taper: (a) excessive roughness, (b) feed and retraction marks, and (c) chatter and retraction marks.

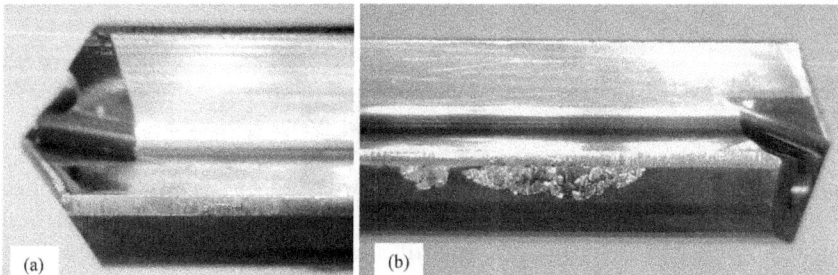

FIGURE 2.88 Showing the appearance of margins with insufficient back taper: (a) leading/major margin and (b) auxiliary/trailing margin.

to as scoring (Astakhov 2010). As can be seen in Figure 2.88b, the auxiliary/trailing margin shows marks of severe rubbing and even some deposit of the work material on its leading edge.

A more sophisticated outcome of insufficient back taper occurs when the back taper is made even smaller than that which caused simple scoring. It is the drill transverse vibrations and the occurrence of increased wear on the margins. Figure 2.89 shows the appearance of chatter marks and wear on the drill margin. It often happens in the automotive industry in drilling various aluminum

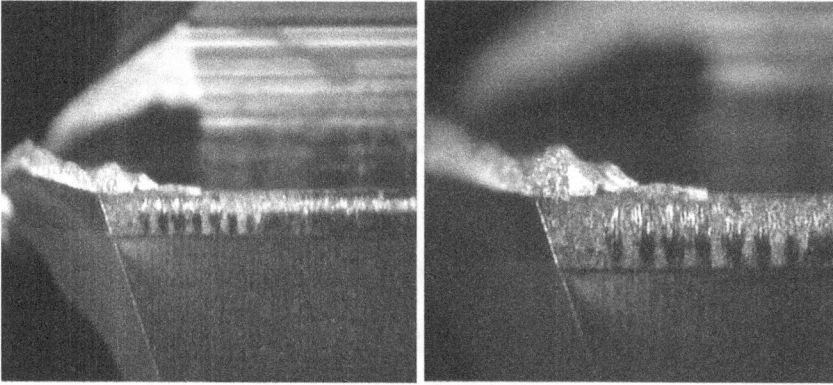

FIGURE 2.89 Chatter marks and wear on the drill margin.

FIGURE 2.90 Wear of the drill major (a) and trailing (b) margins due to high contact pressure caused by the lack of back taper.

alloys having a great thermal expansion/contraction property. The major problem is not only in subpar quality of drilled/reamed holes but also total tool life as to restore drill/reamer margins to their normal working condition, a great amount of tool material should be ground back. This greatly reduces the total number of regrinds, e.g., from commonly intended eight regrinds for drills to 3–4. The matter is even worse for reamers where the margins should form a hole of low roughness and tight diametrical and shape tolerances.

When the back taper is made even smaller, the tool condition deteriorates rapidly. Figure 2.90a and b shows the common wear pattern that occurs on the gundrill leading and trailing margins due to high contact stresses caused by the insufficient back taper. As seen, the margin adjacent to the major cutting edge is ruined due to high contact pressure, and the built-up edge is formed on the auxiliary margin. When this contact pressure becomes high enough, particularly for multi-edge axial tools such as reamers, the tool breaks by an excessive torque (Astakhov 2010, 2014). Often, such a breakage (common examples are shown in Figure 2.91a and b) is wrongly attributed to a misalignment problem not paying attention that the tool is tightly gripped in the hole. Even after the tool breaks, its removal presents a significant problem.

When such a failure happens, the proper tool failure analysis should be carried out. Unfortunately, such analyses are rather exceptions in manufacturing, in general, and in the automotive industry, in particular. Significant unscheduled downtime associated with failed tool and scrap part replacement, restarting and adjusting the production line, and the cost of scrap of almost finished parts are just two examples of the price to pay for misunderstanding of the above-discussed issue.

(a)

(b)

FIGURE 2.91 Appearance of typical torsion-type drill breakage due to insufficient back taper: (a) breakage of a carbide reamer and (b) carbide twist drill.

2.1.3.8.4 Back Taper: Application-Specific Value

As discussed in Section 2.1.8.2, springback depends on the properties of the work material. It is shown that unit elastic recovery of the work material, UER is four times greater for aluminum alloy A380 than that for medium-carbon steels. Moreover, the coefficient of thermal expansion for A380 is almost twice greater for steel AISI 1012, i.e., it expands and contracts in this proportion. As a result, springback of the work material can vary over a great range for various work materials. Therefore, back taper for HP drilling tools should be application-specific, e.g., it should be much greater for titanium and aluminum alloys than that for cabin steels. Unfortunately, this is not nearly the case in modern industry.

Application rule #14: These drill/reamers should have backtaper values that are virtually double those pre-established guidelines: 1–1.5 μm/mm for reamers and 2–2.5 μm/mm for drills. For long drills (length-to-diameter ratio more than 20), back taper should be as high as 3.8–4.2 μm/mm.

2.1.3.8.5 Minor Cutting Edge

As discussed in the Appendix, the minor cutting edge plays a significant role in turning. Because it is a legitimate cutting edge in drilling tools (shown as the edge *1–4* in Figure 2.21), its geometry,

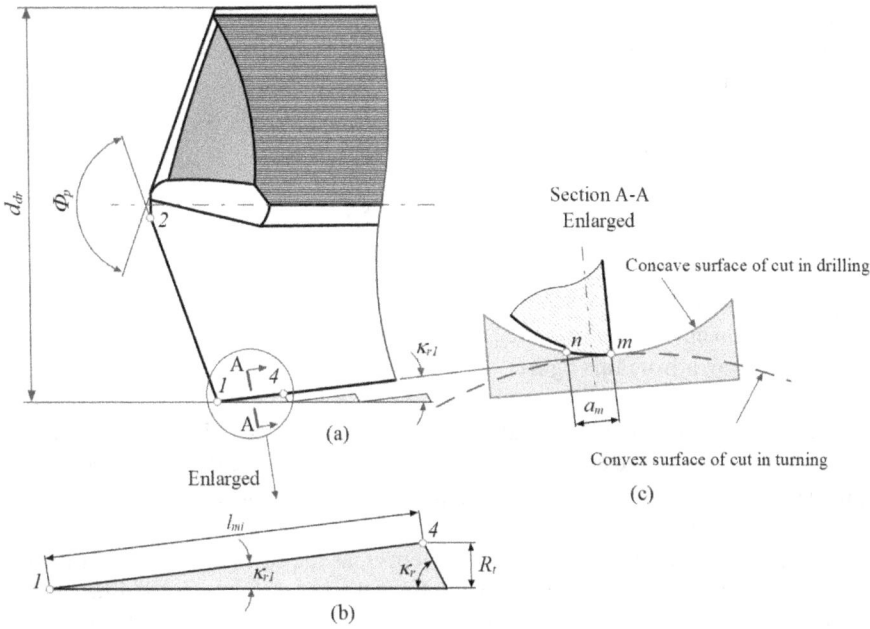

FIGURE 2.92 Visualization and particularities of the minor cutting edge *1-4*.

namely its length, tool cutting edge, rake, and clearance angles should be defined in full analogy with the major and chisel edges. In drilling tools, the minor cutting edge is defined as the line of intersection of the flute face and the margin. Figure 2.92a shows the definition of the tool minor cutting edge angle, κ_1, for a straight-flute drill (in the picture, κ_{r1} is shown significantly exaggerated for clarity). It is defined according to the standard definition provided by ISO 3002-1 standard (see the Appendix) as the acute angle between the projection of the minor (side) cutting edge into the reference plane (a plane that contains the drill longitudinal axis) and the direction set by the vector of the cutting feed.

The drill minor cutting edge angle κ_{r1} is not indicated on the tool drawing. Rather, it is "hidden" in the above-discussed back taper, Δ_{bt} (Eq. 2.17). As such, the tool minor edge angle can be calculated as

$$\kappa_{r1} = \arctan \frac{\Delta_{bt}}{2l_{bt}} \tag{2.18}$$

Figure 2.92b shows an enlarged fragment of the so-called theoretical surface roughness discussed in Section A.8.3 of Appending). In this picture, R_t is the maximum height of surface roughness, which is one of the standard surface roughness parameters. It geometrically follows from Figure 2.92b that

$$R_t = f_z \frac{\cos \kappa_{r1} \cos \kappa_r}{\cos \kappa_{r1} + \cos \kappa_r} \tag{2.19}$$

It follows from this equation that the reduction of the theoretical surface roughness can be achieved if under a given cutting feed f, the number of teeth, z is increased as $f_z = f/z$ and/or the point angle of the drilling tool is reduced because $\kappa_r = \Phi_p/2$. This explains why reamers as finishing tools have a greater number of teeth (3 and more) than drills and smaller point angle (commonly $\Phi_p = 90°$) and even much smaller for special low-surface-roughness gundrills shown in Figure 3.6 in Chapter 3. Moreover, the cutting feed per tooth used in reaming is much smaller than that in drilling. Note that the tool minor edge angle, κ_{r1}, plays a primary role in the formation of surface roughness whereas back taper does not. For example, even when an aggressive back taper $\Delta_{bt}=2\,\mu m/mm$ is used, this angle is only 0.12°. Clearly, such a small angle does not make a noticeable contribution to surface roughness according to Eq. (2.19).

The length of the minor cutting edge *1–4* is rather small. It geometrically follows from Figure 2.92b that this length is

$$l_{mi} = f_z \frac{\cos \kappa_{r1} \cos \kappa_r}{(\cos \kappa_{r1} + \cos \kappa_r)\sin \kappa_{r1}} \tag{2.20}$$

The cutting geometry of the minor cutting edge is shown in Figure 2.92c. It has never been presented in such a way in any literature source on drilling and drilling tools. The following important conclusions can be drawn from this figure:

1. For a straight-flute tool, the rake angle of the minor cutting edge is neutral (equal to zero) at the cutting edge and the actual rake angle is slightly negative due to the location of this edge ahead of the centerline. Note this angle can be altered in special designs of the rake face adjacent to the drill corner.
2. The clearance angle of this edge is zero. Moreover, as the surface of cut in drilling is concave as shown in Figure 2.92c. As a result of machined hole springback after drill corner due to the above-discussed mechanical and thermal factors, the surface of cut tightly grips the margin over its entire width (from point *m* to *n*). Note that this is not the case in turning, where the surface of cut is convex so that even when $\kappa_{r1}=0°$ as in a special case of the so-called viper geometry (Astakhov 2010), it does not present a problem as the surface of cut curves away or escapes from the cutting edge. Moreover, a viper edge is always provided with a clearance angle.

One may argue that the above-discussed hole springback is not always profound so that its extent is not sufficient to consider edge *1–4* as a legitimate cutting edge as its uncut chip thickness can be tiny. The problem, however, is that the hole springback is not a major factor to qualify this edge as the cutting edge. There is a much more serious reason for that based upon the hole generation mechanics.

The role of the minor cutting edge in turning is discussed in Section A.8.4 of the Appendix where a hypothetical cutting tool (Figure A.40a) is used to facilitate understanding of this role. To understand the real significance of the minor cutting edge in drilling, consider a hypothetical drill shown in Figure 2.93. As can be seen, this drill has a single (major) cutting edge and no minor (side) cutting edge.

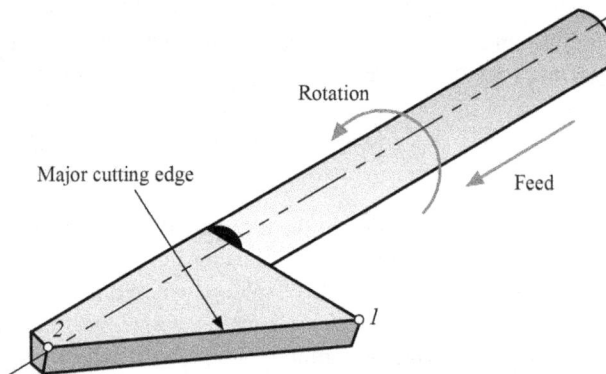

FIGURE 2.93   A hypothetical cutting tool used to explain the role of the minor cutting edge in drilling.

When this drill works, i.e., rotates and translates in the feed direction, it makes a hole. Figure 2.94 shows the axial cross section of this hole. For clarity, the feed per revolution is significantly exaggerated. Although this profile looks odd, Figure 2.95 explains this result. Figure 2.95a shows two successive positions of the discussed drill. As can be seen, because no side cutting edge is provided, a part of the work material represented by triangle ABC forms at each drill revolution that results in the hole profile shown in Figure 2.94. Moreover, if the half-point angle φ increases as shown in Figure 2.95b, the interference of the workpiece and the drill shank takes place. When $\varphi \rightarrow \pi/2$ (an extreme case for the considered hypothetical drill), contour $ABB'C$ representing the materials cut by the major cutting edge changes assumes a rectangular shape having the maximum cross-sectional area.

A real drill is made with the minor (side) cutting edge that cuts the material left by the major cutting edge (represented by area ABC in Figure 2.94). As such, the following is true: the greater the point angle (half-point angle φ) of a drill, the more uncut material is left for the minor (side)

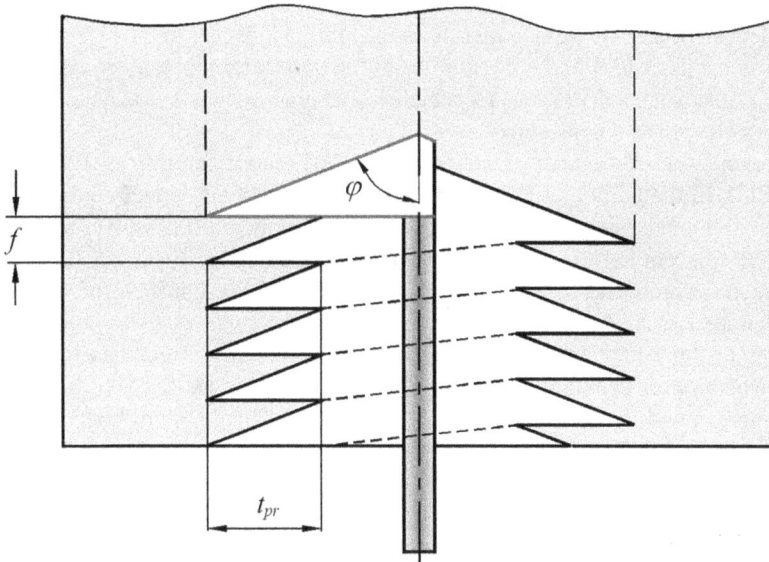

FIGURE 2.94 Profile of the hole drilled by the hypothetical drill shown in 2.93.

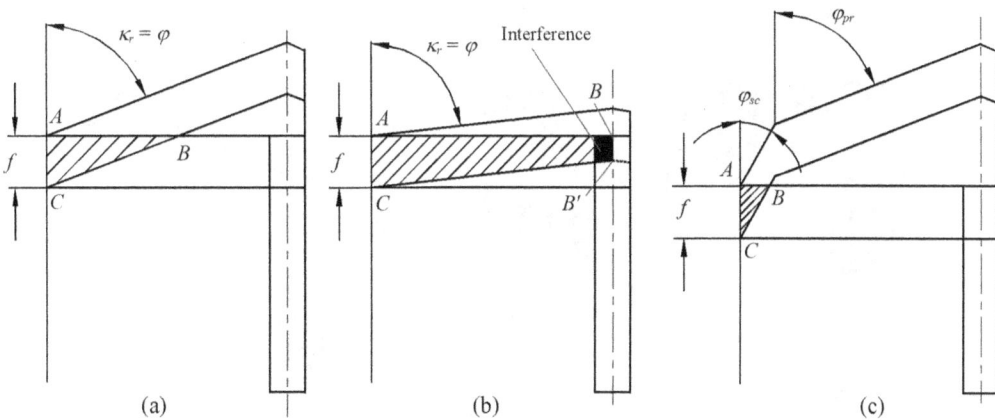

FIGURE 2.95 Explanation of the profile shown in Figure 2.94 and influence of φ on the uncut chip thickness cut by the minor (side margin) cutting edge: (a) for a normal point angle, (b) for a large point angle, and (c) for a double-point angle construction.

cutting edge to cut. Unfortunately, this edge is not meant for cutting according to known designs simply because it does not have the clearance angle (Figure 2.92c), which makes drilling unstable and causes premature wear of the drill periphery corners.

The foregoing analysis also explains a significant improvement in tool life and drilled hole quality in drilling difficult-to-machine materials when a double-point drill shown in Figure 2.31 is used. Figure 2.95c provides a physical explanation. As can be seen, when the major cutting edge of the discussed hypothetical drill is modified by the second point angle having a much smaller half-point angle φ_{sc} compared to point angle φ_{pr}, the area of triangle ABC that represented the work material to be cut by the minor cutting edge is substantially reduced and that significantly improves cutting conditions on the minor cutting edge.

Figure 2.96a and b shows conditions of the margin adjacent to the drill corner, i.e., the flank face of the minor cutting edge in drilling of crankshaft micro-alloy steel RES 10531 Grade A and aluminum alloy A309. As can be seen, the absence of the flank face with a clearance angle causes fracture of the minor cutting edge in drilling of the steel and galling (defined in the drill nomenclature as an adhering deposit of nascent work material on the margin adjacent to the periphery corner of the cutting edge) on the flank face of the margin in drilling of the aluminum alloy. Although it happens each and every time when a drill is used, specialists and practitioners in the field turn a blind eye to the conditions of the margin adjacent to the drill corner.

As it is known, wear of the drill corner is, by far, the greatest and thus is the tool life criterion (Astakhov 2014). In the author's opinion, there are two practically equal contributing causes to this wear. The first one is the combined thermal and mechanical process over a region of the primary flank face adjacent to the drill corner. As known, the temperature and sliding speed in this region are greatest compared to other regions (Astakhov 2014). On the other hand, the extent/ width of the contact area of the primary flank face and the surface of cut is small, particularly when the primary flank face has a sufficient clearance angle. The second cause is high friction/ ploughing/burnishing of the flank face of the minor cutting edge. The width of this contact is equal to the margin width (see width mn in Figure 2.92c). This happens because the minor cutting edge is not provided with a clearance angle that should be the case for any legitimate cutting edge. In the author's opinion, this is a "birth defect" of all known drills that should be finally corrected.

The role and thus the design of drill margins should be re-thought for all drill and drilling tools particularly for high-speed high-penetration rate tools working in modern drilling systems of high accuracy. The major line of thought is providing the minor cutting edge with clearance angle, and thus reducing friction between the margins and the wall of the hole being drilled. The traditional margins are cylindrical. Its width a_m (Figure 2.92c) is selected in the range of 0.3–0.6 mm

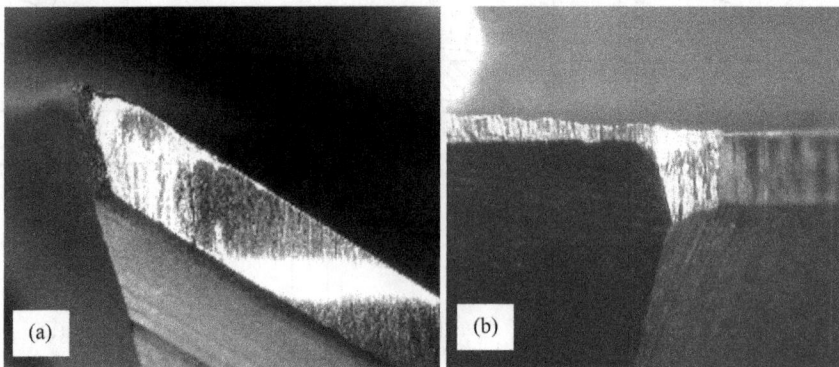

FIGURE 2.96 Corner conditions in drilling: (a) crankshaft micro-alloy steel RES 10531 Grade A and (b) aluminum alloy A309.

depending on the drill diameter. This range was adopted based on experience in the nineteenth century and hardly changed since then.

Various designs of modified flank face of the minor cutting edge are discussed by the authors earlier (Astakhov 2010, 2014). Among these designs, the design, named as the VPA1© design shown in Figure 2.97, is proven to be simple and effective for a wide variety of drilling tools including HP drilling tools. As can be seen, the flank face of the minor cutting edge includes a narrow margin having the width a_{ml} followed by the flank face ground with the clearance angle α_m. The length of the modified portion is l_{mm}. For the general-purpose drill, these parameters are recommended to be as follows:

- $a_{ml}=0.10...0.15\,mm$. For a small drill diameter, it can be $0.05\,mm$ and if a drill is used under conditions with heavy misalignment, e.g., for cored holes, this width can be zero.
- l_{ml} equal to approximately three times greater than the feed per tooth.
- $\alpha_m=5°...7°$.

Application experience shows that the application of the VPA1© design significantly improves the cutting conditions of the minor cutting edge, while the narrow margin still maintains drill stability. Figure 2.98 shows the appearance of the VPA1© design on a drill. The application results show

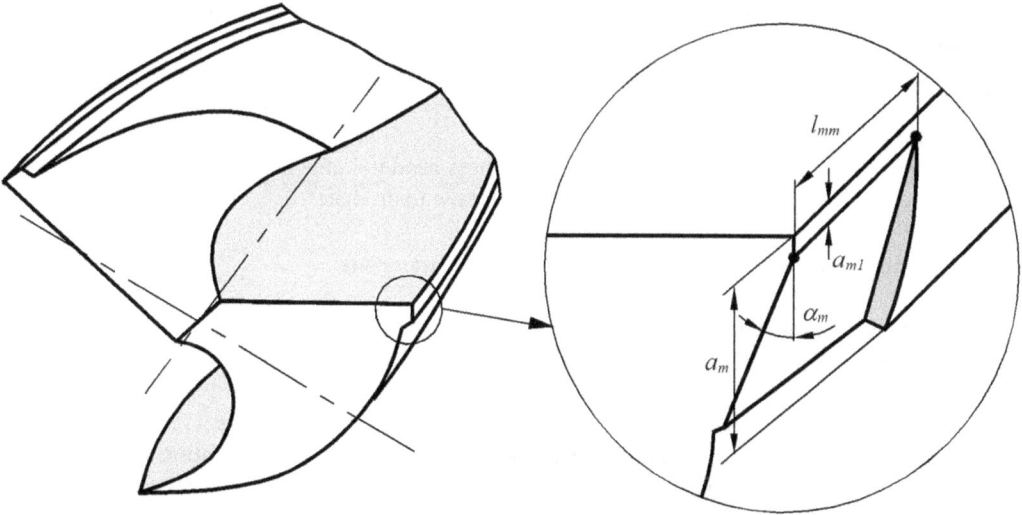

FIGURE 2.97 Margin geometry modification – VPA1© design.

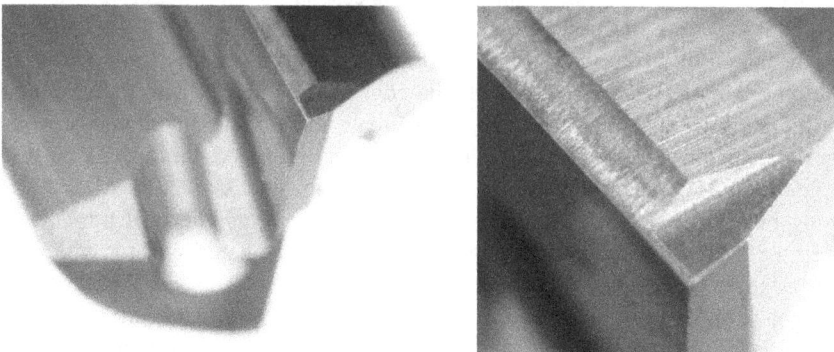

FIGURE 2.98 Appearance of the VPA1© design on a drill.

a significant increase in tool life and quality, both drilled hole shape and surface roughness, of machined hole in drilling and reaming operations. In the author's opinion, this should be a standard feature of common drills and reamers used in industry. The manufacturers of CNC drill point grinding machines should add this design to their standard grinding programs (libraries or galleries) to help end users with wide implementation of the VPA1© design.

2.1.4 MICROGEOMETRY

2.1.4.1 Notion and Basics

So far, the geometry parameters of cutting tools including drilling tools considered at the so-called macrolevel, i.e., at the level at which they are commonly considered and shown in tool drawings. For many-many years, such a consideration was sufficient as corresponding to the tool manufacturing technology level available over these years. Times have changed as higher accuracy and productivity of machining requirements force tool manufacturers to upgrade their technologies to improve the quality of cutting tools and use the modern equipment for their inspection. As such, the considerations of the tool microgeometry, assessment, and assurance of its parameters gained some interest. In the author's opinion, being of vital importance for HP drilling tools, such consideration lacks the correlation of the microgeometry parameters with the metal cutting theory, i.e., with the actual cutting conditions. Therefore, the current section aims to answer the following questions:

What is that microgeometry?
What are the microgeometry parameters?
Why do manufacturing engineers and tool designers need to know them?
Why does one need to measure/inspect them and how to do that?

Microgeometry of cutting tools includes the following parameters:

1. The quality of the cutting edge is often referred to as its sharpness.
2. Surface roughness (a.k.a. surface finish) of the rake and flank contact interfaces.

Being seemingly independent for a casual observer, these two items are correlated directly. Moreover, both have a strong influence on tool life in demanding drilling operations.

As defined in the Appendix, the cutting edge is a line of intersection of the rake and flank surfaces, and thus, it is treated so at the macrolevel. As will be shown further that it may not be the case in reality. To understand how far a common model of chip formation with a perfectly sharp cutting edge is from reality, one first needs to understand the optimal cutting conditions with a perfectly sharp cutting edge. A simple model for this case is shown in Figure 2.99a. The particularities of this model were discussed by the author earlier (Astakhov 1998/1999, 2006, 2010). In this model, the action of the normal force N on the rake face at a certain distance from the cutting edge creates the bending moment ahead of this edge that combines with the compressive load applied by the cutting force. These two together create the so-called combined (complex) state of stress in the zone ahead of the cutting edge that, when optimized, minimizes the resistance of the work material to tool penetration (i.e., cutting) (Astakhov and Xiao 2016, Abushawashi 2017). In machining of not very ductile materials, a visible crack forms ahead of the cutting edge as shown in Figure 2.99b. The sharper the edge, the higher the probability of formation of a visible crack as this edge serves as a stress concentrator.

The springback of the work material δ_1 is at minimum as the minimum cutting force is required at these optimized cutting conditions. Combined with the optimal clearance angle α, the flank contact area (represented by segment AB in Figure 2.99a) is small. This maximizes tool life in terms of reduction of the flank wear.

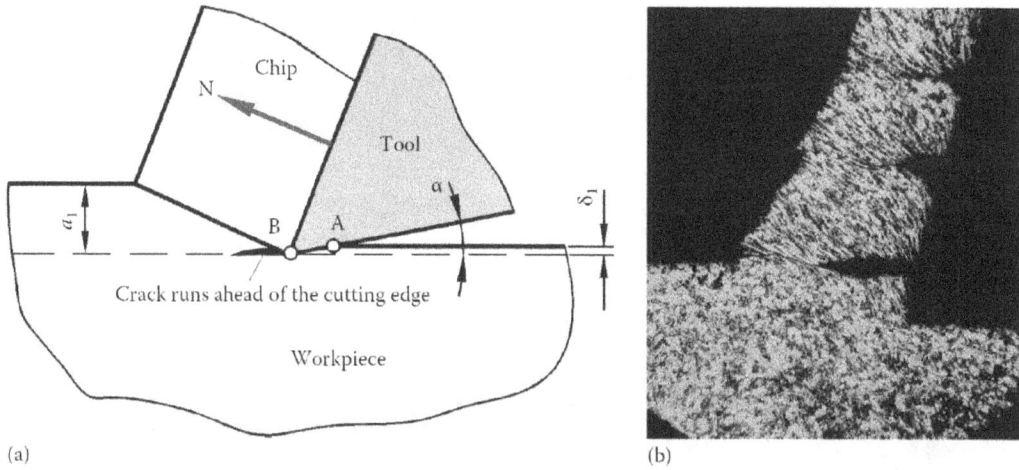

FIGURE 2.99 Cutting with a perfectly sharp cutting edge: (a) model and (b) partially formed chip.

In reality, however, the cutting edge is not a line due to roughness of the rake and flank surfaces. Figure 2.100 shows examples of common roughness of the flank and rake surfaces of drills. Note that these pictures are taken from the top-of-the-line drills made for the most advanced automotive plant. Therefore, it can be safely assumed that surface roughness of other drilling tools is even worse. This explains the earlier statement about the existence of the transition surface between the rake and flank surfaces instead of a sharp cutting edge. Even if it does not exist (visibly) for a new tool, it is formed just after machining of a few holes because of microfracturing of asperities due to rough rake and flank faces. As a result, the model shown in Figure 2.99a is not adequate for many practical drilling operations. In reality, the cutting edge is a highly-irregular transitional surface between the rake and flank faces in any real-world drilling tool as shown in Figure 2.101a. It is convenient for modeling to approximate/represent the irregular surface by a cylindrical surface having radius R_{ce}, termed as the radius of the cutting edge, as shown in Figure 2.101b.

The problem with the rounder cutting edge (in the manner shown in Figure 2.101b) is the variation of the rake angle over its round part. The essence of the problem is that the rake angle becomes highly negative over this radius toward the flank face. It is well established that the cutting, i.e., physical separation of a certain layer chip from the workpiece, stops and burnishing, as pure plastic deformation, takes over at a certain negative rake angle (Zorev 1966, Astakhov 2006). The particular value of this negative rake angle, known as the limiting rake angle, mainly depends on the properties of the work material. Moreover, it varies in a wide range depending on a particular work and tool materials as well as the contact conditions over the tool–chip interface. Outeiro and Astakhov (2005) developed the method of incremental rake angle graphical interpretation shown in Figure 2.102a, which allows to determine the actual rake angle and uncut chip thickness at a given point i of the rounder part of the cutting edge based upon the radius of the cutting edge R_{ce}. In this figure, h_{D-i} is the portion that corresponds to point i of the total uncut chip thickness, h_D, and the intended rake angle of the cutting edge is γ. According to this method, the actual rake angle at point i of the cutting edge, γ_i, is defined as the angle between the tangent to the cutting edge at this point and the vertical line which represents the trace of the reference plane (see Section A3 of the Appendix). The variation over the round part of the cutting edge is represented as

$$\gamma_i = \begin{cases} \arccos\left(\dfrac{h_{D-i}}{R_{ce}} - 1\right) & \text{if } h_{D-i} < R_{ce} \cdot (1 + \sin\gamma) \\ \\ \gamma & \text{if } h_{D-i} \geq R_{ce} \cdot (1 + \sin\gamma) \end{cases} \qquad (2.21)$$

FIGURE 2.100 Examples of common coarse roughness of the flank and rake surfaces: (a) flank face in the region of the chisel edge, (b) the primary flank along the major cutting edge, (c) rake face, and (d) primary flank face adjacent to the drill corner.

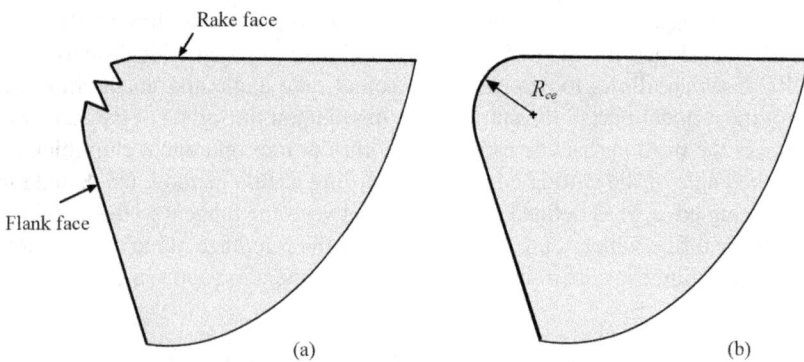

FIGURE 2.101 Cutting edge as the transitional surface between the rake and flank faces: (a) in reality and (b) in practical consideration.

Figure 2.102b shows a model of cutting with a rounded cutting edge. At point A, the rake angle for the given cutting condition reaches its limiting negative value, γ_{lim} at which cutting is ceased and thus burnishing takes over. As a result, the total uncut chip thickness h_D is *physically* separated at point A into two work material flows in its transformation into the chip: the actual uncut chip thickness h_{D-A} and the layer of thickness h_{D-b} to be burnished by the round part adjacent to the tool flank face. In technical terms, physical separation of a material into two or more parts is referred to as fracture, so the fracture stress corresponding to the state of stresses at point A should be achieved to accomplish the discussed separation.

The conditions at point A to achieve such a state of stress are much werse than that at point B of the cutting edge with a sharp cutting tool shown in Figure 2.99a. It is to say that much greater energy required due to a significantly increased amount of plastic deformation of the work material prior fracture. As a result, higher local temperature and cutting force occur. As a result, springback δ_{1-r} shown in Figure 2.102 is much greater than that with a sharp cutting edge, δ_1 shown in Figure 2.99a. This leads to the increase in tool wear/reduction of tool life and possibility of microchipping of the cutting edge. Moreover, burnishing of layer h_{D-b} results in the machining residual stress at the surface layer of the machined part (Outeiro 2020).

Figure 2.102c shows the use of the model shown in Figure 2.102a to determine the location of point i_{lim} and the uncut chip thickness h_{D-r} at this point i when the limiting rake angle γ_{lim} is known.

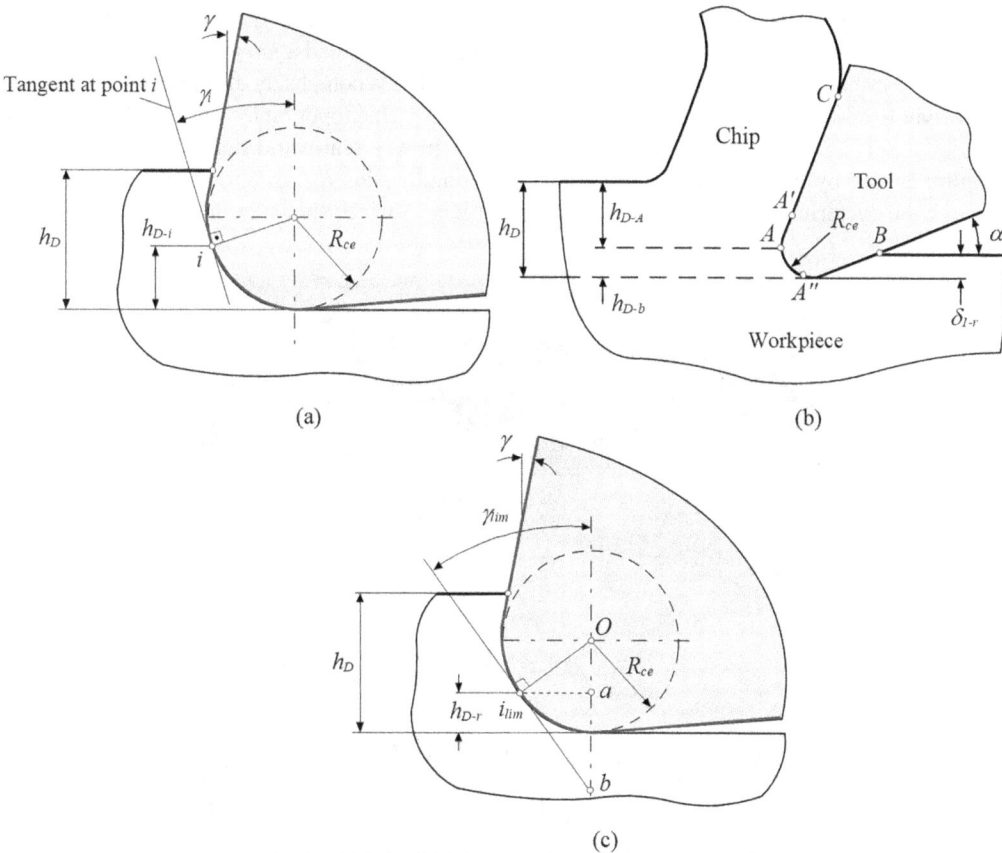

FIGURE 2.102 Rounded cutting edge: (a) graphical interpretation of the method of incremental rake angle, (b) model of cutting with the rounded cutting edge, and (c) model for determining h_{D-r}

2.1.4.2   Cutting Edge Preparation

2.1.4.2.1   Idea

Cutting edge preparation (hereafter, CEP) is a modification of the cutting edge (considered as a theoretical line) into a controlled transitional surface between tool rake and flank surfaces (planes). In other words, CEP is a technology of applying a defined radius to the cutting edge to improve the quality of the transitional surface between the rake and the flank faces and thus improve cutting edge quality.

Although as suggested by the current analysis and confirmed by experimental results (Thiele et al. 2000, Özel et al. 2005), a sharp cutting edge is better than the honed one in terms of cutting forces and temperatures. The problem is such a sharp cutting edge is not normally achieved in modern tool manufacturing (see Figure 2.100). Defects of the cutting edge are present in nearly all cutting tools (Rodriguez 2009). They are mainly the result of finished EDMing, EDG, and grinding. Considered by many as microscopic, these defects are not as small as many practitioners believe. These defects include microcracks, burrs, burns, and serrated cutting edges that lead to great variations in tool performance particularly in demanding applications. In the author's opinion, increased adhesion of the work and tool materials is the major damaging factor. Adhesion between tool and work materials always results in the formation of the BUE, which grows in each cycle of chip formation (Astakhov 1998/1999). The problem is in the size and strength of adhesion bonds. The poorer the topology of the cutting edge, the greater these two factors. When BUE reaches a certain height (depending on the strength of adhesion bonds), it is periodically removed by the moving chip. Such a removal does the damage as BUE takes away pieces of the tool material. Figure 2.103 shows a graphical example of the damaged cutting edge due to this process. This issue is discussed by the author in Chapter 4 in Astakhov (2024) where adhesion tool wear is considered in detail and a *fairy tale* common in metal cutting literature about the so-called *protective action* of BUE is completely dismantled.

The whole idea and objective of CEP is to improve the topography of a highly-irregular transitional surface between the rake and flank faces in any real-world drilling tools as shown in Figure 2.101a by a cylindrical surface having defined radius R_{ce}, as shown in Figure 2.101b. Although the performance of a tool having the transitional surface shown in Figure 2.101b is worse

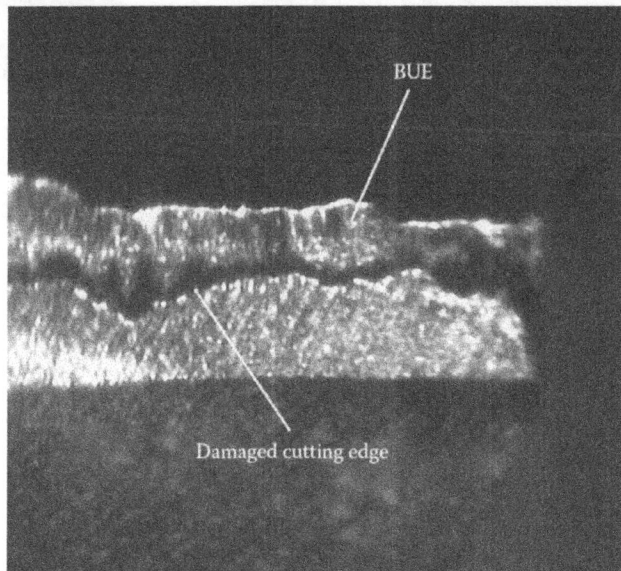

FIGURE 2.103   BUE on the drill cutting major edge.

than that with the sharp cutting edge shown in Figure 2.99a, it is much better than the performance of a tool with a highly-irregular transitional surface shown in Figure 2.101a. The essence of CEP technology is shown in Figure 2.104.

Multiple experimental studies and author's experience reveal the following:

- Tool life of HSS, carbide, PCD, and PCBN tools (single point, twist and straight-flute drills, reamers, milling tools, etc.) increases when the proper (optimal hone radius) CEP is used (Agnew 1973, Mayer and Stauffer 1973, Rech 2006, Biermann and Terwey 2008, Cheung et al. 2008).
- The size and surface finish and the process stability (spiraling, chatter marks) in machining of aluminum alloys are much better when a suitable CEP is used. Even small hand honing by a diamond file can improve these parameters noticeably.

When applied properly, CEP works because it

1. Significantly improves the surface roughness on the tool–chip and tool–workpiece contact interfaces shown as AC and AB, respectively, in Figure 2.102b. This reduces adhesion forces over these interfaces, and thus reduces adhesion-fatigue wear (Astakhov 2006, Astakhov and Shavets 2020).
2. Heals surface microdefects such as cracks and voids in the vicinity of the cutting edge left by the grinding wheel as shown in Figure 2.105. These defects are critical because they cause micro- and then macrochipping of the cutting edge. CEP just *heals* these defects (Komarovsky and Astakhov 2002) as ductile micro-cutting takes place even on superhard tool materials (Jahanmir et al. 1999, Zhong 2003, Kang et al. 2006).
3. Reduces tool vibration (Astakhov 2011b).
4. Improves coatability of the cutting tools. As well known (e.g., Bouzakis et al. 2009), the application of PVD coating on a sharp edge results in a very high internal stress. As a result, such a coating breaks away and peels off very shortly after starting cutting. For this reason, CEP is mandatory for tools to be coated. Figure 2.106 shows an example of coating properly applied on a rounded cutting edge.

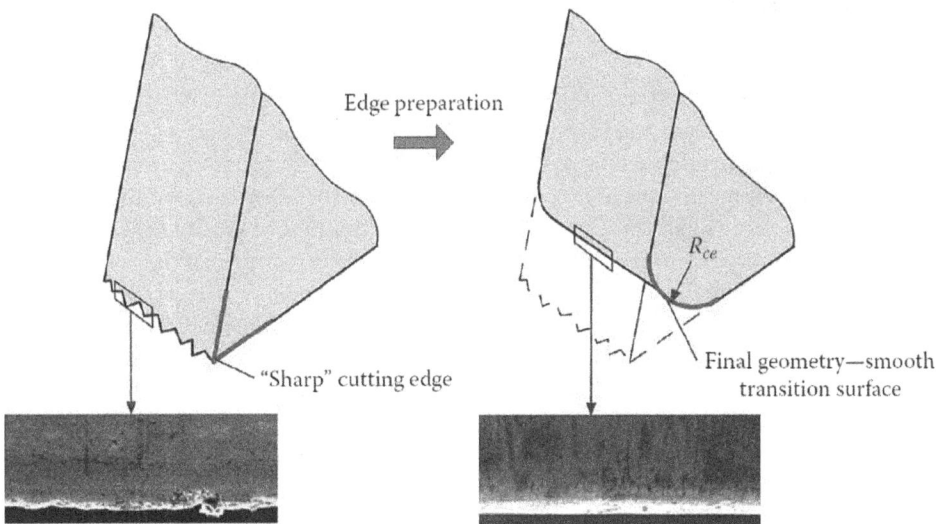

Edge preparation

R_{ce}

"Sharp" cutting edge

Final geometry—smooth transition surface

FIGURE 2.104 Essence of the CEP technology.

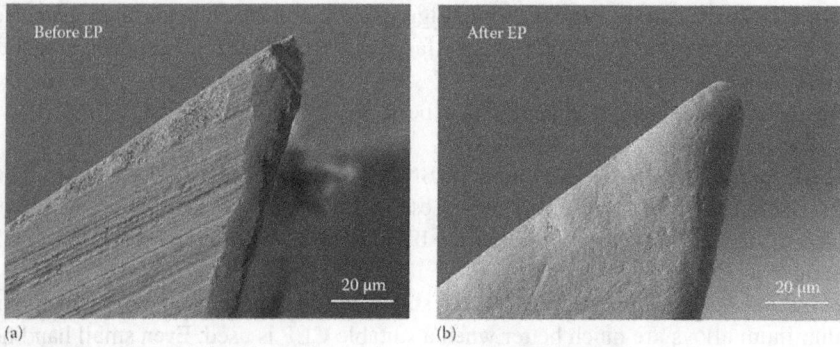

FIGURE 2.105   Example of CEP results: the cutting edge (a) before CEP and (b) after CEP by a drag finishing machine by Otec Co.

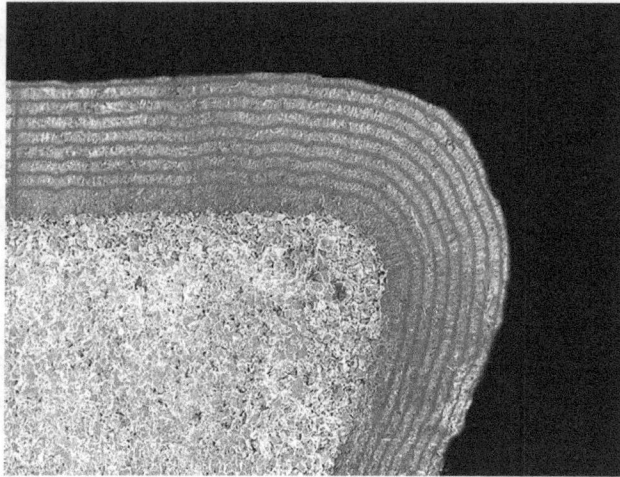

FIGURE 2.106   Tool carbide cutting edge coated by layers of CVD diamond coating.

2.1.4.2.2   Relative Tool Sharpness (RTS): The Existing and New Definition/Interpretation

Reading the foregoing analysis, one may wonder what the radius of the cutting edge R_{ce} should be for a given application. To address this reasonable concern, one should find the proper answer to the equation: What does it mean "sharp"?, thinking how sharp is the sharp. In engineering, any notion should be measurable or quantitative to consider it seriously. Zorev (1966) and Astakhov (2006) suggested that *the radius of the cutting edge R_{ce} should be judged against the uncut chip thickness h_D.* For characterization of cutting tool sharpness, the relative tool sharpness (hereafter, *RTS*) of the cutting edge was introduced as $RTS = h_D/R_{ce}$ (Outeiro and Astakhov 2005, Astakhov 2006). The minimum value of the ratio that corresponds to a negligibly small influence of the cutting edge radius on the cutting process is referred to as the critical relative tool sharpness RTS_{cr}. Zorev suggested the following empirical rule: the radius of the cutting edge does not affect the cutting process if the uncut chip thickness is at least ten times greater than the radius of the cutting edge so *RTS* is equal to or more than 10. In other words, the cutting edge is technically sharp if $RTS \geq 10$.

It was recommended by the author earlier (Astakhov 2014) that for practical *RTS* drilling tools are selected from the range of 6–10. Closer observations of the finishing drilling operations, e.g., reamers, have many teeth so that the feed per tooth is small, and, moreover, micromachining operations that the recommended range of *RTS* is overstated. This is because it was based on the wrong assumption that burnishing begins at point A (Figure 2.102b), i.e., at the point when the rounded part of the transitional surface starts. In reality, the rake angle at this point $\gamma_A = 0°$, which is far from the limiting negative value, $\gamma_{\lim}$, at which cutting is ceased and thus burnishing takes over.

To correct this mistake, Figure 2.102c shows the model for determining h_{D-r}, i.e., the actual part of the uncut chip thickness to be burnished by the rounded transitional surface. For given work and tool materials, $\gamma_{\lim}$ is fixed so that h_{D-r} is a function of the cutting edge radius, R_{ce}. Using this angle to determine the position of point $i_{\lim}$ and then from triangles $Oi_{\lim}b$ and $Oi_{\lim}a$, one can obtain

$$h_{D-r} = R_{ce}\left(1 - \sin|\gamma_{\lim}|\right) \tag{2.22}$$

Note that the absolute values (i.e., with no minus sign) $\gamma_{\lim}$ are used in this equation as it follows from Figure 2.102c.

For example, if it was found that $\gamma_{\lim} = -70°$ for a certain combination of the work and tool materials then when $R_{ce} = 0.3\,\text{mm}$, $h_{D-r} = 0.018\,\text{mm}$, i.e., $18\,\mu\text{m}$; when $R_{ce} = 0.05\,\text{mm}$, $h_{D-r} = 0.003\,\text{mm}$, i.e., $3\,\mu\text{m}$ when $R_{ce} = 0.01\,\text{mm}$, $h_{D-r} = 0.0006\,\text{mm}$, i.e., $0.6\,\mu\text{m}$.

Based on the foregoing analysis, a new definition of RTS is proposed. It is defined as

$$RTS = \frac{h_{d-r}}{h_D} \cdot 100\% \tag{2.23}$$

and for the optimal tool performance should be kept below 10%. When this is the case, the cutting edge is considered to be technically sharp regardless of its actual radius.

It is important to remember that RTS is specific for a given combination of the work and tool materials that results in a specific value of $\gamma_{\lim}$. It is to say that a given cutting edge/cutting tool may be sharp for one material but not sharp for the next having considerably different properties. Application experience shows that RTS should be kept at minimum in machining of high-strength work material of low thermal conductivity and great strain hardening, for example, precipitation hardening (PH) stainless steels and titanium alloys. What is most important, however, is that this optimum CEP lies within a rather narrow range. An example is shown in Figure 2.107a (Cheung et al. 2008) for steel drilling. In drilling of aluminum alloys with PCD drills, the window of opportunity is even smaller as shown in Figure 2.107b. As can be clearly seen, even a small deviation from

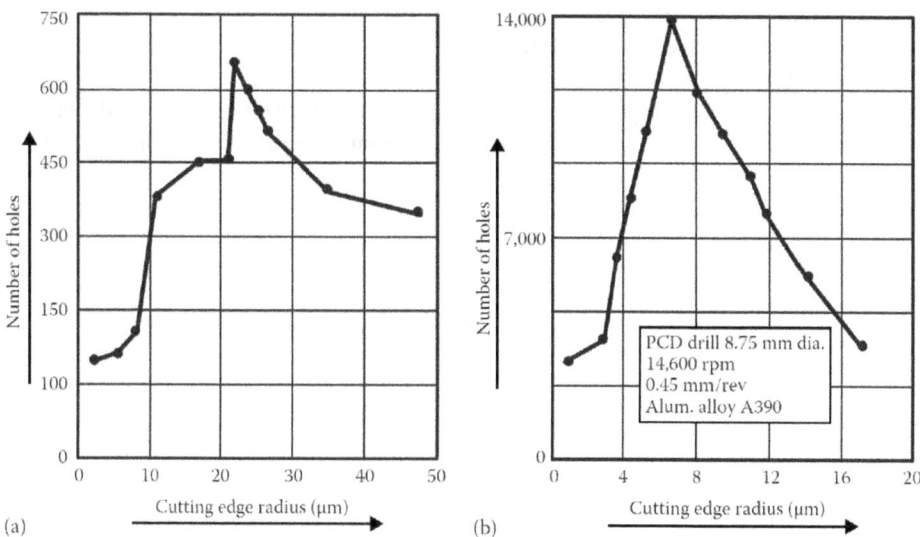

FIGURE 2.107 Influence of the cutting edge radius on the tool life in drilling: (a) twist drills in steel drilling and (b) PCD drills in high-silicon aluminum alloy A390 drilling.

the optimal cutting edge radius to either side results in a steep reduction in tool life. It is explained as follows: initially the rough cutting edge shown in Figure 2.101 is smoothed so tool life is increased to a certain maximum. Then, h_{D-r} gradually becomes excessive that reduces tool life as the cutting force and temperature increase due to excessive burnishing. Obviously, the maximum tool life depends on the initial topography of the cutting edge, i.e., how good it is made by grinding the rake and flank faces. This explains a great scatter in CEP application results as the optimal radius and its allowable deviation range (tolerance) are not normally established and thus are not a part of the tool drawing. Therefore, the optimum radius of CEP should be determined for critical applications.

2.1.4.2.3 Types

ISO 1832: 2004/2005 and ANSI B212-4 both define various types of CEP to be included in the cutting insert designations. The detailed information was presented by the author earlier (Astakhov 2010). Among multiple available forms of CEP, more than 95% of honed drilling tools receive a radius hone (Figure 2.108a), which is centrally located on the cutting corner of the tool, that is, the extent of CEP $L_{ep} = R_{ce}$. Tools with this type of hone are used for general applications. A half-parabolic shape is known as a *waterfall* or *reverse waterfall*, depending on its orientation to the rake and flank surfaces. With a waterfall-shaped hone, CEP is skewed toward the top side of the tool as shown in Figure 2.108b where normally for the waterfall CEP, its size along the rake face is twice greater ($2L_{ep}$) than that along the flank face (L_{ep}). The main benefit of a waterfall-shaped hone is that the honing process leaves more tool material directly under the cutting edge, which further strengthens the corner as suggested by Shaffer (2000). The authors, however, did not find any experimental proof to this statement. The promotion of a waterfall-shaped hone is most probably an excuse due to the fact that modern machines are not capable of applying the perfectly symmetrical radius hone shape shown in Figure 2.108a so that waterfall-shaped hones are actually made in reality.

2.1.4.2.4 Technology

There are three basic methods of EP: mechanical, thermal, and chemical. Each method includes a number of technologies developed. Figure 2.109 lists these methods and associated technologies. The basic description and details of these technologies are discussed by Rodriguez (2009). Each technology developer/CEP machine manufacturer shows various sales presentations, proving that a particular technology/machine is the best. These presentations/brochures/catalogs are generously supported by high-quality SEM images *before and after*, showing significant improvement in the quality of the transition surface and rake and flank contact areas. A brief comparison of these methods was presented earlier (Table C.1 in Astakhov (2014)).

Although the existence of a great number of methods and technologies of CEP, two methods of CEP are mostly used by drilling tool manufacturing companies. The first one is the honing by a nylon-abrasive-filament brush as shown in Figure 2.110. The brush filaments are constructed

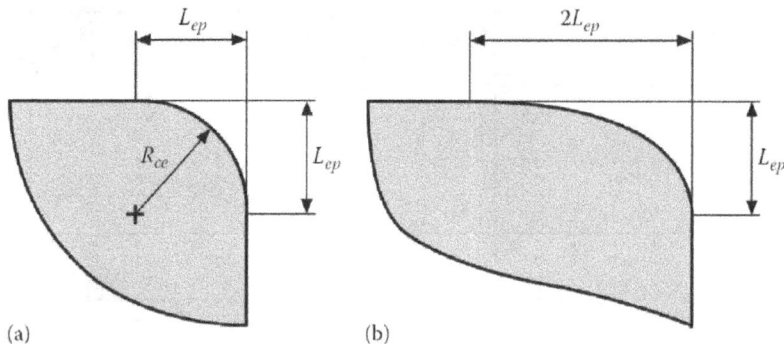

FIGURE 2.108 Common shape of CEP: (a) radius hone shape and (b) waterfall-shaped hone.

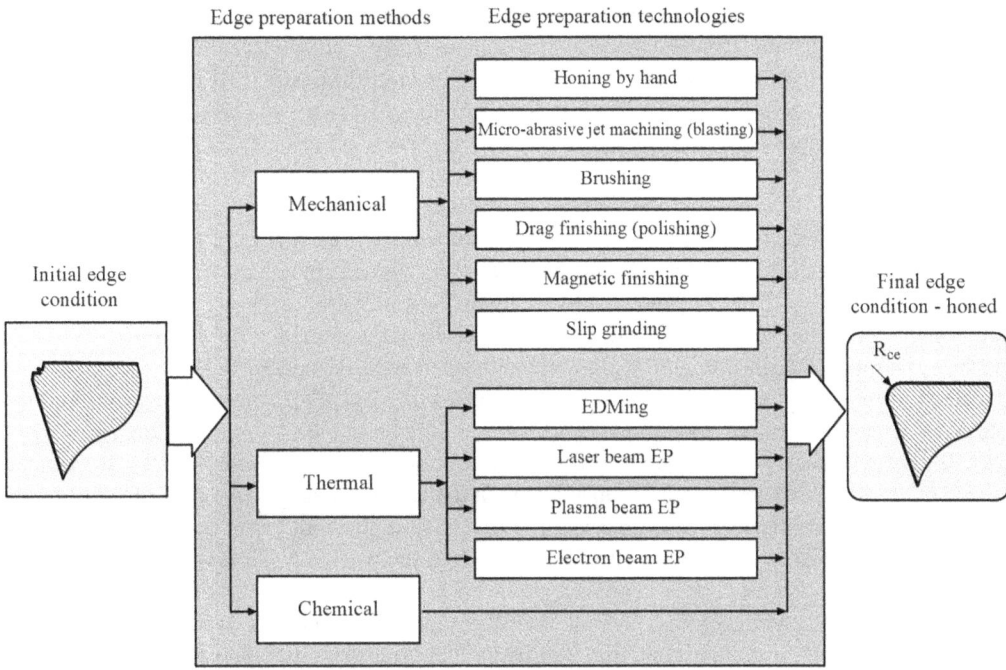

FIGURE 2.109   Basic methods and technologies of CEP.

FIGURE 2.110   CEP by a nylon-abrasive-filament brush.

from a nylon carrier that is co-extruded with abrasive grains. The abrasives used in the brush can also vary greatly including silicon carbide, aluminum oxide, ceramic, and diamond. The filaments can be round or rectangular, straight, or crimped, and brushes are available in various diameters. According to Mutschler Edge Technologies LLC (Cleveland, OH), as the brush wears, new abrasive grains are constantly being exposed to the workpiece. These flexible brush filaments act as "flexible files," wrapping and wiping across all edges evenly. Many factors determine the result, such as speed, direction, cycle time, depth of engagement, and centerline placement.

The second method is drag finishing. Its principle is shown in Figure 2.111a. As can be seen, the tool is fixed in a holder and dragged through granulate/abrasive media, while the tool is additionally rotating on its own axis. The tools follow a complex planetary-like path; this ensures a contact of all areas of the tool with the media. Reportedly, complex geometries can be processed using this method. Real machines have multiple spindle heads as shown in Figure 2.111b so that many tools can be processed simultaneously. The primary process parameters are the process time, speed, immersion depth of the tools, and the abrasive media itself including material, mix, grain size, and so on. These parameters determine the size and shape of CEP and surface roughness. According to various drag finishing machine manufacturers, the process time may vary from as little as a few seconds for removal of droplets from coatings to up to 20 minutes to achieve a 70 μm radius at carbide tools. A great number of good sales pitches, web-based document with "real-life" evidences and success stories convinced many tool manufacturers to acquire such a machine.

The mechanism of drag finishing was presented by Lv et al. (2022). The major drawback of this analysis is the absence of the scale in the consideration of the tool material removal by abrasive particles. Figure 2.111c shows such a mechanism. As can be seen, for the process to work as intended, the size of abrasive particles should be much less than the maximum height of the surface roughness, designated as R_z. The properly made HP drilling tool has the flank and rake faces ground with Ra =0.4 μm (Ra is the arithmetic mean roughness according to standard ISO 4287), which roughly equals to Rz=4.6 μm (Rz is the maximum height of profile according to standard ISO 4287). Therefore, to apply CEP properly, and thus to achieve the intended radius of the cutting edge and better surface roughness of the rake and flank face, the abrasive particle size should be no more than 2 μm. This is not nearly the case in practice where the grid size of abrasive medial is at least 10+ times greater.

The major problem of the common CEP methods is the lack of controllability in terms of CEP shape and place of application. It is to say that the honed surface is applied everywhere in the drilling

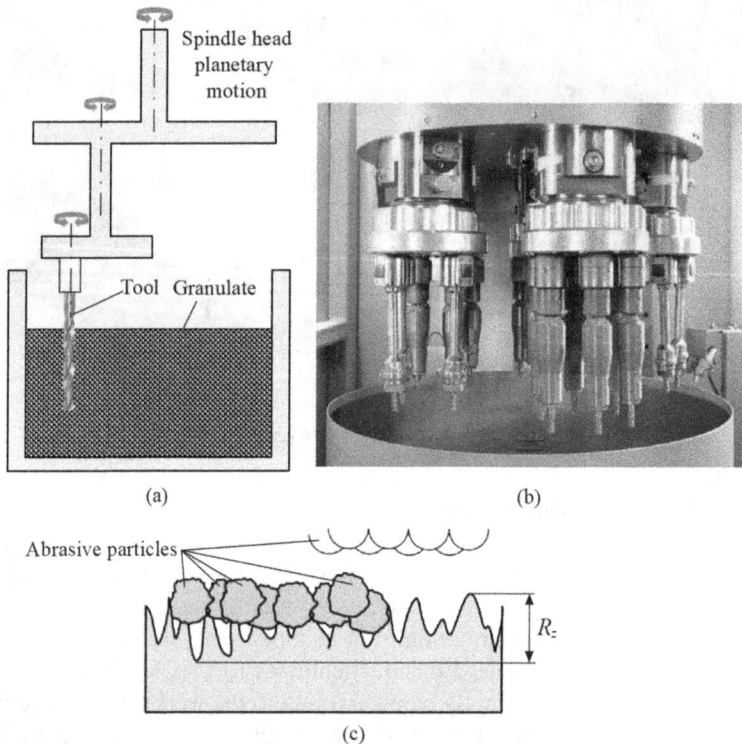

(a)　　　　　　　　　　　　　　　　　　　　　(b)

(c)

FIGURE 2.111　Drag finishing method of CEP: (a) principle, (b) drag finishing machine with planetary driven multiple tool holders for efficient processing, and (c) mechanism of the method.

tool that is incorrect. The most vulnerable part of the major cutting edge is drill corner as it determines the quality of machining and tool life (Astakhov 2014). Therefore, the corners of HP drills, and especially, HP reamers should sharp, i.e., should not be overhoned. The problem is that corner having the smallest amount of "meat" behind them (the cutting wedge) is honed using the above-described common methods of CEP much easily than the cutting edge. As a result, its rounding is the greatest. Figure 2.112 shows the digitally-scanned topography of the whole cutting edge. As clearly seen, CEP applied to the drill corner is much greater than that on the rest of the cutting edge. Unfortunately, this issue cannot be resolved with the above-discussed common methods of CEP. Needless to say, it is never discussed in the literature on CEP. Honing by hand (with diamond file) is the best CEP technology simply because an experienced CEP operator never touches drill corners.

2.1.4.2.5 Cost

Figure 2.113 shows a typical manufacturing cost structure for common drilling tools (Rodriguez 2009).

2.1.4.2.6 Metrology of CEP

The proper metrology has to be an inherent yet most neglected, in practice, part of any application of CEP technology. Such a metrology allows one to

1. Compare the results of CEP, that is, compare the quality of the cutting edge before and after CEP application.
2. Assure that the optimized/intended parameters of CEP are actually applied.
3. Assure the consistency of CEP technology or, alternatively, compare different CEP technologies in order to select the most suitable for a given application.

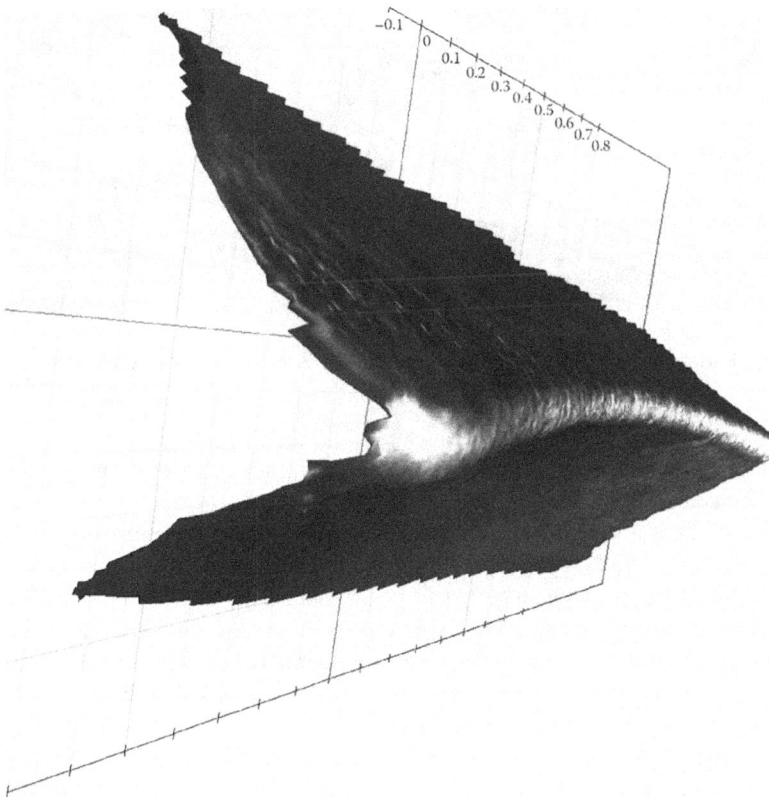

FIGURE 2.112 Topography of the whole cutting edge.

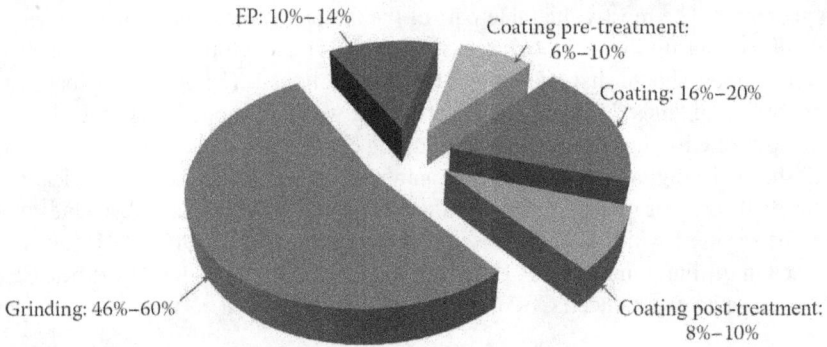

FIGURE 2.113 Manufacturing cost structure for common drilling tools.

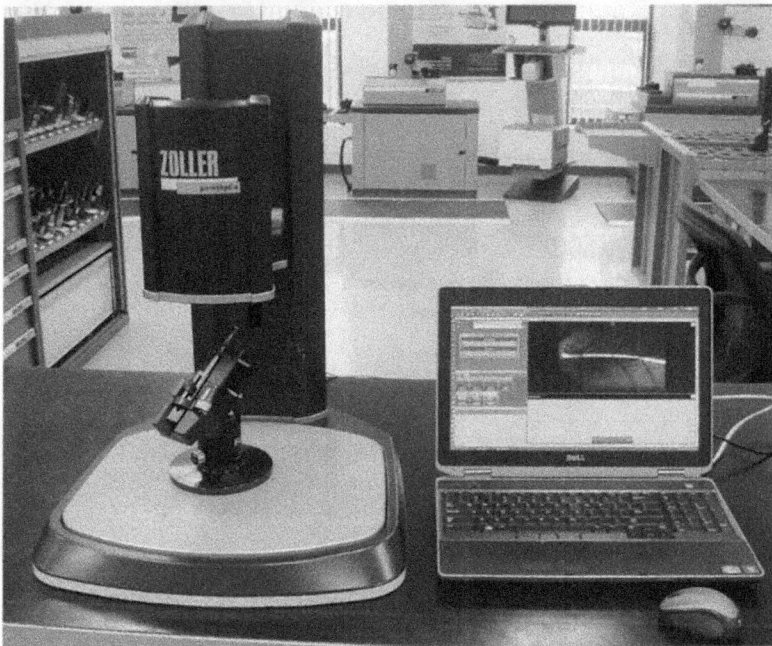

FIGURE 2.114 Portable inspection machine *promSkpGo* by Zoller Co. for the assessment of the quality of the cutting edge.

The rule of thumb, however, is as follows: if a proper metrology of CEP is not available, it is better not to apply CEP at all.

According to the author's experience, the portable inspection machine *promSkpGo* by Zoller Co. (Figure 2.114) is a budget and easy-to-operate system intended for a process-oriented measurement of the cutting edge rounding that can be used for fast and accurate assessment of CEP. It includes a specially designed universal drill holding fixture that allows a fast and proper orientation of the drill's cutting edge to be investigated with respect to the microscope optical axis. A spherical joint and graduated dials allow the final adjustment of the drill position. Using this fixture, any desirable edge of the drill can be properly oriented and inspected.

At the first stage of the quality assessment, a segment of the cutting edge is selected for the assessment of CEP as shown in Figure 2.115. The black shadow lines indicate its boundaries and the middle of the region.

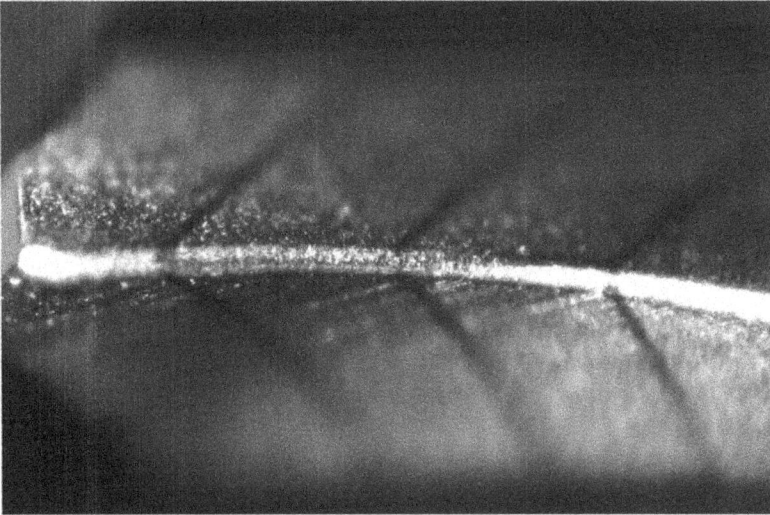

FIGURE 2.115 A segment of the cutting edge selected for measurement.

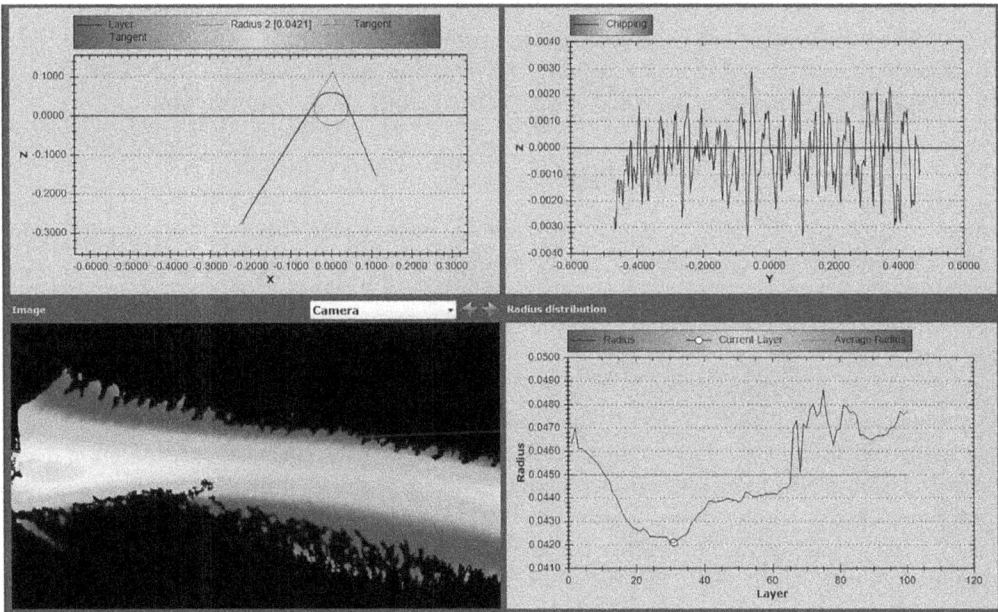

FIGURE 2.116 The results of the cutting edge assessment: (a) before CEP and (b) after CEP was applied.

Figure 2.116a shows the summary report of the quality of the cutting edge before applying CEP. It includes the maximum radius of the cutting edge approximated digitally, the variation of this radius over the length of the selected region, and the roughness of the cutting edge (termed as *chipping*). Using this report, one may conclude that the quality of this cutting edge is very low as it is heavily serrated and the amplitude of the serration is great. Therefore, it was concluded that the application of CEP is needed.

Figure 2.116b shows the result of the application of CEP. As can be seen, the waterfall-type (see Figure 2.108b) CEP is applied. The desirable range of the cutting edge radius (the tolerance on this radius) is set as its lower and upper boundaries. The variation of this radius over the cutting edge is small and just below the mean value preset in the evaluation program. The amplitude of the serration of the cutting edge is much smaller compared to that before CEP.

2.1.4.2.7   Final Recommendation on CEP

Some application recommendations are as follows:

1. As with coating, CEP should not be attempted to use to solve tooling or other drilling-system-related problems. A wrong notion dominating in industry is that when one uses CEP, the drill flank and rake faces can be ground with a coarse-grit grinding wheel as CEP technology then smooths these surfaces so that "high quality" of the cutting edge can be achieved with minimum effort. It never happens in reality.
2. Optimization of the drilling tool geometry, selection of its tool material for a given application, proper MWF application, and so on should be accomplished before considering CEP.
3. CEP machine/technology provider is not responsible for improving tool life and other parameters of a drilling operation. In other words, the parameters of CEP should be defined by the drilling tool manufacturer/user. CEP machine/technology provider is responsible for assuring these parameters are within the tolerance assigned/agreed.
4. As CEP works only in a rather narrow range of EP parameters (i.e., shape, radius), these parameters should be determined experimentally together with allowable ranges of their variation.
5. The tool manufacturer/user should have a reliable means to control CEP parameters.
6. The optimized CEP parameters and their tolerances should be imbedded in the tool drawing properly so that they should be a part of the tool inspection routine/report.

2.1.5   Tool Design/Development/Research Perspective

As pointed out by Webb (1993), although the belief that drilling is simple and well understood may have been valid in the past, i.e., the knowledge of the time satisfied the need of the time, this belief needs a complete revision. Emerging of new work materials, introduction of new and substation improved tool materials, wide use of CNC unmanned drilling machines and manufacturing cells, CNC multi-axis drill grinding machines, grinding techniques and grinding wheel shape/materials, new drilling tool measuring/inspection machines should result in a complete rethink of drill design and geometry. In the author's opinion, the major problem is that normative information, which includes that in various standards, manuals, and handbooks, does not provide help on the matter as it includes obsolete empirical, semi-empirical, and best practice (normally of half of a century ago) formulas and recommendation to select drill's design components, secondary tooling, drilling regime, grinding and inspection practices, and so on.

Another approach is to derive general relationships of a rather complicated nature that hold for various drill designs and geometries (Hsieh and Lin 2002, Fujii et al. 1970a,b, 1971, 1972, Billau and McGoldrick 1979, Tasi and Wu 1979, Tsai and Wu 1979a,b, Kaldor and Lenz 1982, Kaldor and Moore 1983, Hsiau 1985, Watson 1985, Salama and Elsawy 1996). Being useful for academic and development purposes, this approach provides little help in practical drill design/development as a lot of work is yet to be done to bring a general relationship to the level of practical calculation and analysis.

In this chapter, another approach is taken. Rather than developing and analyzing the complete drill geometry, the geometry of essential components is analyzed. Therefore, a researcher/developer/advanced designer can use the analyses as "Lego blocks" to build any drilling tool design consisting of various components or to analyze any new and existing drilling designs. Knowing and understanding this methodology and design components, one should be able to design a drilling tool of any configuration having any number of the cutting edges of any shape and location with respect to the tool coordinate system.

2.1.5.1   Constraints on the Drill Penetration Rate

It was pointed out in Chapter 1 that the prime objective of the HP drilling system is an increase in the drill penetration rate, that is, increase drilling productivity and thus reduce cost per unit. It was explained that this objective is a combined system rather than a component characteristic. The constraints imposed on the penetration rate by the major components of the drilling system are discussed in Chapter 1. The sections to follow introduce the constraints on the drill penetration rate imposed by the drilling tool itself and explain the correlations of the major tool geometry and design parameters with these constraints.

As pointed out in Chapter 1, the feed rate (which is called the penetration rate in drilling) is calculated as the product of the cutting feed (mm/rev or ipr), and the spindle rotational speed (rpm) and can be expressed through the cutting speed v and cutting feed f as (see Eq. (1.16) in Chapter 1)

$$\text{penetration rate} = v_f = k_{dt} f v \qquad (2.24)$$

where $k_{dt} = 1{,}000/(\pi d_{dr})$ is a constant for a given drill.

It directly follows from Eq. (2.24) that the penetration rate/productivity rate can be increased either by increasing the cutting speed or by increasing the cutting feed or both simultaneously. According to the author's experience, a common mistake made in the industry is multiple attempts to increase these parameters for ordinary drilling tools with ungrounded hopes for success. Specialists and practitioners in the field should realize that there are some constraints on each of the listed ways to increase drilling productivity. These constraints should be clearly understood for successful implementation of HP drilling tools, and thus drilling operations.

There are some constraints and limits on each of these ways that should be understood. The major constraint on the rotational speed is the cutting temperature primarily at the drill periphery corners as these have the highest linear (cutting) speed. The maximum allowable temperature is solely the property of the tool material (including its coating), while the maximum allowable speed that causes this temperature is a function of many drill design and geometry variables (Astakhov 2006). This is because they define the state of stress in the deformation zone (the work of plastic deformation) and, thus, drilling force and torque, chip formation, and its sliding direction, as well as the contact conditions on the tool margins and working conditions of the side cutting edges. Moreover, the tool design and geometry define to a large extent the self-centering of the drill and thus affect the drill transverse vibration, which is the prime cause of drill failure in many applications. Unfortunately, the listed factors and their intercorrelations are not well accounted for in the practice of drilling tool design and implementation where the speed for a given tool material is selected based only upon the work material (type and hardness).

Besides the drilling speed, there are many more constraints (attributed to the drill itself) on the allowable cutting feed (feed per revolution). They include the drill buckling stability, excessive deformation, wear, and breakage. The force factors (drilling torque, axial force, and imbalanced forces) constitute this basis. Therefore, it is important in the design of HP drills and drilling systems to understand the concept of the force factors in drilling and the drill-related constraints on the penetration rate imposed by the force factors.

The force constraints on the drill penetration rate are considered in two stages, namely "the external" stage and "internal" stage. The former considers the resultant force factors as the outcomes of the force balance whereas the latter considers the complete force balance in drilling, i.e., the influence of the design/geometry of the drill components on the formation of the resultant force factor. Such an unusual consideration facilitates understanding particularities of the influence of the design/geometry components of drilling tools on the allowable penetration rate. As a result, a clear objective of the optimization of each component can be set as far as the contribution of these components is known.

2.1.5.2   Resistance of a Drill to the Force Factors – "External" Stage

It is well known (Astakhov 2010) that a drill in operation is subjected to two force factors, namely the axial force F_{ax} and drilling torque M_{dr} as shown in Figure 2.117. Two important length dimensions are considered in drill strength calculation: the maximum length L_{dr-1} (section A-A that corresponds to the face of the drill holder) and the maximum flute length L_{dr-s} where the ratio length/cross section is minimal (section B-B location that depends on the flute length and its profile).

2.1.5.2.1 Resistance to the Drilling Torque

In solid mechanics, *torsion* is the twisting of an object due to an applied torque. In drilling, torsion is a result of the drilling torque acting about the longitudinal axis as shown in Figure 2.117. Torsion, like a linear force, will produce both stress and strain. Torsion causes a twisting stress, called shear stress (τ_{st}), and a rotation, called shear strain (γ_{si}). It is important to be able to predict stresses and deformations that occur in the drill in this type of loading. Figure 2.118 shows a simple model of torsion in drilling. In this model, a drill is represented by a cylindrical body, which essentially is a cantilever beam one end of which is rigidly fixed and the drilling torque is applied to the other (free) end.

Let us consider line AB drawn on the surface parallel to the beam longitudinal axis as shown in Figure 2.118a. When the drilling torque (or moment) is applied to the free end of drill represented by a circular beam as shown in Figure 2.118a, line AB becomes helical assuming position AB′, for which the angle of helix is γ_{tw}. The beam twists by an angle θ_{tw}. This angle is a function of the beam length, L, and stiffness represented by the shear modules G of the beam material. The twist angle starts at 0 at the fixed end of the beam and increases linearly as a function of the z-distance from this end. The change of angle, γ_{tw}, is constant along the length.

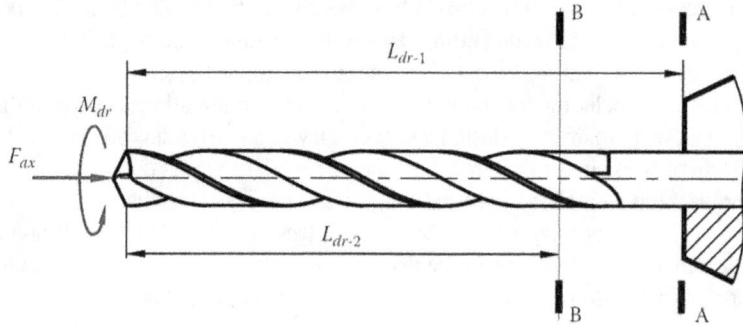

FIGURE 2.117 Drill loaded by the axial force and drilling torque.

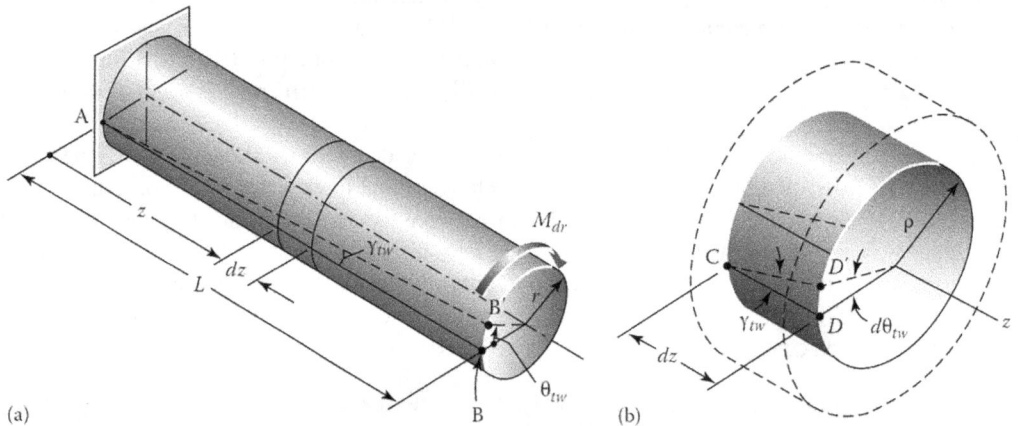

FIGURE 2.118 Model of torsion in drilling: (a) the whole circular beam and (b) a small circular element dz.

A small differential element, dz, is sliced from the beam as shown in Figure 2.118b. Because the cross sections bounded in this element are separated by an infinitesimal distance, the difference in their rotations, denoted by the angle $d\theta_{tw}$, is also infinitesimal. As the cross section undergoes the relative rotation $d\theta_{tw}$, straight line CD deforms into the helix CD'. By observing the distortion of the sliced element, one should recognize that the helix angle γ_{tw} is the *shear strain* of the element.

Two angles γ_{tw} and $d\theta_{tw}$ must be compatible at the outside edge (arc length D-D'). This gives the relationship,

$$\text{Arc Length } D\text{-}D' = \rho \, d\theta_{tw} = \gamma_{tw} dz \qquad (2.25)$$

from which the shear strain γ_{tw} is

$$\gamma_{tw} = \rho \frac{d\theta_{tw}}{dz} \qquad (2.26)$$

The quantity $d\theta_{tw}/dx$ is the *angle of twist per unit length*, where θ_{tw} is expressed in radians. The corresponding shear stress (Figure 2.119a) is determined from Hooke's law as

$$\tau = G\gamma_{tw} = G\rho \frac{d\theta_{tw}}{dz} \qquad (2.27)$$

A simple analysis of Eq. (2.27) reveals that the shear stress varies linearly with the radial distance ρ from the axis of the beam. This variation is shown in Figure 2.119b. As can be seen, the maximum shear stress, denoted by τ_{max}, occurs at the surface of the beam. Note that the previous derivations assume neither a constant internal torque nor a constant cross section along the length of the beam, that is, valid for a general case, for example, for a twist or straight-flute drill.

For practical calculation of τ_{max}, the following formula is normally used (Hibbeler 2010):

$$\tau_{max} = \frac{M_{dr}r}{J} \qquad (2.28)$$

where J is the polar moment of inertia of the beam cross section.

Figure 2.120 shows formulae for calculating the polar moment of inertia for simple cross sections. For complicated cross sections as that of drills (e.g., those shown in Figure 2.16), any modern CAD program, that is, AutoCAD, does this calculation in a single click of a computer mouse. Once one has created a shape from lines and arcs (e.g., the cross section of the twist drill), he or she then

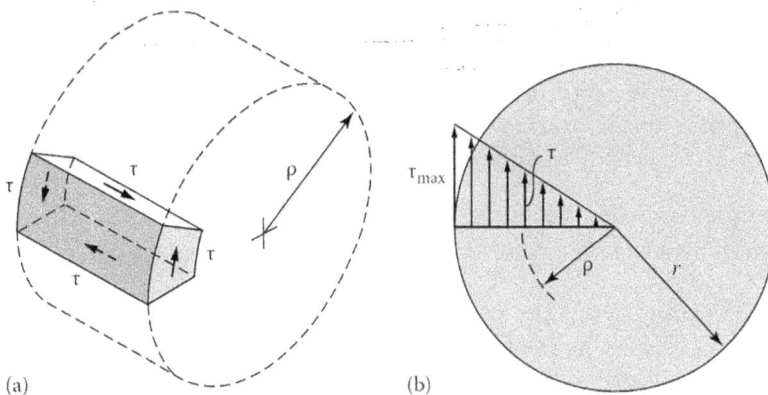

FIGURE 2.119 (a) Shear stress due to torsion and (b) shear stress distribution.

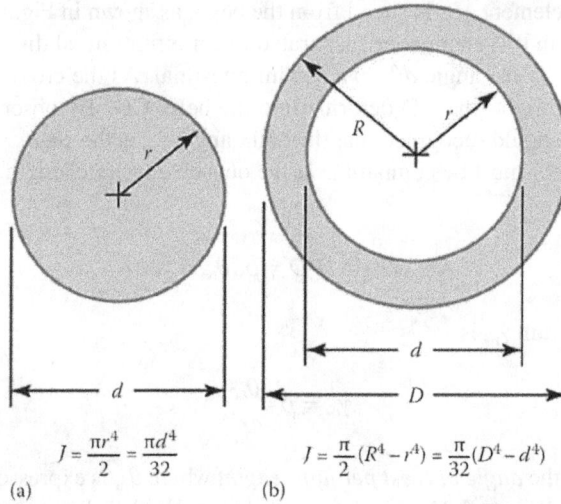

$$J = \frac{\pi r^4}{2} = \frac{\pi d^4}{32}$$

$$J = \frac{\pi}{2}(R^4 - r^4) = \frac{\pi}{32}(D^4 - d^4)$$

(a)   (b)

FIGURE 2.120 Polar moments of inertia for simple cross sections: (a) solid beam and (b) tubular beam.

creates a region from the shape by using the *region* command and selecting all the lines and/or arcs. Then, the position of the coordinate system by location of its origin is defined. Once one has done this, he or she uses the *massprop* command and clicks the region being created. A text box pops up showing the moment of inertia results. When, however, a CAD drawing of the drill profile is not available, the polar moment of inertia J for a drill can be accepted to be half of that for solid beam (see Figure 2.120a).

As follows from Figure 2.119a, elements with faces parallel and perpendicular to the beam axis are subjected to shear stresses only. The normal stresses, shear stresses, or a combination of both may be found for other orientations. In a model shown in Figure 2.121, element a is in pure shear and its maximum shear stress is calculated using Eq. (2.28). Element b shown in Figure 2.121 is located at 45° to the axis. It is subjected to a tensile stress on two faces and compressive stress on the other two. As well known (Hibbeler 2010),

$$\sigma_{max} = \sigma_{45°} = \tau_{max} \tag{2.29}$$

Ductile materials generally fail in shear as many practical shafts. When subjected to torsion, a ductile specimen breaks along a plane of maximum shear, that is, a plane perpendicular to the shaft longitudinal axis. Brittle materials (tool materials) are weaker in tension than shear. When subjected to torsion, a brittle tool breaks along planes perpendicular to the direction in which tension is a maximum, that is, along surfaces at 45° to the drilling tool axis. Figure 2.122 shows an example.

Using the known tensile properties of the tool material, drilling torque, and tool cross-sectional geometrical properties (the polar moment of inertia) as defined by Eq. (2.28), one can make a reasonably accurate assessment if the designed tool can withstand the drilling torque with no fracture.

2.1.5.2.2 Resistance to the Axial Force

A significant axial force in drilling restricts the penetration rate because of the following:

- It affects the buckling stability of the drill. The compromising of this stability causes a number of hole quality problems. It also significantly reduces tool life causing excessive drill corner or even margin wear.
- Some machines used for drilling might have insufficient thrust capacity that limits any increase in the penetration rate with standard drills.

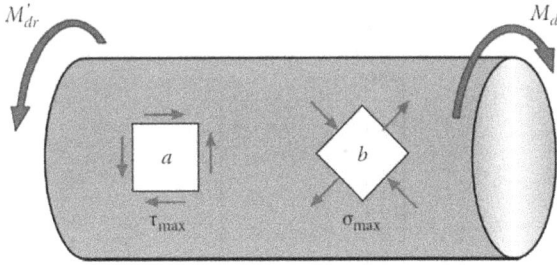

FIGURE 2.121 Maximum shear and normal stresses due to drilling torque.

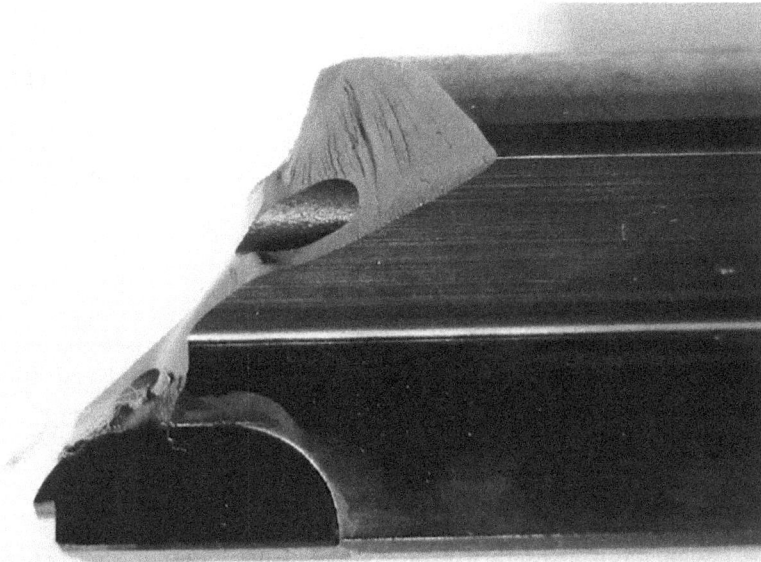

FIGURE 2.122 Typical torsion fracture of a carbide drilling tool.

Therefore, the reduction of the resultant axial force is vitally important when one tries to increase the allowable penetration rate of the drill. As the chisel edge is the major contributor to this axial force, one should (1) reduce the length of this edge and (2) improve the geometry of this edge. These two objectives can be achieved simultaneously as discussed later in this chapter.

The allowable axial force on a drill of a given design is restricted either by its buckling or by its compressive strength. Whenever a structural member is designed, it is necessary that it satisfies specific strength, deflection, and stability requirements. Typically, strength (or in some cases fracture toughness) is used to determine failure, while assuming that the member will always be in static equilibrium. However, when certain structural members, for example, drills, are subjected to compressive loads, they may either fail due to the compressive stress exceeding the yield strength (which in case of brittle tool material is almost the same as the ultimate compressive strength), or they may fail due to lateral deflection (i.e., buckling). The maximum axial load that a drill can withstand when it is on the verge of buckling is called the critical load, F_{z-cr}. Any additional load greater than F_{z-cr} will cause the drill to buckle and therefore to deflect laterally. Buckling is a geometric instability and is related to material stiffness, column length, and column cross-sectional dimensions. Strength does not play a role in buckling but does play a role in compression.

Figure 2.123 shows two basic models for assessments of buckling stability of drills. Figure 2.123a shows a model for a drill as it starts drilling, that is, at hole entrance, while Figure 2.123b shows a

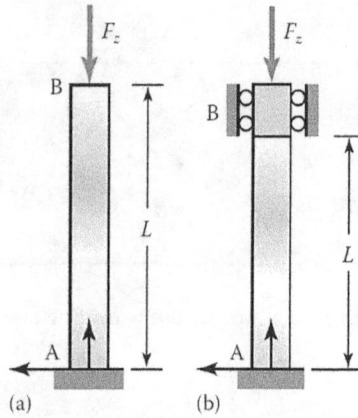

FIGURE 2.123 Two basic models for assessment of the buckling stability of drills: (a) at the hole entrance and (b) in drilling.

model for a fully engaged drill, that is, when its margins are in full contact with the wall of the hole being drilled. The well-known equation (Vable 2012) allows to calculate F_{z-cr} as

$$F_{z-cr} = \frac{\pi EI}{L_{\text{eff}}^2} \tag{2.30}$$

where L_{eff} is the effective length of the drill, E is the modulus of elasticity (sometimes called Young's modulus – see Figure 2.85) of the drill material, and I is the moment of inertia of the drill's cross section. The effective length of the drill in Eq. (2.30) is calculated as

- For the case shown in Figure 2.123a: $L_{\text{eff}}=2L$,
- For the case shown in Figure 2.123b: $L_{\text{eff}}=0.5L$.

For purposes of design, it is more useful to express the critical loading condition in terms of a stress such as

$$\sigma_{cr} = \frac{F_{z-cr}}{A_{d-cs}} = \frac{\pi^2 EI}{A_{d-cs}L_{\text{eff}}^2} \triangleright \sigma_{cr} = \frac{\pi^2 E}{\left(L_{\text{eff}}/k_{gr}\right)^2} \tag{2.31}$$

where A_{d-cs} is the cross-sectional area of the drill and $k_{gr} = \sqrt{I/A_{d-cs}}$ is the smallest radius of gyration determined from the least moment of inertia, I for the cross section. The term L_{eff}/k_{gr} is known as the slenderness ratio and contains information about the length and the cross section of the drill.

To understand the transition from compression failure of a drill under the applied axial force to that due to buckling, consider the case where the applied stress, σ, is equal to the critical buckling stress, σ_{cr}, and the generalized compressive strength, σ_{cm}. At the point where such a transition takes place $\sigma_{cr}=\sigma_{cm}$, i.e.,

$$\sigma = \sigma_{cr} = \sigma_{cm} = \frac{\pi^2 E}{\left(L_{\text{eff}}/k_{gr}\right)^2} \tag{2.32}$$

If Eq. (2.32) is solved for L_{eff}/k_{gr}, the resulting relation marks the combination of length and cross section at which the compressive behavior transitions from compression to buckling. This relation is known as the minimum slenderness ratio:

$$\left(\frac{L_{\text{eff}}}{k_{gr}}\right)_{\text{min}} = \sqrt{\frac{\pi^2 E}{\sigma_{cm}}} \tag{2.33}$$

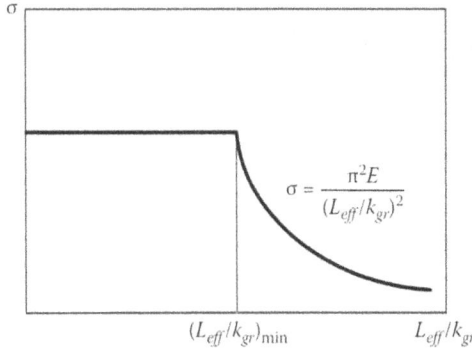

FIGURE 2.124 Stress versus slenderness ratio relation.

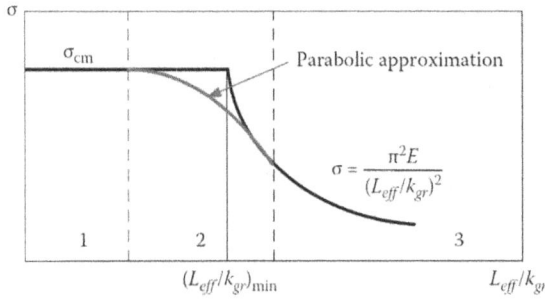

FIGURE 2.125 Stress versus slenderness ratio relation with parabolic approximation.

This transition can be illustrated by plotting the relation between stress, σ, and slenderness ratio, L_{eff}/k_{gr}, as shown in Figure 2.124. Note that in reality, there is no sharply divided transition between yielding and buckling. Instead, the σ versus L_{eff}/k_{gr} curve can be divided into three regions as shown in Figure 2.125. Region 1 is the short-length drill region in which failure occurs due to compression stress when $\sigma = \sigma_{cm}$. Region 2 is the intermediate-length drill region in which the drill may fail either due to compression or may buckle. In this region, empirical relations are used to approximate the resulting curve. Region 3 is for long drills. Buckling is the failure mode in this region.

Calculation of the drill strength in regions 1 and 3 is straightforward (the compression strength and critical stress, respectively), while to do so in region 2, material/geometry/loading-dependent and empirical relations can be used. One common approach is to fit a parabola to the σ versus L_{eff}/k_{gr} curve from $\sigma = \sigma_{cm}$ to $\sigma = \sigma_{cm}/2$ as follows:

$$\sigma = \begin{cases} \sigma_{cm}\left[1 - \dfrac{\left(L_{\text{eff}}/k_{gr}\right)^2}{\left(\left(2L_{\text{eff}}/k_{gr}\right)_{\min}\right)^2}\right] & \text{for } 0 \leq L_{\text{eff}}/k_{gr} > \left(L_{\text{eff}}/k_{gr}\right)_{\min} \\[4mm] \dfrac{\pi^2 E}{\left(L_{\text{eff}}/k_{gr}\right)^2} & \text{for } L_{\text{eff}}/k_{gr} > \left(L_{\text{eff}}/k_{gr}\right)_{\min} \end{cases} \qquad (2.34)$$

Example 2.1

An important example concerning buckling stability is considered next as a part of a drill premature failure analysis. The conditions of the drilling operations considered in the analysis were as follows:

1. Workpiece is a turbine shaft of an automatic six-speed rear-wheel drive transmission. A fragment of the shaft operation drawing is shown in Figure 2.126.
2. Work material is normalized steel SEA 1040 (equivalent AISI 1040) with the following mechanical properties: hardness 170 HB and tensile strength (ultimate 590 MPa, yield 374 MPa).
3. New and CNC-sharpened carbide twist drills having proper geometry, tool material grade, and suitable coating made by one of the most reputable tool companies were used. The drills were preset in drill holders using a Zoller machine so that their total runout did not exceed 25 μm. Figure 2.127 shows a drill preset in a holder.
4. CNC machines (Fanuc Robodrill machines) widely used in the automotive industry were used.
5. The drilling regimes used for the test drills are shown in Table 2.4.
6. MULTAN B-400 MWF supplied at a pressure of 3.1 MPa (450 psi) was used. The coolant concentration, pH, and clearness were kept within the limits assigned by the control plan.

Problem

Tool life of drills of 5.40 mm dia. (for drilling hole 5.500/5.300 mm dia. shown in Figure 2.126) is almost half (380 shafts) compared to that of the drills used for other holes (670 shafts) with no apparent reason.

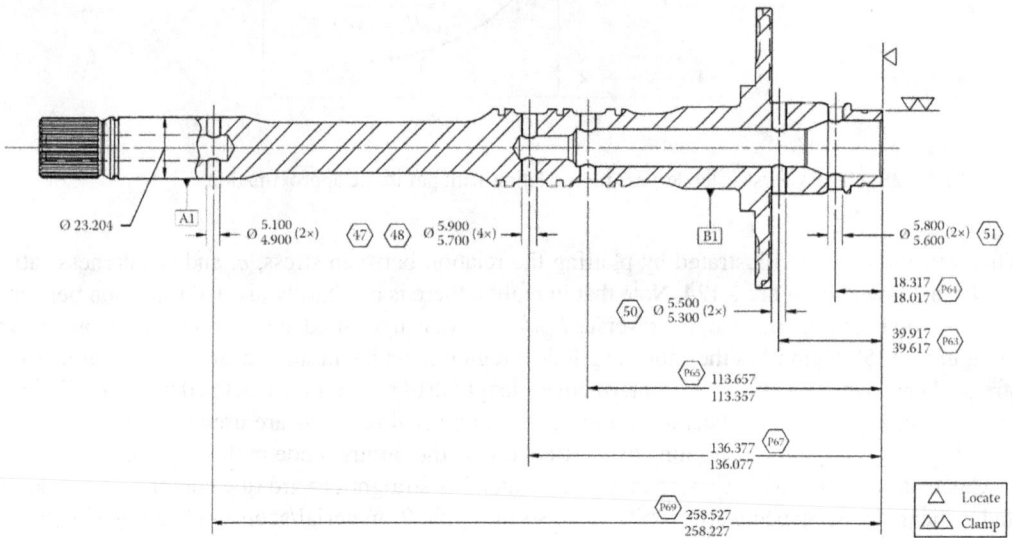

FIGURE 2.126 Turbine shaft with cross holes.

FIGURE 2.127 Drill preset in a holder.

TABLE 2.4
Drilling Regimes Used

Drill diameter, mm	5.00	5.40	5.70	5.80
Rotational speed, rpm	9,315	8,625	8,130	8,013
Cutting speed, m/min	146.3	146.3	145.6	146
Feed, mm/rev	0.13	0.10	0.13	0.13
Feed rate, mm/min	1,211	862.5	1,057	1,042

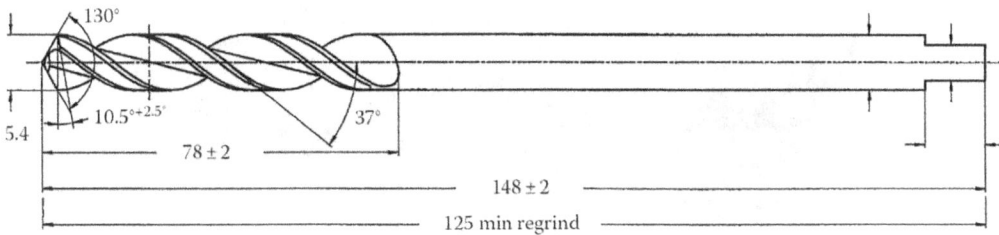

FIGURE 2.128　Length of the drill of 5.4 mm dia. as assigned by its drawing.

Analysis

An analysis of the operational conditions that included all components of the drilling system showed that these components were selected properly including machining regimes shown in Table 2.4. Moreover, they are the same for all drills involved in the consideration. Therefore, these components were excluded from the further analysis. The only difference between the drills is their length. The drill of 5.40 mm dia. (for drilling hole 5.500/5.300 mm dia. shown in Figure 2.126) is a long drill, while other drills are regular-length drills (Jobber-length drills). As can be seen in Figure 2.128, the length-to-diameter ratio for this drill is 25.85 (for a new drill), which well exceeds 10 as with regular-length drills. This is because the hole 5.500/5.300 mm dia. is located near a flange made on the turbine shaft as clearly seen in Figure 2.126.

Analysis of drills of 5.40 mm dia. revealed the following:

- Their geometry (split-point self-started) and flute profile are suitable for the application.
- The overall number of regrind for these drills is almost half that for other drills. This is because of severe wear of drill margins that requires a much longer portion of the drill to be removed on the re-sharpening to restore proper geometry of these margins. Moreover, traces of cutting by the margins were observed as shown in Figure 2.129.

A part autopsy procedure discussed in Chapter 4 in Astakhov (2024) was used to reveal the root cause of the problem. Following the autopsy methodology, a shaft after a completed drilling operation was sectioned as shown in Figure 2.130 and the holes were examined. The examination revealed buckling instability of drills of 5.40 mm dia. Figure 2.131 shows the appearance of a drilled hole when such instability takes place in drilling. As can be seen, the worst instability is observed at the hole entrance as was discussed in the analysis of the model shown in Figure 4.26a. When the drill is fully engaged in the hole being drilled, its deflection reduces as can be seen in Figure 2.131. The reason for this is discussed in the analysis of the model shown in Figure 2.123b.

FIGURE 2.129 Showing an evidence of the cutting by drill margin.

FIGURE 2.130 Showing a sectioned part of the shaft.

Solution

A special thick-web drill (40%) with DIN 1412 Form B point geometry and curved cutting edges (Astakhov 2010) was designed, manufactured, and implemented to solve the problem. Additionally, the cutting feed was reduced by 30% at the hole entrance while a higher feed than listed in Table 2.4 is then applied for the rest of the hole being drilled to meet the requirement of cycle time entrance.

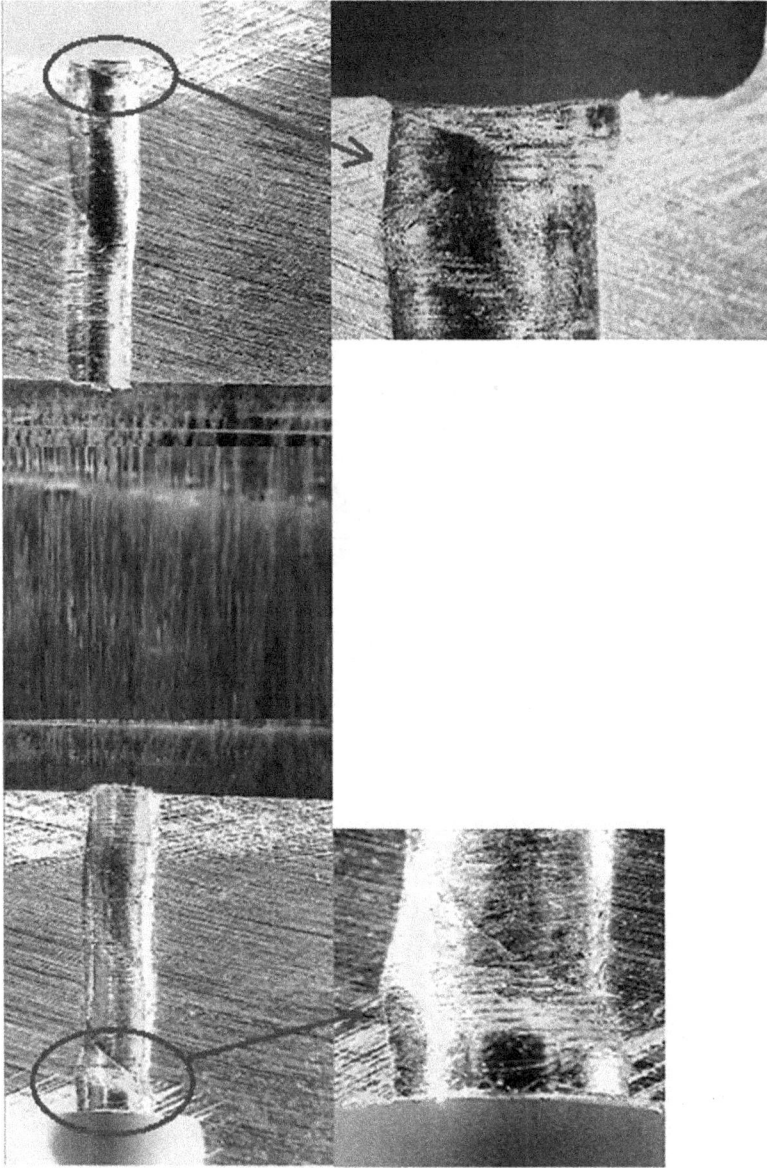

FIGURE 2.131 Showing the results of buckling instability.

2.1.5.3 Drill Design/Process-Related Generalizations/Considerations Related to Resistance to the Force Factors

The foregoing analysis of the drill resistance to the axial force and drilling torque results in the following important drill design/process-related generalizations to be used in the HP drill design:

1. The drilling torque is a function of the work material properties, drill diameter and geometry, and the drilling regime. Of these factors, the drill geometry and drilling regime can be varied to achieve optimal drill performance. As the cutting speed has a weak influence on the cutting force, it also has little influence on the drilling torque so that the cutting feed is the only factor to be considered. In modern production with CNC machines, the drilling

torque is not a limiting factor as these machines are equipped with powerful motors to deliver high torques, for relatively small machines, the drilling torque can be a constraint limited by the power of the drive motor. When the latter is the case, the feed per revolution is lowered, or the hole is drilled in two consecutive drilling operations using first a smaller drill and then a drill to the required hole size.

2. The ultimate tensile strength (S_u in Figure 2.85) is the proper mechanical characteristic of a tool material in its torsional resistance as it is directly correlated with the material shear strength. High-quality HSSs have S_u up to 1,000 MPa while 6%Co cemented carbide has $S_u = 2,200$ MPa. As a result, the allowable torque for the same tool made of cemented carbide is up to two times greater than that made of HSS.

3. The drilling torque is commonly the limiting force factor for regular-length drills (Jobber-length drills). A better accuracy of drill strength calculations is achieved if the Von Mises stress is used as this parameter accounts for the real state of stress in the body of a drill. Because in drilling only the axial force and torque are the force factors and stresses σ_z and τ_{zy} are present (as in combined torsion and bending/axial state of stress), the Von Mises stress is calculated as

$$\sigma_v = \sqrt{\sigma_z^2 + 3\tau_{xy}^2} \tag{2.35}$$

As such, the drill strength equation becomes

$$\frac{S_u}{FS} \leq \sqrt{\left(\frac{F_z}{A_{d-cs}}\right)^2 + 3\left(\frac{M_{dr}d_{dr}}{2J_{d-cs}}\right)^2} \tag{2.36}$$

where FS is the safety factor (Astakhov 2018).

4. Equation (2.33) and Figure 2.124 provide the idea of transition between compression and buckling, i.e., to the region where the axial force is the limiting force factor. Figure 2.125 shows that there is a certain transition range rather than an abrupt transition. Equation (2.34) shows how to calculate the critical stress in this range.

5. The easiest concept to grasp in analyzing buckling stability of a drill is that the maximum allowable axial force F_{z-max} must be less than the critical buckling load F_{z-cr} is given by Eq. (2.30). It follows from this equation that the maximum allowable axial force is calculated as

$$F_{z-max} \leq \frac{F_{z-cr}}{FS} = \frac{\pi^2 EI}{FSL_{eff}^2} \tag{2.37}$$

Therefore, the design equation can be written as

$$EI \geq \frac{\left(FS \cdot F_{z-max}L_{eff}^2\right)}{\pi^2} \tag{2.38}$$

In the design equation, everything on its right-hand side is known or set for the time being. Therefore, the left-hand side can be solved for either E or I, that is, either material or cross-sectional geometry. Usually, the tool material is already known for other reasons so that Eq. (2.38) can be solved for I.

6. Three conclusions important for the drill design/application follow from the design equation (Eq. 2.38):

   a. The maximum allowable axial force F_{z-max} is directly proportional to the modulus of elasticity E. Therefore, for the same drill design (the moment of inertia I), drills made of cemented carbide allow a higher penetration rate than those made of HSS as $E=650\,\text{GPa}$ for cemented carbide while $E=221\,\text{GPa}$ for HSS.

b. The effective length of a drill has the strongest influence. As discussed in the analyses of two models shown in Figure 2.123, this length reduces four times when the drill working end is restricted (Figure 2.123b). *Therefore, the use of starting bushings or starter drills (see Section 2.1.3.4.5) for long drills is the most powerful mean to increase the allowable penetration rate in drilling.* Another less powerful way to increase buckling rigidity is to reduce the feed rate at the hole entrance that, in turn, reduces $F_{z-\text{max}}$.

c. The least effective but commonly used way to increase the drill buckling rigidity is a gradual increase in the moment of inertia from the drill's working end toward its shank by applying web thickness taper as shown in Figure 2.132. The American Society of Mechanical Engineers (ASME B94.11M-1993) and the Aerospace Industries Association of America, Inc. (NAS 907) standards define the conventional web thickness taper rate between 0.60 and 0.76 mm.

2.1.5.4 Improving Drill Rigidity

It follows from the consideration made in the previous section that, for a given drilling torque and axial force, an increase in the allowable penetration rate can be achieved by increasing the resistance of the drill to the axial force and drilling torque. The former can be achieved through increasing the moment of inertia, I, of the drill cross section and/or it is the elasticity modulus, E, of the drill material as follows from Eq. (2.38) provided that the drill length and drill starting conditions are already optimized. The latter can be achieved through increasing the polar moment of inertia, J, of the drill cross section and/or it is the shear modulus, G, of the drill material as follows from Eqs. (2.27) and (2.28).

The simplest way to achieve an increased rigidity is to use a cemented carbide drilling tool instead of tools made of HSS as $E=650\,\text{GPa}$ for cemented carbide while $E=221\,\text{GPa}$ for HSS. The same increase, almost three times, is in the shear modulus, G.

The most common way is to use the wide-web drills shown in Figure 2.65. They are discussed in Section 2.1.3.7.2, and they are more rigid and stronger than conventional drills; thus, higher penetration rates (up to 20%) and length-to-diameter ratio and better accuracy of machined holes are achieved with these drills. Wide-web drills are normally used to produce holes in difficult-to-machine and heat-treated work materials. An apparent drawback to wide-web drills is the reduced cross-sectional area of the chip flutes. One needs to realize that any increase in the web diameter requires the corresponding change in the tool geometry to keep the axial force at a certain low level. Otherwise, any attempt to solve a problem with increasing drill stiffness will lead to another problem with an increase in the axial force that undermines the net result as shown in Figure 2.66.

Another more engineering, but rarely used way to deal with the problem is to select the flute profile intelligently, and thus to optimize I and J of the drill cross section while assuring reliable chip transportation over the chip flute. To apply this way, one needs to consider the flute profile beyond the qualitative representation shown in Figure 2.16 commonly depicted in the literature on drilling and used as generic in the drill grinding machine libraries/galleries. Figure 2.133a shows the profile

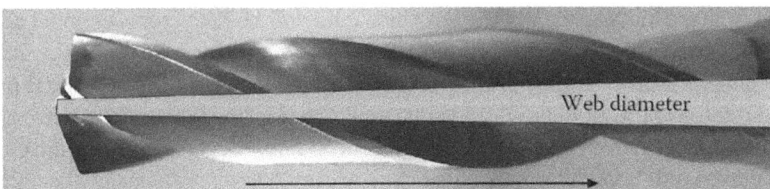

FIGURE 2.132   Web diameter (core thickness) increases from the tip toward the shank.

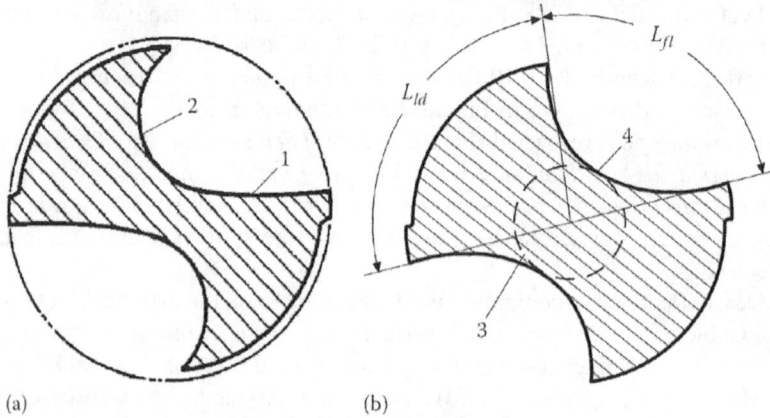

(a)   (b)

FIGURE 2.133 Profile of the (a) standard drill and that of a (b) wide-web drill.

of a standard twist drill. As can be seen, the flute has two distinctive parts, namely the so-called straight part 1, which in its intersection with the flank surface forms the major cutting edge (lip) and concavely curved surface 2 having a relatively large radius of curvature. Figure 2.133b shows the flute profile of a wide-web drill. As the web diameter 3 is great, profile 4 is normally made using a single curve to enhance drill rigidity.

The profile is usually characterized by the so-called flute-width ratio (FWR), which is the ratio of arc length L_{fl} of the flute to arc length L_{ld} of the lands (Figure 2.133b), i.e.,

$$FWR = \frac{L_{fl}}{L_{ld}} \tag{2.39}$$

Figure 2.134 shows the relationships between the drill torsional rigidity and web thickness ratio (web diameter/drill diameter) for two different FWRs according to the data presented in US Patent No. 4,583,888 (1988). As can be seen, an increase in the web thickness ratio and while a decrease in FWR leads to an increase in torsional drill rigidity. The chip evacuation, however, becomes more difficult. Thus, the web thickness ratio and FWR have their limits; generally, the web thickness ratio is set in the range of 15%–23% and FWR is in the range of 1.0–0.76.

According to US Patent Nos. 4,744,705 (1988) and 5,230,593 (1993), the profiles of drills made of HSS and those made of cemented carbide are not the same particularly for heavy-duty drilling operations. For HSS drills having a profile shown in Figure 2.133a, FWR is 0.7 at the forward end of the drill. To increase drill rigidity, FWR can be equal to 1.16 away from the forward end. For carbide drills, FWR is in the range of 0.4–0.9. In the author's opinion, however, FWR should be 1.0 for machining of steels and up to 1.4 for machining of light alloys to improve chip transportation conditions.

A lot has been written on the flute profile and its optimization (Spur and Masuha 1981, Thornley et al. 1987b, Agapiou 1993a,b, Chen 1997, Chen and Ni 1999). Thornley et al. (1987a) discussed that the total area of the chip flute is not the significant factor in chip removal because a typical drill chip is conical in shape and hence only the inscribed circle of the flute cross section it actually occupies is important. In the model shown in Figure 2.135a, the inscribed circles are A and B. In such a model, the radius of these circles r_{fl} was proposed to be considered as the critical parameter. In other words, circles A and B in Figure 2.135a that could be inscribed in the flute space normal to the axis of the drill are considered to be a good index of flute disposal capacity since the curled-conical shape chip occupies only a circular part of the flute space. When the radius of the chip cone tends to be larger than r_{fl}, the chip flute is considered to be clogged. Although it was a great attempt of the body profile optimization, the proposed critical parameter of chip removal is not practical as the

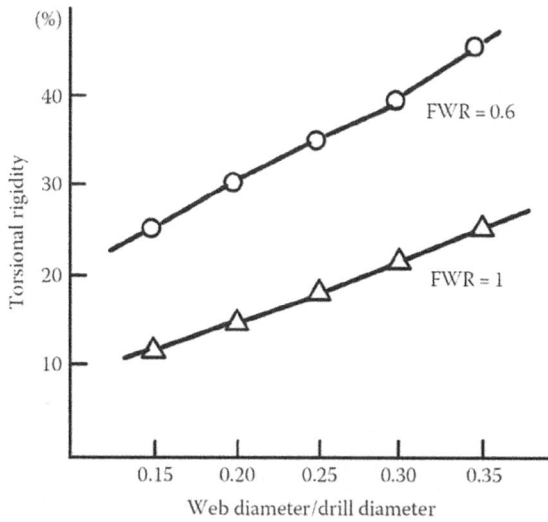

FIGURE 2.134 Drill torsional rigidity as a function of the web thickness ratio.

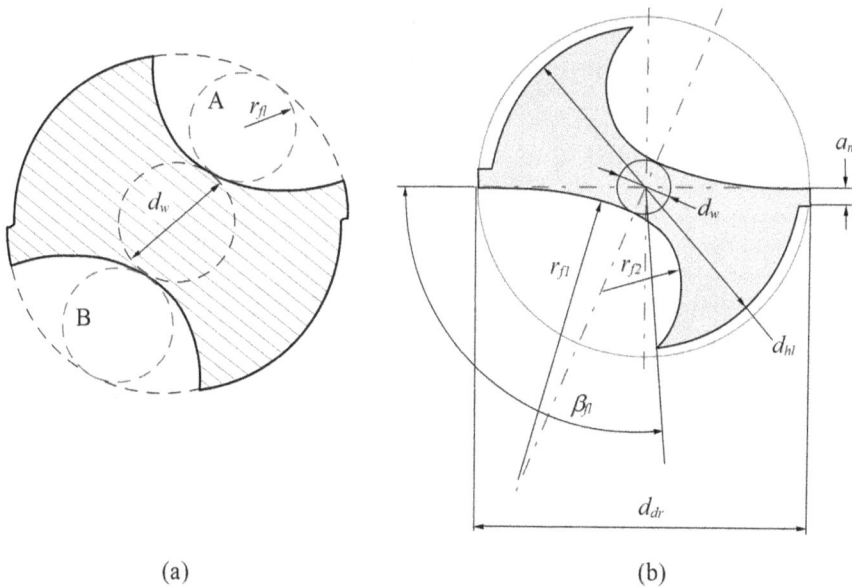

(a)   (b)

FIGURE 2.135 Flute profile: (a) model for the optimization of chip flute suggested by Thornley et al. (1987b) and (b) commonly recommended flute profile.

chip attains various shapes and sizes in drilling depending on the work material, machining regime, tool geometry, and body profile. There were many other attempts to optimize the body profile (its area, polar moment of inertia, and other cross-sectional characteristics).

National and International standard organizations have come up with various recommendations on the forms suitable for given cutting conditions and workpiece materials. One of the common basic forms is shown in Figure 2.135b. In this form, the central angle β_{fl} is selected to be equal to or 2°–3° greater than that of the land. The web diameter is $d_w = (0.125\text{–}0.145)d_{dr}$ and increases toward the shank by up to 1.7 mm per 100 mm drill length; the body clearance diameter is

$d_{cl} = (0.875–0.855)d_{dr}$; profile radii are $r_{f1} = (0.75–0.90)d_{dr}$ and $r_{f2} = (0.22–0.28)d_{dr}$; and margin width is $a_m = (0.2 – 0.5)\sqrt[3]{d_{dr}}$. Back taper of 0.03–0.12 mm per 100 mm length is applied to the margins of HSS drill while that for cemented carbide drill is 0.1–0.2 mm per 100 mm length of the drill.

The foregoing consideration suggests that, besides increasing drill rigidity, the optimization of the flute profile should also aim to improve chip transportation in the flutes. The profile of a drill should be designed in such a way that the flutes provide the maximum space for the chip and facilitate chip removal while ensuring that the drill is capable to adequately withstand the drilling torque and axial force.

2.1.5.5 Axial Force (Thrust)–Torque Coupling

Observations and tests comparing straight-flute and twist drills showed that the latter allows greater critical axial force and torques. This result can easily be explained by the so-called torque–thrust coupling effect in twist drills. The coupling can be explained as follows. The body of a twist drill contains two helical flutes. If a torque is applied as shown in Figure 2.117, this torque tends to *unwind* the helix and thus increases the drill length. Conversely, if an axial force is applied to the end of the drill as shown in Figure 2.117, this force shortens the drill, thus causing its *winding*. Therefore, the effects of the drilling torque and the axial force (thrust) on drill static stability partially compensate each other to a certain extent. This explains the results of observations. It is important to a drill designer/user to understand to which extent this compensation can go and how the design parameters of a twist drill to maximize this compensation.

Narasimha et al. (1987) proposed to assess torque/thrust coupling using the following coupling matrix:

$$\begin{Bmatrix} F_z \\ M_{dr} \end{Bmatrix} = \begin{bmatrix} K_{FF} & K_{MF} \\ K_{FM} & K_{MM} \end{bmatrix} \begin{Bmatrix} \Delta L_{cm} \\ \theta_{tw} \end{Bmatrix} \tag{2.40}$$

where K_{MM} is the axial stiffness under torsional restrain, K_{MF} is the torque-on-thrust coupling stiffness coefficient, K_{FM} is the thrust-on-torque coupling stiffness coefficient, K_{FF} is the torsional stiffness under axial restrain, and ΔL_{cm} and θ_{tw} are linear and angular deformations due to the force factors.

Table 2.5 shows the values of the stiffness coefficients obtained experimentally (Narasimha et al. 1987). Although unconventional, the units of the coefficient are meant for an easy quantitative comprehension of the results, i.e., the coupling effect.

The experimental results of several studies (De Beer 1970, Lorenz 1979, Narasimha et al. 1987) reveal the following:

- The flute helix angle, flute profile, and web thickness significantly affect the axial and torsional stiffness of drills.
- The pure torsional stiffness of drills is maximized for a helix angle of around 28°. Departure from this angle to either side lowers this stiffness significantly (see Table 2.5).
- The torque–thrust interaction (measured by K_{FM} and K_{MF}) has a distinct maximum at a helix angle of about 28°. As the same value of helix angle results in the largest increase in torsional stiffness, this explains a much higher allowable torque and axial force (thrust) for twist drill compared to straight-flute drills. This also provides a theory-based explanation to the fact that general-purpose drills are made with a helix angle of 28°–30°.
- An increase in web thickness decreases the torque–thrust interaction, which means the benefit of the stiffening action of the axial force reduces.

The discussed results indicate that the penetration rate of a properly designed twist drill can be up to 20% higher than that of a straight-flute drill due to the coupling. This advantage is particularly

TABLE 2.5
Stiffness Coefficients

Drill Dia. (mm)	Helix Angle (deg)	Web Thickness (mm)	K_{FF} (N)	K_{MF} (N-m/rad)	K_{FM} (N-m)	K_{MM} (Nm²/ rad)
12.7	32.76	1.905	2.47×10^6	1.91×10^3	2.00×10^3	0.599
	51.42	1.905	2.04×10^6	1.68×10^3	1.56×10^3	0.433
10.3	14.00	1.549	2.22×10^6	514.2	658.4	0.131
	31.94	1.549	2.41×10^6	1.18×10^3	1.42×10^3	0.229
	37.65	1.549	1.88×10^6	978.8	1,359.1	0.165
9.5	12.96	1.473	2.09×10^6	437.9	535.0	0.100
	32.49	1.473	2.47×10^6	959.9	1.29×10^3	0.180
	36.80	1.473	2.47×10^6	765.1	800.8	0.146
6.4	13.67	1.219	2.47×10^6	185.9	151.7	0.019
	31.94	1.219	2.47×10^6	399.5	321.6	0.038
	35.48	1.219	2.47×10^6	339.7	187.8	0.032

important for long drills as the applied torque may significantly improve drill buckling stiffness. In the author's opinion, this is one of the most important advantages of twist drills hardly mentioned in the literature.

2.1.5.6 Force Balance as the Major Prerequisite Feature in HP Drill Design/Manufacturing – "Internal Stage"

This section introduces the concept of the complete force balance in drilling and, based on this concept, discusses the development of VPA2© design. Although the concept is discussed for a twist drill as the most common drill type used in industry, its essence, and the way of assessment are fully applicable for *any type of drilling tools*.

Figure 2.136 shows the complete force model in drilling. For convenience of further considerations, the right-hand Cartesian coordinate system $x_0y_0z_0$, illustrated in Figure 2.136, is set as follows:

1. The z_0-axis is along the longitudinal axis of the drill, with sense as shown in Figure 2.136, toward the drill holder.
2. The y_0-axis is perpendicular to the z_0-axis and passes through point *1* of the major cutting edge in the sense shown in Figure 2.136. The intersection of this axis constitutes the coordinate origin *0* as shown in Figure 2.136.
3. The x_0-axis is perpendicular to the y_0 and z_0 axes as shown in Figure 2.136.

This system should also be regarded as the datum system in drill and drilling machine accessories (drill holder, starting bushing, etc.) design/setting/adjustment.

The complete force model for a two-flute drill shown in Figure 2.136 is partially based on the simplified force balance shown in Figure 2.24. In this diagram, F_{x1} and F_{x2} are resultant power components, F_{y1} and F_{y2} are the resultant radial components, and F_{z1} and F_{z2} are the resultant axial components of the cutting forces acting on the first (*1–2*) and the second (*5–6*) major cutting edges (lips), respectively. These forces are applied at certain points *c* and *d* on the major cutting edge as shown in Figure 2.136.

The power and radial components of the cutting forces that act on the two parts (*2–3* and *3–5*) of the chisel edge are not shown in Figure 2.136 as these are small, while the resultant axial components (F_{z-ce1} and F_{z-ce2}) shown in this figure are significant. The resultant normal forces F_{nm1} and

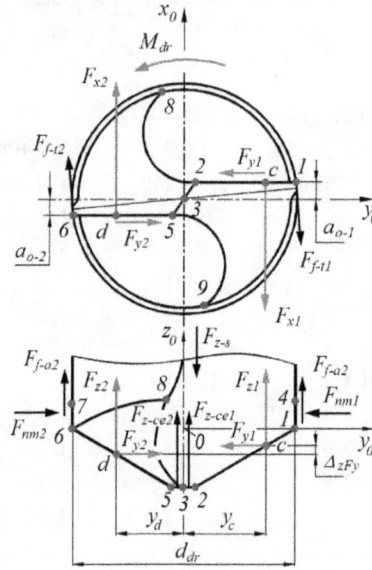

FIGURE 2.136 Complete force model in drilling.

F_{nm2} act on the flank faces of the minor cutting edges *1–4* and *6–7*, respectively. The causes of these forces are discussed in Section 2.1.3.8. When drill/workpiece rotates, these normal forces cause the friction forces F_{f-t1} and F_{f-t2} whereas the feed motion causes the friction forces F_{f-a1} and F_{f-a2} on the margins.

The drilling torque applied through the spindle of the machine is calculated as

$$M_{dr} = F_{x1}y_c + F_{x2}y_d + F_{y1}a_{o-1} + F_{y2}a_{o-2} + F_{f-t1}\left(d_{dr}/2\right) + F_{f-t2}\left(d_{dr}/2\right) \tag{2.41}$$

and the axial force applied by the spindle is

$$F_{z-s} = F_{z1} + F_{z2} + F_{z-ce1} + F_{z-ce2} + F_{f-a1} + F_{f-a2} \tag{2.42}$$

The shown drill is in the static equilibrium in the x_0y_0 and z_0y_0 planes if and only if the following two equilibrium conditions are justified:

In the x_0y_0 plane,

$$F_{x1}y_c + F_{y1}a_{o-1} + F_{f-t1}\left(d_{dr}/2\right) = F_{x2}y_d + F_{y2}a_{o-2} + F_{f-t2}\left(d_{dr}/2\right) \tag{2.43}$$

In the z_0y_0 plane, the arms of forces F_{z-ce1} and F_{z-ce2} are negligibly small so the balance in the plane is represented as

$$F_{z1}y_c + F_{f-a1}\left(d_{dr}/2\right) = F_{z2}y_d + F_{f-a2}\left(d_{dr}/2\right) \tag{2.44}$$

It follows from Eqs. (2.43) and (2.44) that the drilling tool equilibrium can be maintained if only if the discussed forces are equal, i.e., $F_{x1} = F_{x2}$, $F_{y1} = F_{y2}$ and $F_{f-t1} = F_{f-t2}$ and their corresponding arms are the same, i.e., $y_c = y_d$ and $a_{o-1} = a_{o-2}$. These conditions establish the full force balance, known as the theoretical or intended force balance.

Unfortunately, this balance is not the case in practical drilling operations as many additional factors tend to disturb this balance (Astakhov 2014). As shown in Figure 2.136, the radial components/forces may not shear the same line of action because $a^{\cdot}_{o-1} \neq a_{o-2}$ by Δ_{zFy}. This is one of the common cases when a drill has the so-called flute spacing variation (Chapter 1 in Astakhov (2024)). Arms y_c and y_d may not be equal when the drill has a lip-height variation. The power, radial, and axial components on the major cutting edge *1–2* may not be the same as those on the major cutting edge *5–6* due to the difference in the geometry parameters of these edges, conditions of the rake and flank faces, and so on. Assigning tight tolerances on the drill geometry parameters (e.g., lip-height variation, flute spacing, chisel edge centrality, runout of the lands (margins), etc.), one can minimize the deviation from the intended force balance in "normal" drilling. Moreover, standard drills are provided with lends (margins) of excessive width (see Section 2.1.3.8.5) to prevent the side cutting of the drill due to possible force unbalance.

2.1.5.7 Use of the Introduced Complete Force Balance for the Development of a New Design Concept

2.1.5.7.1 Concept

Two extreme cases of drilling, namely drilling of cored holes, which include cored holes, irregular/ curved surface drilling, cross hole and half hole drilling when the entrance into the hole where the face of the workpiece is not perpendicular to the rotation axis of the drills (inclined holes), are considered further to explain the full power/ benefits of understanding and thus proper use of the described force balance.

In the automotive industry, when die castings are used as blanks, cored holes, that is, holes made in castings with some stock to be removed by semi-finishing drilling and then reaming are commonplace today. Figure 2.137a shows a typical example of cored holes made in a die casting of a blank used to machine a part of an automatic transmission. The use of such holes often eliminates the need for solid drilling so that the two- or three-pass hole-making operations (two or three cutting tools) can be replaced by single-pass operations with a finishing reamer having a drill point to break a small flash at the bottom of the hole being drilled. The major problem in drilling such holes is that the tolerances on the location of the axis of cored holes are rather generous (up to 1.8 mm) so that this axis can be significantly shifted from the actual intended axis on the hole in the manufactured part.

The use of cored holes in additive manufacturing (AM) is even more beneficial compared to subtractive manufacturing. Significant reduction of process time and considerable savings on the expensive powder used are of prime importance. Figure 2.137b shows an example of an AM part made on a Renishaw RenAM 500M metal AM machine. A number of cored holes can be seen in this figure. Dimensional inaccuracy of parts produced using AM can be even more problematic (Taheri et al. 2017), particularly when considering a prototype or high-value part where the end use is for a component requiring tight dimensional control (Smith et al. 2016, Farrell and Deering 2018). The layering process used in AM methods can result in rough surfaces and possible deviations from the dimensions and tolerances specified in the CAD model and/or other geometrical anomalies (Zhang et al. 2017). Typically, the CAD model is converted to a stereolithography (*.stl) file format where the designed geometries and surfaces are discretized into geometric meshes. A macrolevel "stair-case" effect can occur on part surfaces due to this discretization (Moroni et al. 2014). In addition, it has been reported that melt pool dynamics have a large influence on sidewall dimensions for AM-made parts (Lee and Farson 2015). The risk of occurrence for curling, waviness, and surface roughness is also all influenced by the previously discussed process and material parameters. To minimize geometrical anomalies, a stable melt pool size/shape is required (Lee and Farson 2015). This is not easy to achieve in practice so the so-called Marangoni effect has a strong influence on melt pool size and shape and can introduce anomalies in deposited layers due to its dependence on composition and the local thermal gradients (Smith et al. 2016).

As above-mentioned, the discussed proper force balance, which assures intended drill working conditions, may not be the case in drilling of cored and angled holes so that the reason for that should be explained. Is schematically depicted in Figure 2.138a that when the axis of the cored hole is shifted from the drill rotation axis, only one major cutting edge (edge *1–2* as shown) is actually

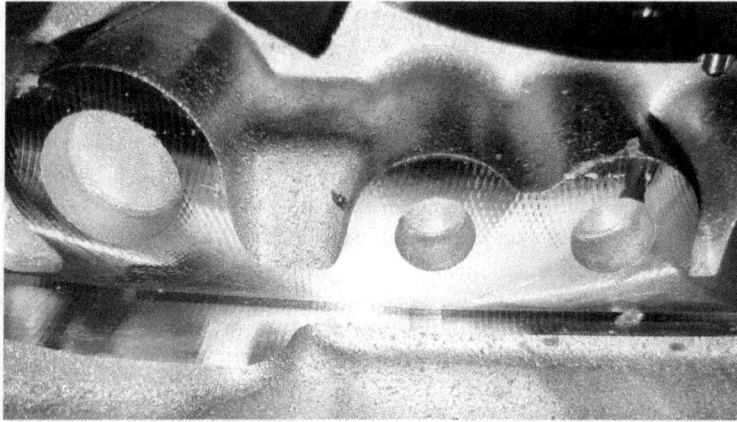

(a)

(b)

FIGURE 2.137 Example of cored holes: (a) made in an aluminum die casting and (b) cored holes made in AM blanks.

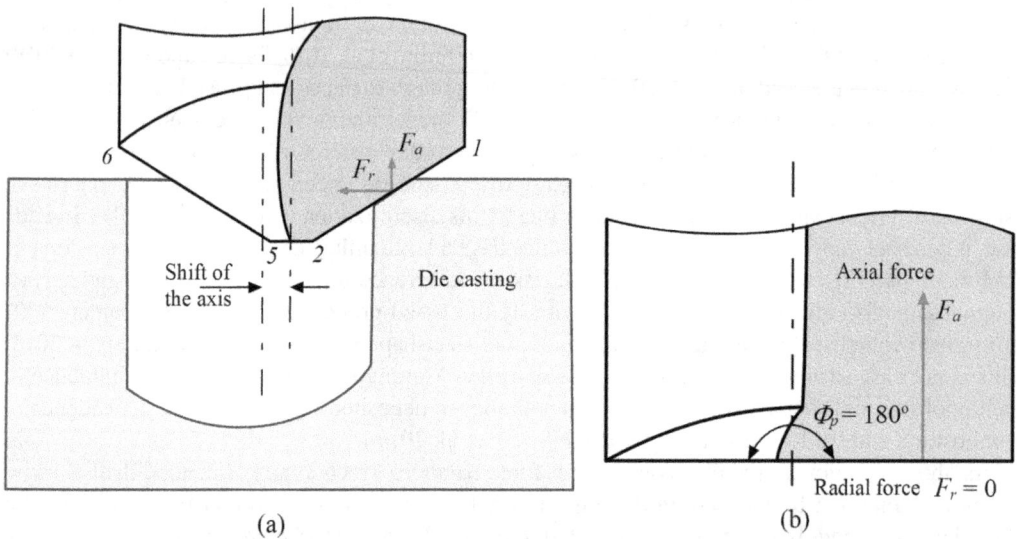

(a)   (b)

FIGURE 2.138 (a) A major problem (as thought) with drilling of a cored hole and (b) its common solution.

engaged in cutting, particularly at the entrance of the hole being machined. As such, the radial force on cutting edge *1–2*, F_r (F_{y1} in Figure 2.136), is not balanced so it tries to bend the drill. When the drill rotates (the most common case in modern machining centers), it is subjected to "whirl" motion. The problem with drill bending in the x_0y_0 plane at the entrance of the hole being drilled has been known for the last 100 years even in drilling of non-cored holes because the technology of drill manufacturing did not allow to grind the drill point perfectly (technically) symmetrical. Particularly, the major cutting edges suffered from a significant lip-height variation. As a result, an excessive bell mouth (the tapered past of the drilled hole at the entrance) (Astakhov 2002), oversize hole and out-of-shape and hole position conditions, and excessive drill wear are reported (Astakhov 2014, 2019). When the shift of the axis of a cored hole is excessive, a carbide drill simply breaks due to its excessive bending.

Analyzing Figure 2.138a, one can conclude that the unbalanced radial force F_r is the root cause of the problem as this force pushes the drill out of balance. Therefore, the common (conventional) wisdom suggests that this force should be reduced (or if possible – eliminated) by changing the drill geometry. As discussed in Section 2.1.3.4.2 and shown in Figure 2.24a, the axial and radial forces are related through the point angle as shown by Eq. (2.4). It follows from this equation that when the point angle $\Phi_p = 180°$ then the radial force $F_r = 0$ as shown in Figure 2.138b, i.e., the radial force on the major cutting edge is zero so that there is no drill benign/distortion on the entrance of the cored hole being drilled even if the axis of this hole is shifted with respect to the drill rotational axis. Therefore, the expectation was that the problem should be solved with the implementation of this drill design once and forever. As a result of this way of thinking, cored holes in the automotive industry are recommended to make using such a drill. Moreover, drills with the point angle $\Phi_p = 180°$ (as shown in Figure 2.139) known today as flat bottom drills have become very popular in the industry as recommended for the drilling of cored and angled holes.

Although drills point with $\Phi_p = 180°$ reduce the severity of the stability problem, particularly at the drill entrance, their application does not solve the problem with drill balance as explained in Astakhov (2019) in great detail. For years, the discussed problem was simply ignored because the diameter of cored holes was rather small compared to that of the drilled holes, tolerances on the drilled and then reamed holes were not tight, drilling speed and the feed rate were not aggressive as restricted by machine capabilities, and primarily relatively low available spindle rpm's and power. In the modern automotive industry, the diameter of the cored holes has increased in an attempt to cut less work material, tolerances on the hole have tightened at least four times in the last 10 years, and rotational speeds and feed rate are quadrupled over the same time period. All these improvements including significantly more rigid machine spindles and part fixtures brought the discussed problem to the forefront.

The reason for this is that the force balance shown in Figure 2.138a considers unbalance only in the z_0y_0 plane (see Figure 2.136), i.e., due to the unbalanced radial force F_r whereas the unbalance in the x_0z_0 plane due to power force component F_{x1} remains. The possible unbalance in the z_0x_0 plane was never considered. Figure 2.140 provides an explanation to the problem. With the help of this figure and referring to Figure 2.136, the problem is explained as follows. The cutting force component F_{x1}, which is normally much greater than axial F_a and radial F_r, is not actually supported. Rather, it has an arm l_{Fx} which length is the distance between points *c* and *9* in their projections into the z_0x_0 plane. Therefore, the force component F_{x1} bends the drill distorting its force balance.

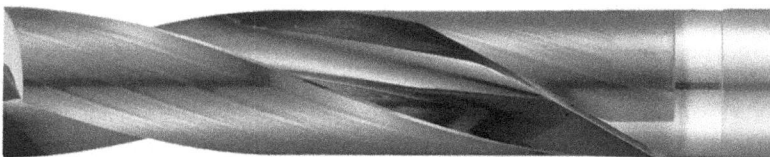

FIGURE 2.139   Modern drill with the point angle $\Phi_p = 180°$.

FIGURE 2.140 Explaining the unbalance problem in the $z_0 x_0$ plane.

The problem, however, is even wider. This is because point *9* (and *8*) is not located on the drill diameter circle so it does not touch the wall of the hole being drilled because, as shown in Figure 2.15, this point is located on the body clearance diameter. As a result, for this point to be a real supporting point, the drill should bend so that point *9* first passes the body clearance gap (Figure 2.15) and only then it can be a real supporting point. With the introduction of new carbide submicron grades having much greater flexural strength, also known as modulus of rupture, or bend strength, or transverse rupture strength, drills with the advanced planar flank face became commonplace (see Figure 2.48). The primary clearance angle applied to these drills is increased to its optimal value ($10°$–$14°$ compared to $6°$–$7°$ used in the past). As a result, the secondary clearance angle was increased up to $35°$, particularly for drills with internal coolant supply to provide better cooling and lubrication conditions at the flank face, that is, where these two actions are mostly needed. Moreover, drills with tertiary flank faces have become common (Astakhov 2014, 2019). The problem with great clearance angles and with application of drills with tertiary flank faces is that the locations of the top of the heels where the auxiliary circular lands (margins) start (points *8* and *9*) were thrown far back in the axial direction (along the z_0 axis) from the cutting edges as shown in Figure 2.141a. Obviously, the arm l_{Fx} (see Figure 2.140) increases significantly that makes significantly increases the severity of the discussed problem.

The problem with drill entrance stability has become even greater when various split-point designs (discussed in Section 2.1.3.7.7) were developed and made as a standard (gallery) feature of modern CNC drill point grinding machines. This advanced design aims at solving the long-standing problems with reduction of the high axial force due to the chisel edge and with improvement of the drill self-centering ability. However, a force balance/entrance stability problem becomes worse because, as shown in Figure 2.141b, the axial location of point *9* becomes even farther from the major cutting edges.

The problem with force balance that causes entrance instability and often affects the diameter, shape, and position of drilled holes particularly for long drills was definitely noticed in practice of drilling. As a result, a great variety of drills design with the auxiliary/secondary circular lands (margins) made on the top of the hills shown in Figure 2.141c was introduced with the intent to prevent drill wandering at the entrance of the hole being drilled. This can significantly reduce the extent of the bellmouth (an enlarged and tapered entrance part of the hole being drilled) and

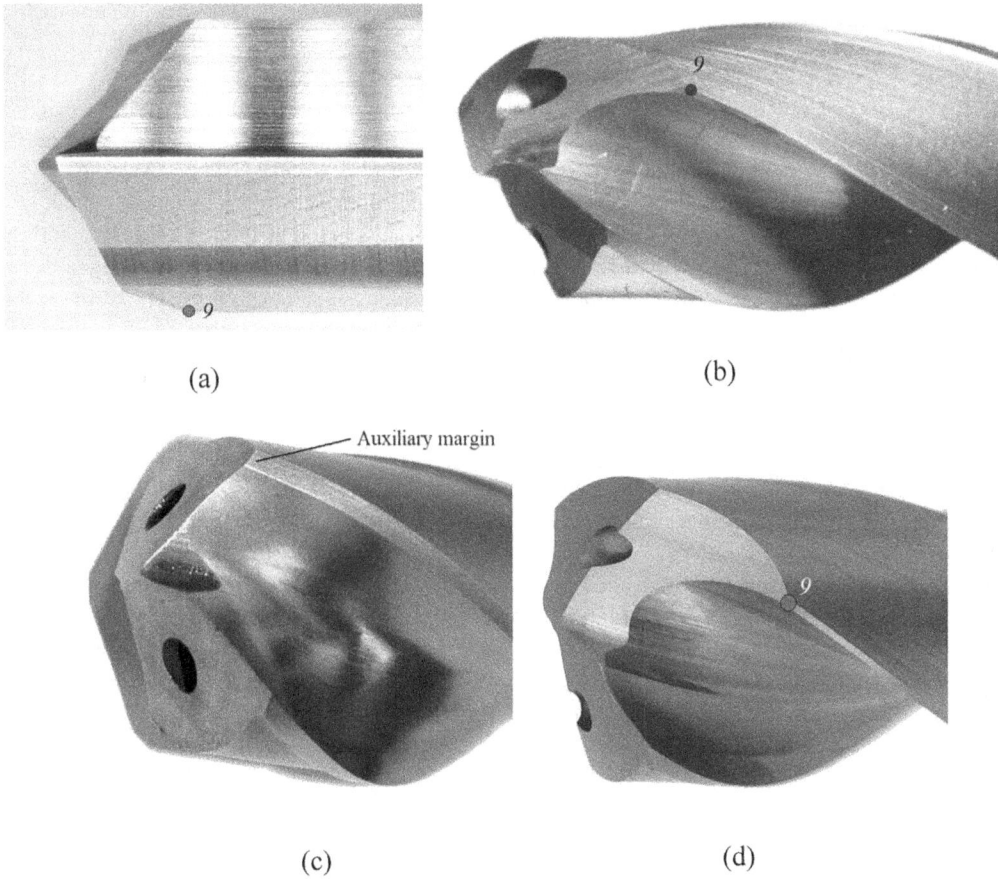

(a)   (b)

Auxiliary margin

(c)   (d)

FIGURE 2.141   Showing the axial location of point 9: (a) for a drill with a tertiary flank face, (b) for the common split-point drill, (c) drill with the auxiliary margins, and (d) a split-point drill with the auxiliary margins.

improve roughness and staginess of drilled holes. Although the problem with bellmouth formation was reduced, it has been never solved. This is because any drill is provided with back taper as discussed above in Section 2.1.3.8. As a result, the diameter of the drill reduces toward the shank so the diameter on which the end of the auxiliary margin is located is smaller than the drill diameter so some gap always exists between the top of this margin and the wall of the hole being drilled. As this gap is normally smaller than the above-discussed body clearance gap, the severity of the problem is reduced. However, when various split-point designs are used, the location of point 9 is often found too far back from the cutting edges as shown in Figure 2.141d. As such, the gap between the auxiliary margin and the wall of the hole being drilled becomes closer to the body clearance gap.

The foregoing analysis reveals that most modern drill designs cannot provide drill force balance, and thus stability at the entrance of the hole being drilled due to incomplete understanding of drill entrance condition. Therefore, a need is felt to develop a new design concept based on the principle of drill entrance stability first suggested for gundrills (Astakhov 1995a, b), i.e., for tool where the quality of machined holes is of prime concern and then formulated as Design Rule #2 (Astakhov 2014) for all drilling tools. Developing this concept further, the author modified this design rule (Astakhov 2019) as Design Rule #2.1 formulated as: Drill-free penetration in the feed direction while assuring the maximum entrance stability is achieved by the location of points 8 and 9 (Figures 2.136 and 2.140) at a distance of $l_{Fx}=f/4$ from the drill corners, where f is the cutting feed, mm/rev. Obviously, due to the drill manufacturing tolerances and setting (in the holder, machine, etc.) inaccuracies, this theoretical distance should be increased. A modeling test result showed that $l_{Fx} \simeq 1.5f$ can be used for practical applications.

The next step was to develop a practical drill design to satisfy Design Rule #2.1. As per Figure 2.140, if a drill is provided with an optimal 10° primary and 25° secondary clearance angles, then the location of point 9 seems to be geometrically fixed. The experience with the development of the flank faces of drills (Astakhov et al. 1995b) shows that it is not so. The development of a new design concept (named as VPA2© design) is shown in Figure 2.142. As shown, the flank face of a drill is designed as consisting of three facets. The first one is the primary flank plane and the second one is the secondary flank plane. Both planes are provided with the optimal flank angles as discussed above so the optimal work performance of the drill is not compromised. The third facet is the tertiary flank face. It starts at a certain distance l_s from the rake face (cutting edge) as shown in Figure 2.142 and it is positioned so that the distance between the drill corner and point 9 $l_s \simeq 1.5 \cdot f$, that is, as shown by the result of the modeling test.

2.1.5.7.2 Feasibility of the Developed VPA2© Design

The next step was to assure that the proposed concept can be realized practically on modern multi-axis CNC drill point grinding machines, that is, to develop a practical procedure of its manufacturing. To do that, modeling of possibility of grinding a special multi-facet flank face using the Helitronic Tool Studio software on a modern Walters Helitronic Vision CNC drill point grinding machine was carried out as shown in Figure 2.143. Figure 2.143a shows a grinding wheel positioning to assure that the second flank lane can be ground independently of the geometry/position of the first and tertiary flank planes. As such, the proper grinding wheel shape and its diameter were also selected to achieve the intended result. Figure 2.143b shows the result of grinding. Therefore, the proposed design concept can be fully realized using modern CNC drill point grinding machines.

2.1.5.7.3 Practical Development and Implementation of the VPA2© Design

A fragment of the process drawing of a drilling operation that includes the worst-case scenario conditions is shown in Figure 2.144. As shown in this figure, a rather tight tolerance (for a cored hole) is assigned on the drilled hole and the position of the hole axis is tightly controlled by two position tolerances. Further analysis of this hole reveals that it is a hole for tapping an NPSF 1/8-27 pipe thread

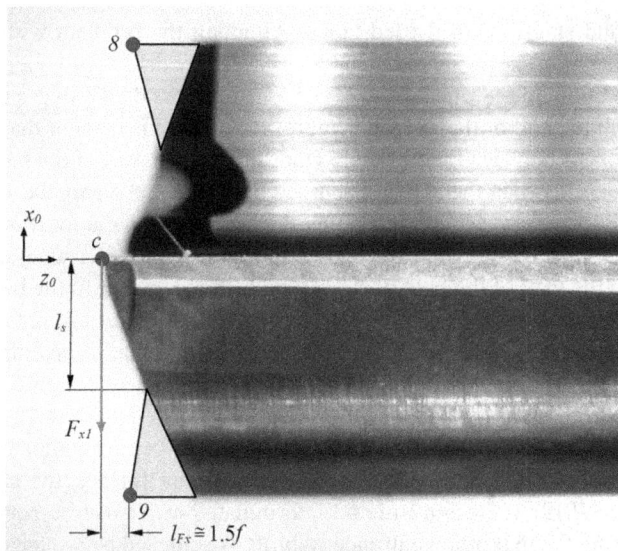

FIGURE 2.142 Development of a new design concept – VPA2© design.

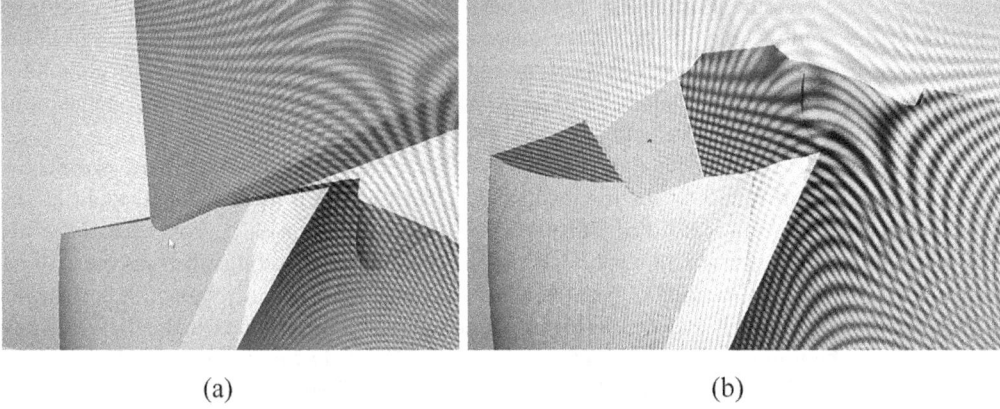

(a)   (b)

FIGURE 2.143 Modeling of the possibility of grinding a special secondary flank plane face using the Helitronic Tool Studio software on a modern Walters Helitronic Vision CNC drill point grinding machine: (a) grading wheel positioning and (b) result of grinding.

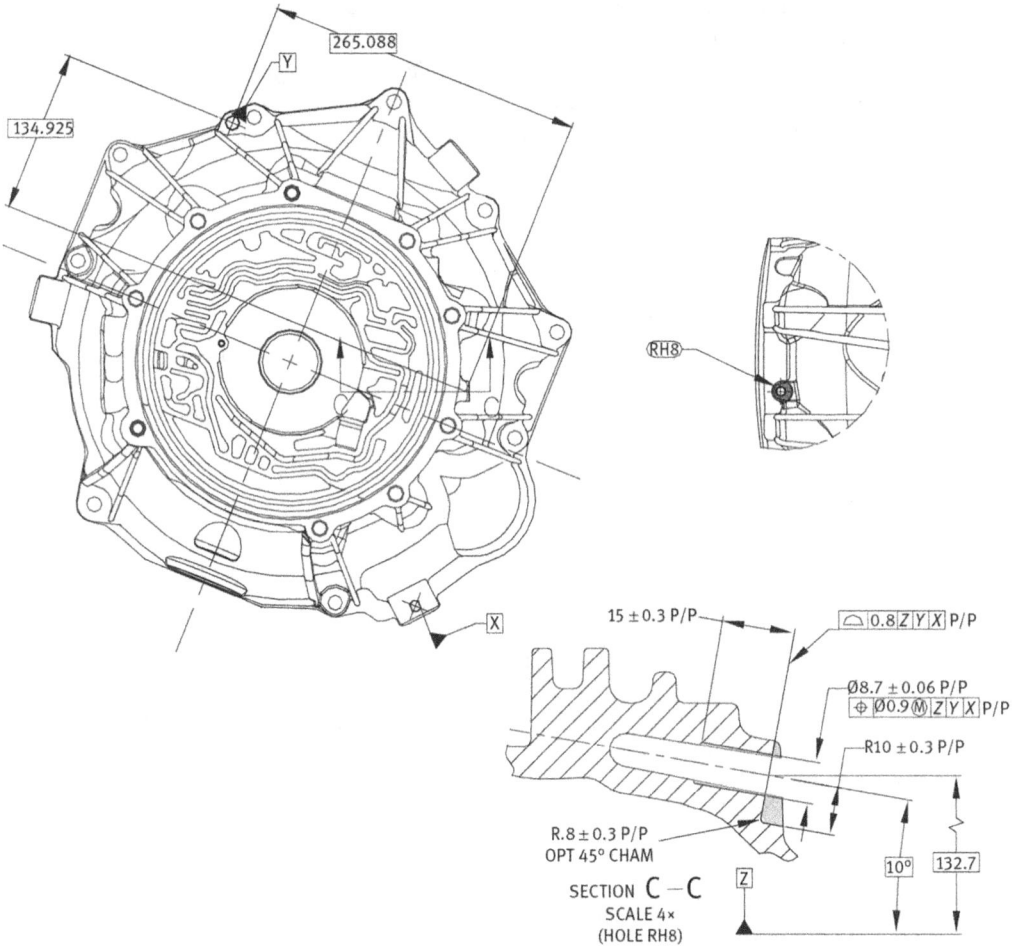

265.088

Y

134.925

RH8

15 ± 0.3 P/P

⌓ 0.8 Z Y X P/P

Ø8.7 ± 0.06 P/P
⊕ Ø0.9 Ⓜ Z Y X P/P

R10 ± 0.3 P/P

X

R.8 ± 0.3 P/P
OPT 45° CHAM

SECTION C – C
SCALE 4×
(HOLE RH8)

Z

10°   132.7

FIGURE 2.144 Fragment of the process drawing.

subjected to a high-pressure leakage test. That is why the tolerance on the drill diameter is tight. For years, the drilling of this hole was one of the major issues due to frequent tap breakage and failing high-pressure leakage test. Massive efforts of the engineering team of one of the largest automotive companies together with major tool manufacturers did not result in a noticeable improvement of this operation as fully described by the author earlier (Astakhov 2019).

A new drill was developed using the proposed VPA2© and above-discussed VPA1© (see Section 2.1.3.6) designs. Figure 2.145 shows the most relevant extract from the developed tool drawing where the proposed concept is realized. Section B-B shows the location of the front ends of the auxiliary circular lands (margins) with respect to the drill corners/major cutting edges (as the drill has an 180° point angle) – the location of point *9* as per Figure 2.142 is actually shown. It is clear that the axial location of point *8* is the same as that of point *9* as the drill is symmetrical. VIEW A shows the geometry of the minor cutting edges (edges *1–4* and *6–7* are shown in Figure 2.136) where the VPA1© design is implemented. As can be seen, the flank face of each of the minor edges includes two parts: the first part (Section E-E in Figure 2.145) is ground as it is usually made on end mills, that is, ground up to a sharp edge; the second part (Section D-D in Figure 2.145) is provided with a thin (0.08 mm) circular land (VPA1© design) for drill stability. The drill contains the auxiliary margins.

The drill was made to the drawing, inspected, and set into a manufacturing cell. Its tool life was tripled compared to the previously used drill, and no tool breakage and problem with passing the leakage test were reported. Figure 2.146 shows the particularities of the new design and wear patterns after 10,000 holes. Figure 2.146a shows the face view of the drill, and Figure 2.146b and c shows two "3D" views where particularities of the rake faces, auxiliary circular lands (margins), and location of the gash forming the rake faces can be seen. Figure 2.146d shows the wear pattern on the face of the auxiliary circular land (margin). As shown, this wear is insignificant as only a small trace of the aluminum

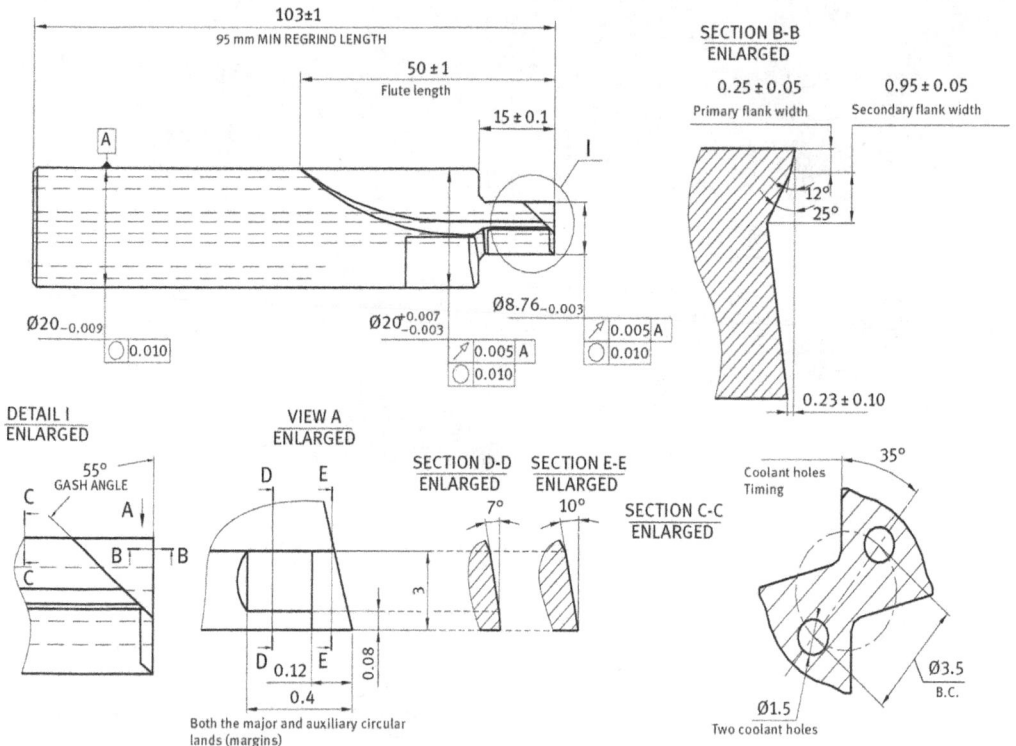

FIGURE 2.145 Relevant fragment of the tool drawing.

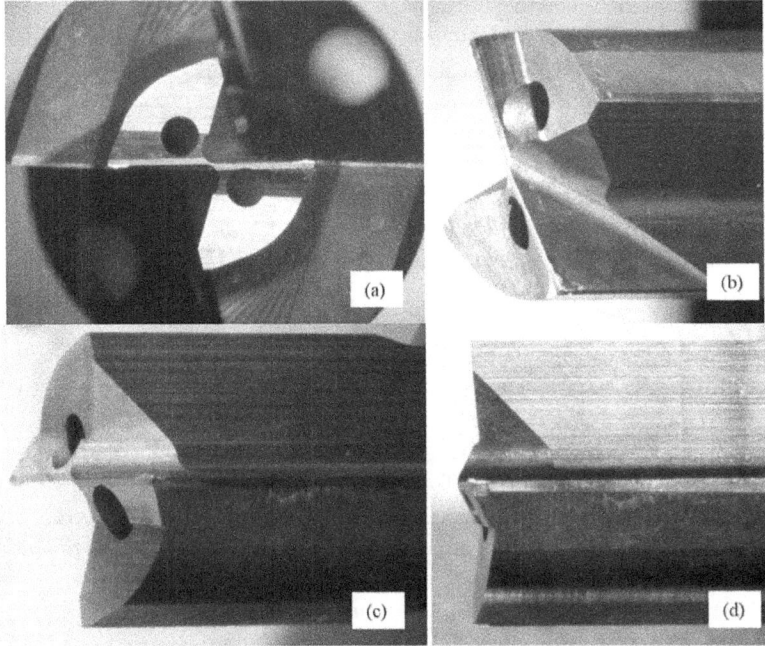

FIGURE 2.146 Particularities of the new drill design and its wear pattern.

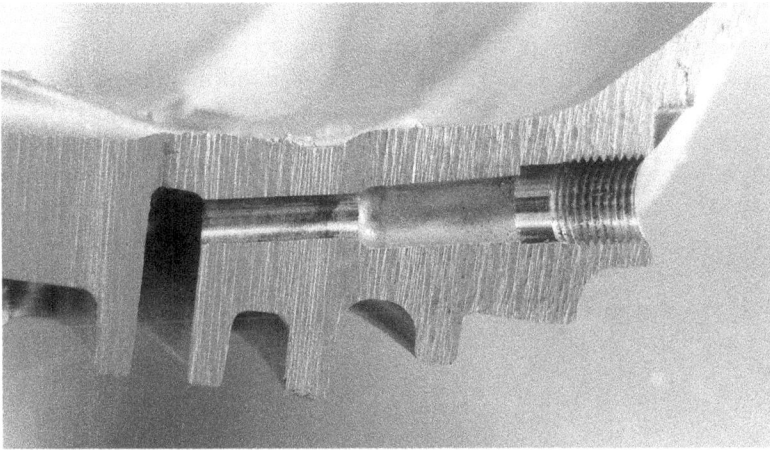

FIGURE 2.147 The proper thread was machined even when the axis of the cored hole is significantly shifted from that of the drill/thread.

built-up edge is observed in the very corner. Such a small wear on the drill corners and the front ends of the major and auxiliary circular lands (margins) suggests that the number of drill regrinds can be substantially increased till min. regrind length (95 mm per Figure 2.145) is reached. Figure 2.147 shows a section of the machined hole with the tapped thread. As can be seen, the proper thread was machined even when the axis of the cored hole is significantly shifted from that of the drill/thread. Moreover, the roughness of the hole drilling by the developed drill is much better than normally obtained in drilling of holes in aluminum alloys by carbide drills. The VPA2© combined with VPA1© designs were implemented for all carbide drills for cored holes in the world's largest automotive transmission plant. Further test and implementation results can be found in Astakhov (2019).

As a conclusion, it can be stated that the basic design of drill for cored hole was developed, tested, and implemented as the combination of the VPA1© and VPA2© designs. Although the development was carried out for cored and angled holes, the proposed designs should be implemented in most of the "standard" straight and twist drills to assure their entrance stability and to improve the quality, i.e., meeting tight tolerances on diametric accuracy, hole shape and straightness, position, and so on. Moreover, the use of the proposed combination improves the quality of the entrance of the hole being drilled so that the proposed designs can be used for drills (both straight and twist) when this quality is of some concern. As such, there is no need to reduce the cutting feed on drill entrance as it is common in drilling, which can result in high drilling productivity.

2.1.5.8 Drilling Tool Geometry in T-hold-S and T-use-S: Clearance Angle

As discussed in the Appendix, there are four systems of consideration of the cutting tool geometry, namely the tool-in-hand system (T-hand-S), the tool-in-holder system (T-hold-S), tool-in-machine system (T-mach-S), and tool-in-use system (T-use-S). As discussed above, the drilling tool geometry parameters in T-hand-S are used in the tool design, and thus assigned in the tool drawing, used in tool manufacturing and its inspection. T-mach-S has a little relevance to drilling tools so it is not considered any further. Therefore, T-use-S and T-hold-S are subjects for further consideration.

The general notion of T-use-S is discussed in the Appendix. According to this notion, T-use-S considers the geometry of the cutting tool accounting for machining kinematics. The parameters of the tool geometry are affected by the actual resultant motion of the cutting tool relative to the workpiece. The velocity of this resultant motion includes the cutting speed and the feed rate (the velocity of the cutting feed). In T-use-S, the so-called kinematic cutting tool angles should be considered. Based upon the consideration made in the Appendix, Figure 2.148 presents the basic idea of the kinematic clearance angle. In this figure, a point of consideration i is selected on the major cutting edge 1–2. The working plane P_f is the same for T-hand-S and T-use-S systems, i.e., $P_f \equiv P_{fe}$. It is drawn through the considered point i and contains the vectors of the cutting speed, v, and that of feed velocity, v_f. The radius of the considered point is r_i. The drill rotates with n (rpm).

As discussed in Chapter 1, as the drill rotates with rotational speed n (rpm) and fed with the cutting feed f (mm/min), the cutting speed for point i is calculated as $v_i = (\pi 2 r_i n)/1,000$ and the feed velocity is calculated as $v_f = nf$. Note that the cutting speed depends on the radius of the point of consideration reducing with this radius, while the feed velocity is the same for any drill point. As discussed in the Appendix, the angle between the resultant cutting velocity $\mathbf{v}_e$ and that of cutting speed $\mathbf{v}_i$ for considered point i is the resultant cutting speed angle η_{e-i}. For a given point i of the cutting edge located at radius r_i, it is calculated as

$$\tan \eta_{e-i} = \frac{v_f}{v_i} = \frac{f}{\pi(2r_i)} \tag{2.45}$$

When the drill is rotated and fed, point i generates a cylinder of radius r_i in its helical motion. Therefore, the most proper consideration of the clearance angle should be in the plane tangent to this cylinder at this point, i.e., in the working plane $P_f \equiv P_{fe}$ as shown in Figure 2.148. As follows from Figure 2.148 and the consideration presented in the Appendix, Section A.7, this angle in a given point i of the cutting edge is calculated as

$$\alpha_{fe-i} = \alpha_{f-i} - \eta_{e-i} \tag{2.46}$$

where α_{f-i} is the clearance angle in T-hold-S.

As discussed in the Appendix, Section A.7.2 in practical turning operations, angle η_e is very small and thus is taken into consideration only for some special cases where excessively high feeds are used, that is, thread cutting. Is it the same in considerations of the drilling tool geometry? To answer this question, let us consider a practical example using a carbide drill of 10 mm dia.

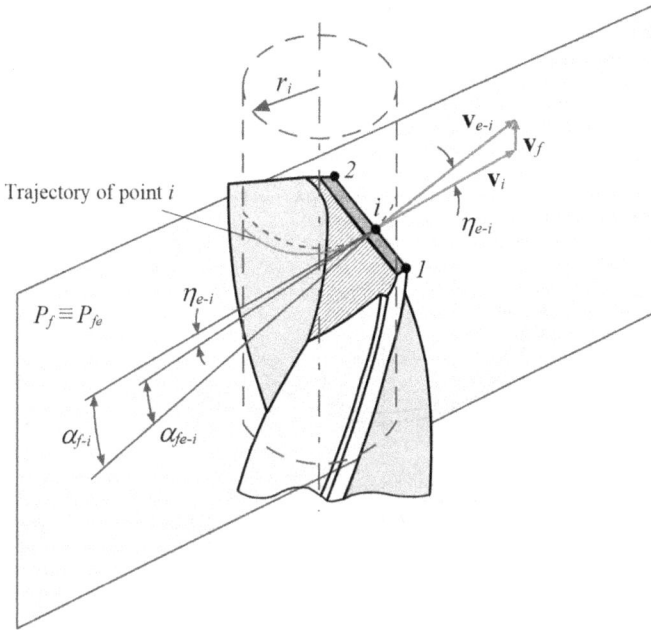

FIGURE 2.148   The basic idea of the kinematic clearance angle.

The major cutting edge dimensions are as follows: radius of point *1* (Figure 2.148) is 5 mm and that of point *2* is 1.5 mm. The cutting speed is chosen to be 100 m/min and the feed per revolution is 0.4 mm/rev. Under these conditions, the drill rotational speed is $n = (1{,}000v)/\pi d_{dr} = (1{,}000 - 100)/(3.141 \cdot 10) = 3{,}183.6 \approx 3{,}184$ rpm. The feed velocity is calculated as $v_f = nf = 3{,}184 \cdot 0.4 = 1{,}273.6$ mm/min ≈ 1.27 m/min. The cutting speed for point *1* is equal to the drill cutting speed, that is, 100 m/min while that for point *2* is calculated as $v_B = (\pi d_B n)/1{,}000 = (3.141 \cdot (2 \cdot 1.5) \cdot 3{,}184)/1{,}000 = 30$ m/min. Using Eq. (2.45), one can calculate angle η_e for points *1* and *2* as $\eta_{e-1} = \arctan(1.27/100) = 0.73°$ and $\eta_{e-2} = \arctan(1.27/30) = 2.42°$. The obtained results show that even in the most inner points of the major cutting edge and rather aggressive cutting feed, angle η_e is not that significant for the major cutting edge. Therefore, it will not be considered in the further analysis of geometry of this edge.

The next step is to establish the clearance angle in the working plane in T-hold-S α_{fh} so its correlation with the corresponding clearance angle, α_f in T-hand-S, should be established. To do this, a model of the T-hold-S geometry based on the model considered in the Appendix is developed as shown in Figure 2.149. As can be seen, the vector of the cutting speed at point *i*, v_i, is perpendicular to the radius, r_i, of this point. Therefore, this actual direction of the cutting speed is located at angle μ_i to the direction of the assumed cutting speed in T-hand-S shown in Figure 2.148. The reference planes in T-hold-S are drawn through point of consideration *i* and defined as follows:

- *Tool reference plane* P_{rh} is perpendicular to the actual direction of primary motion;
- *Working plane* P_{fh} is perpendicular to the reference plane P_r and contains the vectors of the cutting speed and feed velocity;
- *Tool back plane* P_{ph} is defined to be perpendicular to the drill rotational axis. This plane is perpendicular both to the tool reference plane, P_r, and to the tool working plane, P_f.

Figure 2.149 shows the defined reference planes.

The clearance angle α_{fh} is correlated with the normal clearance angle α_n in T-hand-S as follows:

$$\tan \alpha_{fh-i} = \frac{\cos \alpha_n \cos \mu_i + \cos \varphi \sin \mu_i}{\sin \varphi} \tag{2.47}$$

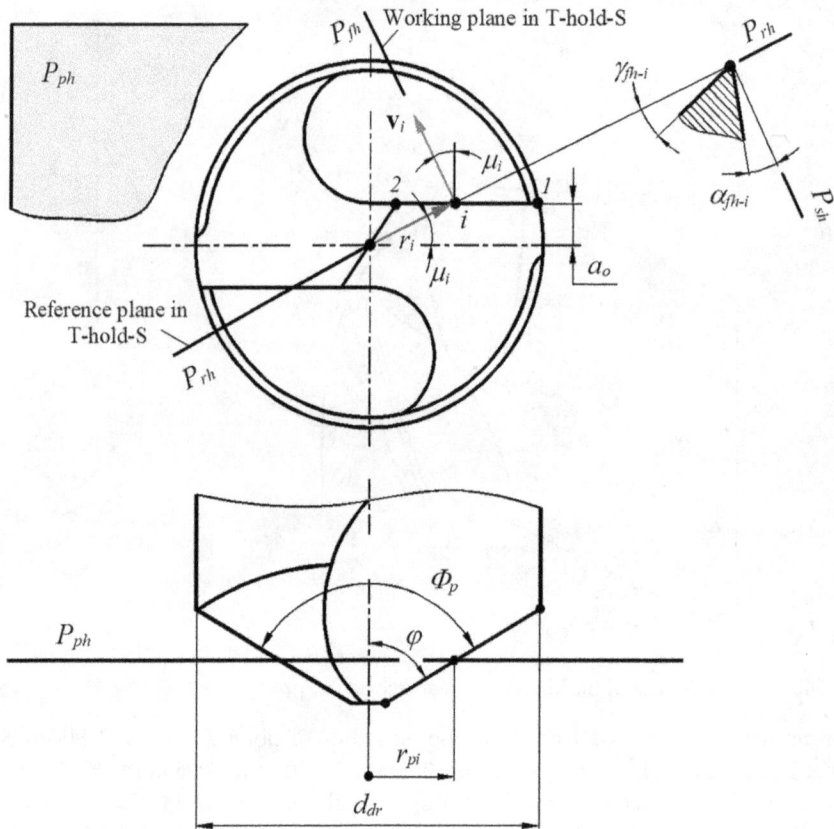

FIGURE 2.149 T-hold-S reference plane and angles.

where the sense of angle μ_i is clearly seen in Figure 2.149 as the angle of the radius of point i.

Equation (2.47) can be rearranged for practical calculations as follows:

$$\alpha_{fh-i} = \arctan\left(\frac{\cos\alpha_n\sqrt{r_i^2 - a_o^2} + a_o\cos\varphi}{r_i\sin\varphi} \right) \tag{2.48}$$

Note also that $r_{pi} = \sqrt{r_i^2 - a_o^2}$.

As angle μ_i varies with the location of the point of consideration i over the major cutting edge 1–2, clearance angle α_{fh} also changes correspondingly even if the normal clearance angle α_n in T-hand-S is kept constant as in the case with a planar flank face. As such, the calculations and thus optimization of α_{fh} do not present a problem. When the primary flank face of a drill is not planar, then the T-hand-S clearance angle α_n also changes with r_i so that Eq. (2.48) should be modified correspondingly

$$\alpha_{fh-i} = \arctan\left(\frac{\cos\alpha_{n-i}\sqrt{r_i^2 - a_o^2} + a_o\cos\varphi}{r_i\sin\varphi} \right) \tag{2.49}$$

where α_{n-i} is the T-hand-S clearance angle at point i of the cutting edge.

An analysis of Eq. (2.48) and Figure 2.150 shows that the clearance angle α_{fh} increases with a_o (Figure 2.149) and decreases with the point angle $\Phi_p = 2\varphi$ particularly in the vicinity of the

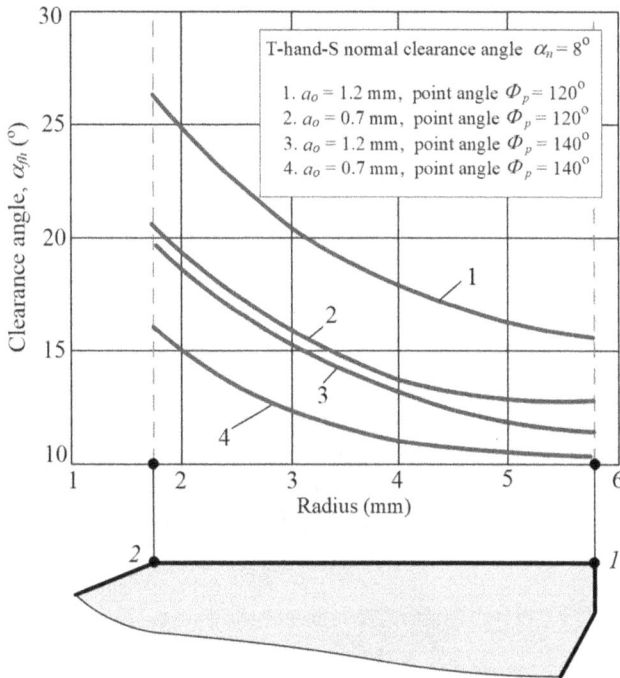

FIGURE 2.150 Variation of the clearance angle in T-hold-S α_{fh} over the major cutting edge *1-2* for a drill of 11.5 mm diameter.

chisel edge. As can be seen, when $a_o = 1.2$ mm, when the point angle Φ_p increases from 120° (old drill standard) to 140° (recent standard for carbide drills), the clearance angle α_{fh} at point *2* decreases from 26.8° to 19.5°, i.e., approximately by 1.4 times. This increases the strength of the cutting wedge in the vicinity of point *2*, which is essential in machining of high-strength steels. This also explains the change of the "old" standard point angle 120° to 140° for carbide drills by practically all major drill manufacturers.

The relevance of the considered clearance angle α_{fh} is that this angle is in the working plane P_{fh} in T-hold-S, i.e., in the line where the sliding velocity over the tool–workpiece interface and the previously discussed springback of the machined surface are the greatest. Unfortunately, these two conditions are seldom fulfilled simultaneously, that is, in one plane of measurement. Therefore, this angle should be considered in any modeling of the conditions over the tool–workpiece interface, including finite element method (modeling) (FEM). It should also be considered in modeling of metal cutting a 2D model of chip formation (e.g., shown in Figure 2.102).

2.1.5.9 Drilling Tool Geometry in T-hold-S and T-use-S: Rake Angle

As discussed in the Appendix, the rake angle in the working plane in T-use-S is calculated as

$$\gamma_{fe} = \gamma_{fh} + \eta_e \qquad (2.50)$$

where γ_{fh} is the rake angle in the working plane in T-hold-S and angle η_e is the angle between the resultant cutting velocity and that of cutting speed.

The determination of the second term of Eq. (2.50) does not present any problem as it is already defined by Eq. (2.45) whereas the determination of the first term is more complicated. Although γ_{fh} can be determined in a pure geometrical way using the model shown in Figure 2.149, γ_{fh} determined using this way requires the justification of its physical meaning in full analogy with the clearance angle in this system. In other words, the actual plane in which this angle is properly defined should be known. There are two principal approaches to the determination of the rake angle in T-hold-S. According to the first approach (Oxford 1955, Galloway 1957, Vinogradov 1985), this angle should be determined in the plane of chip flow so that it is termed the rake angle in the chip flow direction γ_{cf}. According to the second approach, the chip plane is the normal plane P_{nh} ($\equiv P_{ne}$) so the rake angle in this plane γ_{nh} should be considered relevant (Fujii et al. 1970a,b, Astakhov 2010, Takashi and Jürgen 2010). More strictly, however, is to consider the rake angle in the T-use-S angle γ_{ne} determined as

$$\gamma_{ne} = \gamma_{nh} + \eta_e \tag{2.51}$$

2.1.5.9.1 Rake Angle in T-hold-S/T-use-S γ_{cf} Determination According to the First Approach

The first approach determines γ_{cf} as actually the rake angle not only in T-hold-S but also in T-use-S as the chip appears only when the drill actually works. Using the approach discussed in Astakhov (2014), one can determine the chip flow angle η_{cf-i} (the angle between the *theoretical* chip flow direction and the normal to the major cutting edge) for a point of consideration i on the major cutting edge as

$$\sin \eta_{cf-i} = \sin \eta_{e-i} \cos \varphi_i + \cos \eta_{e-i} \sin \varphi_i \sin \mu_i \tag{2.52}$$

Then γ_{cf} is determined as (Vinogradov 1985)

$$\gamma_{cf-i} = \arcsin \left(\cos \lambda_{se-i} \sqrt{1 - \cos^2 \gamma_n \cos^2 \eta_{cf-i}} - \sin \eta_{cf-i} + \sin \eta_{cf-i} \sin \lambda_{se-i} \right) \tag{2.53}$$

where λ_{se-i} is the kinematic inclination angle at a point of consideration i on the major cutting edge. This angle is calculated as

$$\lambda_{se-i} = \arctan \left(\frac{\tan \lambda_{sh-i}}{\cos \eta_e} \right) \tag{2.54}$$

where λ_{sh-i} is the T-hold-S inclination angle for a point of consideration i on the major cutting edge. The sense of this angle is shown in Figure 2.151. It follows from this figure that λ_{sh-i} can be calculated through angle μ_i (Figure 2.149) as

$$\tan \lambda_{sh-i} = \tan \mu_i \sin \varphi \tag{2.55}$$

Figure 2.151 also shows an example of distribution of λ_{sh-i} (and thus λ_{se-i} as angle η_e is small) for a drill with the web diameter equal to $0.2d_{dr} = a_o/2$ and half-point angle $\varphi = 58°$. As can be seen, the smallest $\lambda_{sh-1} = 9.97°$ is at the outer end of the major cutting edge (point 1), while for the inner end 2, this angle is $\lambda_{sh-2} = 50.50°$, that is, the greatest (Granovsky and Granovsky 1985).

According to an extensive experimental study carried out by Galloway (1957), the Stabler rule (Stabler 1964) is applicable for the drill geometry, that is, the chip flow angle $\eta_{cf-i} \approx \lambda_{sh-i}$. As such, Eq. (2.53) can be significantly simplified to

$$\gamma_{cf-i} = \arcsin \left(\cos^2 \lambda_{se-i} \sin \gamma_n + \sin^2 \lambda_{se-i} \right) \tag{2.56}$$

Equation (2.56) establishes the correlation between the T-use-S γ_{cf} and T-hand-S γ_n rake angles through the drill design and geometry parameters. Figure 2.152 shows variation of γ_{cf} over the major cutting edge for a drill of 18 mm dia. with $a_o = 1.5$ mm and point angle $2\varphi = 120°$ for $\gamma_n = 0°$

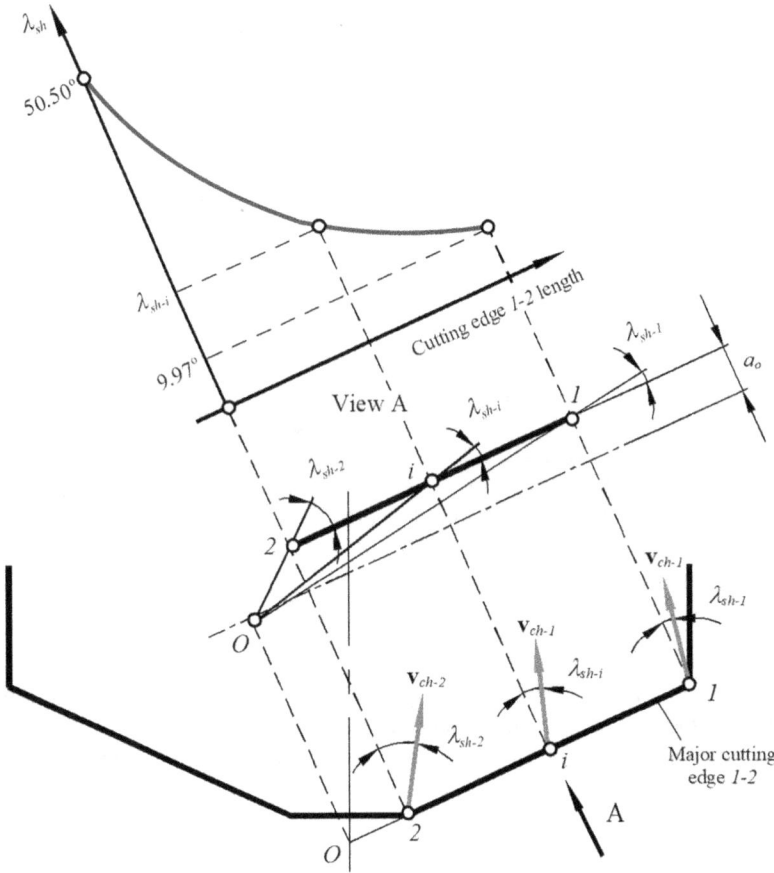

FIGURE 2.151 Inclination angle and the theoretical chip flow directions for the major cutting edge *1–2*.

and $\gamma_n = 10°$ (curves 1 and 2, respectively). As can be seen, the T-use-S γ_{cf} is greater than γ_n over the entire length of the major cutting edge. Moreover, γ_{cf} increases significantly for the region of this edge in the vicinity of its inner end adjacent to the chisel edge.

Equation (2.56) establishes a direct correlation of γ_{cf} with the T-hand-S rake angle γ_n, point angle 2φ, and drill design parameter a_o. It can be assumed for practical calculation that $\lambda_{se-i} \approx \lambda_{sh-i}$ so Eq. (2.56) can be rearranged for convenient use in the drill design as

$$\gamma_{cf-i} = \arcsin \frac{\sin \gamma_n \left(r_i^2 - a_o^2 \right) + a_o^2 \sin^2 \varphi}{r_i^2 - a_o^2 \cos^2 \varphi} \tag{2.57}$$

The analysis of Eq. (2.57) shows that γ_{cf} primarily depends on γ_n, a_o, and r_i while it weakly depends on the point angle $\Phi_p = 2\varphi$. For example, for a drill having $\gamma_n = 6°$, $a_o = 1.5\,\text{mm}$: when $r_i = 4\,\text{mm}$, the rake angle $\gamma_{cf} = 12.33°$ when $\Phi_p = 118°$ and $\gamma_{cf} = 13°$ when $\Phi_p = 140°$. When $r_i = 10\,\text{mm}$, the rake angle $\gamma_{cf} = 7°$ for $\Phi_p = 118°$ and $\gamma_{cf} = 6.86°$ for $\Phi_p = 140°$.

To assess the individual influence of the T-hand-S rake angle γ_n, point angle 2φ, and drill design parameter a_o, the following approximation of Eq. (2.57) can be used (Vinogradov 1985):

$$\gamma_{cf-i} = \frac{0.08 a^{2.13} \varphi^{1.2}}{r_i^{2.08}} + 0.81 \gamma_n \tag{2.58}$$

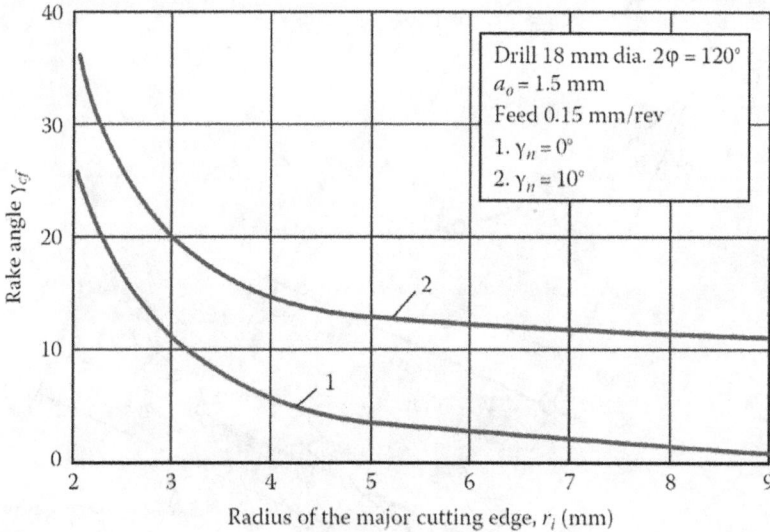

FIGURE 2.152 Variation of γ_{cf} over the major cutting edge.

where $a = 2a_o$.

The second term of Eq. (2.58) includes only parameter γ_n, so by calculating the first term for various r_i, it is possible to determine the values of γ_n for the point of the major cutting edge to keep the same γ_{cf} over the whole major cutting edge, that is, to balance a drill in terms of direction of chip flow. For example, for a drill having the following parameters, $2\varphi = 120°$, $a = 3$ mm, and $\gamma_n = 0°$, the rake angle γ_{cf} is as follows:

r_i (mm)	2	3	5	7	9
γ_{cf} (°)	26.7	11.5	3.9	2.0	1.2

Therefore, to assure the condition $\gamma_{cf} = 0°$ over the entire cutting edge, the T-hand-S rake angle γ_n should vary according to $\gamma_n = -\gamma_{cf}/0.81$ as

r_i (mm)	2	3	5	7	9
γ_n (°)	−33.0	−14.2	−4.9	−2.4	−1.4

and thus the T-hand-S working rake angle as it is actually measured by many drill geometry inspection CNC machines should vary according to Eq. (A.11) (Appendix) as

$$\gamma_f = \arctan\left[\tan\left(\frac{\gamma_{cf}}{0.81} \right) \sin\varphi \right] \qquad (2.59)$$

that is,

R_i, (mm)	2	3	5	7	9
γ_f (°)	−29.3	−12.4	−4.3	−2.1	−1.3

The results of the presented calculations show that even the application of a gash (see Section 2.1.3.6 and Figure 2.54) having rake angle (the working [side] rake angle in T-hand-S) γ_f up to −29° on the drill rake face in the vicinity to its inner end adjacent to the chisel edge assures a neutral γ_{cf} rake angle.

2.1.5.9.2 Rake Angle in T-hold-S/T-use-S γ_{ne} Determination According to the Second Approach

Although the determination of the T-use-S rake angle as γ_{cf} is predominant in the literature, it was noticed a long time ago that it may not be the case (Oxford 1955) as the chip flow direction problem is complicated by the simple fact that the flowing chip is continuous across the major cutting edge. One should realize that in any drill, the length of the chip produced by the periphery point *1* (its nearest vicinity to assume some finite chip width) over one drill revolution is much greater than that by point *2*, which is at the inner end of the major cutting edge where this edge meets with the chisel edge. This is because the path traveled by point *1* in one drill revolution is r_1/r_2 times greater than that by point *2* (where r_1 and r_2 are radii of points *1* and *2*, respectively). For example, for a standard 20 mm dia. drill, $r_1 = 10$ mm and $r_2 = 2$ mm, that is, the path passed by point *1* is fivefold greater than that by point *2*. The chip length is determined as the length of the path divided by a chip compression ratio (CCR) (see the Appendix, Section A.8.2.2.1 and Astakhov (2004, 2006)). Because this ratio is normally 50%–70% higher for point *2* because of its much smaller cutting speed and not favorable tool geometry, the total difference in the chip length produced by point *1* is normally seven to eight times greater than that by point *2*. Because the chip is continuous along the major cutting edge *1–2*, the discussed difference causes chip curving into a cone shape as shown by a model in Figure 2.153a that closely resembles the reality (Figure 2.153b). In other words, the direction of actual chip flow significantly deviates from that set by the major cutting edge inclination angle as shown in Figure 2.154 (Astakhov 2010). As a result, the forming chip *3* flows along straight part *4* of the flute and then over its curved part *5* as shown in Figure 2.154 reaching the walls of the machined hole.

As the exact angle of chip flow is application-specific, the direction of chip flow according to this second approach is assumed to be perpendicular to the cutting edge (Fujii et al. 1970b, Astakhov 2010, Takashi and Jürgen 2010). As such, the T-use-S rake angle considered point *i* of the major cutting edge is calculated as

$$\gamma_{ne-i} = \gamma_{nh-i} + \upsilon_{e-i} \tag{2.60}$$

where γ_{nh-i} is the normal rake angle in T-hold-S and $\upsilon_{e-i} = \arctan(\tan \eta_{e-i} \sin \varphi)$ where η_e is determined according to Eq. (2.45).

As discussed by the author earlier (Astakhov 2010), when the major cutting edge of a drill is made with $\gamma_n = 0$ (the common case for straight-flute drills), then for a given point *i* on this edge, angle γ_{nh-i} is calculated as

$$\gamma_{nh-i} = -\arctan\left(\tan \mu_i \cos \varphi\right) = -\arctan\left[\tan\left(\arcsin \frac{a_o}{r_i}\right) \cos \varphi\right] = -\arctan\left(\frac{a_o}{r_{pi}} \cos \varphi\right) \tag{2.61}$$

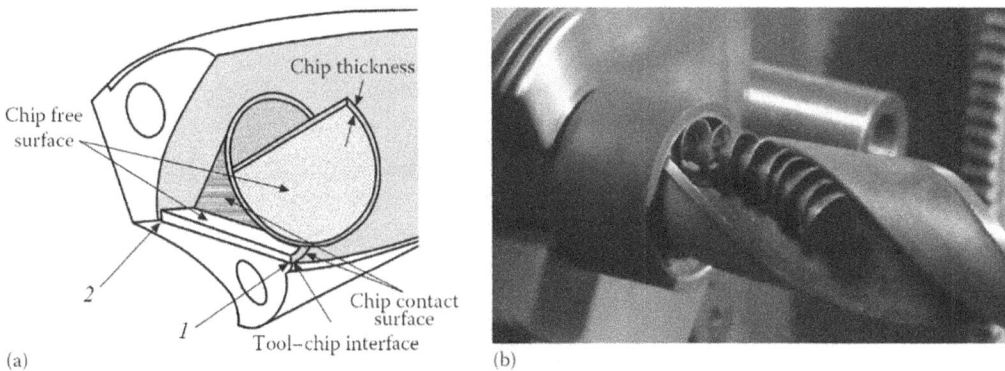

FIGURE 2.153 Chip flow: (a) model and (b) reality.

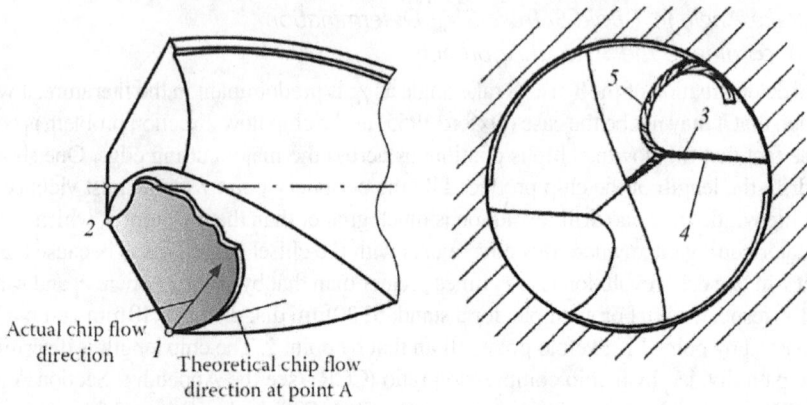

Actual chip flow
direction *1*
        Theoretical chip flow
        direction at point A

FIGURE 2.154 Simple model of chip flow.

Equation (2.61) defines the distribution of rake angles along cutting edge *1–2*, that is, $r_2 \leq r_i \leq r_1$. As can be seen, when $a_o > 0$, that is, when the major cutting edge is located ahead of the centerline, this rake angle is negative for any point of this cutting edge. Because $r_1 > r_2$ and while angle φ and distance a_o are the same for all points of cutting edge *1–2*, the absolute value of the rake angle at point *2* is greater than that at point *1*.

When the major cutting edge of a drill is made with a modified rake surface having the normal T-hand-S rake angle $\gamma_n \neq 0$, for example, modified by applying additional design features such as planar rake face (e.g., shown in Figure 2.56 and Figure 2.58), the distribution of the rake angle over cutting edge *1–2* is calculated as

$$\gamma_{nh}(r_i) = \gamma_n(r_i) - \arctan\left[\tan\left(\arcsin\frac{a_o}{r_i}\right)\cos\varphi\right] \tag{2.62}$$

or

$$\gamma_{nh}(r_{pi}) = \gamma_n(r_{pi}) - \arctan\left(\frac{a_o}{r_{pi}}\cos\varphi\right) \tag{2.63}$$

where $\gamma_n(r_i)$ determines the distribution of the rake face on the modified rake surface. Often, this surface is ground with a constant rake angle in T-hand-S so that this angle does not change over cutting edge *1–2*, i.e., $\gamma_n(r_i) = \text{Const}$.

When the rake face is helical, then calculations of the rake angle distribution over the major cutting edge should follow the following known procedure (Rodin 1971, Astakhov 2010):

$$\tan\gamma_{nh-i}(r_i) = \frac{1 - \sin^2\varphi\sin\mu_i}{\sin\varphi\cos\mu_i}\tan\omega_i - \cos\varphi\tan\mu_i \tag{2.64}$$

As expected, when $\omega_i = 0$, Eq. (2.64) coincides with Eq. (2.61) obtained for a straight flute. Making simple geometrical re-arrangement in Eq. (2.64), one can obtain

$$\tan\gamma_{nh-i}(r_i) = \left(\frac{r_i}{r_{dr}}\tan\omega_d\right)\frac{1 - \left(\dfrac{a_o}{r_i}\right)\sin^2\varphi}{\sin\varphi\cos\left(\arcsin\dfrac{a_o}{r_i}\right)} - \cos\varphi\tan\left(\arcsin\frac{a_o}{r_i}\right) \tag{2.65}$$

An analysis of Eq. (2.65) shows that when the rake face is helical, the T-hold-S normal rake angle of a major cutting edge depends on the point angle Φ_p (as φ is half of the point angle) distance a_o, and the helix angle ω_d. Note that $2a_o$ is often referred to in the literature as the web diameter although in general, the cutting edge may consist of a number of parts with individual a_o's or it can be inclined as per DIN 1214 Type B (Figure 2.56), and on the helix angle ω_d. Figure 2.155 shows the influence of the point angle for twist drill having the following parameters: $\omega_d=30°$, $2a_o=d_w=0.2d_{dr}$ (Rodin 1971) on γ_{nh}. As can be seen, in contrary to γ_{cf}, the point angle has a significant influence on γ_{nh}. Small point angles cause a significant increase in the T-hold-S normal rake angle in the vicinity of the periphery point 1 with a sharp decrease in this angle along the cutting edge toward the drill axis. For a drill with $\Phi_p=180°$, the normal rake angle varies along the cutting edge from 28° to 5°. For a drill with the standard point angle $\Phi_p=120°$, the normal rake angle varies from +30° to −30°. Therefore, an increase in the point angle reduces the spread in the normal rake angle along the cutting edge.

The latter occurs because the point angle affects the shape and thus the curvature of the surface of the cut (the bottom of the hole being drilled). In general, in drilling, this surface is hyperboloid as discussed later, which becomes a plane when $\Phi_p=180°$. When it happens, the normal to the surface of cut does not change its direction along the cutting edge remaining parallel to the drill axis. As such, the location of the major cutting edge with respect to the centerline determined distance a_o ($2a_o=d_w$) has only a weak influence on the normal rake angle. Equation (2.65) can be modified for this case as

$$\tan\gamma_{nh}(r_i) = \left(\frac{r_i}{r_{dr}}\tan\omega_d\right)\cos\left(\arcsin\frac{a_o}{r_i}\right) \tag{2.66}$$

The influence of the distance a_o on the normal rake angle is shown in Figure 2.156 for standard drills with $\Phi_p=120°$. As can be seen, an increase in a_o leads to a decrease in γ_{nh}. If the major cutting

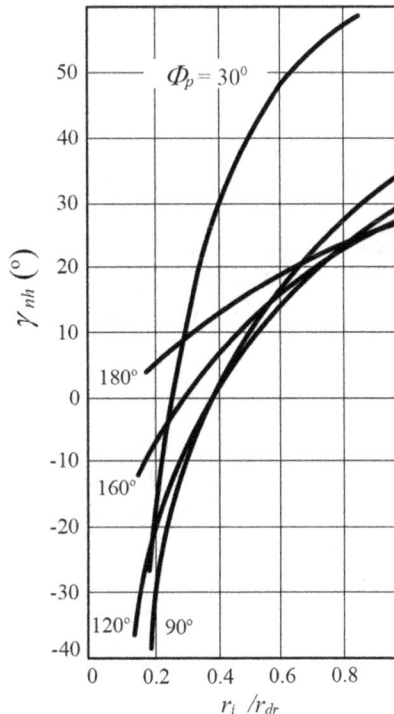

FIGURE 2.155 Influence of the point angle on γ_{nh}.

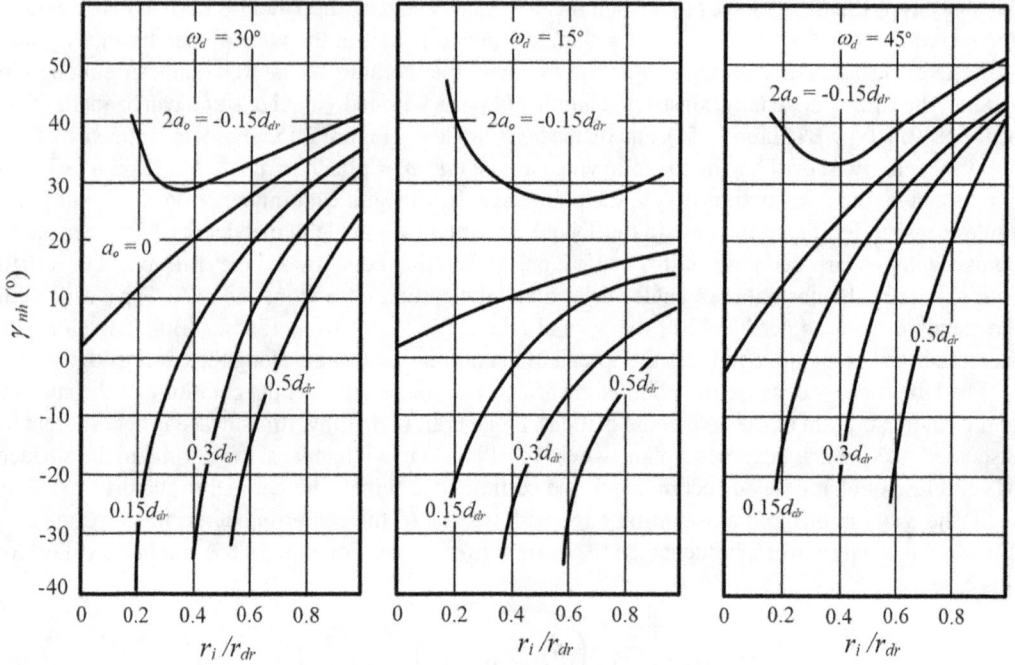

FIGURE 2.156 Influence of the distance a_o on the variation of the normal rake angle.

edge (lip) is located along the drill radius, that is, $a_o=0$, then $\mu_i=0$ for any point of such a cutting edge. For this case, Eq. (2.66) can be modified as

$$\tan \gamma_{nh} (r_i) = \frac{r_i}{r_{dr}} \frac{\tan \omega_d}{\sin \kappa_r} \tag{2.67}$$

Figure 2.157 shows a comparison of the distributions of the T-hold-S normal rake angle along the major cutting edge (lip) of straight-flute and twist drills (Eqs. 2.61 and 2.65, respectively) when drill diameter $d_{dr}=20$ mm, web diameter (thickness) $d_w = 2a_o=4$ mm, and helix angle of the twist drill flute $\omega_d=30°$. As can be seen, the rake angle for the twist drill varies from $+24°$ at the drill periphery (point *1*) to $-37°$ at the inner end of the major cutting edge, while that for the straight-flute drill varies from $-3°$ to $-14°$. Therefore, in terms of cutting conditions, the straight-flute drill has a much more pleasurable rake angle distribution.

2.1.5.9.3 Comparison of the First and Second Approaches

Comparing Figures 2.152 and 2.157, one can conclude that the discussed approaches yield almost directly opposite results. In other words, according to the first approach (Figure 2.152), the T-use-S (T-hold-S) rake angle increases as the radius of the considered point of the major cutting edge decreases, while the opposite is true according to the second approach (Figure 2.157). This is an awkward situation for the one who tries to optimize the rake face drill geometry.

One of the ways to assess the chip flow direction is by analyzing the shape of the chip produced in drilling. Olson et al. (1998) pointed out that currently chip shapes are determined experimentally rather than predicted analytically or using FEM. The analytical or numerical determination of the chip shape and chip flow direction cannot be completed for drilling until all process variables, chip size, and shape have been identified. The experimental program carried out by these authors resulted in the following conclusions:

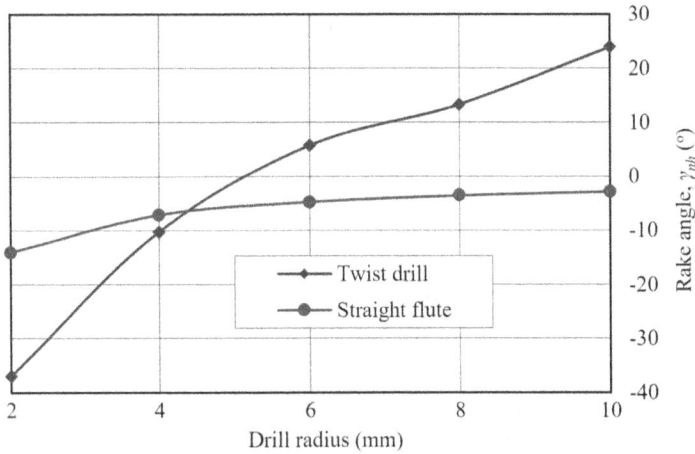

FIGURE 2.157 T-hold-S rake angle variation along the major cutting edge (lip) for twist and straight-flute drills.

- Six basic chip types were identified as produced in drilling. They are (1) conical chips formed when the chip follows a flute shape, (2) fan-shaped chips formed when the chip cannot follow the flute shape, (3) chisel-edged chips formed by the chisel edge, (4) amorphous chips having wrinkled uncurled appearance, (5) needle chip causing severe upcurling, and (6) impacted chip that is not a primary chip form
- In each experiment, more than one chip type existed per sample. However, fan-shaped and corkscrew-shaped chips were the prime shapes.

In the author's opinion, however, the chip flow direction and thus its final shape depend on the properties of the work material, drill geometry and its material, drill manufacturing quality, machining regime, MWF supply conditions, etc. Therefore, there are a number of possibilities in existence.

Figure 2.158 shows the basic chip type in twist drilling at relatively low cutting speeds and feeds. It is a corkscrew-shaped chip. When this chip is the case, the chip flow direction is close to the theoretical chip flow direction provided that the shape of the chip flute is designed properly. To prove that this is the case, a special test was carried out using a pre-drilled workpiece shown in Figure 2.159.

The author carried out a detailed study of the chip flow angle included in the analysis of the chip flow for different work materials, drill geometries, and machining regimes. Examples of the chip shapes obtained in this study are shown in Figure 2.160. The chip structure, its free and contact surfaces, chip deformation, and many other parameters of chip formation were analyzed. The results of the study reveal that, being application-specific, the chip flow direction in HP drills is always close to the direction of the normal to the cutting edge regardless of the particular value of the drill point angle.

The observation of the chip flow direction for various drilling regimes and parts of the major cutting edge as well as for different work materials and drill designs results in the following conclusions:

1. The range of chip flow direction, in general, is often found between the theoretical chip flow direction and the direction of the normal to the cutting edge that differ by the inclination angle λ_{sh-i} as shown in Figure 2.161. The chip flows in the direction of the normal to the cutting edge in HP drilling.
2. In drilling, the majority of steels with a twist drill using small cutting feeds and the chip flow on each individual section of the major cutting edge follows the theoretical direction as shown in Figure 2.162.

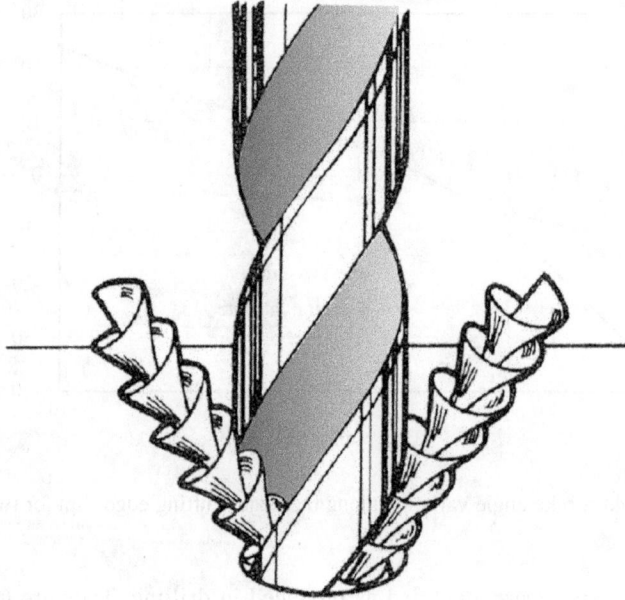

FIGURE 2.158 The basic chip type in twist drilling at relatively low cutting speeds and feeds.

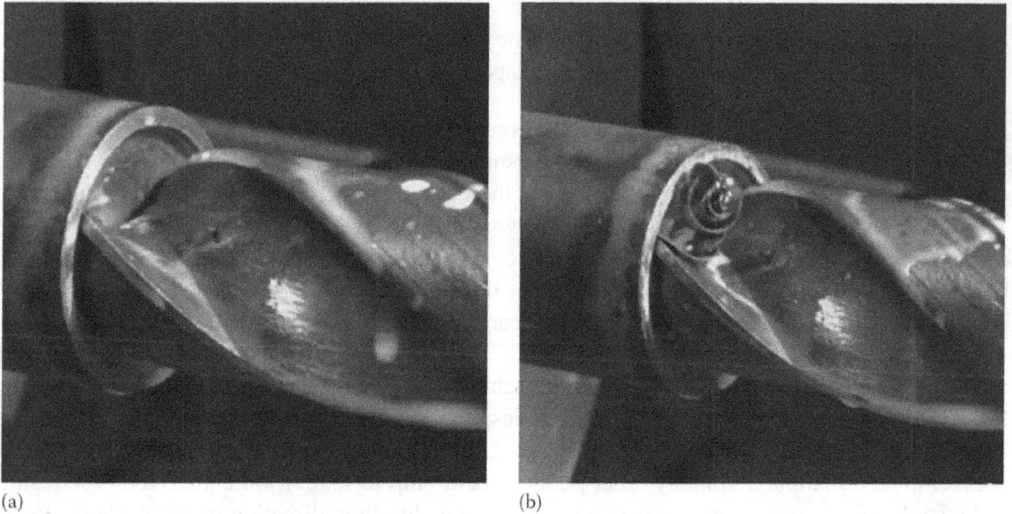

(a)   (b)

FIGURE 2.159 Chip flow direction test: (a) pre-drilled workpiece and (b) observation of the chip flow direction.

3. Although the chip flow on each individual section of the major cutting edge follows the theoretical direction, this is not the case when the whole cutting edge is engaged in cutting as the theoretical chip flow angle for each section is different as defined by Eq. (2.55) and shown in Figure 2.151. As the theoretical chip flow angle varies along the major cutting edge, the unit chip flow from neighboring sections of this edge should cross each other. Moreover, the amount of chip produced by each section of the major cutting edge varies along its length as well as its deformation. As a result, for a wide range of steel, the chip flow direction is between the normal and theoretical directions, while in HP drilling, it is in the normal direction.

FIGURE 2.160   Studying the chip flow direction.

- The geometrical parameters of a common chip fragment shown in Figure 2.163 vary along the length *1–2* (the length of the major cutting edge). For given feed per tooth f_z and the cutting edge geometry along this edge that both define the uncut (undeformed) chip thickness (nominal thickness according to ISO Standard 3002/3) h_D, the chip thickness h_C increased from point *1* to *2*, that is, the chip deformation defined by the CCR ($\zeta = h_C/h_D$) normally increases from the drill periphery corner *1* toward the inner end *2* of the major cutting edge. As such, the radius of curvature of this chip fragment, R_{ch}, decreases from point *1* to *2* due to the difference in the volume of the work material cut by section adjacent to points *1* and 2.

When the feed and/or the cutting speed increase, the corkscrew-shaped chip falls apart into individual cones and small fans.

The foregoing analysis is used in the considerations of the chip breaking in drilling.

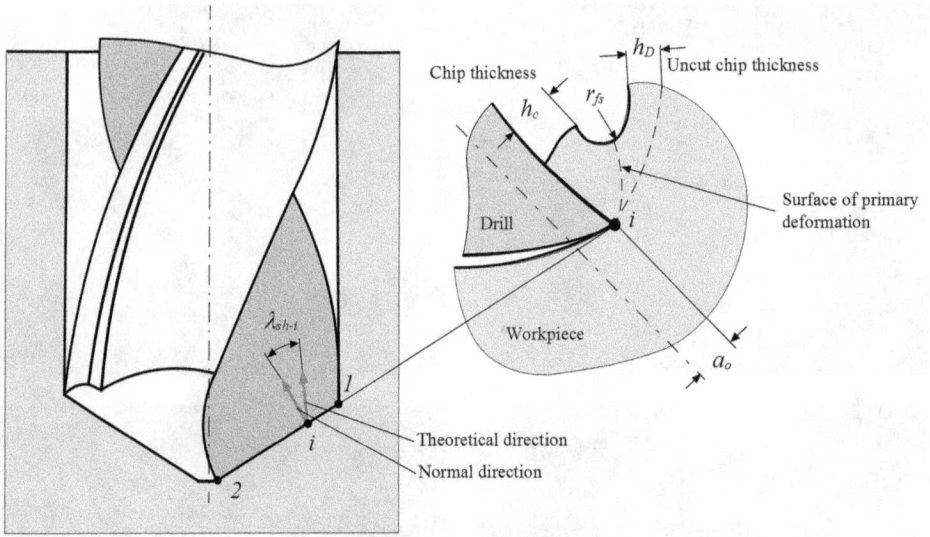

FIGURE 2.161   Possible range of chip flow directions between the theoretical and normal directions.

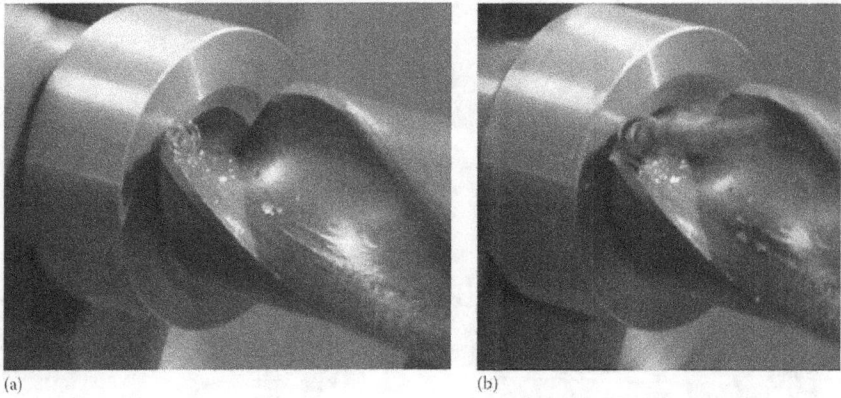

FIGURE 2.162   Observation of the chip flow direction for various drilling regimes and parts of the major cutting edge: (a) chip formation by a segment of the major cutting edge located close to the drill corner and (b) chip formation by a segment of the major cutting edge located in the middle of this edge.

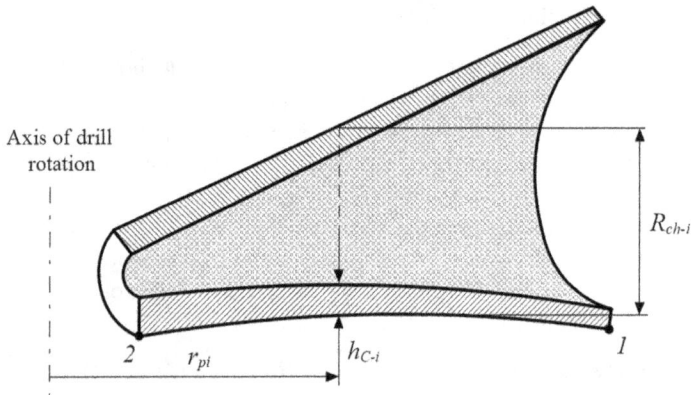

FIGURE 2.163   Parameters of a common chip fragment.

2.1.5.10 Chip Breaking in Drilling

Chip breaking in drilling was always an important issue. This issue, however, has become more and more important with wider acceptance of HP drills, greater variety of highly-ductile and difficult-to-machine materials used in industry, wider adoption of multi-axis sophisticated machines, and many other modern improvements. Under such conditions, there is no time to stop a machine, manufacturing cell, manufacturing line to clean the chip tightly winded over the tool or even the spindle.

The simplest and thus most feasible way to achieve reliable chip breaking in drilling is to apply chip-breaking modification of the rake face of the major cutting edge. Although a number of various designs are known (Astakhov 2010), the plane rake face (PRF) twist drill point geometry design developed by Armarego and Cheng (1972a,b) seems to have a number of proven advantages. This design is shown in Figure 2.164. It has been found that a positive normal rake angle over the entire length of the major cutting edge is achieved by applying a chip-braking, radius-shaped groove. Moreover, the rake face modifications partially split the chisel edge. The cutting mechanics analysis and experimental studies have confirmed the superiority of this drill point design over the conventional twist drills with a significant reduction in the axial force (thrust) and drilling torque and an increase in tool life for both aluminum alloys and difficult-to-machine materials.

A number of modifications to this design have been developed over the last 20 years. An analysis of such modification has been presented by Wang and Zhang (2008). One such modification to the general-purpose drills was attempted by grinding a PRF on the rake face of each lip using a narrow disk-shaped grinding wheel. The essence of the geometry of the PRF drill point design is shown in Figure 2.165. As can be seen, to achieve such geometry, a conventional twist drill is modified by grinding PRF on the rake face of each lip using a disk-shaped grinding wheel.

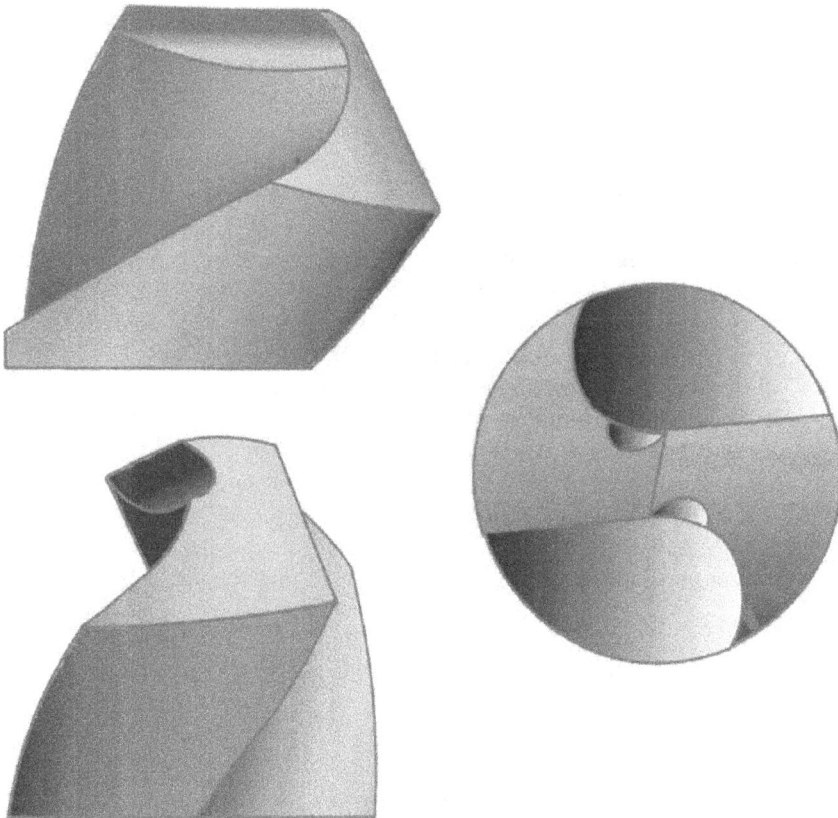

FIGURE 2.164 PRF twist drill point design.

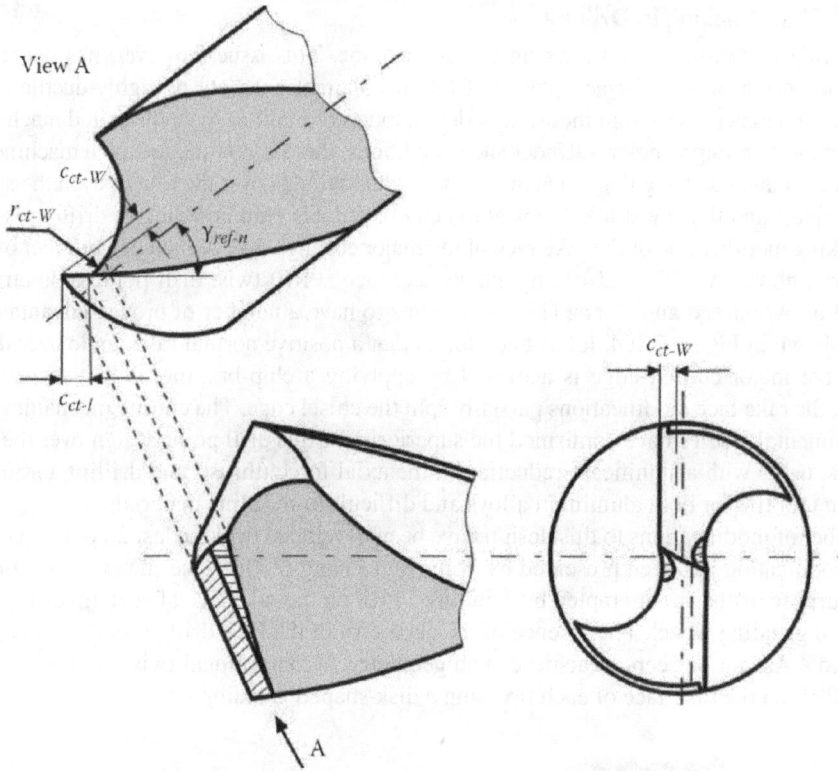

FIGURE 2.165   PRF twist drill developed by Wang and Zhang (2008).

The problem with the described chip breaker is that the parameters of the chip-breaking groove are functions are particular. This makes drills with a chip-breaking groove on the rake face application-specific. The test of such drills has proven that not only particular feed makes them application-specific but also the cutting speed as well as the particular work and tool materials (including coating). As a result, a practical determination of the parameters of the chip-breaking groove turns to be an expensive and time-consuming task. However, high efficiency of this design in terms of achieving reliable chip braking makes it worth to carry out this study for at least the most notorious, in terms of chip-breaking problems, drilling operations.

To simplify the problem, a chip-breaking step can be used as shown in Figure 2.166. To determine the parameters of chip-breaking step, namely, its depth h_{st} and width b_{st} of this step, the following practical procedure can be used. The procedure includes a simple turning test where a single-point cutter made of the same work material as the drill is set to cut the same work material as to be drilled. The cutter has a zero-rake angle and is set with the tool cutting edge angle κ_r (see the Appendix) equal to the half-point angle of the drill φ. The cutting speed is chosen as equal to that in drilling and the cutting feed is equal to the feed per tooth in drilling f_z. The procedure includes the following steps:

1. Determination of the uncut (undeformed) chip thickness h_D as

$$h_D = f_z \sin \kappa_r \qquad (2.68)$$

2. Measuring the CCR (discussed in the Appendix) ζ
3. Determination of the chip thickness h_C as

$$h_c = \zeta h_D \qquad (2.69)$$

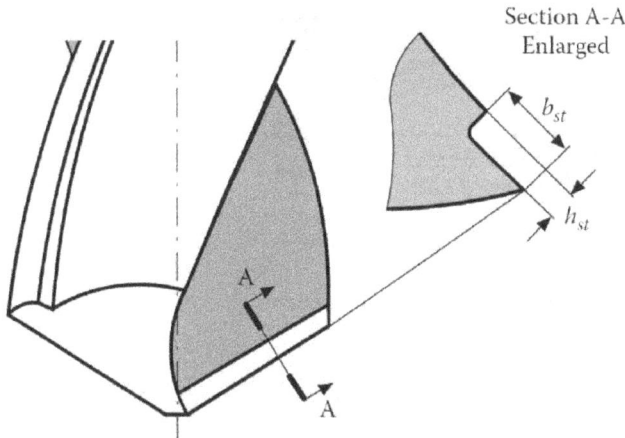

FIGURE 2.166 Chip-breaking step on the drill rake face.

4. Determination of the optimal radius of chip curvature R_{ch-opt}. For a wide range of steels, $R_{ch-opt} = 6.5 h_C$.
5. Calculating the depth h_{st} as $h_{st} \geq h_C$.
6. Calculating the step width b_{st} as

$$b_{st} = \sqrt{h_{st}\left(2R_{ch-opt} - h_{st}\right)} \tag{2.70}$$

The chip breakability can be enhanced if a cutting edge is provided with chip splitting grooves often called nicks. Initially developed for spade drills and evolved into ANSI Standard B94.49-1975, such nicks have been also used for twist drills since the eighteenth century (French and Goodrich 1910, reprinted in 2001). Figure 2.167a shows the drill design according to US Patent No. 1,383,733 (1921). The drill has the major cutting edges 1 and 2 that are provided with chip splitting grooves 3 and 4 extending over the length of flutes 5 and 6. A number of various drill designs with chip splitting nicks were introduced. A detailed analysis of their effect on drill performance was presented by Nakayama and Ogawa (1985). The design of a twist drill having multiple nicks ground as radial grooves on its flank shown in Figure 2.167b was described as early as 1940 by Veremachuck (Figure 244 in Veremachuk 1940) who also extensively studied application specifics of this design in its comparison with the standard twist drill. In this drill, the major cutting edges (lips) 7 and 8 are located so that preferable distributions of the rake and flank angles along these edges are achieved as well as the thinning of the chisel edge 9. Flank surfaces 10 and 11 are provided with radial grooves 12 and 13, respectively, that aim to separate the forming chip into rather narrow strips.

Figure 2.168 shows a drill design according to US Patent No. 5,452,971 (1994). According to Nevilles, the inventor, it is completely a new concept in twist drill theory, design, and method of construction that will revolutionize the rotary end-cutting tool systems. The drill has two curved major cutting edges 1 and 2 formed as intersection lines of flutes 3 and 4 and flank surfaces 5 and 6. These edges are connected by the chisel edge 7. Two series of offset concentric grooves 8 and 9 (deep nicks ground as radial grooves) ground on each flank surface as shown in Figure 2.168. The patent claims that such a drill produces precise, near mirrorlike well-finished holes in various work materials. Besides, its penetration rate is nearly four times faster and tool life seven times longer in comparison to a comparable-sized standard twist drill.

A number of concerns about a drill having radial chip splitting grooves can be identified. The major design concern is the difficulty to assure the flank angle on the sides of the grooves to prevent friction of these sides in drilling. According to the design shown in Figure 2.167b, these grooves

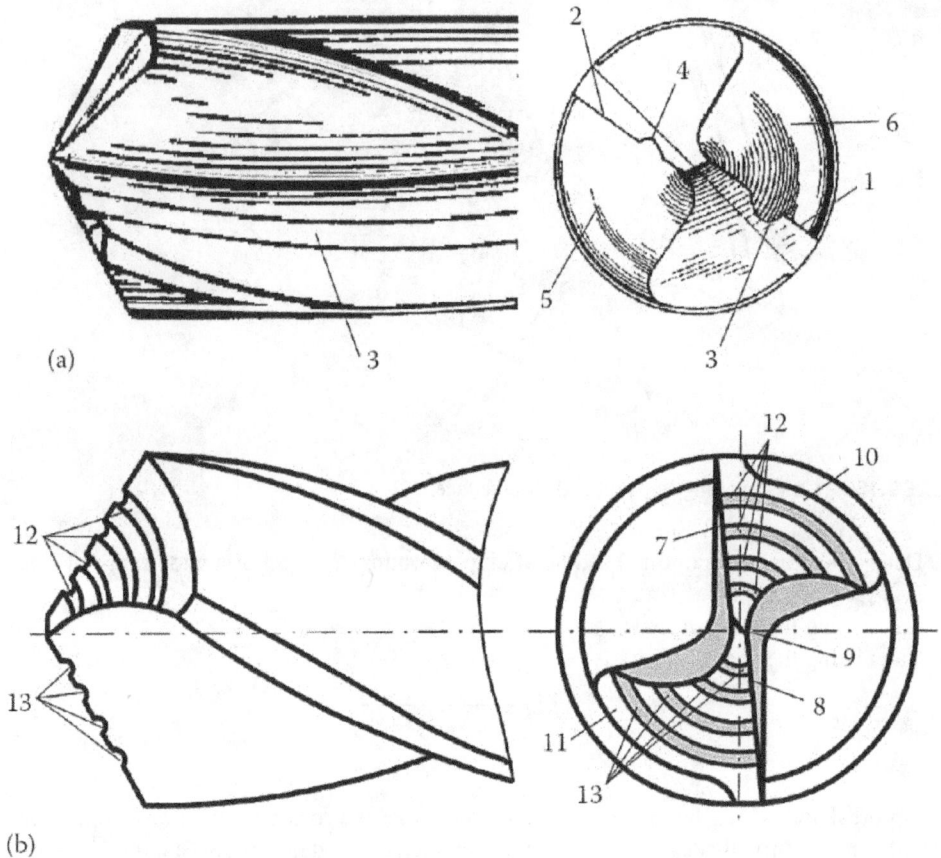

(a)

(b)

FIGURE 2.167 Drills with radial chip splitting grooves: (a) made on the rake face and (b) made on the flank surface.

are rather shallow and have round cross section; thus, severity of the problem is low. In the design shown in Figure 2.168, these grooves are deep and have rectangular cross section; thus, the problem becomes of real concern. To apply the flank angle on the groove sides, their profile should be of a fishtail cross section that is virtually impossible to apply using standard machines and grinding wheels. The application problem is that the radial grooves should be reapplied with relatively high accuracy on each successive re-sharpening. In the author's opinion, these two prime problems prevent practical applications of the discussed designs.

The chip rigidity and thus its enhanced breakability can be achieved if a drill has a means to increase the rigidity of the forming chip. US Patent No. 2,204,030 (1940) offers a drill design where the major cutting edges 1 and 2 are made with ribs (projections) 3 and 4 as shown in Figure 2.169. The formed chip would have a rib of rigidity so its breakability would be much greater. To achieve the same objective, US Patent Nos. 867,639 (1907) and 1,404,546 (1922) offer various combinations of the previously described means located asymmetrically about the axis of rotation. Although the described ideas of chip rigidity enhancement are widely used in modern designs of cutting inserts including those for drills, the formation of additional rigidity means that it is always accompanied by increasing the drilling torque and force.

Often the described measures are not sufficient to obtain the desirable chip shape and to avoid significant force due to chip interaction with the side wall of the flute and with the walls of the machined hole. The latter should be particularly avoided as it results in damage to hole quality and

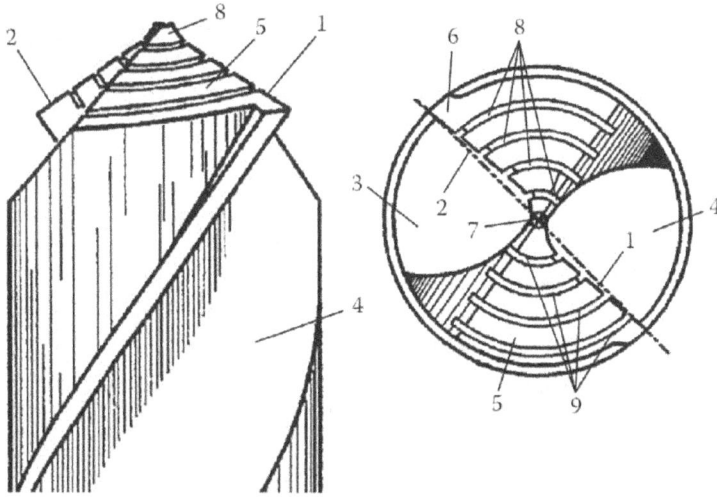

FIGURE 2.168 Drill design according to US Patent No. 5,452,971 (1994).

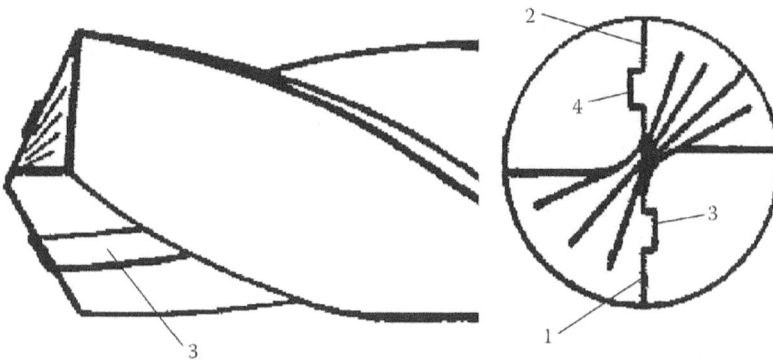

FIGURE 2.169 Enhancing chip rigidity by providing ribs on the drill rake faces.

significant friction. Therefore, the chip flute profile should be made so that it helps to curl the chip into an easy transportable shape and then transport this chip away from the machining zone out of the hole being drilled.

To understand the need for more sophisticated shapes of the chip flute, consider a simple model of chip flow shown in Figure 2.154. As discussed earlier, the length of the chip produced by the periphery point *1* (its nearest vicinity to assume some finite chip width) over one drill revolution is much greater than that by point *2*, which is the inner end of the major cutting edge where this edge meets with the chisel edge. As a result, the forming chip flows along straight part *4* of the flute and then over its curved part *3* as shown in Figure 2.154 reaching the walls of the machined hole.

To deal with the problem, various approaches are used. The first one is to use groove-type and obstruction-type chip breakers as those used in singe-point tool operations (Jawahir and Van Luttervelt 1993, Jawahir and Zhang 1995, Jawahir et al. 1997). Figure 2.170a shows a drill with groove 1 placed on the rake face 2 near the major cutting edge 3 that facilitates chip breaking and prevents chip clogging (Sahu et al. 2003). Figure 2.170b shows a drill according to US Patent No. 2,966,081 (1960). In this drill, the flute profile is of a concave shape 4 and the chip-breaking step (View B) is provided so that the cutting edge 5 is formed.

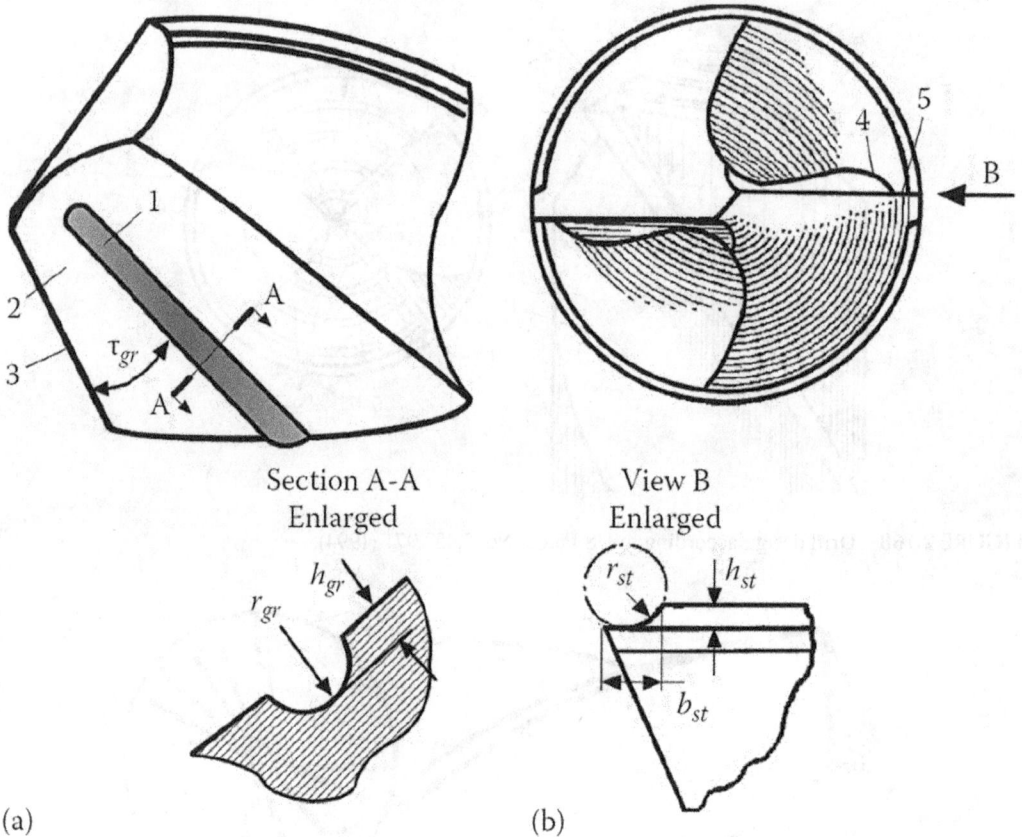

FIGURE 2.170 (a) Groove-type and (b) obstruction-type chip breakers for a drill.

Although these chip breakers may produce broken chips suitable for evacuation from the machining zone, they are rarely used in drilling because of the following:

- Groove geometry (parameters r_{gr} and h_{gr}) and orientation (angle τ_{gr}) as well as chip step geometry (h_{st}, b_{st}, and r_{st}) are application-specific, that is, they work for a rather narrow range of the combination of the work materials properties, feed, speed, MWF, etc. A small departure from one of these parameters may ruin chip breaking.
- Groove-type and obstruction-type chip breakers increase the drilling torque and the axial force.
- The groove and chip step geometry should be reproduced with relatively high accuracy on each drill re-sharpening. Particularly with the groove shown in Figure 2.170a, a great amount of the tool material should be removed to restore the original shape of the drill that significantly lowers the number of possible re-sharpenings.

More intelligent ways to design the flute profile to achieve a suitable shape of the chip can be understood if one considers a simple model of chip shape formation (chip curling during drilling) shown in Figure 2.171a. In this picture, the letters A and B designate the corresponding ends of the cutting edge AB. To curl such a chip into a cone-like shape and then break it when the deformation in its root reaches the strain at fracture or to form long tight curls (Figure 2.171b), a certain force F_{cc} should be applied to the chip at a certain distance h_{cc}. Because the force F_{cc} is the reaction force from some obstacle made on the flute, it depends on the drilling process parameters while the direction of this force and the distance h_{cc} can be varied by flute profile parameters. This is the principle that outlines many patented drill designs that differ only by particular values of these two parameters.

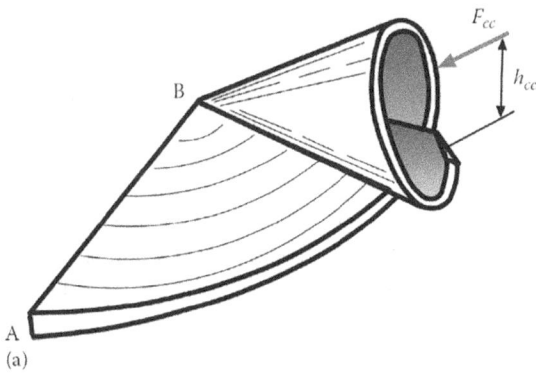

FIGURE 2.171   Model of chip curling: (a) a model and (b) a real chip fragment.

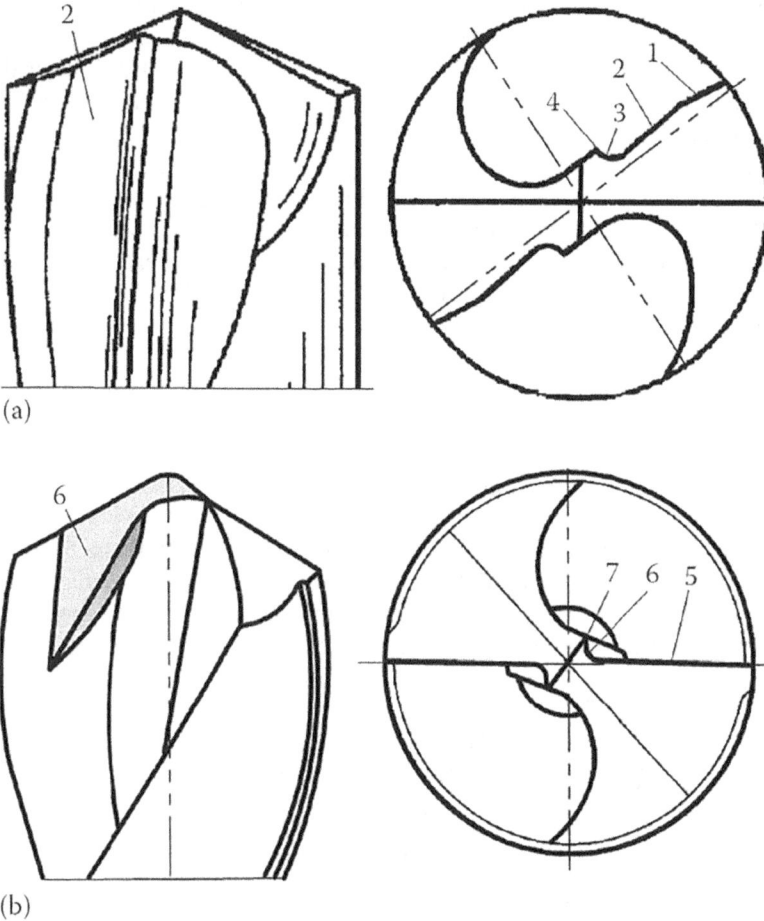

FIGURE 2.172   Two-flute profiles with a mean for chip curling as a part of the major cutting edge: (a) according to US Patent No. 5,622,462 (1997) and (b) according to US Patent No. 5,931,615 (1999).

Figure 2.172a shows the flute profile according to US Patent No. 5,622,462 (1997). As can be seen, each major cutting edge (lip) consists of the outer inclined part 1 and straight inner part 2 having concave portion 3. Apex 4 is formed with the purpose to apply the force F_{cc} shown in Figure 2.171a. The inner part is built-in in the flute profile that assures its consistency over each successive drill regrinding. Figure 2.172b shows a similar flute profile according to US Patent No. 5,931,615 (1999). As can be seen, the major cutting edge 5 is located on the drill transverse axis. It has concave part 6 that ends with apex 7. The location of the major cutting edge on the drill transverse axis improves the distribution of the flank and rake angles over the straight part of the cutting edge and decreases the plastic deformation of the chip that reduces the drilling torque and force. However, complicated concave part 6 should be ground on each successive drill regrind. Note also that at the time of the described inventions, the applying shapes shown in Figure 2.172a and b with the required accuracy to maintain force balance discussed in Section 2.1.5.6 were not yet feasible. Nowadays, with a wide use of sophisticated multi-axis CNC drill point grading machines, this does not present a problem. However, most great designs of the past were forgotten as no one book/paper/article presents and discusses their particularities and advantages.

There are two obvious disadvantages of the profiles shown in Figure 2.172. The first one is that the cutting geometry over the radiused parts and apexes is not favorable so one should expect an increase in the drilling torque and force. The second is that the arm (h_{cc} in Figure 2.171) is small, which may not create the sufficient bending moment at the chip root to break the forming chip into pieces while force F_{cc} is excessive because the radius of the radiused part is too small. However, it has been found that the breaking up of the forming chip into small pieces is not always desirable. Rather, the efficiency of drilling is improved, and less downtime is needed if the drill flute profile is constructed so that the chip coming out from the cutting edge is formed into long sustained curls as shown in Figure 2.158.

Figure 2.173 shows the flute profile according to US Patent No. 4,222,690 (1980) where curved part 1 and apex 2 belong to the flute. Many other known flute profiles originated from the same idea. The difference is in the location of the apex and radius of the curved part. For example, Figure 2.174 shows a twist drill in which a concavely shaped surface 1 extends from the inner end 2 of the major cutting edge 3 toward the outer periphery of the drill body and apex 4 is located on the side wall of flute 5.

In extreme cases, it is worthwhile to consider some special profiles of the chip flute and thus drill cross sections shown in Figure 2.175. In Figure 2.175a, the flute is formed by the cutting edge portion 1 that becomes concave portion 2 extending to apex 3 located almost at the intersection of the flute and the relieved part of the drill body. According to US Patent No. 4,583,888 (1986), the position of apex 3 is selected so the chip contact with the wall of the machined hole is prevented. It can be accomplished by selecting the web thickness $d_w = (0.25...0.35)d_{dr}$ and the FWR (discussed in Section 2.1.5.4) = 0.4...0.8. According to US Patent Nos. 4,983,079 (1991) and 5,088,863 (1992), a significant reduction in the axial force, drilling torque, and thus drilling power as well as improvement in chip transportation from the machining zone in heavy-duty drilling operation is achieved when distance W_{ww} is selected properly from the range of $(0.45...0.65)d_{dr}$.

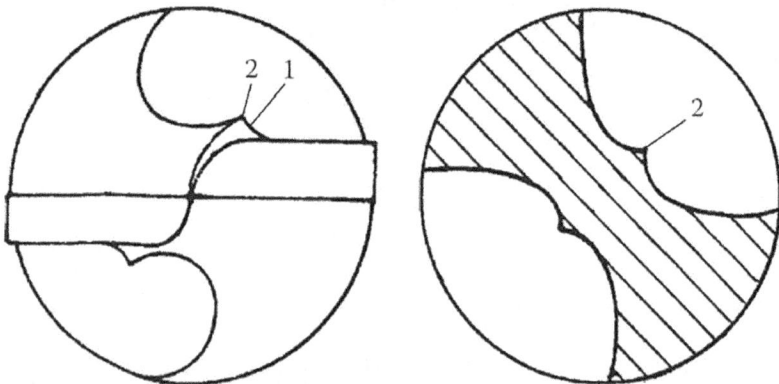

FIGURE 2.173   Drill design according to US Patent No. 4,222,690 (1980).

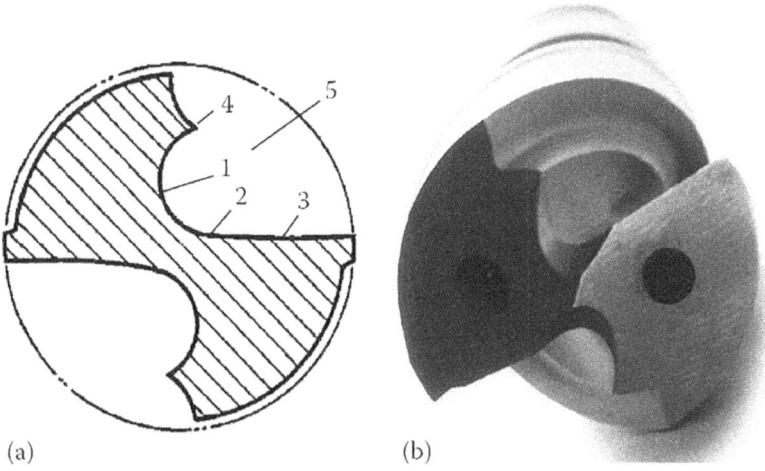

FIGURE 2.174 Drill body and the chip-curling apex located on the side wall of the flute: (a) flute profile and (b) drill appearance.

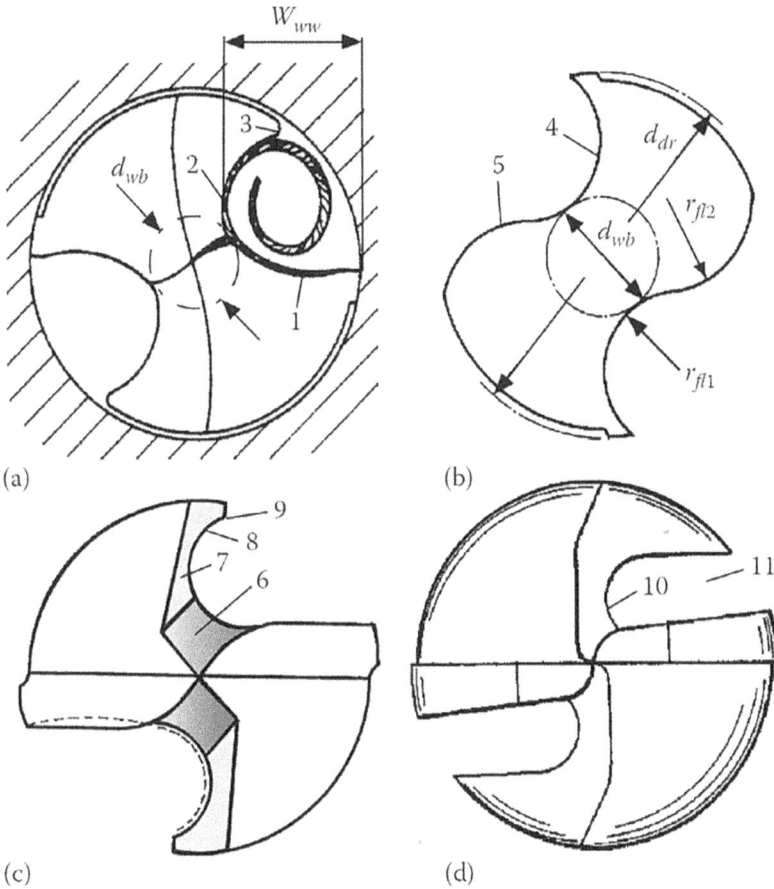

FIGURE 2.175 Some other shapes of the chip flute: (a) profile according to US Patent No. 4,583,888 (1986), (b) double-radius chip flute profile, (c) flute profile according to US Patent No. 5,716,172 (1998), and (d) flute profile for extremely heavy-duty and deep-hole drills.

Figure 2.175b shows the double-radius chip flute profile. According to US Patent No. 4,744,705 (1988), a significant reduction in drilling power and improvement stability of drilling operations are achieved when the web thickness $d_w = (0.25...0.50)d_{dr}$, $r_{fl1} = (0.2...0.3)d_{dr}$, and $r_{fl2} = (0.25...0.40)d_{dr}$. It is understood that the cutting edge formed as the line of intersection of the concave portion 4 of the flute with that flank surface would be concave. Convex part 5 of the flute profile is to curl the forming chip.

Figure 2.175c shows the flute profile according to US Patent No. 5,716,172 (1998) meant for heavy-duty drills. It has the front chip curling means consists of two flat surfaces 6 and 7 that are made to fold and then curl the chip emerging from the cutting edges into the concave flute 8. Flute 8 is provided with apex 9 to prevent the contact of the chip and the wall of the drilled holes.

Figure 2.175d represents the generic idea of extremely heavy-duty and deep-hole drills with a FWR of 0.20...0.02 to increase torsional rigidity to the drill shank (e.g., US Patent Nos. 4,565,473 and 4,975,003). The concave part 10 of the chip flute 11 has a small radius to bend, curl, and break the forming chip or just direct it into the flute without breaking as a small diameter curl or a string.

2.1.5.11   Chisel Edge

The proper definition of the chisel edge, its importance in drill performance, and basics of its geometry are discussed in Section 2.1.3.7, which presents all what any drill user has to know about this edge. However, this knowledge is insufficient at the development, research, and drill advanced design levels. This section aims to discuss particularities of the chisel edge geometry and the root cause of the known problems with this edge in detail.

2.1.5.11.1   The Problem and Its Manifestation in HP Drilling

The problem with the chisel edge in HP drilling has been tackled in Section 2.1.3.7 to a certain extent. However, a much deeper understanding of the problem is required at the considered stage. The problem with many HP drills (HSS, carbide, and PCD) is relatively low tool life and premature tool failures primarily due to oversize conditions, poor roughness of the machined holes, and even breakage of drills. Common appearances of the failed HP drills are shown in Figure 2.176. As can be seen, regardless of a particular drill manufacturer, drill design, quality of surface including roughness and coating, tool material used, and work material, a common denominator for the shown drill is that their central portion adjacent to the chisel edge is covered by a thick layer of aluminum deposit so that the chisel edge cannot be even seen in Figure 2.176a. In machining of steels, the vivid impressions left on the rake and flank surfaces of the chisel edge that indicated the chip flow over this edge in Figure 2.176b as steel does not adhere to these faces. This abnormal chip flow causes drill wandering in drilling and at the hole entrance as such a drill does not possess its self-centering ability as it is intended by its design.

Our multiple closer examinations/microscopic studies of the structure/metallography of the deposit in the vicinity of the chisel edge as well as microhardness tests of the deposit shown in Figure 2.176a revealed that this deposit has the microstructure of heavily-extruded aluminum having much greater hardness than the aluminum in the work material (the matrix material for AA380/AA383 alloys). Moreover, this deposit is strongly adhered to the tool material and thus can be removed only when using strong acids (e.g., muriatic acid).

The sections to follow aim to explain the root cause of the problem, the extent to which the known drill point grind can handle this problem, and proper design of the chisel edge to avoid the problem.

2.1.5.11.2   Common Perception of the Shape of the Bottom of the Hole Being Drilled

Finding the root cause of the problem begins with revealing the inadequacy of a common perception/belief about the shape of the drill point, and thus the shape of the bottom of the hole being drilled with such a point. Figure 2.177 shows the most common perception of the shapes of drill and

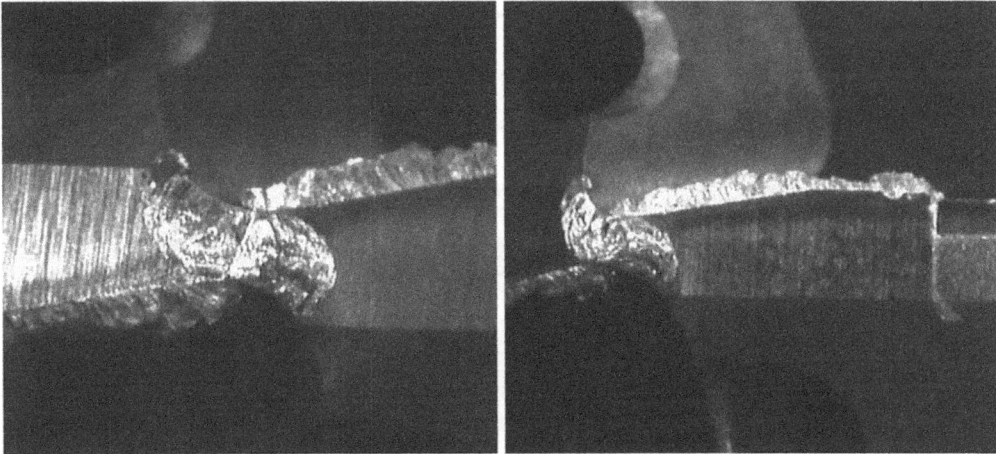

(a)

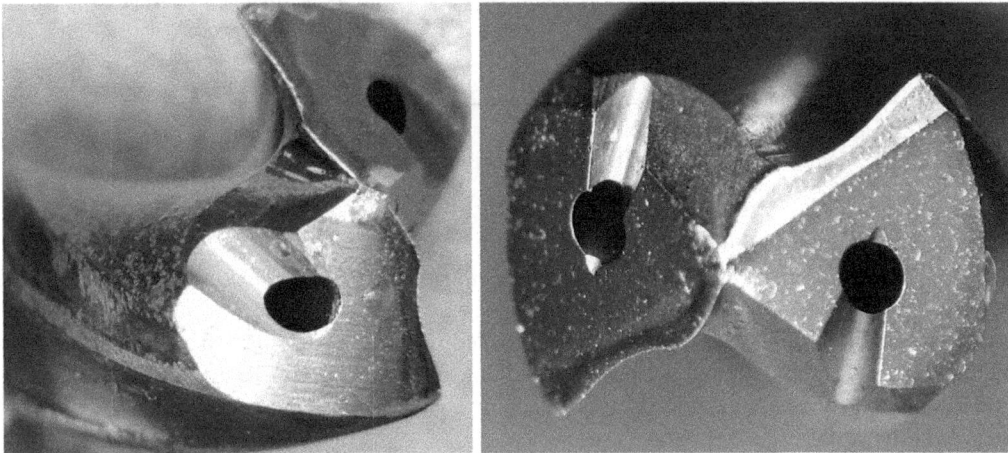

(b)

FIGURE 2.176 Appearance of failed drills and their central portion in the vicinity of the chisel edge (a) in HP drilling of AA380 high-silicon aluminum alloy and (b) in HP drilling of S400/S350-4045 crankshaft steel.

bottom of the hole being drilled. As shown, the drill's major cutting edges meet at the drill center at the axis of rotation so that the bottom of the hole being drilled appears to be a perfect cone surface. This shape is depicted in ALL KNOWN PART DRAWING so that it must be drilled with drills shaped as shown in Figure 2.177. Surprisingly, no one ever questioned the described perception.

To challenge the common perception, let us consider some views of a real drill, namely VIEW A and VIEW B as shown in Figure 2.73. As can be seen, the drill point in VIEW A appears the same as shown in Figure 2.177 for ideal/hypothetic drill. This is the ground of the discussed common belief/perception shown in Figure 2.177 as everything seems to be valid. In reality, it is not true. The real shape of the two chisel edges can be seen only in view B in Figure 2.73 as combined chisel edge 2–3–5. As the drill rotates about point 3 (point on the rotation axis of the drill), it kinematically must make a flat (or whatever the shape of the chisel edge in a given drill) surface of the bottom of the hole being drilled.

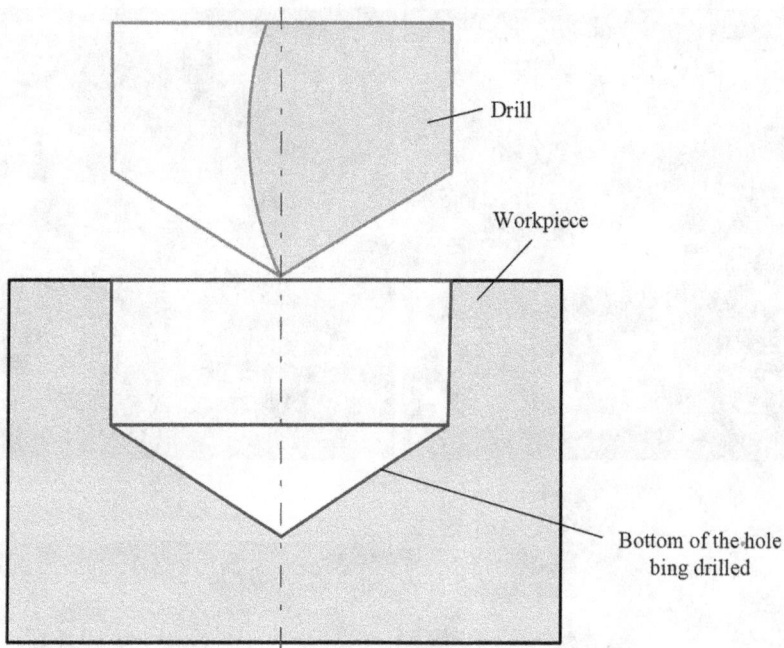

FIGURE 2.177 Common perception of the shapes of drill and bottom of the hole being drilled.

2.1.5.11.3 Real Shape of the Bottom of the Hole Being Drilled

2.1.5.11.3.1 Ruled Surfaces of Revolution: Cone and Hyperboloid   A ruled surface (also called a scroll) can be generated by the motion of a line in space, similar to the way a curve can be generated by the motion of a point. A 3D surface is called ruled if through each of its points passes at least one line that lies entirely on that surface (Prousalidou and Hanna 2007). A surface of revolution is a ruled surface generated by revolving of a plane curve (e.g., $z = f(x)$) about the axis (e.g., Oz) called the axis of the surface of revolution.

2.1.5.11.3.2 Cone as the Shape of the Bottom of the Hole Being Drilled Formed by the Radial Cutting Edge   Let us consider a surface of revolution when the revolving line is a straight line. In its revolving about the z-axis, it generates either the cone, when it intersects the axis or the only other surface of its class known as hyperboloid of revolution. Both cases are considered here as applicable in drilling. Cone is the surface traced by a revolving straight line (the generatrix) that always passes through a fixed point (the vertex) as shown in Figure 2.178. A Cartesian right-hand xyz coordinate system is set as follows. The z-axis is the axis of the cone surface and x-axis and y-axis are drawn from the origin O set the vertex of the cone surface. This coordinate system is termed as the original coordinate system and is used in further consideration. In Figure 2.178, line O-1 revolves about the z-axis. Revolving line O-1 is referred to as generatrix as this line actually generates the cone.

Let us consider whether or not such a cone can be generated (cut) by a drill. Figure 2.179a shows a drill with "radial" (point 3 lies on the axis of rotation, which is the z-axis at the origin of the original coordinate system) cutting edge 1-3 and point angle Φ_p. Obviously, such a drill makes a cone shape of the bottom of the hole being drilled with the cone angle equal to the drill point angle Φ_p as cutting edge 1-3 generates this cone surface. Theoretically, a drill with the radial cutting edges can be made nowadays with the help of modern multi-axis drill point grinding machines.

2.1.5.11.3.3 Shape of the Bottom of the Hole Being Drilled Formed by a Common Drill   Figure 2.180 shows a common drill rotating about the z-axis. Due to its design/cutting edge arrangement, the major cutting edge 1-2 does not intersect the axis of rotation (the z-axis) as clearly

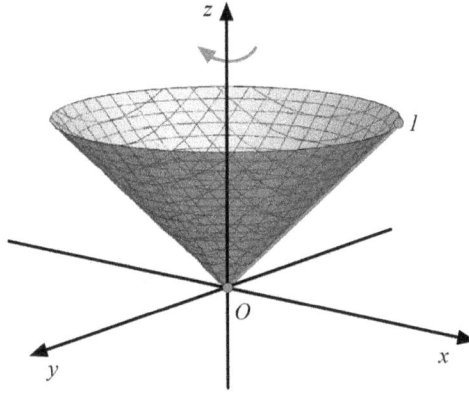

FIGURE 2.178 Formation of a cone surface by line *O-1* revolving about the *z*-axis.

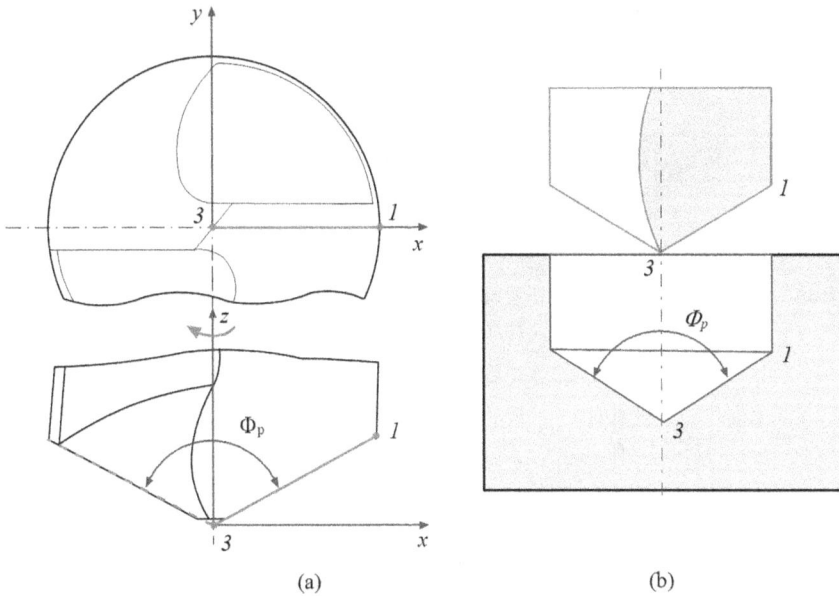

(a)   (b)

FIGURE 2.179 Showing (a) a drill with "radial" (point *3* lies on the axis of rotation, which is the *z*-axis) cutting edge *1-3* and (b) cone shape of the bottom of the hole drilled by this drill.

seen in this figure. Therefore, to cut through the center, the chisel edge *2–3* is needed. As can be seen in the *xy*-plane, the major cutting edge *1–2* is located above the center line at a distance a_o. The inner point (point *2*) of this edge is located at distance z_o from the coordinate origin due to the presence of chisel edge *2–3*.

Let us consider the shape of the bottom of the hole being drilled made by cutting edges *1–2* and *2–3* while revolving about the *z*-axis. Obviously, this shape should consist of two parts as formed by cutting edges *1–2* and *2–3* having different directions. Moreover, both of these surfaces should form the corresponding surfaces of revolution.

As cutting edge *1–2* does not intersect the *z*-axis so it does not form a cone surface. It is known that the surface traced by a revolving straight line that does not intersect the axis of rotation is a hyperboloid of revolution shown in Figure 2.181. It is described by the following equation:

$$\frac{x^2}{a^2} + \frac{y^2}{a^2} - \frac{z^2}{b^2} = 1 \tag{2.71}$$

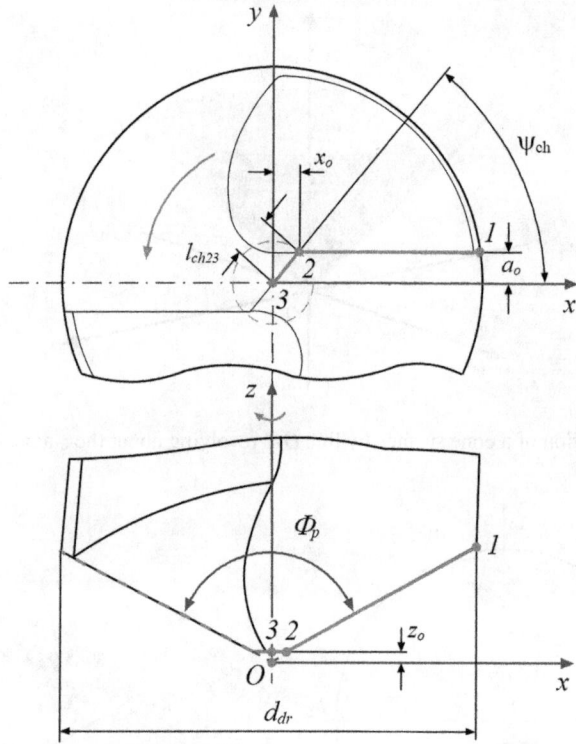

FIGURE 2.180 Arrangement of the cutting edge in a common drill. Note that the major cutting edge *1-2* does not intersect the axis of rotation (the *z*-axis).

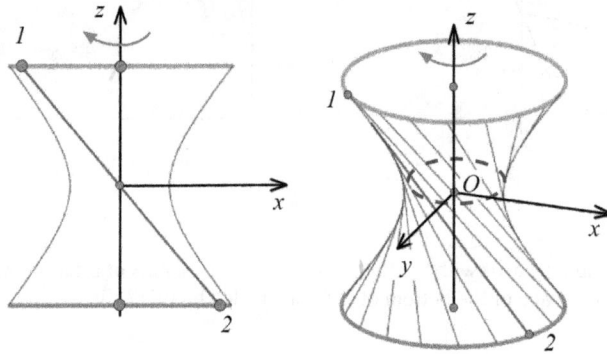

FIGURE 2.181 Hyperboloid traced by revolving straight line *1-2*.

In the *xz*-plane (the section plane), this hyperboloid appears as two branches (known as the east-west branches) of hyperbola described by the following equation:

$$\frac{x^2}{a^2} - \frac{z^2}{b^2} = 1 \tag{2.72}$$

or

$$z = \pm \frac{b}{a} \sqrt{x^2 - a^2} \tag{2.73}$$

Figure 2.182 shows these two branches, their asymptotes, and important geometrical relationships as related to Eq. (2.72). The basic rectangular passing through points *1–9* provides an important help in examining equation of hyperbola. Segment *3–8* is called the transverse axis and segment *1–5* is known as the conjugate axis.

To determine the intersection points of the hyperbola with coordinate axes, one needs to set $z=0$ in Eq. (2.72) that gives $x=\pm a$. Therefore, the intersection points *3* and *8* have corresponding coordinated $(-a, 0)$ and $(a, 0)$, i.e., the distances of segments *1–2*, *4–5*, *5–7*, and *9-1* are equal to a. By setting $x=0$, one obtains $y=\pm zi$ (note that $i = \sqrt{-1}$), i.e., there are no intercepts of hyperbola branches with the *z*-axis.

There are two asymptotes of a hyperbola that coincide with the diagonal of the basic rectangular. Their equations in the *xz*-plane are

$$z = \frac{b}{a}x \quad \text{and} \quad z = -\frac{b}{a}x \tag{2.74}$$

Note that when $x \to \infty$, the branches of a hyperbola become close and close to the asymptotes but never intersect or even touch them.

Let us consider how the above-discussed particularities of a hyperboloid can be used to determine the shape of the bottom of the hole being drilled by cutting edges *1–2* and *2–3* shown in Figure 2.180, i.e., the correlation between coefficients of a and b in the equations of a hyperbola and its asymptotes and the design/geometry parameters of a real drill. To simplify the further considerations, first consider a hypothetical case shown in Figure 2.183 where the fictitious major cutting edge *1–2′* intersects the *z*-axis. As can be seen, the projection of this edge into the *xy*-plane is located above the *x*-axis by distance a_o. Angle Φ_p is the angle between the projection of its two extreme (right-left) positions into the *xz*-plane. Point *2′* of this line lies on the *y*-axis.

Let us consider what happens when line *1–2′* revolves about the *z*-axis as shown in Figure 2.184. In such revolving, line *1–2′* traces a hyperbola in the *xz*-plane with the throat diameter (the diameters of the circle it generated in the *xy*-plane) $d_{th1} = 2a_o$ shown in Figure 2.183. Moreover, the projections of its two extreme (right-left) positions into the *xz*-plane are asymptotes of this hyperbola.

Knowing the established results and referring to the basic rectangular shown in Figure 2.182, one can determine the coefficients a and b of the hyperbola equation (Eq. 2.73) as follows: coefficient $a = a_o$ and coefficient $b = a_0/\tan(\Phi_p/2)$ as the angle between the asymptotes is Φ_p. Therefore, line

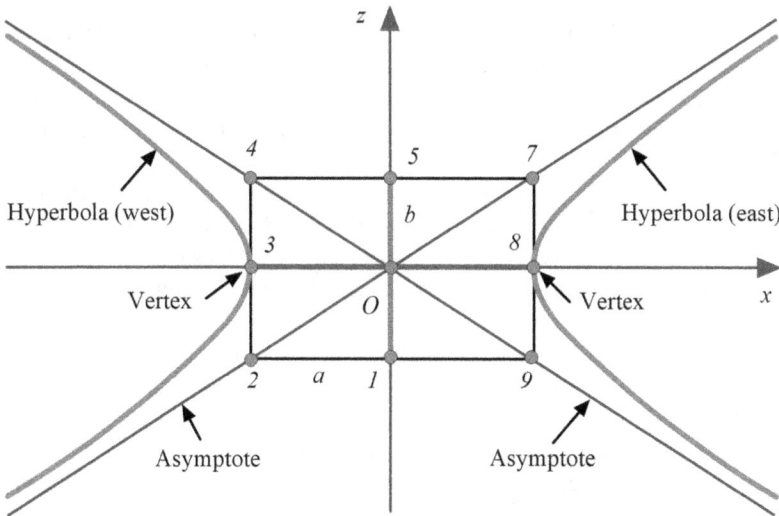

FIGURE 2.182 Branches of hyperbola, asymptotes, and important geometrical relationships.

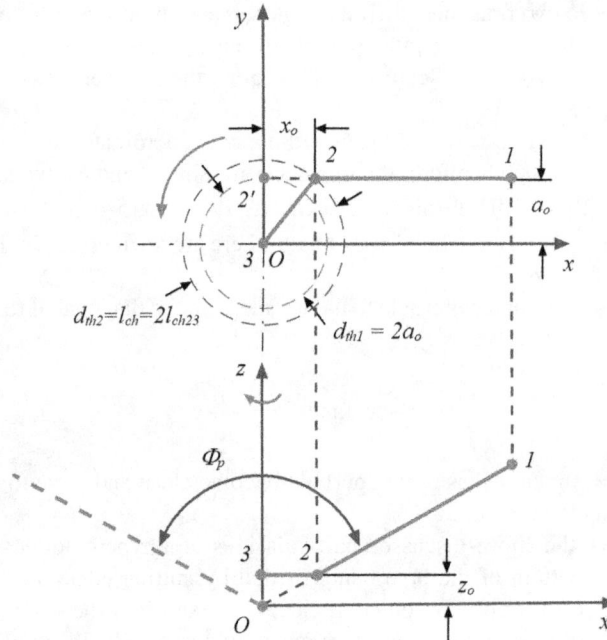

FIGURE 2.183 Model of rotation of cutting edge about the z-axis.

1–2′ in its revolving about the z-axis generates a hyperboloid of revolution, which in the xz-plane appears as two branches of hyperbola having equations:

$$z = \pm \frac{a_o/\tan\left(\Phi_p/2\right)}{a_o}\sqrt{x^2 - a_o^2} = \frac{1}{\tan\left(\Phi_p/2\right)}\sqrt{x^2 - a_o^2} \qquad (2.75)$$

Referring to the arrangement of the cutting edge in a common drill shown in Figure 2.180, one can see that it includes the major cutting edge *1–2* and the chisel edge *2–3* as the drill is meant to cut through the center, i.e., to drill solid holes. The location of the projection of cutting edge *1–2* in the xy-plane with respect to the x-axis is the same as that of line *1–2′* in Figure 2.183. The drill point angle Φ_p (between projections of the cutting edges into the zx plane) is the same as angle Φ_p between the projection of its two extreme (right-left) positions of *1–2′* line into the xz-plane. Therefore, cutting edge *1–2* in its revolving about the z-axis generates a hyperboloid of revolution, which in the xz-plane appears as two branches of hyperbola defined by Eq. (2.75).

The only difference between the location of line *1–2′* (see Figure 2.183) and combined cutting edge *1–2–3* (the major cutting edge *1–2* and the chisel edge *2–3*) is that cutting edge *1–2* does not intersect the z-axis, i.e., the inner point 2 of the cutting edge *1–2* does not lie on this axis. Instead in the xy-plane, it is located at distance x_o from the y-axis. It implies the following:

- It follows from Figure 2.180 $x_o = l_{ch23}\cos\psi_{ch}$, where l_{ch23} is the length of the chisel edge *2–3* and ψ_{ch} is the angle between the chisel edge *2–3* and x-axis known as the angle of the chisel edge (Astakhov 2014).
- The projection of the drill's major cutting edges into the xz-plane does not meet at the coordinate origin O. Rather, point 3 lies on the z-axis and it distance from the origin O is

$$z_o = \left(l_{ch23}\cos\psi_{ch}\right)/\tan\left(\Phi_p/2\right) \qquad (2.76)$$

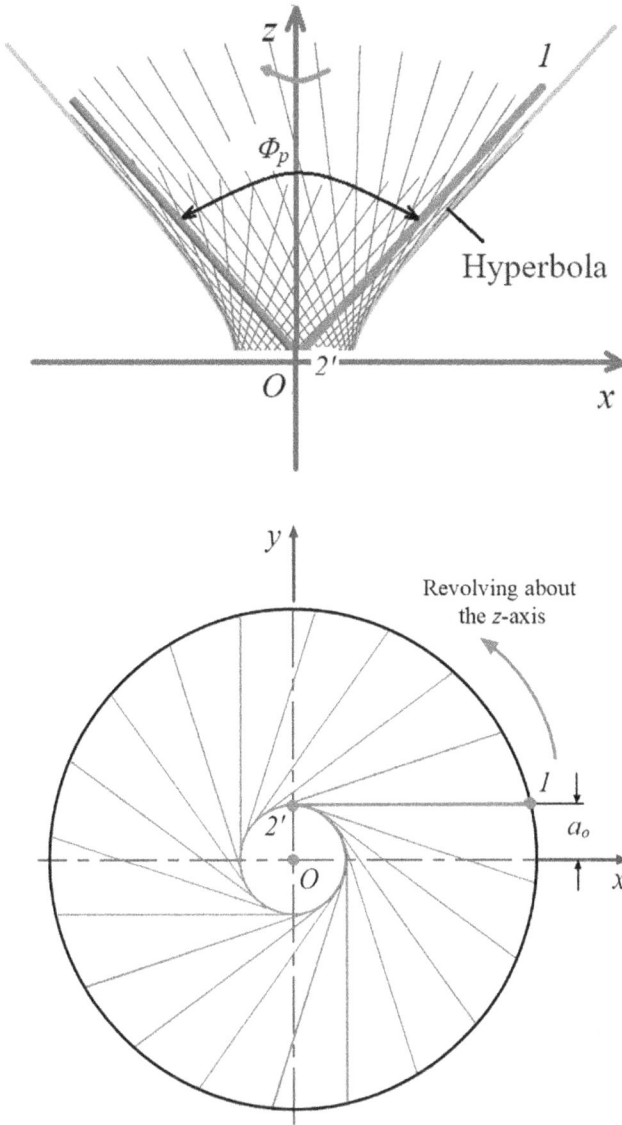

FIGURE 2.184 Revolving line *1-2'* about the *z*-axis.

Figure 2.185 shows a step-by-step consideration of the formation of the shape of the bottom of the hole being drilled. Figure 2.185a shows the shape of the bottom of the hole formed by the major cutting edge *1–2*. As discussed above, a hyperboloid is the shape of the surfaces formed as this edge revolves about the *z*-axis. As a result, the wall of the formed shape in the *zx* plane is hyperbolas and the cutting edge *1–2* in its rotation forms the asymptotes of the corresponding hyperbolas. The central part having a diameter equal to $2l_{ch23}$ is not cut.

Figure 2.185b shows the shape formed by the chisel edge *2–3*. As discussed in Section 2.1.3.7.5 and shown in Figure 2.73, the chisel edge is "flat," it is a straight line crossing the *z*-axis when a standard four-facet drill point grind is considered. The word "flat" in this context means that it is located in a place parallel to the *xy*-plane. In a drill revolving about the *z*-axis, this edge generates a circle having a diameter $d_{th2}=l_{ch}=2l_{ch23}$ as shown in Figure 2.183. When the feed motion is applied, this edge generates the shape shown in Figure 2.185b, which is a hole of diameter d_{th2}.

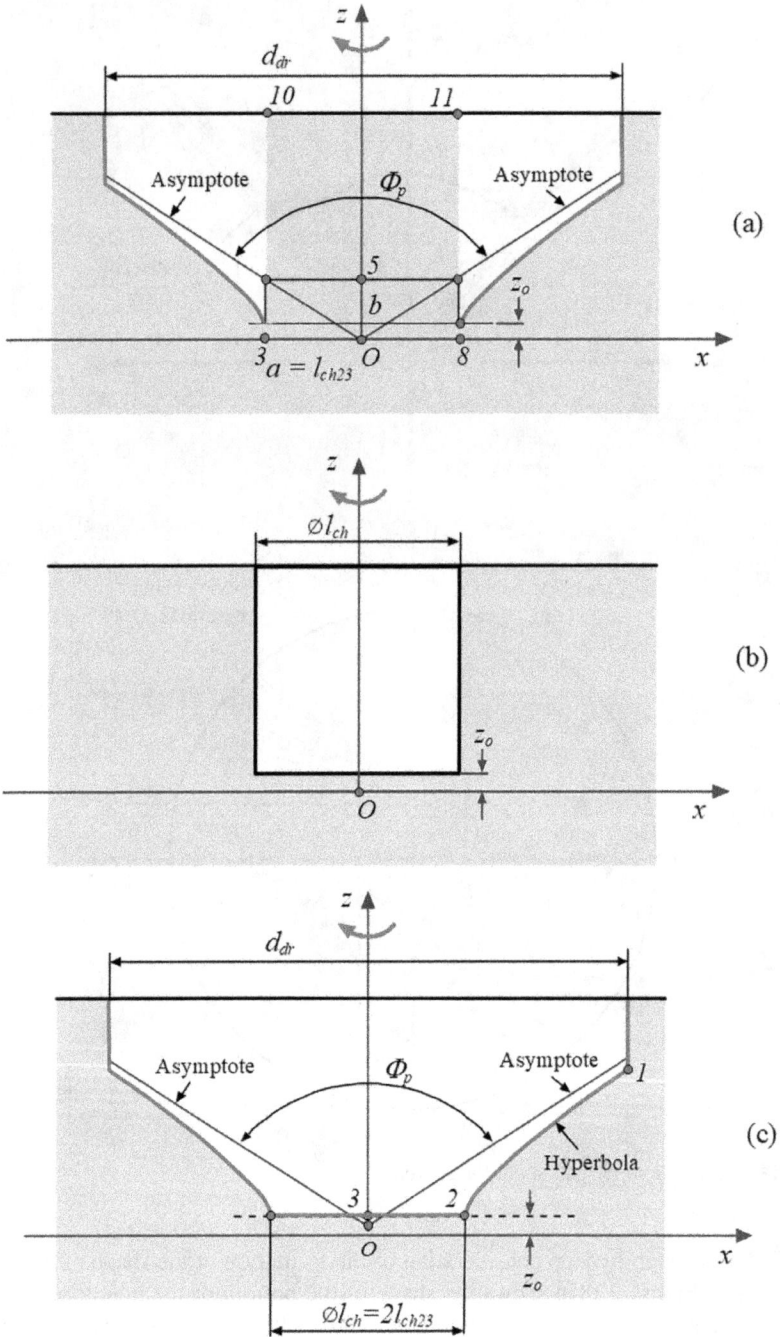

FIGURE 2.185 Formation of the real shape of the bottom of the hole being drilled: (a) shape of the hole formed by the major cutting edge *1-2*, (b) shape of the hole formed by the chisel edge *2–3*, and (c) shape of the hole formed by the cutting edge *1-2-3*.

Figure 2.185c shows the shape of the hole being drilled by the major *1–2* and chisel *2–3* edges, which is a simple combination of the shapes shown in Figure 2.185a and b. It includes the flat part of diameter equal to $2l_{ch23}$ and hyperbolic parts having equation

$$z = \pm \frac{1}{\tan(\Phi_p/2)} \sqrt{x^2 - a_o^2} \tag{2.77}$$

extended from l_{ch23} to $d_{dr}/2$ along the x-axis.

As a practical example, consider a carbide drill of 10 mm dia. with the following geometrical parameters taken from its drawing: point angle, $\Phi_p = 120°$; the location of the major cutting edge ahead of the centerline, $a_o = 0.7$ mm; the angle of the chisel edge, $\psi_{ch} = 60°$; length of the chisel edge $l_{ch23} = 0.8$ mm. The shape of the bottom of the hole being drilled as shown in Figure 2.185c has the following parameters: its flat part has a diameter equal to $2l_{ch23} = 1.6$ mm, the equation of its hyperbolic parts (Eq. 2.77) is

$$z = \pm \frac{1}{\tan(120°/2)} \sqrt{x^2 - 0.7^2} = \pm 0.577 \sqrt{x^2 - 0.49} \tag{2.78}$$

Figure 2.186a shows the summary of the above-obtained results, i.e., the real shape of the bottom of the hole being drilled. A part of the chisel wedge is placed in this figure to show the location of this wedge. This location is important in the further development of the model of chip flow over the rake face of the chisel edge. Figure 2.186b shows the shape of the drilling cups formed at the exit of the drilled holes. Such a cup is formed when the drill pushes a thin wall left to drill at the drill exit. As can be seen, the shape of these caps fully resembles the modeled shape shown in Figures 2.185 and 2.186a.

2.1.5.11.4 Space Considerations

It is of prime importance in considerations of the space available between the chisel edge and the bottom of the hole being drilled to know the distance between a given point of the bottom of the hole being drilled and the rake face of the chisel edge. This section considered only the distance between a given point of the bottom of the hole being drilled and the asymptote of the parabolic surface generated by the major cutting edge *1–2*. Figure 2.187 shows the xz-section of the bottom of the hole being drilled and the asymptote. In this figure, distance Mp' is the axial distance (along the z-axis, i.e., the direction of the cutting feed) and distance Mp is the normal distance of any point of

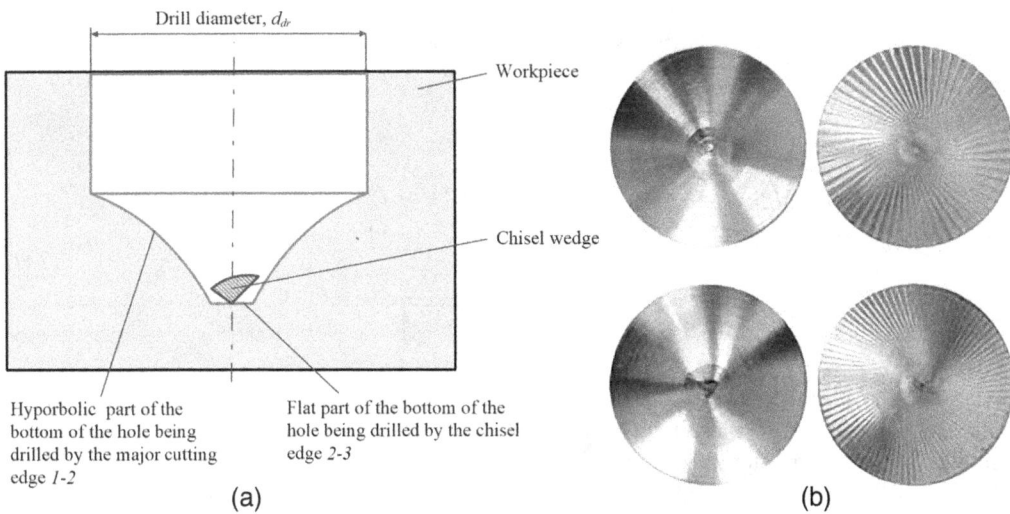

FIGURE 2.186 Showing (a) the real shape of the bottom of the hole being drilled with the chisel wedge placed in it and (b) drilling cups.

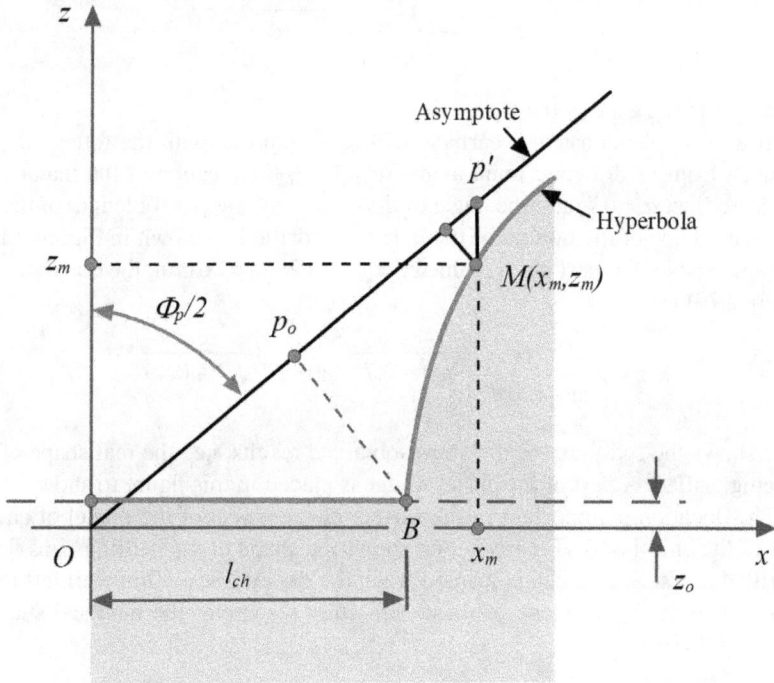

FIGURE 2.187 The xy section of the bottom of the hole being drilled and the asymptote.

parabolic surface having coordinated (x_m, z_m) and its asymptote. Both can be determined as the distance between a point with the known coordinates and a line with the known equation. For a given drill design parameter a_o and a chosen coordinate x_m, asymptote is given by Eq. (2.74) in which $b = a_0/\tan(\Phi_p/2)$, distance Mp' is calculated as

$$Mp'(x_m) = \frac{ab}{x_m + \sqrt{x_m^2 - a^2}} = \frac{a_o^2}{\tan(\Phi_p/2)\left(x_m + \sqrt{x_m^2 - a_o^2}\right)} \tag{2.79}$$

and distance Mp

$$Mp(x_m) = \frac{ab}{x_m + \sqrt{x_m^2 - a^2}}\cos(90° - \Phi_p/2)$$

$$= \frac{a_o^2}{\tan(\Phi_p/2)\left(x_m + \sqrt{x_m^2 - a_o^2}\right)}\cos(90° - \Phi_p/2) \tag{2.80}$$

As can be seen, distance Mp decreases with the x-coordinate eventually becoming infinite close to the asymptote. The decrease rate, however, depends on both a_o and Φ_p. It also follows from Figure 2.187 that the greater distance is Bp_o, where point B is the point of intersection of the hyperbolic and flat parts of the bottom of the hole being drilled. As discussed above, the coordinates of this point depend on the length of the chisel edge, l_{ch}, chisel edge angle ψ_{cl}, and distance a_o. The facts revealed in the last two parameters are of vital importance in the design of HP drills.

As a practical example, let us determine distance Mp at coordinate $x_m = 1.2$ mm for a carbide drill of 10 mm dia. with the following geometrical parameters taken from its drawing: point angle,

$\Phi_p = 120°$; the location of the major cutting edge ahead of the centerline, $a_o = 0.7$ mm; the angle of the chisel edge, $\psi_{ch} = 60°$; length of the chisel edge $l_{ch} = 0.8$ mm. Substituting these data into Eq. (2.80), one can obtain

$$Mp(1.2) = \frac{0.7^2}{\tan 60° \left(1.2 + \sqrt{1.2^2 - 0.7^2}\right)} \cos 30° = 0.112 \text{ mm} \tag{2.81}$$

Using the same way, one can calculate that for $x_m = 1.7$ mm, $Mp(1.7) = 0.075$ mm. The maximum distance when $x_m = l_{ch} = 0.8$ mm is $Mp(0.8) = 0.301$ mm.

Our analysis of Eq. (2.80) and the graphical construction shown in Figure 2.187 reveals the following:

- The asymptote direction does not depend on the distance a_o. Rather is fully set by the drill point angle Φ_p. This is because the equation of the asymptote is $z = (b/a)x$, where $a = a_o$ and $b = a_o/\tan(\Phi_p/2)$ so that the equation of the asymptote is $z = x/\tan(\Phi_p/2)$.
- The greater the length of the chisel edge, l_{ch}, the greater the maximum distance between the bottom of the hole being drilled and the asymptote.

2.1.5.11.5 Conditions of Chip Formation by the Chisel Edge

The following section uses the notions, concepts, and definitions of the CCR and the contact length defined in the Appendix and the definition of the uncut chip thickness in drilling defined and discussed in Chapter 1.

Figure 2.188 shows the sense of the uncut chip thickness, chip thickness, and contact length for the major cutting edge 1–2 and for the chisel edge 2–3. As shown, the drill having point angle Φ_p and diameter d_{dr} (mm) is rotated at n rpm and fed with the feed rate, f_m (mm/min). The axial advance of the drill per one revolution is the cutting feed, f. As the tool is a rigid body, the cutting feed is the same for any point of the drilling tool. As discussed in Chapter 1, the feed per tooth, f_z calculated as the feed per revolution, f over the number of teeth, z. In the considered case $z = 2$ so that $f_z = f/2$.

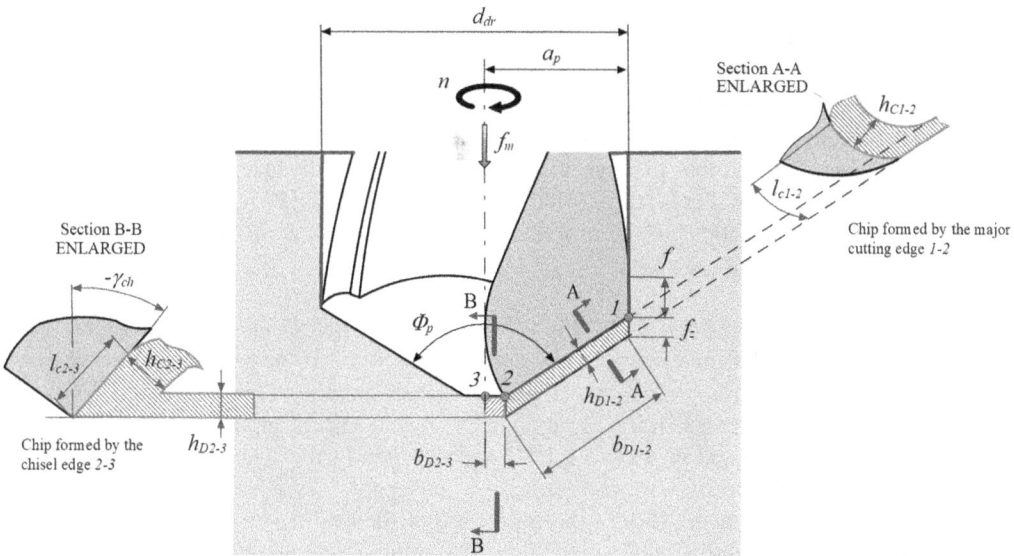

FIGURE 2.188 Showing the sense of the uncut chip thickness, chip thickness, and contact length for the major cutting edge 1-2 and for the chisel edge 2-3.

As the drill rotates with constant n, the linear speed varies over the drill diameter as

$$v_i = \frac{\pi d_i n}{1,000} \text{ (m/minute)} \tag{2.82}$$

Being maximum at point 1 of the major cutting edge where $d_i=d_{dr}$, the cutting speed reduces over the cutting edge toward the drill axis in a linear fashion to 0.

Figure 2.188 visualizes the sense of the uncut chip thickness h_{D1-2} cut by the major cutting edge 1–2. As can be seen, it is calculated as

$$h_{D1-2} = f_z \cos\left(\Phi_p/2\right) \tag{2.83}$$

It follows from Eq. (2.83) that the maximum uncut chip thickness $h_{D\max}=f_z$ when the point angle $\Phi_p=180°$. For the point angle less than 180° (which is normally the case), $h_D<f_z$; it is calculated as $h_{D1-2} = f_z \sin\left(\Phi_p/2\right)$. When the point angle decreases, the length of the cutting edge 1–2 increases so that the chip width, b_{D1-2} increases proportionally as the uncut chip cross section area does not change, i.e., the same volume of the work material is removed per one revolution of the drill regardless of a particular point angle.

Figure 2.188 also visualizes the sense of the uncut chip thickness h_{D2-3} cut by the chisel edge 2–3. In the considered case of the chisel edge geometry, it is equal to f_z as normally the chisel edge has no or very small point angle.

The chip thickness h_C depends on the rake angle, work material, and drilling regime. As discussed in the Appendix, it is calculated as

$$h_C = \zeta h_D \tag{2.84}$$

where ζ is the CCR.

Normally in turning, CCR in cutting of steels, $\zeta=1.8 - 2.5$, i.e., the chip is 1.8–2.5 times thicker than the uncut chip thickness; in machining of ductile materials such as aluminum alloys, CCR can be as great as 10. In other words, the chip thickness can be ten times greater than the uncut chip thickness. The test results show that CCR decreases when the tool rake angle γ and cutting speed increase (Zorev 1966, Astakhov 1998/1999). For a given work material and machining regime, CCR primarily depends on the rake angle and on the cutting speed.

Figure 2.189 shows a generalization of the known results (Astakhov 2010). As can be seen, the rake angle has a significant effect on CCR, which is a direct measure of the amount of work of plastic deformation in metal cutting (Astakhov 2006). The lower the CCR, the lower the cutting force and cutting temperature, and thus the higher tool life. The effect of the rake angle is more profound at low cutting speeds although it is still significant at moderate and high cutting speeds.

Let us consider the variations of the rake angle over the major and chisel edge and the contact length on the rake faces of these edges.

2.1.5.11.5.1 Major Cutting Edge 1–2   The variation of the T-hold-S rake angle along the major cutting edge for straight-flute and twist drills is shown in Figure 2.157. As can be seen, the variation of the T-hold-S rake angle for straight-flute drills is relatively small, i.e., it the T-hold-S rake angle at point 1 is approx. $-3°$, whereas it is $-12°$ at point 2. When a gash is applied to the rake face of a straight-flute drill as discussed in Section 2.1.3.6, the variation of the T-hold-S rake angle is even smaller, i.e., at point 1 is approx. $-3°$, whereas it is $-6°$ at point 2.

The test results show that CCR is in the range of 3–4 (depending on the quality of the rake face and sharpness of the cutting edge) in drilling A380 high-silicon aluminum alloy widely used in the automotive industry. For a common 120° point angle two-flute carbide drill working with the cutting feed $f=0.3$ mm/rev, feed per tooth $f_z=0.3/2=0.15$ mm/rev and uncut chip thickness

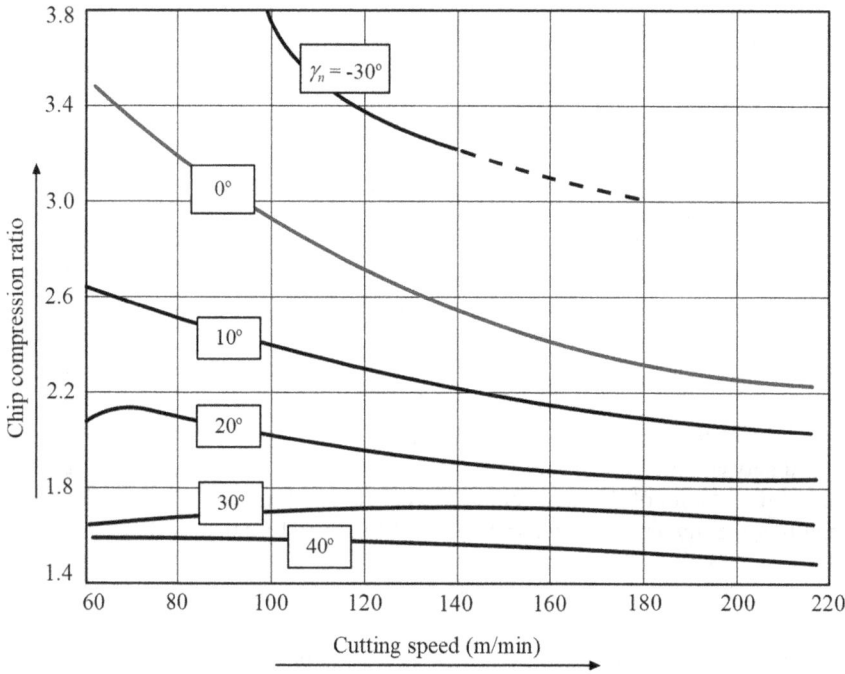

FIGURE 2.189 Influence the rake angle on CCR for the range of the cutting speed used for carbide tools. Free cutting with a_p=6 mm, h_D=0.15 mm. Work material: AISI steel 4130.

h_{D1-2}=0.15·cos (120°/2)=0.075 mm, CCR ζ=4, the chip thickness is calculated as h_{C12}=$h_{D12}\cdot\zeta$= 0.075·4=0.3 mm.

For twist drills, the variation of the rake angle over the major cutting edge is much greater according to Figure 2.157. For a common 120° point angle two-flute carbide twist drill working with the cutting feed f=0.3 mm/rev, feed per tooth f_z=0.3/2=0.15 mm/rev, the uncut chip thickness h_{D12}=0.15·cos (120°/2)=0.075 mm. If CCR ζ=3 then the chip thickness h_{C12}=$h_{D12}\zeta$=0.075·3=0.2 25 mm, which is smaller than that calculated above for a straight-flute drill (0.3 mm) due to greater rake angles when a gash having the proper geometry is applied on the rake face.

The foregoing analysis of chip formation by the major cutting edge *1–2* supported by practical examples reveals that there are no obvious problems with this formation by this edge as chip formation is nearly the same as in turning.

2.1.5.11.5.2 Chisel Edge 2–3 The rake angle of the chisel edge *2–3* depends on a particular point grind of the drill, particularly on the shape of the flank face of the drill and its clearance angle. As discussed earlier, the chisel edge cuts with a constant highly negative rake angle as shown by images Ch1, Ch2, and Ch3 shown in Figure 2.62. As a result, CCR can be as high as 10 in machining of ductile materials in general. In machining of high-silicon aluminum alloys, CCR is smaller reaching 7.

For a common 120° point angle two-flute carbide twist drill working with the cutting feed f=0.3 mm/rev, feed per tooth f_z=0.3/2=0.15 mm/rev, the uncut chip thickness is equal to the feed per tooth as the point angle of the chisel edge is 180°, i.e., h_{D23}=0.15 mm. If CCR ζ=7, then the chip thickness h_{C23}=$h_{D23}\cdot\zeta$=0.15·7=1.05 mm.

2.1.5.11.5.3 Contact Length The sense of the contact length over the major cutting edge of a twist drill is shown in Figure 2.188 (Section A-A). For the above-considered example where h_{D12}=0.075 mm. If CCR ζ=3, the contact length as discussed in the Appendix is calculated as

$l_{c12} = 0.075 \cdot 3^{1.5} = 0.39$ mm. The sense of the contact length over the chisel edge of a drill is shown in Figure 2.188 (Section B-B). For the above-considered example where $h_{D23} = 0.15$ mm. if CCR $\zeta = 7$, the contact length is $l_{c23} = 0.15 \cdot 7^{1.3} = 1.88$ mm.

2.1.5.11.6   *Model of the Flow of the Chip Formed by the Chisel Edge*

Figure 2.190 shows the model of flow of the chip formed by the chisel edge. It "assembles" together all the above-obtained results to quantify what happens when the chisel edge cuts the work material. The chisel wedge rake face *bj* has the rake angle γ_{ch} and the flank face having the clearance angle α_{ch}. The intersection of these faces forms the chisel edge represented by point *b* in Figure 2.190. This edge is located at distance z_o from the coordinate origin *O*. When drilling, the chisel edge cuts the chip having the uncut chip thickness equal to the feed per tooth, $h_{D23} = f_z$. Due to chip plastic deformation, the chip thickness h_{C23} is greater than h_{D23} by CCR, ζ, i.e., $l_{c23} = h_{D23}\zeta^{1.3}$. These data provide a sufficient foundation to analyze all the particularities of the flow of the chip formed by the chisel edge.

The first, and thus the major concern is that the chip having thickness h_{C23} should pass through the gap formed between the rake face of the chisel edge (represented by line *bj* in Figure 2.190) and the side wall of the bottom of the hole being drilled. The width of this gap for a chosen coordinate x_m is the length of the line *mg* as shown in Figure 2.190. It can be calculated as

$$mg = mk \cdot \sin \gamma_{ch} = (mp' + p'k) \sin \gamma_{ch} \qquad (2.85)$$

Distance *mp'* is calculated using Eq. (2.79), distance *p'k* is calculated as

$$p'k = \left(\frac{1}{\tan \gamma_{ch}} - \frac{1}{\tan(\Phi_p/2)} \right) x + z_o \qquad (2.86)$$

Therefore,

$$mg(x_m) = \left[\left(\frac{1}{\tan \gamma_{ch}} - \frac{1}{\tan(\Phi_p/2)} \right) x_m + z_o + \frac{a_o^2}{\tan(\Phi_p/2)\left(x_m + \sqrt{x_m^2 - a_o^2}\right)} \right] \sin \gamma_{ch} \qquad (2.87)$$

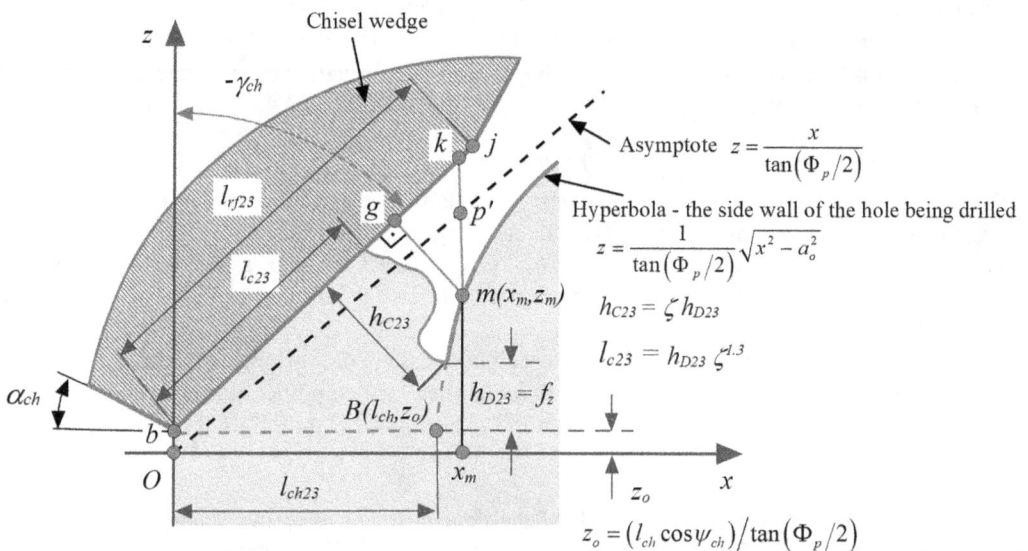

FIGURE 2.190   Model of flow of the chip formed by the chisel edge. Note all the parameters are quantified.

The second issue is the chip formation by the chisel edge. The smaller the length of the chisel edge, l_{ch23}, the closer point B to the rake face of the chisel edge as it is clearly seen in Figure 2.190. This creates a problem as for "normal" chip formation, a certain distance between point B and b (distance Bb) is required to generate the bending moment in the chip so that the chip separation from the rest of the workpiece takes place due to combined action of the compression and bending stresses (Astakhov 1998/1999). When distance Bb becomes critical, the bending moment is not generated so the chip formation takes place due to pure compression. This results in much higher cutting forces, and thus greater energy needed for chip formation. This suggests that the web thinning discussed in Section 2.1.3.7.2 should not go too far, i.e., a certain critical length of the chisel edge is required for the proper chip formation by the chisel edge.

It follows from Eq. (2.87) that the gap which the chip formed by the chisel edge should pass through primarily depends on the rake angle γ_n of this edge so this angle for the above-discussed standard drill point grinds should be determined and analyzed.

2.1.5.11.7 Geometry of the Chisel Edge of the Standard Four-Facet Point Grind

As discussed in Section 2.1.3.7.5, in the standard four-facet point grind, each major cutting edge is provided with the primary flank planar faces called facets having a width equal to or greater than $2a_o$ and the secondary flank faces as shown in Figure 2.191. As such, the clearance angle of the secondary flank face is always greater than that of the primary face. The clearance angle of the first flank face is to assure the proper cutting conditions at the primary cutting edge whereas the clearance angle of the secondary flank face is selected large enough to prevent interference of the drill flank faces with the bottom of the hole being drilled. In Figure 2.191, clearance angle α_{n-pr12} and α_{n-pr45} are primary clearance angles of the major cutting edge *1–2* and *4–5*, respectively, whereas angles α_{n-s12} and α_{n-s45} are secondary clearance angles of the major cutting edge *1–2* and *4–5*, respectively.

As shown by Astakhov (2010), the geometry and working conditions of the chisel edge are defined by the primary clearance angles of the primary flank faces as they serve as the rake faces

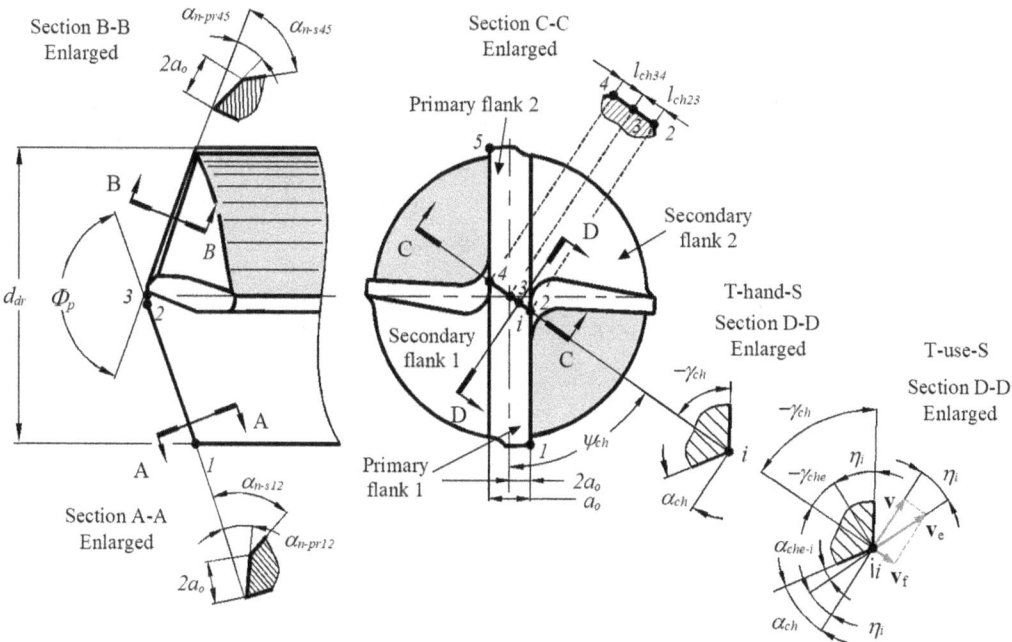

FIGURE 2.191 Particularities of the point geometry of the standard four-facet drill point grind.

for the chisel edge, drill point angle, and distance a_o (the location of the major cutting edge ahead of the centerline) as well as by the tool material and surface conditions (primarily roughness) of the rake face of the chisel edge.

As in this section the concern is the chisel edge, let us consider its geometry parameters shown in Figure 2.191. As discussed in Section 2.1.3.7, there are actually two chisel edges, namely 2–3 and 3–4, each belongs to the corresponding major cutting edges 1–2 and 4–5. Considering Section C-C on the face view of the drill, one can see that these two chisel edges having lengths l_{ch23} and l_{ch45} are aligned creating impression of a single chisel edge. This can be observed under a tool microscope as can be seen in Figure 2.73 (View B). In reality, there are still two chisel edges as shown in Figures 2.63 and 2.64.

Section D-D in Figure 2.191 is the section drawn perpendicular to the projection of the chisel edge in the face view of the drill through a point i selected on the chisel edge 2–3. In this section, the rake, γ_{ch}, and clearance, α_{ch}, angles of the chisel edge are shown. As can be seen, the rake angle of the chisel edge γ_{ch} is highly negative that is the major drawback of this point grind. The same can be observed in View A in Figure 2.73.

Due to drill symmetry, a further consideration will be concentrated only on the major cutting edge 1–2 and chisel edge 2–3. The geometry parameters of this edge are as follows (Astakhov 2010, 2014):

The chisel edge angle is

$$\psi_{ch} = \arctan \frac{\cos\left(\Phi_p/2\right)}{\tan \alpha_{n-pr12}} \qquad (2.88)$$

The length of the chisel edge is

$$l_{ch23} = \frac{a_0}{\sin \psi_{ch}} = \frac{a_0}{\sin\left(\arctan \dfrac{\cos\left(\Phi_p/2\right)}{\tan \alpha_{n-pr12}}\right)} \qquad (2.89)$$

Due to drill symmetry, $l_{ch23} = l_{ch34}$ so that $l_{ch24} = 2\, l_{ch23}$.

Figure 2.192 shows how the chisel edge angle and its total length (l_{ch24}) vary with the point angle and clearance angle of the major cutting edge (8°, 12°, 15°, and 18°). As can be seen, the chisel angle decreases and its total length increases with the point angle. Note that a significant increase in the chisel edge length takes place for drills with the point angle >140° widely used in industry.

2.1.5.11.7.1   The Clearance Angle of the Chisel Edge α_{ch}

$$\alpha_{ch} = \arctan \frac{1}{\tan\left(\Phi_p/2\right)\sin \psi_{ch}} = \arctan \frac{1}{\tan\left(\Phi_p/2\right)\sin\left(\arctan \dfrac{\cos\left(\Phi_p/2\right)}{\tan \alpha_{n-pr12}}\right)} \qquad (2.90)$$

The rake angle of the chisel edge γ_{ch} is

$$\gamma_{ch} = \alpha_{ch} - 90° \qquad (2.91)$$

As the chisel edge passes the axis of rotation (at least, theoretically), the rake and clearance angles of this edge are the same in T-hand-S and T-hold-S, that is, $\gamma_{cl} = \gamma_{cl-h}$ and $\alpha_{cl} = \alpha_{cl-h}$.

Equation (2.90) shows that for a given point angle Φ_p, the normal T-hand-S (T-hold-S) clearance angle of the chisel edge α_{cl} is determined by the T-hand-S normal clearance angle of the

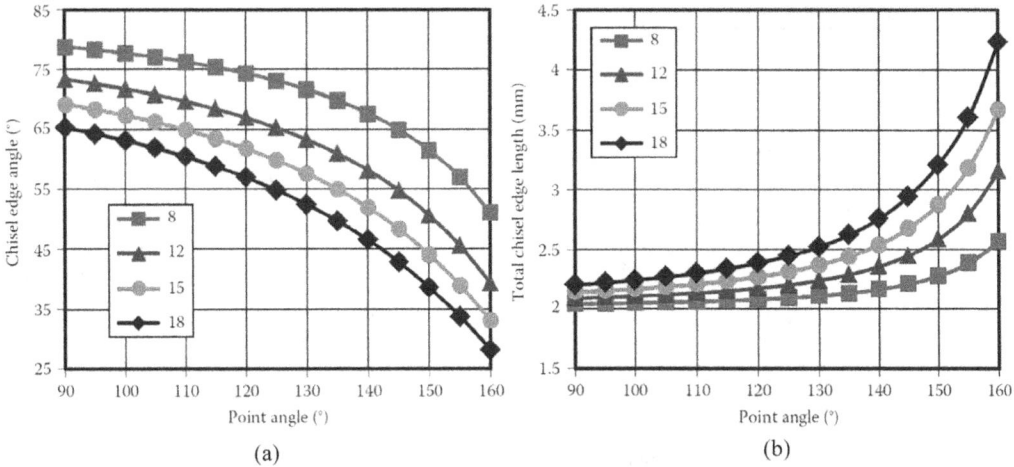

FIGURE 2.192 Chisel edge angle and length: (a) The chisel edge angle as a function of the point angle for four common T-hand-S clearance angles of the major cutting edge (8°, 12°, 15°, and 18°) and (b) the total length of the chisel edge as a function of the point angle for four common T-hand-S clearance angles of the major cutting edge (8°, 12°, 15°, and 18°). A drill of 14.48 mm dia., $a_o = 0.6$ mm.

major cutting edge α_{n-pr}. Because the normal T-hand-S (T-hold-S) rake angle of the chisel edge, γ_{ch}, is fully determined by α_{cl} (Eq. 2.91), this angle is also determined by α_{n-pr}. In other words, the geometry of the chisel edge (for a given point angle Φ_p) is fully determined by the T-hand-S normal clearance angle of the major cutting edge α_{n-pr}. This angle also determines the total length of the chisel edge as follows from Eq. (2.89). Therefore, the normal clearance angle of the primary/first flank face is one of the major design parameters of the drill, and thus, (1) this angle should be clearly indicated on the tool drawing, (2) tight tolerance should be assigned on this angle, (3) this angle should be the same for the major cutting edges to assure drill symmetry and the proper shape of the chisel edge, (4) this angle should be included in the tool inspection procedure and inspection report. The latter, however, with extremely rare exceptions, is not a common practice in the tool industry, and (5) this angle should not be less than 10° for steels and 14° for aluminum alloys work materials for HP drills as discussed in Section 2.1.3.5.

Figure 2.193 shows variations of the normal T-hand-S (T-hold-S) rake angle of the chisel edge with the point angle for four common T-hand-S clearance angles of the major cutting edge (8°, 12°, 15°, and 18°). As can be seen, γ_{cl} is highly negative, for example, for $\Phi_p = 140°$ and $\alpha_{n-pr} = 12°$, it is $\gamma_{cl} = -68°$ while $\alpha_{cl} = 23°$ for the same parameters.

As a practical example, consider conditions of chip formation at the chisel edge and its removal through gap *mg* (Figure 2.190) for a carbide drill of *10* mm dia with the following geometrical parameters taken from its drawing: point angle, $\Phi_p = 140°$; the location of the major cutting edge ahead of the centerline, $a_o = 0.7$ mm; the primary clearance angle $\alpha_{n-prl2} = 10°$. Let us determine distance *mg* (Figure 2.190) at coordinate $x_m = 1.2$ mm. The work material is aluminum alloy A380 and the cutting feed is $f = 0.35$ mm/rev.

Using Eq. (2.88), one can calculate the chisel edge angle as

$$\psi_{ch} = \arctan \frac{\cos\left(\Phi_p/2\right)}{\tan \alpha_{pr12}} = \arctan \frac{\cos\left(140° / 2\right)}{\tan 10°} = 62.7° \tag{2.92}$$

Using Eq. (2.89), one can calculate the length of the chisel edge as

$$l_{ch2-3} = \frac{a_0}{\sin \psi_{ch}} = \frac{0.7}{\sin 62.7°} = 0.79 \text{ mm} \tag{2.93}$$

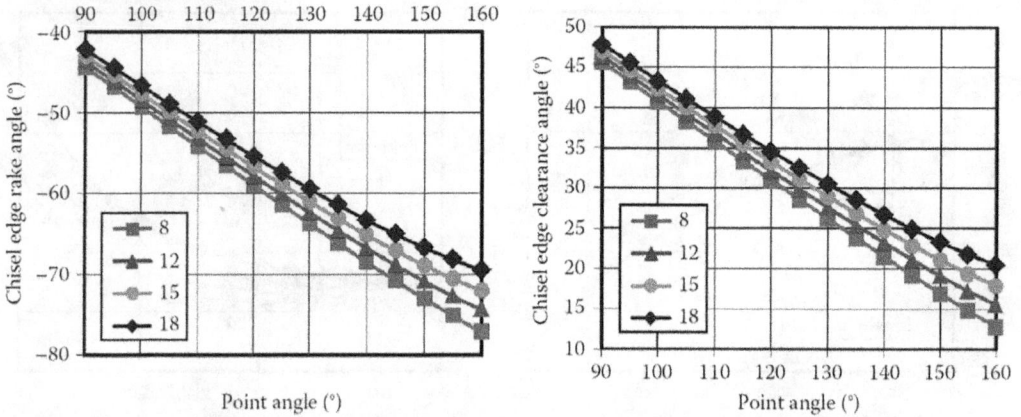

FIGURE 2.193 Variations of the normal T-hand-S (T-hold-S) rake and clearance angles of the chisel edge with the point angle for four common T-hand-S clearance angles of the major cutting edge (8°, 12°, 15°, and 18°).

Using Eqs. (2.90) and (2.91), one can calculate the clearance and rake angles of the chisel edge as

$$\alpha_{ch} = \arctan \frac{1}{\tan\left(\Phi_p/2\right)\sin\psi_{ch}} = \arctan \frac{1}{\tan 70°\sin 62.7°} = 22.3° \tag{2.94}$$

$$\gamma_{ch} = \alpha_{ch} - 90° = 22.3° - 90° = -67.7° \tag{2.95}$$

Using Eq. (2.76), one can calculate distance z_o as

$$z_o = \left(l_{ch}\cos\psi_{ch}\right)/\tan\left(\Phi_p/2\right) = \left(0.7\cos 62.7°\right)/\tan\left(140°/2\right) = 0.117\,\text{mm} \tag{2.96}$$

Distance mg (Figure 2.190) at coordinate $x_m = 1.2\,\text{mm}$ is calculated using Eq. (2.87) as

$$mg(x_m) = mg(1.2) = \left[\begin{array}{c} \left(\dfrac{1}{\tan\gamma_{ch}} - \dfrac{1}{\tan\left(\Phi_p/2\right)}\right)x_m + \\ z_o + \dfrac{a_o^2}{\tan\left(\Phi_p/2\right)\left(x_m + \sqrt{x_m^2 - a_o^2}\right)} \end{array}\right]\sin\gamma_{ch} = \tag{2.97}$$

$$\left[\begin{array}{c} \left(\dfrac{1}{\tan 67.7°} - \dfrac{1}{\tan\left(140°/2\right)}\right)1.2 + 0.117 + \\ \dfrac{0.7^2}{\tan\left(140°/2\right)\left(1.2 + \sqrt{1.2^2 - 0.7^2}\right)} \end{array}\right]\sin 67.7° = 0.24\,\text{mm}$$

For the considered 140°-point angle two-flute carbide twist drill working with the cutting feed $f = 0.35$ mm/rev, feed per tooth is $f_z = 0.35/2 = 0.175$ mm/rev. For the chisel edge, the uncut chip thickness is equal to the feed per tooth, i.e., $h_{D12} = 0.175$ mm. If CCR $\zeta = 6$, then the chip thickness $h_{C12} = h_{D12}.\zeta = 0.175\cdot 6 = 1.05$ mm. Comparing this result with that obtained by Eq. (2.97), one can see that the gap formed between the rake face of the chisel edge and the side wall of the bottom of the hole being drilled (0.24 mm) is not nearly sufficient for the chip having thickness $h_{C12} = 1.05$ mm to pass through.

The obtained result shows that the gaps are small compared to the chip thickness. As such, when the chip formed by the chisel edge attempts to squeeze through the gap smaller than its thickness, the following might happen:

1. When the chip thickness is not much greater than the gap width, further chip plastic deformation takes place. As such, the chip is further deformed by the chisel edge rake face and by non-strain hardened work material. This explains why the axial force due to chisel edge is high as discussed in Section 2.1.3.7.

2. When the chip thickness is much greater than the gap width (a common case for high-penetration rate drills), the chip spreads over the entire rake and flank faces of the chisel edge as shown in Figures 2.74b and 2.176a. Any attempt to increase the drill penetration rate leads to harmful consequences as shown in Figures 2.68 and 2.194.

3. Attempts to apply deep gashes and use of a twist drill instead of a straight-flute drill do not solve the problems as shown in Figure 2.195, which remains of the same severity even if the length of the chisel edge is reduced by application of the web thinning. The observations of failures of thousands of drills of different designs and geometries and made of various tool materials allow to conclude that the known drill transverse (radial) vibration commonly attributed to the lack of drill symmetry occurs due to the described problem

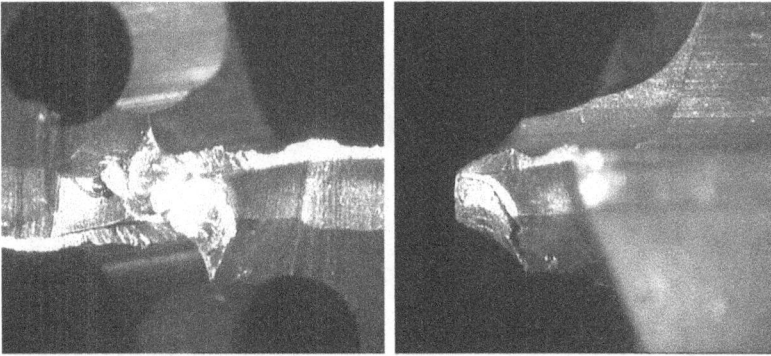

FIGURE 2.194 Consequences of an attempt to increase the drill penetration rate in machining of aluminum alloy A380 – a cemented carbide straight-flute drill.

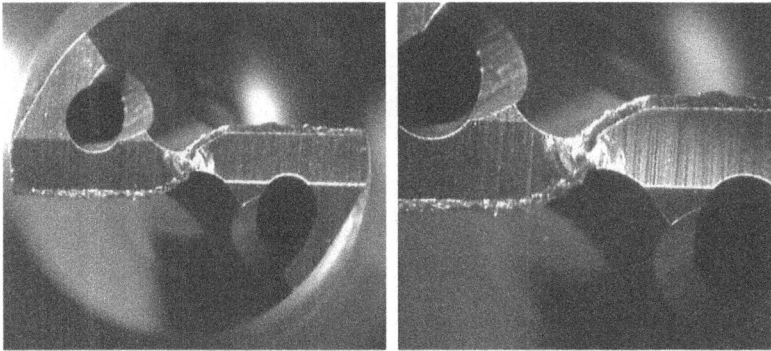

FIGURE 2.195 The chisel edge covered by the aluminum chip/built-up edge of a twist drill having gashes (web thinning) and a plane primary flank.

with the flow of the chip formed by the chisel edge. Figure 2.196 shows side margins with evidences of transverse (radial) vibrations. Such vibrations not only affect the machined hole shape and tolerance (important in drilling precision holes, e.g., in the automotive industry) but also reduce tool life. The harder the tool material, the greater the harmful effect of this vibration.

4. Attempts to change the planar flank face into a conical or helical do not solve the problem. Figure 2.197 shows the consequences of an attempt to increase the drill penetration rate in machining of aluminum alloy A380 using a high-performance cemented carbide twist drill having a helical point grind.

The discussion exemplifies the major problem with the four-facet drill point grind. As a direct result of this problem, the chisel edge accounts for up to 65% of the drilling axial force. Moreover, extrusion of the chip made by the chisel edge creates an impression that this edge does not cut but rather extrudes the work material as discussed in Section 2.1.3.7. The above conclusive results show that this type of point grind should not be used in HP drills although some leading drilling tool manufacturers with the insistence that deserves better application keep manufacturing drill with such a point grind.

The next step is to consider the chisel edge geometry in T-use-S. Section D-D in Figure 2.191 shows the concept of the T-use-S rake and clearance angles for a point of consideration i of the chisel edge. As the point angle of the chisel edge for the considered drill geometry is zero then the

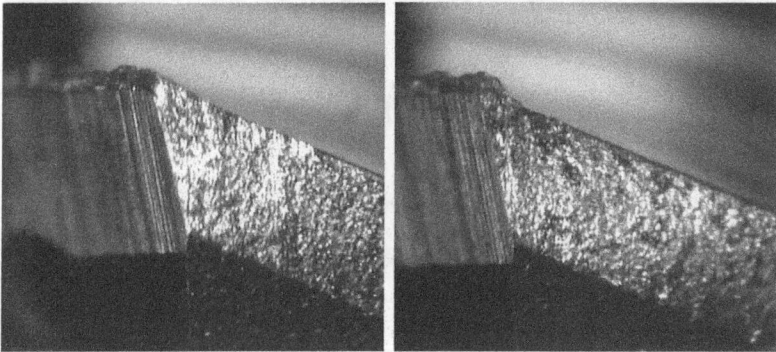

FIGURE 2.196   Evidences of the drill transverse (radial) vibrations.

FIGURE 2.197   Consequences of an attempt to increase the drill penetration rate in machining of aluminum alloy A380 with a high-performance cemented carbide twist drill having a helical point grind.

normal and working angles of this edge are the same as discussed in the Appendix. As can be seen, the rake angle increases while the clearance angle decreases in T-use-S compared to T-hand-S.

As discussed in Section 2.1.5.8, the clearance angle of the chisel edge for a given point i of this edge in T-use-S, α_{cle-i} is calculated as

$$\alpha_{cle-i} = \alpha_{cl} - \eta_{e-i} \tag{2.98}$$

As discussed in Section 2.1.5.9.2, the rake angle of the chisel edge for a given point i of this edge in T-use-S, γ_{cle-i} is calculated as

$$\gamma_{cle-i} = \gamma_{cl} + \eta_{e-i} \tag{2.99}$$

where angle η_{e-i} is defined by Eq. (2.45).

It was discussed in Section 2.1.5.8 that angle η_{e-i} is small for points on the major cutting edge. This is, however, not nearly the case for the chisel edge as the cutting speed is reduced significantly from its point *2* toward point *3* where the cutting speed is zero whereas the speed of feed (feed rate) remains the same. Figure 2.198 shows the variation of the resultant cutting velocity angle η_e over the length of the chisel edge for cutting feeds of 0.2 and 0.35 mm/rev. Note that this graph is "universal" as angle η_e depends only on the cutting feed f and location of point i on the chisel edge (see Eq. 2.45) so it is valid for any drill point grind where the chisel angle is formed. Analyzing this graph, one can conclude that a significant increase (more than 10°) of η_e starts only when a point of consideration locates at a distance of 0.3 mm (for the cutting feed $f=0.35$ mm/rev) and 0.24 mm (for $f=0.2$ mm/rev) from the drill axis of rotation. As a result, the chisel edge normal rake and clearance angles in T-use-S (γ_{cle} and α_{cle}) are defined by Eqs. (2.99) and (2.98), respectively, deviate from their T-hand-S (T-hold-S) values.

Using this result, one can assess the variation of the real (in T-use-S) clearance and rake angles over the length of the chisel edge. Figure 2.199 shows examples for a common drill design. As can be seen, the T-use-S rake angle becomes much less negative compared to that in T-hand-S in the vicinity of the axis of rotation that improves chip formation. The problem, however,

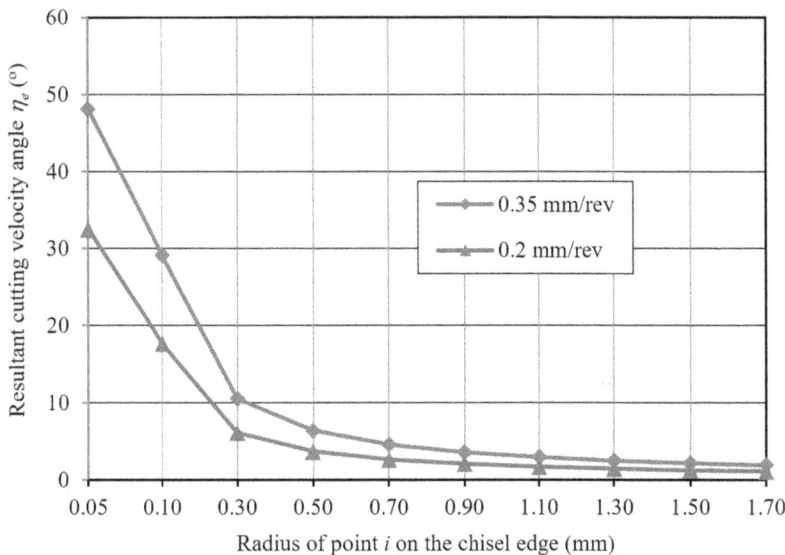

FIGURE 2.198 Variation of the resultant cutting velocity angle η_e over the length of the chisel edge for cutting feeds of 0.2 and 0.35 mm/rev.

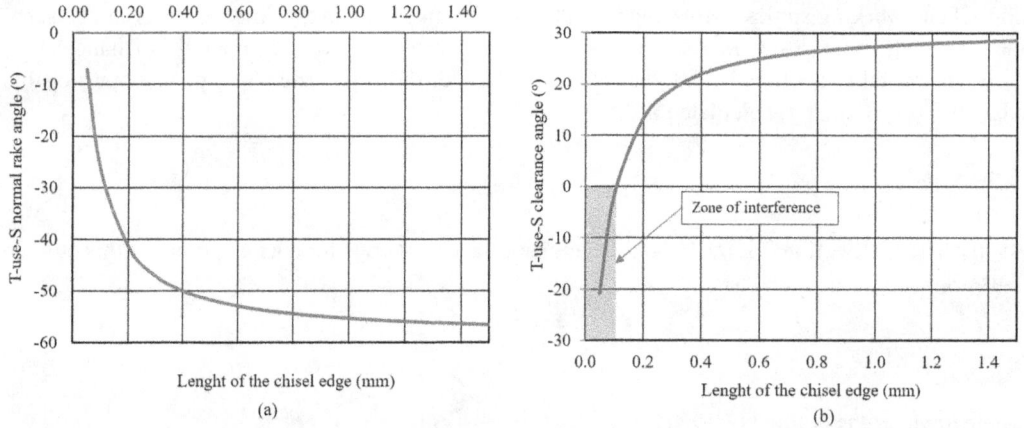

FIGURE 2.199 Variation of the T-use-S rake (a) and clearance (b) angles over the length of the chisel edge (the T-hand-S normal clearance angle $\alpha_{n1-2}=8°$, point angle $\Phi_p=120°$, feed per revolution $f=0.4$ mm/rev, and the length of the chisel edge 1.5 mm).

appears to be with the T-use-S clearance angle. On the one hand, this angle decreases toward the axis of rotation so that the chisel edge does not break as it can with high clearance angles in T-hand-S (Figure 2.193). On the other hand, the interference of the chisel edge flank face and the bottom of the hole being drilled takes place as the T-use-S clearance angle becomes very small and then negative in the vicinity of the axis of rotation as shown in Figure 2.199b as the zone of interference. As follows from Eqs. (2.45) and (2.98), the higher the feed per revolution, the greater the problem. This is because such interference may potentially lead to a significant increase in the axial force in drilling and thus limits the allowable penetration/feed rate. In reality, this problem is not that severe as the location of the chisel edge is not that perfect due to drill runout and tolerance assigned for the chisel edge centrality (see Chapter 1 in Astakhov (2024)). For high-precision, high-penetration rate drill, however, the discussed problem can be dealt with easily by altering the geometry of the flank face of the chisel edge in the vicinity of the drill rotational axis.

2.1.5.11.8   Geometry of the Chisel Edge of the True Four-Facet Point Grind

The particularities at the advanced user level of the true four-facet drill geometry are discussed in Section 2.1.3.7.6 and are shown in Figures 2.75–2.78. Moreover, Figure 2.77 shows a drill with a true four-facet point after drilling soft aluminum alloy. As can be seen in this figure, both chisel edges worked properly so that this figure resembles the modeling of chip formation shown in Figure 2.64. It is to say that this point grind works well even for soft work materials prone to significant plastic deformation in cutting, and thus to the formation of a significant built-up edge. This section aims to explain this result.

Experience shows that significant improvements in drill self-centering and in the geometry of the chisel edge are achieved when the width of the primary flank is equal to a_o, while the secondary flank plane is applied with a normal clearance angle greater than that of the primary clearance angle. As discussed in Section 2.1.3.7.3, this type of drill point grind is known as the true four-facet grind. A model that visualizes the essential geometry parameters of such a drill is shown in Figure 2.200. The same as in the above-discussed four-facet point grind (Figure 2.191), the flank surface of each major cutting edge (lip) (1–2 and 4–5) is planar, i.e., includes two adjacent planes ground at different clearance angles. The major cutting edge 1–2 is provided with the primary flank plane 1 having the clearance angle α_{n-pr12} and secondary flank plane 2 having the clearance angle α_{n-s12}. Respectively, the major cutting edge 4–5 is provided with the primary flank plane 2 having the clearance angle

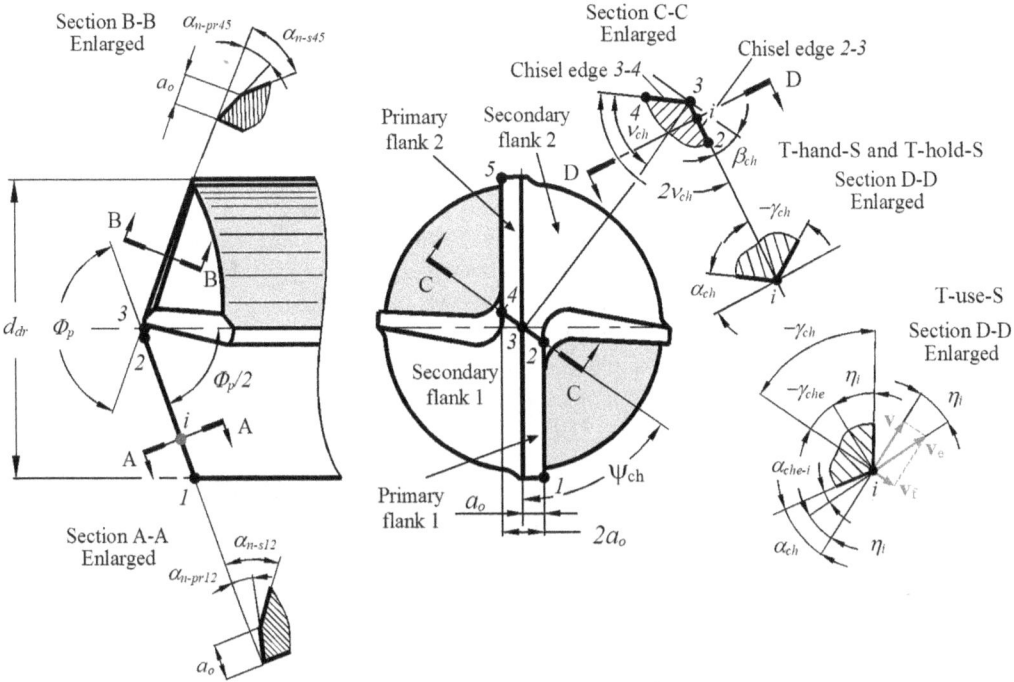

FIGURE 2.200 Particularities of the point geometry of the true four-facet drill point grind.

α_{n-pr45} and secondary flank plane 2 having the clearance angle α_{n-s45}. The difference is in the width of the primary clearance plane: it is equal to or greater than $2a_o$ for a four-facet point grind whereas it is equal to a_o in the true four-facet point grind. As a result, the rake face of the chisel edge *2–3* is the secondary flank plane 2 and the rake face of the chisel edge *4–5* is the secondary flank plane 1. Because the clearance angles of the secondary flank planes are significantly greater than those of the primary flank planes, one should expect greater rake angles of both chisel edges (*2–3* and *3–4*) compared to the four-facet point grind.

Another significant difference of the considered point grind is that the chisel edges *2–3* and *4–5* are not aligned as in the above-discussed four-facet point grind. Rather, each chisel edge makes the so-called approach angle β_{ch} of the chisel edge considered in the tool back plane (the plane perpendicular to the feed motion) as shown in Section C-C in Figure 2.200. As discussed by Astakhov (2010), this angle is calculated as

$$\tan \beta_{ch} = \frac{\left(\tan \alpha_{n-s45} - \tan \alpha_{n-pr12}\right)\sin \psi_{ch}}{2\sin\left(\Phi_p/2\right)} \tag{2.100}$$

where the chisel edge angle ψ_{ch} for this point grind is calculated as (Astakhov 2010)

$$\psi_{ch} = \arctan \frac{2\cos\left(\Phi_p/2\right)}{\tan \alpha_{n-s45} + \tan \alpha_{n-pr12}} \tag{2.101}$$

As can be seen in Figure 2.200, point *3* appears to be the drill apex with the point angle $2v_{ch}$ called the point angle of the chisel edge. This apex first touches the workpiece at the beginning of drilling (provided that this surface is perpendicular to the drill rotating axis) so that it helps to reduce drill wandering and thus reduces drill transverse vibrations at the hole entrance. Therefore, a drill with such a point gains some self-centering ability. It was also found that this shape of the chisel edge makes the chisel wedge stronger and less susceptible to chipping. Figure 2.76 shows the discussed feature of the chisel edge. The point angle of the chisel angle is calculated as

$$2v_{ch} = 180° - 2\beta_{ch} = 180° - 2\frac{(\tan\alpha_{n-s45} - \tan\alpha_{n-pr12})\sin\psi_{ch}}{2\sin(\Phi_p/2)}$$

(2.102)

As follows from Eq. (2.102), $2v_{ch}$ (and thus the drill's self-centering ability) depends on the primary, α_{n-pr12}, and secondary, α_{n-s12}, clearance angles as well as on drill point angle Φ_p. The smaller the chisel edge point angle $2v_{ch}$, the better the self-centering ability of a drill. Figure 2.201 shows the graphical representation of this equation for primary clearance angles $\alpha_{n-pr12} = 8°$, 12°, 15°, and 18°. As can be seen, $2v_{ch}$ increases with both the drill point angle and primary clearance angle whereas it decreases with secondary clearance angle. It should be noticed that for a standard carbide drill with point angle $\Phi_p = 140°$ and secondary clearance angle $\alpha_{n-s12} = 25°$, the chisel edge point angle is about 172° when the primary clearance $\alpha_{n-pr12} = 10°$ whereas when the secondary clearance angle α_{n-s12} is increased to 35°, the chisel edge point angle becomes 163° that increase the drill self-centering ability. *Therefore, from this perspective, an increase in the secondary clearance angle is beneficial.*

Let us now consider how the discussed application of the secondary clearance angle affects the clearance and rake angles of the chisel edge. Using the vector analysis as applied to the cutting tool geometry considerations, the author derived (Astakhov 2010) the following equations for the rake, γ_{ch}, and clearance, α_{ch}, angles of the chisel edge,

$$\gamma_{ch} = \arctan\frac{\tan\alpha_{n-s}\cos\beta_{ch}}{\cos\psi_{cl}}$$

(2.103)

$$\alpha_{ch} = \arctan\left(\frac{1}{\tan(\Phi_p/2)\sin\psi_{cl}}\cos\beta_{cl}\right)$$

(2.104)

Figure 2.202 shows the chisel edge rake angle as a function of drill point angle Φ_p for four common T-hand-S clearance angles of the major cutting edge, α_{n-pr12} (8°, 12°, 15°, and 18°) and for two secondary clearance angles, $\alpha_{n-s45} = 25°$ and 35° angles. If one compares these results with those shown in Figure 2.193 (the chisel edge rake angles for the four-facet drill point), he or she can notice some increase in this angle in the considered case. For example, when the point angle $\Phi_p = 140°$ and the primary clearance angle $\alpha_{n-pr12} = 12°$, the rake angle of the chisel edge for the four-facet drill point $\gamma_{ch} = -66°$ whereas for the true four-facet point $\gamma_{ch} = -57°$ when secondary clearance angles, $\alpha_{n-s45} = 25°$ and $\gamma_{ch} = -48°$ when secondary clearance angles, $\alpha_{n-s45} = 35°$. *This example shows that*

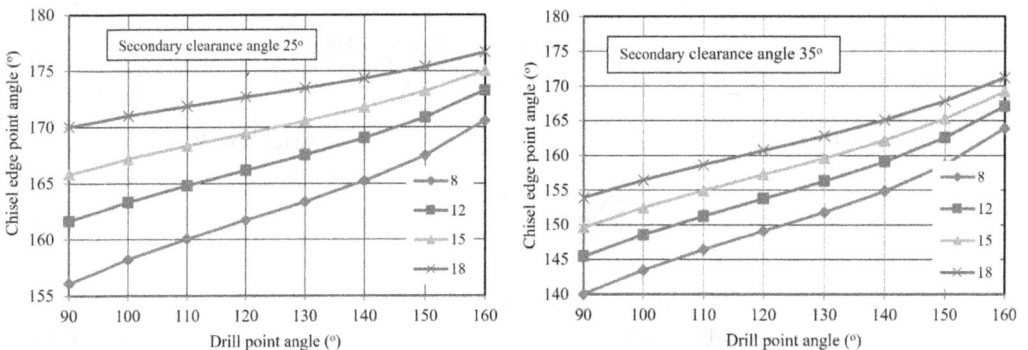

FIGURE 2.201 Chisel edge point angle as a function of the primary, α_{n-pr12} clearance angle and drill point angle Φ_p for two secondary clearance angles, $\alpha_{n-s45} = 25°$ and 35° angles.

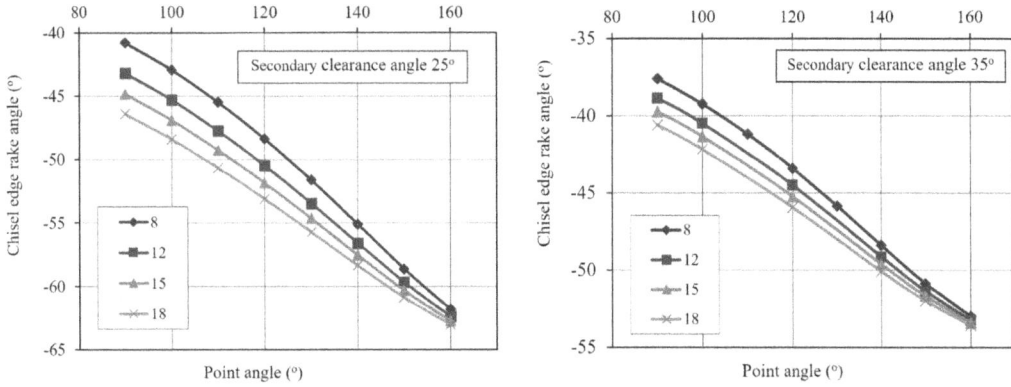

FIGURE 2.202 Chisel edge rake angle as a function of the primary clearance angle, $\alpha_{n\text{-}pr12}$ and drill point angle Φ_p for two secondary clearance angles, $\alpha_{n\text{-}s45}=25°$ and $35°$ angles ($a_o =0.6$ mm).

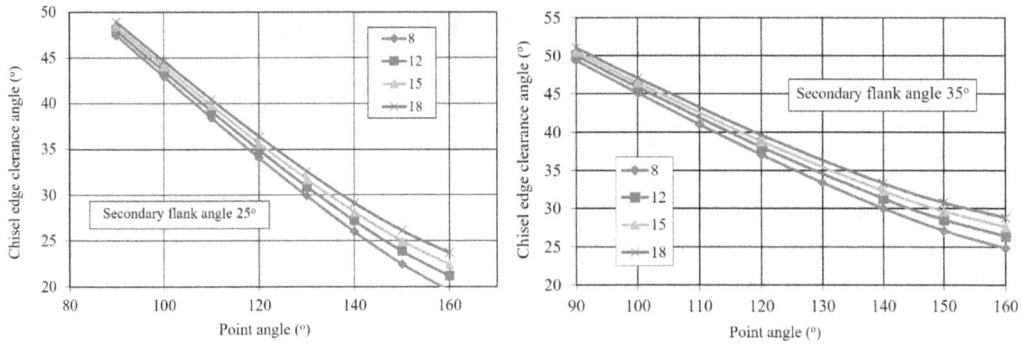

FIGURE 2.203 Chisel edge clearance angle as a function of the primary, $\alpha_{n\text{-}pr12}$ clearance angle and drill point angle Φ_p for two secondary clearance angles, $\alpha_{n\text{-}s45}=25°$ and $35°$ angles ($a_o =0.6$ mm).

any noticeable increase in the chisel edge rake angle can be achieved only when the secondary clearance angle is great.

Figure 2.203 shows the chisel edge clearance angle as a function of the primary clearance angle, $\alpha_{n\text{-}pr12}$ ($8°$, $12°$, $15°$, and $18°$) and drill point angle Φ_p for two secondary clearance angles, $\alpha_{n\text{-}s12}=25°$ and $35°$ angles. If one compares these results with those shown in Figure 2.193 (the chisel edge clearance angles for the four-facet drill point), he or she can notice some increase in this angle for the true four-facet point grind. For example, when the point angle $\Phi_p=140°$ and the primary clearance angle $\alpha_{n\text{-}pr12}=12°$, the clearance angle of the chisel edge for the four-facet drill point $\alpha_{ch}=25°$ whereas for the true four-facet point $\alpha_{ch}=27.5°$ when secondary clearance angles, $\alpha_{n\text{-}s12}=25°$ (a moderate increase) whereas when the secondary clearance angles, $\alpha_{n\text{-}s12}=35°$, the clearance angle of the chisel edge increases to $\alpha_{ch}=32°$. This can be of importance when the kinematic geometry of the chisel edge is considered.

The length of segment *2–3* of the chisel edge l_{ch23} is equal to l_{ch3-4} (distance *4-3*). It is calculated accounting for β_{cl} as

$$l_{cl23} = l_{cl34} = \frac{2a_o}{2 \sin \psi_{cl} \cos \beta_{cl}} \tag{2.105}$$

Figure 2.204 shows the total length of the chisel edge as a function of the primary, $\alpha_{n\text{-}pr12}$ clearance angle and drill point angle Φ_p for two secondary clearance angles, $\alpha_{n\text{-}s12}=25°$ and $35°$ angles. If one

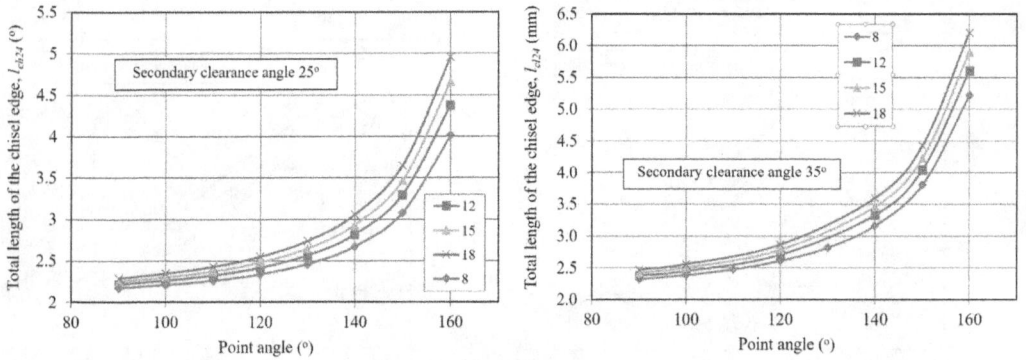

FIGURE 2.204 Total length of the chisel edge as a function of the primary, $\alpha_{n\text{-}prl2}$ clearance angle and drill point angle Φ_p for two secondary clearance angles, $\alpha_{n\text{-}s45}=25°$ and 35° angles ($a_o=0.6\,\text{mm}$).

compares these results with those shown in Figure 2.192b (the total length of the chisel edge angles for the four-facet drill point), he or she can notice a significant increase in this length for the true four-facet point grind. For example, when the point angle $\Phi_p=140°$ and the primary clearance angle $\alpha_{n\text{-}prl2}=12°$, the total length of the chisel edge for the four-facet drill point $l_{ch24}=2.3\,\text{mm}$ whereas for the true four-facet point $l_{ch24}=2.7\,\text{mm}$ when secondary clearance angles, $\alpha_{n\text{-}s12}=25°$ (a moderate increase) whereas when the secondary clearance angles, $\alpha_{n\text{-}s12}=35°$, the total length of the chisel edge increases to $l_{ch24}=3.4\,\text{mm}$.

The rake and clearance angles of the chisel edge in T-use-S are considered accounting for the fact that the half-point angle of this edge is v_{cl}. As such, the T-use-S rake and clearance angles considered point i of the chisel edge are calculated as

$$\gamma_{cle-i} = \gamma_{cl} + \upsilon_{e-i} \tag{2.106}$$

$$\alpha_{cle-i} = \alpha_{cl} - \upsilon_{e-i} \tag{2.107}$$

where $\upsilon_{e-i} = \arctan\left(\tan\eta_{e-i}\sin v_{cl}\right)$, where η_{e-i} is defined by Eq. (2.45).

Figure 2.205 shows examples of the kinematic rake and clearance angles distribution over the chisel edge.

Although the true four-fact point grind is better compared to the standard four-facet point grind, the interference of the chisel flank adjacent to the axis of rotation exists for this point geometry is still there. The following should be taken into consideration:

- To achieve the best drill self-centering ability and to increase the gap between the rake face and the bottom of the hole being drilled (distance mg in Figure 2.190) to assure sufficient space for free flows of the chip formed by the chisel edge, the secondary clearance angle, $\alpha_{n\text{-}s12}$ should be as high as permitted by the strength of the region adjacent to the chisel edge. As such, $\alpha_{n\text{-}s12}=35°$ is a good value to start. As high secondary clearance angles may compromise drill stability as discussed in Section 2.1.5.7, the VPA2© design should be used to avoid any stability problems. The combination of the true four-facet point grind and the VPA2© design in straight and twist drills assures their entrance stability and improves the quality, i.e., meeting tight tolerances on diametric accuracy, hole shape and straightness, position, and so on. Moreover, the use of the proposed combination improves the quality of the entrance of the hole being drilled so that the proposed designs can be used for drills (both straight and twist) when this quality is of some concern.

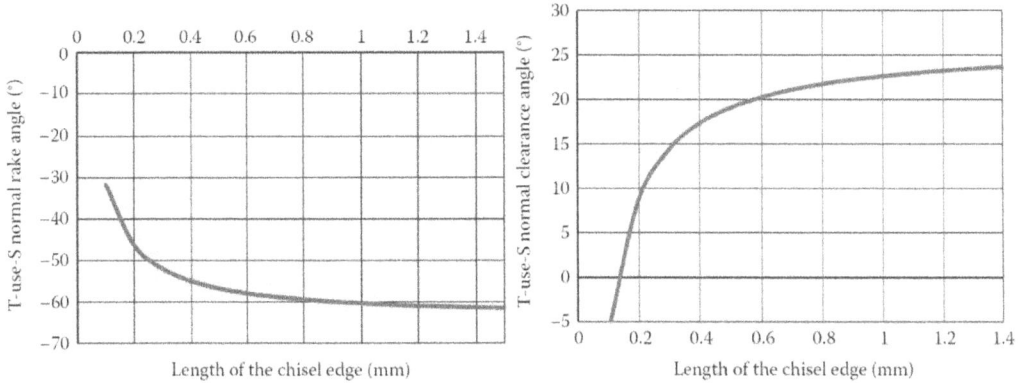

FIGURE 2.205 Variation of the T-use-S rake and clearance angles over the length of the chisel edge (the T-hand-S normal clearance angles: $\alpha_{n-pr12}=10°$ and $\alpha_{n-s45}=25°$, point angle $\Phi_p=140°$, feed per revolution $f=0.4$ mm/rev, and the length of the chisel edge 1.5 mm).

- Because the use of the secondary clearance angle unavoidably increases the length of the chisel edge, the application of gashes (web thinning) as shown in Figures 2.54 and 2.56 should be used to reduce the length of the chisel edge. This is of critical importance for wide-web drills (Figure 2.65), i.e., when the distance a_o (Figure 2.200) is great, as this distance directly affects the length of the chisel edge according to Eq. (2.105).
- To gain the self-centering advantage with less negative rake angle and thus greater allowable penetration rate as the gap between the chisel edge rake face and the wall of the hole being drilled becomes greater, the point grinding machine setup should be very accurate, particularly for small drill diameters (less than 10 mm). Particularly, it is necessary to avoid the grinding flaw shown in Figure 2.78b and discussed in Section 2.1.3.7.6.

2.1.5.11.9 Drill Flank Is Formed by Two Surfaces (Generalization: Tertiary Flank Plane and Split Point)

To further improve the chisel edge geometry, mainly to reduce its negative rake angle, the second flank surface can be angularly located with respect to the primary flank surface. Figure 2.206 shows one of the earlier designs of such a flank face where the primary and secondary flank surfaces are not planes. This drill has two major cutting edges 1 and 2 and two chisel edges 3 and 4. The chisel edges are provided with fully developed lands 5 and 6 that extend to the heels. Such a location of the secondary flanks improves the geometry of the chisel edges and provides better conditions for the chip formed by the chisel edge to pass the gap between the bottom of the hole being drilled and the rake face of the chisel edge.

Although many advantages, including reduction in the axial force, of this and similar point grinds were found in practical applications, this kind of point grinds was not implemented widely in industry. The major reason for that is tight requirements to the location of lands 5 and 6, i.e., to point symmetry. Obviously, such requirements could not be met with hand grinding even using simple grinding fixtures.

In modern designs, the primary and secondary flank surfaces are normally planes. A general objective in forming a two-plane flank can be thought of as the finding of the location of flank planes R and F (Figure 2.207) to assure the desired chisel edge angle, ψ_{cl} (to maintain a certain length of the chisel edge). The location of plane R is uniquely defined by the selected point angle Φ_p and the desired T-hand-S clearance angle α_{n-R} at point 1. Therefore, the problem reduces to finding the location of plane F. The complete solution of such an ambitious problem is considered earlier (Rodin 1971, Astakhov 2010). In this section, a much less ambitious but rather practical model of the geometry parameters is considered.

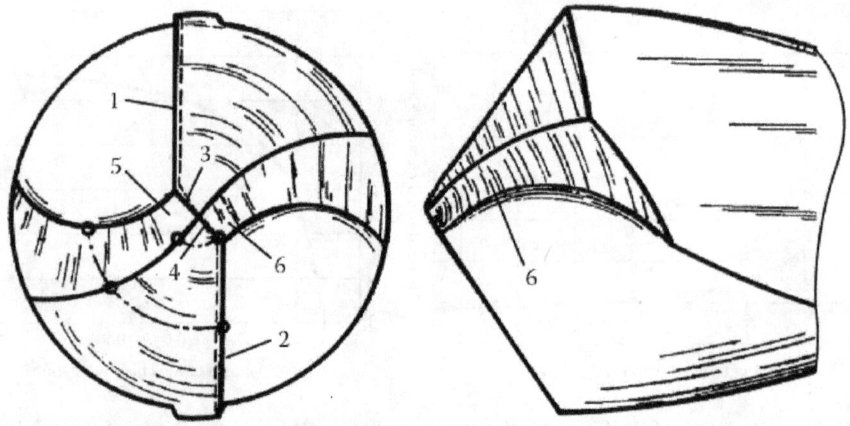

FIGURE 2.206 Drill geometry according to US Patent No. 3,564,947 (1964).

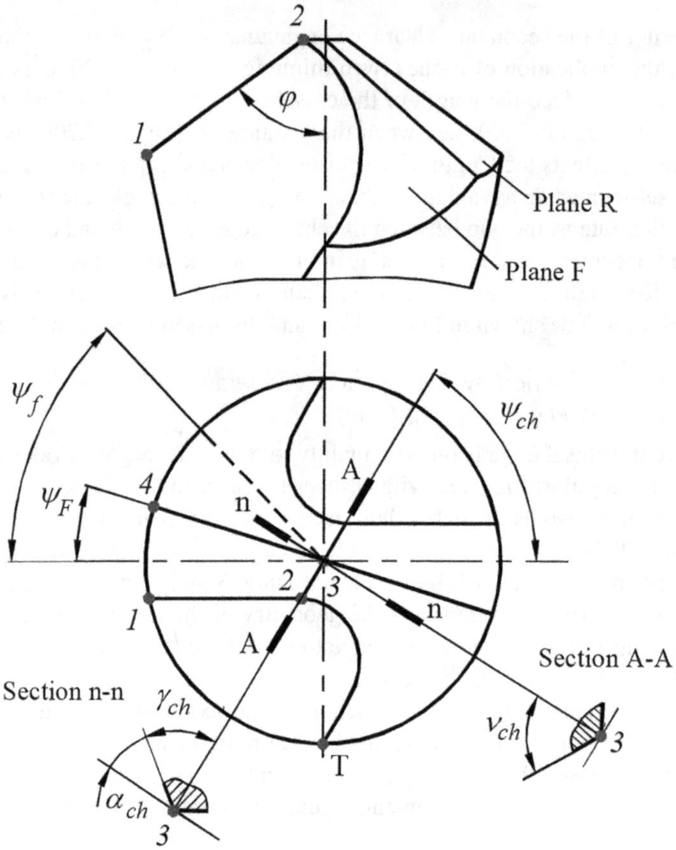

FIGURE 2.207 The flank face formed by two planes – general case.

In practice, plane R having T-hand-S clearance angle α_{n-R} is first ground. Then, the grinding wheel is set with angle α_{n-F} and the drill is revolved by a certain angle ψ_f and the second flank plane F is ground. Angles α_{n-F} and ψ_f together with the half-point angle φ define the geometry of the chisel edge, namely, its inclination angle ψ_{ch}, half-wedge angle v_{ch}, T-hand-S (T-hold-S) rake γ_{ch}, and clearance α_{ch} angles.

2.1.5.11.9.1 Angle ψ_F is Determined as (Vinogradov 1985)

$$\psi_F = \arccos\left(A\big/\sqrt{B+C+D+E}\right) \tag{2.108}$$

where

$$A = \cos\psi_f\left(\sin\alpha_{f-F} + \tan\psi_f\tan^{-1}\varphi\right) - \cos\alpha_{f-F}\tan\alpha_{f-R}$$

$$B = \sin^2\psi_f\left(\frac{1}{\tan^2\varphi} + \sin^2\alpha_{f-F}\right)$$

$$C = \cos^2\alpha_{f-F}\left(\tan\alpha_{f-R} - \cos\psi_f\tan\psi_f\right)^2$$

$$D = \frac{1}{\tan^2\varphi}\left[\frac{\cos\alpha_{f-F}}{\cos\alpha_{f-R}} - \cos\psi_f\right]^2$$

$$E = 2\sin\psi_f\tan^{-1}\varphi\cos\alpha_{f-F}\left[\frac{\cos\alpha_{f-F}}{\cos\alpha_{f-R}} - \tan\alpha_{f-R}\right]$$

where the T-hand-S clearance angles of flank planes F and R in the working plane are

$$\alpha_{f-F} = \arctan\left(\frac{\tan\alpha_{n-F}}{\sin\varphi}\right)$$

$$\alpha_{f-R} = \arctan\left(\frac{\tan\alpha_{n-R}}{\sin\varphi}\right)$$

The chisel edge angle is calculated as

$$\psi_{cl} = \frac{\sin(\varphi_{o1}+\varphi)}{\tan\alpha_{o1}\sin\varphi_{o1}\sin\varphi - \tan^{-1}\psi_R\left(\sin\varphi\cos\varphi_{o1} - \sin(\varphi_{o1}+\varphi)\right)} \tag{2.109}$$

where

$$\varphi_{o1} = \arctan\left[\frac{\cos\psi_F}{\sin\psi_F\tan\alpha_{o1} - \tan\varphi_o}\right]$$

where φ_o is the angle between planes R and F measured in the axial plane perpendicular to the intersection rib *3–4* (Figure 2.207)

$$\varphi_o = \arctan\left(\tan^{-1}\varphi\cos\psi_F + \sin\psi_F\tan\alpha_{o1}\right)$$

$$\alpha_{o1} = \arctan\left(\cos\psi_F\tan\alpha_{oN} + \sin\psi_F\tan\varphi_o\right)$$

$$\alpha_{oN} = \arctan\left[\frac{\tan\alpha_{f-F} - \sin(\psi_F-\psi_f)\tan^{-1}\varphi_o}{\cos(\psi_F-\psi_f)}\right]$$

$$\psi_R = \arctan\left(\frac{\tan \alpha_R}{\cos \varphi}\right)$$

The chisel wedge angle $2v_{ch}$ is calculated as

$$2v_{ch} = 180^o - 2\arctan\left(\sin \psi_{ch} \tan \alpha_{o1} - \cos \psi_{ch} \tan^{-1} \varphi_{o1}\right) \quad (2.110)$$

The T-hand-S (T-hold-S) normal rake angle of the chisel edge is calculated as

$$\gamma_{ch} = -\arctan\left(\frac{\sin \psi_{ch}}{\cos \psi_{ch} \tan v_{ch} + \tan^{-1} \varphi_{o1}}\right) \quad (2.111)$$

The T-hand-S (T-hold-S) normal clearance angle of the chisel edge is calculated as

$$\alpha_{ch-n} = \arctan\left[\frac{\left(\tan^{-1} \varphi - \cos \psi_{ch}\right) \tan v_{ch}}{\sin \psi_{ch}}\right] \quad (2.112)$$

Analysis of the developed model reveals the following features of the discussed point grind:

- It allows to gain the self-centering advantage with less negative rake angle and thus greater allowable penetration rate as the gap between the chisel edge rake face and the wall of the hole being drilled becomes greater. The problem with the chip flow over the rake face of the drill can be solved permanently.
- The rake angle of the chisel edge is mainly determined by the T-hand-S working angle α_{f-F} and is less dependent on ψ_F.
- The clearance angle of the chisel edge depends entirely on the chisel edge angle as this secondary plane F does not affect the primary flank face.
- When the normal clearance angles in T-hand-S of the primary R and secondary F planes are chosen to be small, then interference on top of the secondary flank heel (point T in Figure 2.207) can occur. The fulfillment of the following condition prevents such interference:

$$0.5\left[d_{dr}\left(\tan \alpha_{o1} - \tan^{-1} \varphi\right) - f/4\right] \geq 0.5...2.0 \, \text{mm} \quad (2.113)$$

- The proper use of the discussed geometry of the flank face allows to reduce the negative T-hand-S (T-hold-S) rake angle to $-50°$ that significantly improves chip formation by the chisel edge. The price to pay is an increased length of the chisel edge. For example, when $\psi_f = 15°$ (as such $\psi_F = 16.3°$), $\alpha_{n-R} = 15°$, $\alpha_{n-F} = 25°$, and $2\varphi = 120°$, the chisel edge angle $\psi_{ch} = 43.8°$. For comparison, when the width of the primary flank is equal to or greater than $2a_o$ (case considered in Section 2.1.5.11.7), this angle is equal to 61.8°. The same is for the true four-facet point grind discussed in Section 2.1.5.11.8, and the application of gashes (web thinning) as shown in Figures 2.54 and 2.56 should be used to reduce the length of the chisel edge. This is of critical importance for wide-web drills (Figure 2.65), i.e., when the distance a_o (Figure 2.200) is great, as this distance directly affects the length of the chisel edge according to Eq. (2.105).

The implementation practice of such drills showed that when the T-hand-S (T-hold-S) rake angle is reduced to $-50°$, the drill terminal end becomes weak as the top of the heel point T locates too far from the cutting edge (Vinogradov 1985). It may cause drill vibration at the hole entrance and undermine the use of the additional stabilizing margins. As discussed in Section 2.1.5.7, the VPA2© design can be used to avoid any stability problem. The combination of the true four-facet point grind and the VPA2© design in straight and twist drills assures their entrance stability and to improve the quality, i.e., meeting tight tolerances on diametric accuracy, hole shape and straightness, position, and so on.

In the current practice, however, the flank face consisting of three flank planes is used to ease the problem with the location of point T and significantly improve the rake face geometry of the chisel edge. A drill with such a flank face is shown in Figure 2.208. When the tertiary flank plane is applied, the axial clearance angle α_{f-F} and the location angle ψ_F of this plane should not be great as the secondary flank is meant to gain some space so that the location of point T should be much closer to the cutting edge. Figure 2.208 qualitatively compares the locations of this point when two (point T_2) and three (point T_3) flank planes are applied. According to the author's experience, good results can be achieved when the first two flank planes are ground in the manner for the true four-facet point grind discussed in Section 2.1.5.11.8, and the tertiary plane is applied just to improve the rake angle of the chisel edge while keeping its length unchanged. Any violation of this condition can ruin the drill as discussed by the author earlier (Astakhov 2014).

2.1.5.11.10   Split-Point Grind

Although the split-point geometry seems to solve all the problems with the rake face geometry (a positive rake angle of this edge can be achieved) and the flow of the chip formed by the chisel edge (the chip formed by the chisel edge apparently has plenty of the room between the rake face of the chisel edge and the walls of the hole being drilled) as discussed in Section 2.1.3.7 and widely advertised in trade literature and many professional journals, it is not quite so in reality. Another significant problem remains with the most known designs of the split-point chisel edge. As this problem is not clear for many drill designers, its clarification is needed as this should be solved in any HP drill design.

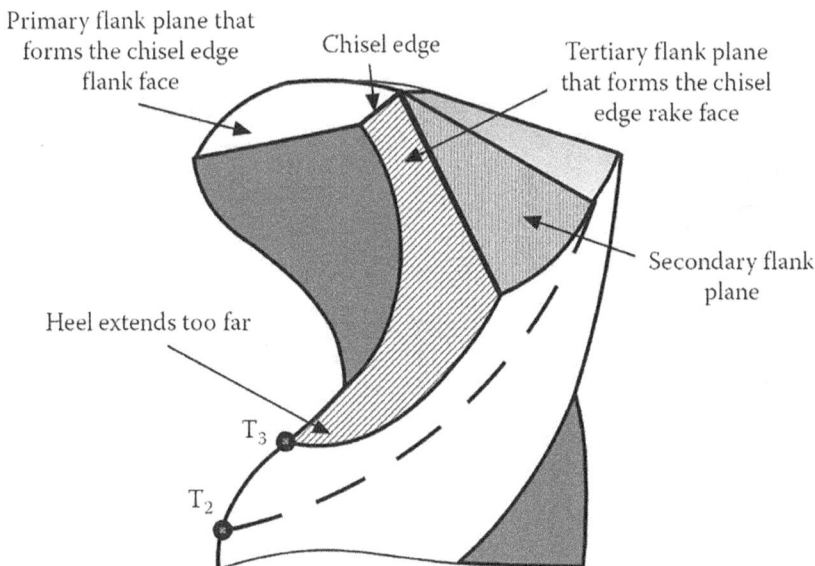

FIGURE 2.208   The flank face consisting of three flank planes.

The essence of the problem is that the rake face of the chisel edge is very small, which becomes of zero length in the vicinity of drill center as can be clearly seen in Figures 2.79, 2.81, and 2.82. As discussed earlier (Section 2.1.5.11.6), the normal chip formation, i.e., with the minimum cutting force and chip plastic deformation takes place when the tool-chip contact length is unrestricted by the vertical wall due to grinding shown in the above-referred figures. This length is calculated as $l_{c23} = h_{D23}\zeta^{1.3}$, where ζ is CCR as discussed in the Appendix and in Section 2.1.5.11.6. As such, the tool–chip contact length is: in drilling an aluminum alloy $l_c = 2.84$ mm and in drilling a high-manganese steel $l_c = 1.95$ mm (Astakhov 2014).

Figure 2.209a shows the appearance of a drill with a real split-point geometry and Figure 2.209b shows the model of the rake face of the chisel edge. As clearly seen in Figure 2.209a, the chisel edge is normally not split over its entire length. This is due to limitation on the strength of the region adjacent to the chisel edge (possibility of fracture) and accuracy of grinding including accuracy of the dressing of the grinding wheel for rake face of the chisel edge particularly for drills of less than 8 mm dia. As a result, a small portion of the chisel edge shown in Figure 2.209b by segment *2–4* remains, which creates a negative chamfer on the chisel edge as shown in Section A-A. Moreover, the rake face of the chisel edge is normally provided with a slightly negative rake angle (up to 15°) due to strength considerations.

Figure 2.209b also shows the length l_{sp} of the rake face of the split-point chisel edge and the transition radius r_{sp} between this face and the third flank plane. According to the traditional split-point geometry, length l_{sp} is at maximum at point 2 and then it rapidly decreases toward the drill center axis becoming virtually zero at this point.

The condition of the chip-free flow can be thought of as $l_{sp} > l_c$ so that the forming chip does not have an obstacle in its way to disturb its flow. Moreover, the transition radius r_{sp} should be rather generous exceeding at least the chip thickness h_C. Unfortunately, these two conditions are not met in the traditional split-point geometry that restricts its use in high-penetration rate drills. However, when the rake face of the chisel edge is designed and made according to the above consideration, i.e., provided for the rake face of the proper length, the built-up edge is uniform on the major cutting edge and on the chisel edge as shown in Figure 2.210 – this is the best test indicator of the properly ground chisel edge geometry.

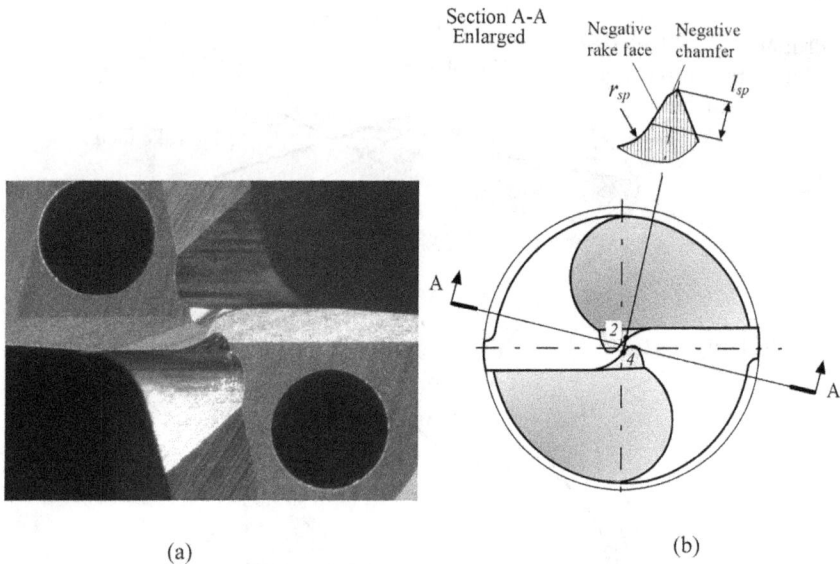

(a)   (b)

FIGURE 2.209   Real split-point geometry (a) and its model revealing the problem (b).

FIGURE 2.210 Showing the test result (on a soft aluminum alloy) of the properly ground rake face of the chiseled edge.

2.1.5.11.11 Altering the Location of the Chisel Edge

Although the split-point grind seems to solve the common problems with the chisel edge, two problems restricting the allowable drill penetration rate remain:

1. Strength of the region adjacent to the chisel edge. For common work materials, this strength is normally sufficient whereas in drilling difficult-to-machine materials, when drilling forces are high, this strength may not be sufficient. As a result, bulk fracture of the region of the chisel edge can occur that often observed in practice.
2. The inherent problem of the chisel edge passing through the axis of rotation is interference of this edge with the bottom of the hole being drilled as shown in Figure 2.199b. Moreover, starting a hole with zero cutting speed at the apex of the chisel angle is not good condition.

The first problem cannot be resolved even in principle if the split-point grind is used. As a result, the true four-facet point grind (instead of the split-point grind) is often used in drilling of difficult-to-machine materials with all the known limitation of this point grind discussed in Section 2.1.5.11.8. This limits the allowable penetration rate (and thus drilling productivity), tool life, and quality of drilled holes.

Considering the second problem, one may argue that the discovered interference is of *theoretical significance* as it is not commonly observed in practice. The simple answer to this *undisputable argument* is that it is not observed due to a significant drill runout observed in practice. For years, the discussed problem was not as profound as drills, drill holders, machine spindles, etc. resulted in a drill significant runout that *automatically* solved the discussed problems. Improvements in the accuracy of the components of the drilling system achieved for the last 10–15 years significantly reduced the total runout of drills so that the discussed problem became noticeable. As discussed by the author (Chapter 1 in Astakhov (2024)), drill runout significantly reduces drill life and causes oversized holes so it has to be limited in drilling. Therefore, the second problem should be dealt with in the design and applications of high-accuracy drills having great penetration rates.

The foregoing considerations suggest that the following requirements should be kept in mind/met in the design and development of new drill designs/geometries:

1. No one point of the chisel edge should have zero cutting speed.
2. Self-centering ability without compromising the strength of the drill tip, which is important in machining difficult-to-machine heat-treated work materials.

In the author's opinion, any new drill design should be checked against these requirements to prove that this design is better than the existing one.

The first known by author attempt to solve the discussed problems was presented in US Patent No. 3,028,773 "Drill" (1962, Borneman G.H., assignor to General Electric Co.). The author pointed out that in drilling, there is a tendency for the cutting drill to "walk." By "walking" means the axis of the tool wanders with respect to the axis of the hole being drilled. This causes the shape of the hole to be either out of round, or to be of a size larger than desired, or both. The major reason for this walking is the drilling conditions at the region of the chisel edge, particularly at the beginning of drilling.

The object of the invention is forming a projection in the hole being drilled, which acts as a journal for the drill as it passes through the workpiece. This journal and bearing arrangement support the drill at a point adjacent to the cutting surfaces of the drill thereby assuring that the hole and the drill are concentric. It is realized by providing an improved drill which forms a protuberance in the hole cut in the workpiece having an annular surface concentric with the hole upon which the drill is journalled as it passes through the workpiece.

One basic realization of the stated objective (out of four in the patent) is shown in Figure 2.211a. Referring to this figure, there is shown a twist drill 1 having a general cylindrical shape with a tip portion. This tip portion includes the flank faces 2 and 3. The twist drill 1 includes means defining flutes 4 and 5 extending helically along the cylindrical surface of the drill. The cutting major edges of the drill are formed as the intersection lines of flutes 4 and 5 with flank faces 2 and 3. In Figure 2.211a, the flute 4 is shown intersecting flank face 3 to form the cutting edge 6. This portion of the twist drill is typical for most conventional drills.

In operation, rotation of the twist drill in a clockwise direction will cause the cutting edge 6 to engage the workpiece and remove material in the form of chips. These chips are directed away from the cutting edge by being urged through the flutes associated with the cutting edge and away from the workpiece.

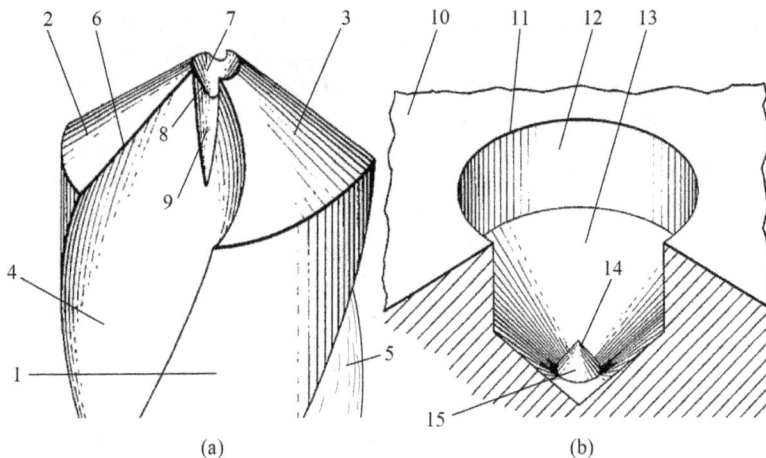

FIGURE 2.211 Drill design according to US Patent No. 3,028,773 (1962): (a) the basic design and (b) the shape of the bottom of the hole being drilled.

In order to provide guide means located substantially adjacent to the cutting area of the tool, a substantially conical opening is formed in the center of the tool concentric with the axis of the tool. This substantially conical surface 7 performs a bearing function against a protuberance formed upon initial entry of the drill into the workpiece. The protuberance achieves a maximum height and, depending on its shape, it is necessary to cut away a portion of the complementary bearing surface of the protuberance so that the tool may advance. To perform this function, the cutting edge 8 is provided with a suitable secondary flute 9. It can be seen in Figure 2.211a that as the tool advances the protuberance in the workpiece has its profile shaved by the cutting edge 8. The material cut from the protuberance is discharged into the secondary flute 9, which passes the material into the flute 4.

Figure 2.211b shows a section of hole being drilled in workpiece 12 drilled by the twist drill 1 shown in Figure 2.211a. The hole 11 in the work piece 10 has a cylindrical portion 12 and conical portion located below the cylindrical portion. Due to the particularities of the drill design, protuberance 14 having bearing surface 15 is formed at the very bottom of the hole to support the drill and thus prevent it wondering.

It is cut by cutting edge 8, which can be considered as the chisel edge shifted from its traditional location. Because conical surface 7 made in the drill performs a bearing function against mating surface 15 of protuberance 14, the clearance angle of cutting edge 8 in T-hand-S is zero. However, it is positive as considered in T-use-S, i.e., when the direction of the resultant cutting speed is considered.

Although the idea of self-supporting the discussed drill is valuable and can be successfully implemented, one obvious drawback of the proposed design and geometry is that it is rather difficult to grind the proposed drill as a specially designed and dressed grinding wheel together with grinding fixture and metrological equipment are needed. Resharpening of this drill presents an even greater challenge.

One of the most promising directions in the development of new drill design/geometry meeting the above-listed requirements to the design/geometry of the chisel edge is the drill point design with non-central chisel edge (hereafter, NCCE drill). As first suggested by Vinogradov (1985), the shift of the chisel edge from the drill rotational axis even over some small distance changes the kinematics and dynamics of the cutting by the chisel edge. Unfortunately, the above-cited work by Vinogradov contains a number of unclear explanations and graphic and designation errors so that it was never understood and thus followed by other researchers including his co-workers. In the author's opinion, however, this suggestion has a great potential so it is worth to revisit it taking a closer look at the essence of the proposal. Obviously, modern designations, principles, and notions should be applied in such an endeavor.

The shift of the chisel edge from the drill axis can be accomplished using various drill point designs. The first one considered in this book is one meant for drilling of work materials of high strength. It is designated as the VPA3-strength© design. Due to the unusual point geometry, the basic components of this geometry are considered in a step-by-step fashion. The development of such a point grind starts with the consideration of the standard drill point grind as shown in Figure 2.212a. As can be seen, it has two major, right $1–2$ and left $4–5$, cutting edges made symmetrical with respect to the drill axis, i.e., their half-point angles, primary clearance angles, and lengths are the same. The shown drill has a central chisel edge $2–4$, which, as discussed above, consists actually of two chisel edges $2–3$ and $3–4$. Point 3 lies of the drill axis so that the lengths of these edges are equal. It is called as central because it passes through the drill axis. As discussed above, the speed at point 3 is zero.

According to the VPA3-strength© design, the shift of the chisel edge from its central location is accomplished by grinding of a portion of the major cutting edge $1–2$ (its length is l_{12}) with the increased half-point angle, which is greater than the original point angle by $\Delta\varphi$ so that the point angle of this ground portion is $\varphi_1=\varphi+\Delta\varphi$. As shown in Figure 2.212b (Section E-E), this portion starts at distance l_{16} from the drill corner (point 1). As a result, the chisel edge assumes a new position $7–9$. Its central point 8 does not belong to the drill axis. Rather it is shifted from this axis by a distance $3–8$ as shown in Figure 2.212b. The chisel edge angle of so formed chisel edge $7–9$, ψ_{ch79} is smaller than the chisel edge angle ψ_{ch24} of the original chisel edge $2–4$. Moreover, the chisel edge

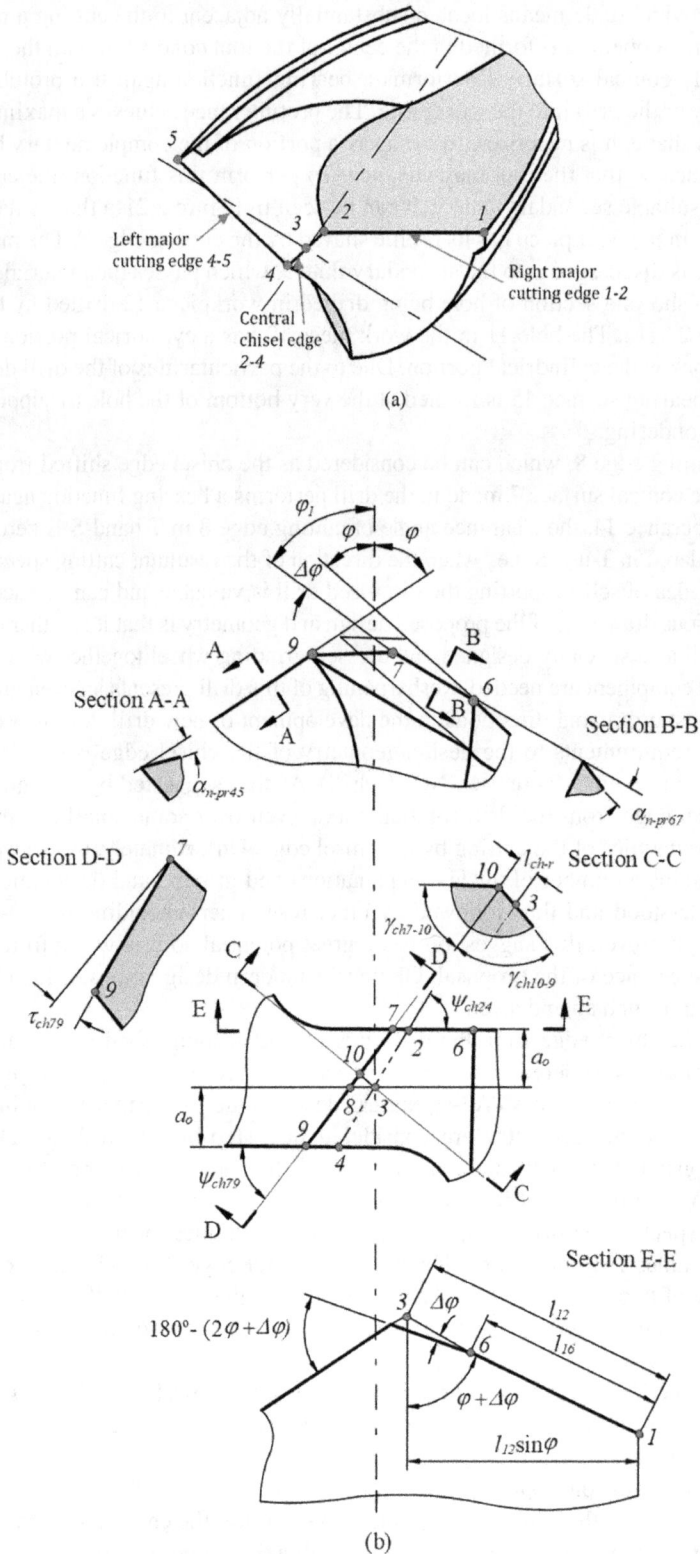

FIGURE 2.212 Geometry parameters of NCCE drill – the VPA3-strength© design: (a) original drill and (b) NCCE drill.

7–9 is inclined at angle τ_{ch79} to the plane perpendicular to the drill axis (considered above and in the Appendix as the back plane) – this angle is referred to as the point angle of this chisel edge. This is one of the major advantages of NCCE drills.

When the primary clearance angles of edges 1–2 and 6–7 are kept the same, i.e., when $\alpha_{n-pr12}=\alpha_{n-pr67}=\alpha_{n-pr}$, the chisel edge angle ψ_{ch79} is calculated as

$$\tan \psi_{ch79} = \frac{\sin(\varphi_1+\varphi)}{\tan \alpha_{n-pr}(\sin \varphi_1+\sin \varphi)} \tag{2.114}$$

When the primary clearance angles of edges 1–2 and 6–7 are not the same, i.e., when $\alpha_{n-pr12}\neq \alpha_{n-pr67}$, the chisel edge angle ψ_{ch79} is calculated as

$$\tan \psi_{ch79} = \frac{\sin(\varphi_1+\varphi)}{\left(\tan \alpha_{n-pr79}\sin \varphi_1+\tan \alpha_{n-pr12}\sin \varphi\right)} \tag{2.115}$$

The point angle of the chisel edge 7–9 is calculated as

$$\tan \tau_{ch79} = \frac{\cos \varphi_1 \cos \varphi - \tan \alpha_{n-pr}\sin \psi_{ch79}}{\sin \varphi} \tag{2.116}$$

The shift δ=distance $(3$–$8)$ of the chisel edge 7–9 relative to the drill axis in the back plane is the most important parameter defining the proper working condition of NCCE drills. It is calculated as

$$\delta = a_o \cot \psi_{ch79} - \left(2a_o \cot \psi_{ch24}\left(\sin(2\varphi+\Delta\varphi)+\sin \Delta\varphi\right) - \frac{d_{dr}-2l_{16}\sin \varphi}{2\sin(2\varphi+\Delta\varphi)} \right) \tag{2.117}$$

This shift depends on three parameters, namely $\Delta\varphi$, l_{16}, and α_{n-pr67}. Note that δ and l_{16} should be a part of the tool drawing subject to close inspection in drill manufacturing.

The distance l_{ch-r} is the distance between points 3 and 10 in the radial direction perpendicular to the chisel edge 7–9, it is the shortest distance between the drill axis and the chisel edge. It is calculated as

$$l_{ch-r} = \delta \sin \psi_{ch79} \tag{2.118}$$

The model for visualizing the clearance and rake angle of the chisel edge is shown in Figure 2.213. The T-hand-S clearance angle of the part 7–10 of the chisel edge 7–9 is calculated as

$$\tan \alpha_{ch7-10} = \left(\sin \psi_{ch79}\cos \varphi_1+\cos \psi_{ch79}\tan \alpha_{n-pr67}\right)\sin \varphi_1 \tag{2.119}$$

The T-hand-S clearance angle of the part 10–9 of the chisel edge 7–9 is calculated as

$$\tan \alpha_{ch10-9} = \left(\sin \psi_{ch79}\cos \varphi+\cos \psi_{ch79}\tan \alpha_{n-pr12}\right)\sin \varphi \tag{2.120}$$

The T-hand-S rake angle of the part 7–10 of the chisel edge 7–9 is calculated as

$$\gamma_{ch7-10} = -(90-\alpha_{ch10-9}) \tag{2.121}$$

The T-hand-S rake angle of the part 10–9 of the chisel edge 7–9 is calculated as

$$\gamma_{ch10-9} = -(90-\alpha_{ch7-10}) \tag{2.122}$$

TABLE 2.6

T-hand-S Geometry Parameters of the Chisel Edge 7–9

$\psi_{ch79}(°)$	$\tau_{ch79}(°)$	$\alpha_{ch7-10}(°)$	$\alpha_{ch10-9}(°)$	$\gamma_{ch7-10}(°)$	$\gamma_{ch7-10}(°)$	δ(mm)	t_{ch-r} (mm)
59.83	1.33	28.83	33.17	−56.83	−61.17	0.58	0.5

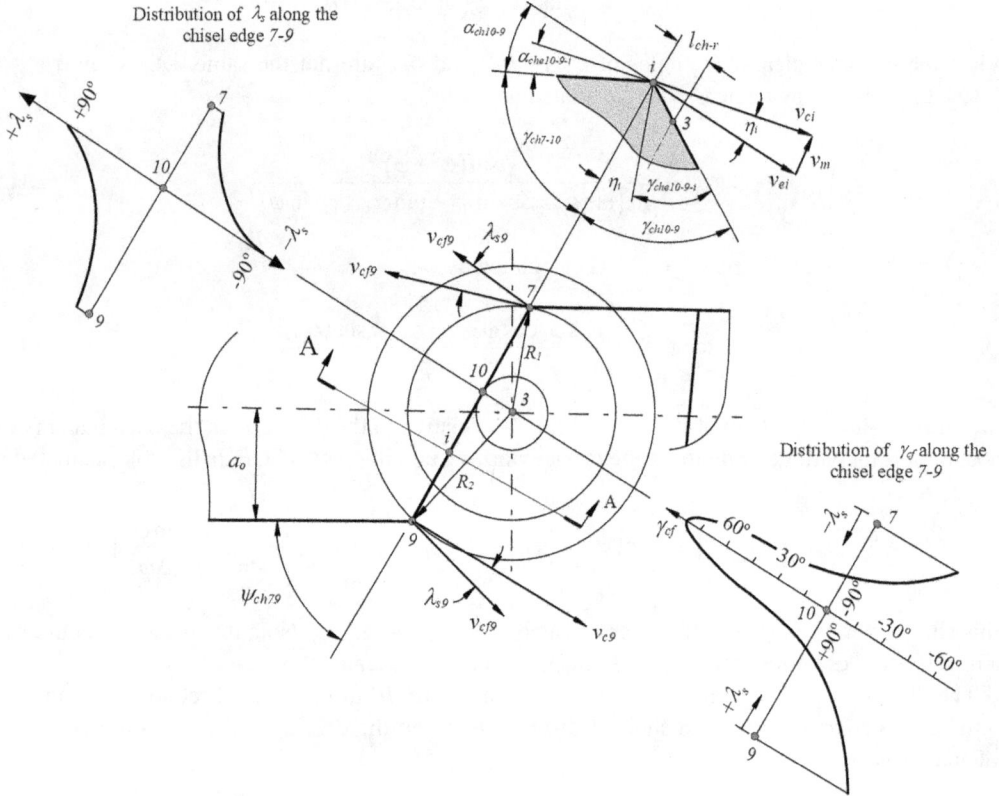

FIGURE 2.213   Geometry of the chisel edge of the NCCE drill.

Table 2.6 shows the T-hand-S clearance and rake angles of the of parts *7–10* and *10–9* of the chisel edge *7–9* calculated for a drill of 20 mm dia., $2\varphi = 120°$, $\Delta\varphi = 5°$, $\alpha_{n-pr12} = \alpha_{n-pr45} = 15°$, $l_{1-6} = 5$ mm, $a_o = 1.5$ mm.

When the drill rotates, end points *7* and *9* of the chisel edge trace circles have R_1 and R_2 radii, respectively. As can be seen in Figure 2.213, $R_2 > R_1$. These radii are calculated as

$$R_1 = \frac{\sqrt{\left(a_o - \delta \sin \psi_{ch79} \cos \psi_{ch79}\right)^2 + \delta^2 \sin \psi_{ch79}}}{\sin \psi_{ch79}} \tag{2.123}$$

$$R_2 = \frac{\sqrt{\left(a_o + \delta \sin \psi_{ch79} \cos \psi_{ch79}\right)^2 + \delta^2 \sin \psi_{ch79}}}{\sin \psi_{ch79}} \tag{2.124}$$

For example, for a drill of 20 mm dia., $a_o = 1.5$ mm, $2\varphi = 120°$, $\Delta\varphi = 5°$, $l_{16} = 5.8$ mm, these radii are $R_1 = 1.54$ mm and $R_2 = 2.06$ mm.

It follows from this consideration that part *10–9* of the chisel edge *7–9* does much more cutting work than part *7–10*. As a result, the load due to cutting forces, particular radial force, is much greater on this part. As a result, a non-balanced radial force due to the action of the chisel edge occurs. Moreover, as the major cutting edge is not anymore symmetrical with respect to the drill axis, the radial forces acting on these edges are not the same that results in another unbalanced radial force. Therefore, the force balance considered in Section 2.1.5.6 and as a prerequisite feature for drill normal performance is obviously violated in the discussed NCCE drill. As a result, the unbalanced radial forces on the chisel and major cutting edges should deflect the drill in the radial direction causing a number of problems from hole oversize and up to drill breakage.

However, it is not the case in the discussed NCCE drill. The reason for that is in unique shape of the bottom of the hole being drilled by this drill. Figure 2.214 shows a section of the bottom of the hole being drilled by an axial plane and enlarged section of the bottom of the hole being drilled cut by the chisel edge to reveal the shape of this bottom formed by the chisel edge *7–9*. At drill entrance, the chisel edge *7–9* first touches the face of the workpiece by point *7* as this point is the farthest point of the drill in the axial direction. This point has the radius R_1. The entrance of the chisel edge ends when point *9* touches the face of the workpiece. This point has radius R_2. As discussed in Section 2.1.5.11.3 in great detail, as the cutting edge *7–9* does not intersect the *z*-axis, which is the axis of rotation, it forms a hyperboloid of revolution shown in Figure 2.181. Because the chisel edge *7–9* forms a hyperboloid, the line formed by this edge in the axial section shown in Figure 2.214 is a hyperbola having equation

$$\frac{x^2}{l_{16}^2} - \frac{z^2}{\left(l_{ch-r}/\cot \tau_{79}\right)^2} = 1 \tag{2.125}$$

Due to the non-central location of the chisel edge, a cone surface forms in the center of the hole being drilled as shown in Figure 2.214. The diameter of the base of the cone is

$$d_c = 2l_{ch-r} = 2\delta \sin \psi_{ch24} \tag{2.126}$$

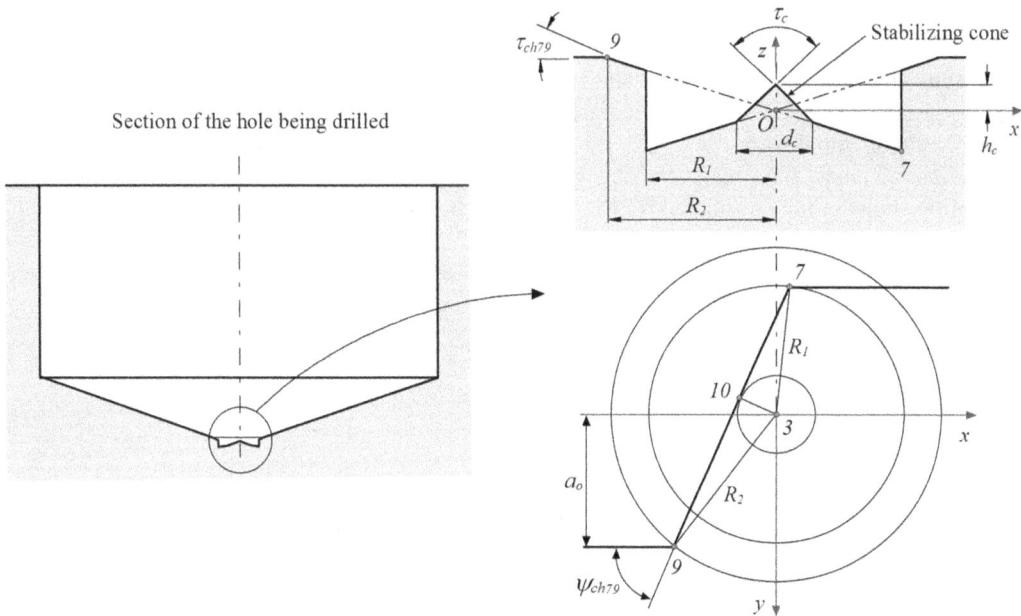

FIGURE 2.214 Showing a section of the bottom of the hole being drilled by an axial plane to reveal the shape of this bottom formed by the chisel edge *7–9*.

the cone angle is

$$\tau_c = 180° - 2\alpha_{n-pr67} \tag{2.127}$$

and its height is

$$h_c = l_{ch-r} \tan \alpha_{n-pr67} \tag{2.128}$$

The shift of the chisel edge from the drill axis results in that this edge cuts the so-called stabilizing cone, which prevents any deflection of the drill in the radial direction due to the above-discussed unbalanced radial force. As a result, the drill enters the hole smoothly with no wandering and/or radial vibrations as the farthest point 7 of the chisel edge has non-zero cutting velocity that assures self-centering of the drill and close position tolerance of the drilled hole. This principle is the same as used in gundrilling (discussed in Chapter 3) for years but has never been utilized in usual drilling. This constitutes one of the important advantages of NCCE drills.

To comprehend the second most important advantage of NCCE drill, one needs to recall the above-described working conditions of the chisel edge. As discussed in the Appendix and used in Section 2.1.5.11.5, the chip deformation is determined by CCR, which defines both the chip thickness, h_C, and the contact length, l_c, on the rake face. It was shown in Section 2.1.5.11.5 that both h_C and l_c present significant problems with the flow of the chip formed by the chisel edge in the known drill point grinds. Moreover, the interference of the chisel edge flank face and the bottom of the hole being drilled takes place as the T-use-S clearance angle becomes very small and then negative in the vicinity of the axis of rotation as shown in Figure 2.199b as the zone of interference. As follows from Eqs. (2.45) and (2.98), the higher the feed per revolution, the greater the problem. This is because such interference may potentially lead to a significant increase in the axial force in drilling and thus limits the allowable penetration/feed rate. This problem is inherent in the known drill point grinds used today and cannot be resolved even in principle.

Fortunately, this is not nearly the case with NCCE drills due to unique kinematics of cutting by the shifted chisel edge. The shift of the chisel edge from the drill axis changes the cutting kinematics from orthogonal, the case in various drills with central chisel edges, into oblique cutting that significantly improves the cutting conditions on this edge. Such a change is due to the presence of the cutting edge inclination angle λ_s which, as discussed above, is zero in any drill with a central chisel edge. As discussed in the Appendix (Figure A.4) and in Section 2.1.5.9.1, this angle changes the direction of chip flow characterized by the chip flow angle according to Eq. (2.53). The greater the inclination angle, the greater the change. Therefore, it is important to assess the cutting edge inclination angle for the considered NCCE drill in order to understand the significance of the influence of this angle.

Following the consideration made in Section 2.1.5.9 and Figure 2.151, one can define the cutting edge inclination angle for the considered chisel edge (see Figure 2.213) as

For the points of the part 7–10 of the chisel edge

$$+\sin \lambda_{si} = \frac{\delta \sin \psi_{ch2-4}}{R_{i(7-10)}} \tag{2.129}$$

For point of the part 10–9 of the chisel angle

$$-\sin \lambda_{si} = \frac{\delta \sin \psi_{ch2-4}}{R_{i(10-9)}} \tag{2.130}$$

It is shown in Figure 2.213 that angle λ_s is relatively small at the end points 7 and 9 of the chisel edge. For example, for the above-considered conditions for a drill of 20 mm dia., $a_o = 1.5$ mm, $2\varphi = 120°$,

$\Delta\varphi=5°$, $l_{16}=5.8\,\text{mm}$, these radii are $R_1=1.54\,\text{mm}$ and $R_2=2.06\,\text{mm}$, $\lambda_{s7}\approx-15°$, and $\lambda_{s9}\approx+11°$. Then, the inclination angle on both parts of the chisel edge increases rapidly reaching $\pm90°$.

It is known that the cutting edge inclination angle significantly changes the whole cutting process reducing the chip deformation, cutting force, direction of chip flow, tool wear, etc. (Bobrov 1962, Zorev 1966, Astakhov 2010). Chip deformation determined by CCR defines both the chip thickness, h_C, and the contact length, l_c, on the rake face, that significantly reduces when λ_s is high. At a certain high λ_s, the built-up edge does not form on the rake face of any cutting edge including the chisel which assures the free flow over this rake face. Moreover, λ_s changes the direction of chip flow as discussed in the Appendix (Figure A.4). When $\lambda_s=90°$, the chip flows parallel to the cutting edge with practically no deformation.

The variation of the chip flow angle, γ_{cf} over the chisel edge 7–9 according to Eq. (2.53) is shown in Figure 2.213. As can be seen, this angle is positive at the central part of the chisel edge (over approx. 1/3 of the length of this edge). At periphery points of the chisel edge 9 and 7, γ_{cf} is equal to $-51°$ and $-46.5°$, correspondingly. As such, the T-hand-S rake angle is the same over the chisel edge and equal to 56.3°. A positive fact for the considered NCCE assuring its stable performance is that the cutting feed does not affect the chip flow angle and working clearance angle. Moreover, a relatively thin chip and small contact length over the rake face due to high λ_s combined with preferable chip flow direction and a great area between the rake face of the chisel edge and the wall of the hole being drilled assure reliable removal of the chip formed by the chisel edge.

For drilling ductile work materials that do not have high strength, another design of NCCE drill is recommended. This design is shown in Figure 2.215 and it is designated as the VPA3-tough© point design. As shown, the shift of the chisel edge is accomplished by grinding step having depth k_{23} on the right major cutting edge at a certain distance 1–2 from the drill periphery corner 1. As such, the half-point angle φ of the cutting edge 3–4 is kept the same as that for the cutting edge 1–2. If the normal clearance angle of the primary flank face of the cutting edges 3–4 and 1–2 is the same, i.e., $\alpha_{n-pr34}=\alpha_{n-pr12}$ then the shift $\delta=$distance (7–9) of the chisel edge 4–5 relative to the drill axis in the back plane depends only on k_{23} where other T-hand-S parameters of the chisel edge remain the same as for the drill with the central chisel edge. When $\alpha_{n-pr34}>\alpha_{n-pr12}$, the chisel edge is no more located in the back plane. Rather it has the point angle τ_{4-5} and the chisel edge angle ψ_{ch45} becomes smaller than that of the drill with the central chisel edge. The geometrical parameters of such a drill are calculated as follows:

$$\delta = \frac{k_{23}+\left(a_o-f_{pr34}\right)\left(\tan\alpha_{n-pr34}-\tan\alpha_{n-pr12}\right)}{2\cos\varphi} \tag{2.131}$$

$$\tan\tau_{34} = \frac{\cos\psi_{sym}\sin\left(\alpha_{n-pr34}-\alpha_{n-pr12}\right)}{\tan\varphi\sin\left(\alpha_{n-pr34}+\alpha_{n-pr12}\right)} \tag{2.132}$$

$$\psi_{ch45} = \frac{2\cos\varphi\cos\alpha_{n-pr12}\cos\alpha_{n-pr34}}{\sin\left(\alpha_{n-pr34}+\alpha_{n-pr12}\right)} \tag{2.133}$$

$$\tan\gamma_{cl45} = \frac{\sin\varphi\cos\alpha_{n-pr34}}{\sqrt{\sin^2\alpha_{n-pr34}+\cos^2\alpha_{n-pr12}\cos^2\varphi}} \tag{2.134}$$

To simplify considerations, this section presents only two basic VPA3© point designs. In general, the VPA3© point design may include more complicated designs with the modification of the chisel edge geometry, particularly the value and distribution of the rake angle, which is developed to be tailored for a particular application of HP drill including the properties of the work material and required accuracy of drilled hole and their position tolerances.

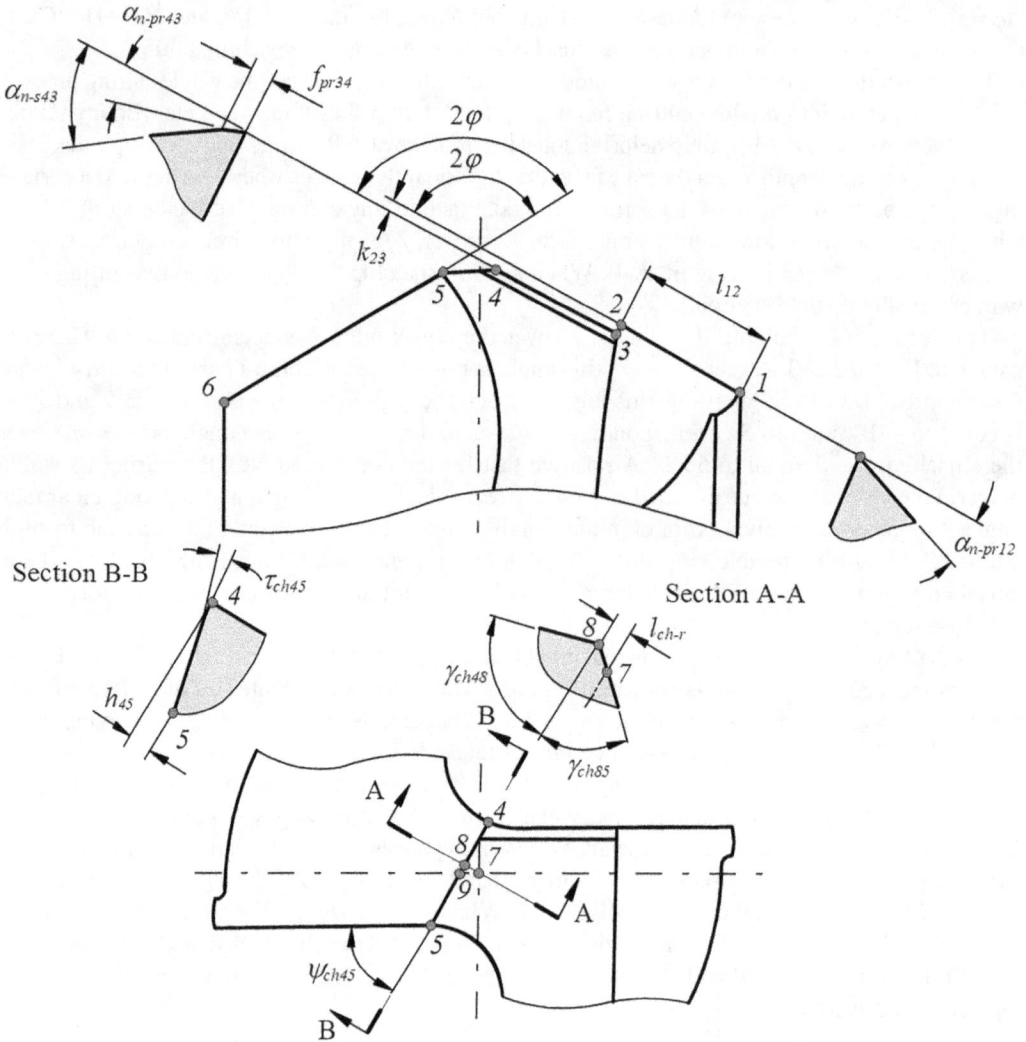

FIGURE 2.215 Geometry of the VPA3-tough© point design.

2.1.5.11.12 Eliminating the Chisel Edge

The problems created by the chisel edge were always at the center of attention in the drill design (Thornley et al. 1987a). The most radical solution to these problems is the total elimination of this edge from the drill point design. As early as 1911, Mather patented a drill with no-chisel edge (US Patent No. 989,379 [1911]). In the proposed drill shown in Figure 2.216, a common twist drill 1 is provided at the apex of the web with the slot 2. This slot having a width equal to the web diameter extends upward into the drill body and terminates at the upper end in the inclined surface 3. When drill 1 penetrated into workpiece 4 drilling hole 5, core 6 forms. This core either breaks due to bending when its front end comes into contact with inclined surface 3 or, when the work material is ductile, bends into the chip flute 7 and thus is removed at the end of drilling.

The subsequent developments, for example, described in US Patent Nos. 4,143,723 (1979) and 4,342,368 (1982) and well summarized in US Patent No. 4,373,839 (1983) did not offer new ideas. Rather, various design applications of the ideas described earlier particularly to various drill configurations were attempted.

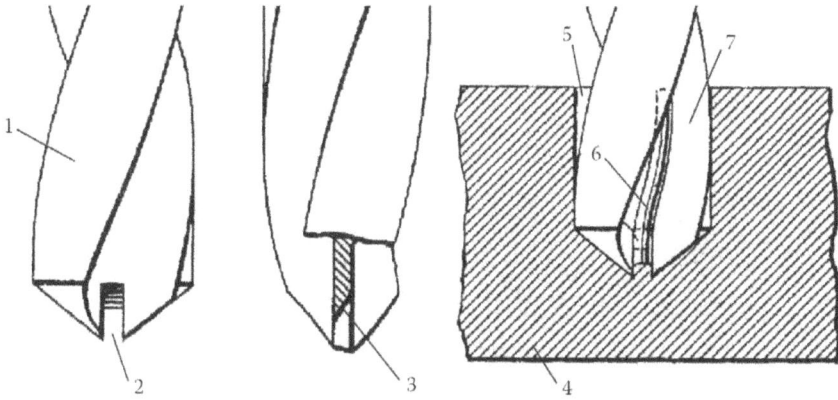

FIGURE 2.216 Drill design according to US Patent No. 989,379 (1911).

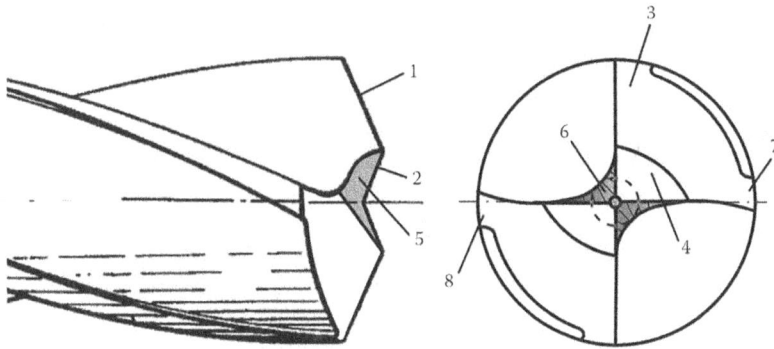

FIGURE 2.217 Drill geometry according to US Patent No. 2,334,089 (1939).

A special drill point design shown in Figure 2.217 was developed by Hallden (US Patent No. 2,334,089 [1939]) that, in the author's opinion, opened a new line of truly self-stabilizing drills. These can be referred to as double-apex drills (widely known as W-point drills). Each major cutting edge of this drill has two portions, namely, the outer 1 and the inner 2 cutting edges and corresponding flank surfaces 3 and 4. Flank surface 4 is shaped as an inverted conical surface as shown in Figure 2.217. Rake face 5 of the region of inner cutting edge 2 adjacent to the drill center is formed by providing (grinding) gash 6.

There are a number of advantages of the proposed design not realized by the inventor and apparently by the subsequent specialists who did not notice such a great leap ahead in the design of drills:

- Because the cutting edges located on the axis of the drill, there are no variations of the rake and clearance angles along these edges over the drill radius as in the traditional drill design.
- While drilling, inner cutting edge 2 forms a conical surface on the drill bottom that definitely stabilizes the drill preventing its wandering and thus improving the drilling stability and quality of the hole being drilled. This stabilizing cone 1 formed at the bottom of the hole being drilled is shown in Figure 2.218. In the presence of cone, any unbalanced radial force does not act on the wall 2 of the hole being drilled through the side margins, thus preventing drill radial vibration and deterioration of the drilled hole. Rather, it acts on the cone forcing one of the inner cutting edges to cut deeper to compensate for this unbalanced

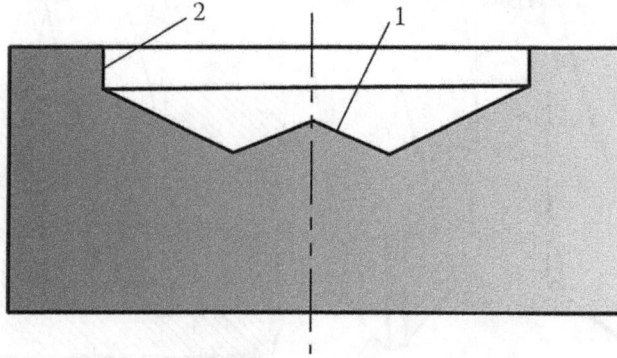

FIGURE 2.218   The bottom of the hole being drilled by the drill shown in Figure 2.217.

force stabilizing in this manner drill performance. Additional supporting lands 7 and 8 (Figure 2.217) shown in the patent would interfere with drill stability as any redundant supports in mechanical systems.

Regardless of the above-listed advantages, the drills made using the design shown in Figure 2.217 often show non-stable performance that limits their wide use in industry. This is because of a few severe drawbacks of this design:

- The inner ends of the major cutting edges must meet exactly at the axis of rotation. Otherwise, a pin forms at the bottom of the hole being drilled or interference takes place as explained in Chapter 3. It is not feasible to meet such a condition due to a variety of reasons including manufacturing quality, drill inspection, and accuracy of tool holdings.
- To achieve "no-chisel edge" conditions, deep gashes should be ground on the rake faces of the major cutting edge that weakens the drill point causing its chipping in drilling of high-strength work materials.

One of the modern designs of a double-apex drill is shown in Figure 2.219. A slot having length l_{sl} and width b_{sl} at the intersection of inner cutting edges is to compensate the manufacturing and installation inaccuracies. As shown, the inner cutting edges are not straight. Their curvature depends on the lead of the helical flute. A number of tests of such a drill on steels and on the automotive aluminum alloy revealed that the roundness of hole drilled is almost two times better than that with the best standard drills. The design of this drill requires further optimization for application-specific conditions to gain its maximum potential.

2.1.5.12   Drill Design Optimization Based on the Load Over the Drill Cutting Edge

The degree of optimality of the geometry of any cutting tool can be assessed through optimality of the geometry of the cutting edge, on the one hand, and by uniformity of the load over this edge. The latter is particularly true for cutting tools having long cutting edges over which the cutting conditions change substantially as in drills. For given work material and tool design, the load at point i of the cutting edge is characterized by the cutting speed, v_i, and uncut (undeformed) chip thickness, h_{D-i}. This load determines the tool wear rate (Astakhov 2004) and thus tool life.

Often for tools with long cutting edges, the cutting speed and uncut chip thickness may vary over the cutting edge. Therefore, accounting for these variations, one should establish a criterion of the load on a given part of the cutting edge to be able to determine the most loaded portions of this edge. As pointed out by Rodin (1971), one of the prime reserves in cutting tool improvements is to make the load evenly distributed. As a criterion of optimality, the wear rate (in the sense as it was

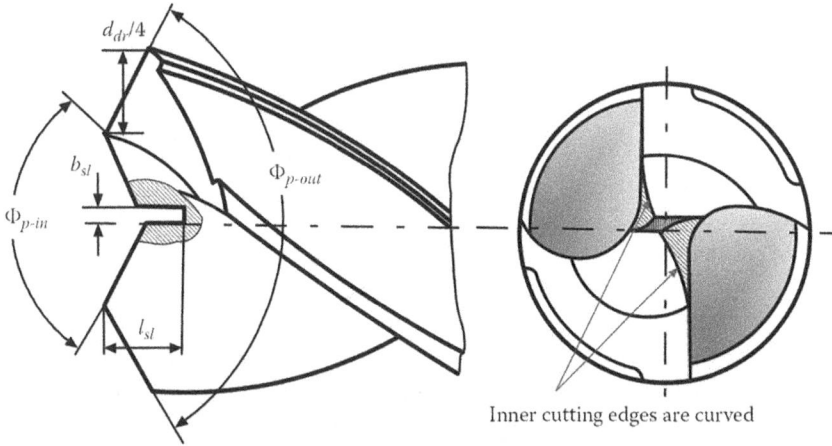

FIGURE 2.219 One of the modern designs of double-apex drill.

introduced by the author Astakhov 2004) or tool life (if the tool life of the considered portion of the cutting edge is greater compared to other parts) can be used.

In the simplest case, the following empirical formula is used to correlate the cutting speed with tool life and cutting parameters:

$$v = \frac{C_T C_w C_{cc}}{T^{m_v} h_D^{m_t} b_D^{m_b}} \tag{2.135}$$

where C_T, C_w, and C_{cc} are constant depending upon the tool material, work material, and cutting conditions, respectively, T is tool life, h_D and b_D are the uncut chip thickness and its width, respectively, m_v, m_t, and m_b are powers to be determined experimentally using cutting tests (e.g., as discussed by Astakhov (Chapter 5 in Astakhov (2006)).

To analyze the distribution of the load over a given cutting edge, a point i on this edge is selected to be the base point. For this point, the cutting speed is v_i, uncut chip thickness is h_{D-i}, and tool life is T_i. For any other point of the cutting edge, for example, point p, tool life T_p can be calculated using Eq. (2.135). The load coefficient at this point can be calculated as $k_{N-p} = T_i/T_p$. If $k_{N-p} > 1$, then this point p is loaded more than the base point i.

The load coefficient k_N can also be determined as the ratio of the uncut (undeformed) chip thickness h_{D-p} at the considered points to that h_{D-i} corresponding to tool life T_i determined using Eq. (2.135), that is,

$$k_{N-p} = \frac{h_{D-p}}{h_{D-i}} \tag{2.136}$$

If k_N's for various points of the cutting edge are known, then the uniformity of the load over this edge can be assessed.

2.1.5.12.1 Uncut Chip Thickness in Drilling

Although the uncut (undeformed) chip thickness can be easily determined using the vector analysis (Astakhov 2010), it was found instructive to visualize this important characteristic of the cutting process graphically to help a tool designer/optimizer to develop its material sense and only then to derive an equation for its calculation. Figure 2.220 shows a model for visualization of the uncut chip thickness. In this model, a drill having two lips is shown. A right-hand $x_o y_o z_o$ coordinate system is set as shown in the figure, i.e., its origin O is on the drill axis, the z_o-axis is along the drill axis, the

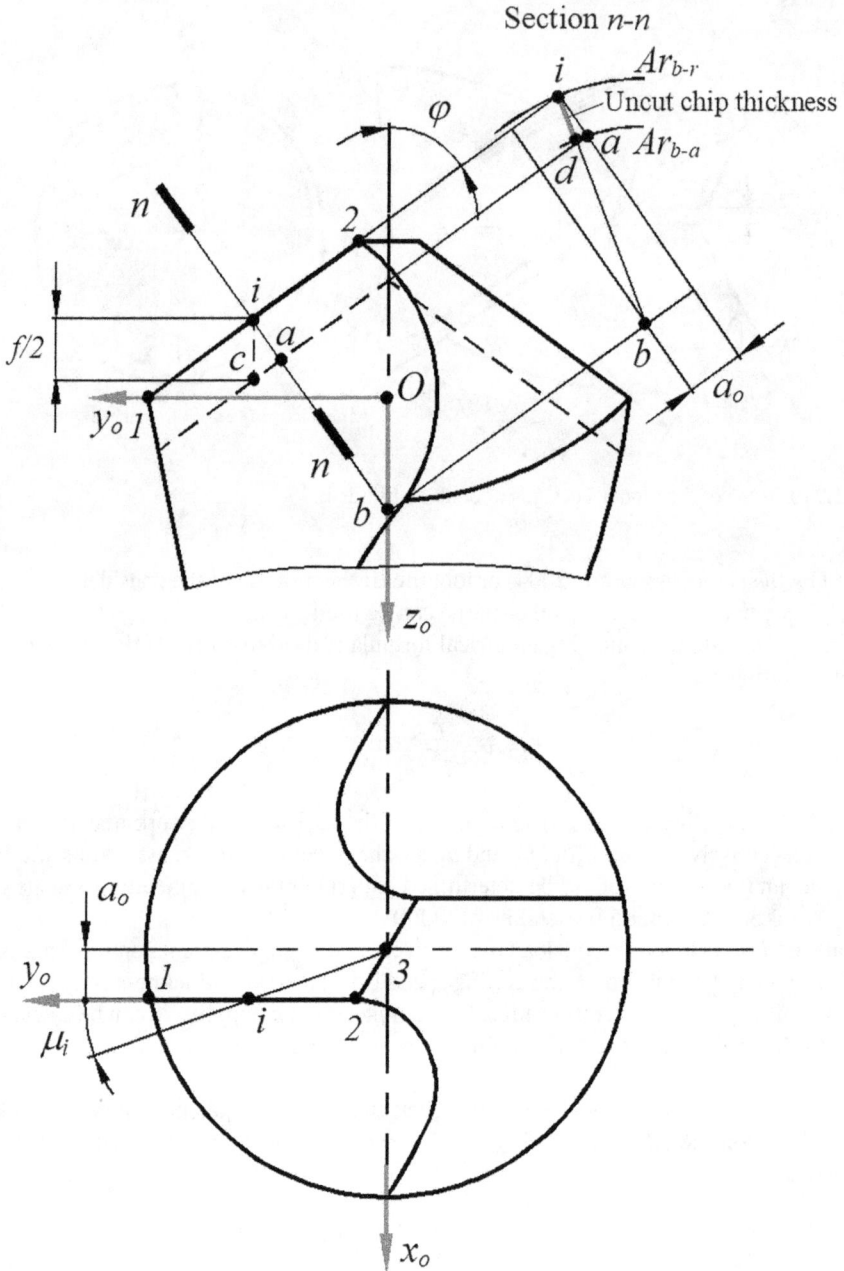

FIGURE 2.220 Model for graphical determination of the uncut chip thickness.

y_o axis passes through the drill corner *1* and the x_o-axis is perpendicular to the other two axes to complete the introduced coordinate system. There are two surfaces of the cut in the axial (the y_0z_0 plane) section apart from each other by $f/2$.

In the model shown in Figure 2.220, the uncut chip thickness h_D is determined for point *i* of the cutting edge as follows. The normal to the surface of cut lays in the plane *n-n*, which is normal to the cutting edge at point *i*. As this plane crosses the z_0-axis (the longitudinal axis of the drill) at point *b*, line *ib* is the normal to the surface of cut at point *i*. The normal section plane *n-n* crosses the two consecutive surfaces of cut located in the axial direction by distance $f/2$ (for a two-flute drill), so curves Ar_{b-r} and Ar_{b-a} are intersection lines. For the sake of simplicity, these curves can be replaced by circular arcs having point b as their center. Segment id of the normal to the point of cut is between point i and point d formed at the intersection of arc Ar_{b-a} with this normal (line ib). This segment id is a graphical interpretation of the uncut (undeformed) chip thickness.

Following the procedure presented by the author earlier (Astakhov 2010) for the determination of the uncut chip thickness analytically, let us consider a vector $\mathbf{F}$ along the direction of the cutting feed f and the normal to the surface of cut $\mathbf{N}_p$. If the angle between these two vectors is designated as ε_{NF}, then the uncut chip thickness can be represented as

$$h_D = \frac{1}{2} f \cos \varepsilon_{NF} \qquad (2.137)$$

The angle between two vectors $\mathbf{F}$ along the direction of the cutting feed f and the normal to the surface of cut $\mathbf{N}_p$ is calculated as

$$\cos \varepsilon_{NF} = \frac{\mathbf{N}_p \cdot \mathbf{F}}{\|\mathbf{N}_p\| \|\mathbf{F}\|} \qquad (2.138)$$

A unit vector of the normal to the surface of cut in the $x_0 y_0 z_0$ coordinate system is

$$\mathbf{N}_p = -\mathbf{i} \sin \mu_i \cos \varphi + \mathbf{j} \cos \mu_i \cos \varphi - \mathbf{k} \cos \mu_i \sin \kappa \qquad (2.139)$$

so its modulus is

$$\|\mathbf{N}_p\| = \sqrt{\sin^2 \mu_i \cos^2 \varphi + \cos^2 \mu_i \cos^2 \varphi + \cos^2 \mu_i \sin^2 \varphi} = $$
$$\sqrt{\cos^2 \mu_i + \sin^2 \mu_i \cos^2 \varphi} \qquad (2.140)$$

Unit vector $\mathbf{F}$ in the feed direction is

$$\mathbf{F} = -\mathbf{k} f \qquad (2.141)$$

Substituting Eqs. (2.138)–(2.141) into Eq. (2.137), one can obtain

$$h_D = \frac{f}{2} \frac{1}{\sqrt{1 + \left(\cot \varphi / \cos \mu_i \right)^2}} = \frac{f}{2} \frac{1}{\sqrt{1 + \left(\cot \varphi / \cos \left(\arcsin \left(a_o / r_i \right) \right) \right)^2}} \qquad (2.142)$$

2.1.5.12.2 Load Distribution Over the Cutting Edge

The load distribution over the drill cutting edge is determined for a drill that rotates with the angular velocity ω and fed with the feed f. The cutting speed at point i of the cutting edge located at radius r_i is calculated as

$$v_i = \omega r_i \qquad (2.143)$$

The cutting speed changes over the cutting edge 1–2 in proportion to the radius from the inner point 2 to the periphery point 1. It follows from Eq. (2.142) that the uncut chip thickness also varies over cutting edge 1–2 as angle μ_i depends on the location of a considered point on this edge.

Calculations showed (Rodin 1971) that for standard twist drills with the web diameter $d_w = 0.15 d_{dr}$ having the point angle $\Phi_p = 120°$ and chisel edge angle $\psi_{cl} = 55°$, the uncut chip thickness h_D changes from $0.43 f$ at the periphery point 1 to $0.35 f$ at point 2 where the major cutting edge intersects the chisel edge. Thus, h_D at point 2 is 18% lower than that at point 1.

When only the influences of the uncut chip thickness and cutting speed on tool life are considered, the condition when tool life is constant ($T = $ Const) can be represented as

$$v = \frac{c_{vt}}{h_D^{m_{vt}}} \tag{2.144}$$

Substituting Eq. (2.143) into Eq. (2.144) and accounting for the fact that the angular velocity is constant, one can obtain

$$r_i = \frac{c_{vt-\omega}}{h_{D-i}^{m_{vt}}} \tag{2.145}$$

where $c_{vt-\omega} = c_{vt}/\omega$.

If the uncut chip thickness corresponding to constant tool life for point i is h_{D-i}, then that point p on this edge can be represented as

$$\frac{r_p}{r_i} = \left(\frac{h_{D-i}}{h_{D-p}}\right)^{m_{vt}} \tag{2.146}$$

Experiments showed (Rodin 1971) that for medium-carbon steels, $m_{vt} \cong 0.5$. If it is assumed that the uncut chip thickness is equal to 1 at the periphery point 1, then the theoretical uncut chip thickness assuring the equality of tool life for other points of the cutting edge 1–2 can be determined by the following formula:

$$h_{D-i} = \left(\frac{r_{dr}}{r_i}\right)^2 \tag{2.147}$$

Table 2.7 shows theoretical and real (for a drill having point angle $\Phi_p = 120°$ and the web diameter $d_{wb} = 0.15 d_{dr}$) uncut chip thickness calculated for various points of the cutting edge 1–2. The load coefficients k_N for the same points are also shown. As can be seen, the load coefficient for the periphery point 1 is many folds greater than that for the central part of the cutting edge 1–2. This is the prime cause for drill non-uniform wear observed in practice. Observations showed that tool wear at the drill corner is normally much greater than that of central parts of the cutting edge.

2.1.5.12.3 Drills with Curved Cutting Edges

The twist drill used in industry for more than 140 years for the most widely employed metalworking operation (drilling) can be regarded as an unperfected cutting tool. As discussed in this chapter, severe problems associated with this cutting tool are great variations of the load coefficient and tool geometry parameters over the cutting edge. For example, the normal rake angle varies from +30° at drill periphery part of the cutting edge (point 1 in Figure 2.220) to −30° at the inner end of this edge

---

TABLE 2.7

Theoretical and Real Uncut Chip Thickness and Load Coefficients for Points of Cutting Edge 1–2

r_i/r_{dr}	Theoretical Uncut Chip Thickness	Real Uncut Chip Thickness	Load Coefficient
1	1	0.43f	1
0.6	2.8	0.43f	0.36
0.2	25	0.38f	0.035

(point 2 in Figure 2.220); the load coefficient varies from 1 at point *1* to 0.035 at point *2*. These are the greatest variations in terms of tool geometry, and thus load distributions among general-purpose cutting tools used in industry.

Although researchers, tool engineers, and professionals in the field are well aware about the importance of the region of the major cutting edge adjacent to periphery point *1* in tool life consideration, the influence of other parts of the major cutting edge (lip) as well as the chisel edge on tool life is not well understood. For example, improving conditions of the chisel edge (web thinning or splitting) cause improvement not only in self-centering ability but also in tool life defined by the wear of the drill corners (point *1*). This influence also follows from the results of the tests carried out with a standard drill of 30 mm dia. used for enlarging pre-drilled holes in gray cast iron (Rodin 1971). Series of tests with pre-drilled holes of 26, 17.5, and 11.5 mm were carried out. It was found that tool life in machining of pre-drilled holes of 26 mm dia. was 29.5 min while that in machining pre-drilled holes of 11.5 mm dia. was threefold lower. This result exemplifies the interinfluence of various parts of the cutting edge on tool life. Unfortunately, this interinfluence was not a subject of extensive research studies.

For simplicity, consider a drill with the so-called diametral cutting edges, that is, the major cutting edge (lip) is along the drill radius as in the drill design shown in Figure 2.217. For such a drill, $\mu_i = 0$ for any point *i* of the major cutting edge as follows from Figure 2.220 so that, as follows from Eq. (2.142), the uncut (undeformed) chip thickness is constant along this cutting edge. The cutting speed changes significantly along the cutting edge causing non-uniform wear of the drill along this edge. The objective of further considerations is to alter the shape of the cutting edge to achieve uniform drill wear.

According to Eq. (2.144), the cutting speed and the uncut chip thickness under $T = \text{Const}$ for any point *i* of the cutting edge *1–2* correlate as

$$v_i = \frac{C_{vt}}{h_{D-i}^{m_{vt}}} \tag{2.148}$$

For the diametral cutting edge, the uncut chip thickness for point *i* of this edge is calculated as

$$h_{D-i} = \frac{f}{2}\sin\varphi_i \tag{2.149}$$

where φ_i is the half-point angle for point *i* of the major cutting edge as, in general, this angle can vary along the major cutting edge.

As discussed in Chapter 1, the cutting speed at point *i* is calculated as

$$v_i = \frac{2\pi n}{1000}r_i \tag{2.150}$$

where $\pi = 3.141$, r_i is in millimeters, and *n* is the rotational speed in r.p.m. or rev/min.

Substituting Eqs. (2.149) and (2.150) into Eq. (2.148), and after some rearrangements, one can obtain

$$\tan\varphi_i = \frac{\left(A_i/r_i\right)^{1/m_{vt}}}{\sqrt{1 - \left(A_i/r_i\right)^{2/m_{vt}}}} = \frac{dr_i}{dz_0} \tag{2.151}$$

Differential Eq. (2.151) defines the shape of the uniformly loaded cutting edge in the y_0z_0 plane. Its numerical differentiation, however, shows that the length of the drill point under the accepted conditions is too long for practical applications. The closest known shape of the cutting edge to the obtained result is the drill design with the ellipsoidal point (Figure 4.60d in Astakhov (2014)). The obtained result shows why even though complicated and difficult to grind as the grinding accessories and programs are not fully developed, this drill point is still in use showing remarkable results when applied properly. The problem is in grinding of such a complicated profile. The calculation of a suitable grinding wheel profile used to generate the flank face of a twist drill involves complicated numerical methods, and this is of major concern. The available methods for modeling grinding wheel profiles are based on the theory of enveloping or theory of conjugations and computer simulation methods. So far, only particular solutions to the problem are found, for example, the toroidal grinding method for curved cutting edges of twist drills (Fetecau et al. 2009). Although Ivanov et al. (1998) proposed a generalized analytical method for profiling all types of rotation tools for forming helical surfaces, it was not noticed by the developers of CNC drill grinders.

Another feasible possibility is varying angle μ_i, that is, the cutting edge can be curved not only in the $y_0 z_0$ plane but also in the $x_0 y_0$ plane. The advantages of such geometry were found out by the trial-and-error method as early as the beginning of the twentieth century. For example, US Patent No. 1,309,706 (1917) describes a drill design shown in Figure 2.221 where the major cutting edges (lips) 1 and 2 are curved in both the $y_0 z_0$ and $x_0 y_0$ planes. The advantages as a remarkable increase in tool life and drilling *smoothness* are explained using an intuitive but very precise perception as

> By this constriction the work of removing metal, that is, the cutting, is distributed in such a manner that a unit of length of edge at the periphery does no more work than a unit of length of edge nearer the axis and therefore that the amount of heat produced in removing the metal is more nearly uniform for each unit length of the cutting edge...

If this heat results in the uniform temperatures equal to the optimal cutting temperature along this edge (Astakhov 2019), then the maximum tool life and uniform tool wear along the cutting edge can be achieved.

The model of optimization of the cutting edge shape for the considered case was discussed by the author earlier (Astakhov 2010). It was shown that because the uncut chip thickness varies over the cutting edge due to variation of angle μ_i, the distribution of this angle should be as follows:

$$\cos \mu_i = \frac{f}{2} \frac{\cot \varphi_i}{\sqrt{\dfrac{f^2}{4r_i^2} - 1}} \tag{2.152}$$

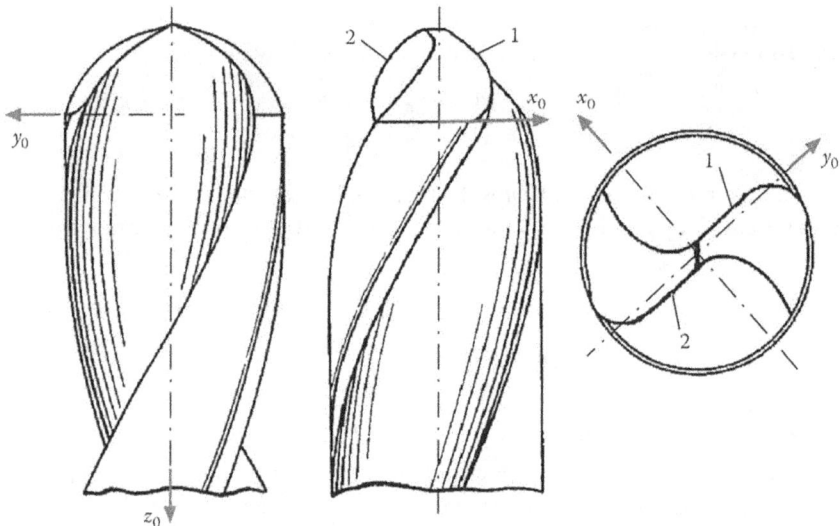

FIGURE 2.221   Drill geometry according to US Patent No. 1,309,706 (1917).

Figure 2.222 shows a drill geometry where the cutting edge is constructed using angles μ_i calculated by Eq. (2.152). As can be seen, the uncut chip thickness varies along the cutting edge following a linear fashion with the minimum at the drill periphery. The obtained results constitute the background for drills with curved cutting edges.

Another way to improve the performance of a twist drill is to achieve a constant rake angle over the entire length of the major cutting edges (lips). The major benefit of such a goal is *the balanced drill* where the unit chip flows from the unit length of the cutting edge do not cross each other.

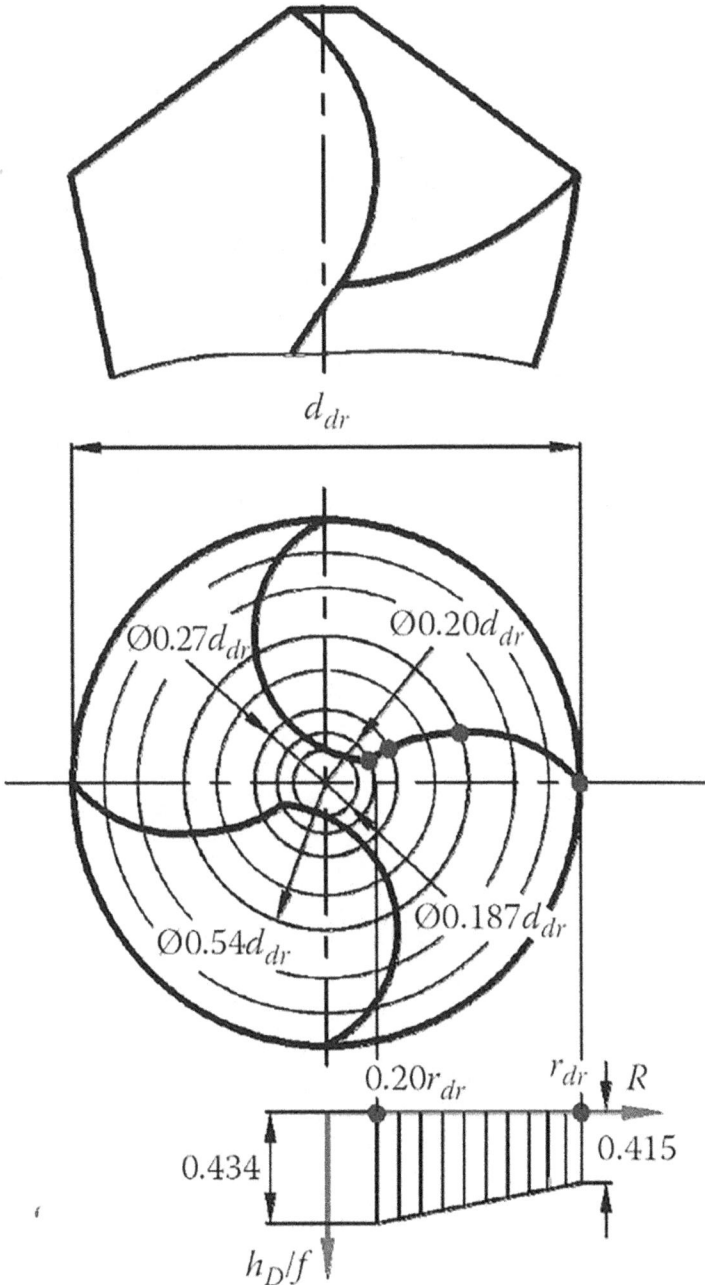

FIGURE 2.222   Calculated curvilinear cutting edge.

It follows from Figure 2.155 that the suitable distribution of the normal rake angle along the cutting edge can be achieved by continuous variation of the tool cutting edge angle (the half-point angle) φ along this cutting edge. Schematically, such a point grind of a twist drill is shown in Figure 2.223. Each point of the cutting edge, for example, point i, is characterized by its own point angle φ_i, which is the angle between the tangent to the curved cutting edge at i and the drill longitudinal axis.

To determine the shape of the cutting edge *1–2* that results in a constant rake angle along this edge, a graph similar to that shown in Figure 2.155 is used. For a given rake angle, say 29°, one may draw a horizontal line in Figure 2.155 that corresponds to this angle. Points of intersection of this line with corresponding curves define point angles Φ_p $(=2\varphi)$. Table 2.8 shows the calculated half-point angles for the considered case.

A drill shown in Figure 2.224a is described in US Patent No. 5,273,380 (1993). As seen, it is similar to that shown in Figure 2.223. According to this patent, the drill is provided having a novel drill point that includes the concave cutting edges. As claimed, this drill is characterized by a much greater tool life due to improved heat dissipation along the concave cutting edge.

A great disadvantage of the drill shown in Figure 2.224a is that the uncut chip thickness is the greatest drill periphery that lowers the tool life. This issue can be easily resolved if a second point angle similar to that shown in Figure 2.31 and discussed in Section 2.1.3.4.4 is applied as shown in Figure 2.224b. ICS Cutting Tools (Casco, WI) manufactures such a drill calling it the Perfect Point drill (Figure 2.225). As claimed by ICS Cutting Tools, this point geometry offers advantages not found with other conventional or specialty point styles. Most notably, these include an improved cutting action that generates superior surface finishes with excellent drilling precision and accuracy. This design also provides excellent stability while substantially reducing the chatter and vibration so common with other drill point shapes. The drill combines the best features of metal and wood cutting drills and is well suited for drilling holes in a variety of materials. It may be used for drilling ferrous and non-ferrous metals, natural and composite wood products, and a range of plastics and

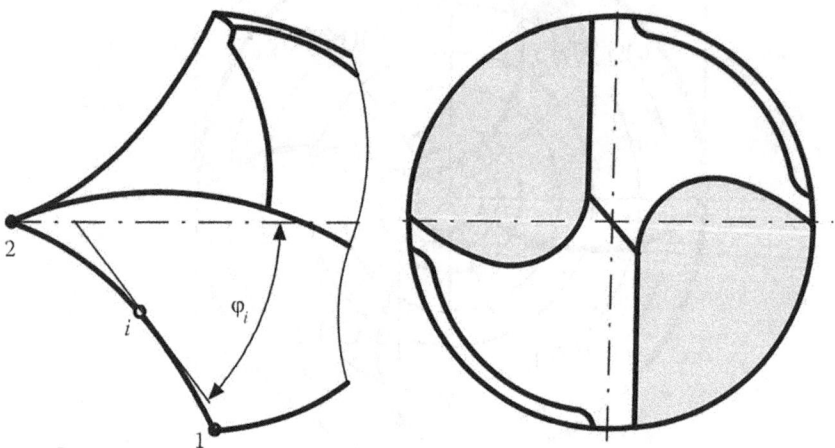

FIGURE 2.223 A point grind of a twist drill with the variable half-point angle.

TABLE 2.8
Tool Cutting Tool Angles for the Curved Cutting Edge 1–2

r_i/r_{dr}	0.40	0.69	0.84	0.95	1.00
Φ	15°	30°	45°	60°	90°

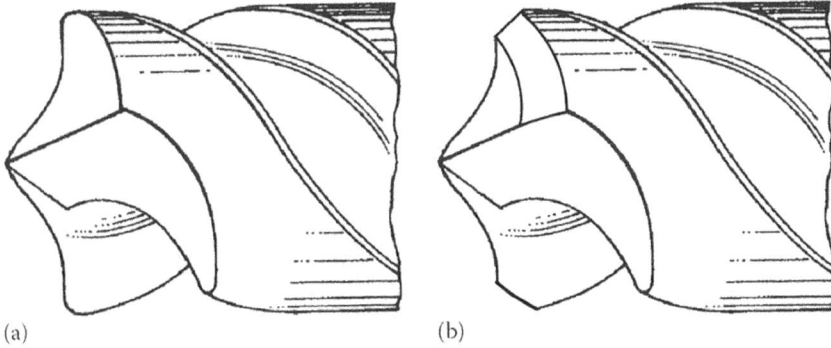

FIGURE 2.224 Practical realization of the concept geometry: (a) drill design according to US Patent No. 5,273,380 (1993) and (b) its proposed modification.

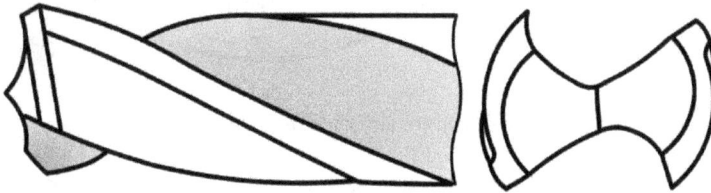

FIGURE 2.225 The Perfect Point drill.

TABLE 2.9
Distances a_o's along the Cutting Edge to Achieve a 20° Invariable Rake Angle

r_i/r_{dr}	0.43	0.50	0.73	0.85	0.95
a_o	$-0.07d_{dr}$	0	$0.15d_{dr}$	$0.30d_{dr}$	$0.50d_{dr}$

similar manufactured materials. Unlike other point designs, the Perfect Point geometry consistently produces perfectly round and uniform holes with no metal burr breakthrough or wood grain tears and splinters in virtually any material thickness.

Yet another possibility to achieve the constant rake angle over the entire length of the major cutting edges (lips) and thus to design *the balanced drill* can be thought of as a drill geometry where the cutting edge is constructed using an alternation of μ_i through the distance a_o because, as discussed earlier, the distance a_o directly affects the rake angle. (see Eqs. 2.61 and 2.65). In other words, the major cutting edge of a drill is curved in the x_0y_0 plane so that each point of this edge has its own a_o. Therefore, the curvature of the cutting edge can be selected so that a uniform rake angle along the cutting edge *1–2* is achieved. If the tool designer knows what the rake angle is to be achieved, then he can draw a horizontal line on the corresponding graph similar to those shown in Figure 2.155. The intersection points of this line with corresponding curves indicate a_o's along the cutting edge are used to achieve the objective.

For example, for a twist drill having point angle $\Phi_p=120°$ and helix angle of the drill flute $\omega_d=30°$ when a 20° invariable rake angle is to be achieved along the major cutting edge, a_o's for points of the major cutting edge as shown in Table 2.9. Figure 2.226 shows a graphical representation of these results. The experimental study of a drill with this geometry showed that the tool life increases while the drilling torque decreases.

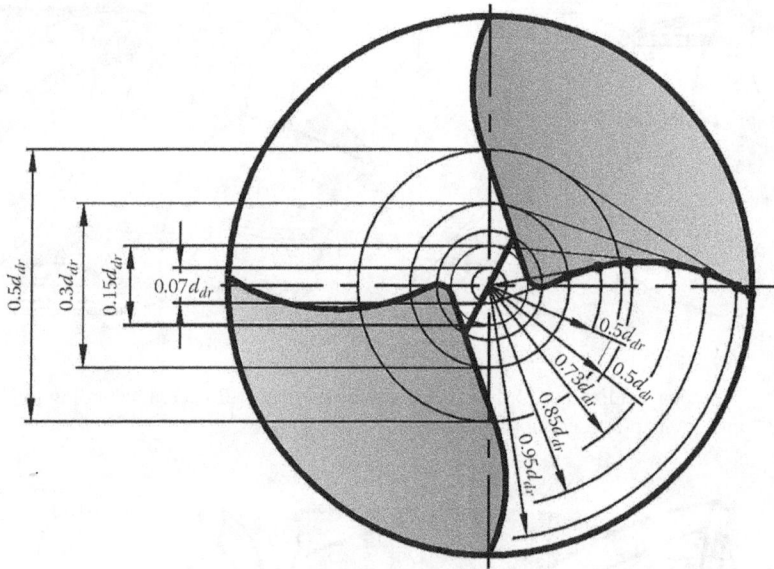

FIGURE 2.226   The curved cutting edge constricted using the data of Table 2.9.

It is interesting to mention that the similar profile shown in Figure 2.226 was developed as early as 1882 empirically. Drill design according to US Patent No. 262,588 (1882) is shown in Figure 2.227: (a) drill, (b) normal cross section, and (c) face view. This drill has a drill body 1 and two straight or helical flutes (or grooves as in the text of the patent) 2 and 3 situated opposite to each other (Figure 2.227a and b). As can be seen in Figure 2.227b, the side of each flute that makes the cutting edge is of convex curvature along its entire length, so that when the forward end of the drill is ground away to form the point by providing the flank surfaces 4 and 5, the corresponding lips or the major cutting edges 6 and 7 (which form as the intersection lines between the flutes and the flank surfaces as shown in Figure 2.227c) of convex shape are produced. According to the patent, such lips' shape results in a significant reduction of the drilling torque. It was appreciated and thus became popular over a century later.

It is not a surprise then that Sandvik Coromant Co. has developed a similar drill CoroDrill Delta-C R846, which fully resembles the discussed drill geometry. Reportedly, CoroDrill Delta-C R846 shows great results in drilling of Ni/Co-based heat-resistant alloys as well as titanium alloys and stainless steels.

2.1.6   GENERALIZATION

Reading this chapter, one may ask a logical question: "Why do we need to know particularities of the drill geometry and the means of its optimization as drills are standard on-shelf items that can be readily selected/bought using colorful catalogs of numerous drill manufacturers?" In fact, drill diameters, lengths, shank types, and sizes are standardized by many ISO and DIN standards, for example, ISO 235:1980 Parallel shank jobber and stub series drills and Morse taper shank drills; ISO 494:2009 Cylindrical shank twist drill (long series); ISO 2306: 1972 Drills for use prior to tapping screw threads; and DIN 6539:1991 Continuous parallel shank solid hardmetal twist drills and dimensions. Therefore, a need is felt to provide a clear answer to this question.

According to various estimates, 75% or more of the tools used in certain aerospace manufacturing operations are special. In the automotive industry, the use of special drills exceeds 80%. What exactly determines if a tool is special is a matter for discussion. Although even when produced in large quantities, custom tools lack the economy of scale characteristic of standard tools, there are a

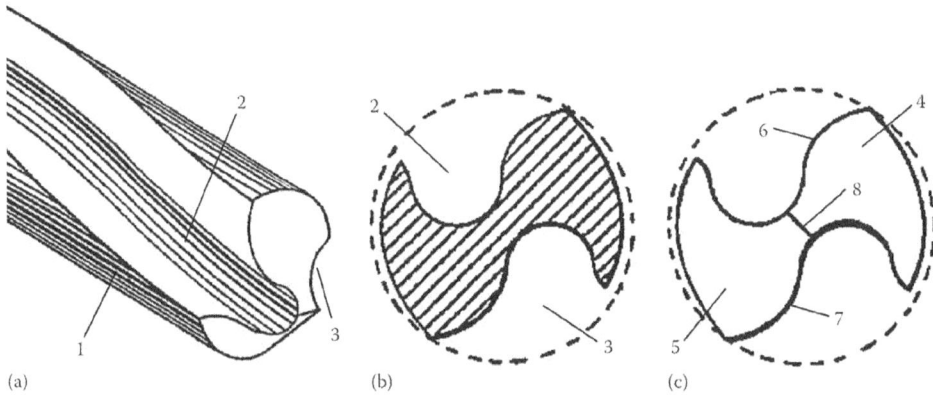

FIGURE 2.227 Drill design according to US Patent No. 262,588 (1882): (a) drill, (b) normal cross section, and (c) face view.

number of advantages in the use of custom drills in these industries so that part makers' willingness to pay a premium and take advantage of special drills can be readily explained. Standard drills are cheaper, but there is value in the more expensive custom HP drills; they offer more capability. As explained in Chapter 1, the cost of machining time is way greater than that of the drill so that even if a special HP tool cost twice compared to a standard, its productivity saves ten times more due to saving the machining time. If a shop uses the custom HP drills properly, their productivity can go way up and their profits can go way up too. Because many manufacturing companies are reluctant to make large capital purchases, a custom HP tool may enable a shop to increase productivity without purchasing new equipment.

The other advantage of custom HP drills is that they are application-specific and can be designed and manufactured according to the recent trends in the development of new tool materials, coatings, manufacturing, control, and inspection equipment (both in drill manufacturing and its use). Unfortunately, many standards on drills lag far behind these trends.

The matter discussed in this chapter provides a clear way for optimization of drill parameters and explains the rationale behind the known high-performance drills. It argues that a special HP drill can be "*assembled*" out of design/geometrical components (e.g., VPA1©, VPA2©, and VPA3© design features discussed in this chapter) needed to achieve the goal of optimization. Such an assembly is significantly facilitated when a circular diagram shown in Figure 2.228 is constructed. This diagram visualizes the requirement to drilling operation and thus makes it easy to select the design/geometry parameters of a HP drill for a given application.

It should be clear that

- High-penetration-rate, high-efficient drilling operation cannot be achieved by just doubling the penetration rate with the standard drills as they are not designed and made for such conditions. Moreover, the tolerances on the important (for HP drills) design/geometrical parameters of standard drills, for example, on runout, lip-height variation, and back taper are too wide and thus are not acceptable for HP drills. A common mistake in the designing of HP drills is to assign the same tolerances as those for standard drills on the *secondary* features, as, for example, on the back taper. This chapter explains why such a practice should not be used.
- The optimization of the drill design/geometry parameters makes sense if and only if the drilling system supports such an optimization (see Chapter 1).
- No one particular drill design/geometry can maximize all the parameters shown in Figure 2.228 simultaneously.

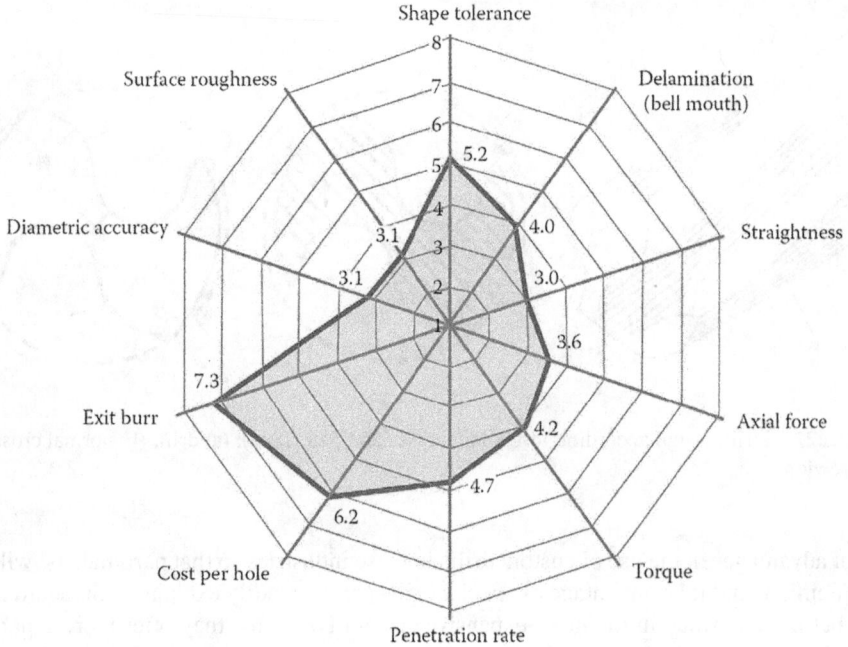

FIGURE 2.228 Circular diagram showing relative importance (scale 1–8) of the parameters of a drilling operation.

- Adding any additional parameter of the tool geometry may increase the cost of the tool as additional programming, grinding wheel, grinding time, etc. may be required.
- The geometry/design parameters that affect the goal for a given application/case are of prime importance, and thus should be selected first and then adjusted by going down through other parameters. In the example shown in Figure 2.228, the point angle that results in the tolerable exit burr is selected. Then, as the cost per drilling hole is the second important objective, the double-point drill should be selected to minimize the cycle time and improve the quality of machining holes.
- The design/geometry parameters optimized for one tool material or even tool material grade may not be suitable for the other as the friction/adhesion, achievable contact surface conditions, sharpness of the cutting edge, temperature-affected properties, and many other parameters hidden from the view of manufacturing specialists may change the tool performance dramatically.

2.2 REAMERS

2.2.1 DEFINITION AND CURRENT ISO STANDARDS

Reaming is a finishing hole-making operation carried out with a cutting tool called a reamer. The major objectives of using this operation are to achieve high diametric accuracy, low surface roughness, and close tolerance of the hole shape in the transverse direction.

Reamer is a rotary cutting tool with one or more cutting elements, which is used for enlarging to size and contour a previously formed hole. Its principal support during the cutting action is obtained from the workpiece as defined by *Metal Cutting Tool Handbook* (1989). This book also provides the complete classification of reamers. Figure 2.229 shows the three most common types of reamers. Figure 2.229a shows a solid reamer, i.e., reamer made of one piece of tool material. It is used for

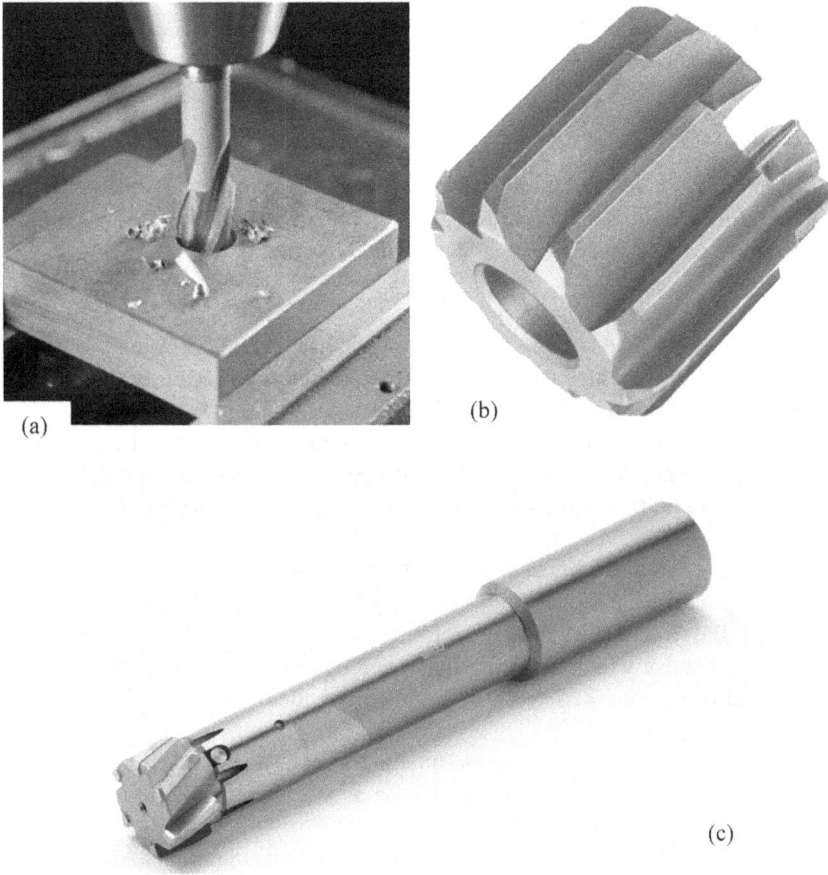

FIGURE 2.229 Most common types of reamers: (a) solid reamer, (b) shell reamer, and (c) modular head indexable reamer.

high-precision holes. Figure 2.229b shows a shell reamer. Normally, these reamers made of HSS are multi-fluted, end-cutting tools used to enlarge previously formed holes to a precise diameter with a smoother finish. They are used with a special shell reamer arbor, which is tapered to fit the holes in the reamer. Figure 2.229c is a modular head indexable reamer used for less precision holes. Such a reamer has a number of advantages in terms of cost per reamed hole, tool life, and flexibility.

A general impression in industry is that reamers are well-developed and investigated tools. Their types, geometry (the cutting geometry and length items), involved terms, and tolerances are extensively covered by multiple ISO standards. In year 2022, the following ISO standards were related to reamers:

ISO 236-1:1976: Hand reamers.
ISO 236-2:2013: Reamers – Part 2: Long-fluted machine reamers with Morse taper shanks.
ISO 521:2011: Machine chucking reamers with cylindrical shanks and Morse taper shanks.
ISO 522:1975: Special tolerances for reamers.
ISO 2238:2018: Machine bridge reamers.
ISO 2250:2017: Finishing reamers for Morse and metric tapers, with cylindrical shanks and Morse taper shanks.
ISO 2402:1972: Shell reamers with taper bore (taper bore 1: 30 (included)) with slot drive and arbors for shell reamers.

ISO 3465:1975: Hand taper pin reamers.
ISO 3466:2016: Machine taper pin reamers with parallel shanks.
ISO 5420:1983: Reamers – Terms, definitions, and types.

2.2.2 General Classification

ISO standards do not provide any classification of reamers. The best classification found is given in Metal Cutting *Tool Handbook* (1989). Using this classification as the starting point, the author modified some definitions accounting for modern manufacturing practice.

   A: Classification Based on Constriction
   Solid Reamers: those made of one piece of tool material (Figure 2.229a).
   Tipped Reamers: those which have a body of one material with cutting inserts of another material brazed or otherwise bounded in the body.
   Inserted Blade Reamers: those that have replaceable mechanically retained cutting inserts. These inserts may be solid or tipped (cartridges) and are usually adjustable.
   Expansion Reamers: those whose size can be increased by deflecting or bending of the reamer body.
   Adjustable Reamers: those whose size may be changed by sliding, or otherwise moving, the cutting inserts/cartridges toward or away from the reamer axis.
   B: Classification Based on the Method of Holding, or Driving
   Hand Reamers: those which are ordinarily used by hand. A driving square is provided at the end of the shank. The cutting end is provided with a starting taper for easy entry.
   Machine Reamers: those having shanks suitable for mounting in machines.
   Shell Reamers: machine reamers mountable on arbors specifically designed for the purpose, called "Shell Reamer Arbors" (Figure 2.229b).
   Modular head indexable reamer: those having a reamer head mountable on the specially designed body (Figure 2.229c). It is the most modern trend in reamer designs and applications.

The terms "right hand" and "left hand" are used to describe both the direction of rotation of reamers and the direction of flute helix for helical reamers. Therefore, a need is felt to clarify the difference and encourage specialists (reamer manufacturers and end users) to use the proper terms as "hand of rotation" and "hand of flute helix" for those two articles as suggested by *Metal Cutting Tool Handbook* (1989).
   A - Hand of Rotation or *Hand of Cut*
   Right-Hand Rotation or *Right-Hand Cut* is a counterclockwise rotation of a reamer revolving so as to make a cut when viewed from the cutting end.
   Left-Hand Rotation or *Left-Hand Cut* is a clockwise rotation of a reamer revolving so as to make a cut when viewed at the cutting end. This type of reamers is used rarely, only in very special cases.
   B – Hand of Flute Helix
   Reamers with their cutting edges and their rake faces in a plane parallel to the reamer axis are described as having "Straight Flutes" as shown in Figure 2.230a. Reamers with every other flute of opposite (right and left) helix are called "Alternative Helix Reamers." Reamers with flute helix in one direction only are described as having Right-Hand or Left-Hand helix. A reamer has Right-Hand helix when the flutes twist away from the observer in a clockwise direction waved from either end of the reamer as shown in Figure 2.230b. A reamer has Left-Hand helix when the flutes twist away from the observer in a counterclockwise direction waved from either end of the reamer as shown in Figure 2.230c. According to Fullerton Tool Co., straight-flute reamers were designed

(a)

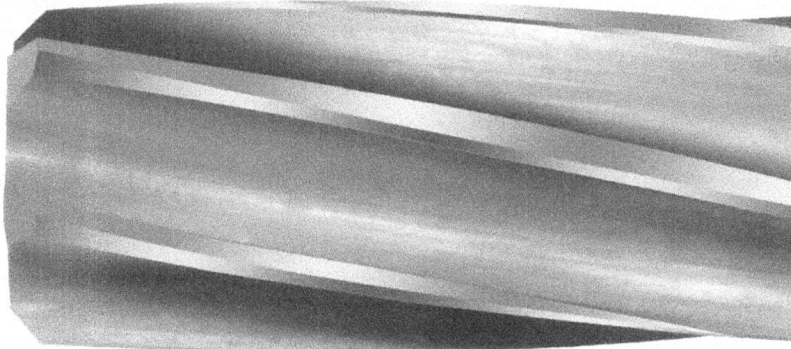

(b)

(c)

FIGURE 2.230   Showing (a) straight-flute reamer, (b) right-hand helix reamer, and (c) left-hand helix reamer.

for all types of general reaming applications. Right-hand helix flute reamers are recommended for blind holes and abrasive, ductile materials. Left-hand helix flute reamers produce better finishes on hard materials such as heat-treated steels.

2.2.3   Nomenclature/Terminology of Reamer Elements and Associated Terms

Reamer Diameter: the diameter over the margins of the reamers measured at the periphery corners, i.e., at the leading end of the chamfered section.

Arbor Hole: the central mounting hole in a shall reamer.

Axis: the imaginary straight line that forms the longitudinal centerline of a reamer, usually established by rotating the reamer between centers.

Bach Taper: a slight decrease in diameter, from front to back, over the flute length of reamers.

Bell Mouth: an entrance portion of the hole that is larger in diameter at the start of the hole then at some distance beyond.

Bevel: an unrelieved tapered surface of revolution located (mainly in hand reamers) between the front face and chamfer. Its angle of taper measured in an axial plane is greater than the chamfer angle.

Blade: a tooth or cutting element inserted in a reamer body. It may be adjustable and/or replaceable.

Blending Radius: a relieved radius joining the chamfer and the periphery.

Body: (1) the fluted full diameter portion of a reamer, inclusive of the chamfer, starting taper, and bevel. (2) The principal supporting member for a set of reamer blades, usually including shank.

Burnishing Reamer: a finishing reamer intended to take a light scraping cut and impart a fine finish.

Chamfer Angle: the angle between the reamer's axis and the cutting edge of the chamfer measured in the axial/reference plane at the cutting edge.

Chamfer Length: the length of the chamfer measured parallel to the axis.

Chamfer Cutting Geometry: it includes the rake and clearance faces. Their definitions and relevant angles in various systems of considerations are the same as for drills – see Sections 2.1.3.5 and 2.1.3.6.

Chucking Reamer: a type of machine reamer with relatively short straight or helical flutes on which the peripheral lends are relieved.

Clearance: the space created by the flank face ground behind the cutting edge and the bottom of the hole has been reamed.

Core: the central portion of a reamer below the flutes which joints the lands.

Core Diameter: the diameter at a given point along the axis of the largest circle which does not project into the flutes.

Cutter Sweep: the section removed by the milling cutter or grinding wheel in entering or leaving a flute.

Cutting edge: the edge on the chamfer formed as the intersection line of the rake and flank faces of a reamer.

Cutting speed: the peripheral linear speed resulting from rotation. As discussed in Chapter 1 (see Eq. 1.1), it is calculated in m/min as

$$v = \frac{\pi d_1 n}{1,000} \tag{2.153}$$

where $\pi = 3.141$, d_1 is the reamer diameter in millimeters (the designation d_1 is according to standard ISO 521), and n is the rotational speed in rpm or rev/min no matter which rotates, the reamer or the workpiece. If both the reamer and the workpiece rotate in opposite directions (the so-called counterrotation), then n is the sum of the rotational speeds of the drill, n_{rm}, and the workpiece, n_w, that is, $n = n_{rm} + n_w$.

Cutting feed: the axial advance in millimeters per revolution of the reamer with respect to the workpiece.

Flutes: longitudinal channels formed in the body of the reamer to provide cutting edges, permit passage of chips, and allow MWF to reach the cutting edges.

Helical Flute: a flute that is formed in a helical path around the axis of a reamer (see Figure 2.230b and c).

Straight Flute: a flute that is formed with the margin edges parallel to the axis of a reamer (see Figure 2.230a).

Flute Length: the length of the flutes not including the cutter sweep, i.e., the length of the full flute profile.

Heel: the trailing edge of the land in the direction of rotation for cutting.

Helix Angle: the angle at which helical margin edges of a reamer make with the axis views in the projection of these edges in the working plane.

Irregular Spacing: a deliberate variation from a uniform spacing of the reamer cutting edges as viewed in the back plane, i.e., the plane perpendicular to the axis of a reamer.

Land: the section of a reamer between adjacent flutes.

Land Width: the distance between the leading edge of the land and the heal measured in the back plane at a right angle to the leading edge.

Margin: the unrelieved part of the periphery of the land adjacent to the cutting edge.

Nominal Size: the designed basic size of the reamer. It is normally equal to the basic size of the hole to be/been reamed.

Coolant (MWF) Holes: holes through which MWF (coolant) is fed into machining zone.

Overall Length: the extreme length of the complete reamer from end to end, but not including external centers and/or expansion screws (if any).

Pilot: a cylindrical portion preceding the entering end of the reamer body to maintain alignment.

Pull Reamer: reamers that are designed to be pulled through long holes (such as gun barrels) while the reamer or workpiece is rotated.

Runout: (1) *Radial Runout* – The radial variation from a true circle which lies in the back plane and is concentric with the reamer axis. See term Total Indicator Variation. (2) *Axial Runout* – The axial variation from a true circle that lies in the back plane and is concentric with the reamer axis.

Secondary Chamfer: a slightly relieved chamfer adjacent to and following the initial chamfer on a reamer.

Shank: the portion of the reamer by which it is held and driven.

Step Reamer: a multiple-diameter reamer with all lands in each step ground to the same diameter.

Total Indicator Variation: the difference between the maximum and minimum educator reading obtained in a checking cycle.

Figure 2.231 illustrates some important terms applying to reamers.

2.2.4 Basic Reamers Geometry and Its Relevant Particularities

In the author's opinion, the proper geometry of reamers and its particularities are not well presented/ explained/illustrated in literature sources including the above-listed ISO standards. Therefore, this section aims to clarify many previous misunderstanding and beliefs widely spread in the theory and practice of the reamer design, manufacturing, and application.

To understand the reamer geometry properly, one should correlate the results/information related to drilling discussed in Section 2.2 and that related to reaming simply because these two tools are very similar, i.e., a reamer is a drill aimed to enlarge a pre-existing hole. The same as the design of a drill, the reamer design corresponds to its dual purpose: (1) Cut cylindrical holes in a hollow workpiece, and (2) transport the chip formed in such cutting outside the hole being cut. Therefore, the

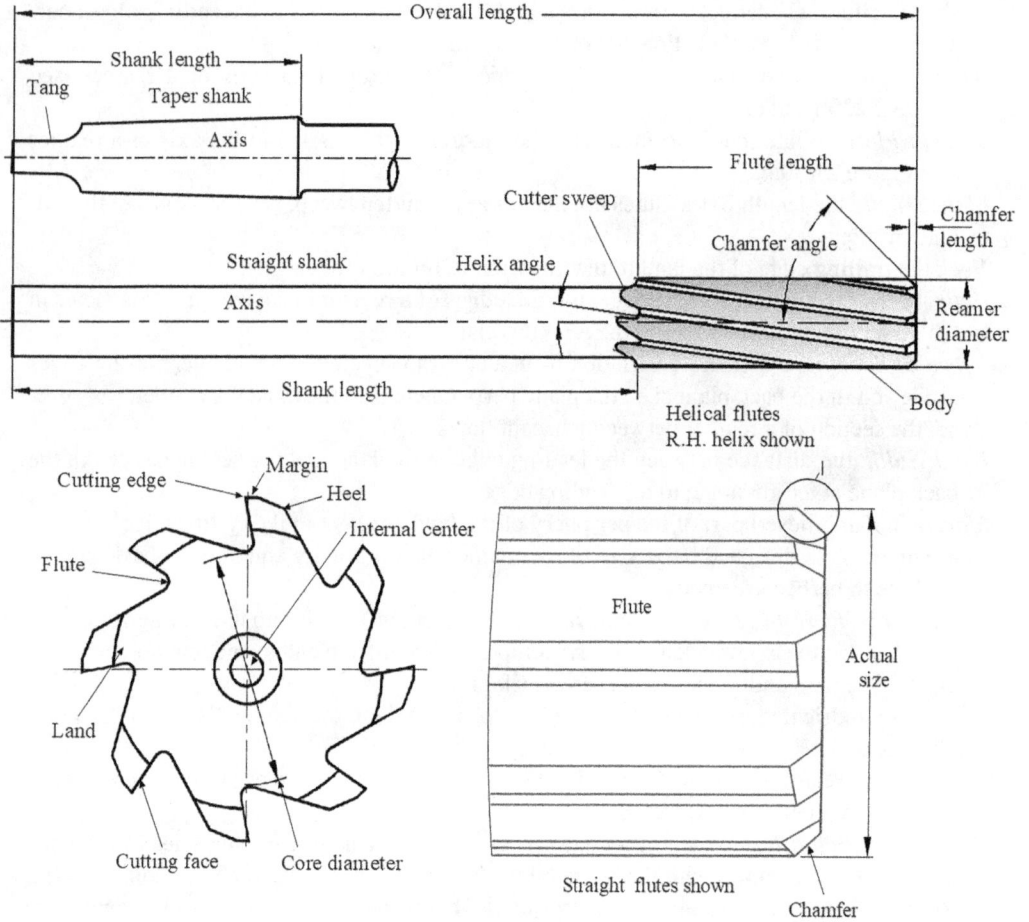

FIGURE 2.231 Illustration of some terms applying to reamers.

optimality of the reamer design for a given reaming condition should always be considered keeping in mind how well this reamer fulfills the stated purposes.

2.2.4.1 Number of the Cutting Edges in a Common Reamer

This section considers the number of the cutting edges of a common reamer in full analogy with Section 2.1.3.3 which considered the number of the cutting edge of a common drill. The cutting portion of a reamer should be called as *the reamer point* in analogy with drills although it is called the chamfer in the corresponding standards, in scientific and trade literature all over the world.

Figure 2.232 shows a common reamer point of a multi-flute reamer. Each cutting flute has two cutting edges, namely:

1. Major cutting edge *1–2*, routinely called the chamfer edge.
2. Minor cutting edge *1–3*, not mentioned in the literature.

As discussed in the Appendix, a cutting edge is actually a line of intersection of the rake and flank faces so that its shape is entirely defined by the shape of these faces. In metal cutting, the geometry of the cutting edge is understood as a set of certain angles, namely the rake, clearance, tool cutting edge, and inclination angles as discussed in the Appendix. Therefore, these angles in terms of their meaning and significance to reamer performance should be considered, and thus unambiguously

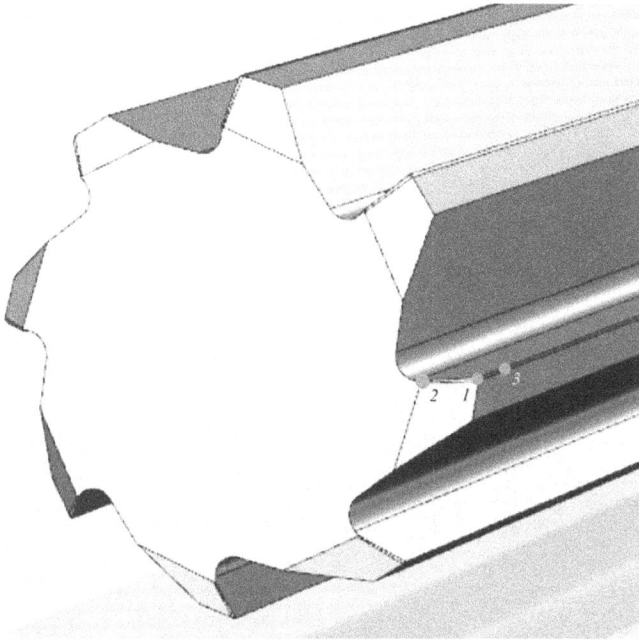

FIGURE 2.232 Cutting edges of a flute of a common reamer.

understood by specialists for the generic reamer point shown in Figure 2.232. Besides geometry parameters, the parameters of the uncut chip geometry, namely the uncut chip thickness (a.k.a. the chip load in industry) and its width should also be determined as functions of the reaming feed. The latter is of prime importance in the consideration of drill microgeometry discussed in Section 2.1.4. The next sections define the listed parameters for the drill cutting edges shown in Figure 2.232.

2.2.4.2 Major Cutting Edge

2.2.4.2.1 Chamfer Angle

The particularities of the geometry of the major cutting edge *1–2* are shown in Figure 2.233a. As discussed in the Appendix, the tool cutting edge angle κ_r is the most important geometrical parameter of this tool. It is called as the chamfer angle, φ, in reaming. This angle is defined as the angle between projection of the major cutting edge *1–2* into the reference plane P_r and the direction of the cutting feed (as explained in the Appendix).

In full analogy with modern drills manufactured with point angles of 118° for HSS and of 140° for carbide drills (see Section 2.1.3.4), the 45° chamfer angle is the industry standard for reamers to the extent that many leading cutting tool manufacturers do not include its angle in their catalogs. Moreover, no information on other reamers geometry parameters is provided. It mysteriously suits many users who do not want to deal with the application-specific reamer geometry thinking that it is up to the tool manufacturers to suggest the optimum reamer geometry, its material and coating for a given application. Only very few reamers manufacturers, e.g., RTS Cutting Tools, offer special chamfer angles.

The author's many-year experience, however, shows otherwise. The same as with drills, reamer manufacturers always try to sell the so-called on-the-shelf products or stock items, which they can be produced in mass quantity at low manufacturing cost and with decent quality. With such reamers, the lead time (the time between a purchase order and actual tool delivery) is minimal and the tool is relatively inexpensive. These real-world conveniences often overshadow the potential gains in efficiency (tool life, productivity, etc.) that can be achieved with application-specific reamers.

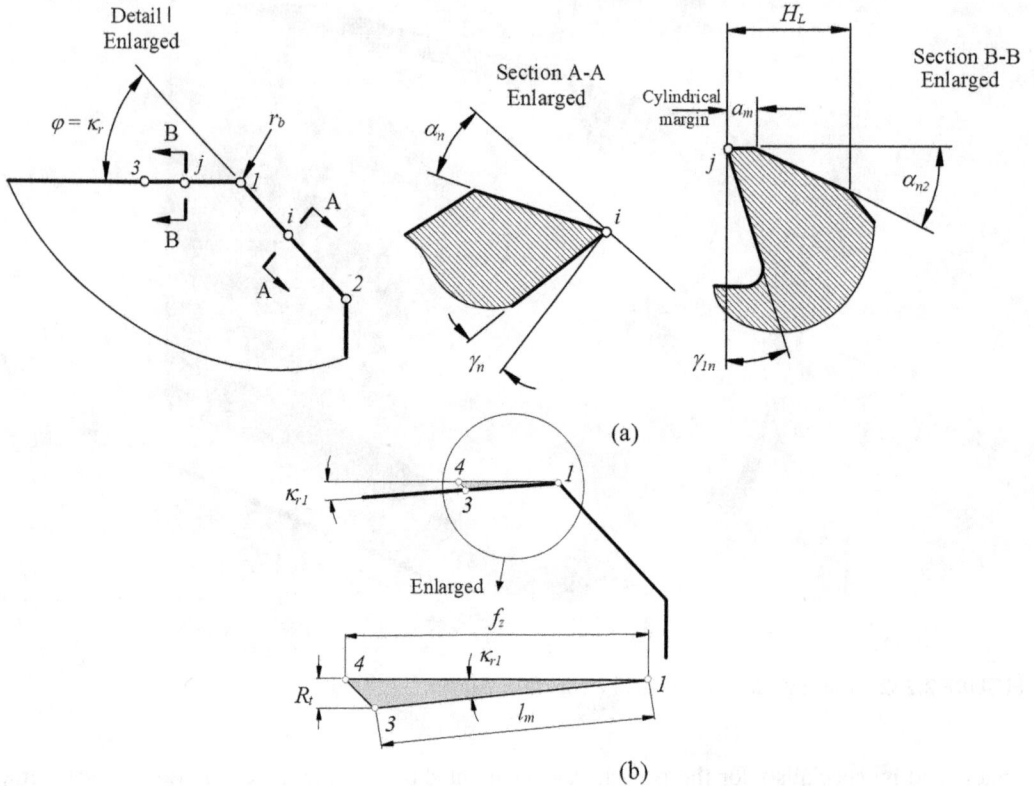

FIGURE 2.233 Particularities of the cutting geometry: (a) of the major *1-2* and minor *1-3* cutting edges (Detail I from Figure 2.231) and (b) model of the theoretical surface roughness formation in reaming.

It is important to note that standard ISO 521 lists the chamfer angles to be 5°, 15°, and 45°. These angles are used to ream the holes of "standard" quality, i.e., corresponding IT7–IT9 grades. When holes of higher grades to be reamed particularly in difficult-to-machine work materials, then the triple-section chamfer design is used. The first section is made with the chamfer angle of 45°, the second – of 15°, and the third – of 2–3°. As such, the length of the third section is 0.8–2 mm as it blends this chamfer to the minor edge. Moreover, according to standard ISO 2238, the so-called bridge reamers are provided with 5°45' angle of the taper over the chamfer length that corresponds to 2°22' chamfer angle. As a result, the chamfer length is extended correspondingly. For example, for a reamer of 10 mm dia., this length is 38 mm.

The chamfer angle φ affects the force balance in reaming the same as in drilling as discussed in Section 2.1.3.4.2. The same as in drilling, it is customary to represent the cutting force, normal to the cutting edge, F_n as applied at a certain point m of this edge. The same as in drilling (see Figure 2.24), for a given cutting condition including work and tool materials, cutting speed and feed, and so on, the magnitude and direction of F_n as well as the point of its application m on the cutting edge *1–2* does not depend on the point angle. However, because the direction of the major cutting edge *1–2* varies with the chamfer angle, φ, the direction of F_n also varies to keep it perpendicular to the major cutting edge. Figure 2.234 visualizes these considerations. As the direction of F_n varies with φ, the ratio of the axial and radial forces also varies as

$$\frac{F_a}{F_r} = \tan \varphi \tag{2.154}$$

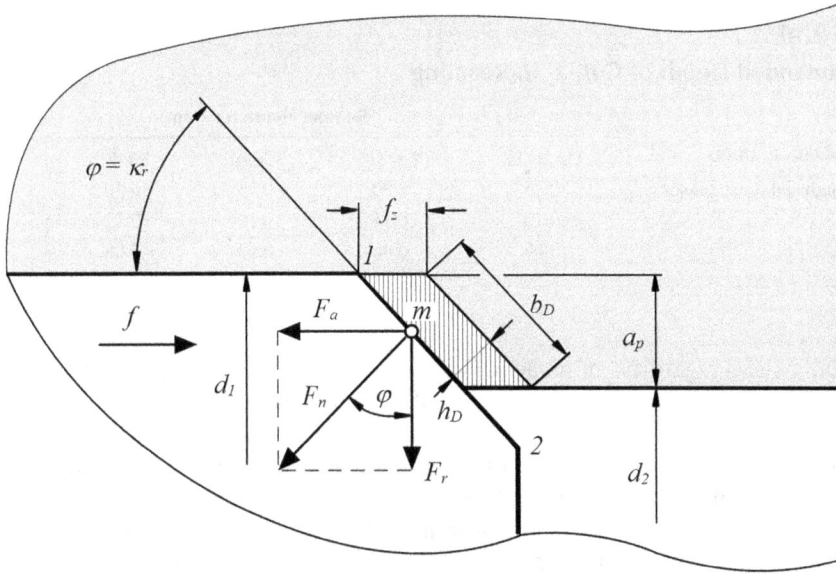

FIGURE 2.234   Uncut chip thickness parameters and force balance in reaming.

As follows from this equation, the radial component increases and the axial component decreases when the point angle decreases. This can be a valuable means at tool/process designer disposal to increase the allowable reamer penetration rate normally restricted by the axial force for long reamers.

As discussed in Chapter 1 and in Section 2.1.3.4.3, the nominal thickness of cut known in the literature as the uncut (undeformed) chip thickness or chip load, h_D, and the nominal width of cut known in the literature as the uncut (deformed) chip width, b_D, are two important factors in metal cutting affecting its major outputs as cutting force and tool life. In reaming, they are calculated as follows from Figure 2.234 as

$$h_D = f_z \sin \varphi \tag{2.155}$$

$$b_D = \frac{a_p}{\sin \varphi} \tag{2.156}$$

where f_z is the feed per tooth and a_p is the depth of cut in reaming.

As discussed in Chapter 1, $f_z = f/z$, where f is the feed per reamer revolution, $a_p = (d_1 - d_2)/2$, where d_2 is the hole diameter prior reaming (see Figure 2.234). Table 2.10 gives some guidelines for the selection of a_p for reaming medium-carbon steels.

A critical item in reaming is the undeformed/uncut chip thickness, h_D, as the cutting feed in reaming is rather shallow and the number of teeth of reamers is great. The smaller the cutting feed and the greater the number of teeth, the smaller the uncut chip thickness. As a result, according to Eq. (2.155), h_D can be very small, particularly for multi-tooth reamers that can present a major problem. To understand the extent of the problem, one needs to read and understand the whole concept of microgeometry presented in Section 2.1.4. It is discussed here that *the radius of the cutting edge R_{ce} should be judged against the uncut chip thickness h_D.* As discussed in Section 2.1.4, for characterization of cutting tool sharpness, the *RTS* is defined in Section 2.1.4.2.2.

For proper drilling tool performance, *RTS* defined by Eq. (2.23) should be kept below 10% for optimal performance of drilling tools. When this is the case, the cutting edge is considered to

TABLE 2.10

Recommended Depth of Cut, a_p in Reaming

Depth of cut, a_p (mm)	Reamer diameter (mm)				
	Up to 18	18–30	30–50	50–75	75–100
Total, rough and finish passes	0.15	0.20	0.25	0.30	0.40
Rough pass	0.10	0.14	0.18	0.22	0.30
Finish pass	0.05	0.06	0.07	0.08	0.10

be technically sharp regardless of its actual radius. Unfortunately, it is often not nearly the case in reaming. In the author's opinion, this is one of the major issues in reaming. It is important to remember that *RTS* is specific for a given combination of the work and tool materials that results in a specific value of γ_{lim}. It is to say that a given cutting edge/cutting tool may be sharp for one material but not sharp for the next having considerably different properties. Application experience shows that *RTS* should be kept at minimum in machining of high-strength work material of low thermoconductivity and great strain hardening, for example, PH stainless steels and titanium alloys.

To improve the working condition of the cutting edge in reaming, CEP is widely used. CEP, its basics, ranges, and methods are extensively covered in Section 2.1.4.2. What is the most important, however, is that this optimum CEP in reaming lies within a range, which is much narrower compared to drilling tools because the undeformed/uncut chip thickness in reaming is significantly smaller than that in drilling. Unfortunately, this important fact is totally ignored in the design of reamers, their manufacturing and applications. It is to say that the cutting edges of reamer (characterized by *RTS*) should be made and kept sharp for optimal and even for normal performance of reamers.

2.2.4.2.1 Rake and Clearance Angles

The normal clearance and rake angles of the cutting part of a reamer are shown in Section A-A in Figure 2.233a. As discussed in Section 2.1.3.5 and the Appendix, the clearance angle is the only distinguished feature of *the cutting edge* of any type of cutting tool. Its sole purpose is to clear the motion of the cutting edge, i.e., to prevent any unwanted contact between the tool and machined surface. Consequently, whereas the other angles of this edge can be positive, neutral, or negative, the clearance angle *MUST* always be positive to clear the motion of the cutting edge. When it comes to drilling tools, the clearance angle is probably most important yet least understood, and thus not properly assigned parameters of the tool geometry.

In most modern reamers, the flank surface is planar (flat). As such, regardless of a particular reamer type, the primary clearance angle is the same for all points of the major cutting edge *1–2* in T-hand-S. In the most common case, the rake face of a reamer lies in the reference plane or slightly deviates from this plane by the rake angle. However, the length of the major cutting edge is small so that the rake angle in T-hand-S is practically equal to that in T-hold-S.

The recommended clearance angles for HP drills and reamers are given in Table 2.3. Application experience shows that these values should be considered as good starting values in the optimization of the reamer geometry parameters. Unfortunately, many reamer manufacturers and users do not follow these conclusions/recommendations, which results in lower tool life, inferior quality of drilled holes, and, sometimes, reamer breakage. The clearance angle should be an inherent part of the reamer drawing. It has to be indicated in the manner shown in Figure 2.233a, i.e., as α_n.

Although the considerations/results/recommendations of the rake angle of the major cutting edge of a reamer are exactly the same as for the major cutting edge of drills presented in Section 2.1.3.6, such considerations are significantly simplified accounting for a short length of this edge in reamers. Therefore, for most straight-flute reamers, the rake angle is zero and it is not indicated in tool drawings. In some special case, when one optimizes the performance of reamer for special materials, a small positive rake angle (5°–10°) can be applied that is particularly beneficial for difficult-to-machine materials. Note that this angle is application-specific so its particular value should be determined experimentally. In this case, it should be indicated in tool drawing in the manner shown in Figure 2.233a.

As discussed in Section 2.1.3.6 for drills, when a reamer with helical flute is used, the rake angle of the major cutting edge in the working plane is equal to the angle of helix (Figure 2.55) as the major cutting edge is short. Obviously, for right-hand-helix reamers, this rake angle is positive, whereas it is negative for left-hand-helix reamers. In any case, it is not indicated in the tool drawing because the angle of helix is normally shown.

2.2.4.3 Minor Cutting Edge and Back Taper

2.2.4.3.1 Back Taper

Back taper on the reamer margin(s) is applied as discussed in Section 2.1.3.8.3. The purpose of back taper is to reduce the heat due to friction while the tool is engaged in the workpiece thus to prevent binding of a reamer in the hole being drilled. If no back taper is applied to the reamer's margins, the reamer will be gripped by the contracted wall of the machined hole in the manner shown in Figure 2.91a. The purpose of back taper is to reduce the heat due to friction while the tool is engaged in the workpiece thus to prevent binding of a reamer in the hole being reamed.

In drilling tools as drills and reamers, the back taper Δ_{bt} (included) is assigned on the drawing as a diameter decrease per certain length as shown in Figure 2.86. It is calculated using Eq. (2.17). As discussed in Section 2.1.3.8.3, for years, rather small back taper was applied to the drills and reamers (for reamers 0.01–0.08 mm/100 mm). This "standard" range was not of any concern as the runout of reamers, reamer holders, and spindles of machines (the so-called system runout discussed in Chapter 1 in Astakhov (2024)) was enormous from the modern standpoint resulting in the diameter of drilled hole much greater than that of drills.

As discussed in Section 2.1.3.8.3, times have changed and so the components of drilling system as discussed in Chapter 1. It is to say that better reamers with small runouts, better tool holders, and machines were introduced in industry over relatively recent times. Moreover, the use of high cutting speeds in reaming, particularly with carbide and PCD drills and reamers, became commonplace. What was not changed is the amount of back taper applied to standard drilling tools.

When the back taper is insufficient, the result depends on how small an actual back taper is and many other particularities of the hole-making operations. A common indication of insufficient back taper is poor quality of reamed holes. Figure 2.87b shows feed and retraction marks on the hole surface. Figure 2.87c shows chatter marks (known as "tiger strips" in industry) and retraction marks on the hole surface.

Figure 2.235 shows an example of abnormal reamer wear when back taper is insufficient. The shown reamer is a helical-flue (5° right-hand helix) carbide reamer having six equally spaced flutes. The work material is SAE 1020 steel of hardness 10–12 HRC. The reamer diameter is 17.93/17.90 mm. The reamer has double chamfer: the first chamfer angle is 45° and its length is 0.120 mm. The second chamfer angle is 20°, and its length is 0.33 mm. Back taper applied is 0.0002 inch/inch, i.e., 0.2 μm/mm. MWF is a water-based (or water-mixable) coolant having 8% concentration. Tool life is 200 parts (the spider deck). The tool life criterion is an undersized hole. These conditions are common in industry.

Figure 2.235a shows significant built-up edge and tool wear over some distance after the chamfer. This distance constitutes the minor cutting edge. After this, the whole edge of the flute margin shows scoring. Attempts were made to solve the problem by increasing the blending radius r_b (Figure 2.233). Figure 2.235b shows that increasing the blending radius led to even worse results. The problem was solved and the tool life was significantly increased only when the proper recommendation by *Application rule #14* presented in Section 2.1.3.8.4 is applied.

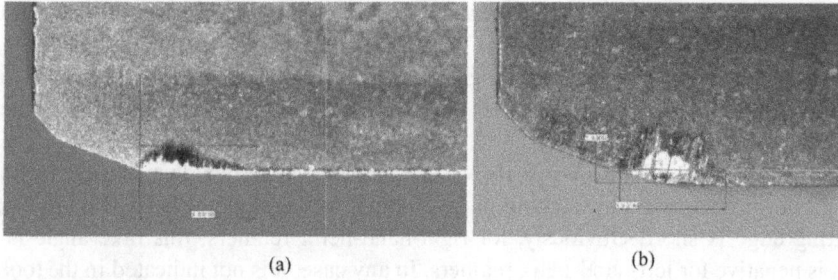

FIGURE 2.235 Common wear patterns on reamers when (a) back taper is insufficient and (b) blend radius is too great.

2.2.4.3.2 Minor Cutting Edge

The geometry of the minor cutting edge *1–3* is presented in Figure 2.233a, Section B-B. If there is no back taper, then the length of this edge would be equal to that of the margin. In reality, back taper on the reamer margin(s) is applied that forms the minor cutting edge as discussed in Section 2.1.3.8.5. As discussed in Section 2.1.3.8.5, the minor cutting edge plays a significant role in drilling. Even more significant role it plays in reaming. In drilling tools, the minor cutting edge is defined as the line of intersection of the flute face and the margin.

Figure 2.233b shows the definition of the tool minor cutting edge angle, κ_{r1} for a straight-flute drill (in the picture, κ_{r1} is shown significantly exaggerated for clarity). It is defined according to the standard definition provided by ISO 3002-1 standard (see the Appendix) as the acute angle between the projection of the minor (side) cutting edge into the reference plane (a plane that contains the drill longitudinal axis) and the direction set by the vector of the cutting feed. Equation (2.18) shows how this angle is calculated as this equation is universal for all drilling tools.

Figure 2.233b also shows an enlarged fragment of the so-called theoretical surface roughness (see detailed explanations in Section 2.1.3.8.5). In this picture, R_t is the maximum height of surface roughness which is one of the standard surface roughness parameters. It geometrically follows from this figure that it is calculated as

$$R_t = \frac{f}{z} \frac{\cos \kappa_{r1} \cos \kappa_r}{\cos \kappa_{r1} + \cos \kappa_r} \tag{2.157}$$

It follows from this equation that, for a given cutting feed in reaming, R_t, and thus roughness of reamed holes proportionally decreases when the number of teeth, z, of a reamer increases. That is why multi-tooth reamers are used in practice. Moreover, it also follows from Eq. (2.157) that roughness of the machined hole decreases with the chamfer angle $\varphi = \kappa_r$. This is well known in practice of reaming where two- or three-stage chamfer designs with progressively decreased chamfer angles are often used to achieve better quality reamed holes. Unfortunately, this is totally ignored by many tool manufacturers as discussed above.

The length of the minor cutting edge *1–3* is rather small. It geometrically follows from Figure 2.233b that this length is calculated using Eq. (2.20) as

$$l_m = f_z \frac{\cos \kappa_{r1} \cos \kappa_r}{(\cos \kappa_{r1} + \cos \kappa_r) \sin \kappa_{r1}} \tag{2.158}$$

Important particularities shown in Section B-B in Figure 2.233a are as follows:

1. The edges of the cylindrical part of reamers (including the minor cutting edge) are provided with thin (compared to drills) cylindrical margins, i.e., there is no clearance angle applied. The margin width, a_m, is in the range of 0.05–0.40 mm for reamers of 3–50 mm dia.: for

reamers of 3–5.5 mm dia. a_m=0.08–0.15 mm; for 5.5–10 mm dia. a_m=0.10–0.20 mm; and for 10–20 mm dia. a_m=0.10–0.25 mm. The surface roughness on the margins should not be worse than 0.2 µm Ra. For finishing reamers, runout of the margins should not be worse than 0.003 mm, whereas for semi-finishing reamers and for reamers used to ream the holes of "standard" quality, i.e., corresponding IT7 – IT9 grades, this runout is 0.005 mm.

3. Teeth on the cylindrical part of reamers is provided with secondary clearance angle α_{n2}=10°–15°. As such, smaller α_{n2}'s are used for finishing reamers whereas greater α_{n2}'s are used for semi-finishing reamers.

4. For a straight-flute reamer, the rake angle of the minor cutting edge, γ_{1n}, is neutral (equal to zero). If, however, the chamfer stage is provided with a rake angle, γ_n then γ_{1n} is calculated as $\gamma_{n1} = \arctan(\tan\gamma_n \cos\kappa_r)$ as discussed in the Appendix.

5. Land width, H_L, is a design parameter which depends on the reamer diameter and the number of teeth of a reamer. The common recommendations are as follows:

- For 6-flute reamers of 3–5.5 mm dia., H_L is in the range of 0.25–0.40 mm. The tolerance on H_L is 0.2 mm.
- For 6-flute reamers of 5.5–10 mm dia., H_L is in the range of 0.5–0.7 mm. The tolerance on H_L is 0.25 mm.
- For 8-flute reamers of 10–20 mm dia., H_L is in the range of 0.6–1.0 mm. The tolerance on H_L is 0.3 mm.

2.2.4.4 Number of Teeth and Their Spacing

The number of teeth, z for HSS and carbide inserts depends on the reamer diameter, ductility of the work material (affects thickness, shape, and the conditions of its transportation of chips over the chip flute), and reamer design. It is calculated (with rounding to the nearest integer) as follows:

$z = 1.5\sqrt{d} + 2$ for ductile work materials;
$z = 1.5\sqrt{d} + 4$ for brittle work materials;
$z = 1.2\sqrt{d}$ for indexable blade reamers.

Special reamers can have a smaller or greater number of teeth. For example, PCD reamers often have z=4 to accommodate PCD inserts.

It is strongly recommended to use an irregular spacing of the flutes, which is a deliberate variation from uniform spacing of the reamer cutting edges as viewed in the back plane, i.e., the plane perpendicular to the axis of a reamer. This prevents chatter in reaming, and thus leads to improvements in the circularity and surface roughness of reaming holes. Although a great number of papers and books were published to address the issues with chatter in reaming by optimizing flute spacing, no general or even practical model is available today to calculate the optimal spacing for given reaming conditions including work material, reamer geometry and material, machining regime and so on. Standard GOST 7722-71 includes the recommendation on flute spacing for reamers with 4, 6, 8, 10, and 12 teeth. Figure 2.236 shows an example a reamer with 8 teeth. Note also that the reamer's catalogs of even leading tool manufacturers do not mention, and thus list this parameter although it is of vital importance in reaming.

2.2.5 TRENDS IN THE DEVELOPMENT

As can be seen, the existing standards, particularly related to tolerances, were developed a long time ago so one may be of impression that nothing has changed since then. In the author's opinion, it is not quite so.

It is true that the design and geometry of reaming tool do not change significantly in recent years. However, the requirements for the accuracy and quality of reamed holes as well as to productivity of reaming, particularly in the automotive industry, have been increasing significantly. To comply

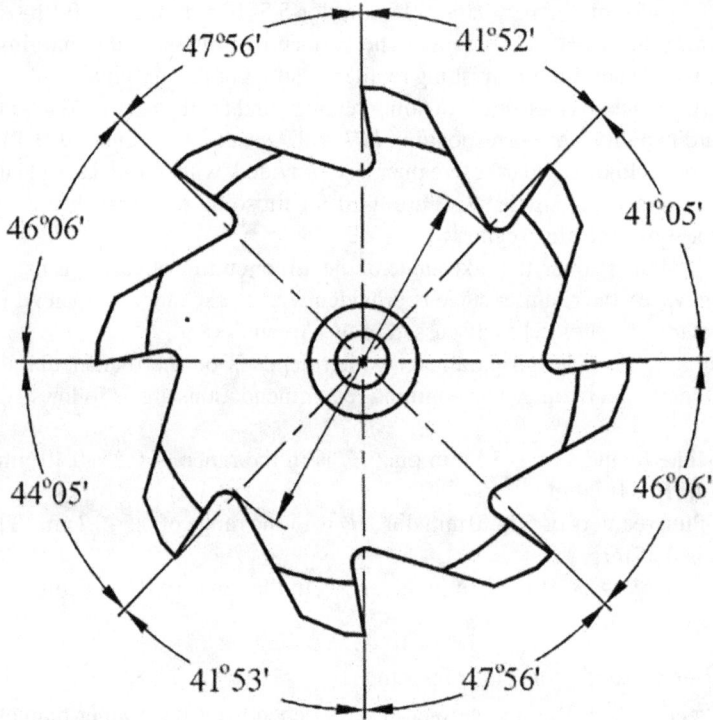

FIGURE 2.236   Flute spacing for a reamer with 8 teeth according to standard GOST 7722-77.

with these tough requirements, a number of features related to tool quality (see the concept of tool quality presented in Chapter 1) as tool manufacturing quality and tool materials/coating have been improved. Unfortunately, these changes are not covered by ISO standards.

Probably, the most noticeable change in the automotive industry is a wide use of PCD reamers. Figure 2.237 shows two examples. The use of these tools allowed a significant increase in both productivity and quality of reaming operation. The latter is particularly noticeable in surface roughness – mirror shiny holes are obtained in various aluminum alloys that were not ever possible with HSS and carbide reamers. As PCD tools allow a three–ten times increase in the cutting speed, the productivity of drilling operations increases accordingly (see Chapter 1). These tools however require special attention and equipment for their presetting, rigid spindle/low runout, high-speed machine to runs, and tool holders of high quality. The major problem, however, is in their manufacturing quality, particularly brazing of PCD inserts and their finishing – these issues are discussed in Chapter 4 in great detail.

Another important trend in advanced industries is the increasing popularity of monoblock reamers.

Achieving high accuracy of reaming tool is restricted by the excluding inaccuracies occurring on tool clamping in the holder and its presetting. These inaccuracies occur because of inherent tool holder runout and the ability/proper calibration of even the most advanced tool presetting machines. With growing requirement to the diametric and shape tolerances to reamed holes, the combined error (i.e., runout) due to the tool clamping/presetting can exceed the tolerances set by the part drawing. Another aspect is a tool back taper, which for reamers is much smaller than that for drills and other drilling tools so that the actual back taper after tool setting may be insufficient for optimal tool performance. When this is the case, the surface roughness of the machining hole deteriorates, tool life reduces, and the possibility of tool frictional chatter known as spiraling increases. As the accuracy of tool holders, even the most advanced ones, as, for example, hydraulic tool holders, and

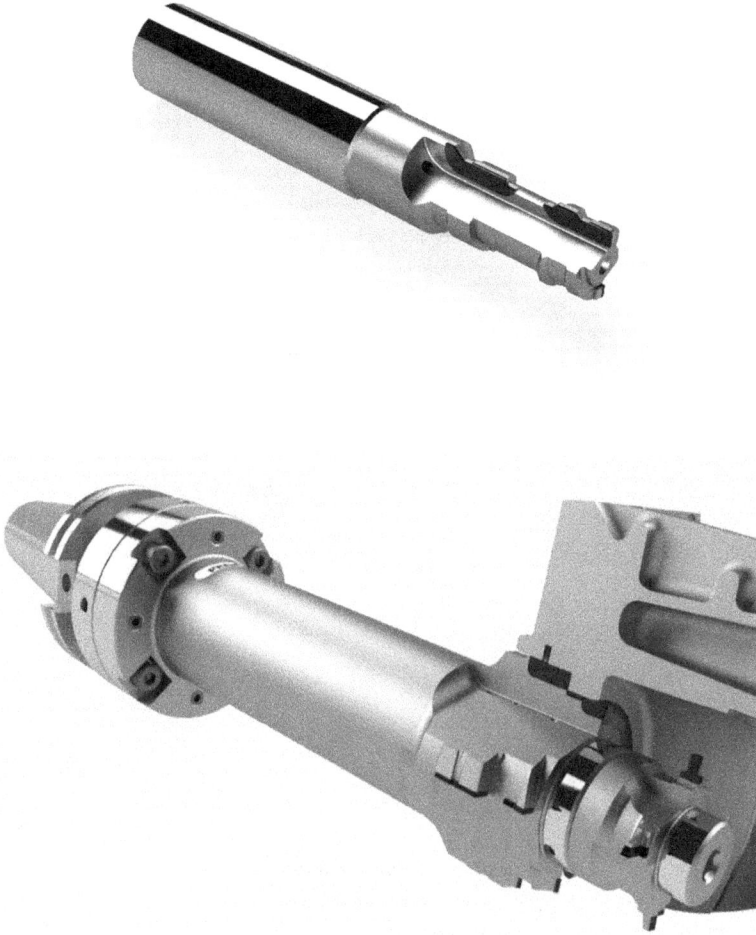

FIGURE 2.237   PCD multi-stage reamer by Premier Tooling Systems Co (MI, USA).

tool presetting machines reached its possible maximum, another way to deal with the tool clamping issues was developed.

The concept of monoblock drilling tool is simple – a semi-finished tool is mounted into a tool holder and permanently clamped till its next reconditioning. Then, the tool is finished using the tool holder as the datum. As a result, a monoblock, i.e., inseparable assembly (holder and tool) is formed. This automatically illuminates inaccuracies of the tool mounting and setting as the holder is directly clamped in a machine. Obviously, no presetting for such a tool is required. An example is shown in Figure 2.238.

2.2.6   PROCESS AND DESIGN PARAMETERS TO CONSIDER

When performing reaming operations, there are several parameters that directly affect the quality of reamed holes and reamer tool life to consider:

- Reamer design/geometry (including the number of flutes) and tool length
- Tool material
- Depth of cut understood as $a_p = (d_{dr} - d_1)/2$, where d_1 is the diameter of the pilot (core) hole (see Chapter 1)

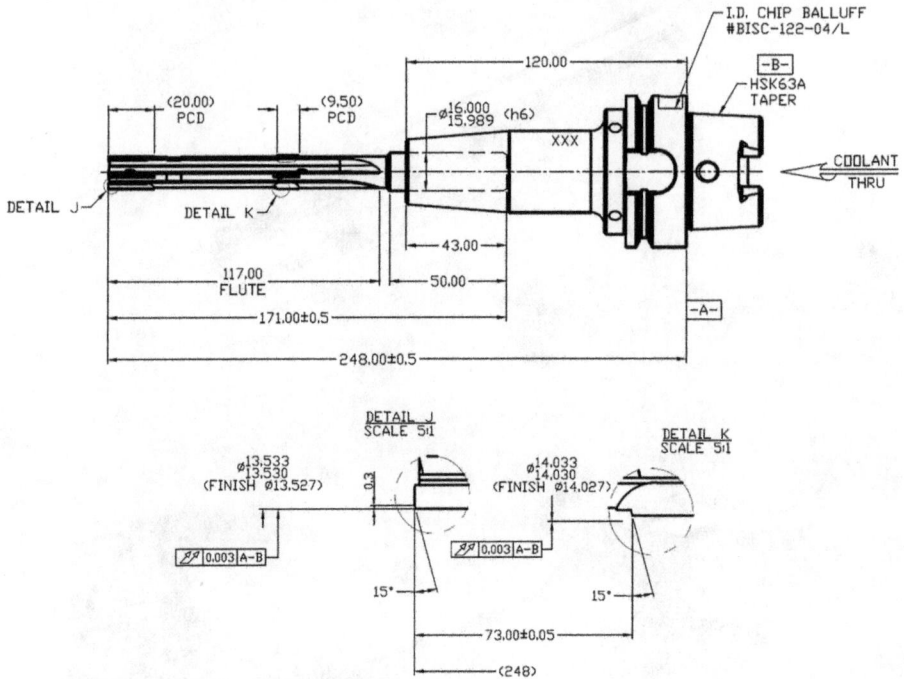

FIGURE 2.238 Example of monoblock reamers.

- Cutting speed, m/min and cutting feed, mm/rev
- Workpiece material
- Reamer setting accuracy including the system runout, offset, etc.
- MWF parameters: method of supply, flow rate, concentration, clearness, etc.
- System rigidity/stiffness including workpiece clamping
- Interrupted cuts.

It is important to point out that the listed considerations are the same as for HP drills. Moreover, they are conceptually the same for all modern HP and high-efficiency cutting tools.

2.3 TAPS

2.3.1 Introduction Words

Among many other cutting tools, the tap is the most regulated by a number of ISO standards. For example, the basic terminology of taps is specified in ISO 5967 "Taps and thread cutting – Nomenclature of main types and terminology" reviewed and confirmed in 2020; ISO 529 specifies the general dimensions of short and hand taps; ISO 2857 "Ground thread taps for ISO metric thread of tolerances 4H to 8H and 4G to 6G coarse and fine pitches – Manufacturing tolerances on the treaded portion, defines the tolerances on the tap profile"; ISO 8830 "High-speed steel machine taps with ground threads – Technical specification"; and so on. Many tap manufacturing companies such as Walter, OSG, and EMUGE provide detailed catalogs with important features of their products with technical information/recommendations for their proper use. Being far more detailed and disruptive compared to others, Walter's technical guides provide the basics of tapping theory and its utilization in practical; tap design and applications. When it comes to scientific literature, there is not even a relatively recent paper on taps as a general understanding is that there is nothing new can be developed in this field besides tool materials and coatings to improve tool life. Therefore, this section presents only material that advanced tool users and tap designers should know, particularly information not well explained by the standards and guide materials.

2.3.2 DEFINITION OF TAPPING AND TAP

Tapping: a drilling operation for producing internal threads with a drilling tool called a tap. Such a tool can be used in a great variety of machine tools. Figure 2.239a shows a common tapping arrangement on a lathe. The workpiece is clamped in a three-jaw self-centering lathe chuck mounted on the lathe spindle. This spindle provides the rotary motion. The tap is clamped in a chuck mounted on the tailstock of the machine, which is engaged with the carriage to provide the feed motion. Figure 2.239b shows a common tapping arrangement on a modern machining center. The workpiece is clamped in a part fixture mounted on the table of the machine. The tap is clamped in a specialized tapping chuck installed on the machine spindle, which provides both the rotary and feed motions.

Tap: a cylindrical or conical threading tool with one or more cutting or forming elements having screw threads of a desired form on the periphery. By a combination of rotary and axial motions, the leading end produces an *internal thread*.

Although this standard definition includes taps with forming elements, known as forming taps, roll taps, cold-forming taps, etc., the author agrees with Walter Co. who named it thread formers to distinguish them clearly from thread taps. This is because thread forming is a chipless process accomplished by plastic deformation of the work material, i.e., a thread-forming tool is not a cutting tool. Nevertheless, some important features of this type of tap are discussed in Section 2.3.9.

2.3.3 CLASSIFICATION

The complete tap classification is covered by standard ISO 5967 "Taps and thread cutting – Nomenclature of main types and terminology" reviewed and confirmed in 2020. The standard serves as a reference for tap users and manufacturers. The diagrams given, however, are only to illustrate the terminology as tap design can vary according to the manufacturer. They include nomenclature of the main types of taps, styles of threaded portion, sets of taps, dimensional characteristics, and thread profiles.

2.3.4 NOMENCLATURE

Figure 2.240 shows a combined sketch that helps to visualize the tap basic terminology. ISO 5967 does not define the terms. In the author's opinion, however, the definitions of the basic terms

(a)   (b)

FIGURE 2.239

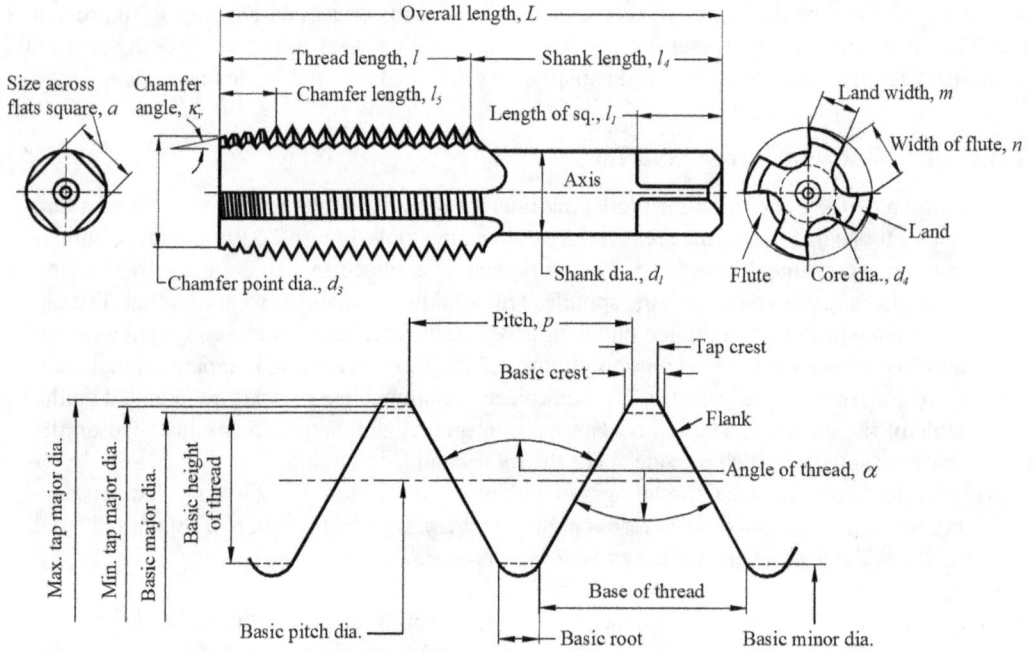

FIGURE 2.240   Visualization of tap basic terminology.

facilitate their better understanding and proper usage. Therefore, it was found necessary to present such definitions following guidelines established by the United States Cutting Tool Institute (1989).
Terminology

Axis: the imaginary straight line that forms the longitudinal line of the tool or threaded part.

Back Taper: a gradual decrease in the diameter of the thread forms on a tap from the chamfer end toward the back, which creates a slight relief on the threads.

Basic Profile of Thread: the cylindrical outline, in the axial plane, of the permanently established boundary between the provinces of the external and internal threads.

Body: the threaded full diameter portion of a solid tap, inclusive of the chamfer.

Bottoming Tap: a tap having a chamfer length of 1–2 threads.

Chamfer: the tapering of the threads at the front end of each land of a tap by cutting away and relieving the crest of the first few teeth to distribute the cutting action over several teeth.

Chamfer Angle: the angle formed between the chamfer and the axis of the tap, measured in an axial plane at the cutting edge.

Chamfer Length: the chamfer length is measured at the cutting edge and is the axial length from the point diameter to the theoretical intersection of the tap major diameter and the chamfer angle.

Chamfer Clearance Angle: the complement of the angle formed between the line tangent to the relieved surface at the cutting edge and radial line to the same point on the cutting edge.

Class of Thread: an alphanumeric designation to indicate the standard grade of tolerance and allowance specified for a thread.

Clearance: any space provided to prevent undesirable contact of the toll and the workpiece.

Concentric: having a common center.

Concentric Margin: a portion of the treaded land, adjacent to the cutting edge, that has concentric threads.

Core: the central portion of the tap below the flutes that joints the lands.

Core Diameter: the diameter of a circle that is tangent to the bottom of the flutes at a given point on the axis.

Core Taper: the taper in the core of a tap.

Crest: that surface of the thread that joints the flanks of the thread and is farthest from the cylinder or cone from which the thread projects.

Crest Clearance: the radial distance between the root of the internal thread and the crest of the internal thread of the coaxially assembled design forms of mating threads.

Cutting Edge: the leading edge of the land in the direction of the rotation that does the actual cutting.

Cutting face: the leading side of the land in the direction of rotation for cutting on which the chip impinges.

First Full Thread: the first full thread on the cutting edge behind the chamfer. It is at this position that rake, hook, and thread elements are measured.

Flank: the part of a helical thread surface that connects the crest and roots and that it theoretically a straight line in an axial plane section.

Flute: the longitudinal channel formed in a tap to create cutting edges on the thread profile and to provide chip space and cutting fluid passages.

Flute, Helical: a flute with uniform axial lead and constant helix in a helical path around the axis of c cylindrical tap.

Flute Lead: the axial advance of a helical cutting edge in one turn about the axis.

Flute Lead Angle is the angle that a helical cutting edge as a given point of the cutting edge makes with the axial plane.

Flute, Straight: a flute that forms a cutting edge lying in an axial plane.

Flute Length: the full axial length of a flute including the cutters sweep.

Hand of Cut: rotation for cutting viewed from the chamfered end of a tap is clockwise for *Left-Hand Cut* and couterclockwise for *Right-Hand Cut.*

Hand of Flutes: flutes, when viewed axially, twist in a counterclockwise direction for *Left-Hand Flutes* and in a clockwise direction for *Right-Hand Flutes.*

Hand of Threads: a thread, when viewed axially, winds in a clockwise and receding direction for *Left-Hand Threads* and counterclockwise and receding direction for *Right-Hand Threads.*

Heel: the edge of the land opposite the cutting edge.

Land: one of the threaded sections between the flutes of a tap.

Land Width: the chordal width of the land between the cutting edge and the heel measured normal to the cutting edge.

Lead: the distance of a helix advances axially in one complete turn.

Lead of Thread: the distance a screw thread advances axially in one complete turn. On a single start tap, the lead and pitch are identical. On a multiple start tap, the lead is the multiple of the pitch.

Pitch Diameter: the diameter of an imaginary cylinder or cone at a given point on the axis, of such a diameter and location of its axis that its surface would pass through the thread in such a manner as to make the thread ridge and the thread groove equal, and, therefore, is located equidistant between the sharp major and minor cylinders or cones of a given thread form. On the theoretical perfect thread, these widths are equal to one half of the basic pitch (measured to the axis).

Point Diameter: the diameter at the cutting edge of the leading end of the chamfered section.

Rake: the angular relationship of the straight cutting edge of a tooth with respect to a radial line through the crest of the tooth at the cutting edge. *Positive Rake* means that the crest of the cutting face is angularly ahead of the balance of the cutting face of the tooth. *Negative Rake* means that the crest of the cutting face is angularly behind the balance of the cutting face of the tooth. *Zero Rake* means that the cutting face is directly on a radial line.

Relief: the removal of metal behind the cutting edge to provide clearance between the part being threaded and threaded land.

Root: that surface of the thread that joints the flanks of adjacent thread forms and is immediately adjacent to the cylinder or cone from which the tread projects.

Runout: the radial variation from the true circle that lies in a diametral plane and is concentric with the tool axis.

Shaving: the excessive removal of material from the product thread profile by the tool thread flanks is caused by an axial advance per revolution less than or more than the actual lead of the tool. In tapping, this results in an increase in product pitch diameter without an increase in product major diameter.

Square: four driving flats parallel to the axis on a tap forming a square or square with round corners.

Taper Tap: a tap having a chamfer length of 7–10 threads.

Tread Angle: the angle formed by two adjacent flanks in an axial plane.

Total Indicator Variation [tiv]: the difference between maximum and minimum indicator readings obtained during a checking cycle. It is actually runout.

2.3.5 TAPS CUTTING GEOMETRY

Although standard ISO 5967, multiple tap manufacturers catalogs, and technical guides provide some help in their tapping guides, the cutting tool geometry of the tap is not covered in the listed materials. Therefore, a need is felt to explain the parameters of this geometry in order to understand tap performance and use introduced angles in the optimization of the tap performance knowing the influence of these angles (Astakhov 2010, 2014).

As discussed above, the analysis of the tool geometry of any drilling tool should start with the proper identification of the cutting edges involved in cutting. After that, the uncut chip thickness for each identified cutting edge should be determined as well as the tool cutting edge, rake, clearance, and inclination angles (see Section 2.1.3).

2.3.5.1 Uncut Chip Thickness

Figure 2.241 shows a model of cutting with a common tap. As follows from this figure, the cutting edge of the tap is not straight but consists of three segments, namely the major segment *1–2* and two side segments *1–3* and *2–4*. When the tap works, this cutting edge cuts the chip having the uncut (undeformed) chip cross section 3124. As can be seen, being the greatest for the first tooth, this cross-sectional area reduced toward the end of the chamfer as the uncut chip width b_z reduces whereas the uncut chip thickness, h_{D_z} is the same for all cutting teeth.

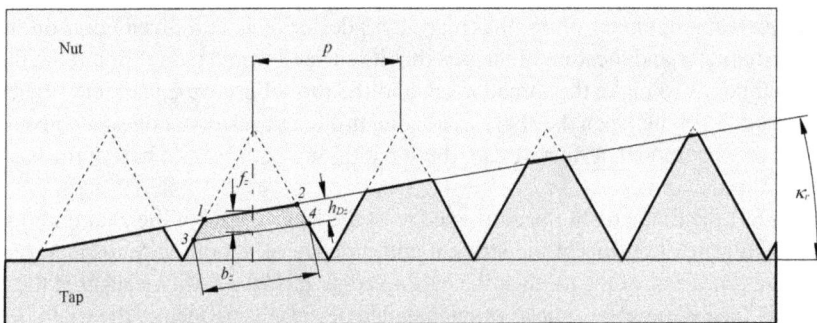

FIGURE 2.241 Model of cutting with a tap.

As mentioned above, for a one-start tap, the cutting feed, f, is equal to the pitch p. Therefore, the feed per tooth is determined the same as for any drilling tool as

$$f_z = p/N \tag{2.159}$$

where N is the number of flutes (ISO 5667). The uncut chip thickness is calculated as (Chapter 1)

$$h_{Dz} = f_z \cos \kappa_r = \frac{f}{N} \cos \kappa_r \tag{2.160}$$

where κ_r is the tool cutting edge angle of the major segment *1–2*. This angle is known as the chamfer angle (ISO 5667).

2.3.5.2   Chamfer Angle and the Chamfer Length

The chamfer angle κ_r plays the same role as in any drilling tool (see Section 2.1.3.4) as it affects the uncut chip thickness, i.e., the load on the cutting teeth, and the axial force. Tool life of taps increases with a reduction of κ_r, whereas the axial force in tapping decreases. The latter is important in hand tapping. On the other hand, small chamfer angles lead to the corresponding increase in the chamfer length, and thus in the cycle time in tapping. Kyreev (2008) discussed the recommended uncut chip thickness in tapping and the chamber length for various work materials.

Walter Co. offers five chamfer forms of taps. Form A has a chamfer length of 6–8 threads and is recommended for straight-flute taps for through holes in medium- and long-chipping materials. Form B has a chamfer length of 3.5–5 threads and is recommended for straight-fluted with spiral (helical) point taps for through holes in medium- and long-chipping materials. Form C has a chamfer length of 2–3 threads and is recommended for straight- and helical-fluted taps for blind holes in long- and medium-chipping materials and through holes in short-chipping materials. Form D has a chamfer length of 3.5–5 threads and is recommended for straight-fluted or helical-fluted taps for holes with long-chipping materials. Form E has a chamfer length of 1.5–2 threads and is recommended for straight-fluted or helical-fluted taps for bind holes with very short thread runout.

2.3.5.3   Shape and Direction of the Chip Flutes

The chip flute is an important design component of any tap. Its part makes the tap's rake face, which in its intersection with the flank faces, forms the cutting edges of the tap's cutting teeth. The shape and size of such a flute should be sufficient for allocating the chip formed in tapping and its successful transportation over this flute. The flute should have a shape convenient for manufacturing, e.g., to dress of a grinding wheel for fluting. The profile of the flute should be smooth to avoid possible cracking or other defects on tap heat treatment or sintering. According to Kyreev (2008), a double-radius profile of the tap flute shown in Figure 2.242 satisfies the listed requirements. Its geometrical parameters shown in this figure, namely, X_1, X_2, Y_1, Y_2, Y_3, R_1, R_2, d_4, and θ are given by Kyreev for taps of diameter $d = 6$–50 mm in Table 1.13 of his book (Kyreev 2008).

2.3.5.4   Inclination, Rake, and Clearance Angles

The definition of the cutting edge inclination angle λ_s is discussed in the Appendix. Obviously, this angle is zero in straight-flute taps as the cutting edges lie in the radial (reference) plane. The Appendix also points out that when this angle is applied to the cutting edge, i.e., $\lambda_s \neq 0$ it determines the direction of chip flow. This is widely used in the design and application of taps.

Figure 2.243a shows the geometry of the chamfer with a positive λ_s as a modification of straight-flute taps, called in industry the taps with a spiral point. In the author's opinion, a helical point term should be used as there are no spirals (a spiral is a flat curve) in the tap shown in

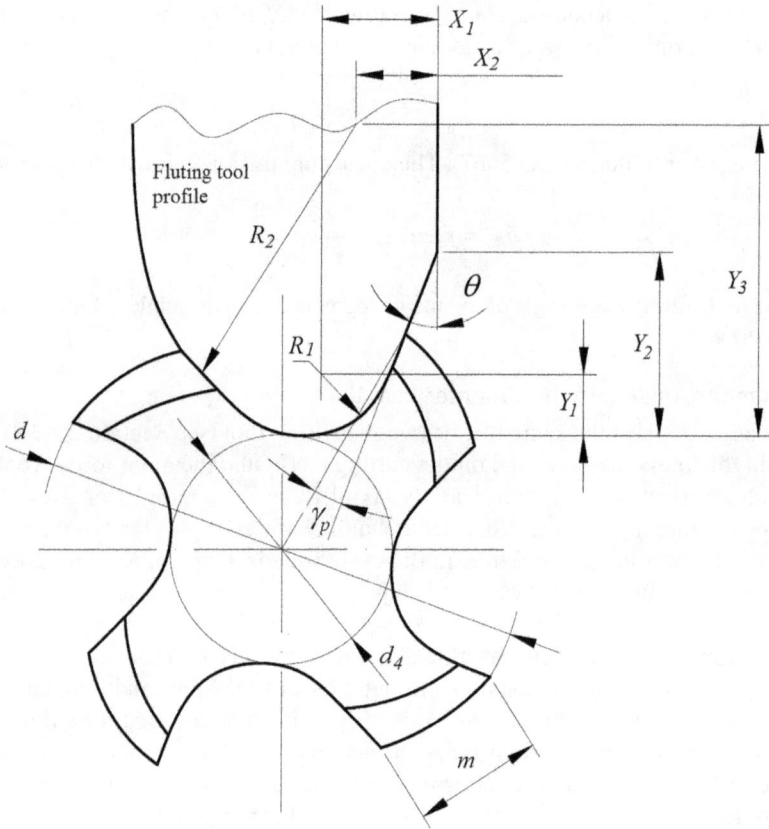

FIGURE 2.242 Recommended shape of the chip flute.

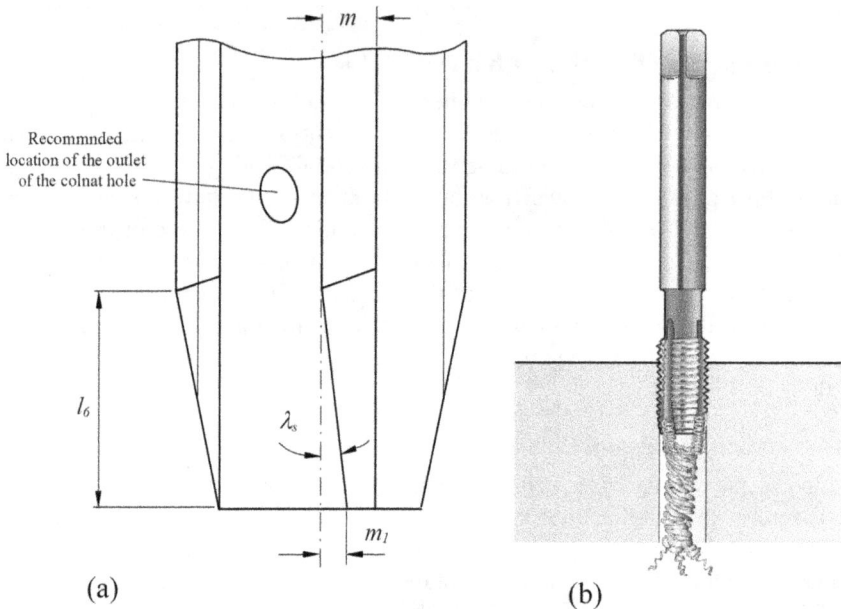

FIGURE 2.243 Helical point modification of the rake face of a straight-flute tap: (a) definition of the cutting edge inclination angle λ_s and (b) chip flow pattern.

Figure 2.243a. Taps with such a modification are used for tapping through holes. As follows from Figure 2.243a, the cutting edge inclination angle λ_s is calculated as

$$\lambda_s = \arctan \frac{m_1}{l_6} \qquad (2.161)$$

where $m_1 = (0.5-0.7)m$.

When the internal high-pressure MWF supply is used, the outlet of the coolant hole should be located as shown in Figure 2.243a to facilitate chip removal. Figure 2.243b shows the direction of chip flow in a tap with a spiral point according to Walter Co.

When the type of the hole (blind or through) is known, then a helical-fluted tap should be used. Left-hand helical flute taps are used for tapping through holes whereas right-hand helical flute taps are used for tapping blind holes (bottoming taps) as shown in Figure 2.244. As it follows from the definition of the cutting edge inclination angle, this angle is positive in the former case and equal $\lambda_s = \omega$ (ω is the angle of helix), whereas in the latter case, this angle is negative and equal $\lambda_s = -\omega$. The angle of helix depends on the work material. For low- and medium-carbon steels is in the range of $\omega = 10°-16°$, for long-chip materials $\omega = 25°-30°$.

The rake angle of segment $1-2$ of the cutting edge (see Figure 2.241) is assigned for the first full-form tooth (at the angle of the chamfer) as the rake angle in the back plane, γ_p (see the Appendix) as shown in Figure 2.242. This angle should be in the range of $5°-30°$ depending on the properties of the work material. This angle is not the same for all the cutting teeth. Rather, it slightly increases toward the first cutting tooth as

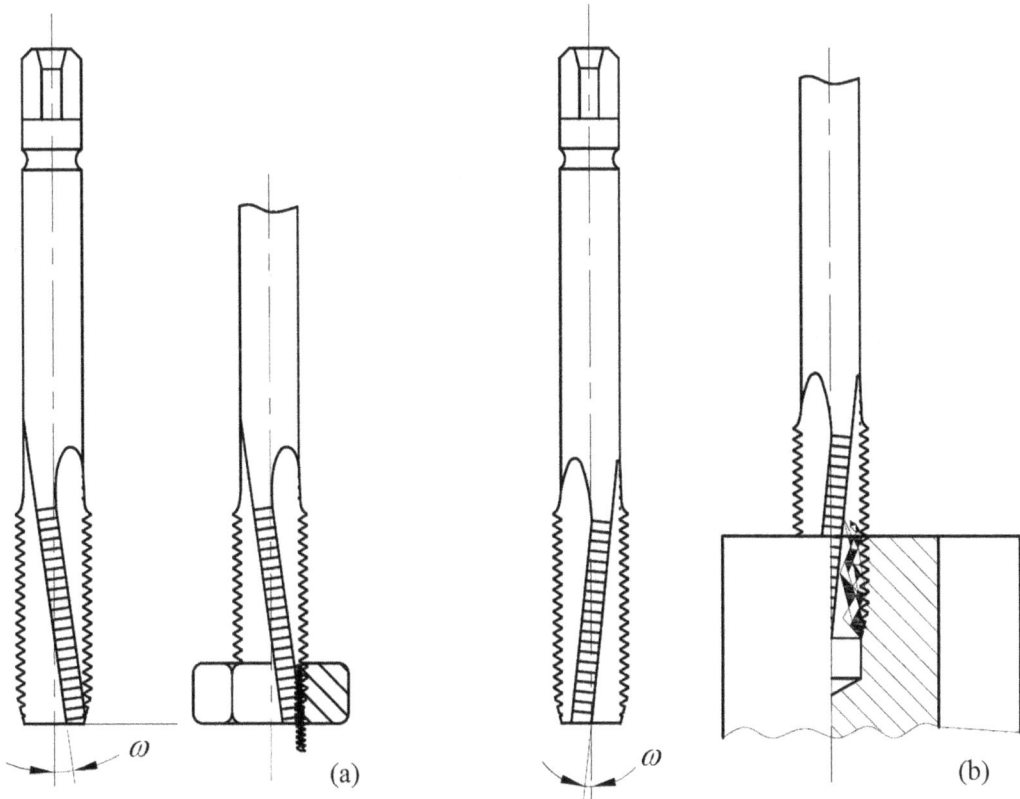

FIGURE 2.244 Taps with helical flutes: (a) left-handed helical tap for through holes and (b) right-handed helical tap for blind hole.

$$\gamma_{p-i} = \arctan\left(\frac{d}{d_i}\tan\gamma_p\right) \qquad (2.162)$$

where d_i is the diameter of the center of segment 1–2 of the considered cutting tooth i.

Being of great importance, the tap rake angle and the shape of the chip flute are not normally listed in catalogs of tap manufacturers. According to the author's experience, zero or even negative rake angle taps are often supplied to manufacturing companies that result in lower tool life and problems with chip removal. Therefore, if one wants to improve/optimize a tapping operation(s), he or she needs to request these two important parameters of the tap geometry from the tap supplier.

The flank face of the cutting teeth of a tap is formed by a reliving operation, i.e., its eccentric grinding having the path of an Archimedes spiral as shown in Figure 2.245. The clearance angle for segment 1–2 of the cutting edge (see Figure 2.241) is assigned for the first full-form tooth (at the angle of the chamfer) as the clearance angle of cutting edge in the back plane, α_p (see the Appendix) as shown in Figure 2.245. It is determined by the setting parameter K as

$$K = \frac{\pi d}{z}\tan\alpha_p \qquad (2.163)$$

As well as the rake angle, the clearance angle α_p slightly increases toward the first cutting tooth as

$$\alpha_{p-i} = \arctan\left(\frac{d}{d_i}\tan\alpha_p\right) \qquad (2.164)$$

where d_i is the diameter of the center of segment 1–2 of the considered cutting tooth i.

Table 2.11 provides the recommended rake γ_p and clearance angle α_p for some common work materials (Kyreev 2008).

The T-use-S clearance angle α_{pe} and thus angles α_{pe-i} on the cutting teeth of the chamfer can significantly differ from its T-hand-S values defined by Eqs. (2.162) and (2.164) due to high cutting feed equal to the pitch of the thread (see the Appendix). These angles are calculated as

$$\alpha_{pe-i} = \arctan\left(\tan\alpha_{pi} - \frac{p}{\pi d_i}\tan\kappa_r\right) \qquad (2.165)$$

When it was found that α_{pe-1} is too small (it commonly happens when the pitch p is great), the T-hand-S angle α_p should be increased.

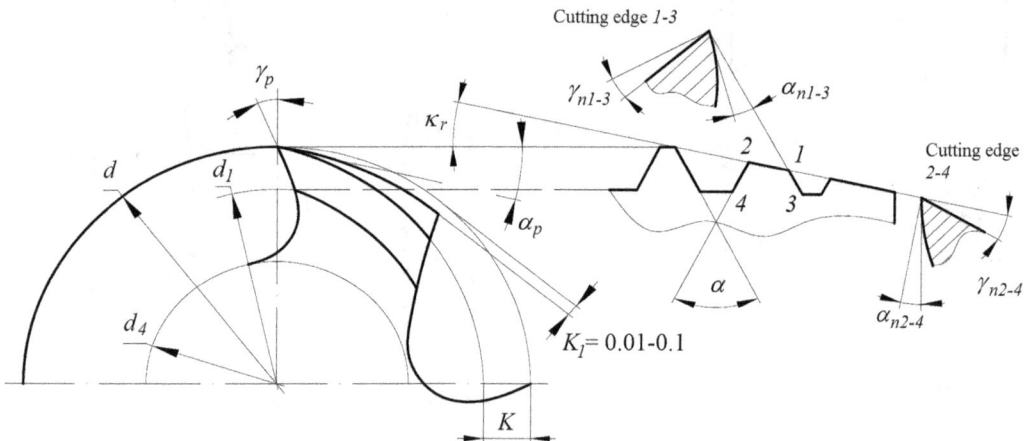

FIGURE 2.245 Geometry of the cutting teeth.

TABLE 2.11

Recommended Rake γ_p and Clearance Angle α_p for Some Common Work Materials

Work Material	γ_p, (°)	α_p, (°)
Carbon steel, strength MPa to 400 MPa	12–15	10–12
to 700	8–12	10–12
to 900	6–8	10–12
Steel casting and forging	6–10	5–7
Alloy steel, strength MPa to 900	6–9	5–7
over 900	3–6	5–7
Stainless steels	6–12	6–8
Cast iron: HB < 180	2–3	5–7
HB > 180	0–2	5–7
Ductile cast iron	6–8	5–7
Cupper	15–20	16–20
Bronze	6–8	6–8
Zink	15–20	16–20
Al-Si alloys	8–12	10–12

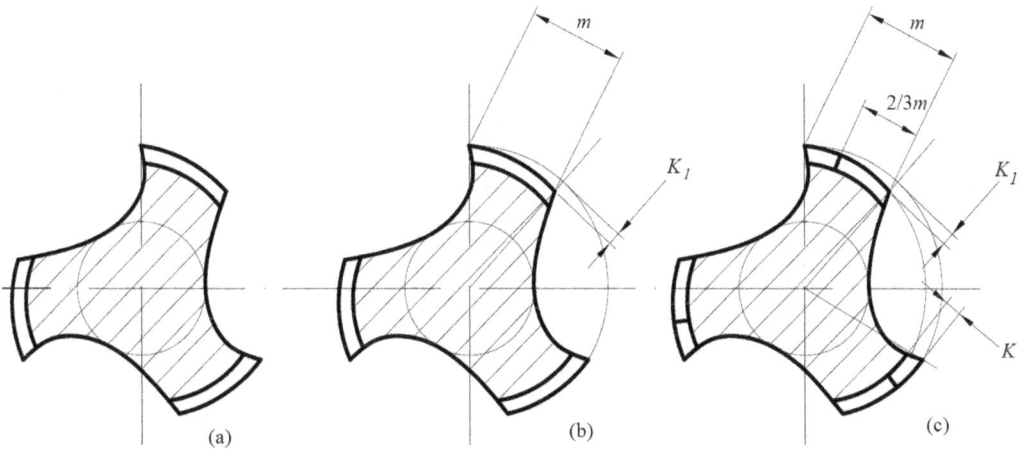

FIGURE 2.246 Methods of forming the flank faces of segments *1–3* and *2–4* of the cutting edge.

There are three methods of forming the flank faces of segments *1–3* and *2–4* of the cutting edge:

1. The flank faces of segments *1–3* and *2–4* of the cutting edge are not relieved (Figure 2.246a). This is common for taps of $d < 3$mm. As such, the flank faces have a full contact with the formed thread that increases tapping torque.
2. Taps of diameters 3–52 mm are relived over the whole land width with $K_1 = 0.01$–0.1 mm as shown in Figure 2.246b. As a result, the clearance angles are formed on segments *1–3* and *2–4* of the cutting edge. This leads to reduced tapping torque due to reduced friction on the corresponding flank faces and much smaller deposits of the work material adhered to the flank faces. This is the most common method.

3. Reliving only a part (2/3) of the land width as shown in Figure 2.246c. Experience shows that this method is the best for tapping stability and quality of tapped thread. Due to its technological complicity, this method is used only for some special cases.

The normal clearance angles of segments *1–3* and *2–4*, namely α_{n1-3} and α_{n2-4} are calculated as

$$\alpha_{n1-3-i} = \arctan\left(\tan\alpha_{p-i}\sin\frac{\alpha_{p-i}}{2} \right) \tag{2.166}$$

$$\alpha_{n2-4-i} = \arctan\left(\tan\alpha_{p-i}\sin\left(90-\kappa_r\right) \right) \tag{2.167}$$

The normal rake angles of segments *1–3* and *2–4*, namely γ_{n1-3} and γ_{n2-4}, are calculated as

$$\gamma_{n1-3-i} = \arctan\left(\tan\gamma_{p-i}\sin\frac{\gamma_{p-i}}{2} \right) \tag{2.168}$$

$$\gamma_{n2-4-i} = \arctan\left(\tan\gamma_{p-i}\sin\left(90-\kappa_r\right) \right) \tag{2.169}$$

2.3.6 THREAD PROFILE AND TOLERANCES

2.3.6.1 Thread Profile

There are a great number of screw threads used in industry covered in numerous ISO and national standards. Among them, the ISO metric screw threads are the world-wide most commonly used type of general-purpose screw thread. Due to high usage, these threads were one of the first international standards agreed when the International Organization for Standardization was set up in 1947. The "M" designation for metric screws indicates the nominal outer diameter of the screw, in millimeters, e.g., an M12 screw has a nominal outer diameter of 12 mm.

Threaded joints are defined as separable joints which are used to hold to machine parts together by means of threaded fastening such as bolt and nut. Figure 2.247 shows such a joint. When considering

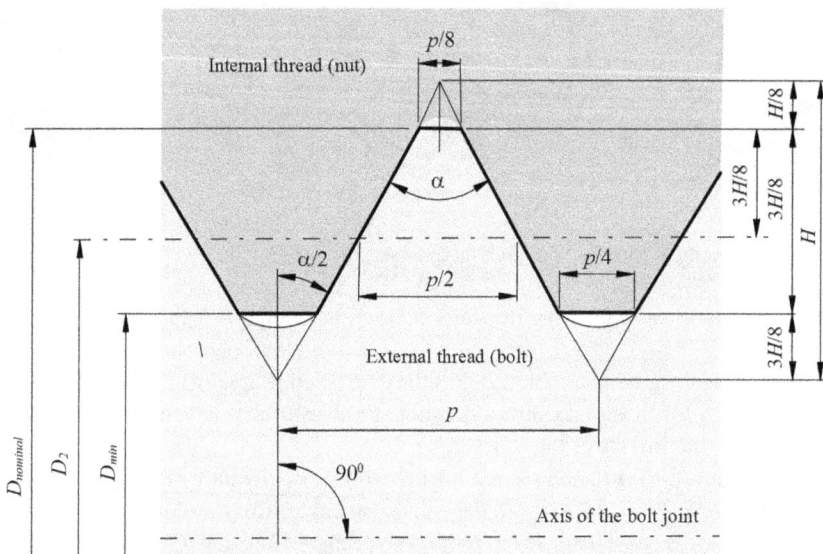

FIGURE 2.247   Basic dimensions of a thread joint.

a joint, one needs to know the characteristic dimensions of both the external thread and internal thread. The thread dimensions shown in Figure 2.247 for coarse thread and fine thread can be found in corresponding standards (e.g., ISO/R 724, ISO general-purpose metric screw threads – Basic dimensions) and tables located in any standard machine design handbook, e.g., Oberg et al. (2020).

The prime parameter of many threads is the included angle of thread, α. Its sense is shown in Figure 2.247 for metric threads $\alpha=60°$.

A thread size is specified based on a nominal (major) diameter, $D_{nominal}$, and either the number of threads per inch, TPI (for unified inch threads) or the pitch, p (for metric threads). The pitch, p, is the distance between the threads. TPI and p are related as

$$TPI = 1/p \qquad (2.170)$$

An important dimension of thread is the pitch diameter, D_2 related to pitch p as

$$D_2 = D_{nominal} - 0.75H = D_{nominal} - 0.64951905p \qquad (2.171)$$

The minor diameter of thread, D_{min}, is related to pitch p as

$$D_{min} = D_{nominal} - 1.25H = D_{nominal} - 1.08253175p \qquad (2.172)$$

2.3.6.2 Thread and Tap Tolerances

There are a number of ISO standards for specification of thread tolerances. Two of them are mostly used in metalworking industries: ISO/R 965/I, ISO general-purpose metric threads – Tolerances – P rinciples and basic data and ISO/R 965/II, ISO general-purpose metric threads – Tolerances – Limits of sized for commensal bolt and nut threads – Medium quality. The thread profile of nuts with relevant tolerance zones visualization important in the tap design/manufacturing is shown in Figure 2.248. In this figure,

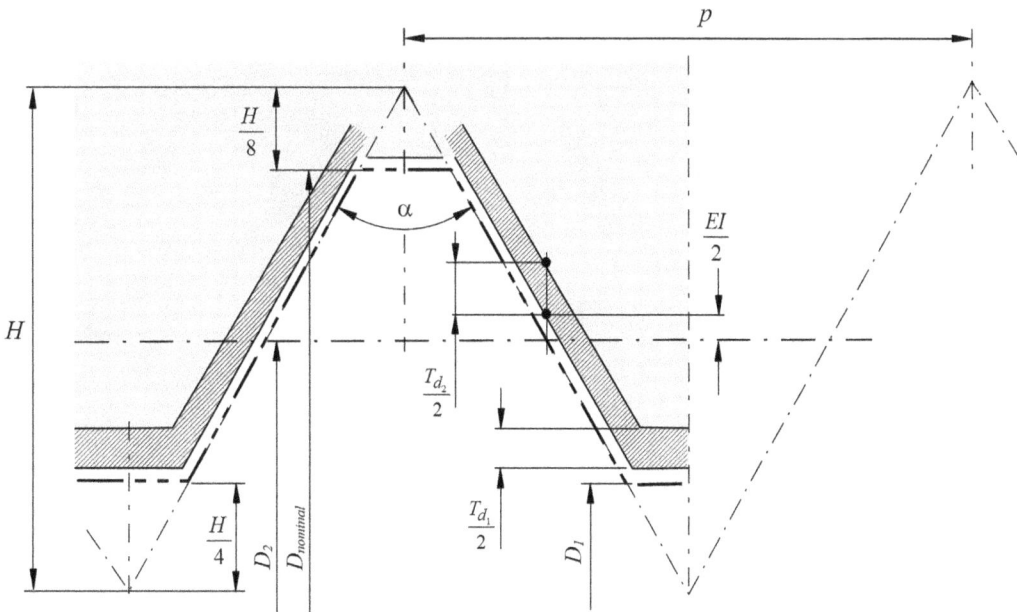

FIGURE 2.248 Thread profile of nuts.

T_{D_1} =minor diameter tolerance,
T_{D_2} =pitch diameter tolerance,
EI=minimum clearance (zero for H deviation, positive for G deviation).

These tolerances are defined by the above-mentioned standards.
To meet the tolerances on nut defined by the standards, the tolerances on the tap profile are defined by ISO standards. For the considered standards thread, standard ISO 2857, Ground thread taps for ISO metric thread of tolerances 4H to 8H and 4G to 6G coarse and fine pitches – Manufacturing tolerances on the treaded portion, defines the tolerances on the tap profile. The thread profile of tap with relevant tolerance zones is shown in Figure 2.249.

In this figure,

$d=D_{nom}$ =nominal diameter,
d_{min} =permissible minimum major diameter,
J_s =minimum clearance on the major diameter,
$d_2=D_2$ =pitch diameter,
d_{2min} =minimum pitch diameter,
$d_{2\,max}$ =maximum pitch diameter,
Es =upper deviation of pitch diameter,
Em =lower deviation of pitch diameter,
T_{d_2} =tolerance on pitch diameter.

For the production of the following nut classes
4H – 5H – 6H – 7H – 8H with zero minimum clearance,
4G – 5G – 6G with positive minimum clearance,
Three tolerance classes of taps have been accepted:
Class 1 – Class 2 – Class 3

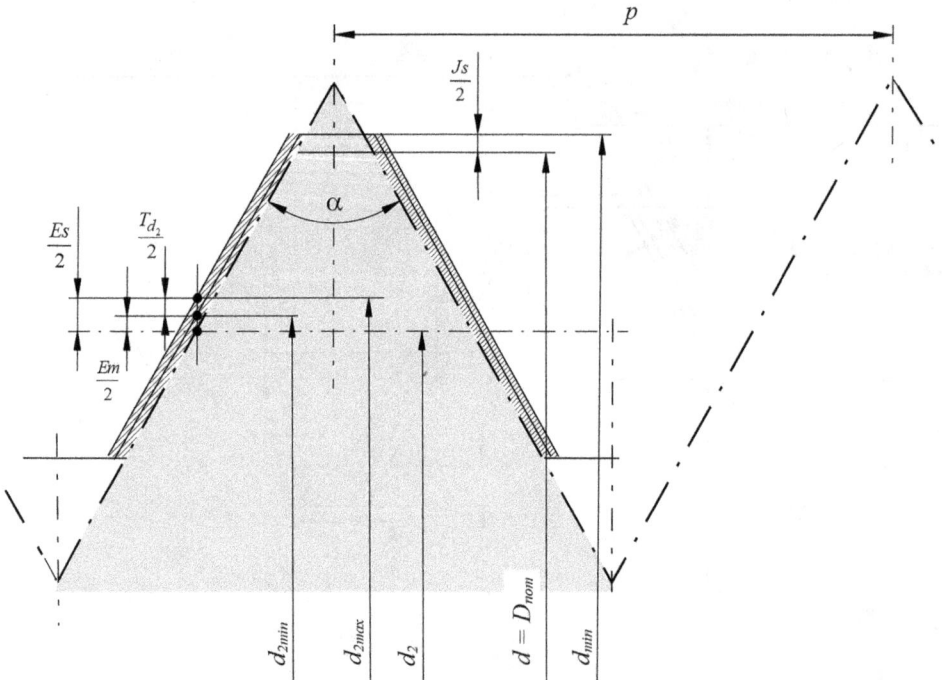

FIGURE 2.249   Thread profile of tap.

The tolerances of these three classes are determined in terms of a tolerance unit t, the value of which is equal to the pitch tolerance value T_{D_2}, grade 5 of the nut (extrapolated up to pitch 0.2 mm), i.e.,

$$t = T_{D_2} \text{ grade 5 of the nut} \tag{2.173}$$

The particular values of T_{D_2} are given in ISO/R 965/I, section 9. The value for the tap pitch diameter tolerance T_{d_2} is the same for all three classes 1, 2, and 3: it is equal to 20% of t (Eq. 2.173).

The position of the tolerance of the tap with respect to the basic pitch diameter results from the lower deviation Em, the values of which are (see Figure 2.249)

for tap class 1: $+0.1 \cdot t$
for tap class 2: $+0.3 \cdot t$
for tap class 3: $+0.5 \cdot t$

The taps of classes 1–3 are generally used for the manufacture of nuts of the following classes:
Class 1: for nuts of classes 4H and 5H
Class 2: for nuts of classes 6H and also 4G and 5G
Class 3: for nuts of classes 7H–8H and also 6G

It is shown graphically in Figure 2.250.

Standard ISO 2857 presents calculations (with examples) of tap thread dimensions and tolerances based upon tolerance unit t per Eq. (2.173). For example, the tolerance on tap pitch diameter d_2 (see Figure 2.249) for ISO class 1 tap is calculated as

$$Em = 0.1t \tag{2.174}$$

$$Es = 0.3t \tag{2.175}$$

$$d_{2\min} = d_2 + Em; \quad d_{2\max} = d_2 + Es \tag{2.176}$$

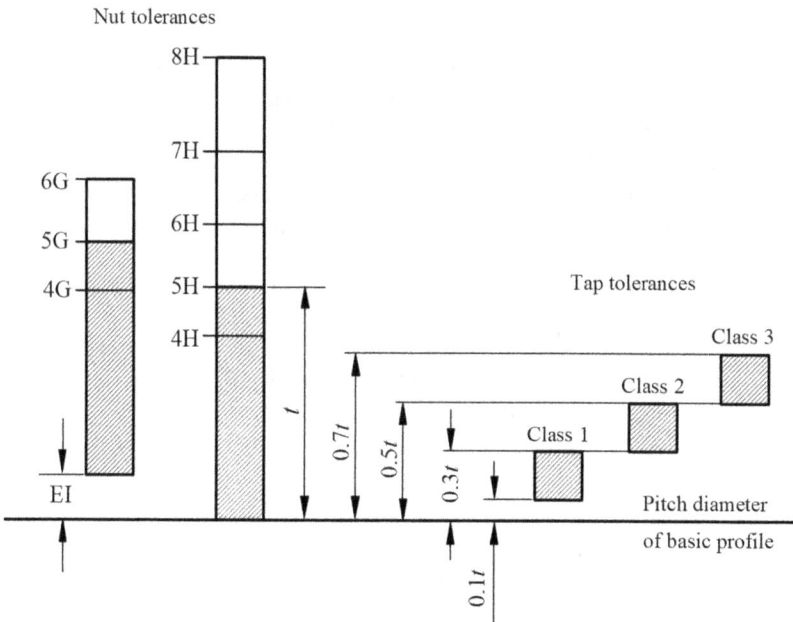

FIGURE 2.250 Graphical representation of the tolerance of the taps of classes 1–3.

where d_2 values correspond to the values of pitch diameter D_2 of the nut in conformity with ISO/R 724.

According to the standard ISO 2857, the taps shall bear, after their dimensional designation (as indicated in ISO/R 529), the nominal diameter and, if necessary, the pitch of the thread, and the symbol ISO followed by the class of the tap, a dash being placed before the ISO symbol.

Examples:

For an M6 coarse pitch tap of class 2: M 6- ISO 2

For an M 20 tap with a pitch of 2 of class 1: M 20 × 2- ISO 1

2.3.6.3 Runout Tolerances

The circular runout is checked when the tap is mounted between centers at the following locations as shown in the following:

- in the middle of the chafer length (t_1),
- on the first complete thread after the point on the flank (t_2),
- on the shank, at twice the driving square length (t_3).

The circular runout tolerances t_1, t_2, and t_3 for HSS taps with ground thread according to standard ISO 8830 are given in Table 2.12 in relation to the nominal diameter of the tap, d (Figure 2.251).

2.3.7 METHODOLOGY TO CALCULATE THE DRILLING TOOL DIAMETER FOR TAP HOLES

A screw thread complying with the requirements of the International Standards for ISO general-purpose metric screw threads in accordance with ISO 68-1, ISO 261, ISO 262, ISO 724, ISO 965-2, and ISO965-3 is designated by the letter M followed by the value of the nominal diameter and of the pitch, expressed in millimeters and separated by the sign "×". For coarse pitch threads listed in ISO 261, the pitch may be omitted. The tolerances class designation comprises a class designation for the pitch diameter tolerance followed by a class designation for the crest diameter tolerance. Each class designation consists of a letter indicating the tolerance position, capital for internal threads, and small for external threads. If the two class designations for the pitch diameter and crest diameter (major or minor diameter for internal and external threads, respectively) are the same, it is not necessary to repeat the symbols. Figure 2.252 shows an example of standard assigning tolerance on the internal (minor) diameter of thread. Therefore, the tolerance on the internal (minor) thread diameter is known as shown in the part drawing. This tolerance should be used in the selection of the proper tolerance on the drill for tap holes (the tap drill).

This section discusses a methodology of the selection of the proper diameter and tolerances on a tap drill (used to drill a hole for further tapping with a threading tap). Figure 2.253 shows

TABLE 2.12

The Circular Runout Tolerances t_1, t_2, and t_3

D	t_1	t_2	t_3
Mm		µm	
$d < 10$	18	18	30
$10 \leq d < 18$	22		
$18 \leq d < 30$	26	22	40
$30 \leq d < 40$	30		
$40 \leq d$	36	26	

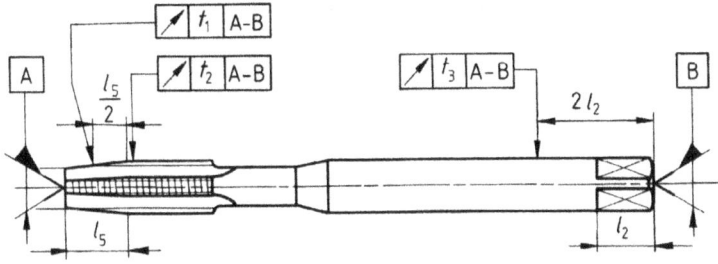

FIGURE 2.251 Assigning runout tolerances on the tap according to standard ISO 8830.

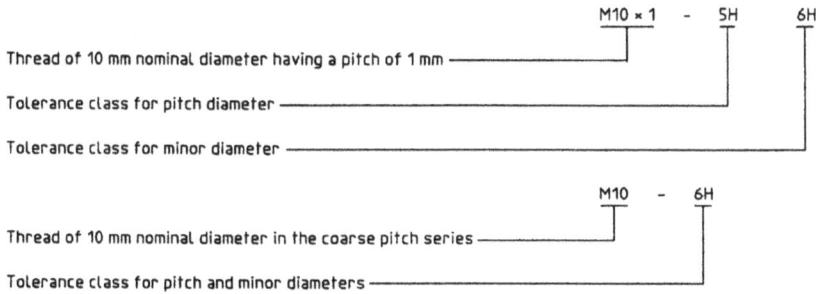

FIGURE 2.252 Example of standard assigning tolerance on the internal (minor) diameter of thread.

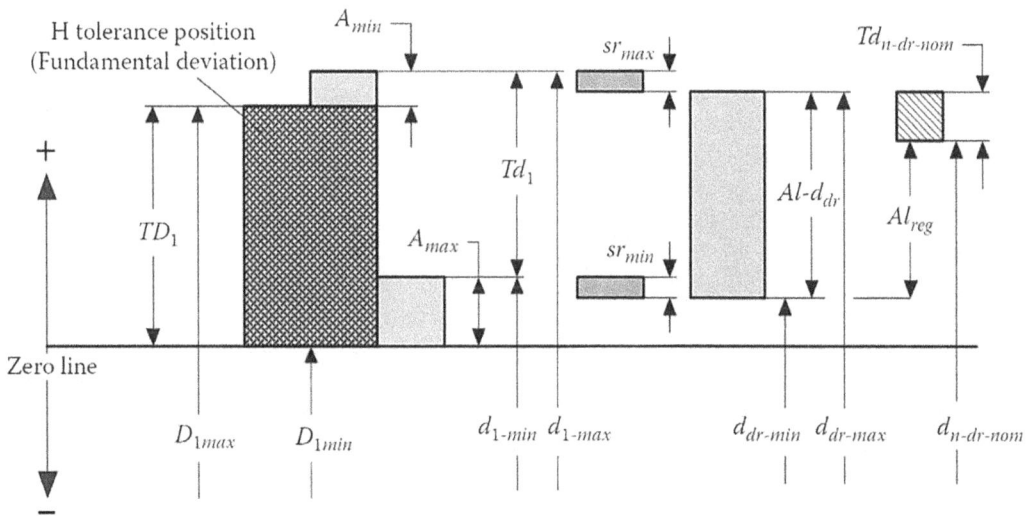

FIGURE 2.253 The model to calculate tap drill diameter including metric tolerance system for screw threads with H tolerance position.

the model to calculate tap drill diameter, its tolerance, and minimum allowable diameter after regrind for metric threads with H tolerance position. In the model shown in Figure 2.253, the following applies

$D_{1\,min}$ is the minimum minor diameter of the internal thread (Table 2.13).
$D_{1\,max}$ is the maximum minor diameter of the internal thread (Table 2.13).

TABLE 2.13

Internal Metric Thread – M Profile Limiting Dimensions, ANSI/ASME B1.13M-1983 (R1995) for M6, M8, and M10 Threads

Basic Thread Designation	Tolerance Class	Minor Diameter D_1(mm)	
		Min.	Max.
M5×0.75	6H	4.134	4.334
M6×1	6H	4.917	5.153
M8×1.12	6H	6.647	6.912
M10×1.5	6H	8.376	8.676

TABLE 2.14

Maximum and Minimum Growth (recovery) of the Minor Diameter on Tapping for 8%Si Aluminum Alloys

Pitch (mm)	0.8	1	1.25	1.5
A_{max} (mm)	0.064	0.080	0.100	0.120

A_{max}, A_{min} are the maximum and minimum growth (springback, see Section 2.1.3.8) of the minor diameter on tapping. These depend on the properties of the work material. Table 2.14 shows A_{max} for 8%Si aluminum alloys (e.g., A380), while A_{min} is approximately equal to $A_{max}/2$ (GOST 19257-73).

d_{1-max} and d_{1-min} are maximum and minimum diameters of the tap hole, respectively.

$TD_1 = D_{1-max} - D_{1-min}$ is the tolerance on the tap hole.

sr_{max} is the maximum drill setting runout.

sr_{min} is the minimum drill setting runout.

$d_{dr-min} = d_{1-min} - sr_{min}$ is the minimum diameter of the tap drill (cutoff diameter for regrinds) (to be shown in the drawing).

$d_{dr-max} = d_{1-min} - sr_{max}$ is the maximum diameter of the tap drill.

$Al\text{-}d_{dr} = d_{1-max} - d_{1-min}$ is the total allowance for the tap drill diameter variation.

$d_{n\text{-}dr\text{-}nom}$ is the chosen nominal diameter of a new tap drill (to be shown in the drawing).

$Td_{n\text{-}dr\text{-}nom} = d_{dr\text{-}max} - d_{n\text{-}dr\text{-}nom}$ is the tolerance on a new drill (to be shown in the drawing).

Al_{reg} is the allowance for regrinding.

The methodology allows determining (1) the diameter of a new drill and its tolerance and (2) the minimum diameter of the tap drill (cutoff diameter for regrinds). It includes the following simple steps:

1. *Determination of D_{1min} and D_{1max} using the tolerance zone assigned by the part drawing.* For a given thread size, these are determined from Table 2.13. For example, for M6×1 thread, $D_{1min} = 4.917$ mm and $D_{1max} = 5.153$ mm. Graphical representation of the obtained diameters shown in Figure 2.254 (based on the model shown in Figure 2.253) makes this and further steps much more transparent helping to avoid numerical errors.

2. *Determination of the maximum and minimum diameter of the machined hole prior springback.* As follows from Figure 2.253, $d_{1-max} = D_{1max} + A_{max}/2$ and $= D_{1-min} + A_{max}$. For example, for M6×1 thread, $A_{max} = 0.08$ mm (Table 2.14); therefore, $d_{1-max} = 5.153 + 0.08/2 = 5.193$ mm and $d_{1-min} = 4.917 + 0.08 = 4.997$ mm as can be seen in Figure 2.254.

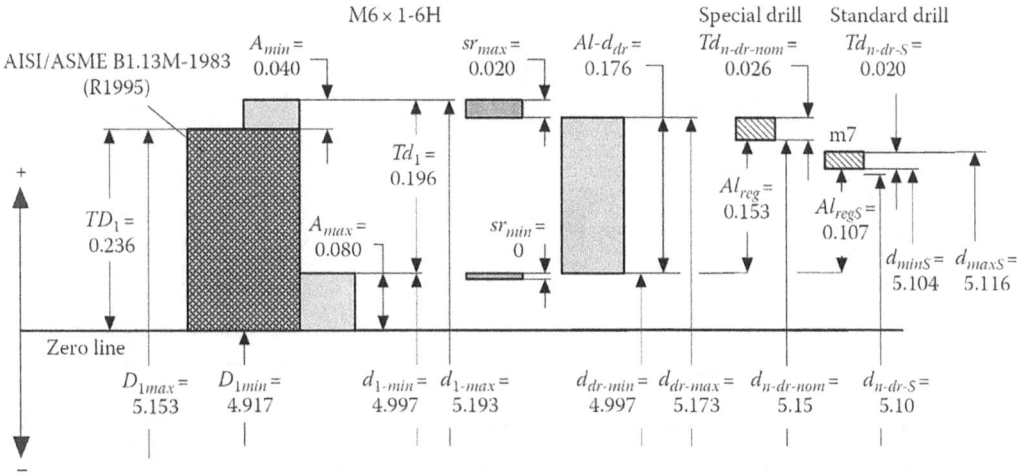

FIGURE 2.254 Graphical representation of tap drill diameter calculations for M6×1-6H tap drill.

3. *Establishing the drilling system runout depending upon the particular holder and setting practice used.* For example, for the tap drill for M6×1 thread, $sr_{max}=20\,\mu m=0.02\,mm$ for a standard drill setting.

4. *Calculating the minimum tool diameter.* In the considered example, the minimum diameter of the tap drill (cutoff diameter for regrinds) is $d_{dr-min}=d_{1-min}-sr_{min}$. The minimum runout can be zero in the most conservative case. For M6×1 thread, the minimum diameter is then calculated as $d_{dr-min}=4.997-0.000=4.997\,mm$ as indicated in Figure 2.254.

5. *Calculating the maximum tool diameter.* In the considered example, the maximum diameter of the tap drill is calculated as $d_{dr-max}=d_{1-max}-sr_{max}$ as follows from Figure 2.253. In the considered example for M6×1 thread, $d_{dr-max}=5.193-0.02=5.173\,mm$ as indicated in Figure 2.254.

6. *Calculating the tolerance on a new tool.* For the tap drill, it is calculated as $Td_{n-dr-nom}=d_{dr-max}-d_{n-dr-nom}$ (Figure 2.253). In the considered example for M6×1 thread, $Td_{n-dr-nom}=5.173\,mm-4.997=0.176\,mm$ as indicated in Figure 2.254.

7. *Selecting the nominal tool diameter.* For the tap drill, it should be selected in the range of $d_{dr-min}-d_{dr-max}$ (Figure 2.253). In the considered example for M6×1 thread, special and standard tools are considered. The nominal drill diameter of 5.15 mm is selected for a special tool while that of 5.10 mm – for standard.

8. *Assigning tolerance zone and deviations for the drilling tool.* For the considered example for M6×1 thread, the tolerance zone of the diameter of the special drill is assigned to be 0.026 mm whereas it is 0.020 mm for the standard drill as shown in Figure 2.254.

The suggested tap drill diameter and tolerance are ⌀5.15+0.026/−0 mm, and the minimum diameter of the tap drill (cutoff diameter for regrinds) is ⌀4.977 mm. According to the old rule, established by manufacturing practice a long time ego, the tap drill diameter is selected as the tap size minus the pitch. This rule somehow made its way into ISO 2306-72 (revision date February 28, 2008), according to which the drill diameter for M6×1 is 5 mm with no tolerances assigned. Accounting for this standard, all reference books on tool selection and some tool manufacturers' catalogs recommend this diameter. If one uses a standard 5 mm drill with the standard tolerance and if this drill is made of a carbide tool material, then according to Table 2.13, the tolerance limits are +0.016/+0.04, that is, at the low end of the acceptable limit according to Figure 2.254. Due to back taper, such a drill will be out of the acceptable tolerance after a few regrinds. If, however, the drill is made of HSS, its tolerance is h8, that is, 0/−0.018 mm (Table 7.9 in Astakhov (2014)). As can be seen, the drill

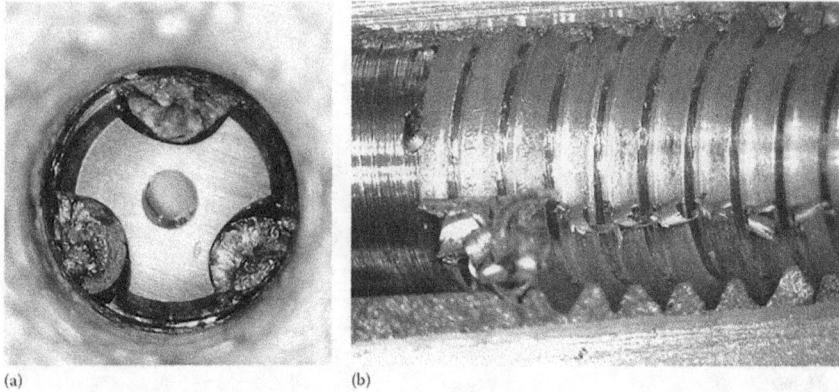

FIGURE 2.255 Tap can break due to the significant amount of chips accumulated in the flute (a) practically all threads are involved in cutting due to a severe synchronization problem and insufficient hole diameter (b).

diameter can be well below the acceptable minimum according to Figure 2.254. Moreover, when such a drill is re-sharpened, its diameter becomes even smaller.

Figure 2.255 shows what happens on tapping when a standard carbide tap drill subjected to five re-sharpenings is used. As can be seen, the tap flute is full of chips (Figure 2.255a). Not only cutting threads (two to three first relieved threads) but also practically all the threads participate in cutting (Figure 2.255b), which creates high cutting torque and leads to tap breakage. Even when the tap can manage to complete the full thread, it breaks on retraction as the hole becomes smaller due to bore springback after tapping. Moreover, the chip left in the tap flute causes its recutting and severely damages the thread as can be clearly seen in Figure 2.256a.

The problems with tap breakage and poor quality of the threads were solved in the setting of the powertrain plants of one of the world's largest automotive companies. The tap drills having the diameters calculated using the proposed methodology were used for threading holes M5, M6, M8, M10, and M12. A great reduction in the tap breakage and part scrap and a significant improvement in thread quality were achieved.

The foregoing considerations suggest that the tolerance on the drill should be calculated carefully rather than relying on multiple tables and recommendations found in various literature reference sources. Many of these recommendations and even standards were developed a long time ago (although recently revised) so they do not reflect the changes that have occurred in industry over the past decade. In the considered example, the tap drill of 5 mm served its objective when tap drills were used in the old machining system characterized by a great system runout (spindle runout + tool holder runout). As such, the system runout can reach 0.15 mm that overlaps the lack of tap drill diameters. In modern manufacturing with high-speed spindles and precision shrink-fit and hydraulic holders, the system runout normally does not exceed 5–10 μm (0.005–0.010 mm) that creates the problems discussed previously with the diameter of tap drills. Figure 2.256b shows the quality of thread cut when the calculated diameter of the tap drill was used.

2.3.8 Synchronization Issue in Tapping

Although the section to follow is more related to the tapping process than to tap design, it is important to be aware about one of the common problems in the tapping process as it directly affects tap performance. Moreover, the known consequences of this problem (e.g., poor thread quality and tap breakage) are often unfairly blamed on the tap although they have nothing to do with the tap itself.

The problem is explained as follows. It is well known that, in theory, the feed per revolution set on the machine used for tapping should be exactly equal to the tap pitch. However, the problem

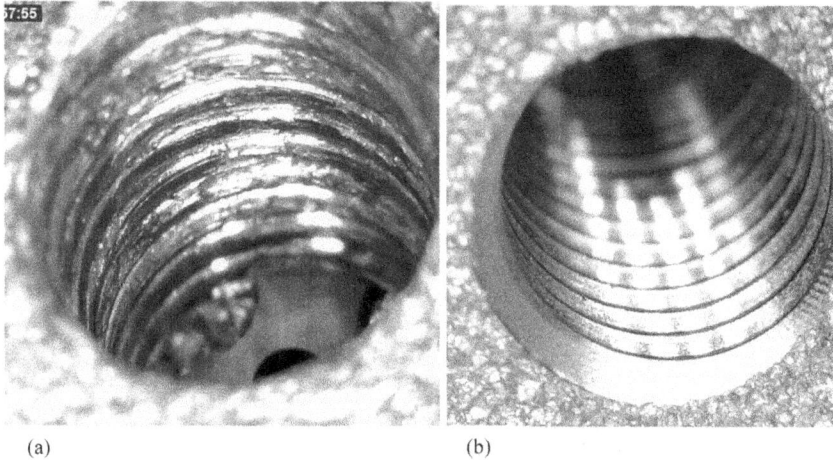

(a) (b)

FIGURE 2.256 Showing (a) poor quality of the machined thread and the tap broken on its retraction and (b) thread of good quality cut when the calculated diameter of the tap drill was used.

occurring in practice is that it is impossible for the machine to exactly match the pitch of the specific tap being used due to specific machine particularities and tap pitch variation within the tolerance. As a result, there is always a slight discrepancy between what the machine is doing and the actual pitch of the tap. The tap wants to move down into the hole at a rate determined by how fast it is fed versus its actual thread pitch. Such a rate can be called as the natural rate. If the tap is moved down into the hole slower than the natural rate, the pull on the threads gets greater and greater as the tap falls behind the place the threads dictate it should be. If it is moved down the hole faster than the natural rate, it starts to "push" the threads faster than they want to go. When the discrepancy between the natural and actual rate of a tap becomes too great, significant contact stresses occur on the tap flanks that increase both the tapping torque and axial force, tear the thread, or even break the tap. Obviously, both are undesirable outcomes.

Although there are a number of approaches to regulate the feed rate, the tap bringing is closer to the proper rate. Among them, two are most common in industry:

1. *Tapping with tension-compression tap holders known as synchro tapping holders*: A synchro tap holder is a tap holder constructed to allow a slight amount of axial movement to compensate for the discrepancy between the natural and actual feed rates that are unavoidable in rigid tapping. If designed and applied properly, its micro tension-compression float eliminates the extra-axial forces on the tap. As pointed out by Big Kaiser Co, it leads to longer tool life, consistent tapping depth, and improved thread quality. This is because the integrated float unit absorbs the loads in the thrust direction, significantly reducing thrust forces that occur during reverse rotation.
2. *Rigid tapping*: With this approach, the tapping holder is rigid, meaning it has no tension/compression play. Therefore, the motion of the spindle and the axis moving down the hole have to be precisely synchronized. Delivering this synchronization is one reason rigid tapping requires a more sophisticated controller capable of such a synchronization.

As discussed by Warfield (2020), various modern controllers approach rigid tapping using different methods. The most common method, used on a majority of controllers, is to let the spindle go, monitor its actual speed with some sort of encoder, and then vary the feed rate of the synchronized axis to mirror the right ratio of spindle rpm to feed rate for the thread being cut. Lately, the smaller machines, especially the "tapping and drilling" centers, use a type of rigid tapping often called "synchronous tapping." In this method, both the spindle and the axis speeds are dynamically controlled as servos to get the best synchronized result. This is only possible on machines whose spindles have low enough mass and inertia that their rotation can be dynamically modified quickly enough, hence on relatively smaller machines.

2.3.9   Some Notes on Thread Formers/Forming Taps

Thread formers, also called as cold forming taps and roll taps, create threads by plastically displaced/extruding the work material in the drilled hole up into the thread form instead of removing material. Obviously, thread forming is chipless process since threads are not cut but formed. Thread forming has a number of advantages and limitations which both should be clearly understood. This section aims to explain some of the most common.

2.3.9.1   Common Explanation of the Major Advantage

Commonly, the improved strength of the treads made by cutting and forming taps is explained qualitatively using a picture similar to that shown in Figure 2.257. Figure 2.257a shows cutting tap having four cutting flutes (in the section shown). The tap cuts the tread profile shown in Figure 2.257b. The weakness of this thread is explained by the direction of the deformation bands, which are parallel to the axial force in the tread joint. As shown, such a thread has the weakest shear resistance in the direction of the axial force. As a result, the applied axial force can easily shear the profile when the tress in the axial direction reaches the yield shear strength

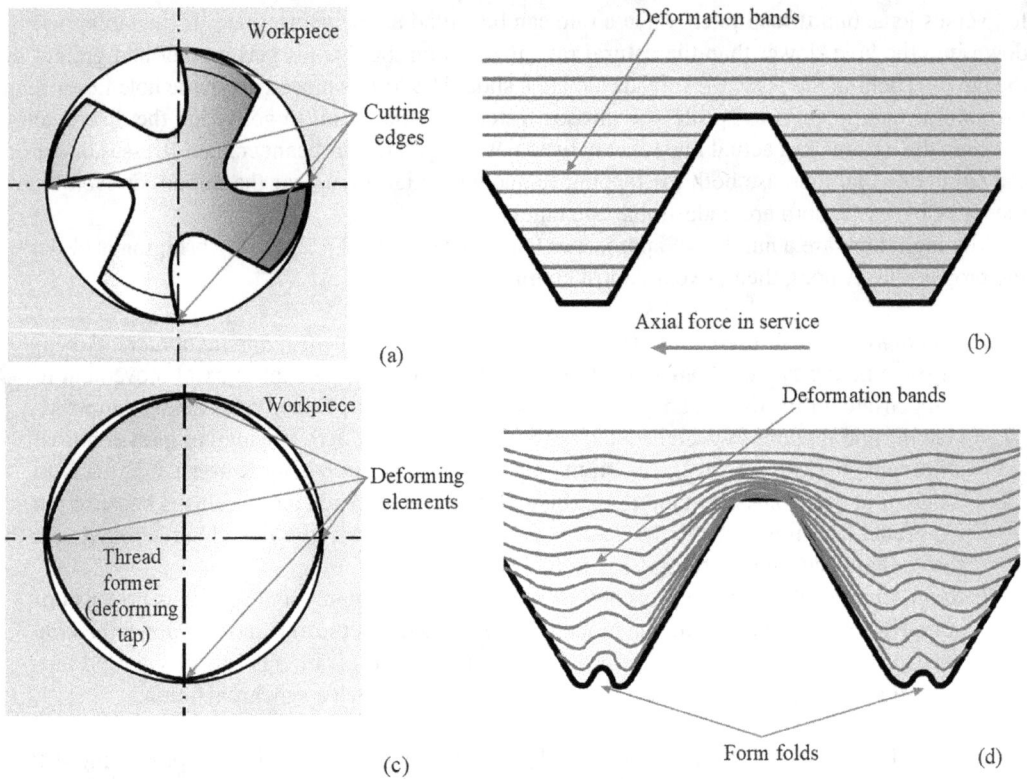

FIGURE 2.257   Common explanation of the improved strength of the thread made by forming taps: (a) profile of a cutting tap, (b) profile of the tread and deformation bands after cutting, (c) profile of the forming tap, and (d) profile of the tread and deformation bands after form tapping.

of the work material. On the contrary, the forming tap has the deforming elements shown in Figure 2.257c, which plastically deform/extrude the work material to form the thread profile. Due to three-dimensional plastic deformation, the shapes of the resulting deformation bands are shown in Figure 2.257d. As can be seen, they have a complex curved shape not parallel to the direction of the axial force. As a result, the shear strength of such a thread should increase. Moreover, as significant strain hardening of the work material takes place on the discussed plastic deformation, the flank of the thread becomes a greater strength than those in thread cutting. Note also that similar qualitative pictures are used in discussing the advantages of various operations of thread rolling.

A critical analysis of the thread profile shown in Figure 2.257b shows that this profile has little to do with reality as the shown deformation/shear bands are not parallel to the direction of the axial force in reality. The shown shape of the shear bands can be found only after severe cold extrusion of a highly-ductile material. In reality, modern workpiece/blanks are made by casting/die casting, forging, etc. with no deformation/shear band. Even in hot rolling/extrusion, recrystallization of the work material takes place (2008) so that no deformation band weakening the work material shear resistance in the direction of the axial force is found. The second important issue is the nature of the cutting process where the cut surface of the work material is strained to the maximum, i.e., to the strain at fracture as discussed in Section 2.1.3.8.3 (Figure 2.85). As such, cold working (i.e., strain hardening) of the machined surface takes place. The greater the ductility of the work material, the greater the degree of coldworking. The latter is the ratio of the harnesses of the machined surface and that of the work material. Particularly in thread tapping where the clearance angle of the cutting teeth is small, the springback of the work material is high and the degree of coldworking is great.

2.3.9.2   Advantages

The following major advantages of forming tap can be listed as follows:

1. Forming taps offer an extended tool life, which is up to 20x longer compared to cutting taps. Even though forming taps is generally more expensive than cutting taps, the savings due to increased tool life significantly offsets this cost difference.
2. It is feasible to make deep threads down to four times diameter as standard because there are no problems associated with chip removal.
3. Threads produced are stronger than those made by cutting tap-tapping threads. Reportedly, formed threads have more than double the fatigue strength under dynamic load due to strain hardening.
4. Greater productivity due to increased cutting speed. Form taps operate most efficiently at spindle speeds 1.5–2 times faster than those recommended for conventional cutting taps, especially in softer materials and/or with fine-pitch forming taps.

Knowing these advantages and the results of some pure laboratory studies and some researchers (e.g., Nardi et al. 2012) suggested that "The above advantages make this technique recommendable in a lot of applications, especially in automatic operations where HP rates are required as it is the case of automotive applications." In the author's opinion, however, it is also important to understand the known limitation of forming taps before making any decision on their implementation in a particular case.

2.3.9.3   Limitations

The limitations of the forming tap can be described as follows.

2.3.9.3.1   Improved Hole Quality for Thread Forming Is Required

A limiting factor to form tapping is the drilled hole which has to be much more precise (diameter, shape, straightness, and surface roughness) than for cutting tapping. If the drilled hole is too large, there will not be enough material to displace to completely form the thread. If the hole is too small, there will be too much material to displace, causing high torque and tap failure/breakage. This is particularly true when the tap design is customized to achieve the minimum size of folds for a particular work material. As a result, the tolerance on the diameters of the drilled hole for tapping should be at least two times tighter than that for thread-cutting taps. As stated by EMUGE Co., "Every lack of precision, every kind of surface roughness will be mirrored in the finished internal thread and its minor diameter." It may be necessary to find the correct diameter by actual testing for a given threading operation (work material, machine, tool holder, MWF, etc.). Obviously, the drill allowable runout and wear criterion should be tighter accordingly. All these set the additional strain on the system of drills quality/cost, presetting, handling, and wear control.

2.3.9.3.2 Problem to Control Springback of the Work Material

Although it is university claimed that form taps make threads of better quality than cut taps, it is much more difficult in reality to control the size of the thread with a form tap than with a cutting tap. This is because formed threads are more susceptible to material expansion and shrinkage (see springback discussed in Section 2.1.3.8) due to inherent variation of work material ductility even within the tolerance on its machinal properties. This makes it harder to hold tight tolerances with form taps. According to EMUGE Co., the produced thread is always smaller than the thread part of the forming tap. One will never be able to screw the forming tap back into the thread manually after form tapping, as would be possible without any problem with a cut thread and a cutting tap. For this, the thread part of forming taps is made closer to the upper tolerance limit of the internal thread.

2.3.9.3.3 Only Certain Materials Are Recommended for Thread Forming

Not all materials are suitable for thread forming. Unfortunately, the list of materials and their acceptable properties suitable for form tapping are significantly overstated in the catalogs and guide materials for form tapping by leading manufacturers. For example, EMUGE Co. states that suitable materials usually have a tensile strength of less than 1,400 MPa and the minimum strain at a fracture of 5%; Walter Co. states that materials having a tensile strength in the range of 1,200–1,400 MPa and strain at fracture lower than 7% are not suitable for thread forming. The best materials for form tapping include aluminum and Al-Si alloys (with silicon content less than 8%), brass, copper, 300 series austenitic stainless steel, and lead steel. In other words, any material having a significant elongation (strain) at fracture and strength less than approx. 1,000 MPa, i.e., materials cutting of which results in stringy chips are a good candidate for forming threads. The so-called short-chipping materials such as cast iron, Cu-Zn alloys, and similar have poor deformation properties so they are not suitable for thread forming.

2.3.9.3.4 MWF (the Coolant) Specifics

Three aspects of MWF should be considered in thread forming, namely the type of MWF, its clearness, and application technique. As thread forming involves heavy plastic deformation of the work material that involves high friction/severe contact conditions at the interfaces of the forming elements and the workpiece. These severe contact conditions combined with the process high speed result in high temperatures over these interfaces. To reduce the forming tap wear and improve the quality of the produced threads, it is best to use a heavy-duty drawing MWF recommended by the MWF companies for similar applications involving cold extrusion or metal drawing processes. This type of MWF normally contains extreme-pressure additives, such as sulfur, chlorine, and some non-chlorine additives. This is not a very attractive option for many manufacturing companies where components of MWF such as sulfur and chlorine are prohibited. Moreover, it is not feasible to have a specific type of MWFs for one given operation, particularly when a central MWF supply system is used. Therefore, a water-mixable high concentration (approx. 10%) MWF is normally used to keep tread forming operation efficient, and this concentration should be controlled/marinated on the hourly basis.

Cleanliness of MWF is important since a chip shaving and other debris found in MWF cause premature tap failure or other poor performance problems if the chip becomes wedged between the tap and the wall of the hole. A suitable filtration system normally recommended for HP drilling tools (see Chapter 3 in Astakhov (2024)) is normally suitable for thread forming.

To improve MWF access to the deformation zones of the forming element, modern designs of forming taps are provided with lubrication grooves ground along the length of the tap working part. Walters Co. makes forming taps with internal MWF supply and radial outlets to deliver MWF in the critical place (in terms of pressure and temperature) of taps. Taps with lubrication grooves on the shank are used although they are not as efficient as radial coolant holes with an internal MWF supply.

2.3.9.3.5 Acceptance in Some Industries

An important limitation on form tapping is that formed threads are not accepted in some industries. The major reason is folds (see Figure 2.257) formed inside the thread crest that can hold contaminants. Industries such as medical and food have strict regulations regarding cleanliness, so they cannot accept threads with those folds. The dynamics of folds formation are presented by EMUGE Co. as an instructional animation showing how the thread-forming process works. The video demonstrates how the material is displaced rather than cut and the slight "fish mouth" that is formed at the crest of each thread. Figure 2.258 provides a graphical explanation. The size of folds can be reduced to minimum through optimizing the deformation process and tap design. Walter Prototyp reported that the company is able to design special tools in which the form folds can be closed under specific conditions.

Aerospace is another industry that may restrict acceptance of form threads due to tight-tolerance requirements. Although it is universally claimed that form taps make threads of better quality than cut taps, it is much more difficult in reality to control the size of the thread with a form tap than with a cut tap. This is because formed threads are more susceptible to material expansion and shrinkage (see springback discussed in Section 2.1.3.8.1) due to inherent variation of work material ductility even within the acceptable tolerance of machinal properties. This makes it harder to hold tight tolerances with form taps. One of the possible ways to improve form tap performance in this respect is to tighten a tolerance on the pitch diameter on one IT grade. For example, if H6 (IT6 grade) tolerance on the pitch diameter is required for a given tread then IT5 (IT grade 5) is assigned to forming taps. This makes the grinding, inspection, and presetting (allowable runout and required high-precision

FIGURE 2.258 Showing formation of the folds in the forming process.

tool holder) of forming taps more difficult and expensive. Besides tighter tolerances, the aerospace industry does not accept form threads due to the resulting small gap at the tip of the minor thread diameter, which can hold contamination.

2.3.9.3.6 Miscellaneous

Two miscellaneous items should also be pointed out. The first one is a necessity of generous chamfer at the beginning and end of the hole to be form tapped. This is because a forming tap may slightly raise the area surrounding the top and bottom surfaces of the hole. Larger chamfers help eliminate this raised area resulting in a flatter, more even surface after form tapping. As a result, if a part is not designed for thread forming, these chamfers should be added to its design as well as special tools/ operations to make these chamfers.

There is also a small potential problem when a blind hole is thread formed using the so-called bottom style forming tap. The chamfer on this tap is approximately 2 threads long and requires a drilled hole depth of 3–4 pitches beyond the full thread required. It lowers the tool life of the tap and requires a deeper hole to form tapping compared to that made for a cutting tap.

With cutting taps, small diameter internal threads can be cut by hand, using a tap holder/wrench. It is virtually impossible with forming taps as the alignment of the tap and the hole is critical.

REFERENCES

"Roll Form Tapping Process." from https://www.emuge.com/videos/emuge-roll-form-tap.

(1989). *Metal Cutting Tool Handbook*. New York, The Metal Cutting Tool Institute by Industrial Press. https://www.freertool.com/products/metal-cutting-tool-handbook-7th-edition-reference

F. C. Campbell (2008). Deformation processing. *Elements of Metallurgy and Engineering Alloys*. Novelty, OH, ASM International.

Abrao, A. M., Aspinwall, D. K. (1996). "The surface integrity of turned and ground hardened bearing steel." *Wear* **196**: 279–284.

Abushawashi, Y., Xiao, X., Astakhov, V. (2017). "Practical applications of the "energy-triaxiality" state relationship in metal cutting." *Machining Science and Technology* **21**: 1–18.

Agapiou, J. S. (1993a). "Design characteristics of new types of drill and evaluation of their performance drilling cast iron. - I. Drills with four major cutting edges." *The International Journal of Machine Tools and Manufacture* **33**: 321–341.

Agapiou, J. S. (1993b). "Design characteristics of new types of drill and evaluation of their performance drilling cast iron. - II. Drills with three major cutting edges." *The International Journal of Machine Tools and Manufacture* **33**: 343–365.

Agnew, J. (1973). "The importance and methods of carbide edge preparation." SME paper MR73-905.

Ahmad, Z., Ed. (2012). *Aluminium Alloys - New Trends in Fabrication and Applications*. Rijeka, Croatia, InTech.

Armarego, E. J. A., Cheng, C. Y. (1972a). "Drilling with flat rake face and conventional twist drills-I. Theoretical investigation." *Journal of Machine Tools and Manufacture* **12**: 17–35.

Armarego, E. J. A., Cheng, C. Y. (1972b). "Drilling with flat rake face and conventional twist drills-II. Experimental investigation." *International Journal of Machine Tools and Manufacture* **12**: 37–54.

Astakhov, V. P. (1998/1999). *Metal Cutting Mechanics*. Boca Raton, FL, CRC Press.

Astakhov, V. P. (2002). "The mechanisms of bell mouth formation in gundrilling when the drill rotates and the workpiece is stationary. Part 1: The first stage of drill entrance." *International Journal of Machine Tools and Manufacture* **42**: 1135–1144.

Astakhov, V. P. (2004). "The assessment of cutting tool wear." *International Journal of Machine Tools and Manufacture* **44**: 637–647.

Astakhov, V. P. (2006). *Tribology of Metal Cutting*. London, Elsevier.

Astakhov, V. P. (2010). *Geometry of Single-Point Turning Tools and Drills. Fundamentals and Practical Applications*. London, Springer.

Astakhov, V. P. (2011a). Drilling. *Modern Machining Technology: A Practical Guide*. P. Davim. Oxford (UK), Woodhead-Chandos: 79–212.

Astakhov, V. P. (2011b). Turning. *Modern Machining Technology: A Practical Guide*. P. Davim. Oxford, Woodhead-Chandos: 1–78.

Astakhov, V. P. (2014). *Drills: Science and Technology of Advanced Operations*. Boca Raton, FL, CRC Press.

Astakhov, V. P. (2018). Mechanical properties of engineering materials: Relevance in design and manufacturing. *Introduction to Mechanical Engineering*. P. Davim. Cham, Switzerland, Springer International Publishing AG: 3–41.

Astakhov, V. P. (2024). *High-Productivity Drilling Tools: Materials, Metrology and Failure Analysis*. Boca Raton, FL, CRC Press.

Astakhov, V. P., Galitsky, V. V., Osman, M. O. M. (1995a). "An investigation of the static stability in self-piloting drilling." *International Journal of Production Research* **33**: 1617–1634.

Astakhov, V. P., Galitsky, V. V., Osman, M. O. M. (1995b). "A novel approach to the design of self-piloting drills. Part 1. Geometry of the cutting tip and grinding process." *ASME Journal of Engeneering for Industry* **117**: 453–463.

V. P. Astakhov, S. Joksch, Ed. (2012). *Metal Working Fluids for Cutting and Grinding: Fundametals and Recent Advances*. London, Woodhead.

Astakhov, V. P., Outeiro, J. (2019). Importance of temperature in metal cutting and its proper measurement/modeling. *Measurement in Machining and Tribology*. P. J. Davim. London, Springer: 1–47.

Astakhov, V. P., Patel, S. (2019). Development of the basic drill design for cored holes in additive and subtractive manufacturing. *Additive and Subtractive Manufacturing*. J. P. Davim. London, De Gruyter: 113–148.

Astakhov, V. P., Shvets, S. (2004). "The assessment of plastic deformation in metal cutting." *Journal of Materials Processing Technology* **146**: 193–202.

Astakhov, V. P., Shavets, S. V. (2020). "Technical resource of the cutting wedge is the foundation of the machining regime determination." *International Journal of Manufacturing, Materials, and Mechanical Engineering* **10**: 1–17.

Astakhov, V. P., Xiao, X. (2016). The principle of minimum strain energy to fracture of the work material and its application in modern cutting technologies. *Metal Cutting Technology*. P. Davim. Boston, MA, De Gruyter Publishers: 1–35.

Atkins, A. G. (2003). "Modelling metal cutting using modern ductile fracture mechanics: quantitative explanations for some longstanding problems." *International Journal of Mechanical Science* **43**: 373–396.

Atkins, T. (2009). *The Science and Engineering of Cutting*. Amsterdam, Butterworth-Heinemann.

Biermann, D., Terwey, I. (2008). "Cutting edge preparation to improve drilling tools for HPC processes." *CIRP Journal of Manufacturing Science and Technology* **1**: 76–80.

Billau, D. J., McGoldrick, P. F. (1979). "An analysis of the geometry of the periphery of the flank face of twist drills ground with cylindrical and conical forms." *International Journal of Machine Tool Design and Research* **19**: 69–86.

Bobrov, V. F. (1962). *Inlcuence of the Cutting Edge Inclination Angle on the Metal Cutting Process*. Moscow, Mashgiz.

Bouzakis, K.-D., Gerardisa, S., Katirtzogloua, G., Makrimallakisa, S., Bouzakisa, A., Cremerd, R., Fussd, H.-G. (2009). "Application in milling of coated tools with rounded cutting edges after the film deposition." *CIRP Annals-Manufacturing Technology* **58**(1): 76–80.

Chen, W.-C. (1997). "Applying the finite element method to drill design based on drill deformations." *Finite Elements in Analysis and Design* **26**: 57–81.

Chen, Y. R., Ni, J. (1999). "Analysis and optimization of drill cross-sectional geometry." SME Paper MR99-162.

Cheung, F. Y., Zhou, Z. F., Geddam, A., Li, K. Y. (2008). "Cutting edge preparation using magnetic polishing and its influence on the performance of high-speed steel drills." *Journal of Materials Processing Technology* **208**: 196–204.

Davim, J. P., Reis, P. (2003). "Study of delamination in drilling carbon fiber reinforced plastics (CFRP) using design of experiments." *Composite Structures* **59**: 481–487.

Davim, J. P., Reis, P., Antonio, C. C. (2004a). "Drilling fiber reinforced plastics (FRPs) manufactured by hand lay-up: influence of matrix (Viapal VUP 9731 and ATLAC 382-05)." *Journal of Materials Processing Technology* **155-156**: 1828–1833.

Davim, J. P., Reis, P., Antonio, C. C. (2004b). "Experimental study of drilling glass fiber reinforced plastics (GFRP) manufactured by hand lay-up." *Composites Science and Technology* **64**: 289–297.

De Beer, C. (1970). "The web thickness of twist drills." *Annals of the CIRP* **18**: 81–85.

Ernst, H., Haggerty, W. A. (1958). "The spiral point drill-a new concept in drill point geometry." *Transactions of the ASME* **80**: 1059–1072.

Farrell, S. P., Deering, J. (2018). "Analysis of seeded defects in laser additive manufactured 300M steel." *Materials Performance and Characterization* **7**(1): 300–315.

Fetecau, C., Stan, F., Oancea, N. (2009). "Toroidal grinding method for curved cutting edge twist drills." *Journal of Materials Processing Technology* **209**: 3460–3468.

Fiesselmann, F. (1993). "An option for doubling drilling productivity." *The Fabricator* **23**(4): 36–38.

French, L. G., Goodrich, C. I. (1910, reprinted in 2001). Principles of deep hole drilling. Deep Hole Drilling. Machinery's Reference Series No. 25. Bradley, IL, Lindsay Publication Inc.

Fujii, S., DeVries, M. F., Wu, S. M. (1970a). "An analysis of drill geometry for optimum drill design by computer-II: Computer aided design." *ASME Journal of Engineering for Industry* **92**: 657–666.

Fujii, S., DeVries, M. F., Wu, S. M. (1970b). "An analysis of drill geometry for optimum drill design by computer-I: Drill geometry analysis." *ASME Journal of Engineering for Industry* **92**: 647–656.

Fujii, S., DeVries, M. F., Wu, S. M. (1971). "An analysis of the chisel edge and the effect of the d-theta relationship on drill point geometry." *ASME Journal of Engineering for Industry* **93**: 1093–1105.

Fujii, S., Devries, M. F., Wu, S. M. (1972). "Analysis and design of a drill grinder and evaluation of grinding parameter." *ASME Journal of Engineering for Industry* **94**: 1157–1163.

Galloway, D. F. (1957). "Some experiments on the influence of various factors on drill performance." *ASME Transactions* **79**: 191–231.

Granovsky, G. E., Granovsky, V. G. (1985). *Metal Cutting (in Russian)*. Moscow, Vishaya Shkola.

Heisela, U., Pfeifroth, T. (2012). "Influence of point angle on drill hole quality and machining forces when drilling CFRP." *Procedia CIRP* **1**: 471–476.

Hibbeler, R. C. (2010). *Mechanics of Materials*. Boston, MA, Prentice Hall.

Hsiau, J. S. (1985). "Computer aided design of multifacet drills." *Transactions of NAMRI/SME* **13**: 490–496.

Hsieh, J.-F., Lin, P. L. (2002). "Mathematical model of multi-flute drill point." *International Journal of Machine Tools & Manufacture* **42**: 1181–1193.

Ivanov, V., Nankov, G., Kirov, V. (1998). "CAD-orientated mathematical model for determination of profile helical surfaces." *International Journal of Machine Tools and Manufacture* **38**: 1001–1015.

Jawahir, I. S., Balaji, A. K., Stevenson, R., van Luttervelt, C. A. (1997). "Towards predictive modeling and optimization of machining operations. Manufacturing Science and Engineering." *Proceedings of 1997 ASME International Mechanical Engineering Congress and Exposition*, Dallas, TX, ASME.

Jawahir, I. S., Van Luttervelt, C. A. (1993). "Recent developments in chip control research and applications." *Annals of the CIRP* **42**(2): 659–693.

Jawahir, I. S., Zhang, J. P. (1995). "An analysis of chip curl development, chip deformation and chip breaking in orthogonal machining." *Transactions of NAMRI/SME* **XXIII**: 109–114.

Kaldor, S., Lenz, E. (1982). "Drill point geometry and optimization." *ASME Journal of Engineering for Industry* **104**: 84–90.

Kaldor, S., Moore, K. (1983). "Drill point designing by computer." *Annals of the CIRP* **32**: 27–31.

Kang, X., Junichi, T., Akihiko, T. (2006). "Effect of cutting edge truncation on ductile-regime grinding of hard and brittle materials." *International Journal of Manufacturing Technology and Management* **9**(1-2): 183–200.

Khashaba, U. A. (2012). "Drilling of polymer matrix composites: A review." *Journal of Composite Materials*, 1–16. doi: 10.1177/0021998312451609.

Kim, J., Dornfeld, D. (2002). "Development of an analytical model for drilling burr formation in ductile materials." *ASME Journal of Engineering Materials and Technology* **124**(2): 192–198.

Ko, S.-L., Chang, J.-E. (2003). "Development of drill geometry for burr minimization in drilling." *CIRP Annals - Manufacturing Technology* **52**: 45–48.

Ko, S.-L., Changa, J.-E., Yang, G.-E. (2003). "Burr minimizing scheme in drilling." *Journal of Materials Processing Technology* **140**: 237–242.

Komarovsky, A. A., Astakhov, V. P. (2002). *Physics of Strength and Fracture Control: Fundamentals of the Adaptation of Engineering Materials and Structures*. Boca Raton, FL, CRC Press.

Kyreev, G. I. (2008). *Desing of Threading Taps and Dies (in Russian)*. Ulanovsk, UlGTU.

Lee, Y. S., Farson, D. F. (2015). "Surface tension-powered build dimension control in laser additive manufacturing process." *International Journal of Advanced Manufacturing Technology* **85**(5): 1035–1044.

Lorenz, G. (1979). "Helix angle and drill performance." *Annals of the CIRP* **28**: 83–86.

Lv, D., Wang, Y., Yu, X., Chen, H., Gao, Y. (2022). "Analysis of abrasives on cutting edge preparation by drag finishing." *International Journal of Advanced Manufacturing Technology* **119**: 3583–3594.

Mayer, J. E. J., Stauffer, D. G. (1973). "Effects of tool hone and chamfer on wear life." ASME Technical paper MR73-907.

Moroni, G., Syam, W. P., Petro, S. (2014). "Towards early estimation of part accuracy in additive manufacturing." *Procedia CIRP* **21**: 300–305.

Nakayama, K., Ogawa, M. (1985). "Effect of chip splitting nicks in drilling." *Annals of the CIRP* **34**(1): 101–104.

Narasimha, K., Osman, M. O. M., Chandrashekhar, S., Frazao, J. (1987). "An investigation into the influence of helix angle on the torque-thrust coupling effect in twist drills." *The International Journal of Advanced Manufacturing Technology* **2**(4): 91–105.

Nardi, D., De Lacalle, L. N. L., Lamikiz, A. (2012). "Friction drilling of dual phase steels." *Revista de Metalurgia* **48**: 13–23.

E. Oberg, F. D. Jones, H. L. Horton, H. H. Ryffel, C. McCauley, Ed. (2020). *Machinery's Handbook*. South Norwalk, Connecticut, Industrial Press, Inc.

Olson, W. W., Batzer, S. A., Sutherland, J. W. (1998). Modeling of chip dynamics in drilling. *Proceedings of the CIRP 1st International Workshop on Modeling of Machining*, Atlanta, GA.

Outeiro, J. C. (2020). Residual stresses in machining. *Mechanics of Materials in Modern Manufacturing Methods and Processing Techniques*. V. V. Silberschmidt. Amsterdam, Elsevier: 297–360.

Outeiro, J. C., Astakhov, V. P. (2005). The role of the relative tool sharpness in modelling of the cutting process. *8th CIRP International Workshop on Modeling of Machining Operations*, Chemnitz University of Technology, Chemnitz, Germany.

Oxford Jr., C. J. (1955). "On the drilling of metals-I. Basic mechanics of the process." *Transactions of the ASME* **77**: 103–114.

Özel, T., Hsu, T.-K., Zeren, E. (2005). "Effects of cutting edge geometry, workpiece hardness, feed rate and cutting speed on surface roughness and forces in finish turning of hardened AISI H13 steel." *The International Journal of Advanced Manufacturing Technology* **25**(3–4): 262–269.

Pilny, L., De Chiffre, L., Pıska, M., Villumsen, M. F. (2012). "Hole quality and burr reduction in drilling aluminium sheets." *CIRP Journal of Manufacturing Science and Technology* **5**: 102–107.

Pres, P., Skoczynski, W., Jaskiewicz, K. (2014). "Research and modeling workpiece edge formation process during orthogonal cutting." *Archives of Civil and Mechanical Engineering* **14**: 622–635.

Prousalidou, E., Hanna, S. (2007). A parametric representation of ruled surfaces. *Computer-Aided Architectural Design Futures (CAADFutures)*. A. Dong, A. V. Moere, J. S. Gero. Dordrecht, Springer.

Rech, J. (2006). "Influence of cutting edge preparation on the wear resistance in high speed dry gear hobbing." *Wear* **261**: 505–512.

Ren, K., Ni, J. (1999). "Analyses of drill flute and cutting angles." *International Journal of Advanced Manufacturing Technology* **18**: 546–553.

Rodin, P. R. (1971). *Cutting Geometry of Twist Drills (in Russian)*. Kyiv (Ukraine), Technika.

Rodriguez, C. J. C. (2009). Cutting edge preparation of precision cutting tools by applying micro-abrasive jet machining and brushing, Kassel University.

Jahanmir, S., Ramulu, M., Koshy, P., Ed. (1999). *Machining of Ceramics and Composites Manufacturing Engineering and Materials Processing*. Boca Raton, FL, CRC Press.

Sahu, S. K., Ozdoganlar, O. B., DeVor, R. E., Kapoor, S. G. (2003). "Effect of groove-type chip breakers on twist drill performance." *International Journal of Machine Tools & Manufacture* **43**: 617–627.

Salama, A. S., Elsawy, A. H. (1996). "The dynamic geometry of a twist drill point." *Journal of Materials Processing Technology* **56**: 45–53.

Shaffer, W. R. (2000). "Cutting tool edge preparation." SME Paper TP99PUD68.

Smith, C. J., Derguti, F., Hernandez Nava, E., Thomas, M., Tammas-Williams, S., Gulizia, S., Fraser, D., Todd, I. (2016). "Dimensional accuracy of electron beam melting (EBM) additive manufacture with regard to weight optimized truss structures." *Journal of Materials Processing Technology* **229**: 128–138.

Spur, G., Masuha, J. R. (1981). "Drilling with twist drills of different cross section profiles." *CIRP Annals - Manufacturing Technology* **30**: 31–35.

Sreenivasulu, R., SrinivasaRao, C. (2019). "Review on investigations carried out on burr formation in drilling during 1975 to 2020." *Technological Engineering* **16**(1): 43–57.

Stabler, V. G. (1964). The chip flow law and its consequence. *5th International Machine Tool Design and Research Conference*, Birmingham, UK.

Taheri, H., Shoaib, M. R. B. M., Koester, L. W., Bigelow, T. A. (2017). "Powder-based additive manufacturing - a review of types of defects, generation mechanisms, detection, property evaluation and metrology." *International Journal of Additive and Subtractive Materials Manufacturing* **1**(2): 172–209.

Takashi, M., Jürgen, L. (2010). "Simulation of drilling process for control of burr formation." *Journal of Advanced Mechanical Design, Systems, and Manufacturing* **4**(5): 966–975.

Tasi, W. D., Wu, S. M. (1979). "Measure and control of the drill point grinding process." *International Journal of Machine Tool Design and Research* **19**: 109–120.

Thiele, J. D., Melkote, S. N., Peascoe, R. A., Watkins, T. R. (2000). "Effect of cutting-edge geometry and workpiece hardness on surface residual stresses in finish hard turning of AISI 52100 steel." *Journal of Manufacturing Science and Engineering* **122**(4): 642–649.

Thornley, R. H., Wahab, A. B. I. El., Maiden, J. D. (1987a). "A new approach to eliminate twist drill chisel edge. Part 1. Asymmetrical configuration." *International Journal of Production Research* **25**(4): 589–602.

Thornley, R. H., Wahab, A. B. I. El., Maiden, J. D. (1987b). "Some aspects of twist drill design." *International Journal of Machine Tools Manufacturing* **27**: 393–397.

Tsai, W. D., Wu, S. M. (1979a). "A mathematical model for drill point design and grinding." *ASME Journal of Engineering for Industry* **101**: 333–340.

Tsai, W. D., Wu, S. M. (1979b). "Computer analysis of drill point geometry." *International Journal of Machine Tool Design and Research* **19**: 95–108.

Vable, M. (2012). *Mechanics of Materials*. Houghton, MI, Michigan Technological University.

Veremachuk, E. S. (1940). *Deep Hole Drilling (in Russian)*. Moscow, State Publishing House of Defence Industry.

Vinogradov, A. A. (1985). *Physical Foundation of the Drilling of Difficult-to-Machine Materials with Carbide Drills (in Russian)*. Kyiv (Ukraine), Naukova Dumka.

Wang, J., Zhang, Q. (2008). "A study of high-performance plane rake faced twist drills. Part I: Geometrical analysis and experimental investigation." *International Journal of Machine Tools & Manufacture* **48**: 1276–1285.

Warfield, B. (2020). "Advantages and pitfalls of rigid tapping." CNC Cookbook.

Watson, A. R. (1985). "Geometry of drill elements." *International Journal of Machine Tool Design and Research* **25**: 209–227.

Webb, P. M. (1993). "The three-dimensional problem of twist drilling." *International Journal of Production Research* **31**(5): 1247–1254.

Zhang, B., Yongtao, L., Bai, Q. (2017). "Defect formation mechanisms in selective laser melting: A review." *Chinese Journal of Mechanical Engineering* **30**: 515–527.

Zhong, Z. W. (2003). "Ductile or partial ductile mode machining of brittle materials." *The International Journal of Advanced Manufacturing Technology* **21**(8): 579–585.

Zorev, N. N., Ed. (1966). *Metal Cutting Mechanics*. Oxford, Pergamon Press.

3 Deep-Hole Drilling Tools

Don't design for everyone. It's impossible. All you end up doing is designing something that makes everyone unhappy.

Leisa Reichelt (UK), a user experience consultant (Freelance)

3.1 INTRODUCTION

The term deep-hole machining (DHM) relates to a machining operation of a hole with an excessive length-to-diameter ratio (known as *L/D*). For more than 100 years, it was considered that DHM range begins when $L/D > 5$. With the introduction of some modern machinery and tooling, *L/D* greater than 15 eventually was accepted as a reasonable indicator. However, modern high-pressure through tool metal working fluid (MWF) (coolant) supply solid-carbide straight-flute and twist drills are fully capable to cover $L/D \leq 40$ range showing much greater productivity and efficiency than traditional DHM drills. When design/geometries such as VPA1©, VPA2©, VPA3© (Chapter 2), VPA2d©, and VPA4© are used, this range can be extended even to $L/D \leq 60$. As such, the tolerances on the drilled hole straightness and position can be kept the same as with traditional DHM drills. It is to say that the essence of the traditional term "deep-hole machining drills" should be re-defined in terms of the working principle of such drills and the achieved accuracy of drilled holes. In this book, the term "self-piloting tool (SPT)" is used to distinguish the traditional deep-hole drills from the other drills capable of drilling deep holes, e.g., long twist drills developed to drill deep holes.

The use of the *L/D* criterion, however, is not nearly sufficient to make the proper decision about the tool when high machined hole specifications, for example, diametric accuracy, hole straightness and shape, and surface roughness, are required. This is because when properly used, SPTs produce superior holes with close tolerances. While most applications of SPTs involve hole depths varying from 10 to 30 diameters, it is common to encounter drills with depth-to-diameter ratios of 100 to 1. Furthermore, holes with depth-to-diameter ratios of 300 to 1 have been successfully drilled (Swinehart 1967). While such features may be accomplished with a twist drill, the extra problems involved in getting MWF (the coolant) in and the chips out of the hole make it quite difficult to drill beyond an *L/D* 50 to 1.

3.2 COMMON CLASSIFICATION OF DEEP-HOLE MACHINING OPERATIONS

In general, DHM operations are classified by the method of MWF supply and chip/swarf transportation from the machining zone. It can be gundrill (gun)-type, STS-type (known also as Boring and Trepanning Association (BTA)-type), and ejector-type machining.

The principle of gundrilling is shown in Figure 3.1a. MWF (the coolant) is supplied under high pressure to the tool holder and then to the drill shank. Then, it flows through a kidney-shaped internal passage made in the shank and then through the coolant passage made in the tip. After cooling and lubricating the machining zone, MWF carries away the formed chips as the mixture (often referred to as the swarf) over the external V-shaped flute made on the shank. The swarf ends up in the chip box from where it passes through the chip separator and flows into the MWF tank. After cleaning and cooling down, MWF is pumped back to the gundrill.

DOI 10.1201/9781003263296-3

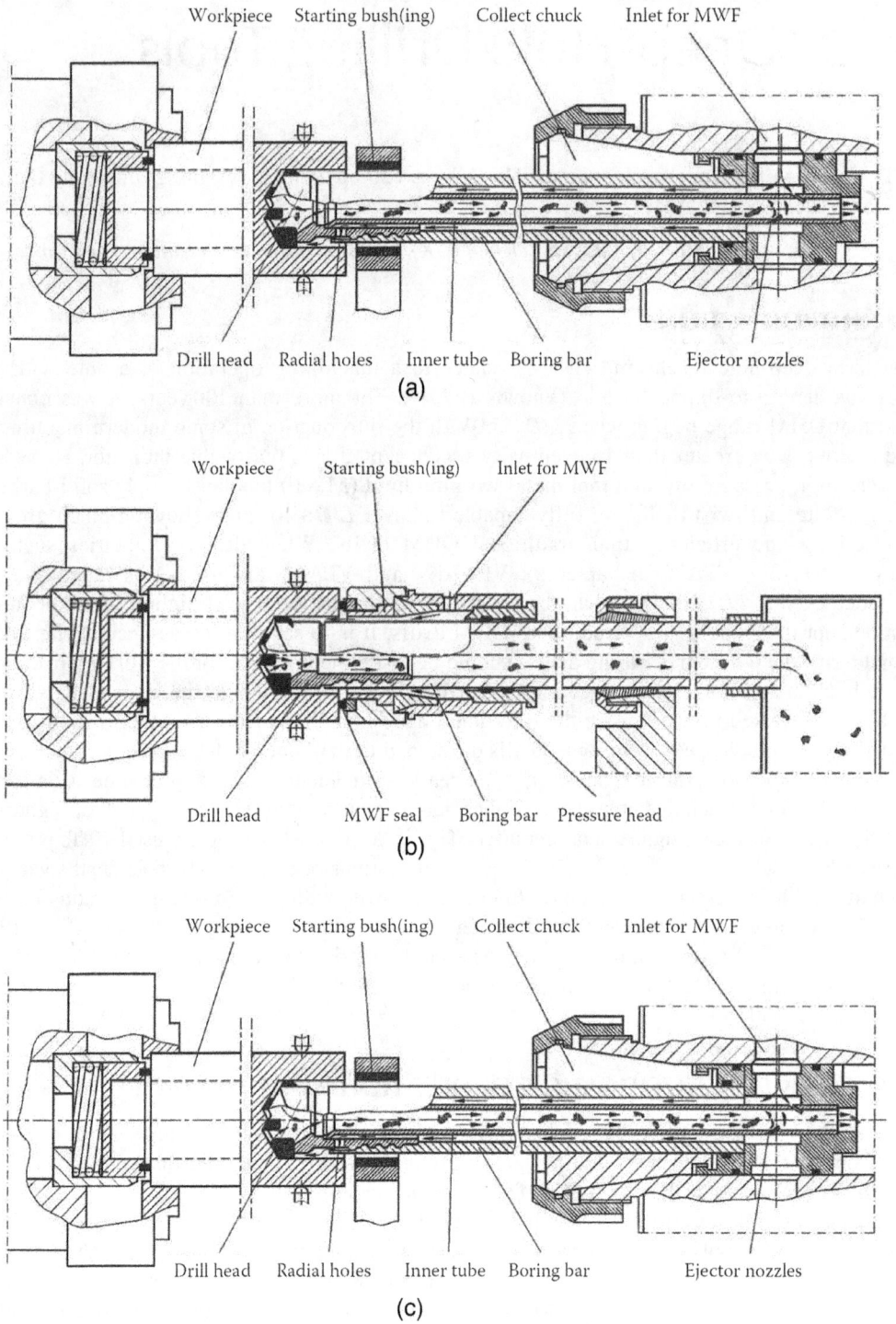

Workpiece   Starting bush(ing)   Collect chuck   Inlet for MWF

Drill head   Radial holes   Inner tube   Boring bar   Ejector nozzles

(a)

Workpiece   Starting bush(ing)   Inlet for MWF

Drill head   MWF seal   Boring bar   Pressure head

(b)

Workpiece   Starting bush(ing)   Collect chuck   Inlet for MWF

Drill head   Radial holes   Inner tube   Boring bar   Ejector nozzles

(c)

FIGURE 3.1   Deep-hole machining classification by method of MWF supply: (a) Gundrilling, (b) STS drilling, (c) ejector drilling.

Figure 3.1b shows the principle of single tube system (STS) drilling. The STS tool assembly consists of a boring bar and a single- or a multi-edged drill head secured at its terminal end. MWF is supplied under high pressure through the inlet of the pressure head and flows through the annular channel between the boring bar and the bore wall toward the drill head. After cooling and lubricating the machining zone, MWF carries away swarf through the interior of the drill head and boring bar. Contrary to gundrilling, the returning chips do not come in contact with the bore wall; thus, a better surface finish can be achieved.

The tubular cross section of the boring bar possesses greater buckling stability compared to the gundrill's shank; thus, greater feed rates can be achieved in STS drilling. As such, the annular chip removal channel is of much greater cross section compared to the V-flute of gundrill; thus, more chips per unit time can be transported without clogging. Moreover, even when chip clogging occurs due to poor chip breaking, the inlet pressure increases helping to push forward the chip cluster that clogs the chip removal channel. A requirement for reliable sealing between the face of the workpiece and the pressure head is a price to pay for this advantage.

Figure 3.1c shows the principle of ejector drilling. MWF is supplied to the inlet of the collet chuck. Then, this fluid separates into two portions. The first portion flows in the annular channel formed by the boring bar, called the outer tube, and the inner tube. In the drill head, MWF flows outside through the radial holes made in the drill head; thus, this part reaches the machining zone where it cools and lubricates the cutting elements and the bearing areas. Then, the MWF–chip mixture (swarf) goes into the inner tube. The second portion flows through the ejector nozzle(s) made on the inner tube. As a result, this flow creates a partial vacuum (the ejector effect) in the inner tube that sucks the swarf into the inner tube as a vacuum cleaner.

The great advantages of ejector drilling are as follows: (1) there is no need to have a non-reliable seal between the face of the workpiece and the starting bushing and (2) much lower inlet MWF pressure is needed. As a result, ejector drilling does not always require a special drilling machine so this type of drilling can be used in many general-purpose machines or even machining centers as one of the common drilling operations.

However, this versatility comes at certain costs/restrictions. First, it is not available for hole diameters below 20 mm. Second, it is suitable only for the work materials that generate easy-to-control chips. That eliminates most nickel-based and many non-ferrous alloys. This is because if the chip even slightly clogs the chip removal channels including the inner tube, there is no pressure available to push the formed chip cluster through as the maximum pressure created by the ejector is much less than 0.1 MPa. As a result, MWF leaves the radial coolant holes and then flows between the boring bar and the bore outside the drilling coolant circuit. When this happens, the drill normally breaks.

3.3   GUNDRILLS: BASIC DESIGN AND GEOMETRY

A generic gundrill shown in Figure 3.2 consists of a drill body having a shank and a tip (a.k.a. the drill head). The tip is made up of a hard wear-resistant material such as a sintered carbide. The other end of the shank incorporates an enlarged driver used to set up the drill in the drill holder.

The driver can have a flat surface for holding the drill. The design of the driver is machine specific. The shank is of tubular shape having a V-shaped flute on its surface. The flute terminates in an inclined crease formed adjacent to the driver. The tip is larger in diameter than the shank and also has the V-flute, which is similar in shape to the flute on the shank and extends along the full length of the tip. These two flutes are longitudinally aligned. On the flank face of the tip, an orifice as an outlet for the MWF (coolant) is located. Gundrill manufacturers have adopted various shapes for this orifice including one or two circular holes or a single kidney-shaped hole.

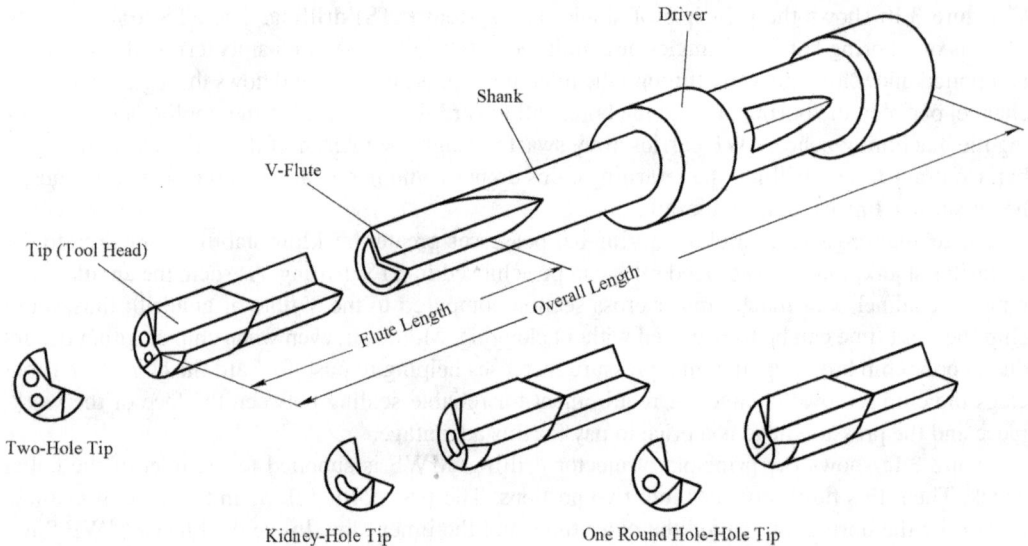

FIGURE 3.2 Common gundrill and its components.

3.3.1 HISTORY

Present-day deep gundrilling has its origin in the firearm industry of the eighteenth century. During the period between years 1500 and 1750, the town of Suhl in Germany was known as the center of deep-hole drilling. In such a drilling, a water mill was used as the machine tool and a spade drill bit was the cutting tool. Two barrels could be drilled simultaneously by two parallel boring spindles. The feed and the thrust were provided by the operator actuating a lever, which supplied certain force amplification. Jean Maritz developed an advanced horizontal boring machine in 1754. A drilling head had been mounted on the end of a boring bar that was rotated by animal power and a downward feed motion was applied to the gun barrel. The first boring machine in which the workpiece was rotated and feed motion was given to the drilling tool appears to have been used in about 1758 by J. Verbruggen in Germany.

A major survey of the methods and equipment available for making guns was published in 1794 by G. Monge in his book, *L'Art de Fabriquer les Canons* at the investigation of the Committee of Public Safety of the French Revolution. This survey shows that these earlier-date SPTs were in wide use, albeit primarily for finishing gun barrels.

An article published in 1886 by Landis describing gun-barrel manufacture shows that by this time, not only was there a variety of SPTs available but the refinement of pressurized cutting oil systems was employed (Landis 1886). SPTs illustrated in this paper are remarkably similar in design to present-day tools. He even showed a tool with replaceable supporting pads and an indexable cutting edge. This very detailed and comprehensive article shows that the advantages of SPTs had been recognized, and they were employed to improve the efficiency of gun manufacturing for centuries. Although a number of improvements have been made since then, for example, the application of tungsten carbide and specially blended MWF (coolants), it remains a fact that the foundation of modern DHM was laid by the turn of the twentieth century. Another milestone publication on DHM is a book edited by Swinehart (1967) that summarized the tool and machine design for deep-hole machining as well as their application practice for a great variety of work materials. Unfortunately, this book is practically unknown to many specialists including researchers in the field. Should patent experts have read Landis' paper and book edited by Swinehart, well more than a half of the existing patents on "new" designs of gundrills would not have been granted in the author's opinion. Moreover, practical manufacturing engineers and application specialists would not repeat the common mistakes from the past in the selection of drills, machine tools, and drilling parameters.

3.3.2 Common Design and Geometry: Advanced Users' Perspective

There is no ISO standard on gundrill design and geometry. However, it is covered by the German Technical Rule VDI 3208:2014-4: Deep-hole drilling with gundrills available in German and English (VDI – The Association of German Engineers). It describes the general advantages of gundrilling, the specifics of MWFs application, and lists the properties of the suitable machines. It includes the guidelines for the tool design and provides recommendations on the selection of the machining regime parameters for various work materials.

3.3.2.1 Defining the Parameters

The basic design and geometry parameters of a commonly used gundrill are shown in Figure 3.3. The gundrill consists of a drill body having a shank (1) and a tip (2). The tip is made up of a hard wear-resistant material such as tungsten carbide. The other end of the shank incorporates an enlarged driver (3) having a machine-specific design. The shank is of tubular shape having an elongated passage (4) extending over its entire length and connecting to the MWF supply passage (5) in the driver. The shank has a V-shaped flute (6) on its surface that serves as the chip removal passage. The shank length depends mainly on the depth of the drilled hole as well as on the lengths of the bushing and its holder, chip box, etc., so it is determined from the tool layout as discussed later.

The tip is larger in diameter than the shank that prevents the shank from coming into contact with the walls of the hole being drilled. Flute (7) on the tip, which is similar in shape to flute (6), extends along the full length of the tip. This flute is bounded by side faces (8 and 9) known as the cutting face and side face, respectively. The depth of this flute is such that the cutting face (8) extends past the axis (distance c_{ax}) of the tip, which is also the axis of the drill body. The angle ψ_v between the side and cutting faces is known as the profile angle of the tip, which is usually equal or close to the V-flute profile of the shank.

The terminal end of the tip is formed with the approach cutting edge angles φ_{a1} and φ_{a2} of the outer (10) and inner (11) cutting edges, respectively. These cutting edges meet at the drill point P (a.k.a. the apex). The location of the drill point (defined by the distance m_d in Figure 3.3) can be varied for optimum performance depending on the work material and the finished hole specifications. One common point grind calls for the outer angle, (φ_{a1}), to be 30° and the inner angle, (φ_{a2}), to be 20°. The geometry of the terminal end largely determines the shape of the chips and the effectiveness of MWF, the lubrication of the tool, and the removal of the chips. The process of chip formation is also governed by other cutting parameters such as the cutting speed, feed rate, and work material.

The primary flank surface (12) having a normal primary clearance angle α_{n1-p} of 7°–10° is applied to the other cutting edge (10). To assure drill-free penetration, the secondary flank surface (13) having normal clearance angle α_{n1-s} of 12°–20° is applied as shown in Figure 3.3. Flank surface (14) having a normal clearance angle α_{n2} of 8°–12° is applied to the inner cutting edge (11). To assure drill-free penetration, that is, to prevent the interference of the drill's flanks with the bottom of the hole being drilled, the auxiliary flank (15) (normal clearance angle α_{n3}) and shoulder dub-off (16) (point angle φ_{a3} and clearance angle α_4) are provided. Their location and geometry are uniquely defined for a given gundrill. The tip 2 is provided with a coolant hole (17) to supply MWF (coolant) into the machining zone.

Another common shape of the flank surface is a helical surface rather than a planar surface. The helical flank surface is normally applied to the flank of the outer cutting edge, while other flanks are planar. Different manufacturers have different standards on the lead and generating diameter of the helical flank surface, depending upon the drill diameter and design of the grinding fixture.

Modern point grinds use great lead and generating diameters so that the flank surface of the outer cutting edge does not affect the planar flanks of the inner cutting edge or the shoulder dub-off (angle φ_4) as shown in Figure 3.4a (Zhang et al. 2004). The lead and generating diameter of this surface is relatively large, so the line, or rib, of intersection (1) of outer (2) and inner (3) flank surfaces does not extend too far from the vertical axis. Therefore, shoulder dub-off (4) is not affected by the helical surface. In older designs (still in use in some manufacturing facilities), for example, US Patent No.

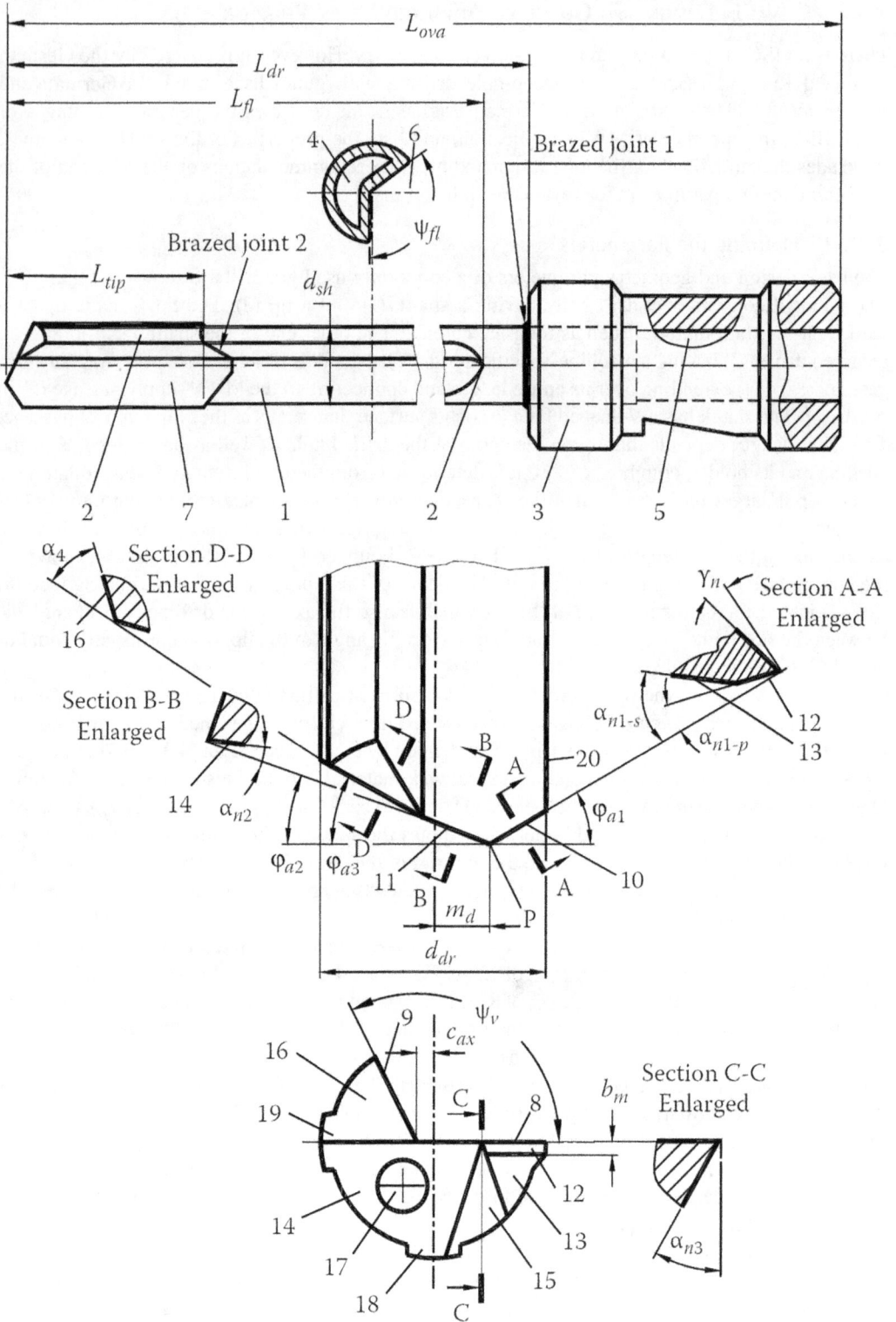

FIGURE 3.3   Geometry of a common gundrill.

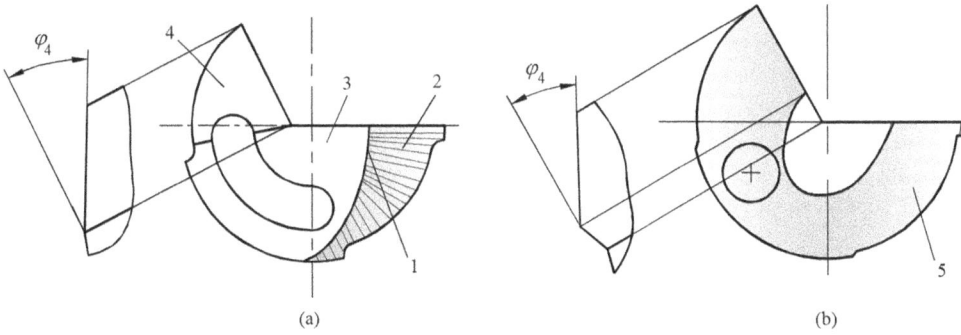

FIGURE 3.4 Alternative shapes of the flank surfaces: (a) combined helical and planar and (b) helical.

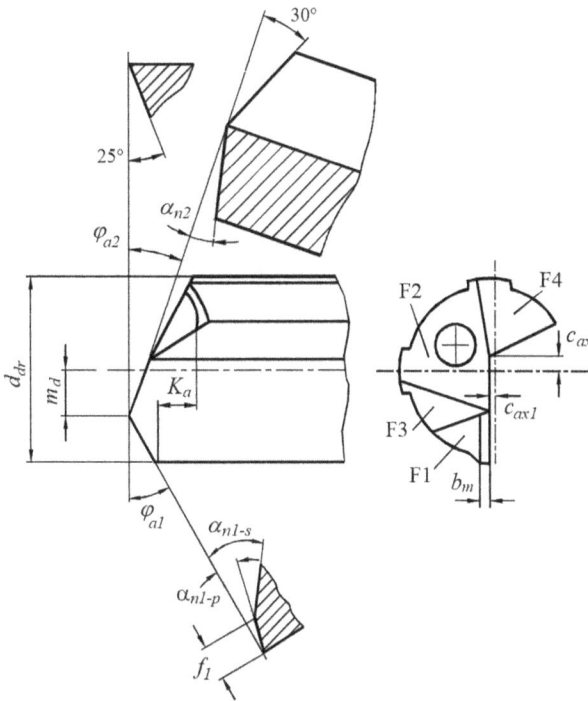

FIGURE 3.5 Parameters of a gundrill with planar flanks as they should appear in design/manufacturing drawings.

2,325,535 (1943), which is still in use, a relatively small lead and generating diameter of the helix surface is used so that this surface passes through the outer flank and shoulder dub-off as shown in Figure 3.4b. In this figure, a helical convex surface (5) is applied to the outer cutting edge. This helical convex surface has a small lead and generating diameter so that it passes through a flank surface of the inner cutting edge.

3.3.2.2 Common Recommendations on the Selection of Tool Geometry/Design Parameters

Gundrills with the planar flank faces (Figure 3.3) became the most popular in modern manufacturing for drilling holes of 3–30 mm diameter. As shown in Figure 3.5, the flank face of such a drill consists of four planar flank faces:

- The flank face F1 of the outer cutting edge ground with the primary, α_{n1-p} and secondary, α_{n1-s} normal clearance angles. These angles are work material specific.
- The flank face F2 of the inner cutting edge, ground with the normal clearance angle α_{n2}. This angle is work material specific.
- The first auxiliary flank face F3. As shown, this plane begins just on the drill apex (shifted by the distance m_d from the drill axis) having a 25° normal clearance angle.
- The second auxiliary flank face F4, ground with a clearance angle of 30°.

Note that the purpose of auxiliary faces F3 and F4 is to assure gundrill-free penetration with no interference of the flank faces F1 and F2 with the bottom of the hole being drilled. In other words, these conditions assure that only the outer and inner cutting edges are in contact with the workpiece.

The recommended parameters of the geometry of a common gundrill shown in Figure 3.5 are shown in Table 3.1. Other parameters are as follows: $c_{ax1} = 0.02 d_{dr}$, $c_{ax} = 0.05 d_{dr}$, $b_m = (0.02...0.04) d_{dr}$, $f_1 = (0.03...0.05) d_{dr}$, and $K_a \geq 0.5$ mm.

Figure 3.6 shows the geometry of a gundrill for drilling low-surface-roughness (Ra=0.63–0.32) holes. Tables 3.2 and 3.3 list the values of the parameters shown in this figure.

TABLE 3.1

Standard Parameters of Gundrills with Planar Flanks

Work Material	Hardness HB	m_d (mm)	φ_{a1} (°)	φ_{a2} (°)	α_{n1-p}(°)	α_{n1-s} (°)	α_{n2} (°)
Steel	To 240	$0.20 d_{dr}$	45	20	7	15	20
	240–320	$0.25 d_{dr}$	35	25	7	15	20
Cast iron	120–300	$0.25 d_{dr}$	30	30	6	8	15

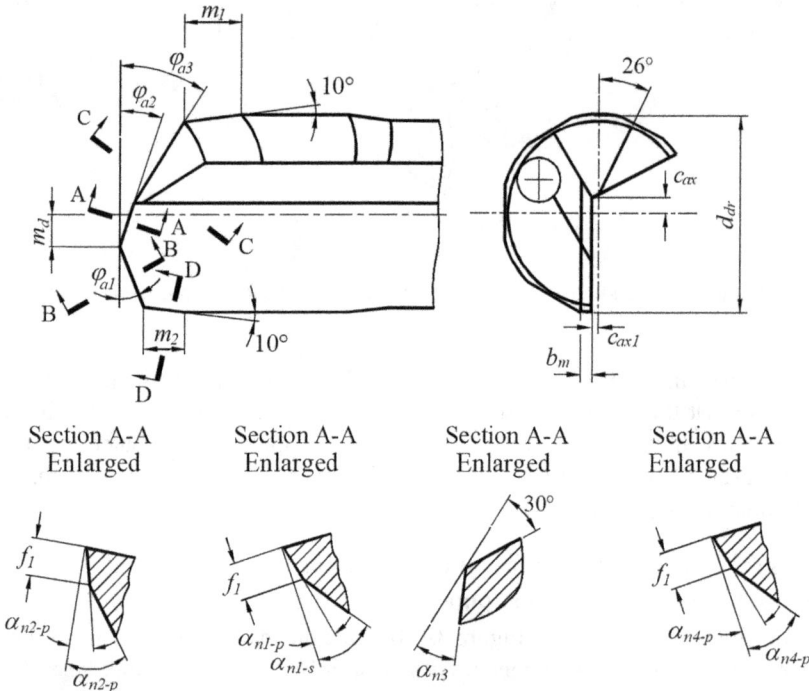

FIGURE 3.6 Parameters of a gundrill for drilling low-surface-roughness (Ra=0.63–0.32) holes.

TABLE 3.2

Dimensional Parameters of Gundrills with Planar Flanks for Holes of Ra=0.63–0.32 μm

Drill Diameter	m_d (mm)	m_1 (mm)	m_2 (mm)	c_{ax} (mm)	c_{ax1} (mm)	b_m (mm)	f_1 (mm)
From 3 to –5	0.7	0.5	–	0.4	0.08	0.12	–
Over 5–8	1.2	1.0	1.0	0.6	0.10	0.15	0.3
Over 8–12	2.0	1.3	1.3	0.7	0.15	0.20	0.5
Over 12–15	2.4	1.6	1.6	0.8	0.20	0.25	0.5
Over 15–20	2.8	2.0	2.0	1.0	0.25	0.30	0.6

TABLE 3.3

Geometrical Parameters of Gundrills with Planar Flanks for Holes of Ra=0.63–0.32 μm

Work Material	Hardness HB	φ_{a1} (°)	φ_{a2} (°)	$\alpha_{n1-p}, \alpha_{n2-p}, \alpha_{n4-p}$ (°)	$\alpha_{n1-s}, \alpha_{n2-s}, \alpha_{n4-p}$ (°)	α_{n3} (°)
Steel	To 240	35	15	12	18	20
	240–320	30	20	7	16	20
Cast iron	120–300	30	20	6	8	15

3.3.2.3 Particularities of the Rake Face Geometry

The rake face shown in Figure 3.5 is used in approx. 99% of gundrills due to its apparent simplicity. As such, the T-hand-S rake angle of any point of a gundrill is considered to be zero, and it does not change with re-sharpening. Moreover, because the shift of the outer and inner cutting edges with respect to the centerline (CL) (distance c_{ax1} in Figures 3.5 and 3.6) is small, these angles are considered to be diametral, that is, the rake angles in T-hand-S and T-hold-S are the same because the assumed, as well as the actual, vector of the cutting speed is approximately perpendicular to the cutting edge. This simplicity, however, is only apparent because there is a not well-understood issue in the region near the axis of rotation.

Although the gundrill does not have the chisel edge, an *Achilles' heel* of straight-flute and twist drills, the problem with cutting in the region close to the axis of rotation does not magically disappear. In basic designs (similar to that shown in Figure 3.3) presented in many literature sources, the cutting edge is shown as passing through the axis of rotation, so there should be no problem with cutting in the region close to the axis of rotation. One should realize, however, that a certain tolerance on the cutting edge location with respect to the CL should be allowed.

Figure 3.7 shows the possible locations of the cutting edge with respect to CL (the y_0-axis). Figure 3.7a shows the theoretical location of the cutting edge where its projection into the $x_0 y_0$ plane coincides with the projection of the y_0-axis, that is, the cutting edge passes through the center of rotation. In reality, a certain shift (due to a certain manufacturing tolerance) from its ideal location is the case. The rake face can be made so that the location of the cutting edge is as shown in Figure 3.7b, i.e., above CL. This is the worst-case scenario because of interference between the flank face of the drill and the bottom of the hole being drilled as clearly seen in this figure. In other words, when the cutting edge is shifted above CL by a certain distance c_{ax1} up with respect to the y_0-axis, the drill has no means to remove the cylindrical core having a radius equal to c_{ax1} as shown in Figure 3.7b. When this distance is relatively small, the drill bends to compensate for this core. When c_{ax1} exceeds a certain threshold (depending upon drill particular drill design and parameters of the gundrilling system), the tip simply breaks.

To avoid interference, the rake face should be always located so that the projection of the cutting edge into the $x_0 y_0$ plane occupies a position right on or slightly below CL (the y_0-axis) as shown in

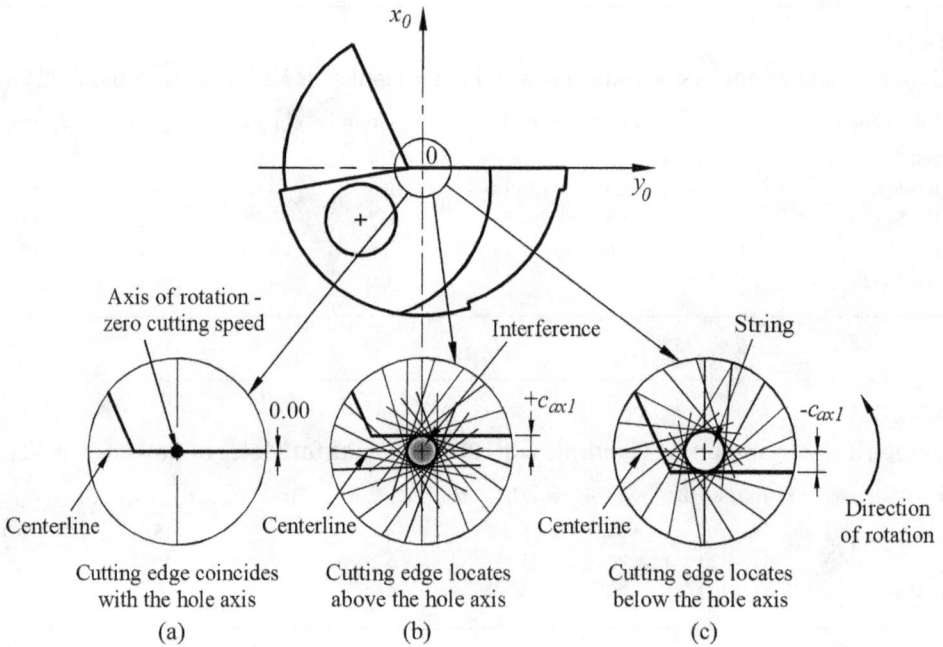

FIGURE 3.7 Possible locations of the cutting edge with respect to the centerline (CL) (the y_0-axis): (a) on the axis, (b) above the axis, (c) below the axis.

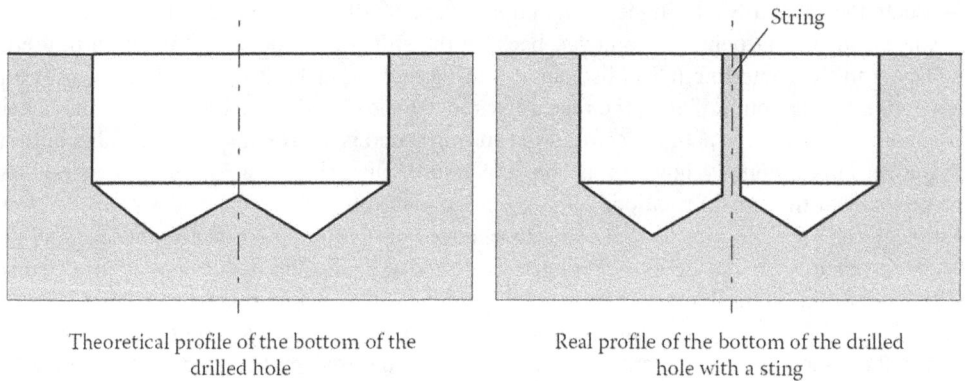

FIGURE 3.8 Theoretical and real profile of the bottom of the drilled hole.

Figure 3.7c and in the above-discussed recommendations (see Figures 3.5 and 3.6). When the cutting edge locates below CL, a string is formed as the result of such a location. The theoretical string diameter would be equal to $2(-c_{axl})$. The string will be attached to the bottom of the hole being drilled as shown in Figure 3.8. Figure 3.9 shows examples of the strings. It is clear that this string is undesirable particularly when blind holes are drilled.

It is also worthwhile to explain the role of c_{axl} at the entrance of the hole being drilled. Figure 3.10a shows the ideal location of a drill tip in a starting bushing (no misalignment). In this figure, Δ_{sb} is the radial clearance (gap) between the tip and the bushing and c_{axl} is the location distance of the cutting edge behind CL. It is clear that the intended (theoretical) string diameter is calculated as $d_{st} = 2c_{axl}$. When drilling begins, the cutting force applied to the cutting edge shifts the tip toward the bushing walls (Astakhov 2002a). As such, when the drill rotates, the string changes its location at the entrance with respect to the center of the drilled hole (Figure 3.10b). Eventually, when the supporting part enters the hole, the drill occupies a new location and its longitudinal axis coincides

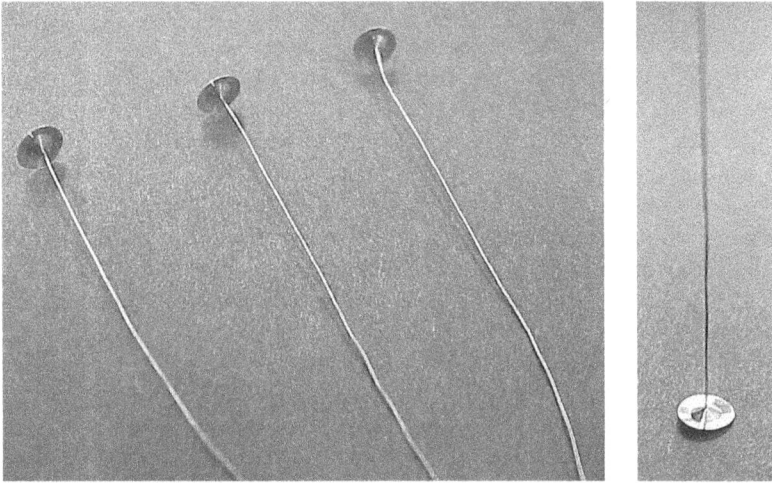

FIGURE 3.9 Strings.

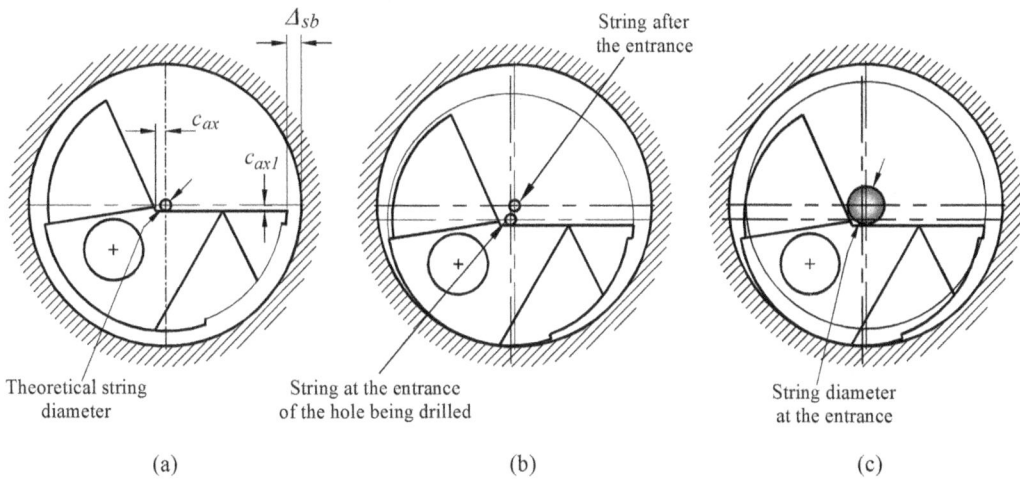

FIGURE 3.10 (a) Theoretical string diameter, (b) string diameter of the hole entrance when the drill rotates, and (c) string diameter of the hole entrance when the workpiece rotates.

with the axis of rotation. The string, however, does not change its diameter. It shifts together with the tip after the full tip entrance into the hole being drilled. A small additional force due to this shift does not normally present any problem. This is not the case, however, when the workpiece rotates and the gundrill is stationary. This case is shown in Figure 3.10c. As such, the maximum diameter of the string increases becoming equal to $d_{st} = 2(c_{axl} + \Delta_{sb})$. If Δ_{sb} is significant (due to improper diameter of the starting bushing or due to bushing wear), the large additional force acts on the rake face that often causes tip breakage. Besides, the interference between the string and the sidewall of the V-flute may take place that affects the position of the gundrill with respect to the axis of rotation. To avoid this, the distance c_{ax} (Figure 3.10a) should not be more than $c_{ax} = 0.05d_{dr}$.

In practice, however, the string presents some problems only when blind holes are being drilled in special high-alloy steels. Otherwise, the sting usually breaks into relatively small portions and does not create any problems. Moreover, multiple testings of gundrills have shown that proper selection of distance c_{axl} results in lower cutting forces and greater drill stability. The author's application experience allows to suggest that this distance should be assigned within the range of 0 to $-cax_1 = 0.02d_{dr}$. This should be included in an inspection report and clearly indicated on any gundrill drawing.

The discussed string can provide some useful benefits in gundrilling if its diameter is deliberately increased to a rod (Figure 3.11). As can be seen, the gundrill has the same components as the usual gundrill, namely, the driver, shank, supporting pads, and MWF supply passage made in the shank. The difference is in the circular opening that extends from the front of the gundrill to its opposite end passing through the shank as a hole. In the operation of the drill, the presence of the circular opening causes the formation of a rod. This rod extends from the bottom of the hole being drilled throughout the length of the workpiece, thus forming a kind of support around which the drill rotates. The rod is removed after the drilling operation is over. Technically, the discussed drill is a trepanning tool as it does not drill the entire work material. Such a drill can be used only for through holes as it was intended for rifle barrels, camshafts, etc.

The discussed design pioneered a number of variations of gundrill designs using the same or similar idea adjusting it to the needs of improved gundrill manufacturing quality and requirement to make more versatile tools. Figure 3.12 shows a gundrill design according to US Patent No. 3.089,359 (1963). As claimed, the significant opening is made for the sake of using replicable cutting inserts in which location accuracy on the shank does not affect the conditions of cutting in the center as the rod diameter is great, so only a small variation of this diameter can be caused by the differences in the cutting insert position. The circular opening on the tip requires a special cross-sectional configuration of a tubular shank.

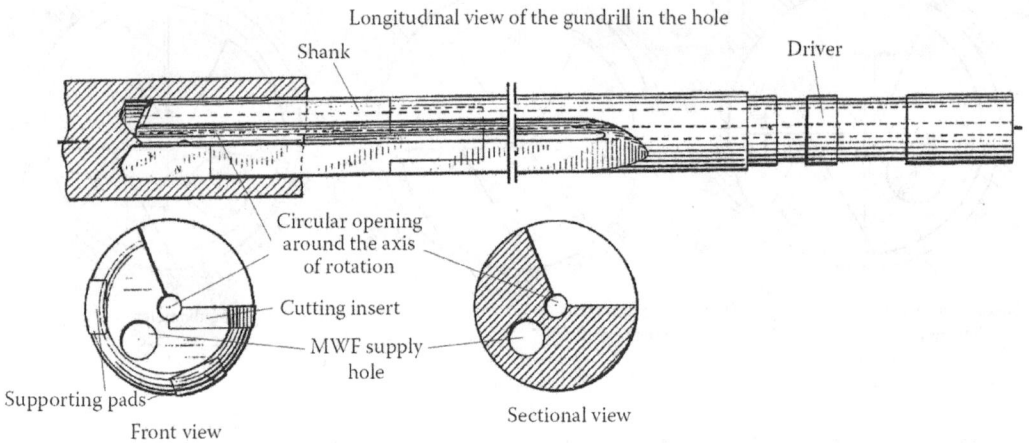

FIGURE 3.11   Gundrill design according to US Patent No. 2,418,021 (1947).

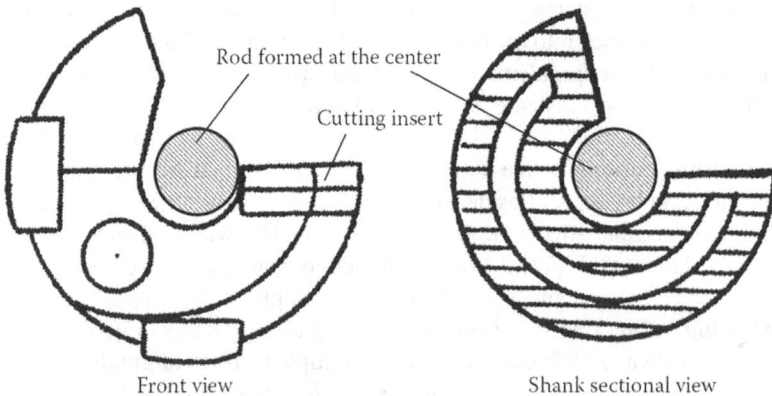

FIGURE 3.12   Gundrill design according to US Patent No. 3.089,359 (1963).

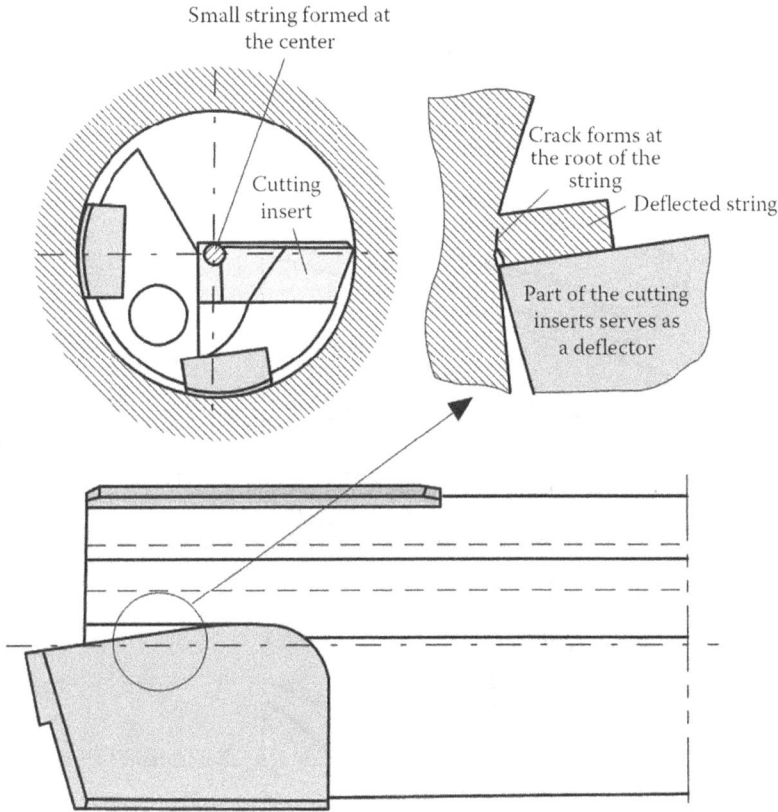

FIGURE 3.13 Dealing with the string in modern gundrills with indexable inserts.

In modern gundrills of small-diameter ground on CNC grinders, the location of the cutting edge is precisely controlled so that this distance is small, and thus, the resultant small-diameter string does not present problems as it breaks itself into small pieces. In modern gundrills with indexable inserts, STS, and ejector drills, a universal mean to deal with a partially formed string is shown in Figure 3.13, where a general string deflector is a part of the insert design with a negative rake angle (e.g., US Patent No. 4,565,471, 1986). As can be seen, when a partially formed string contacts the deflector, it bends and then fractures from the bottom of the hole being drilled. The design and location of the chip deflector depends on the type and design of a particular drill and thus can be used with any type of drill when one tries to solve the problems that unavoidably occur for any drill in the region adjacent to the axis of rotation.

3.3.2.4 Particularities of the Flank Face Geometry

The geometry of major and auxiliary flank faces of standard gundrills having planar flank faces is simple as the major cutting edges are located practically on CL. As a result, the clearance angles in T-hand-S and T-hold-S are the same because the assumed, as well as the actual, vector of the cutting speed is approximately perpendicular to the cutting edge. Therefore, the formulae correlating clearance angle in different standard reference planes are the same as for straight-flute drills in T-hand-S considered in Chapter 2. Particularities of more complicated and rarely used flank geometries are considered by the author earlier (Astakhov 2010).

3.3.2.5 Shank

Shanks consist of tubing with a flute rolled or swaged to the required shape (Figure 3.14). The most common flute form is a V-shape that matches the flute on the tip although other shapes are still

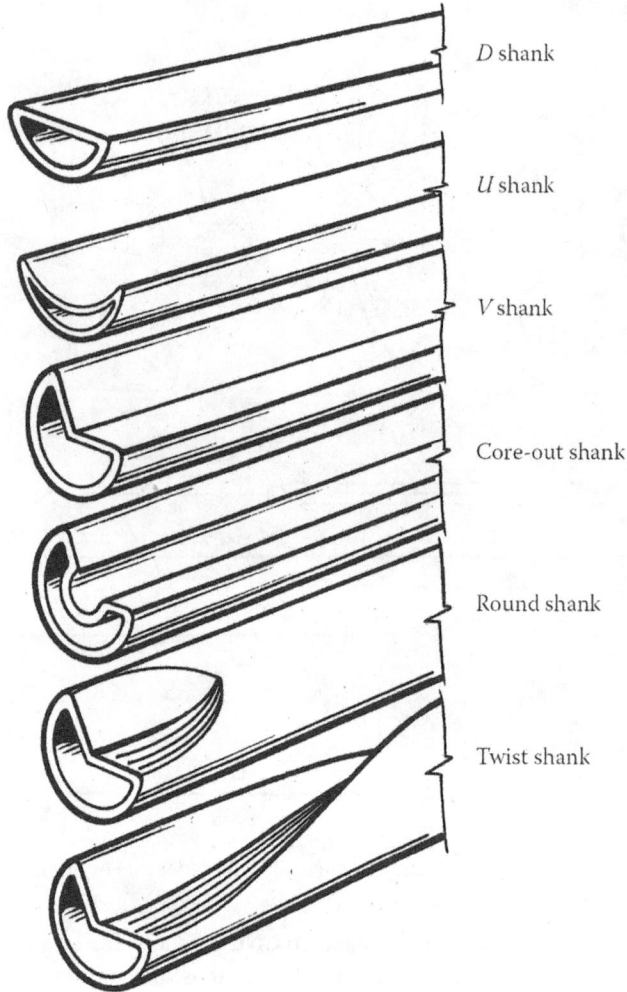

FIGURE 3.14   Various styles of gundrill shanks.

in use for some specific applications as each of the listed shapes has its own advantages. These may include simplicity of manufacturing, chip removal space, torsional and buckling stiffness, and MWF supply cross-sectional area. Although the shank is as critical as the tip and driver, practically no information on the shank design, material, and manufacturing requirements is available in the published literature.

The major dilemma in the shank design is the proper selection of its external diameter and wall thickness (or the internal diameter). On the one hand, the MWF (coolant) channel through the length of the shank must be large enough to fully supply MWF to the tip. This leaves a limited wall thickness for the shank and explains why the shank is probably the most critical component of the gundrill. On the other hand, the wall thickness should be large enough to withstand the axial force and drilling torque. This wall thickness combined with the shank length is a major constraint on the allowable feed/penetration rate and thus productivity of drilling. The better the shank design and its material, the higher the allowable feed per revolution and thus the penetration rate of gundrilling.

The shank length L_{sh} is another critical parameter of the shank. Needless to say, this length should be as short as possible. As discussed in Chapter 1, the proper way to determine this length

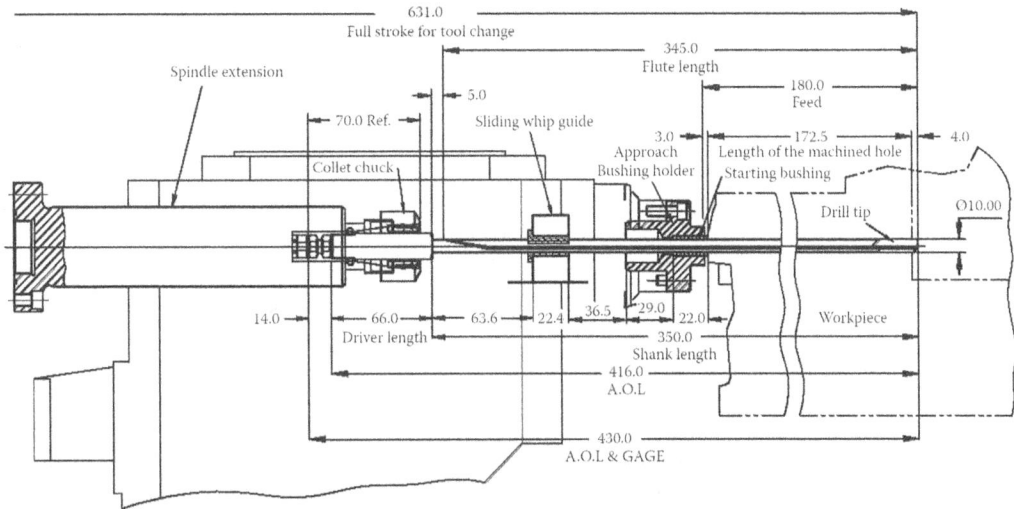

FIGURE 3.15 A typical tool layout to determine the length of the shank, flute, and drill.

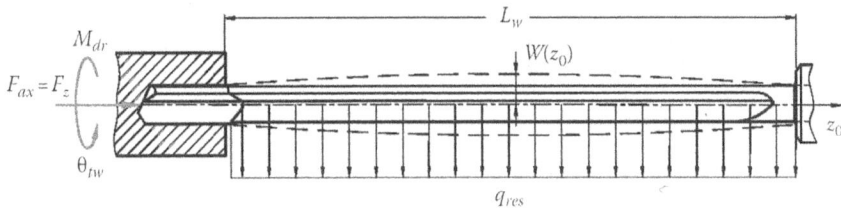

FIGURE 3.16 Force factors acting on a gundrill.

is to design the tool layout for a given drilling operation (see Figures 1.16 and 1.18). An example of the tool layout for a gundrilling operation is shown in Figure 3.15. The following sequence is recommended in determining the shank length: start with the length of the machined hole then add the approached and overshoot distances and then the length of the bushing including the bushing holder and chip box.

Once the shank length is determined, an important decision is to be made about the use of steady rest(s) often called whip guides and their number to restrict whipping of the shank. Although whipping is mentioned in many gundrill-/gundrilling-related technical and trade literature sources, its nature is not revealed. A need is felt to clarify the issue.

Because a gundrill is a drill, the resultant force factors the shank must withstand are the same as discussed in Chapter 2 (Figure 2.114), that is, the axial force F_{ax} and the drilling torque M_{dr} as shown in Figure 3.16. The drilling torque causes drill twisting, while the axial force causes its buckling due to an extensive drill length. It may be logically assumed that whipping (shown in Figure 3.16 as $w(z_0)$) is a result of shank buckling due to the action of the axial force. The problem, however, is that whipping is observed with no drilling when a gundrill is just installed into the starting bushing and rotates. It was noticed that the higher the rotation speed, the greater the whipping.

Our understanding of gundrill whipping based on observation of many gundrilling systems allows to represent whipping as consisting of two components, namely, forced and *natural*. The forced whipping occurs when a gundrill rotates, and it is due to the action of the axial force as considered in Chapter 2. When the drill is stationary, some static bending of the shank is observed due to the action of the axial force. Natural whipping is caused by a distributed load q_{res} acting on the gundrill shank as shown in Figure 3.17a. This distributed load is vectorial load consisting of two components, namely,

$$\mathbf{q}_{res} = \mathbf{q}_c + \mathbf{q}_g \qquad (3.1)$$

where $\mathbf{q}_c$ is the centrifugal component arising because the center of rotation of the shank does not coincide with its center of gravity and $\mathbf{q}_g$ is the load due to shank weight. The magnitudes of these distributed loads are calculated as

$$q_c = A_{sh}\rho_{sh}y_c\omega_{sh}^2 \quad (N/m) \qquad (3.2)$$

where A_{sh} is the shank cross-sectional area (m²), ρ_{sh} is the density of the shank material (kg/m³), y_c is the distance between the center of shank rotation and its center of gravity as shown in Figure 3.17a, and ω_{sh} is the shank angular velocity (s⁻¹). Figure 3.17b shows an example of q_c for some common shank diameters (for shank having flute angle $\psi_c = 110°$ and wall thickness $\delta_{sh} = 0.12d_{sh}$).

$$q_g = A_{sh}\rho_{sh}g \quad (N/m) \qquad (3.3)$$

where g is the acceleration due to gravity (m/s²). When gundrilling on the Earth, $g = 9.81$ m/s².

The rotating coordinate system $x_r y_r$ is introduced to obtain the final equation for the magnitude of distributed load q_{res} shown in Figure 3.17a. As such, the angle between the stationary and the rotating axes is a function of time, t as $\psi_r = \omega_{sh}t$. Therefore,

$$q_{res} = \sqrt{q_c^2 + q_g^2 - 2q_c q_g \cos \omega_h t} \qquad (3.4)$$

It follows from Eq. (3.4) that the resultant distributed load q_{res} changes with the rotation angle ψ_r and its maximum is achieved when vectors $\mathbf{q}_c$ and $\mathbf{q}_g$ have the same direction.

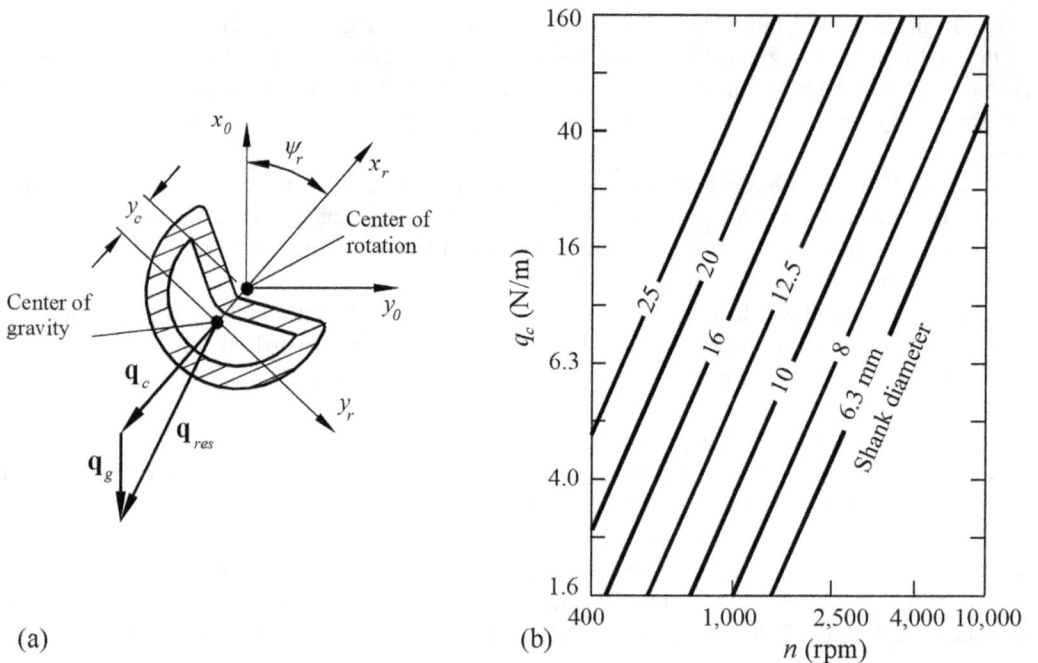

(a)   (b)

FIGURE 3.17   Showing the nature of (a) the distributed load on a gundrill and (b) its centrifugal component.

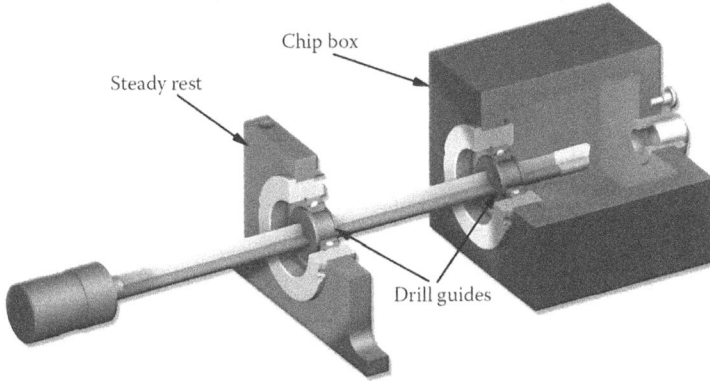

FIGURE 3.18   Steady rest for a gundrill by International Drill Guide Co.

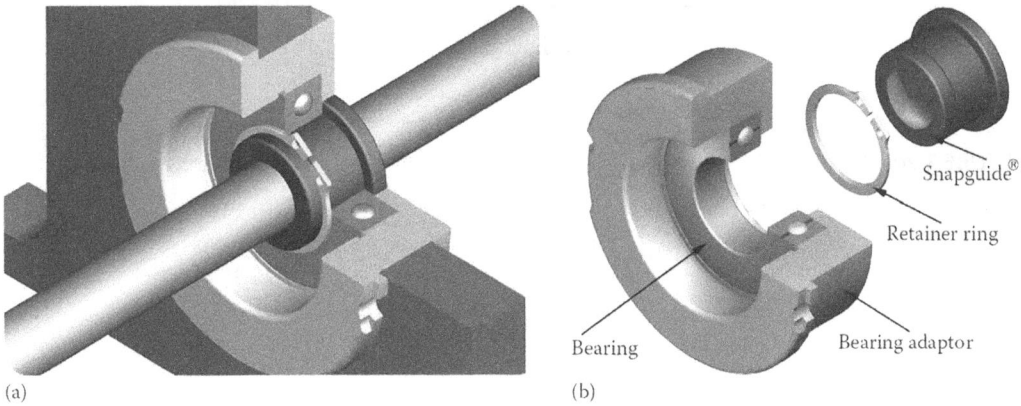

(a)   (b)

FIGURE 3.19   Snapguide® steady rest bushing: (a) assembly and (b) its components. (Courtesy of International Drill Guide Co., Willoughby, OH.)

A steady rest is installed on the gundrilling machine as shown in Figure 3.18 to limit whipping (both its components – forced and natural). Figure 3.19 shows a simple and reliable design by International Drill Guide Co. that provides significant help in damping vibration and drill whipping. These flexible plastic bushings support the shank with a slight compression fit and are mounted in a bearing on the steady rest.

Figure 3.20 helps to make the decision about the use of steady rests. For example, if the shank length of an 8 mm drill rotating at 3,000 rpm is 400 mm, then a whip guide is needed because the maximum allowable distance between supports for these conditions is 350 mm. Information similar to that shown in Figure 3.20 should be always requested from gundrill manufacturers for the design of the layout for a given gundrilling operation.

The foregoing analysis suggests the same as with twist and straight-flute drills, a gundrill shank should withstand buckling and torsion (see Chapter 2, Section 2.1.5.2). The buckling stability is enhanced by the proper location of steady rest according to Figure 3.20 so that resistance to torsion is of prime concern in the selection of shank material and its cross-sectional shape. Figure 3.21 shows how the torsion rigidity GJ (G is the shear modulus of the shank material and J is the polar moments of inertia – see Chapter 2, Section 2.1.5.2.1) varies with the shank diameter for the annular and V-shaped cross sections. As can be seen, the torsion rigidity is always significantly greater for the annular cross section. Heat treatment of the shank improves the torsion rigidity but such an improvement is rather marginal (Nikolov 1986).

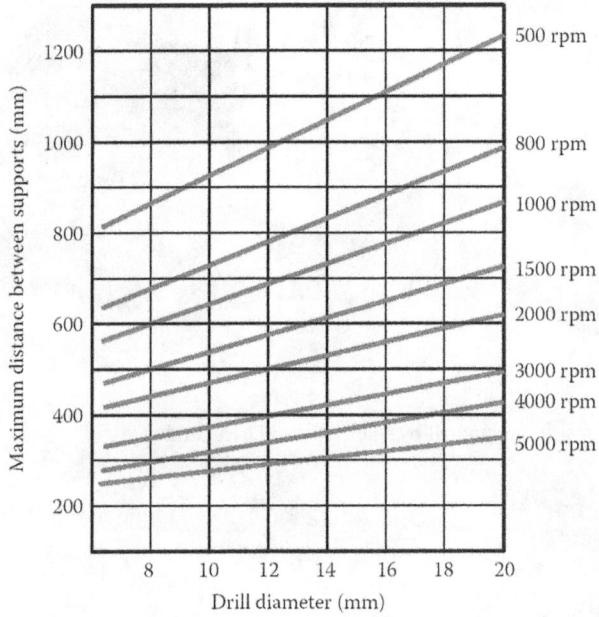

FIGURE 3.20 Maximum allowable distance between supports for high-penetration-rate gundrills.

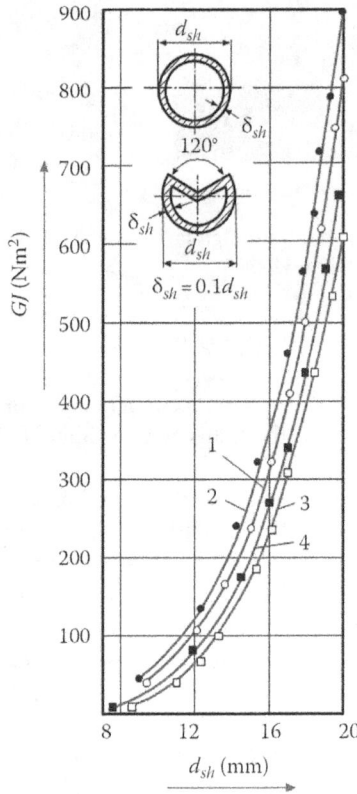

FIGURE 3.21 Torsion rigidity *GJ* versus shank diameter for the annular and V-shaped cross sections: The data shown for shanks made of AISI steel 1040 ($G = 0.8\,\text{MPa}$): (1 and 2) annular cross section, non-heat-treated and heat-treated, respectively, and (3 and 4) V-shaped cross section, non-heat-treated and heat-treated, respectively.

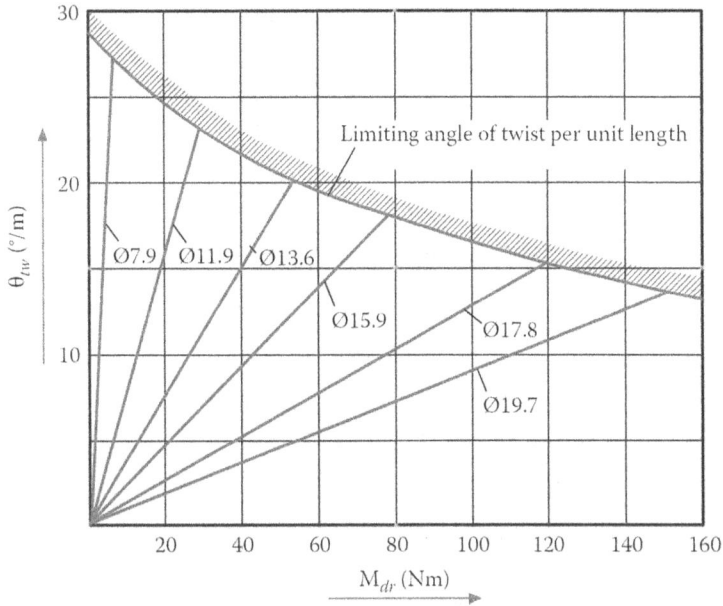

FIGURE 3.22 The limiting angle of twist as a function of the drill diameter and applied torque.

As discussed in Chapter 2, Section 2.1.5.2.1, torsion deformation defined by the limiting angle of twist is the limiting factor. This angle is calculated as

$$\theta_{tw} = \frac{M_{dr} \cdot 180°}{GJ} \tag{3.5}$$

Heat treatment of shanks improves the limiting angle of twist (Figure 3.22) allowing greater torques and thus penetration rates. The use of twist shanks results in a favorable distribution of the load q_c over the shank length and due to the axial force (thrust)–torque coupling (see Chapter 2. Section 2.1.5.5) allows greater penetration rates.

3.3.2.6 Drivers

Drivers are mounted at the end of the shank. There are a number of various styles of drivers. The basic styles are shown in Figure 3.23. Drivers 1–4 are plane drivers that are most common. The plane cylindrical driver (driver 1) is mostly used in HP drills as it is installed into a precision shrink-fit or hydraulic holder. The standard tolerance h6 is used for its diameter. Plane drivers 2, 3, and 4 have means to accommodate a setscrew to transmit the drilling torque: driver 2 has flats, driver 3 has tapered undercut, and driver 4 has tapered flat. They are used in normal precision gundrilling.

Drivers 5 and 6 are preset drivers as they have the adjustments for the depth of presetting. Driver 5 has the front adjustment, while driver 6 has the rear adjustment by means of a locking hollow screw. Some drivers have been made with key, pin, or tang drive or with locking thread (driver 8). Locking into the socket can be accomplished by using a threaded cap over the front of the socket or a collet closure in the socket.

Normally, MWF (the coolant) is fed through the center of the driver from its rear end. This is possible when the spindle has a central hole through it to supply MWF. When the spindle has no hole through it, a special rotary union can be mounted in front of the spindle or around the driver where crossholes shown in driver 7 carry MWF at right angle to the central coolant of the gundrill.

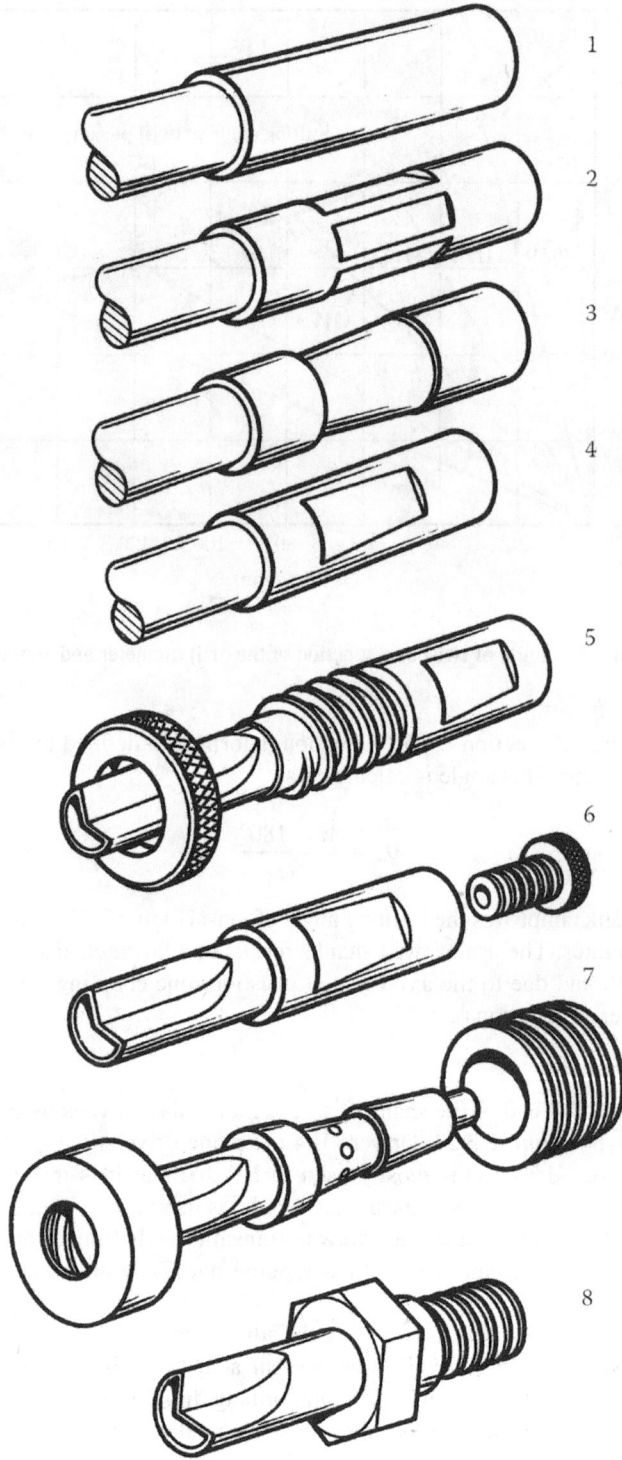

FIGURE 3.23   Types of drivers for gundrilling.

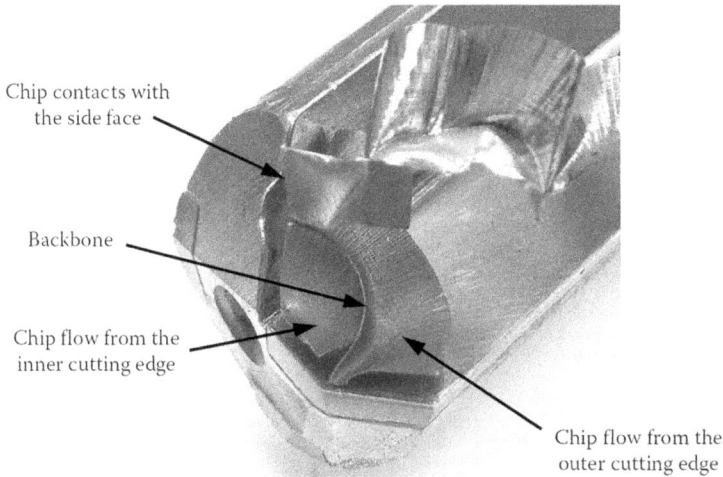

Chip contacts with the side face

Backbone

Chip flow from the inner cutting edge

Chip flow from the outer cutting edge

FIGURE 3.24 Interaction of the chip flows from the outer and inner cutting edges.

3.3.2.7 Chip Breaking

Among deep-hole drills, gundrills are the only drills without chip-breaking steps ground on the rake face. This makes re-sharpening of gundrills much simpler and faster using standard grinding wheels and fixtures often even by a gundrilling machine operator. This simplicity, however, comes at some cost. The rake face with no chip-breaking step is possible because the chip formed at the inner cutting edge should impinge on the chip formed by the outer cutting edge and thus should serve as an obstacle chip breaker. In other words, the collision of these two chip flows should result in the formation of the so-called backbone at their interface which, colliding with the rotating bottom of the hole being drilled (or, at worst, the sidewalls of the hole being drilled), causes the breakage of such a combined chip. This point is illustrated in Figure 3.24.

To achieve reliable chip breaking, the following parameters can be varied: distance m_d (Figure 3.5) that defines the width of the chips formed by the outer and inner cutting edges; the approach cutting edge angles φ_{a1} and φ_{a2} (Figure 3.5) that define the chip thickness (under a given cutting feed) and the interaction of the chips formed by the outer and inner cutting edges; the cutting feed (feed per revolution) that defines the chip thickness; the cutting speed that defines the temperature of the formed chip; and the shoulder dub-off angle φ_4 that defines the interaction conditions of the chips and MWF (Astakhov 2010). In the author's experience, reliable chip breaking can be practically always achieved (besides a few exceptions) if one understands the interinfluence of the listed parameters and thus varies them accordingly. For example, marginal chip breakage is achieved with the drill shown in Figure 3.24 because the chip flow from the outer cutting edge is much stronger so that the intersection rib collides with the side of the chip removal flute. To solve the problem, distance m_d was increased so that reliable chip breaking was achieved due to the interaction of the discussed rib with the bottom of the hole being drilled as shown in Figure 3.25.

Unfortunately, this is not always the case so some means enhancing chip breaking/separation is added to the basic gundrill design. These can be broadly divided into two groups. The chip-breaking/ separation features in the design and geometry of drills of the first group are applied on drill rake face on its manufacturing so that there is no need to reproduce them on drill re-sharpening. Figure 3.26 shows an example of the drill design developed in the 1930s with the chip-separating steps applied to the rake face of the outer cutting edge. All the design/geometries of the drills of the first group are just variations of this principle.

FIGURE 3.25 Improved tool geometry to achieve better chip breaking.

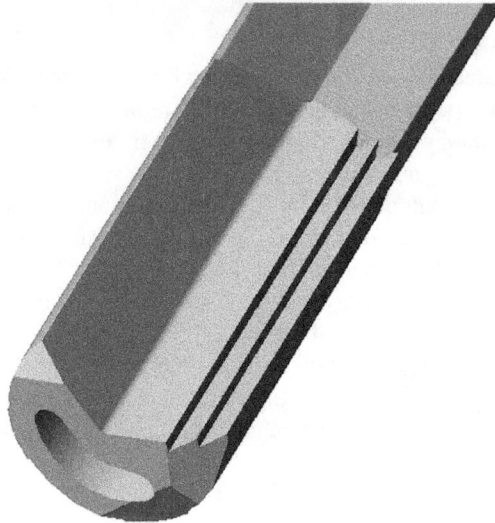

FIGURE 3.26 Chip-separating steps applied to the rake face of the outer cutting edge.

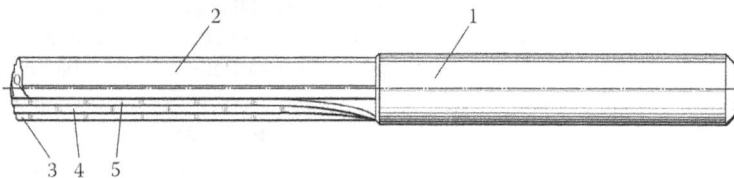

FIGURE 3.27 Design of gundrill according to Japanese Patent Laid-Open N 2007-50477.

For example, Figure 3.27 shows the so-called nicked gundrill according to Japanese Patent Laid-Open No. 2007-50477. As can be seen, the drill has the driver (1) and the shank (2). This nicked drill is configured so that each cutting edge has a plurality of concave groove-like nicks (3, 4, and 5 as shown in Figure 3.27) arranged, so on drilling they divide the chip in the width direction. The disadvantages of such a design, for example, possible undesirable interaction of chip flows from nicks, were listed

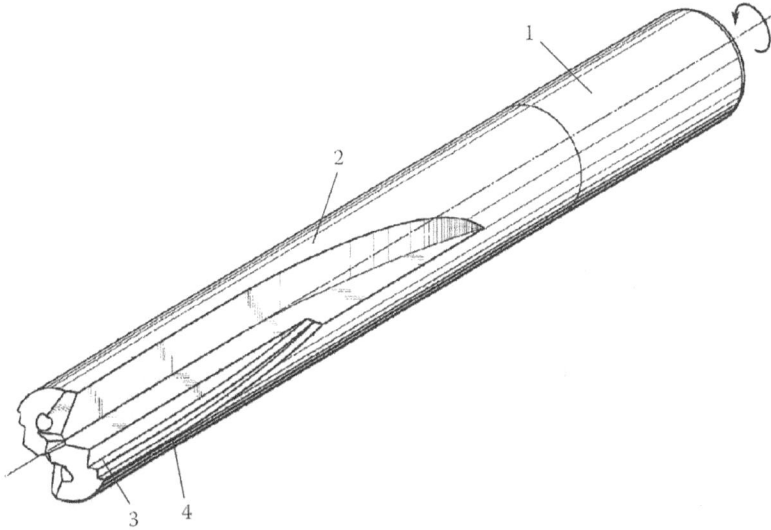

FIGURE 3.28 Drill design according to US Patent Application Publ. No. US 2012/0063860.

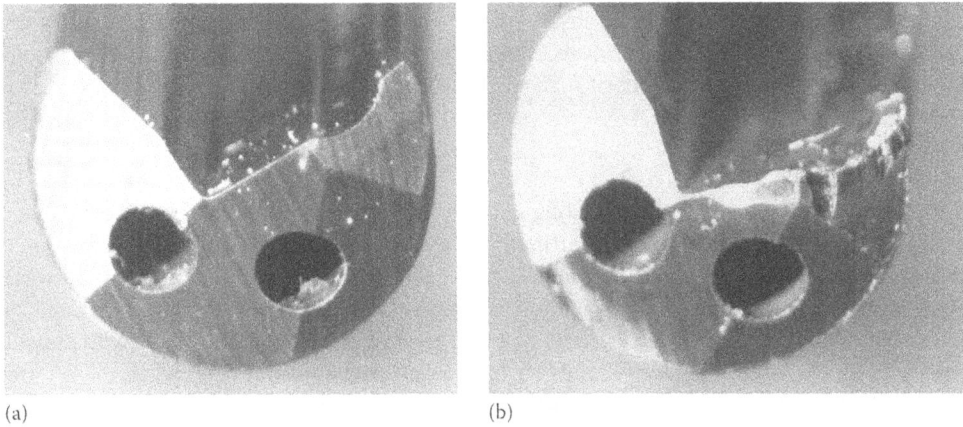

(a)   (b)

FIGURE 3.29 Chip-separating groove on the gundrill's rake face: (a) new drill and (b) wear pattern.

in the description of US Patent Application Publ. No. US 2012/0063860, which offers a modified design shown in Figure 3.28. The gundrill in this figure has the driver (1) and shank (2). The tool has a plurality of rake faces (e.g., 3 and 4 shown in Figure 3.28) arranged in a steplike manner and spaced in the front and rear direction of the tool rotation direction. The inventors believe that the chip is segmented into strips having a narrow width that improves chip controllability.

According to the author's experience, however, application of the grooves on the gundrill's rake face is not that effective as such grooves only separate the formed chip in a number of strips providing no means for their breakage. Moreover, in machining of difficult-to-machine work materials, grooves on the rake face lower tool life. As an example, Figure 3.29a shows a chip-separating groove on the gundrill's rake face. As can be seen in Figure 3.29b, deep wear notches form at the edges of the groove lowering tool life.

The chip-breaking/separation features in the design and geometry of drills of the second group are applied on drill manufacturing as well as on each re-sharpening so that the gundrill loses its

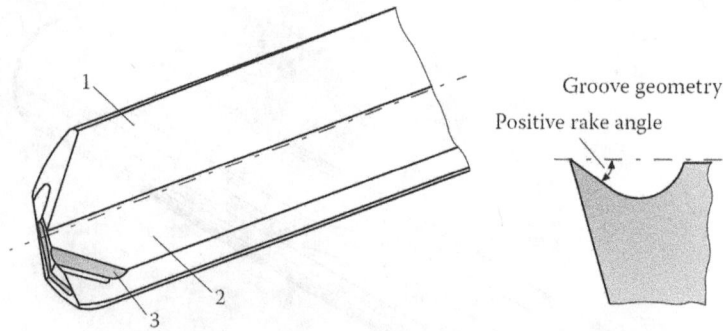

FIGURE 3.30 Gundrill tip geometry according to US Patent No. 7,753,627 (2010).

re-sharpening simplicity. As an example, Figure 3.30 shows the gundrill tip geometry according to US Patent No. 7,753,627 (2010) where the tip (1) having the rake face (2) is provided with a chip-breaking groove (3). The advantages of such gundrills are reliable chip breaking and cutting with a positive rake angle. The latter enhances tool life by reducing the cutting force/power and also makes it suitable for near dry machining (NDM). The price to pay is a necessity to restore the grove geometry with high accuracy on each successive tool regrind so that a special grinding fixture and wheel is needed. This is not a problem, however, when a CNC point grinding machine is used (which actually should be the case with HP gundrills). Another drawback is that a much greater amount of the tool material has to be removed on regrinding to restore the chip-breaking flute, which lowers the total number of regrinds. This should be compensated with increased tool life between regrinds.

3.3.2.8 Alignment

The alignment of the spindle and the starting bushing in gundrilling significantly affects the deviation of the hole axis (position error) of drilled hole as shown in Figure 3.31a. The most common method of gundrilling is one where the gundrill rotates and the workpiece is stationary. This method imposes special requirements on the accuracy of gundrilling machines and their components. The alignment of the gundrill components should be next to perfect when drilling holes with diameters less than 10 mm (0.4 in.) in light materials such as aluminum alloys, that is, when the rotational speed (6,000–15,000 rpm) and the feed rate (600–1,000 mm/min [24–40 in/min]) are high. The clearance in the starting bushing, drill holder-starting bushing, and whip guide alignments, and the accuracy of the feed motion are key factors in using this method.

Although awareness of the importance of machine alignment grows in industry, there are at least three important issues that remain unaddressed. First, there is no simple way to check the discussed alignment in many gundrilling machines built in production lines. Normally, it takes many hours to clean up the space for such an inspection. Second, the discussed gundrilling machines do not have any means to correct alignment when needed. Commonly, shims are used to adjust this alignment that reduces the machines dynamic stability. Third, the alignment is normally checked between the starting bushing holder and the spindle of the machine. Although it is an important parameter, it is not sufficient. It should be clearly understood that the alignment in the system *actual gundrill holder–actual starting hushing* should be examined although it is not that easy to account for the current method and accessories used for misalignment inspection. For HP gundrilling application, the discussed misalignment should not exceed 0.004 mm. It should be checked between the actual gundrill holder and the actual starting bushing. The use of modern laser alignment systems with digital targets significantly simplifies the verification of this parameter.

Yet another important issue in the consideration of alignment is the steady rest (whip guide) (Figure 3.18). Unfortunately, there is not much data on the influence of this alignment on the drill

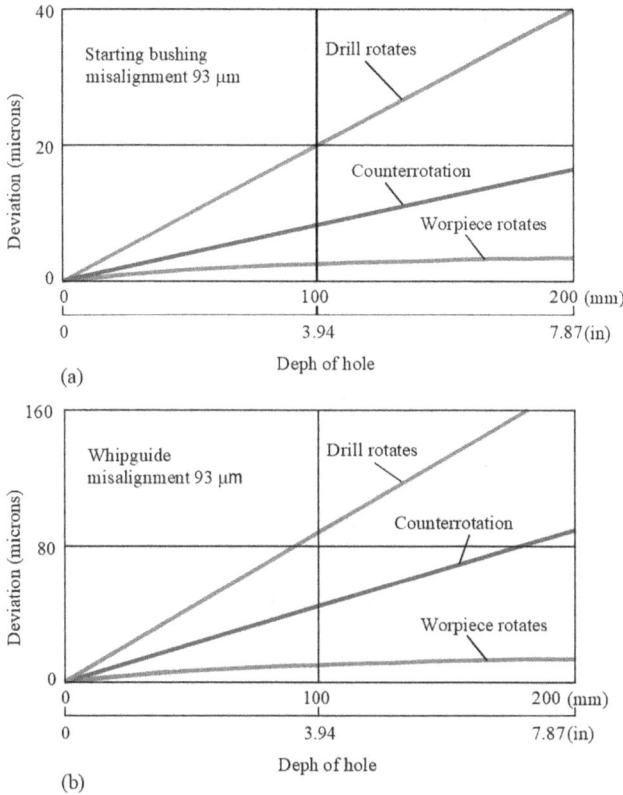

FIGURE 3.31 Deviation of the hole axis (position error) for different gundrilling methods: (a) influence of the misalignment of the starting bushing and (b) influence of the misalignment of the whip guide. Gundrill 10 mm (0.394 in.) diameter, $n = 1600$ rpm, $f = 0.04$ mm/rev (0.0016 in./rev).

performance. An extremely important and unknown fact to end users follows from the comparison of data presented in Figure 3.31: the whip guide alignment affects the deviation of the hole axis much more than that of the starting bushing. Unfortunately, there is no simple way to check and to correct the alignment of the whip guide(s) on many machines used in the industry.

The clearance between the gundrill tip and the starting bushing is another important but often ignored system parameter. Ideally, this clearance should be next to zero. However, it is not possible in any practical situation due to a number of different factors (drill-free rotation and penetration, tip back taper, wear of the tool and starting bushing, etc.). The excessive clearance in the starting bushing is the prime cause for entrance instability in gundrilling when the rotational speed and feed rate are high. This instability causes the formation of a bell-shaped part at the hole entrance known as the bell mouth (Astakhov 2002a, b). According to the author's experience, excessive clearance in the starting bushing caused by the use of non-specialized for gundrilling or excessively worn starting bushing is the prime cause for the so-called unpredicted drills' failures (Astakhov 2001). Therefore, when one uses HP gundrilling, only specialized, preferably carbide gundrilling bushings should be used and gundrills should not be re-sharpened beyond approx. 1/3 of the original tip length.

3.3.2.9 MWF: Flow Rate and Filtration

High-pressure MWF delivery is necessary to cool the workpiece and the tool, to provide lubrication between tool and workpiece, as well as to carry away chips from the cutting area along the flute to the chip box. Cooling action dissipates both the external heat of friction and the internal heat of plastic deformation due to cutting and burnishing. Lubrication between the workpiece and the drill contact areas reduces contact stresses and amount of the thermal energy generated in these areas, so it reduces adhesion and/or diffusion wear of gundrills. To effectively carry chips away, MWF should have a sufficient combination of viscosity and velocity. Improper selection of this combination causes chip clogging in the flute that leads to an increase in torque and probable drill breakage.

In simple terms, the MWF flow rate needed for reliable chip removal in gundrilling can be calculated using the assessment of the work needed to be done by the MWF flow. This work can, for horizontally located gundrill, be represented as

$$W_{fl-\text{req}} = (F_{r-a} + F_{r-t}) L_{dr} \tag{3.6}$$

where
F_{r-a} is the resistance force due to atmospheric pressure (N)
F_{r-t} is the resistance force due to chip transportation (N)
L_{dr} is the length of drilling (m).

The resistance force due to atmospheric pressure is calculated as

$$F_{r-a} = p_{\text{atm}} A_{v-fl} \tag{3.7}$$

where
p_{atm} is the atmospheric pressure
$p_{\text{atm}} \approx 0.1\,\text{MPa}$
A_{v-fl} is the cross-sectional area of the chip flute (V-flute) (m^2).

The resistance force due to chip transportation is calculated as (Astakhov and Joksch 2012)

$$F_{r-t} = K_{ch} G_{ch} \tag{3.8}$$

where
$K_{ch} = 0.15$ is the proportionality coefficient
G_{ch} is the weight of the chip produced by the drill (N)

$$G_{ch} = 0.785 d_{dr}^2 fn\gamma_w \tag{3.9}$$

where
d_{dr} is the drill diameter (m)
f is the feed per revolution (m/rev)
n is the spindle rotational speed (rev/s)
γ_w is the specific weight of the work material (N/m^3)

$$\gamma_w = \rho_w g \tag{3.10}$$

where

ρ_w is the density of the work material (kg/m³)
g is the acceleration due to gravity (m/s²).

The work done by the MWF flow is calculated as (Astakhov and Joksch 2012)

$$W_{fl} = \gamma_{fl} \frac{Q_{fl}^3}{A_{v-fl}^2 2g} \tag{3.11}$$

where

Q_{fl} is the flow rate of MWF (m³/s)
γ_{fl} is the specific weight of MWF (N/m³).

The condition of the reliable chip transportation is

$$W_{fl-req} \geq W_{fl} \tag{3.12}$$

which, accounting for Eqs. (3.6) and (3.12), yields an expression for the minimum required MWF flow rate (m³/s)

$$Q_{fl-min} \geq \sqrt[3]{\frac{W_{fl-reg} A_{v-fl}^2 2g}{\gamma_{fl}}} \tag{3.13}$$

There are three basic types of MWFs (coolants) used in gundrilling:

1. Oil-based MWFs known as *straight oils* are generally used for gundrilling of alloy steels on stand-alone machines having their own MWF supply system. Compared to water-mixable MWFs, oil-based MWFs significantly reduce tool wear, yield better surface finishes, and generally improve the accuracy of drilling. However, it happens only when such MWFs contain extreme pressure additives of sulfur (2.5%–3.5%) for high-alloy steels and heat-treated cast irons, chlorine (3.5%–5%) for light ferrous materials, and 10%–14% of fat. Low coolant viscosity aids in good heat dissipation and good load-carrying capacity. It also reduces the risk of pump starving when cold starting, improves the efficiency of filters, and reduces the amount of oil carried off with the chips. The kinematic viscosity should generally not exceed 20–30 cSt/20°C. In exceptional cases, 45 cSt/20°C can be used.
2. Water-based MWFs known as *water-mixable oils* are used for machining aluminum alloys on in-line gundrilling machines where MWF is supplied from a central coolant pump station. These MWFs are generally more economical and can be used for non-ferrous metals and high machinability steels under light-cutting conditions. Extreme pressure additives contained in water-mixable MWFs prevent work material adhesion to the tip. These MWFs also contain film strength enhancers (animal and vegetable fats) to reduce friction and wear. Dilution (concentration) 8 to 1 is normally used, while dilution greater than 10 to 1 significantly reduces film strength and creates vapor pockets at high tool load areas in the gundrilling zone that reduces tool life dramatically.
3. Synthetics are water-based MWFs that are easier on the environment but harder on the gundrill. It is used, however, in gundrilling cast irons with great success.

MWF filtration is essential to system performance. Because MWF collects and circulates considerable quantities of both coarse and fine chips, it must be carefully purified in the interests of both tool life and hole quality. Poor filtration leads to increased MWF temperatures and rapid failure of the coolant pump. It also causes premature failure of solenoid valves, leaking servo valves, and bearing failure in the rotating coupling. Cartridge filters of size in the range of 5–10 µm for high-precision holes should be used. Filtration of particles of 15–20 µm for precision holes and in the range of 20–30 µm for normal holes should be targeted. Drilling cast iron requires rough filters, magnetic drums, or rolled media, followed by a bag-type or woven media polishing filter.

The MWF temperature defines to a large extent its cooling, lubricating, and transportation abilities. This is particularly important with oil-based MWFs. About 40°C–50°C (100°F–120°F) is generally recommended as the maximum temperature of MWF. It can often be maintained by circulation through a heat exchanger or even by installing a fan to blow across the surface of the coolant reservoir. When precise holes are to be drilled, refrigeration systems may be necessary.

MWF pump(s) plays a more important role than one may think. Unfortunately, most of the gundrilling coolant supply systems have the inferior type of pumps, called variable-displacement pumps. A variable-displacement pump is designed to maintain *set* pressure. If an obstruction is encountered by the MWF flow, the *set* pressure (the pressure seen on the gauge by the operator) will be maintained but the flow rate supplied to the tool will actually decrease because MWF will be diverted through the pump's internal relief valve. As a result, the obstruction (in the case of chip clogging in the flute of the tool) can, in fact, be worsened and quickly lead to drill failure. Experienced practitioners in the industry change these pumps with fixed-displacement pumps.

3.3.3 Tool Design/Development/Research Perspective

The following section is probably the most important in terms of comprehending the systemic design considerations for any kind of drill. Although the understanding of the rake and flank geometries and their adjustments for a particular work material are important stages in the design of any drill including gundrills, it is only the tip of the iceberg. At the tool research and development stage, a specialist should analyze the objective functions (goals) and constraints on these functions so that he or she can solve or reduce severity of the problems associated with these constraints. After this, a consideration of a reasonable balance in justifying a number of objective functions of the tool design can be carried out. The proposed strategy, i.e., the consideration of the objective functions and constraints on their achievement, is of methodological significance as they can be used for any kind of drills or other cutting tools. In the author's opinion, it is the only professional way to deal with the development of any advanced cutting tool.

The following objective functions must be considered in the gundrill design:

1. Meeting hole quality requirements, which in gundrilling includes hole diametric and shape (in both radial and axial directions) accuracies and minimum surface roughness of gundrilled holes, minimum bell mouth at the hole entrance (assured by tool entrance stability), and hole straightness.
2. Achieving maximum tool life under the desirable productivity (the penetration rate).
3. Producing chip shapes suitable for reliable removal.

Although these objective functions seem to be independent, it was shown (Astakhov 2010) that they are just different facets of the proper tip design/geometry as the tip does the actual cutting and burnishing. In other words, one can start the design with practically any feature and then develop others using iteration procedures to balance the final design. In the author's opinion, however, it is easy and more logical to start with the consideration of the influence of the cutting force components and typical gundrill wear pattern.

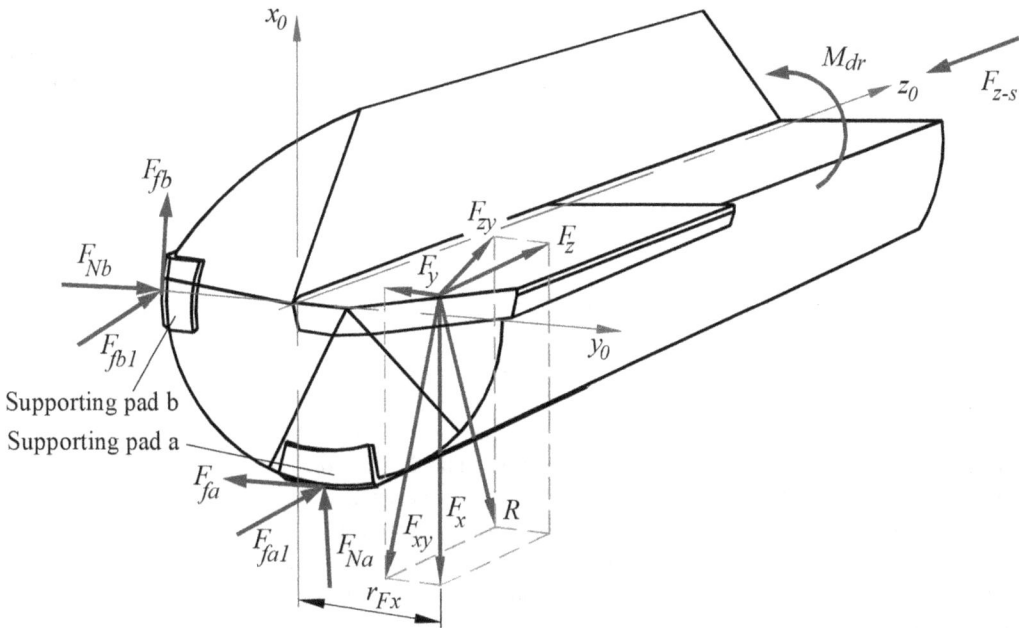

FIGURE 3.32   Force balance in gundrill.

3.3.3.1   Force Balance and the Meaning of the Term *Self-Piloting Tool*

3.3.3.1.1   *Theoretical (Intended) Force Balance*

Although there are a great number of designs of SPTs including gundrills, two important features, namely, the theoretical force balance and the locating principle, are the same. Therefore, they should be considered before particular designs of gundrills are analyzed further in the subsequent sections.

To comprehend the concept of SPTs, one should first consider the force balance of such a tool. Following the methodology of constructing the force balance diagram used in Chapter 2 (Section 2.1.5.6), one can construct a force balance diagram for the tool shown in Figure 3.32. For convenience and simplification of further considerations, the right-hand $x_0y_0z_0$ coordinate system, illustrated in Figure 3.32, is set as follows:

1. The z_0-axis along the longitudinal axis of the drill, with sense as shown in Figure 3.32, toward the drill holder.
2. The y_0-axis passes through the periphery corner and is perpendicular to the z_0-axis. The intersection of these axes constitutes the coordinate origin as shown in Figure 3.32.
3. The x_0-axis is perpendicular to the y_0 and z_0 axes as shown in Figure 3.32.

Figure 3.32 shows the force balance in the T-hold-S $x_0y_0z_0$ coordinate system (Astakhov and Osman 1996a, b). When a gundrill works, the cutting force generated is due to the resistance of the workpiece material to cutting. This force is a 3D vector that can be thought of as applied at a certain point of the drill rake face as discussed in Chapter 2, Section 2.1.3.4.2, and Figure 2.23. The cutting force R (or the resultant cutting force for multi-edge tools) can be resolved into three components, namely, the power (tangential) F_x, axial F_z, and radial F_y components, respectively.

As discussed in Chapter 2, Section 2.1.3.4.2, and Figure 2.23, these components are commonly referred to as the tangential, radial, and axial forces. The axial force is balanced (equal in magnitude and oppositely directed) by the axial force F_{z-s} of the feed mechanism of the deep-hole machine, while the tangential and radial forces sum to create force F_{xy} (acts in the x_0y_0 plane) that (in contrast to other axial tools such as twist drills, reamers, and milling tools) generally is not balanced, regardless of the number of the cutting edges used. To prevent drill bending due to this unbalanced force, some special measures are taken. The term *deep-hole drilling* has grown to mean that the unbalanced cutting force F_{xy} generated in the cutting process is balanced by the equal and opposite force due to the normal forces F_{Na} and F_{Nb} on the leading a and trailing b supporting pads, which bear against the wall of the hole being drilled. When the discussed drill rotates about the z_0-axis, burnishing forces F_{fa} and F_{fb} acting in the direction opposite to the corresponding supporting pads rotation occur. When the discussed drill translates due to feed motion along the z_0-axis, burnishing forces F_{fa1} and F_{fb2} acting in the direction opposing the feed motion occur.

3.3.3.1.2   Prevailing Concept of Self-Piloting

According to the prevailing notion, due to the action of the mentioned forces on the supporting pads, these pads ensure balancing of the tool in the x_0y_0 plane while simultaneously providing for a unique additional machining operation known as burnishing. The fact that the pads bear against the wall of the hole being drilled behind the side cutting edge effectively means that the tool is provided with its own guide. The concept of self-piloting (sometimes referred to as self-guidance), meaning the tool guiding or steering itself along the bore, has been recognized as the major underlying principle of the design of SPTs (Veremachuk 1940, Swinehart 1967, Sakuma et al. 1980, 1981, Griffiths and Grieve 1985, Griffiths 1993).

Nowadays, multi-edge and multi-supporting pad SPTs are used; thus, the described self-piloting feature may not be obvious. As discussed by the author earlier (Astakhov 2010), no matter how many cutting edges an SPT has, as long as there is an unbalanced radial force acting on the supporting pads of the tool, it remains self-piloting. Alternatively, no matter how many supporting pads a hole-making tool has, it should not be regarded as self-piloting if there is no (at least, theoretically) unbalanced radial force to cause self-guiding.

Sakuma et al. (1981) suggested that not only does the self-piloting design make it possible to machine deep holes, but it also provides stable cutting conditions for the cutting insert or cutting portion of the tool by eliminating the radial vibrations that often cause drill failure in ordinary drills. When machining conditions are selected properly, a machining operation with SPT is very stable and consistently produces holes of high quality. Although the use of SPTs requires a number of additional accessories such as starting bushings, high-pressure coolant system, etc., the benefits gained with the use of these tools are greater as SPTs proved capable of maintaining close size control and producing holes of good surface finish that meets the output requirements of honing. Thus, the use of SPTs is considered whenever one or more of the following conditions exist:

- The $L/D > 15$;
- High-precision requirements difficult to attain by conventional bore machining operations;
- The tight position tolerance on the longitudinal axis of the machined hole;
- The tight shape tolerance on the machined hole.

3.3.3.1.3   Original Concept of Self-Piloting

The original concept of self-piloting was brought by the Pratt & Whitney Co. in the 1900s, principally for use in connection with their gun-barrel drilling machines (1910 (reprinted in 2001)). Since then, it has not even been mentioned in the literature on the subject. The concept was explained using a newly-developed tool (the Pratt & Whitney Co. in the 1900s) shown in Figure 3.33a. According to the description provided, A is the cutting edge (the outer cutting edge in the modern terms). It is explained that the rake face of the drill is located on the drill center line that provides "very-free cutting as compared to the ordinary two-lip twist drill which has a central web." Continue to refer to Figure 3.33a, B is the oil duct (the coolant hole in the modern terms), C is the chip groove (V-shaped flute), E is the high point of the drill, and D is a cross section of the work being drilled. It is stated that "grinding the drill in this manner makes possible its running trues or straight, the teat F on the

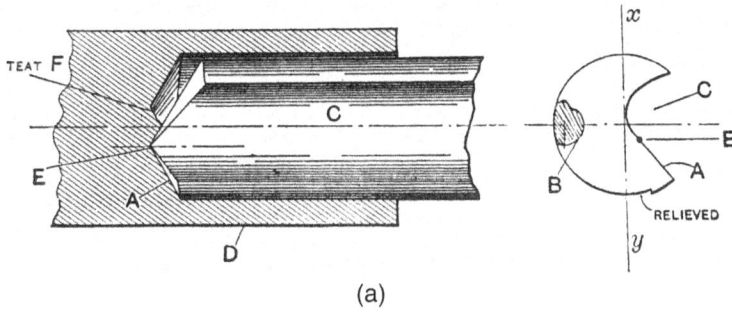

(a)

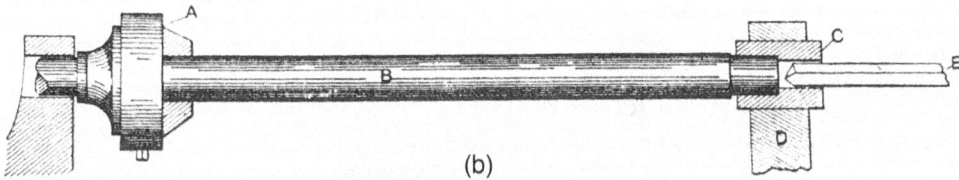

(b)

FIGURE 3.33   Pratt & Whitney Co. deep-hole drill (a) and gundrilling arrangement (b).

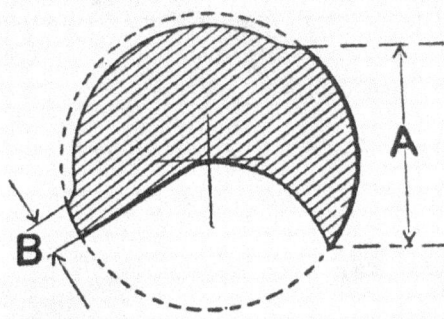

FIGURE 3.34   Relieving practice of large drills.

work acting as a support to the drill, which, owing to its periphery being partly relieved, would have a tendency to travel in a curve away from its cutting side."

Because teat F cannot be cut with no significant tip deflection at the hole entrance, a special gundrilling arrangement shown in Figure 3.33b is needed. As shown, it includes the chuck, A, the workpiece, B, the bushing, C, and the support, D, for holding the bushing C and the drill, E. The workpiece being drilled is held and revolved at one end by a suitable chuck on the live spindle of the machine, while the other end, which should be turned perfectly true, runs in a stationary bushing. In other words, the drill tip starts to drill when its tip is located in the starting bushing to prevent its deflection.

The proposed concept of self-piloting resembles that described in Chapter 2, Section 2.1.11.11, for non-central chisel edge drills. As such, the tout is termed as the stabilizing cone as shown in Figure 2.210.

It is also discussed in (1910 (reprinted in 2001)) that the body of large drills is relieved as shown in Figure 3.34. The straight, or radial, edge is the cutting edge and the distance B is about 1/8 inch (approx. 3.2 mm) on a one-inch (25.4 mm) drill. The surface A is left of the full radius of the drill and makes a good backrest. Therefore, it can be concluded that the action of teat F as a stabilizing cone may not be sufficient for self-piloting for larger drills so that additional support is needed.

3.3.3.1.4   Analysis of the Common and Original Concepts of Self-Piloting

In the author's opinion, both concepts are correct and, moreover, are fully applicable in various SPTs. Depending upon the geometry of a given SPT, the common or original concept is prevailing whereas the second one is supplemental. This was intuitively used in the SPT's design practice. For gundrills having smaller diameters (compared to STS and ejector drills), the teat, or stabilizing cone is made deliberately large as the distance m_d (see Figure 3.3) is normally equal to $0.25d_{dr}$ so that the self-piloting mostly relies on the stabilizing cone whereas the supporting pads provide drill stabilization and formation of the surface of the hole being drilled. That is why, the formation of this surface including its roughness and roundness only weakly depends on the design and location of the supporting pads. In STS and ejector drill, this distance is normally equal to $0.1d_{dr}$ so the self-piloting of the drill mostly relies on the action of the supporting pads. As such, design/geometry, location, and material of the supporting pads play a significant role in the formation of the quality characteristics of the machined hole surface (Astakhov 2010).

3.3.3.1.5   Complete Force Balance

In full analogy with the complete force balance for a two-flute drill discussed in Chapter 2, Section 2.1.5.6, this section considers the complete force balance in gundrilling in order to use this balance in the practice of the tool design in the manner discussed in Chapter 2, Section 2.1.5.7, i.e., to apply the VPA2© design to gundrills. This design is termed as VPA2d© design.

Following the introduced cutting force and its components in a gundrill (Figure 3.16), one can represent the components of the cutting force as the projections of this force on the axes of the original (T-hold-S) coordinate system for cutting edges 1–2 and 2–3 as shown in Figure 3.35. Besides the previously discussed bending moments due to the axial components of the cutting force, the following bending moments act in the y_0z_0 plane:

$$M_{y_0z_0} = F_{y12}z_{b12} - F_{y23}z_{b23} - F_{z12}y_{12} + F_{z23}y_{23} \qquad (3.14)$$

and in the x_0z_0 plane

$$M_{x_0z_0} = F_{x12}z_{a12} + F_{x23}z_{a23} \qquad (3.15)$$

It follows from Eq. (3.14) that the bending moment $M_{y_0z_0}$ is not significant and, moreover, can be even negative under a certain combination of approach angles φ_{p12} and φ_{p23}. On the contrary, the

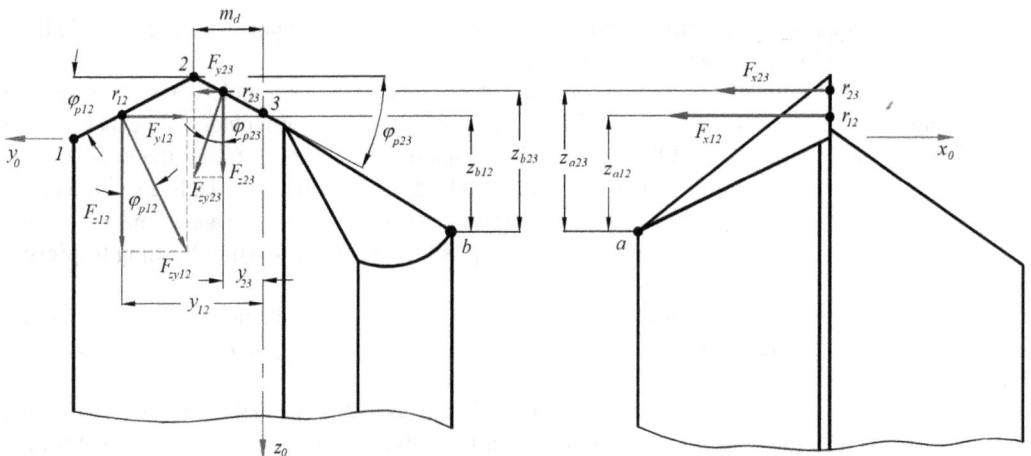

FIGURE 3.35   Components of the cutting force causing additional bending moments.

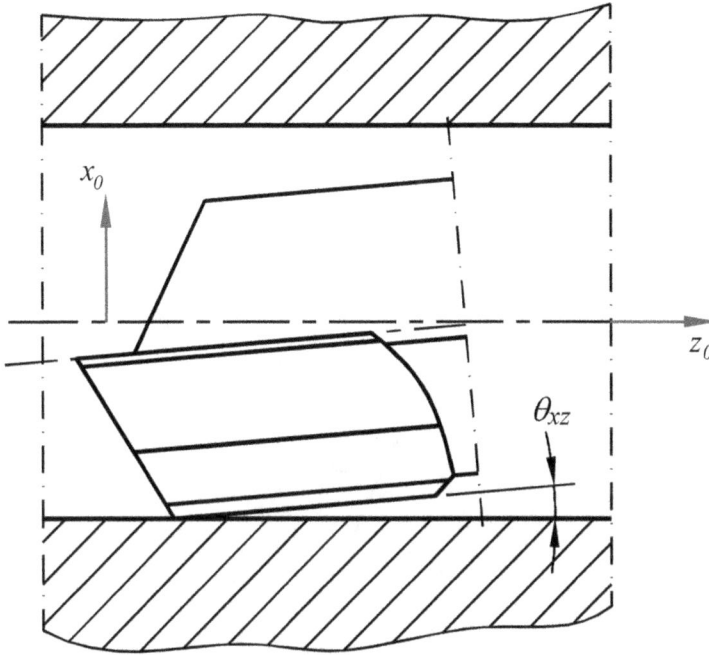

FIGURE 3.36   Bending of the gundrill in the x_0z_0 plane.

bending moment in the x_0z_0 plane is great as the components of the cutting force in the x_0 direction are the highest. The bending moment in the x_0z_0 plane is calculated at the sum of the moments due to F_{x12} and F_{x23} so that the bending of the gundrill in this plane by angle θ_{xz} (Figure 3.36) is normally of prime concern. The greater the angle θ_{xz}, the lower the quality (surface finish and diametric accuracy) of the machined holes. As discussed by the author earlier (Astakhov 2014), the bending moment in the x_0z_0 plane is the *Achilles' heel* of any SPT, it is more profound in gundrilling due to small drill diameters and thus weak shanks susceptible to bending. The problem is greatest in the industry-standard facet point grind shown in Figures 3.3 and 3.5. To assure drill-free penetration, that is, to prevent the interference of the drill's flanks with the bottom of the hole being drilled, the auxiliary flank (15) (normal flank angle α_{n3}) in Figure 3.3 or flank face F3 in Figure 3.5 is applied that pushes the location of point "a" far behind the cutting edge so that the arms z_{a12} and z_{a23} (Figure 3.35) become significant. This explains why the facet point grind often shows no advantages over the CAM point grind (see Figure 3.4) although the former is much superior in terms of drilling itself. The mentioned arms are much smaller when the CAM point grind is used. Therefore, the tool geometry parameters should be optimized to minimize the bending moment in the x_0z_0 plane – this constitutes *Rule No.* 1 in the HP SPT drill tip geometry/design. This rule directly follows from the VPA2© design discussed in Chapter 2, Section 2.1.5.7.

3.3.3.1.6   Practical Realization of the VPA2d© Design in Gundrilling

To start the consideration, one should ask a logical question: how many flank planes are really needed for free penetration of a gundrill? As discussed above, normally, four flank surfaces (planes) are provided. Figure 3.5 shows FI, F2, F3, and F4 flank planes. Planes FI and F2 are flank planes of the outer and inner cutting edges, while planes F3 and F4 are meant to assure drill-free penetration into the hole being drilled. Because there are no recommendations on the selection of the location of planes F3 and F4, most of gundrill manufacturers and users follow the common pattern established more than 50 years ago. Obviously, the concept of the VPA2© design cannot be applied under such a design/geometry.

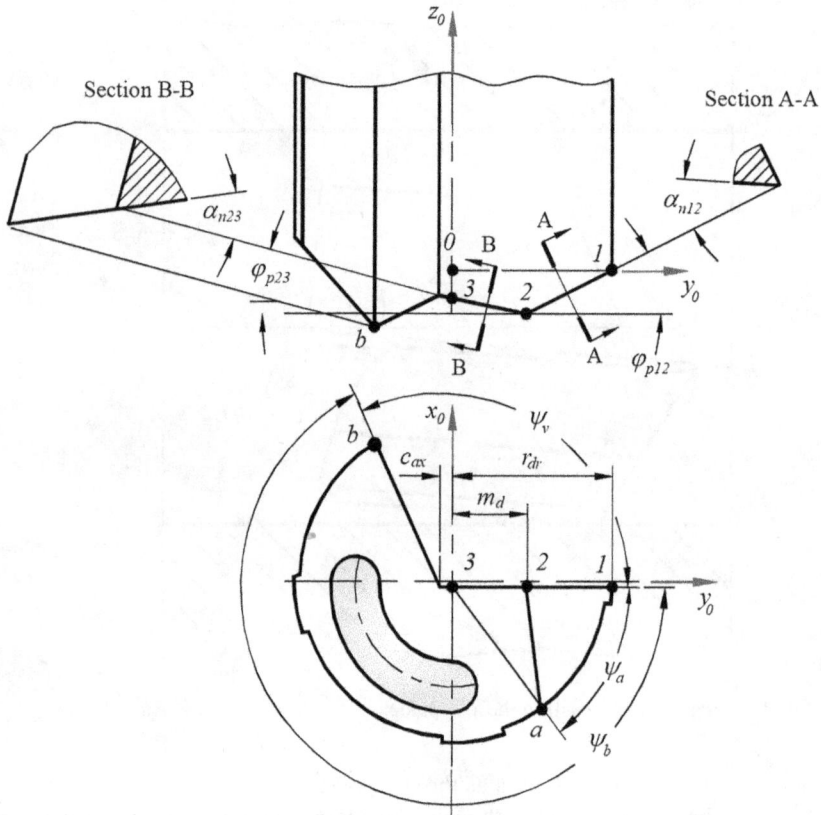

FIGURE 3.37   A gundrill is made with just two flank planes.

To challenge this status quo, Figure 3.37 shows a gundrill made with just two flank planes. The free penetration of the drill's flank can be verified by a set of clearances δ_{cl} between the flank points and the bottom of the hole being drilled (Astakhov 2010). The gundrill's flank touches the bottom when the value of this clearance is equal to zero. The value of δ_{cl} varies continuously along the flank and the graphical analysis shows that the most *dangerous* points of the drill flank (the possibility of interference with the bottom) are located at the periphery points a and b shown in Figure 3.37. Figure 3.38 shows visualization of these points on a solid model of a gundrill.

For the realization of the VPA2© design, designated for gundrilling as the VPA2d© design, point a should be located as close to the bottom of the hole being drilled. Therefore, at this stage, the limiting location of this point should be determined. It was shown by the author earlier (Astakhov 1995a) that the condition of the gundrill-free penetration into the hole (no interference with the bottom of the hole being drilled) as related to point a with designations shown in Figure 3.39 is as follows:

$$\xi_a \geq \xi_l \tag{3.16}$$

where

$$\xi_l = \arctan\left(\tan\varphi_{p12} - \frac{m_d}{r_{dr}}\left(\tan\varphi_{p12} - \tan\varphi_{p23}\right)\right) \tag{3.17}$$

$$\xi_a = \arctan\left[\left(\cos\psi_a + \tan\alpha_{n23}\frac{\sin\psi_a}{\sin\varphi_{p23}}\right)\tan\varphi_{p23}\right] \tag{3.18}$$

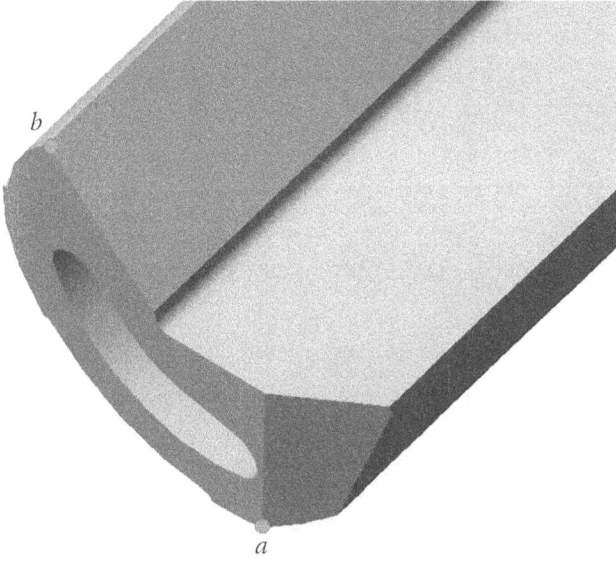

FIGURE 3.38 Visualization of the most *dangerous* points *a* and *b* of the drill flank.

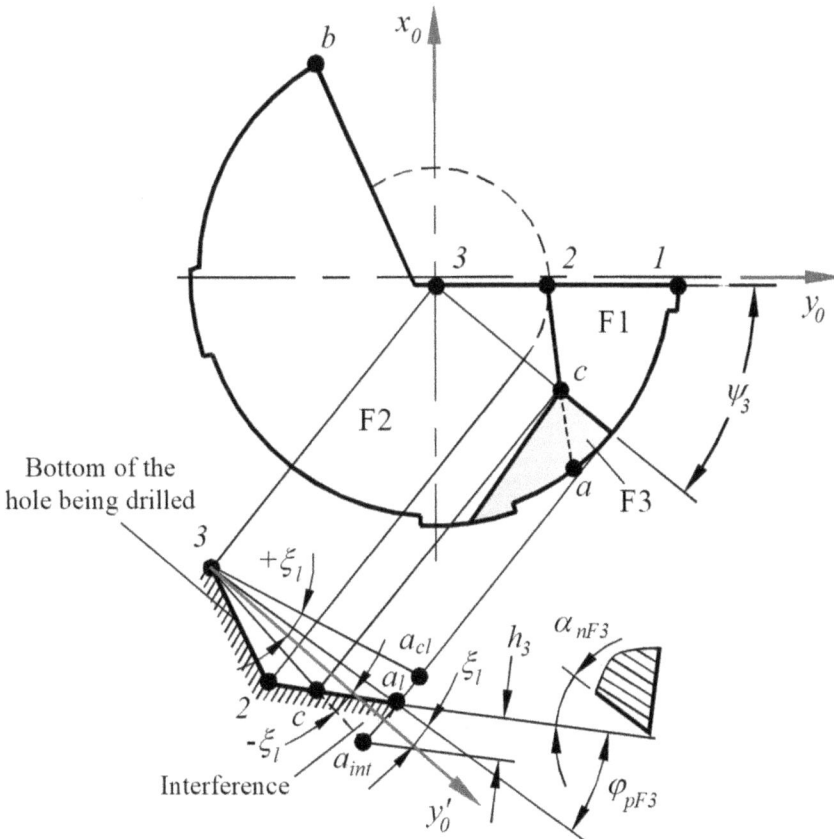

FIGURE 3.39 Model of the condition of gundrill-free penetration as related to point *a*.

where

$$\psi_a = 180 - \alpha_a - \arcsin\left(\frac{m_d}{r_{dr}}\sin\alpha_a\right) \tag{3.19}$$

$$\alpha_a = \frac{\sin\left(\varphi_{p12} - \varphi_{p23}\right)}{\tan\alpha_{n12}\cos\varphi_{p23} - \tan\alpha_{n2}\cos\varphi_{p12}} \tag{3.20}$$

If $\alpha_a > 0$ then its value calculated using Eq. (3.20) is substituted into Eq. (3.19); otherwise,

$$\psi_a = \alpha_a - \arcsin\left(\frac{m_d}{r_{dr}}\sin\left(180 - \alpha_a\right)\right) \tag{3.21}$$

In practical design of a gundrill, first parameters m_d, φ_{a1} (φ_{p12}), φ_{a2} (φ_{p23}), α_{n1-p}, α_{n1-s}, and α_{n2} are selected from Table 3.1 or 3.2 depending upon a given work material and required accuracy of machined hole. Then, conditions set by Eq. (3.16) are verified. When this condition is justified, there is no need to make flank surfaces F3 to assure drill-free penetration so that the VPA2d© design is applied automatically. However, for practical parameters of the tool geometry, flank surface F3 is most often needed. Therefore, the next question to be answered is about what the geometry of flank plane F3 should be.

There are four geometry parameters needed to make addition flank face F3, namely, the location angle, ψ_3, the approach angle, φ_{pF3}, the depth, h_3, and the clearance angle, α_{nF3}. Analytical determination of the introduced geometrical parameters of F3 resulted in the following (Astakhov 1995a). It is obvious as it directly follows from Figure 3.39 that $\varphi_{pF3} = \varphi_{pF1}$ (φ_{p12} in Figure 3.37) as the considered part of the bottom of the hole being drilled is formed by the outer cutting edges 1–2.

The location angle ψ_3 is calculated as

$$\psi_3 = 2\arctan\frac{\tan\alpha_{n12}}{\sin\varphi_{p23}} \tag{3.22}$$

The normal clearance angle of flank F3 is calculated as

$$\alpha_{nF3} = \arctan\left(\sin\varphi_{pF3}\tan\left(\psi_m - \psi_3\right)\right) \tag{3.23}$$

where

$$\psi_m = \arcsin\frac{\cos\psi_3 + \sin\left(\tau_{13} + \alpha_{p2}\right)}{2\sin\frac{c_m}{2}} \tag{3.24}$$

$$c_m = 90 + \tau_{13} + \alpha_{p2} - \psi_3 \tag{3.25}$$

$$\tau_{13} = -\arcsin\left(\frac{\left(r_{dr} - m_d\right)\tan\varphi_{p12} + m_d\tan\varphi_{p12}}{r_{dr}}\right)\frac{\cos\alpha_{p2}}{\tan\varphi_{p23}} \tag{3.26}$$

$$\alpha_{p2} = \arctan\frac{\tan\alpha_{n23}}{\sin\varphi_{p23}} \tag{3.27}$$

and the depth h_3 is calculated as

$$h_3 = r_{dr}\sin\theta_{\psi2}\tan\left(\alpha_{n3} + \theta_{\psi3}\right) + \frac{f}{360}\left(90 + \tau_{13} + \theta_{2b}\right)\frac{\cos\alpha_{nF3}}{\cos\varphi_{pF3}} \tag{3.28}$$

where

$$\theta_{\psi3} = \arctan \frac{\tan \alpha_{n12} \cos \psi_3}{\tan \varphi_{pF3} \sin \psi_3 + 1} \qquad (3.29)$$

$$\alpha_{p1} = \arctan \frac{\tan \alpha_{n12}}{\sin \varphi_{p12}} \qquad (3.30)$$

3.3.3.2   Gundrill Wear and Its Correlation with the Tool Geometry

3.3.3.2.1   Typical Wear Pattern

As well known, after a cutting operation has progressed for some time, wear takes place in two different regions on the cutting tool (Ramakrishana and Shunmugam 1988). Wear appears on the flanks of the tool below the cutting edge(s) forming a wear land and also appears on the tool face forming a characteristic cavity at a certain distance from the cutting edge on the rake face. Like with other cutting tools, tool wear is of prime concern in gundrilling as it defines tool life. In gundrilling, two characteristics of tool life should always be considered:

1. *Tool Life* is the time over which a tool performs cutting between two successive re-sharpenings. It can be also measured by a number of drilled holes per one re-sharpening providing that these holes are the same. It can also be measured by the overall length of the drilled holes produced by the drill between two re-sharpenings. Therefore, tool life can be measured in the units of time and length or by non-dimension numbers of the holes. To measure tool life properly, a clear and accurate criterion of tool life should be established (Astakhov 2004a).
2. *Total Tool Life* calculates as tool life times the number of re-sharpenings. This number varies depending upon the wear pattern achieved within tool life. Unfortunately, this is seldom understood in gundrilling where the topography of the wear pattern has never been considered as a factor limiting overall tool life. Normally, a gundrill is ground back on a re-sharpening by 0.4–0.6 mm to remove visible wear marks. As such, an average gundrill normally allows 7–10 re-sharpenings.

To establish a clear criterion of tool life, the regions of wear on a gundrill and wear patterns should be known. Figure 3.40 shows typical regions of wear on a gundrill which are as follows:

- *Wear on the flank surfaces*: of the inner and outer cutting edges. This is known as flank wear.
- *Wear on the rake face*: This wear is due to high-temperature conditions and high contact pressures at the tool–chip interface. Normally, it appears as a groove (called the crater) along the cutting edges.
- *Wear of the side cutting edge*: It was discussed in Chapter 2 that this cutting edge actually cuts in any kind of drilling even if all the parameters of the machining system are perfect. Another common cause of premature wear of the side cutting edge is a small backtaper normally applied to gundrills as these tools are normally finishing tools. This type of wear is the least desirable because a substantial amount of tool material should be ground back on re-sharpening to restore the drill. As a drill is provided with a backtaper, its diameter decreases on drill sharpening so that the total tool life reduces significantly.
- *Wear of the supporting pads and their faces*: Normally in gundrilling, the leading supporting pad is loaded much more than the trailing supporting pad. This is due to a higher force acting on this part (compared to the trailing pad) causing high plastic deformation of the work material due to burnishing compared to the trailing pad.

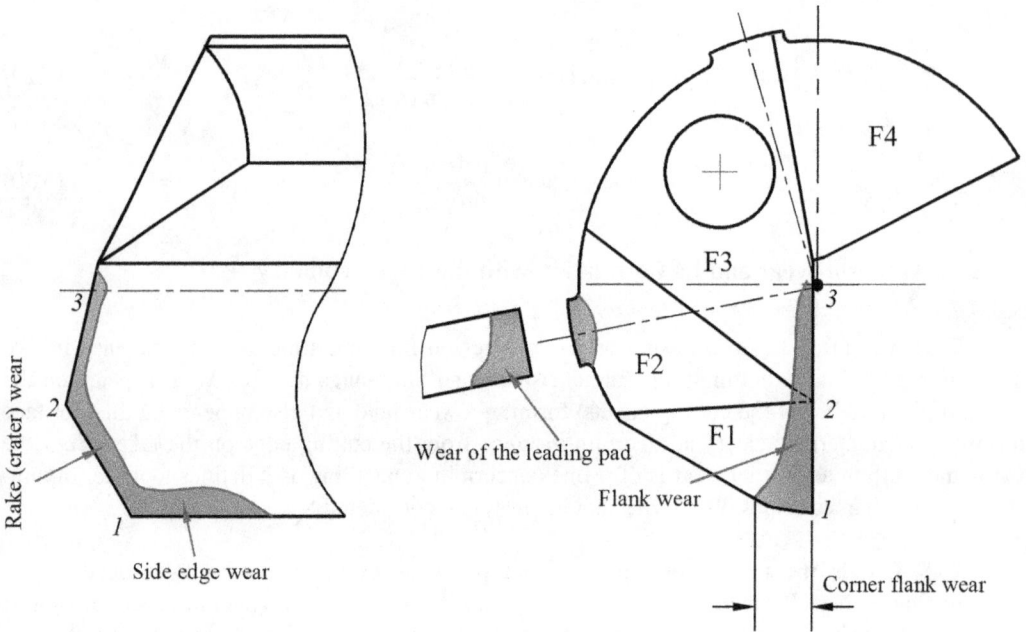

FIGURE 3.40 Typical regions of wear on a gundrill.

An analysis of the wear pattern shown in Figure 3.40 shows that the flank and rake face of the outer cutting edge *1–2* experience much greater wear compared to the flank and rake faces of the inner cutting edge *2–3*. It is commonly explained by the variation in cutting speed along the cutting edges with the maximum at the periphery point *1* of the drill. Moreover, the amount of material removed by the part of the cutting edge adjacent to the drill periphery point is much greater than that by the regions of the inner cutting edge. The third feature is that the distance passed by the periphery point of a gundrill per one drill revolution is greater than that by other points of the cutting edge. As a result, the wear on the flank and rake faces of the outer cutting edge is much greater than that on the corresponding faces of the inner cutting edge and naturally, wear is maximum in the region of the drill periphery point *1*.

The problem is that being correct, this plausible explanation "freezes" the wear pattern shown in Figure 3.40 as something given, or unavoidable as the variation of the cutting speed along the cutting edges, and thus the amount of the work material removed by these edges cannot be changed. Although it is true, the conditions of cooling and lubricating of the flank face F1 can be improved to reduce the temperature on this flank face as this temperature defines its wear (Astakhov and Outeiro 2019).

This is the major objective of the VPA4© design.

3.3.3.2.2   Bottom Clearance Space Definition

While drilling, the gundrill geometry results in the formation of the sculptured surface 1 known as the bottom of the hole being drilled (Figure 3.41). This bottom, from one side, and the drill's flanks (2, 3, 4, and 5), from the other side, form a limited space (6) named as the bottom clearance space. The topology of the bottom clearance space can be appreciated in different cross sections as shown in Figure 3.41. MWF (the coolant) is supplied into the bottom clearance space (6) under pressure through the internal passage (7) of the shank and coolant hole (9) made in the tip (8).

The topology of the bottom clearance is rather complicated as it is formed by two sculptured surfaces: the bottom of the hole being drilled and the flank faces of the gundrill. The bottom of the hole being drilled in general case is formed by two adjacent hyperboloids of revolution. If the

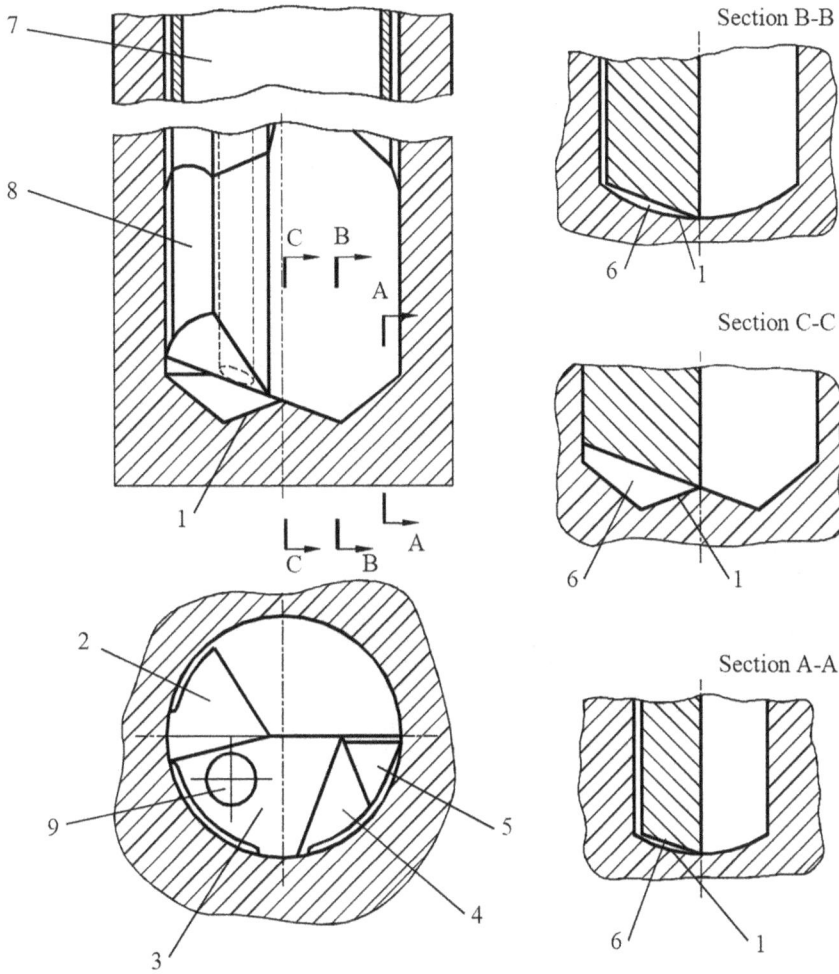

FIGURE 3.41   Concept of the bottom clearance space.

cutting edges are not straight, this bottom has an even more complicated shape. In particular case, however, when the cutting edges are straight, horizontal, and located on the y_0-axis, the bottom (1) (Figure 3.41) consists of two adjacent conical surfaces (one is reverse due to the outer cutting edge and the other is direct as formed by the inner cutting edge). A section of the bottom of the hole being drilled is shown in Figure 3.42. The complicated shapes of the drill flanks and the bottom make the shape of the bottom clearance space very irregular and complex.

One may ask a logical question: why is it important to know the topology of this space? There are two prime reasons for that. First is to assure drill-free penetration into the hole being drilled with no interference. Such a condition is already considered in Section 3.3.3.1.6. Second is to assure the preferable MWF flows in this space to reduce tool wear.

3.3.3.2.3   *MWF Pressure Management in the Bottom Clearance Space*

The MWF pressure in the bottom clearance space directly affects tool life as it has a major influence on the cooling and lubrication conditions on the tool flank (of the inner, outer, and side cutting edges) faces as well as on the supporting pads. The higher the MWF pressure, the higher the tool life. This constitutes *Rule No. 2* in the HP drill tip geometry/design. Moreover, high MWF in the bottom clearance space significantly improves the drill stability and quality of drilled holes (Astakhov 2010). High MWF pressure in the bottom clearance space provides a better penetration

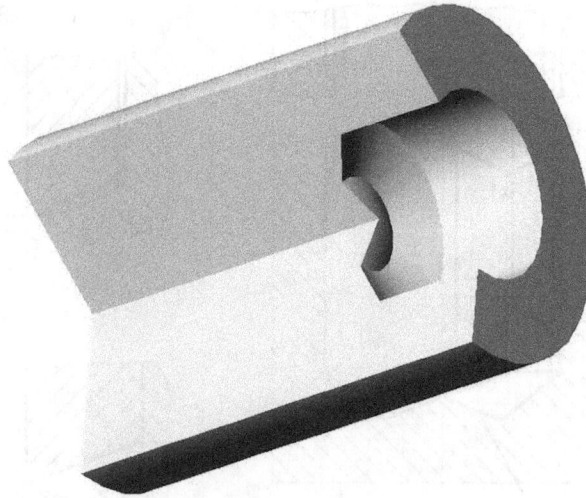

FIGURE 3.42   Section of the bottom of the hole being drilled.

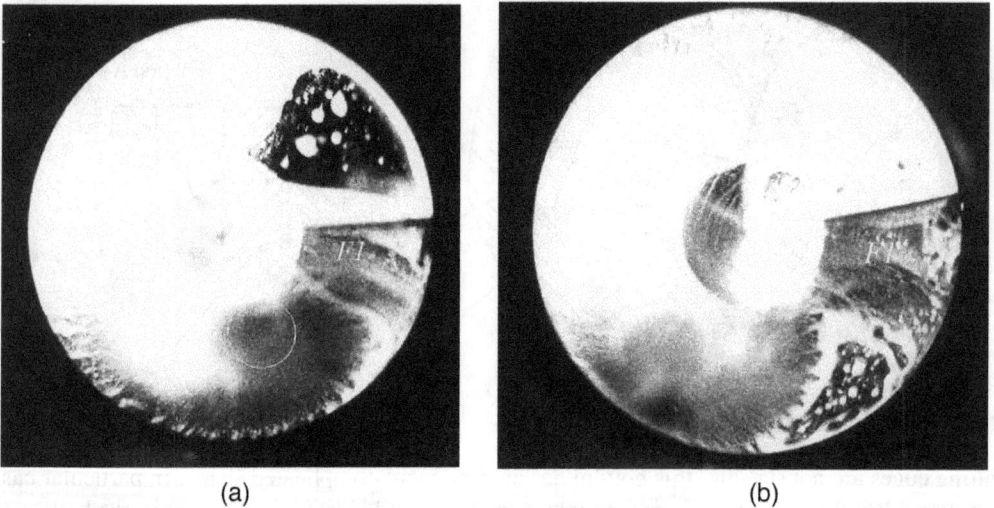

(a)   (b)

FIGURE 3.43   Visualization of the MWF elementary flows in the bottom clearance space when a common gundrill is used.

of the MWF into the extremely narrow passages (see Sections A-A and B-B in Figure 3.41) between the tool flanks and the bottom of the hole being drilled, that is, better conditions for lubricating and cooling of the flank contact areas over the outer and side cutting edges. This is particularly important for MWF penetration in the regions adjacent to the drill periphery region where this MWF is mostly needed, i.e., where tool wear is highest (see Figure 3.40).

Reading this introduced *Rule No. 2*, one may wonder what seems to be the problem? The pressure of MWF supplied to the bottom clearance space is very high in gundrilling operation so it can be set at the level sufficient for MWF penetration into even the smallest gaps between the flank face and the bottom of the hole being drilled. In other words, the problem does not exist. The author's experimental studies of MWF flows in the bottom clearance space have shown that this is not the case in reality.

The experiments carried out using a transparent workpiece and a stroboscopic light having variable flashing frequency allowed observation of the drilling fluid flow distribution in the bottom clearance space. Figure 3.43 shows examples of the observation taken at two different rotational frequencies of the gundrill. As can be clearly seen, the region of the outer flank surface adjacent

FIGURE 3.44   Visualization of the outlet section of the side passage *12345*.

to the drill periphery point (flank face F1) has the poorest cooling and lubricating conditions. This explains why a common gundrill has the greatest wear in this region.

The reason for the result shown in Figure 3.43 should be explained, i.e., why regardless of the high pressure of MWF supplied into the bottom clearance space, the poor MWF flow over flank face F1 is observed resulting in its high wear. The reason for that is shown in Figure 3.44. As can be seen, the bottom clearance space has an opening of a significant cross-sectional area *12345* into the chip flute. No matter how high the pressure of MWF is supplied into the bottom clearance space, the pressure of MWF in this space cannot be kept high as the most of MWF flows from the bottom clearance space into the chip flute trough opening *12345* so that not much MWF left for flank face F1. Therefore, one of the feasible ways to increase MWF pressure in the bottom clearance space and thus improve the cooling condition of the outer and side cutting edges is to reduce the cross-sectional area *12345* formed by flank face F4, i.e., to change the geometry of this flank face.

3.3.3.2.4   VPA4© Design

Figure 3.45 visualizes the essence of the VPA4© design. In Section 3.3.3.1.6, a gundrill made with just two flank planes is considered the most *dangerous* points of the drill flank (the possibility of interference with the bottom) are located at the periphery points *a* and *b* shown in Figure 3.37. Figure 3.38 shows visualization of these points on a solid model of a gundrill. In the practical realization of the VPA4d© design, point *b* is considered. Normally, when the clearance angle α_{n23} (Figure 3.37) is more than 6°, which is the case in practical applications (see Table 3.1), point *b* comes into contact with the bottom of the hole being drilled much before the outer and inner cutting edges, i.e., the interference of this point takes place as shown in Figure 3.45a. To avoid this interference, flank face F4 should be applied. Figure 3.45b shows the limiting case when the flank plane F4 is applied to avoid interference and minimize the area of the outlet section of the side passage, and thus to maximize MWF pressure in the bottom clearance space.

3.3.3.2.5   Practical Realization of the VPA2© Design in Gundrilling

For practical realization of the VPA4© design, the limiting location of this point *b* should be determined and, should the flank plane F4 be applied, then the geometrical parameter of this plane should be determined in full analogy with the flank plane F3 in the manner discussed in Section 3.3.3.1.6. It was shown by the author earlier (Astakhov 1995a) that the condition of the gundrill-free penetration into the hole (no interference with the bottom of the hole being billed) as related to point *b* with designations shown in Figure 3.46 is as follows:

$$\xi_b \geq \xi_l \tag{3.31}$$

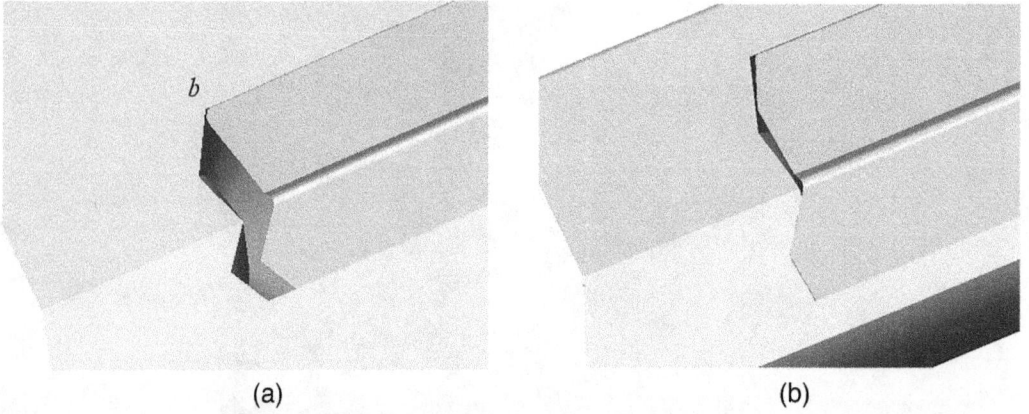

FIGURE 3.45 Visualizing the essence of the VPA4© design: (a) interference of point b, (b) applying the flank plane F4 to avoid interference and minimize the area of the outlet section of the side passage.

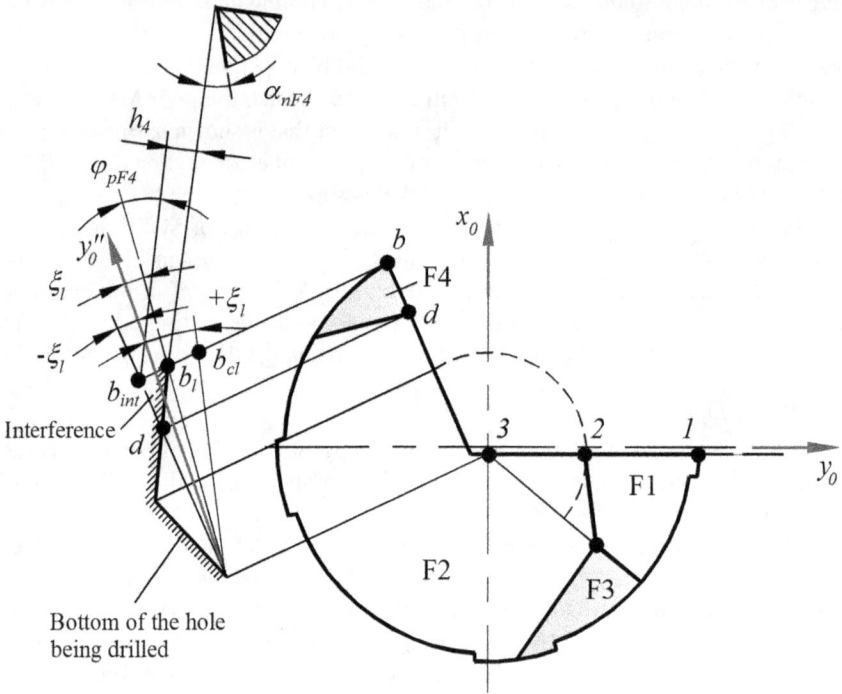

FIGURE 3.46 Model of the condition of gundrill-free penetration as related to point b.

where

$$\xi_l = \arctan \frac{(r_{dr} - m_d)\tan\varphi_{p12} - (m_d + c_{ax})\tan\varphi_{p23}}{c_{ax}\cos\psi_v + \sqrt{r_{dr}^2 - c_{ax}^2\sin^2\psi_v}} \tag{3.32}$$

$$\xi_b = \arctan\left[\left(\cos\psi_b + \tan\alpha_{2n}\frac{\sin\psi_b}{\sin\varphi_{p23}}\right)\tan\varphi_{p23}\right] \tag{3.33}$$

where

$$\psi_b = \psi_{1b} + \arcsin\left(\frac{c_{ax}}{r_{dr}}\sin\psi_{1b}\right) \tag{3.34}$$

where ψ_{1b} $(= 360° - \psi_v)$ is the angle of the sector the $x_0 y_0$ plane corresponding to the drill body as ψ_v is the chip flute profile angle shown in Figure 3.37.

When condition set by Eq. (3.31) is justified, there is no need to make flank surface F4 to assure drill-free penetration. However, for practical parameters of the tool geometry flank surface F4 is often needed. Therefore, the next question to be answered is about what the geometry of flank F4 should be.

There are three geometry parameters needed to make addition flank face F4, namely, the approach angle, φ_{pF4}, the depth, h_4, and the clearance angle, α_{nF4} as the location angle, $\psi_4 = \psi_b$, which is known. Analytical determination of the introduced geometrical parameters of F4 resulted in the following (Astakhov 1995a). It is obvious as it directly follows from Figure 3.46 that $\varphi_{pF4} = \varphi_{pF1}$ (φ_{p12} in Figure 3.37) as the considered part of the bottom of the hole being drilled is formed by the outer cutting edges 1–2.

The normal clearance angle of flank F4 is calculated as

$$\alpha_{n4} = \arcsin \frac{1 - \cos\left(\theta_{\psi4} - \psi_v\right)}{r_{dr}\sin\left(\theta_{\psi4} - \psi_v\right) - c_{ax}\sin\psi_v} r_{dr}\sin\varphi_{p12} \tag{3.35}$$

where

$$\theta_{\psi4} = \arccos\left(\frac{\tan\xi_l\cos\varphi_{p23}\sin\varphi_{p23} - \tan\alpha_{n12}\sqrt{\tan^2\alpha_{n12} + \sin^2\varphi_{p23} - \tan^2\xi_l\cos^2\varphi_{p23}}}{\sin^2\varphi_{p23} + \tan^2\alpha_{n23}}\right) \tag{3.36}$$

and the depth h_4 is calculated as

$$h_4 = \cos\varphi_{p12}\left[\begin{array}{l}r_{dr}\left(\tan\varphi_{p12} - \tan\xi_b\right) - c_{ax}\cos\psi_a\left(\tan\varphi_{p23} - \tan\xi_b\right)\\ -m_d\left(\tan\varphi_{p12} - \tan\varphi_{p23}\right)\end{array}\right]$$

$$+ \frac{f\psi_v}{360}\frac{\cos\alpha_{n4}}{\cos\varphi_{pF4}} \tag{3.37}$$

3.3.3.3 Practical Example

As an example, let us consider a gundrill having the parameters shown in Table 3.4. For these parameters, one needs to determine using the introduced model if flank planes F3 and F4 are needed. Table 3.5 shows the results of the calculations. Because $\xi_a \geq \xi_l$ (the condition set by Eq. (3.16)), flank plane F3 is not needed. On the contrary, as $\xi_b < \xi_l$, so that flank F4 is needed. Its geometry is calculated and shown in Table 3.5. Taking into account the grinding, measurements, and installation tolerances, the parameters $\alpha_{n4} = 7°$ and $h_4 = 1.2$ mm were accepted. A gundrill with these parameters is shown in Figure 3.47a.

Let gundrill 2 have the same parameters as shown in Table 3.4 except the approach angles of the outer and inner cutting edges, which are now $\varphi_{p12} = 30°$ and $\varphi_{p23} = -20°$.

For these parameters, one needs to determine using the introduced model if flank planes F3 and F4 are needed. Table 3.6 shows the results of the calculations. As seen in this table, both conditions

TABLE 3.4
Parameters of Gundrill 1

Drill diameter	$2r_{dr}$	31.75 mm
Approach cutting edge angle of the outer cutting edge	φ_{p12}	25°
Main cutting angle of the inner cutting edge	φ_{p23}	−15°
Normal flank angle of the outer cutting edge	α_{n12}	20°
Normal flank angle of the inner cutting edge	α_{n23}	8°
Size of V-shaped flute extension	c_{ax}	1.8 mm
Angle of the sector corresponds to the V-shaped flute	ψ_v	116°
Angle of the sector corresponds to the drill body	ψ_b	244°
Offset of the point P of the cutting edge	m_d	7.94 mm
Maximum cutting feed	f	0.2 mm/rev

TABLE 3.5
Grinding Parameters of Gundrill 1

ξ_l	ξ_a	ξ_b	φ_{F4}	α_{n4}	h_4 (mm)
5.65°	5.84°	1.01°	25°	6.37°	1.065

of free penetration (Eqs. 3.16 and 3.31) are not justified. Therefore, to assure gundrill-free penetration into the workpiece, flanks F3 and F4 are needed.

Based on the calculated results shown in Table 3.6, parameters $\psi_3=72°$, $\alpha_{n3}=3°$, $h_3=1.8$ mm, $\alpha_{n4}=6.5°$, and $h_4=0.7$ mm were accepted. A drill with these parameters is shown in Figure 3.47b.

Figure 3.48 shows some results of the comparison of the best conventional gundrills used in the automotive industry (light bars) with the gundrills made according to the VPA2© and VPA4d© designs (dark bars). It was also found that an increase in tool life is even greater when difficult-to-machine materials, for example, nickel- and chromium-based PH alloys, were drilled. In this case, up to a 4 times increase in tool life was recorded. Moreover, the quality of drilled holes, including their diametric accuracy, shape, straightness, and surface roughness, was significantly greater when the gundrills were made according to the VPA2d© and VPA4© designs.

3.3.3.4 Shank and Driver

The shank must be designed and made properly. Although there are a number of issues that affect shank performance, the excessive corner radii and shank material-related considerations are of prime concern in HP gundrilling (Astakhov 2004b).

Gundrill shanks must be made of a high-yield-strength material and properly heat treated. Unfortunately, these issues are not always followed by gundrill manufacturers. First, high-yield-strength materials present problems (such as excessive warping, wrinkling, and cracking) when the V-flute is formed (crimped or even swaged) using old tube crimping technology. So tubular products made of 4130 and 34Cr6Mo steels having moderate strength are common in the gundrilling industry. Second, very few gundrill manufacturers understand the proper heat treatment procedure for shanks and thus the fact that it must include a thermomechanical rather than pure thermal relief of the stresses formed on producing the V-flute. The best structure of the shank for short gundrills is a tempered martensitic structure, while for normal and long gundrills, the upper bainitic structure is the best choice. This is the only structure that possesses a very unique combination of high hardness, increased toughness, and great wear resistance suitable for gundrill shanks. Unfortunately, no one shank of gundrills produced today has this structure.

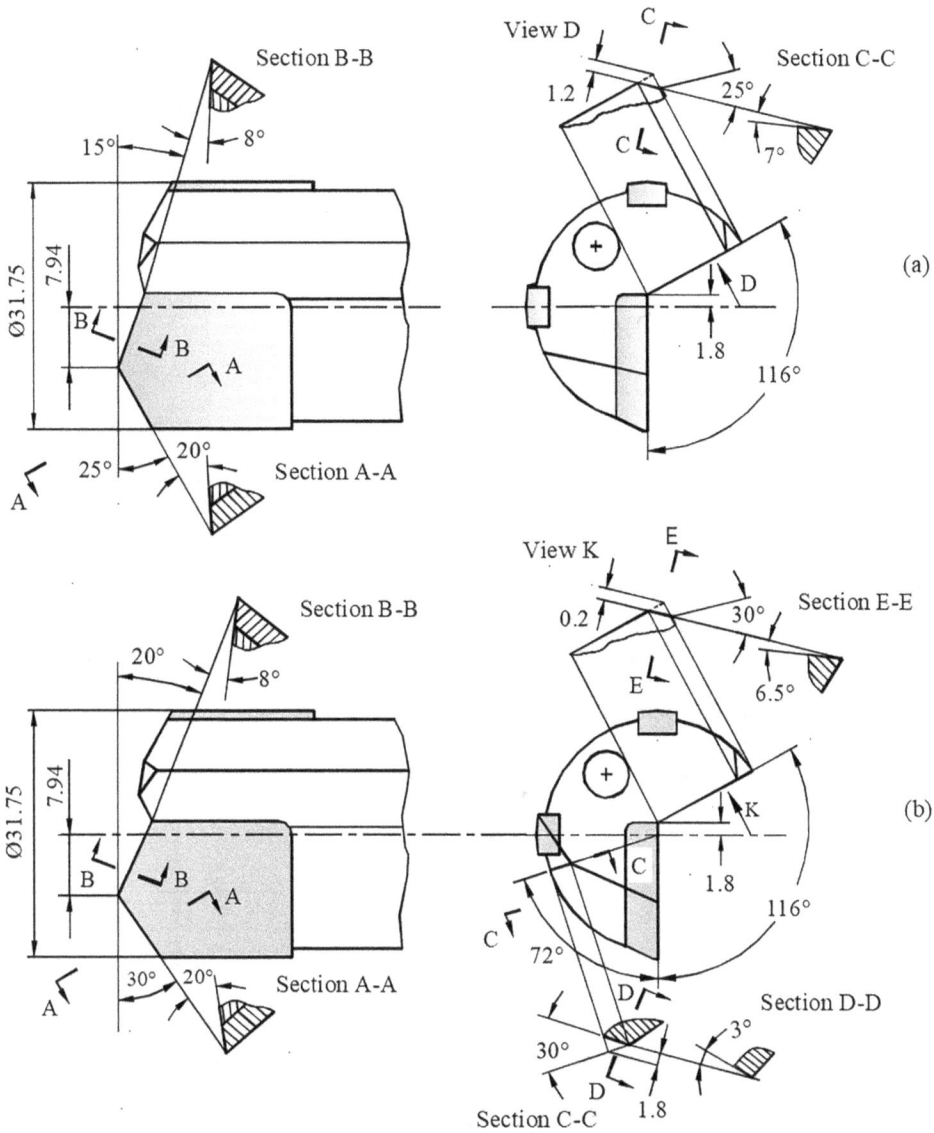

FIGURE 3.47 Gundrills with the calculated parameters: (a) gundrill 1 and (b) gundrill 2. (US Patents US Patents 7,147,411 (2006) and 7,195,428 (2007).)

TABLE 3.6

Grinding Parameters of Gundrill 2

ξ_l	ξ_a	ξ_b	φ_3	ψ_3	α_{n3}	h_3 (mm)	φ_4	α_{n4}	h_4 (mm)
6.09°	3.78°	3.71°	30°	72.1°	2.45°	1.43	30°	6.15°	0.44

When the shank is brazed to the tip and to the driver, the excessive heat from this brazing often ruins the results of the heat treatment at the brazed joints. This heat may cause high residual thermal stresses hidden in the tip. When an increased drilling torque occurs due to, for example, chip clogging or tool wear, the tip fails (Astakhov 2004b). Therefore, the use of low-temperature,

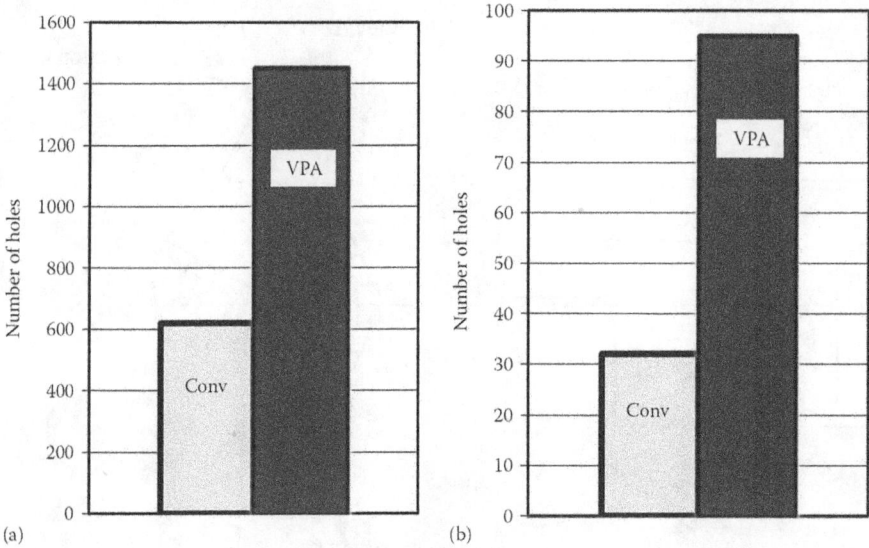

FIGURE 3.48 Tool life comparison of the conventional with the gundrills made according to the VPA2d© and VPA4© designs: (a) drill diameter 5.5 mm, length 200 mm; drilling regime: rotational speed 5100 rpm, feed rate 155 mm/min; machine, Excello NC; work material, nodular cast iron of HB240 hardness and (b) drill diameter 5 mm, length 900 mm; drilling regime: rotational speed 2800 rpm, feed rate 32.5 mm/min; machine, Technidrill; work material, SS15-15LC Mod. of HB300 hardness HB300.

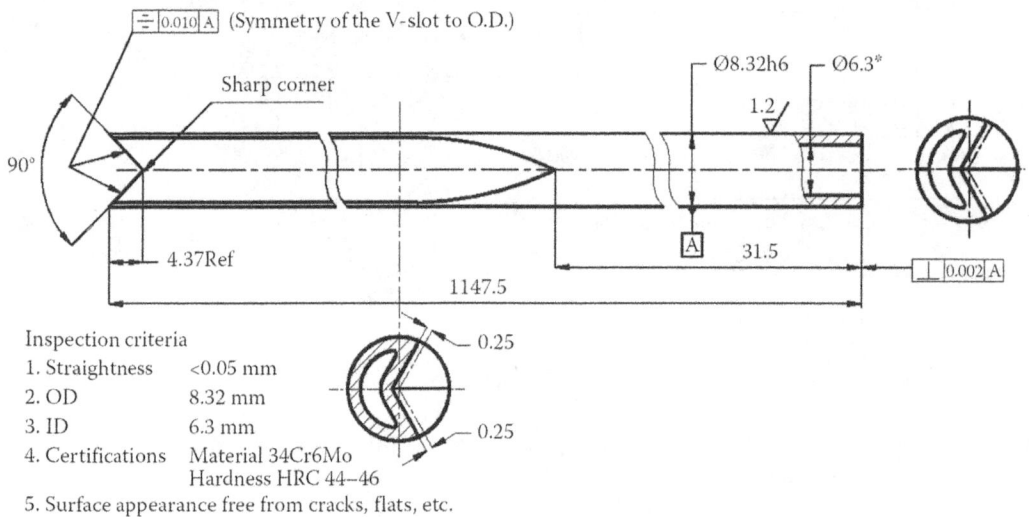

FIGURE 3.49 Example of the shank design for HP rotating gundrill.

high-strength brazing filler materials combined with infrared in-process temperature control followed by a 100% torque test is mandatory for gundrill brazing operations.

Figure 3.49 shows an example of the shank design for HP rotating gundrill. The following should be noted:

1. The shank is made of a high-quality steel and heat treated to high hardness.
2. The proper datum (datum A) is specified.
3. The shank is ground to meet the requirements on its OD and surface roughness.
4. The V-groove for the tip is properly dimensioned with the corresponding form tolerance.

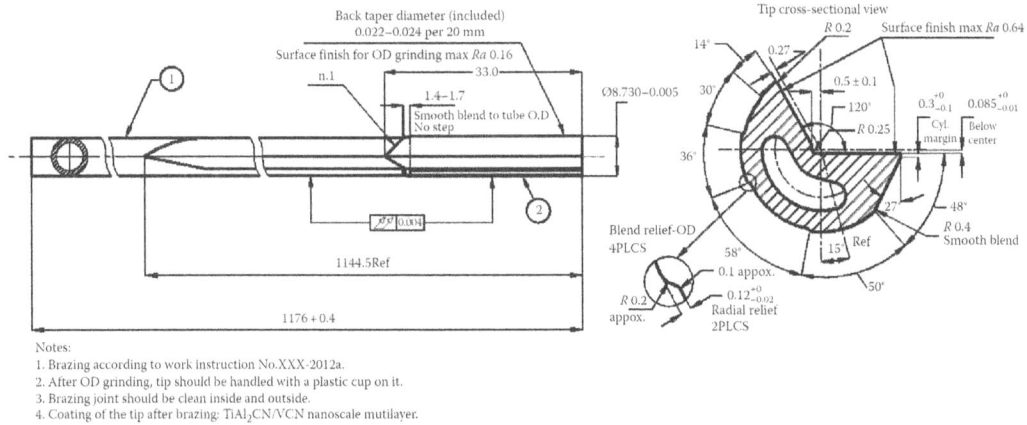

FIGURE 3.50   Example of the tip ground on the shank for HP rotating gundrill.

Figure 3.50 shows an example of the tip ground on the shank for HP rotating gundrill that should always be the case for HP gundrills. The following should be noted:

1. Even small features/details of the tip profile to be ground are clearly indicated with tolerances.
2. High requirements for surface roughness of the tip.
3. Clear indication of the place of brazing and reference on the corresponding work instruction where the procedure, materials, and operating regime are clearly specified.

Figure 3.51 shows the final assembly requirements for HP rotating gundrill. The following should be noted:

1. Clear indication of the place of brazing and reference on the corresponding work instruction where the procedure, materials, and operating regime are clearly specified. The contact of the shank face and the shoulder of the driver is assured by applying the axial force on brazing.
2. Torque and flow rate requirements with reference to the corresponding work instruction where the procedures and equipment used are specified.

Experience shows that when the shank is made of high-yield-strength material, properly heat treated to achieve small grain size bainitic structure, and properly connected to the driver (using a low-temperature brazing filler metal), the increase in the gundrill penetration rate can be as high as twice compared to gundrills commonly used today.

3.4   STS DRILLS

There is no ISO standard on STS tools design and geometry. However, it is covered by the German Technical Rule VDI 3209: 2019-3: Deep-hole boring systems with an external supply of coolant. It provides guidelines for both the prospective and the experienced users to design the STS process with a comprehensive and easily manageable information base. Machining parameters (starting values) for the various process types (e.g., solid, rough, and core drilling) are given for all relevant work material groups. The guide values are of a significant aid, especially for work planning and for the machine operator.

For example, initial values can be determined for machining materials. The deep-hole drilling methods BTA and ejector deep-hole drilling as well as other methods related to deep-hole drilling

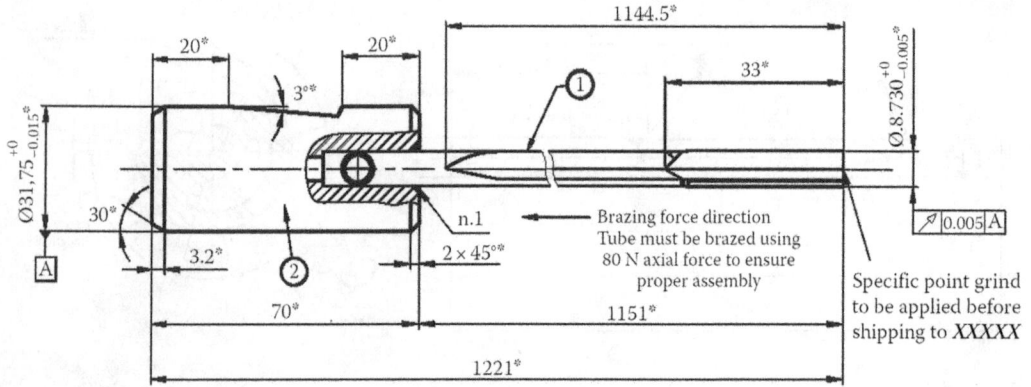

FIGURE 3.51 Final assembly requirements for HP rotating gundrill.

(e.g., chambering) are dealt with. The practical design of the tools, processing machines, and necessary accessories is discussed. In addition, the guideline provides diagrams for the amount of cooling lubricant and the cooling lubricant pressure, which depend on the drilling diameter, as well as the machine drive power. A comprehensive list of useful notes completes the guideline table.

3.4.1 SHORT HISTORY

Many literature sources including German books (e.g. König and Klocke 2008) named STS drills as BTA drills and maintained that the process was developed in Germany in the late 1930s. According to an archive search attempted by the author, it is not exactly so. This method of MWF supply through the annular gap between the boring bar and the bore wall using the pressure head and swarf removal through the interior of the drill head and boring bar was widely used as an industrial technology as early as in the mid-1930s. Moreover, drilling heads with carbide inserts secured on the boring bar using rectangular threads were used. The cutting speed used was as high as 112 m/min. A two-spindle Fritz Werner horizontal deep-hole drilling/boring machine was used in the production of gun and artillery barrels (Veremachuk 1940). The method of mounting these drilling heads on the boring bar with a square multi-start thread and two centering shoulders is the same as used today.

In 1942, Beisner developed a better design of the drilling head. The major improvement was bringing the chip mouth closer to the cutting edges so that more efficient cooling and lubrication of the contact areas can be achieved due to increasing MWF velocity over these areas. Moreover, better conditions for chip removal over the chip mouth can be achieved under conditions of relatively small (compared to modern drilling conditions) flow rate. The additional stabilizing pad is added behind the cutting edge in the axial direction to limit drill vibrations.

The BTA was founded in 1945 to commercialize the "BTA" deep-hole products made in Germany, particularly in France, the UK, and the USA. The central role was played by Gebruder Heller in Bremen, which resulted in the formation of the BTA group founded in the US and UK subsidiaries. A US federal trademark registration was filed for BTA by Gebruder Heller GmbH (Bremen-Mahndorf) on December 2, 1959 and USPTO has given the BTA trademark serial number 72087104. The BTA trademark was filed in the category of Musical Instrument Products.

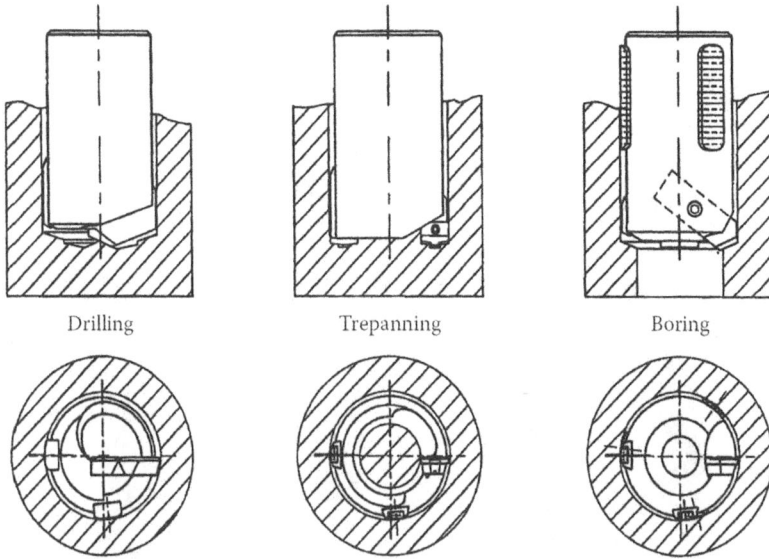

FIGURE 3.52   Basic STS hole-manufacturing operations: drilling, trepanning, and boring.

The description provided to the USPTO for BTA is Industrial Oils and Greases. It was canceled on September 21, 2001. Currently, American Heller Corporation owns this trademark. The description of Goods and Services is Boring Tools and Machines for Solid Boring, Trepanning Counter Boring, and Finish Boring. Sandvik Coromant Co. developed this method for years and termed this process as the STS drilling and thus STS drills.

3.4.2   BASIC OPERATIONS

Although the STS method was initially developed as drilling deep-hole technology, its implementation in modern manufacturing significantly broadened over the years since its introduction. Today, it includes the following basic operations (Figure 3.52):

- STS solid drilling
- STS counterboring
- Trepanning.

Figure 3.53 shows STS drills with a single cutting insert and with three cutting inserts brazed in the drill body. These drills are used for small hole diameters starting with 11 mm. Figure 3.54 shows the drilling heads with indexable cutting inserts and supporting pads. The minimum diameter of the drill with two cutting inserts is 24 mm (Figure 3.54a), while that with multiple cutting inserts is used in the diameter range of 150–350 mm.

Instead of making a hole by cutting all the metal into chips, trepanning (pronounced TREE-panning or treh-PAN-ing) removes a solid core of material by cutting around it. This is an advantage when cutting expensive alloys, as the solid core can be used to make other parts, or, if it is recycled, is more valuable than chips. The trepanning head is completely hollow as can be seen in Figure 3.55, and the cutting process is similar to the solid STS drilling, but it requires less spindle power, as it cuts less material at each revolution. Normally, the indexable trepanning tools cover a range of diameters from 90 to 600 mm and provided with either an internal or external fast lead thread. The carbide inserts of multiple carbide grades, coatings, and chip breakers can be used although precision ground inserts provide improved chip control when trepanning specialty materials.

FIGURE 3.53 STS drills with single cutting insert (a) and with three cutting inserts (b) brazed in the drill body by BTA Heller, Inc.

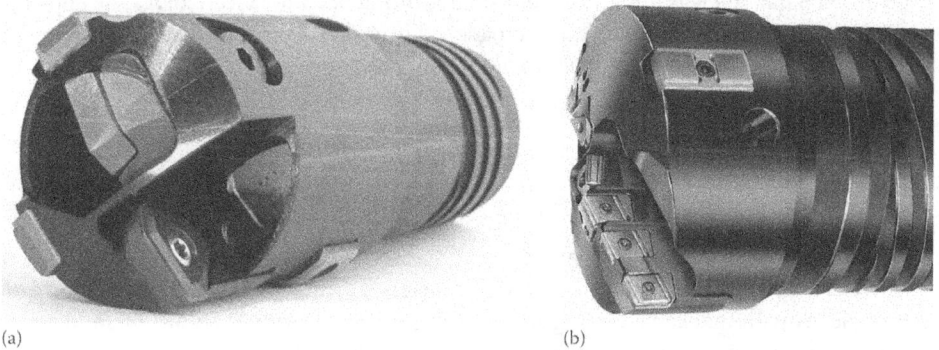

FIGURE 3.54 STS drill with indexable cutting inserts and supporting pads: (a) drill for smaller hole diameters (by BTA Heller Inc.) and (b) drill for large drill diameters (type 43 drill head by Botek Co.).

Skiving and roller burnishing are performed when close diameter, roundness, and surface finish tolerances are required, often for hydraulic cylinder applications. For a hydraulic cylinder to operate effectively, the cylinder's diameter must be precisely round and have a mirrorlike surface finish to ensure a tight seal between it and the mating internal piston. This is commonly achieved through skiving and subsequent roller burnishing inside a tubular workpiece. Skiving

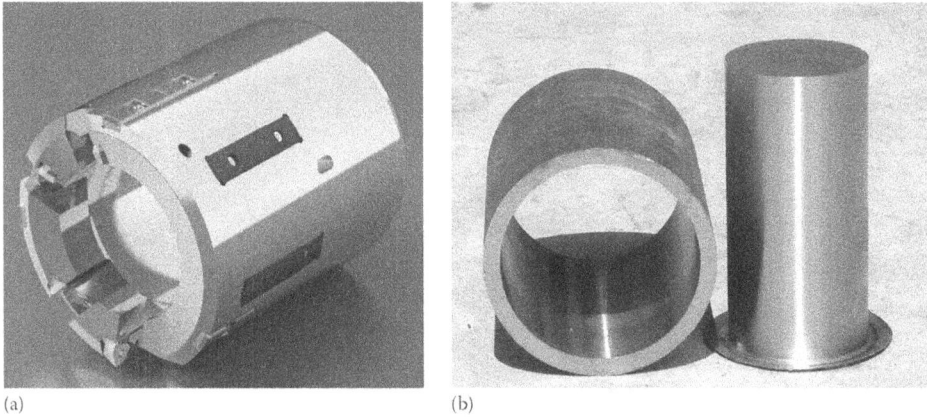

(a)   (b)

FIGURE 3.55   Trepanning tool (a) and the trepanned workpiece and its core (b).

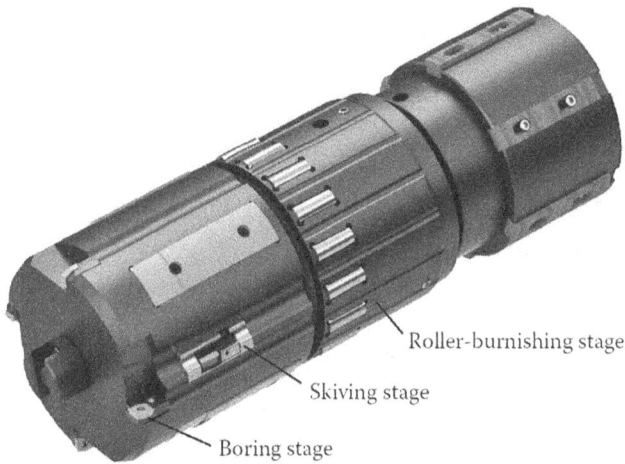

Roller-burnishing stage

Skiving stage

Boring stage

FIGURE 3.56   Combined three-stage boring, skiving, and roller-burnishing tool.

uses a set of carbide blades positioned around the diameter of a tool to slice away chips and create a geometrically round bore. Roller burnishing, a cold-working process, uses multiple rollers to compress the peaks of material left behind after skiving to generate an extremely smooth surface finish. Burnishing also introduces a residual stress layer into the cylinder wall, which improves cylinder fatigue life.

A skiving tool is essentially a modified floating reamer, using multiple knives in a rapid stock removal process. This utilizes high penetration rates with low radial engagements. Roller burnishing uses one or more rollers to cold work the surface of the bore. Rollers are pressed against the bore, plastically deforming the top layer of the metal, compressing peaks, and filling in valleys. Roller burnishing can produce surface finishes of 1 μm Rz. Skiving and roller burnishing are often combined into one tool to complete both operations in a single pass in these DHM operations. Figure 3.56 shows a combined three-stage boring, skiving, and roller-burnishing tool.

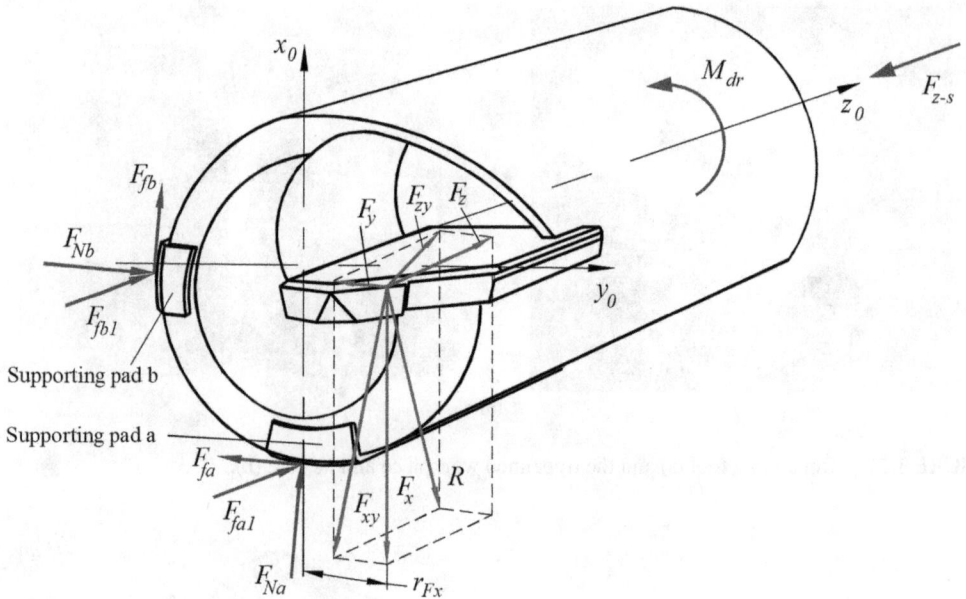

FIGURE 3.57 Simple STS drill head force balance.

3.4.3 FORCE BALANCE

3.4.3.1 Theoretical (Intended) Force Balance

Although there are a great number of designs of STS tools, two important features, namely, the theoretical force balance and the locating principle, are the same. Therefore, they should be considered before particular designs of STS tools are analyzed further in the subsequent sections.

Following the methodology of constructing the force balance diagram used in Section 3.3.3, one can construct a force balance diagram for the tool shown in Figure 3.57. For convenience and simplification of further considerations, the right-hand $x_0 y_0 z_0$ coordinate system, illustrated in Figure 3.57, is set as follows:

1. The z_0-axis along the longitudinal axis of the drill, with sense as shown in Figure 3.57, toward the drill holder.
2. The y_0-axis passes through the periphery corner and is perpendicular to the z_0-axis. The intersection of these axes constitutes the coordinate origin as shown in Figure 3.57.
3. The x_0-axis is perpendicular to the y_0 and z_0 axes as shown in Figure 3.57.

Figure 3.57 shows the force balance in the T-hold-S $x_0 y_0 z_0$ coordinate system (Astakhov 1996). When an STS tool works, the cutting force generated is due to the resistance of the workpiece material to cutting. This force is a 3D vector that can be thought of as applied at a certain point of the drill rake face. The cutting force R (or the resultant cutting force for multi-edge tools) can be resolved into three components, namely, the power (tangential) F_x, axial F_z, and radial F_y components, respectively. These components are commonly referred to as the tangential, radial, and axial forces (Sakuma et al. 1980). The axial force is balanced (equal in magnitude and oppositely directed) by the axial force F_{z-s} of the feed mechanism of the deep-hole machine, while the tangential and radial forces sum to create force F_{xy} (acts in the $x_0 y_0$ plane) that (in contrast to other axial tools such as twist drills, reamers, and milling tools) generally is not balanced, regardless of the number of the cutting edges used. To prevent drill bending due to this unbalanced force, some special measures are taken. The term *deep-hole drilling* initially has grown to mean that the unbalanced cutting force F_{xy} generated in the cutting process is balanced by the equal and opposite force due to supporting pads, which bear against the wall of the hole being drilled as the shift of the drill apex from the tool axis is small compared to gundrills so the stabilizing cone (see Section 3.3.3.3.3) is small and thus makes a relatively small contribution to the tool self-piloting action.

Due to the action of the mentioned forces on the supporting pads, these pads ensure balancing of the tool in the x_0y_0 plane while simultaneously providing for a unique additional machining operation known as burnishing. The fact that the pads bear against the wall of the hole being drilled behind the side cutting edge effectively means that the tool is provided with its own guide. The concept of self-piloting (sometimes referred to as self-guidance), meaning the tool guiding or steering itself along the bore, has been recognized as the major underlying principle of the design of SPTs (Griffiths and Grieve 1985, Griffiths 1993).

Nowadays, multi-edge and multi-supporting pad STS tools are used; thus, the described self-piloting feature may not be obvious. As discussed by the author earlier (Astakhov 2010), no matter how many cutting edges an STS tool has, as long as there is an unbalanced radial force acting on the supporting pads of the tool, it remains self-piloting. Alternatively, no matter how many supporting pads a hole-making tool has, it should not be regarded as self-piloting if there is no (at least, theoretically) unbalanced radial force to cause self-guiding.

Not only does the self-piloting design make it possible to machine deep holes, but it also provides stable cutting conditions for the cutting insert or cutting portion of the tool by eliminating the radial vibrations that often cause drill failure in ordinary drills (Sakuma et al. 1981). When machining conditions are selected properly, a machining operation with STS tool is very stable and consistently produces holes of high quality. Although the use of STS tools requires a number of additional accessories such as starting bushings, high-pressure coolant system, etc., the benefits gained with the use of these tools are greater as these tools are capable of maintaining close size control and producing holes of good surface finish that meets the output requirements of honing. The elimination of the whole sequence of standard hole-making operations makes the use of STS tools appealing even for machining shallow holes.

3.4.3.2 Additional Force Factors in Real Tools

The additional bending moments similar to those discussed for gundrills in Section 3.3.3.1.5 are considered in this section. One important feature of the force balance shown in Figure 3.57 is that the tangential and radial forces (or their resultant F_{xy}) are normally fully balanced by the normal (F_{Na} and F_{Nb}) and tangential (F_{fa} and F_{fb}) reactions acting on the supporting pads (Griffiths 1993, Richardson and Bhatti 2001). In the design of STS and ejector heads, the arms a_{zx-ub} and a_{zy-ub} (Figure 3.58) (the distances from the point of total force application A to the corresponding apexes B and C on the supporting pads) cause the unbalanced moments due to F_x and F_y forces. The axial force F_z is balanced by F_{z-s}. The problem, however, is that the arm e_{zy-ub} is significant so the additional bending moment due to the axial force presents a problem.

The unbalanced moment in the z_0y_0 plane is given in the following equation:

$$M_B = F_z e_{zy-ub} - F_y a_{zy-ub} \tag{3.38}$$

This moment tends to bend the drill in the counterclockwise direction in the z_0y_0 plane because the axial force F_z is approximately three times greater than the radial force F_y and the arm e_{zt-ub} is greater than the arm a_{zy-ub} for the standard STS design.

Similar to gundrills (see Section 3.3.3.1.5), the additional bending moment in the z_0x_0 plane is given in the following equation:

$$M_{C-F_x} = F_x a_{zx-ub} \tag{3.39}$$

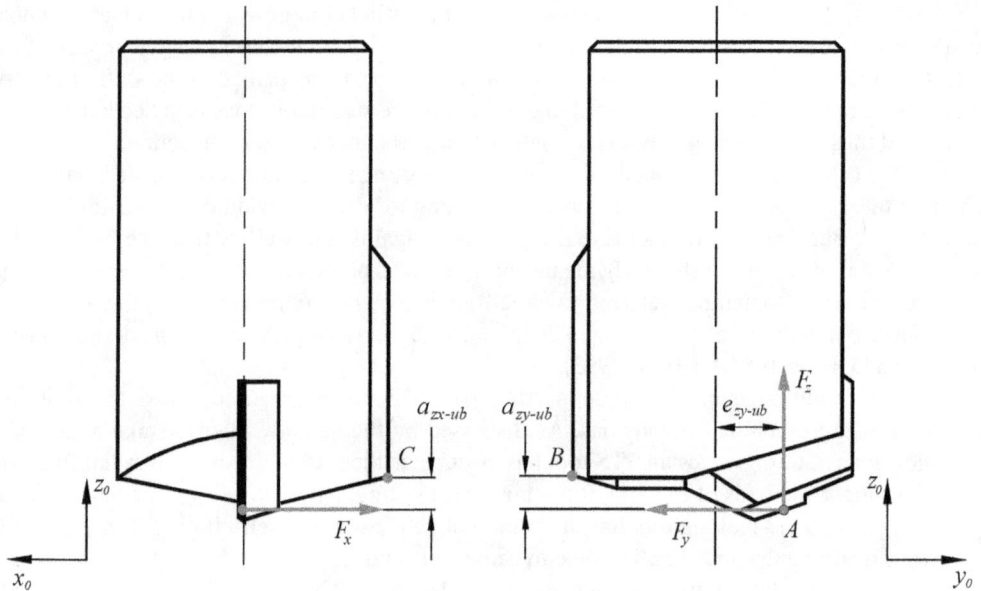

FIGURE 3.58   System of unbalanced moments.

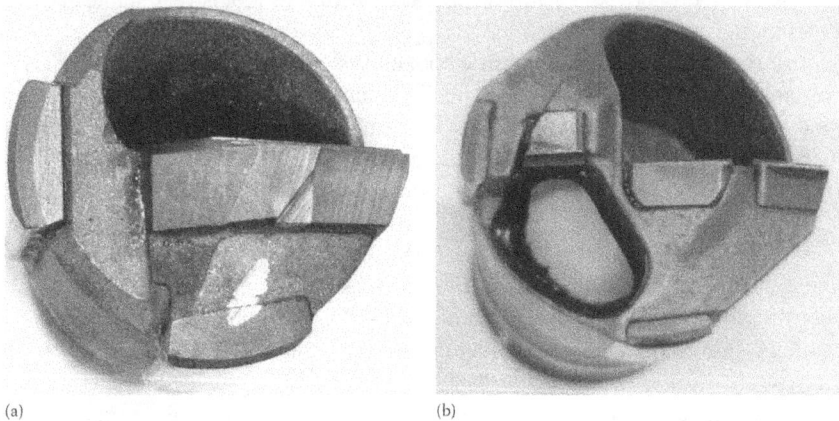

(a)   (b)

FIGURE 3.59   Drill designs: (a) with single cutting insert and (b) with partitioned cutting edge made by three cutting inserts.

A number of measures have been undertaken to reduce harmful consequences of these moments. The most common is to use very shallow feed rates, thus reducing the power (tangential) and axial forces. This measure, however, results in low productivity in STS machining. Another common measure is to introduce additional array of the supporting pads at the rear end of the drill head without any justification.

The problem with the mentioned force balance in the $z_0 y_0$ plane has eased when STS drills with the partitioned cutting edge were introduced. In such drills, the cutting inserts are located on both sides of the x_0 axis. Figure 3.59a shows a traditional drill head and Figure 3.59b shows a head with the partitioned cutting edge. A simplified force model in the $y_0 z_0$ plane for a drill with the partitioned cutting edge is shown in Figure 3.60. As can be seen, the unbalanced moment in the $z_0 y_0$ plane in this case is given in the following equation:

$$M_B = \left(F_{z1}r_{in1} + F_{z2}r_{in2}\right) - \left(F_{z3}r_{in3} + F_y a_{zy-ub}\right) \tag{3.40}$$

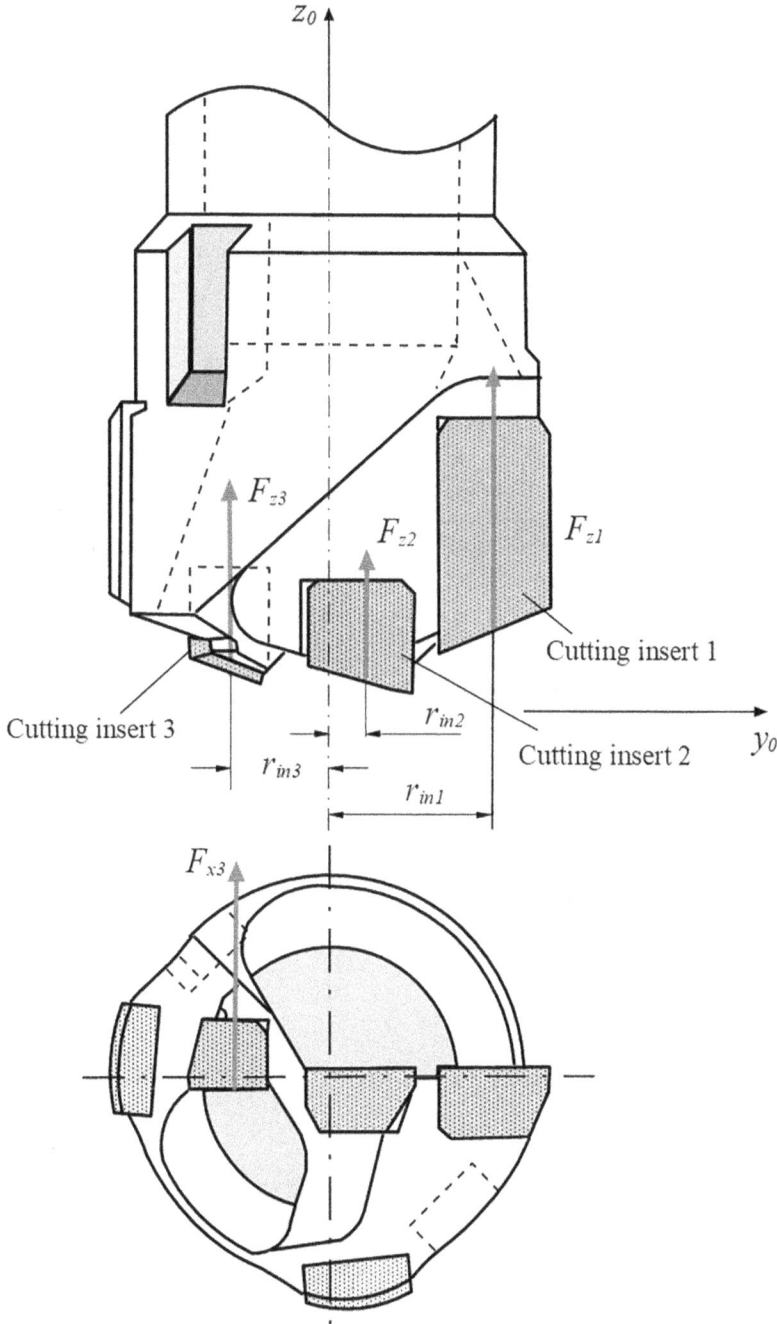

FIGURE 3.60 Simplified force balance in the $y_0 z_0$ plane.

Because the sum $\left(F_{z1} r_{in1} + F_{z2} r_{in2}\right)$ is smaller than $F_z e_{zy-ub}$ and the sum $\left(F_{z3} r_{in3} + F_y a_{zy-ub}\right)$ is greater than $F_y a_{zy-ub}$, the unbalanced moment M_B defined by Eq. (3.40) is much smaller than that defined by Eq. (3.38). This is a significant advantage of the drill design shown in Figure 3.59.

Unfortunately, the problem with the additional bending moment of the $z_0 x_0$ plane defined by Eq. (3.39) is not solved although the power (tangential) force F_{x3} (shown in Figure 3.60) reduces its severity. Moreover, it is never even discussed in the literature although it is a serious problem in terms of achieving high accuracy of the drilled holes as was discussed in Chapter 2, Section 2.1.5.7. In the author's opinion, this problem should be resolved using the VPA2© design (see Figures 2.139–2.143). In other words, practically all STS drills should be re-designed according to this concept. Note that it is also used in the design of gundrills as discussed in Section 3.3.3.1.6.

3.4.4 BASIC GEOMETRY OF STS DRILLS

Figure 3.61 shows the particularities of the tool geometry of the simplest STS (BTA) drill. Although one may think that visually it is different from that for a gundrill (Figure 3.3), it is practically the same in reality. The shown drill has a carbide cutting insert (1) brazed into the tool body (2). The cutting insert includes the outer (3) and the inner (4) cutting edges. The cutting insert has side margin (5) ground as a circular land as in any other drills. The leading (6) and trailing (7) supporting pads are mounted on the drill body to balance the drill. However, this drill has the following particularities compared to the gundrill shown in Figure 3.61:

1. The geometry of the outer cutting edge:
   a. The chip-breaking step is provided.
   b. The chip-separating step of 0.6 mm divides the outer cutting edge into two parts with different approach angles (12° and 18°).

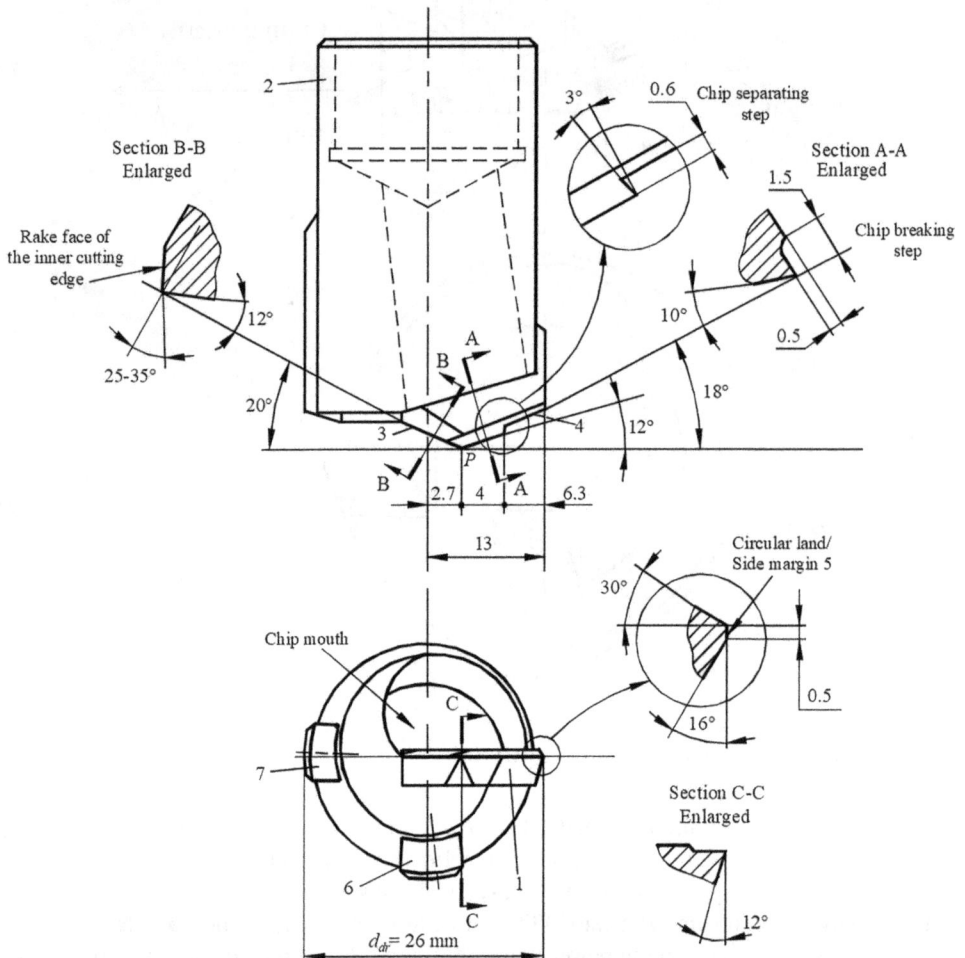

FIGURE 3.61 Particularities of the tool geometry of the simplest STS (BTA) drill.

2. The geometry of the inner cutting edge:
   a. The offset of the drill point P equal to 2.7 mm is much smaller compared to the gun-drill geometry. This offset is approx. 15% of the drill radius, while in gundrill, it is normally 50%. It is compensated, however, by a relatively large approach angle (20°) of the inner cutting edge compared to the outer.
   b. The inner edge rake face is ground with a high negative rake angle of 25°–35° (the reason will be explained later in this chapter).
3. The drill body has the chip mouth.

The presence of the chip-breaking and chip-separating steps as well as the different approach angles of the portions of the outer cutting edge is explained by the necessity of reliable chip breaking. The chip should be broken into small pieces to be able to pass through the chip mouth and then transported without clogging as swarf through the interior of the drill head and boring bar.

Figure 3.62 shows particularities of the tool geometry of a three-insert STS drill with indexable inserts and supporting pads. Although such a drill is more complicated and thus expensive compared to that shown in Figure 3.61, its major advantages (tested by the author) are as follows:

- Up to 30% higher allowable feed per revolution that results in proportionally higher productivity.
- The design with indexable cutting inserts and replaceable supporting pads allows restoring a worn drill in a short time without sending the drill for re-sharpening.
- Different grades of carbide of the cutting edges, different coatings, and different chip breaker parameters can be used on the same drill body that allows using the same body for drilling various work materials.

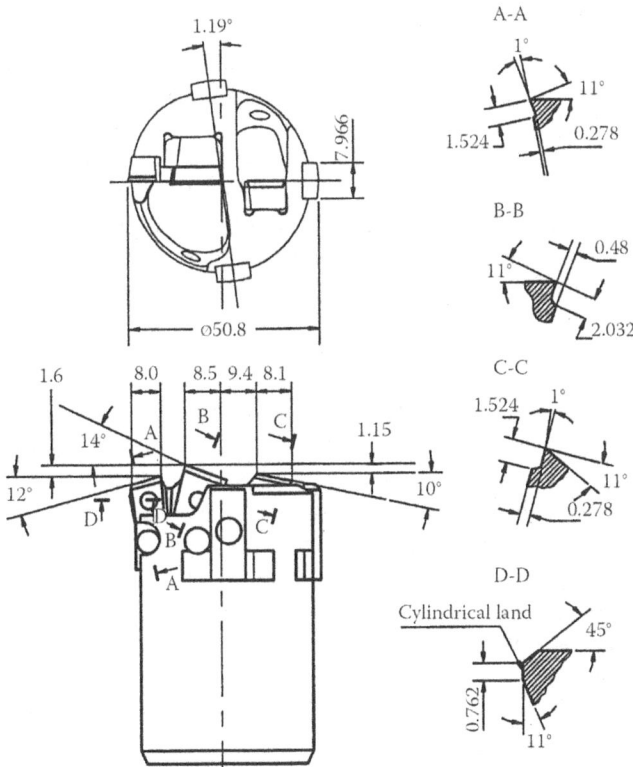

FIGURE 3.62 Particularities of the tool geometry of a three-insert STS drill with indexable inserts and supporting pads.

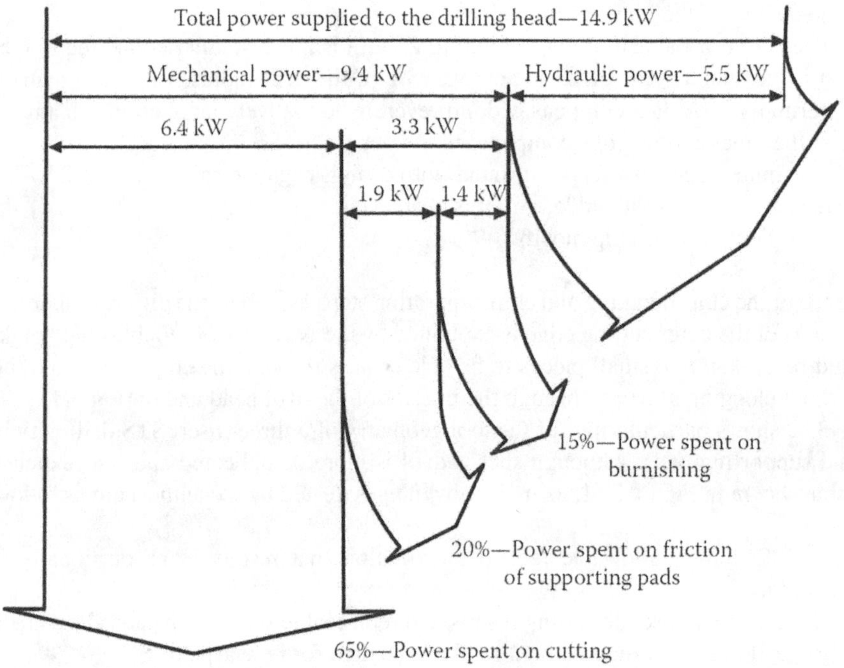

FIGURE 3.63 Power balance for the following conditions: solid STS drill of 50 mm diameter; cutting speed $v=90$ m/min, cutting feed $f=0.14$ mm/rev; and work material AISI steel 1060, MWF flow rate $Q_{fl}=227$ L/min.

3.4.5 POWER AND FORCE

Understanding the partition of the total power supplied to an STS drill allows the proper assessment of drilling efficiency and the optimization of the drilling operation including the drill design optimization. The method of assessment of the energy partition was proposed by Stockert and Weber (1977). Knowing the rotational speed and torque records, one can assess the power spent in drilling and its partition. Figure 3.63 shows the power balance. In this balance, the total power spent by the supporting pads is divided into two parts. The first part is the power spent in pure friction of the supporting pad against the workpiece, and the second part is the energy spent in burnishing the drilled surface by these pads. This became possible as a series of tests with pre-burnished holes was used to assess the power spent in pure friction. The diameter of these holes was adjusted so that no change in the surface roughness occurred, so there was no burnishing.

To verify that the obtained result on the power balance is not drill diameter/design specific, three other series of tests were carried out with STS drill having 25.5, 35, and 50 mm diameter. The same work material AISI steel 1060 and the cutting speed $v=90$ m/min were used in all the tests. For the test carried out with a drill of 25.5 mm diameter, the cutting feed $f=0.08$ mm/rev; MWF flow rate $Q_{fl}=108$ L/min. For a drill of 35 mm diameter, the cutting feed $f=0.10$ mm/rev; MWF flow rate $Q_{fl}=154$ L/min. For a drill of 50 mm diameter, the cutting feed $f=0.14$ mm/rev; MWF flow rate $Q_{fl}=227$ L/min. The obtained results are shown in Figure 3.64. As can be seen, the power partition is practically the same as presented in Figure 3.63.

The analysis of the power balance reveals the following:

1. The highest portion of the power (approx. 65%) is spent on chip formation, so it should be *considered* first in the drilling optimization. Such optimization may include the use of the application-specific chip breakers (both geometry and dimensions) and optimal clearance angles. In the author's opinion, however, the cutting conditions near the axis of rotation where the rake angle of $-30°$ is used (Figure 3.61) should be optimized first.

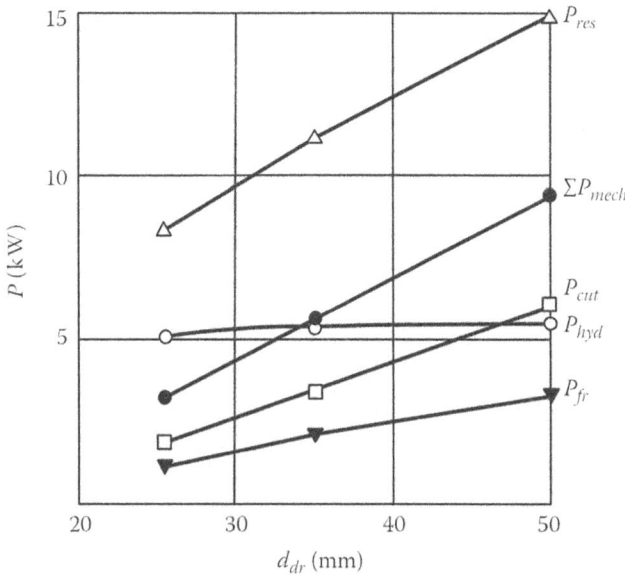

FIGURE 3.64   The total (resultant) power (P_{res}) and its components: ΣP_{mech}, total mechanical power; P_{cut}, cutting power; P_{fr}, total power spent on friction and burnishing of the supporting pads; and P_{hyd}, hydraulic power.

2. The energy spent on friction is greater than that on burnishing. It should be clear that burnishing is a useful function of the tool as it improves integrity of the drilled surface while energy spent on friction is wasted. Although it cannot be avoided, it certainly can be minimized by the optimum pad location and pad geometry and by the proper selection of the combination of the pad and work materials. Pad coating and polishing can also help.

3. The hydraulic power constitutes about 37% of the total power spent in drilling. In reality in modern machines, the portion of the hydraulic power is much greater than the power of secondary equipment as, for example, chillers (for maintaining MWF temperature), automated chip crushing, and centrifuge system should also be taken into consideration.

Optimization of the cutting tool geometry/design and the proper selection of the drilling regime may require the determination of the cutting force components and the location of the point of application of the resultant force. There are two principal ways of determining these parameters, namely, experimental and theoretical (Astakhov 1996, Astakhov and Galitsky 2005, 2006). The former is more accurate but valid only for the test conditions. The latter is much less accurate including too many unknown variables. However, in the author's opinion, a tool designer/process planner/manufacturing engineer should be aware of both, particularly in the way of representing the results obtained using each method.

As an example of experimental determination of the cutting forces, consider an STS tool shown in Figure 3.61 having the geometry shown in Figure 3.65. In order to obtain detailed information on the cutting force components and drilling torque, the tests were carried out for each part of the cutting insert (shown as t_1, t_2, t_3, and t_4 stages in Figure 3.65) (Bescrovny 1984). In the tests, the drilling head was located in the starting bushing mounted in a dynamometer. The dynamometer design and the methodology of the tests were described by Astakhov and Shvets (2001) and Astakhov and Galitsky (2006).

Briefly, the test conditions were as follows:

• Machine – a special CNC deep-hole drilling machine with a high-pressure MWF supply system capable of delivering a flow of up to 220 L/min and generating a pressure of 4.5 MPa was used.

• The stationary workpiece-rotating tool working method was used in the experiments.

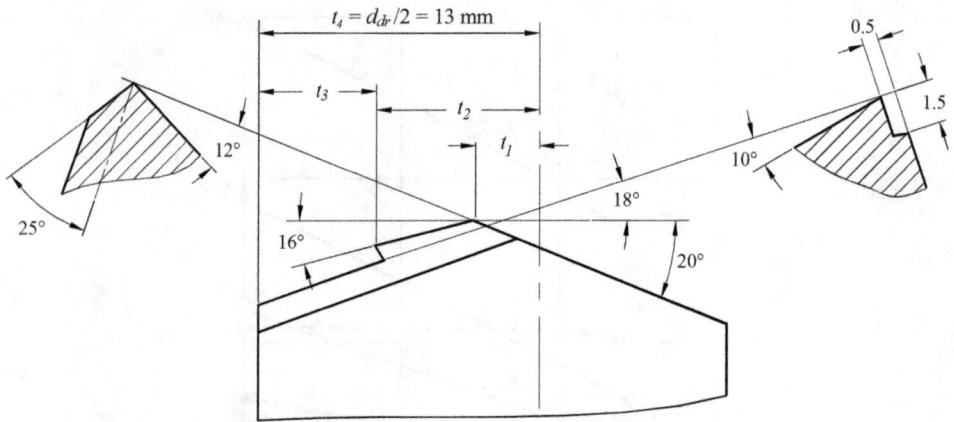

FIGURE 3.65 Geometry parameters used in the tests.

- Workpiece material was AISI 1040, having an ultimate tensile strength of 580 MPa. Test bar of 150 mm length and 60 mm diameter was used.
- Cutting tool – specially designed STS heads of 26 mm diameter were used. The tool material was carbide ISO P10. As shown in Figure 3.65, the cutting edge is divided into three sections: the outer, middle, and inner cutting edges. The outer and middle edges have 18° and 16° approach angles, respectively. These edges are separated by a step. The inner cutting edge is reversed. The offset is approx. 15% of the drill radius. The rake angle for the inner cutting edge is −25° in comparison to 0° for the outer and middle cutting edges. Tolerances for all angles were ±0.5°. The roughness Ra of the face and flank of the drills was less than 0.45 μm. Each cutting edge was examined at a magnification of 25× for visual defects such as chips or cracks.
- Parameters of experiments – the following parameters were selected as the cutting conditions: $t_1=2.7–6.5$ mm; $t_2=1.8–6.3$ mm; $t_3=3.0–6.3$ mm; $t_4=d_{dr}/2=13$ mm; $f=0.062–0.125$ mm/rev; and $v=40–90$ m/min.
- Statistical analysis of the results – the 22 factorial, complete block type of the design of experiments was used to establish the experimental force relationship (Astakhov 2012). The mathematical models obtained as regression equations have been statistically analyzed. Such an analysis included the examinations of variance homogeneity, die significance of the model coefficients, and the model adequateness (Astakhov 2012).

The experimental results are shown in Table 3.7.

The experimental results showed the following:

1. The width of the stage (the width of cut) t_i and the cutting feed f have the prime influence on the force components, while the influence of the cutting speed is negligibly small.
2. The data obtained allow us to find out not only the force magnitudes but also the point of the resultant force application. It was proven in the test that the force components can be considered as applied in the middle of the corresponding stages of the cutting edges.
3. For practical calculation, components F_x and F_y can be determined as

$$F_z = (0.60...0.65) F_x \tag{3.41}$$

and

$$F_y = (0.30...0.33) F_x \tag{3.42}$$

Similar to gundrills, the problem with cutting in the region close to the axis exists in BTA drills.

TABLE 3.7
Experimental Results

Cutting Edge Number (Figure 3.65)	Statistical Relationships for Cutting Force Components		
	F_x (N)	F_y (N)	F_z (N)
1	$1497t^{0.98}f^{0.81}$	$450t^1 f^{0.81}$	$594t^{0.94}f^{0.61}$
2	$1560t^{0.99}f^{0.78}$	$585t^{1.07}f^{0.96}$	$636t^{0.93}f^{0.66}$
3	$1620t^{0.94}f^{0.77}$	$770t^{0.92}f^{0.90}$	$728t^{0.93}f^{0.63}$

3.4.6 PROBLEM WITH THE CORE

It is discussed in Section 3.3.2.3 that although the gundrill does not have the chisel edge, the problem with cutting in the region close to the axis of rotation does not magically disappear. To avoid interference with the bottom of the hole being drilled, the rake face of gundrills is located below CL allowing formation of a string. A number of ways to deal with this string were also discussed.

It should be not a surprise that the problem with cutting in the region close to the axis of rotation exists in STS and ejector drilling. Traditionally, the stage of the cutting edge adjacent to the center of rotation is ground with a highly negative rake angle (25°–30°, see Section B-B in Figure 3.61), so the formed core at the center of the hole being drilled is bent and thus fractured by this rake face. In the author's opinion, this problem is not normally discussed in the literature on the subject because when the basic geometry of STS drills was introduced, the cutting feed used for drilling was relatively low so the core fracture contribution was not profound. Moreover, other problems, for example, chip breakage, spindle-starting bushing alignment, quality and accuracy of shanks (tubes), and supplying of the sufficient flow rate of MWF, greatly overshadowed the problem with the core fracturing. When the listed problems were eventually fixed and when higher feeds were attempted, the problem with core fracturing became important.

One can understand the problems easily if he or she analyzes the data presented in Table 3.7. As can be seen, the contribution to the total cutting force of stage 1 of the cutting edge is practically the same as that of stage 3 although stage 3 removes approximately five times greater volume of the work material per one revolution. It means that the cutting by stage 1 of the cutting edge requires fivefold more energy per unit volume of the work material removed.

A universal approach to deal with a partially formed core is shown in Figure 3.66. In this design, the core deflector is a part of the drill body (US Patent No. 4,565,471 (1986)). As can be seen, when a partially formed string contacts the deflector, it bends and then fractures from the bottom of the hole being drilled. The design and location of the chip deflector depends on the type and design of a particular drill and thus can be used with any type of drill when one tries to solve the problems that unavoidably occur for any drill in the region adjacent to the axis of rotation.

Figure 3.67 shows the design according to US Patent No. 4,616,964 (1986) where the side of the cutting insert deflects the partially formed core. As can be seen, the area around the axis of rotation is made a non-machining zone by shifting the cutting inserts from this axis. As a result, the cutting edge is deliberately shifted from the zone of very low cutting speeds reducing the axial force and thus allowing greater penetration rates. The distance between the axis of rotation and the cutting insert corner of the insert defines the radius of the forming core. To deflect the core, this distance is made larger than that at the rear end of the insert in the axial direction. As a result, the core is deflected and then fractured.

Figure 3.68 shows the modification of the previously described idea for the drilling head design according to US Patent Application No. 2011/0067927 (2011). This drilling head includes screw-clamped carbide inserts. As can be seen, the central cutting insert forms an angle in a specific range with respect to the insert tip (or the axis of the drill). As a result, lateral displacement of the partially formed core takes place by the side surface of the cutting insert. To protect the edge of the

Partially formed
cylindrical core

Tool

Axis of rotation

Core deflector made in
the tool body

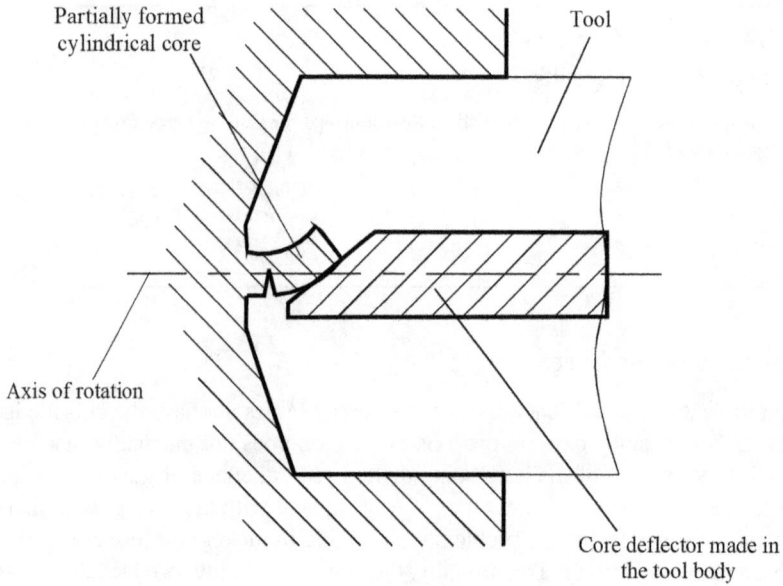

FIGURE 3.66 Core deflector as a part of the drill body (US Patent No. 4,565,471 (1986)).

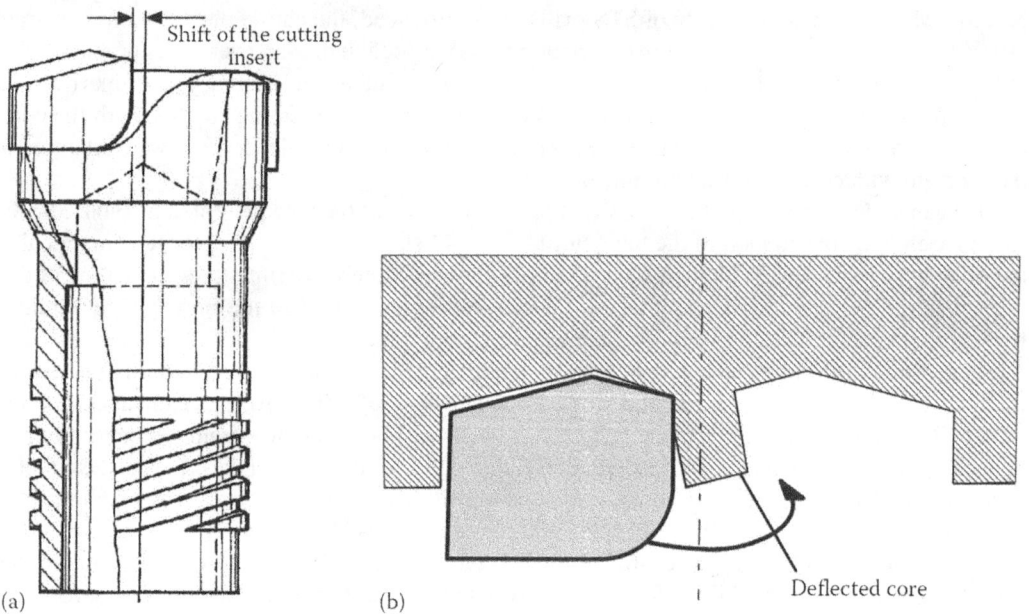

Shift of the cutting
insert

(a)

(b)

Deflected core

FIGURE 3.67 Drilling head design according to US Patent No. 4,616,964 (1986): (a) drill design and (b) core deflection in drilling.

insert from chipping, the central cutting insert is located well above CL so only the side of this insert participates in the core deflection and fracture.

In the author's opinion, practically all the known designs utilize the same idea – deflection and the fracture of the partially formed core. This still results in an increase in cutting force and in destabilization of drill smooth performance.

Tip shifted from the
rotational axis

Core

Crack forms at the root of the core

Deflected core

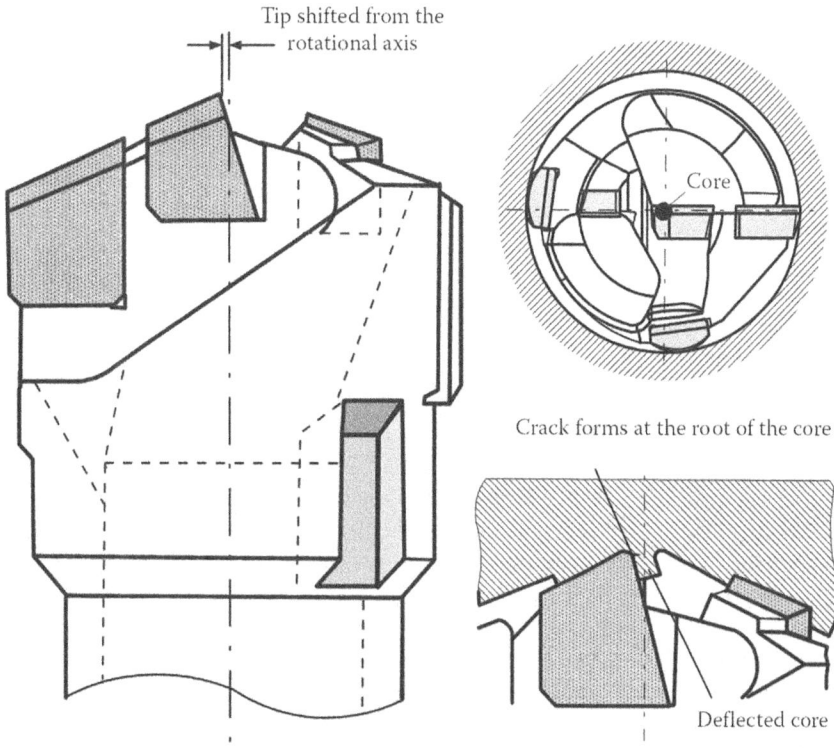

FIGURE 3.68   Drilling head design according to US Patent Application No. 2011/0067927 (2011).

3.4.7   PROBLEM WITH THE PRESSURE DISTRIBUTION

The architecture of the drilling head developed by Beisner addressed the conditions of MWF delivery parameters (relatively low flow rate) assuring smooth MWF flow in the machining zone and proper removal of the chips generated by the cutting insert. The introduction of the indexable cutting inserts and replicable supporting pads gave drilling heads some new *squared/boxed* style, while not much care about the MWF conditions around the supporting pad and in machining zone was taken as the modified Bernoulli equation (see Appendix C in Astakhov (2024)) seems to be forgotten.

The modified Bernoulli equation relates the static pressure in the MWF flow with the velocity of this flow. The higher the velocity, the lower the static pressure under a given flow rate. Because the diameter of the drilling head is made rather close to that of the hole being drilled and because a relatively high flow rate is to be delivered into the machining zone of STS drills, the velocity of MWF should be high so that the static pressure of the flow around the supporting pads and the side of the cutting insert can be rather low. Having noticed the problem with insufficiently lubricated supporting pads, the tool manufacturers tried to alter the pad design instead of addressing the problem using the modified Bernoulli equation.

Figure 3.69 shows an example of the modified pad design according to US Patent No. 6,602,028 (2003). The full contact surface of such a pad is reduced to two triangular areas (1 and 2 in Figure 3.69), while a countersink (3) is formed between these two contact areas in the hope to improve pad's cooling and lubricating by the enhanced flow of MWF.

The special geometry of the drill head results in the formation of a special shape of the bottom of the hole during drilling. The space enclosed between this bottom from one side and the flanks of the drill head from the other is called the bottom clearance space as was discussed earlier in the

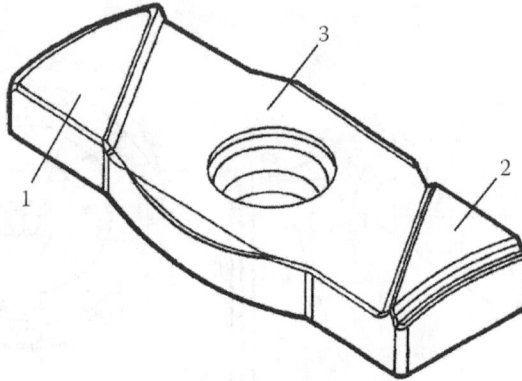

FIGURE 3.69   The improved pad design according to US Patent No. 6,602,028 (2003).

consideration of gundrilling. The MWF pressure in the bottom clearance space has a major influence on the cooling and lubricating conditions of the flank and rake contact areas. Increasing the cutting fluid pressure in the bottom clearance space provides better penetration of MWF to the narrow passages between the tool flanks and the bottom, that is, better conditions for lubrication and cooling of the flank contact areas. Therefore, the pressure in the regions close to the cutting edge(s) should be as high as possible under a given flow rate. This leads to a considerable reduction in flank wear and therefore increases the tool life.

Unfortunately, little attention has been paid to the influence of the drill design parameters on the MWF pressure distribution in the bottom clearance. The *smooth* designs of the STS drilling head with a single and multiple brazed cutting insert (Figure 3.53) were gradually replaced with body designs of STS drilling head with partitioned brazed and indexable cutting inserts that should have changed considerably the architecture of the flank faces (e.g., as shown in Figures 3.54 and 3.62). No adjustments to the architecture of the flanks were made. To comprehend the difference in the MWF fluid pressure distribution in the bottom clearance space between the STS drilling heads with a single and with indexable cutting inserts, a series of tests were carried out for the STS drill having a 50.8 mm diameter. The static (the drill just brought to close contact with the bottom of the hole) and dynamic (the drill was rotating) pressure distributions were measured (Astakhov 1995a, Astakhov et al. 1998). Figures 3.70 and 3.71 show examples of the experimental results. A subsequent analysis of these results revealed the following:

1. The static and dynamic pressure distributions in the bottom clearance space were not uniform, contrary to the currently held view. Both distributions change significantly in the bottom clearance, and there exist regions with zero pressure and even negative pressure under static and dynamic conditions.
2. The rotation of the tool leads to a significant change in the pressure distribution in the bottom clearance space. The rotation makes this distribution more uniform in the case of the STS drill with a single cutting insert and less uniform in the case of the STS drill with indexable inserts. This seems to be due to the swirling effect produced by the "rough" features of the drill head.
3. In deep-hole drilling, the tool life is defined by the flank wear of the cutting edge(s). It is known that flank wear is influenced by the rate of penetration of MWF to the contact areas between the bottom of the hole being drilled and the flank(s). This rate depends upon both the properties of MWF and its static pressure in the region adjacent to the flank contact areas. The experiment results show that the STS drill with indexable inserts has a higher

MWF pressure in these areas. For the STS drill with a single cutting insert, the zone of maximum pressure is shifted to the direction of the second supporting pad and plays little role in improving the cooling and lubrication of the cutting edge. However, a number of zones with negative (compared to atmospheric) pressure and with recirculating stagnating were found for the STS drill with indexable inserts.

4. The architecture of the STS drill with a single cutting insert provided better conditions for the chip removal process due to much more uniform pressure distribution at the chip mouth of the drilling head.

5. The results also showed that the architecture of the drill head can be changed to achieve a better pressure distribution, avoid negative pressure regions in the bottom clearance space, and provide better conditions for cooling, lubrication, and chip removal.

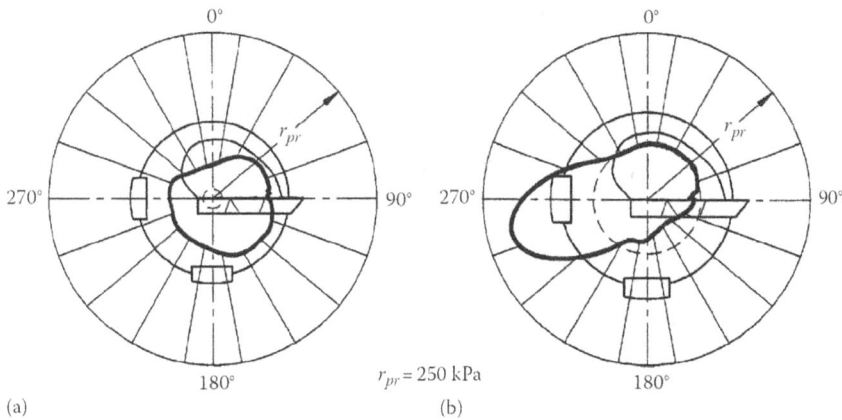

FIGURE 3.70 Examples of the dynamic (the drill rotates) MWF pressure distribution in the bottom clearance space for the STS drill with a single cutting insert: (a) position of the pressure transducer corresponds to the radius $r=3.1$ mm and (b) position of the pressure transducer corresponds to the radius $r=13.1$ mm.

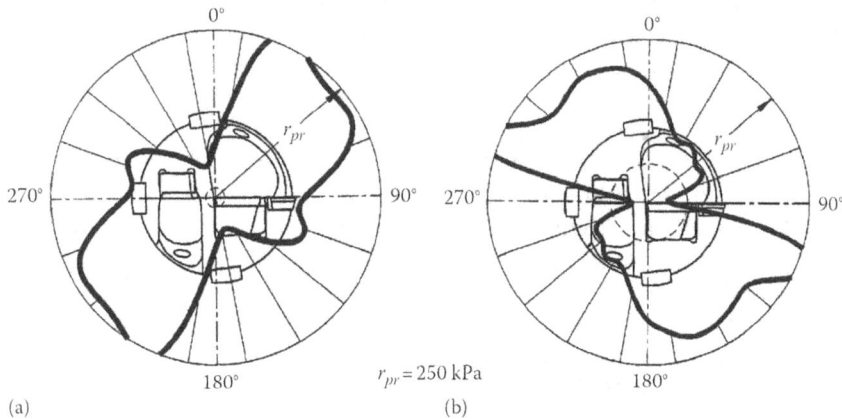

FIGURE 3.71 Examples of the dynamic (the drill rotates) MWF pressure distribution in the bottom clearance space for the STS drill with indexable inserts: (a) position of the pressure transducer corresponds to the radius $r=3.1$ mm and (b) position of the pressure transducer corresponds to the radius $r=13.1$ mm.

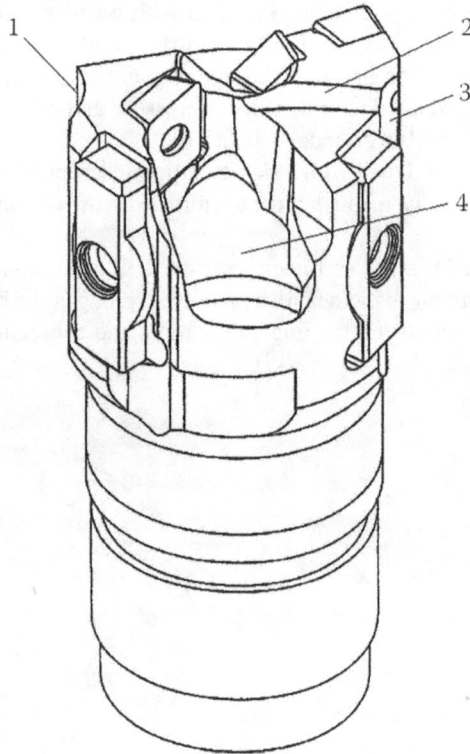

FIGURE 3.72   Improved architecture of the drilling head according to US Patent No. 6,682,275 (2004).

3.4.8   ADDRESSING THE PROBLEMS WITH THE PRESSURE DISTRIBUTION AND THE CORE

Figure 3.72 shows an improved architecture of the drilling head according to US Patent No. 6,682,275 (2004). As can be seen, additional MWF channels 1, 2, and 3 are made to provide better MWF delivery to the supporting pads and to the flanks of the cutting inserts. The major difference, however, is in the location of the chip mouth (4). As can be seen, it is located much further down in the axial direction from the cutting inserts that definitely improves chip removal and MWF flows in the bottom clearance space. Still intuitive and not complete, this architecture is a step forward in the right direction.

A new STS drill was developed that solves the problems with the cutting of the core (see VPA3© design discussed in Chapter 2, Section 2.1.11.11), drill stability (VPA2© design discussed in Chapter 2, Section 2.1.5.7), and MWF pressure distribution. Moreover, its implementation practice shows that practically all parameters of drill hole quality improve, drilling productivity and tool life increase particularly in drilling difficult-to-drill work materials. Figure 3.73 shows an example of an optimized STS drilling head for a range of bore diameters from 20 to 60 mm. It is based on the drill design originally proposed by Vinogradov (1985) so that it includes the point design with non-central chisel edge. As can be seen in Figure 3.73, it includes only one supporting pad to balance the additional moment according to VPA2© design.

The distinctive feature of this drill is the presence of the chisel edge, which is not normally used in SPTs. This chisel edge is shifted from the center of rotation and is similar to that discussed in Chapter 2 (Section 2.1.11.11, Figure 2.208) where the advantages of such a chisel edge are listed. Among them, two are most relevant: (1) no one point of the chisel edge has zero cutting speed and (2) self-centering ability without compromising the strength of the drill tip. The problem with the core is permanently solved. Moreover, the architecture of the drill includes the optimized MWF flow passes that assure proper cooling and lubrication of the contact areas.

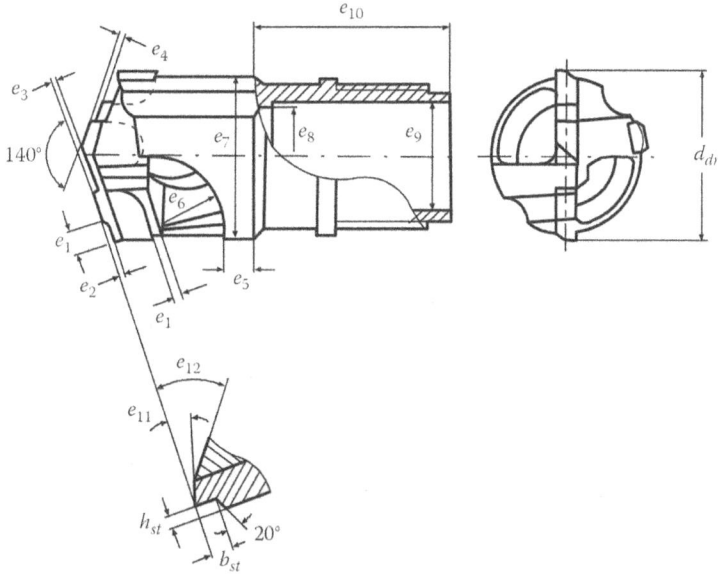

FIGURE 3.73 An example of the optimized design of the STS drilling head.

The shift of the chisel edge from the drill axis results in the edge cuts the so-called stabilizing cone, which prevents any deflection of the drill in the radial direction due to the above-discussed unbalanced radial force (discussed in Section 3.3.3.1.2). As a result, the drill enters the hole smoothly with no wandering and/or radial vibrations.

The proper accounting for the force balance allowed using one supporting pad, which should be the target for any STS and ejector drilling head designs. The following particularities of the drill designs should also be noticed:

1. Diameter e_7 is selected to be equal to $d_{dr}-(1–3)$ mm.
2. The bore diameter e_8 is made so that the distance between its left shoulder and the end of the cutting insert does not exceed 3 mm.
3. The radius of the z0y0 projection of the chip mouth $e_6=(0.4–0.6)$ e_9.
4. Size $e_5=(4–8)$ mm.
5. Size e_{10} and thread parameters are standard for STS drills.
6. The point angle is 140° for a wide range of difficult-to-machine materials.
7. Size e_3 depends on the drill diameter and the cutting feed. It is 0.2–0.5 mm.
8. The step on the outer cutting insert having size e_2 is ground for chip separation, and it should be approximately $2f$ but should not be greater than $0.2b_{st}$.
9. The step size is $e_4 \geq e_2$.
10. The rake angle of all the cutting inserts is zero. However, it can be increased to 5° in drilling of difficult-to-machine materials.
11. The primary clearance angle $e_{11}=10°–12°$, while secondary clearance angle $e_{12}=25°–20°$.

3.5 EJECTOR DRILLS

There is no ISO standard on ejector drills including tool design and geometry. However, it is covered by the German Technical Rule VDI 3209: 2019-3: Deep-hole boring systems with an external supply of coolant. It covers, at least partially, the practical design of the tools, the processing machines, and the necessary accessories.

The principle of ejector drilling was invented by Kurt Faber in 1963. US Patent No. 3,304,815 (1967) fully describes this principle. The prime objective of this invention was "to provide a

For manual tool change   For automatic tool change   With ISO/Morse taper

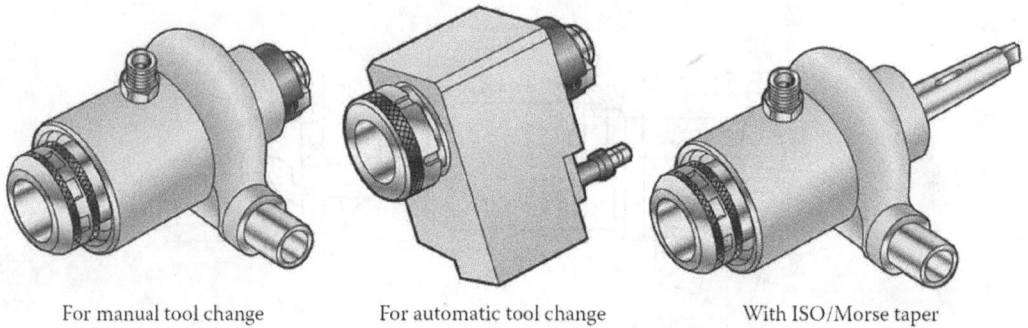

FIGURE 3.74 Adapters for ejector drills offered by Sandvik Coromant Co.

separate rearward action on the return flow of the flushing medium thereby lessening the pressure of this medium at the front of the drill and also lessening the tendency of breakage." As a result, ejector drilling does not always require a special drilling machine as it can be used in many general-purpose machines or even machining centers as one of the common drilling operations. To make it possible, Sandvik Coromant Co. offers a variety of ejector adapters (examples are shown in Figure 3.74) designed for different machines and their spindle particularities.

There are three basic myths about ejector drilling circulated in the trade (and unfortunately sometimes in scientific) literature and textbooks:

1. This type of deep-hole drilling is an alternative to STS drilling.
2. The pressure and flow rate needed for ejector drilling are much lower than those for STS drilling.
3. Ejector drilling is suitable only for a group of relatively easy-to-machine work materials.

None of these myths are true.

As discussed in Chapter 3 in Astakhov (2024), in practice of MWF systems, machine tools, and cutting tool design and implementation, the pressure of MWF is always considered to be of prime concern, while its flow rate is totally neglected. In the authors' opinion, this is one of the most severe misconceptions in MWF applications as it affects various facets of machining efficiency, productivity, and quality. The MWF flow rate is of prime concern, while its pressure is only a means to assure the flow rate needed to assure that MWF can perform its intended actions in machining (Astakhov 2012). As such, four basic actions of MWFs are commonly considered: (1) cooling, (2) lubricating, (3) chip transportation from the machining zone (or even from the working zone), and sometimes (4) chip control. All of these actions are directly defined by the MWF flow rate.

Although all of these actions are important, reliable chip removal is one of the first and foremost requirements to any deep-hole drilling application. This is particularly true for STS and ejector drilling. The MWF (coolant) is supplied to a deep-hole drill and then, after performing its cooling and lubricating actions in the machining zone, it carries away the chips in the form of swarf through the interior of the drill head and boring bar or inner tube. As a result, after the machining zone, a two-phase flow, that is, the chip–coolant mixture, should be considered. When the MWF flow rate is insufficient to transport the formed chip, the chip clogs the chip removal passages. Therefore, it is of prime concern to determine the sufficient flow rate.

The flow of the chip–coolant mixture through the chip removal passage may occur in different transportation modes. Figure 3.75 shows the influence of the MWF velocity on the chip transportation and the velocity profiles of the mixture in a tube located horizontally. When the MWF velocity is low (zone 1 in Figure 3.75), the MWF does not have any effect on the chip accumulated in the tube. As such, the velocity profile shows that the MWF moves, while the chip does not. Increasing the MWF

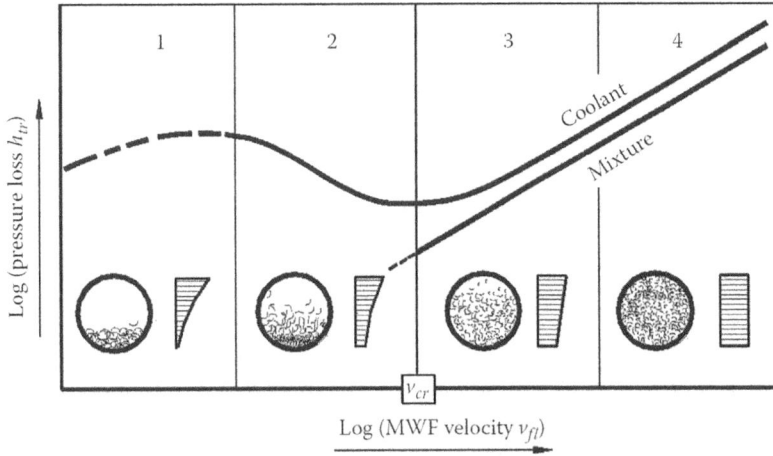

FIGURE 3.75   Flow modes and velocity profiles in a tube with chips.

velocity leads to its interaction with the chip (zone 2 in Figure 3.75). When it happens, a part of the chip layer moves with the MWF, whereas the other part forms a slow-moving (gliding) layer at the bottom. This transportation mode is called heterogeneous with a gliding layer. Because a lot of chips concentrate at the tube bottom, there is a possibility of formation of chip clogging. Such a mode, however, is considered to be acceptable in gundrilling where achieving the middle of the second zone is desirable. This is because high MWF pressure is needed in gundrilling due to small cross-sectional areas of the coolant channels to deliver the MWF flow rate corresponding to this mode.

Even further increase in the MWF velocity leads to the heterogeneous transportation mode where the coolant–chip mixture forms (zone 3 in Figure 3.75). When the MWF velocity becomes sufficient to reach this transportation mode, all the chip fragments flow in the mixture although the concentration profile is still not fully uniform as shown in Figure 3.75. The coolant–chip mixture velocity corresponding to the beginning of the heterogeneous mode is referred to as the critical velocity, v_{cr}. It is clear that the flow rate corresponding to this critical velocity called the critical MWF flow rate is calculated through the tube diameters and v_{cr} as discussed in Chapter 3 in (Astakhov 2024).

Although the further increase in the MWF flow rate leads to the pseudohomogeneous flow mode (zone 4 in Figure 3.75), the pressure losses in swarf transportation become significant. Moreover, achieving this regime requires a great coolant flow rate that is not always feasible in deep-hole drilling.

The foregoing analysis implies that no matter what type of drill, STS or ejector, is used, at least the minimum MWF flow rate should be delivered to achieve the critical velocity, v_{cr}, in the boring bar or inner tube to assure reliable chip transportation.

Figure 3.76 presents a side-by-side comparison of the STS and ejector drilling principles in terms of MWF supply parameters provided that the designs of the drilling heads, work materials, and machining parameters (the cutting speed and feed) used are the same. In other words, the shape and amounts of chip generated and thus to be transported out are the same for both drills. Being reasonable and practical, these assumptions simplify a MWF flow analysis in the drill shown in Figure 3.76.

The pressure and flow balance for the STS drill shown in Figure 3.76, namely, the outlet MWF flow rate Q_{fl-out}, are almost equal (except for possible minor leaks) to the inlet MWF flow rate Q_{fl-in}. As MWF from the boring bar flows into a tank with atmospheric pressure, the outlet pressure is equal to the atmospheric pressure and the inlet pressure is equal to the sum of the major and minor pressure losses calculated from the inlet hole of the pressure head to the exit of the rear end of the

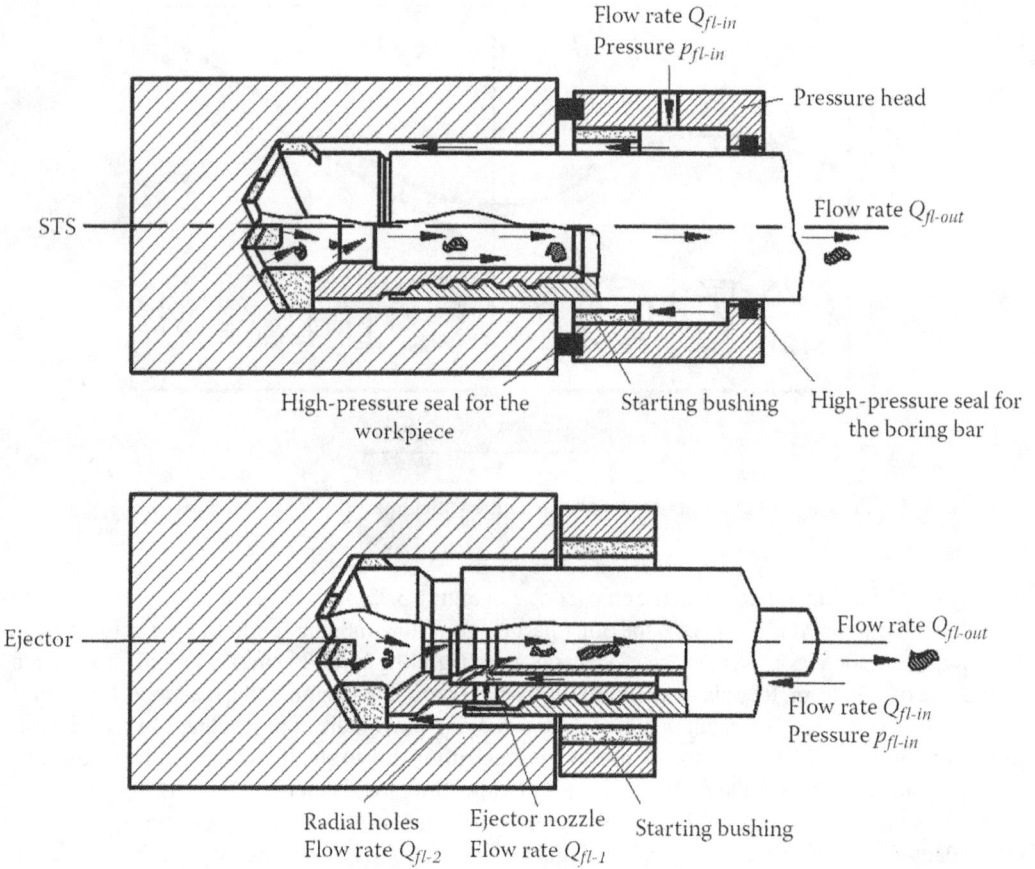

FIGURE 3.76 Side-by-side comparison of the STS and ejector drilling principles.

boring bar as discussed in Chapter 3 in Astakhov (2024) in great detail. The experimental results and calculations show that the MWF pressure losses in the annular gap between the boring bar and the wall of the hole being drilled are normally the highest out of major pressure losses (Astakhov 1995b). Moreover, this portion of the total pressure loss increases gradually as the drill goes deeper because the length of the annular gap increases proportionally. Because in high-penetration-rate STS drilling a great amount of chip is generated, the pressure losses in the mixture in its flow inside the boring bar are significant.

A device called the pressure head is used to deliver the MWF flow into the annular clearance between the boring bar and starting bushing and then the wall of the hole being drilled as drilling progresses. Although in the literature it is often depicted in the manner shown in Figure 3.76, the functions and the design of the pressure head are wider and more complicated.

The pressure head is a unit of the drilling machine. This unit

- Directs MWF into the drill;
- Mounts drill bushing;
- Contains locating bore and pilot for work holding components to provide support for the workpiece;
- Contains precision rotating spindle with high thrust and radial load capacity (for rotating workpieces);
- Contains boring bar seal packing glands.

Figure 3.77 shows an example of the pressure head design capable of delivering 10 MPa MWF pressure (Shertladse et al. 2006). Rotating spindle (1) is installed on the bearings mounted in the head housing (2). The workpiece locator (3) mounted on the front end of the spindle has a tapered surface to locate the workpiece (4). The taper contact between the workpiece and the locator assures reliable seal under the action of surface springs (5) and due to MWF pressure acting on the face of the locator. The labyrinth seal package (6) prevents MWF leaks between the spindle and the stationary unit (7) having the starting bushing (8) in it. MWF is supplied through the connector (9) and then through tangential holes (10). The directing flange (11) is to unify the coolant flow and thus prevent any MWF flow fluctuations. The seal (12) is for the boring bar.

Figure 3.78 shows the required MWF pressure and flow rate depending upon BTA (STS) drill diameter. Because of significant flow rates of MWF involved in STS drilling and high requirement to MWF clearness, a typical MWF supply system includes a high-volume tank having the volume equal

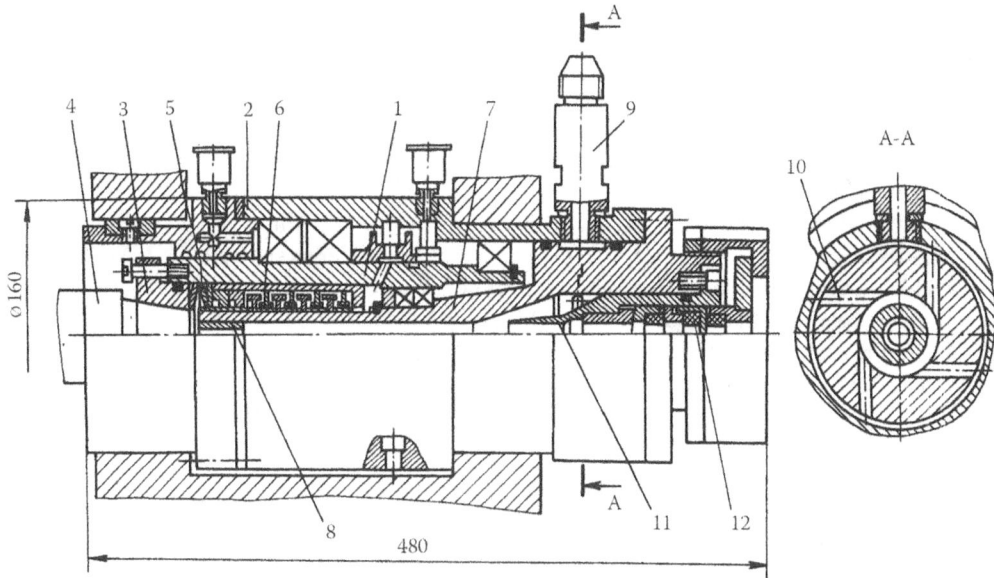

FIGURE 3.77   Pressure head.

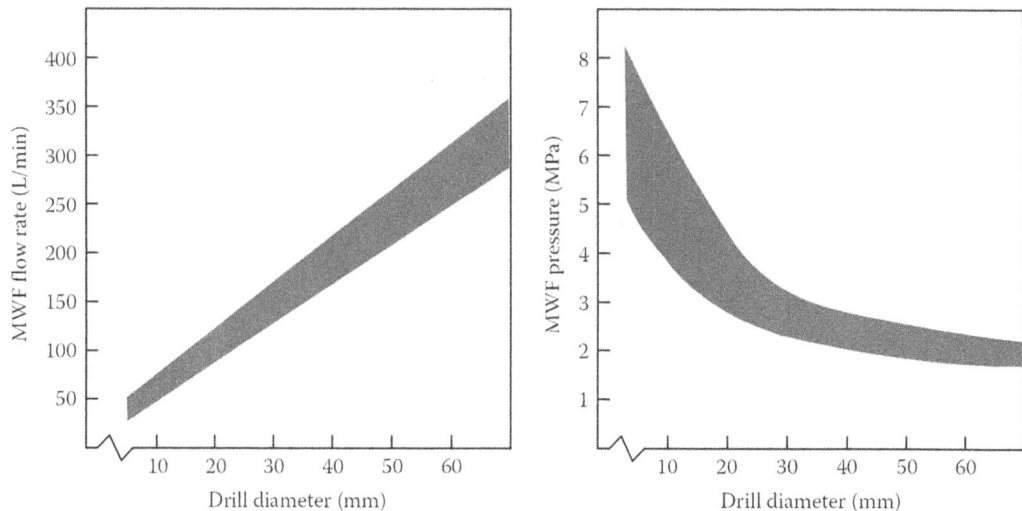

FIGURE 3.78   Pressure and flow rate of MWF as functions of STS drill diameter.

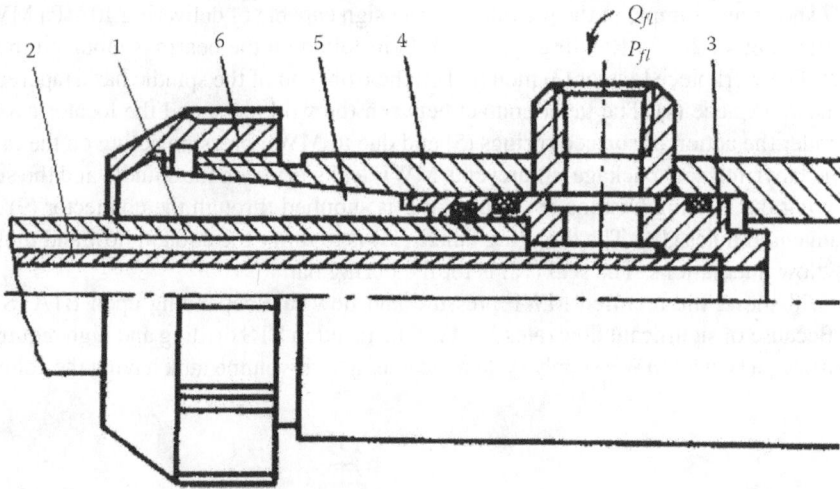

FIGURE 3.79 Basic design of the adapter for non-rotating ejector drills.

to the maximum MWF minute flow rate per 10 minutes (e.g., if the maximum flow rate is 200 L/min, then the tank volume should not be less than 2,000 L); array of MWF filters that commonly includes magnetic, centrifugal, and paper-cartridge filters; refrigerant chiller to stabilize the MWF temperature; and high-pressure pumps. As a result, the shop-floor area occupied by such a system often is the same or even larger than that occupied by the STS deep-hole machine itself (Astakhov 2012).

The pressure and flow balance for the ejector drill are shown in Figure 3.76. Similar to the STS system, the outlet MWF flow rate Q_{fl-out} is almost equal (except for possible minor leaks) to the inlet MWF flow rate Q_{fl-in}. This flow rate needed for chip transportation as well as for cooling and lubrication of the cutting and bearing contact areas is passed to the drill head by means of a two-tube system using an adapter (Figure 3.74).

Figure 3.79 shows the basic design of the adapter for non-rotating ejector drills. As can be seen, the outer (1) and inner (2) tubes are mounted into the sealing sleeve (3) installed into the adapter body (4). The precision collet (5) is tightened by the nut (5) to secure the outer tube in the adapter.

The design is simple and reliable and the whole unit is not expensive.

The outer tube referred to as the boring bar takes up the drilling torque and axial force, while the inner tube is rather thin. The rear end of the inner tube is installed in the drilling head. The front end is slightly conical forming an annular slit termed as the ejector nozzle. In the case shown in Figure 3.76, the ejector is the annular nozzle located in the drilling head as was in the original patent by Kurt Faber (US Patent No. 3,304,815 (1967)).

Once the flow of MWF supplied in the annular clearance between the boring bar and the inner tube reaches the drilling head, it is divided into two flows as indicated by arrows in Figure 3.76. The first flow termed as the ejector flow having the flow rate Q_{fl-1} passes through the ejector nozzle, and thus, it is directed rearwardly by this nozzle. The second flow having the flow rate Q_{fl-2} is often termed as working flow, passes through the radial holes in the drilling head, and then goes to the machining zone as indicated by arrows in Figure 3.76. The MWF flow around the ejector drill head as thought of by Sandvik Coromant Co. is shown in Figure 3.80. As can be seen, this second flow then returns with the formed chips rearwardly from the machining zone inside the drilling head and joins the flow ejected through the ejector nozzle. This happens because the first flow passing through the ejector nozzle lessens the pressure (compared to the atmospheric pressure) in the inner tube in front of the nozzle. After these two flows join again, they transport the chip over the inner tube similar to that in STS drilling.

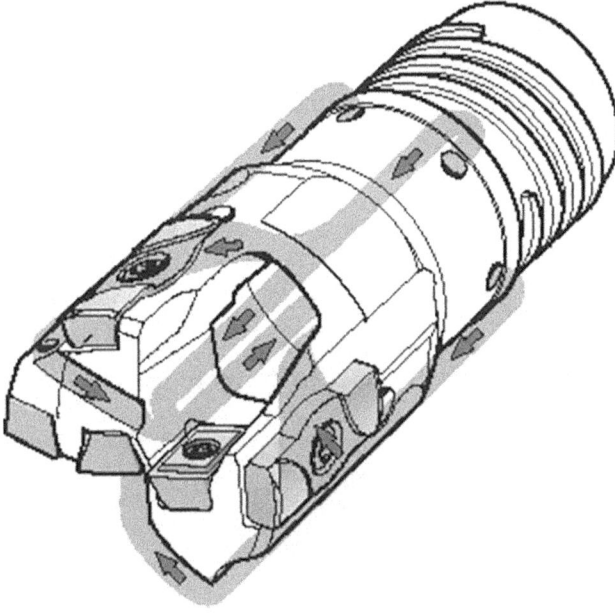

FIGURE 3.80 MWF flows around the ejector drill head.

The lessening of the pressure by an ejector is known as the ejector effect (Astakhov 1995b). When it is sufficient, no seal is needed between the face of the workpiece and the starting bushing. As a result, with simple adapters, the ejector drilling can be used on most standard machine tools and even on CNC machines as a hole-making operation, and moreover, it can potentially substitute STS drilling as was widely advertised when ejector drills became available. Although it is partially true, the practice of ejector drilling showed its limitations and the proper application range/conditions were gradually established in industry. To understand the limitations of ejector drilling and thus to establish its proper implementation conditions, one needs to understand the basic particularities of the ejector as a hydraulic apparatus.

As discussed by the author earlier (Astakhov et al. 1991, Astakhov 1995b, 1996), the ejector is characterized by three similarity numbers:

Relative Flow Rate

$$q_e = \frac{Q_{fl-2}}{Q_{fl-1}} \tag{3.43}$$

As explained previously, the flow rate Q_{fl-1} passes through the ejector nozzle and the flow rate Q_{fl-2} passes through the radial holes in the drilling head flowing to the machining zone and then sucked into the inner tube due to the ejector effect.

Relative Hydraulic Head

$$h_e = \frac{H_{fl-1}}{H_{fl-2}} \tag{3.44}$$

where

H_{fl-1} is the hydraulic head (see Chapter 3 in (Astakhov 2024) for the definition) of the flow that passes through the ejector nozzle;

H_{fl-1} is the hydraulic head of the flow just in front of the ejector nozzle(s).

Ejector Modulus

$$m_e = \frac{A_{en}}{A_{mc}} \tag{3.45}$$

where

A_{en} is the cross-sectional area of the ejector nozzle(s);

A_{mc} is the cross-sectional area of the mixing chamber where two flows are mixed.

In the considered case, it is the cross-sectional area of the inner tube, and the three similarity numbers define the hydraulic efficiency of the ejector as

$$\eta_e = q\frac{h}{1-h} \tag{3.46}$$

The experimental result show (Astakhov et al. 1991, Astakhov 1996) that the partial vacuum (the ejector effect) is reliably formed in the machining zone when the MWF pressure in front of the ejector nozzle is at least $0.3\ldots0.4\,\text{MPa}$. As such, the maximum vacuum even with the optimal ejector can be as high as $-0.06\,\text{MPa}$. Therefore, the inlet MWF pressure should be high enough to achieve this target accounting for the hydraulic resistance of MWF circuit from the adapter inlet to the ejector nozzle. Moreover, the ejector modulus defines the following relationship:

$$q_e = 0.205m_e - 0.007m_e^2 - 0.041 \tag{3.47}$$

That is, knowing m_e (which is the design parameter – see Eq. 3.45) and flow rate Q_{fl-1} passed through the ejector nozzle, the flow rate Q_{fl-2} can be determined. If the separation of Q_{fl} into Q_{fl-1} and Q_{fl-2} in the drilling head does not correspond to q_e according to Eq. (3.47), the following may happen:

- The flow rate Q_{fl-2} is greater than that determined by q_e according to Eq. (3.47). In this case, the excess of MWF would flow outside as the gap between the starting bushing and face of the workpiece is not sealed. As far as the flow rate that is sucked by the ejector is sufficient to transport the chip from the machining zone to the ejector nozzle region, there is no problem with drilling besides the waste of energy
- The flow rate Q_{fl-2} is smaller than that determined by q_e according to Eq. (3.47). In this case, the air will be sucked into machining through the same gap to compensate the difference. Note that the flow rate of the sucked air is much greater than the difference because the density of the air is much smaller than that of MWF.

The normal working regime of an ejector drill should be considered when *light* suction of the air takes place.

Figure 3.81 shows the efficiency of ejectors with annular nozzles as a function of the relative flow rate q_e and modulus m_e. As can be seen, the maximum efficiency is about 20% so as the conservation law implies no energy gain is made by installing the ejector in the drill. This efficiency is the price to pay for advertised advantages of ejector drills. It can be seen in Figure 3.81 that maximum efficiency is achieved when $m_e = 3.5$–4.0, while for commercial drills, $m_e > 10$.

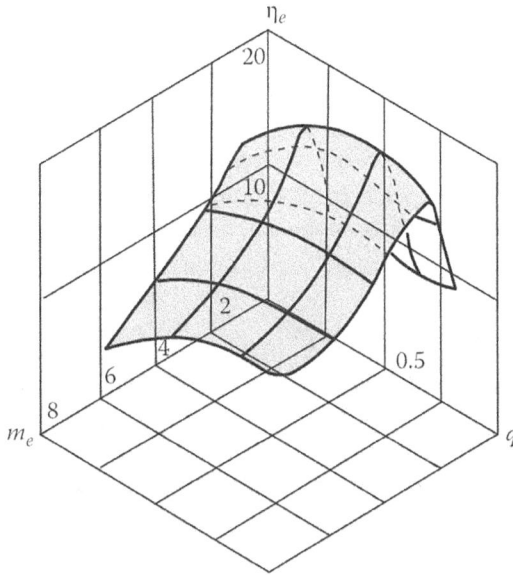

FIGURE 3.81 Influence of the relative flow rate q_e and modulus m_e on the energy efficiency of the ejector.

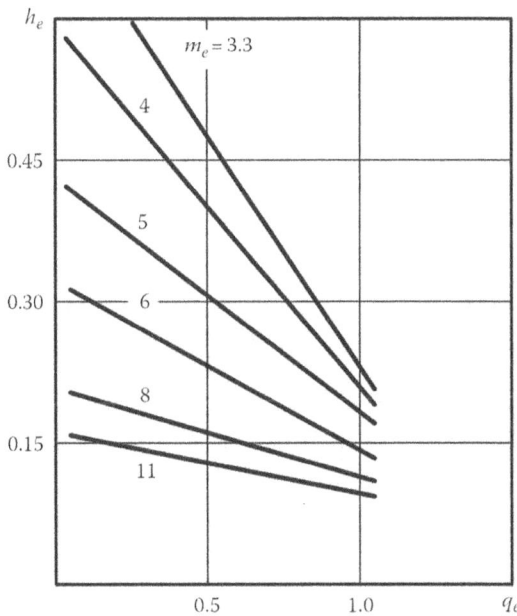

FIGURE 3.82 Generalized energy characteristics of ejectors with annular nozzles under maximum energy efficiency.

Based upon maximum efficiency, the general characteristic $h_e - f(q_e, m_e)$ is obtained as shown in Figure 3.82. As can be seen, the greater the modulus m_e, the less hydraulic head h_e can be generated by the ejector. For a given ejector drill, that is, for the given hydraulic head before the ejector nozzle, H_{ej} and its m_e, one can determine the hydraulic head H_{mx} of the chip–MWF mixture in the inner tube just after the ejector nozzle as

$$H_{ej} = hH_{mx} \tag{3.48}$$

or in terms of more convenient manometric pressure

$$\frac{p_{mx}}{\rho_{fl}} = h\frac{p_{ej}}{\rho_{mx}}$$

(3.49)

where ρ_{fl} and ρ_{mx} are the densities of MWF and MWF–chip mixture, respectively.

Knowing the hydraulic head H_{mx}, one can determine the maximum length of the inner tube under the condition of reliable chip transportation in a given ejector drill based upon the following energy condition:

$$H_{mx} \geq \pm\Delta h_{ver} + \Delta h_L + \Delta h_m$$

(3.50)

where Δh_{ver} is the difference in elevations between the outlet of the inner tube and the ejector nozzle accounted for only in vertical or inclined drills. As such, the sign "–" is used when the ejector nozzle is located below the outlet of the inner tube (the common case) and the sign "+" when the opposite is true; Δh_L is the major hydraulic loss as discussed in Chapter 3 in Astakhov (2024). One can calculate this loss as

$$\Delta h_L = \lambda\frac{l_{in-ej}}{d_{in}}\frac{Q^2_{mx}}{2gA^2_{in}}$$

(3.51)

where
   l_{in-ej} is the length of the inner tube from the ejector nozzle to its outlet;
   d_{in} is the diameter of the inner tube;
   Q_{mx} is the volumetric flow rate of the MWF–chip mixture (can be thought of as equal to the total flow rate of MWF supplied to the ejector drill, that is, $Q_{mx}=Q_{mx}$ can be accepted to the first approximation);
   A_{in} is the cross-sectional area of the inner tube.

λ is the D'Arcy coefficient often referred to as the friction factor. The values of this coefficient depending upon the flow conditions were presented by the author earlier (Astakhov 1996).

Δh_m is the major hydraulic loss as discussed in Chapter 3 in Astakhov (2024). One can calculate this loss as

$$\Delta h_m = K_{fr}\frac{Q^2_{mx}}{2gA^2_{in}}$$

(3.52)

where K_{fr} is the flow resistance coefficient accounted for only in rotating ejector drills. The values of this coefficient depending upon the flow conditions were presented by the author earlier (Astakhov 1996).

The hydraulic calculation has shown that the length l_{in-ej} cannot exceed approx. 2 m when the ejector nozzle is located in the drilling head as shown in Figure 3.76. This limits the allowable length of ejector drill to an approx. 1.8 m, while STS drills do not have this limitation. Therefore, the notion that the ejector drill can be considered as an alternative to STS drills does not have any ground.

The MWF flow rate delivered to the ejector drill Q_{fl} should be greater than that for the STS drill for normal functioning of the ejector and reliable chip removal from the machining zone to the ejector nozzle. Because the cross-sectional area of the annular clearance between the inner and outer tubes used for MWF delivery into the drilling head is much smaller than the cross-sectional area of the annular clearance between the outer tube (the boring bar) and the wall of the hole being drilled, the pressure needed to deliver even the same as for STS drill flow rate to the ejector drill is higher for the same drill length when the ejector nozzle is located in the drilling head.

The notion that the ejector drill required less inlet pressure stems from the alternative (to that shown in Figure 3.76 and originally patented by Faber) ejector nozzle(s) location in some designs of modern ejector drills, for example, by Sandvik Coromant Co. This location in the rear end of the inner tube in standard ejector drills is shown in Figure 3.83. Figure 3.84 shows such an ejector drill in operation. The same as discussed previously, the workpiece (1) is drilled by the drilling head (2) located at the end of the outer tube (3). The inner (4) tubes are installed in the drilling head and in the connector (5) to form the annular MWF channel. The difference is, however, that the total flow rate Q_{fl} is divided into two, not in the drilling head but in the adapter. The first flow termed as the ejector flow having the flow rate Q_{fl-1} passes through the ejector nozzles located at the rear end of the inner tube, and thus, it is directed rearwardly by these nozzles as indicated by the arrows in the figure. The second flow having the flow rate Q_{fl-2} passes through annular clearance between the outer and inner tubes to the drilling head, then it passes through the radial holes in the drilling head and then goes to the machining zone as indicated by the arrows in Figure 3.84. Due to the ejector effect created by the ejector nozzle in the inner tube, this flow rate is sucked from the machining zone into the inner tube providing the transportation of the chip as shown in the picture.

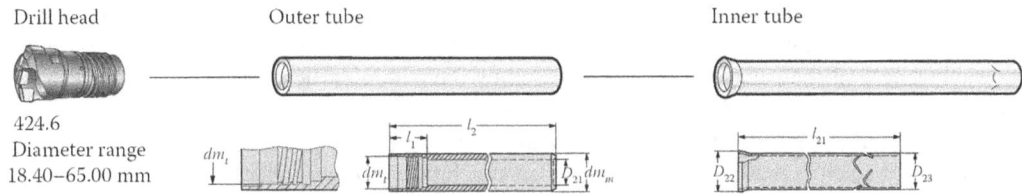

FIGURE 3.83 Ejector nozzles location in the rear end of the inner tube in standard ejector drills by Sandvik Coromant Co.

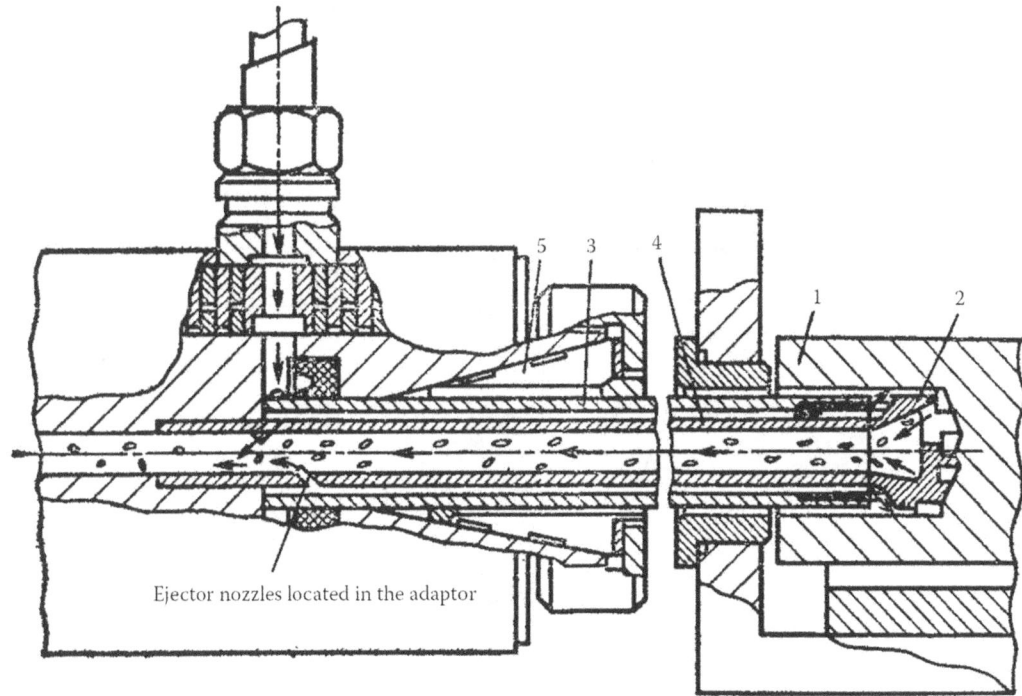

FIGURE 3.84 The ejector nozzles are located in the adapter according to the design of the inner tube shown in Figure 3.83.

The obvious advertised advantage of such nozzle location is that only small portion Q_{fl-2} of the total flow rate should pass through really narrow clearance between the outer and inner tubes that require a much smaller inlet pressure for the normal operation of ejector drills. Moreover, as the inlet pressure is not a concern, the relative flow rate q_e can be made optimal so the hydraulic head produced by the ejector can be maximized according to Figure 3.82. However, this great hydraulic head is not needed anymore after the adapter as the inner tube ends so that there is no more hydraulic resistance to overcome.

The second concern with such a design is far more important. As discussed previously, the maximum vacuum even with the optimal ejector can be as high as −0.06 MPa. In other words, only 0.06 MPa times the density of MWF–chip mixture is available to overcome the pressure loss in the inner tube over the distance from the machining zone to the ejector nozzles. This significantly restricts the maximum allowable length of the ejector drills of such a design. This also explains why this length does not exceed one meter for the standard ejector drills.

In the description of the original patent on the ejector drill (US Patent No. 3,304,815 (1967)), Kurt Faber, the inventor, suggested that "for improving the ejector effect, especially in long drills, one or more extra ejector devised may be provided along the drill." The foregoing analysis, however, suggests that as far as the flow rate for an additional ejector is taken from that supplied to the ejector drill, no gains can be achieved in terms of increasing the ejector effect and thus increasing the allowable length of the ejector drill. As discussed by the author earlier, some improvement in the chip removal from the ejector drills is achieved when an additional ejector with an independent supply of MWF is used (Astakhov 1995b). Such an additional ejector can be embedded in the design of the adapter for ejectors drill or can be made as an attachment to the standard adapters as discussed by the author earlier in Chapter 5 in Astakhov (2014).

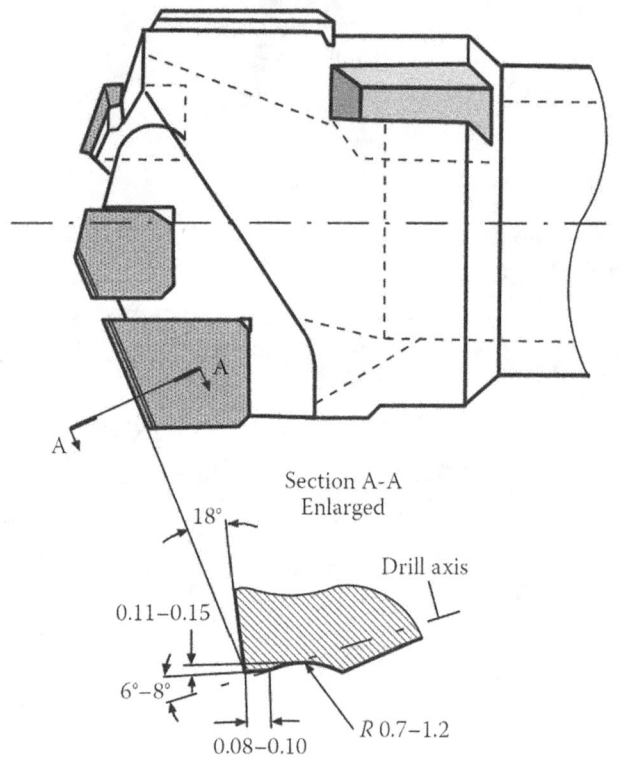

FIGURE 3.85   Chip breaker for ejector drill for drilling difficult-to-machine materials.

The last myth that ejector drilling is suitable only for a group of relatively easy-to-machine work materials stems from the use of the standard chip breakers initially developed for STS drilling (shown as chip-breaking steps in Figures 3.61, 3.62, and 3.65). Such a chip breaker works well for STS drill where the MWF flow rate through the machining zone is great, and if, by any chance, a small chip clogging occurs say in the chip mouth, the external pressure increases to push the clogged chip through. As this is not the case in ejector drilling, much more intelligent handling of chip breaking is needed for ejector drills. The author has experience with successful productive ejector drilling (20 and 40 mm hole diameter and 1.2 m length) of soft austenitic stainless steel (AISI 321) using the chip breaker design shown in Figure 3.85.

REFERENCES

(1910 (Reprinted in 2001)). *Deep Hole Drilling*. Kankakee, IL, Reprinted by Lindsay Publications Inc.
Astakhov, V. P. (2001). "Gundrilling know how." *Cutting Tool Engineering* **52**: 34–38.
Astakhov, V. P. (2002a). "The mechanisms of bell mouth formation in gundrilling when the drill rotates and the workpiece is stationary. Part 1: The first stage of drill entrance." *International Journal of Machine Tools and Manufacture* **42**: 1135–1144.
Astakhov, V. P. (2002b). "The mechanisms of bell mouth formation in gundrilling when the drill rotates and the workpiece is stationary. Part 2: The second stage of drill entrance." *International Journal of Machine Tools and Manufacture* **42**: 145–152.
Astakhov, V. P. (2004a). "The assessment of cutting tool wear." *International Journal of Machine Tools and Manufacture* **44**: 637–647.
Astakhov, V. P. (2004b). "High-penetration rate gundrilling for the automotive industry: System outlook." SME Paper TPO4PUB249: 1–20.
Astakhov, V. P. (2010). *Geometry of Single-Point Turning Tools and Drills: Fundamentals and Practical Applications*. London, Springer.
Astakhov, V. P. (2012). Chapter 1. Design of experiment methods in manufacturing: Basics and practical applications. *Statistical and Computational Techniques in Manufacturing*. J. P. Davim. London, Springer: 1–54.
Astakhov, V. P. (2014). *Drills: Science and Technology of Advanced Operations*. Boca Raton, FL, CRC Press.
Astakhov, V. P. (2024). *High-Productivity Drilling Tools: Materials, Metrology and Failure Analysis*. Boca Raton, FL, CRC Press.
Astakhov, V. P., Abi Karam, S., Osman, M.O.M. (1998). "The static and dynamic pressure distribution in the machining zone of BTA drills." *Journal of Manufacturing Science and Engineering* **120**(4): 820–822.
Astakhov, V. P., Frazao, J., Osman, M.O.M. (1991). "On the design of deep-hole drills with non-traditional ejectors." *International Journal of Production Research* **28**(11): 2297–2311.
Astakhov, V. P., Galitsky, V. (2005). "Tool life testing in gundrilling: An application of the group method of data handling (GMDH)." *International Journal of Machine Tools and Manufacture* **45**: 509–517.
Astakhov, V. P., Galitsky, V.V. (2006). "The combined influence of various design and process parameters of gundrilling on tool life: Experimental analysis and optimization." *The International Journal of Advanced Manufacturing Technology* **36**(9–10): 852–864.
Astakhov, V. P., Galitsky, V.V., Osman, M.O.M. (1995a). "A novel approach to the design of self-piloting drills. Part 1. Geometry of the cutting tip and grinding process." *ASME Journal of Engeneering for Industry* **117**: 453–463.
Astakhov, V.P., Joksch, S. (2012). *Metal Working Fluids: Fundamentals and Recent Advances*. Cambridge, UK, Woodhead Publishing Limited.
Astakhov, V. P., Osman, M.O.M. (1996a). "An analytical evaluation of the cutting forces in self-piloting drilling using the model of shear zone with parallel boundaries. Part 1: Theory." *International Journal of Machine Tools and Manufacture* **36**(11): 1187–1200.
Astakhov, V. P., Osman, M.O.M. (1996b). "An analytical evaluation of the cutting forces in self-piloting drilling using the model of shear zone with parallel boundaries. Part 2: Application." *International Journal of Machine Tools and Manufacture* **36**(12): 1335–1345.
Astakhov, V. P., Outeiro, J. (2019). Importance of temperature in metal cutting and its proper measurement/modeling. *Measurement in Machining and Tribology*. P. J. Davim. London, Springer: 1–47.
Astakhov, V. P., Shvets, S.V. (2001). "A novel approach to operating force evaluation in high strain rate metal-deforming technological processes." *Journal of Materials Processing Technology* **117**(1–2): 226–237.

Astakhov, V. P., Subramanya, P. S., Osman, M.O.M. (1995b). "Theoretical and experimental investigations of coolant flow in inlet channels of the BTA and Ejector drills." *Journal of Engineering Manufacture, Part B, Proceedings of the Institution of Mechanical Engineers* **209**: 211–220.

Astakhov, V. P., Subramanya, P.S., Osman, M.O.M. (1995c). "An investigation of the cutting fluid flow in self-piloting drills." *International Journal of Machine Tools and Manufacture* **35**(4): 547–563.

Astakhov, V.P., Subramanya, P. S., Osman, M.O.M. (1996). "On the design of deep hole drills with ejectors." *International Journal of Machine Tools and Manufacture* **36**(2): 155–171.

Bescrovny, A. M. (1984). "On the location of the supporting pads in deep-hole tools (in Russian)." *Rezanie i Instrument* **31**: 99–103.

Griffiths, B. J. (1993). "Modeling complex force system, Part 1: The cutting and pad forces in deep drilling." *ASME Transactions, Journal of Engineering for Industry* **115**: 169–176.

Griffiths, B. J., Grieve, R.J. (1985). "The role of the burnishing pads in the mechanics of the deep drilling process." *International Journal of Production Research* **23**: 647–655.

König, W., Klocke, F. (2008). *Fertigungsverfahren Band 1: Drehen, Fräsen, Bohren.* Berlin, Springer.

Landis, A. B. (1886). "Deep holes by continuous boring." *American Machinist* **27**: 4.

Nikolov, S. (1986). "Improving quality with deep-hole tools (in Bulgarian)." *Mashinostroenie* (4): 1986.

Ramakrishana, R. K., Shunmugam, M.S. (1988). "Wear studies in boring trepanning association drilling." *Wear* **124**: 33–43.

Richardson, R., Bhatti, R. (2001). "A review of research into the role of guide pads in bta deep-hole machining." *Journal of Materials Processing Technology* **110**(1): 61–69.

Sakuma, K., Taguchi, K., Katsuki, A. (1980). "Study on deep-hole-drilling with solid-boring tool: The burnishing action of guide pads and their influence on hole accuracies." *Bulletin of the JSME* **23**(185): 1921–1928.

Sakuma, K., Taguchi, K., Katsuki, A., Takeyama, H. (1981). "Self-guiding action of deep hole drilling tools." *Annals of the CIRP* **30**(1): 311–315.

Shertladse, A. G., Ivanov, V.I., Kareev, V.N. (2006). *Hydraulic and Pneumatic Systems (in Russian).* Moskow, Vishyja Shkola.

Stockert, R., Weber, U. (1977). "Untersuchung der energie-verhaltnisse beim tiefbohren mit einschneidigen BTA-Vollbohropfen." *Industrie-Anzeiger* **99**: 26.

Swinehart, H. J., Ed. (1967). *Gundrilling, Trepanning, and Deep Hole Machning.* Dearborn, MI, SME.

Veremachuk, E. S. (1940). Deep Hole Drilling (in Russian). Moscow, State Publishing House of Defence Industry.

Vinogradov, A. A. (1985). *Physical Foundation of the Drilling of Difficult-to-Machine Materials with Carbide Drills (in Russian).* Kyiv (Ukraine), Naukova Dumka.

Zhang, W., He, F., Xiong, D. (2004). "Gundrill life improvement for deep-hole drilling on manganese steel." *International Journal of Machine Tools and Manufacture* **44**(2–3): 237–331.

4 PCD Drilling Tools

Diamonds are not forever.

An experienced user

No pressure, no diamonds.

Thomas Carlyle (1795–1881) a Scottish essayist, historian and philosopher

4.1 CHALLENGES OF WORK MATERIALS – WHY DIAMOND AS A TOOL MATERIAL IS NEEDED?

Many work materials with enhanced special properties have been developed in modern industry to meet the ever-growing application requirements/conditions. Among these materials, composite materials occupy one of the leading places. A composite material is a combination of two materials with different physical and chemical properties, which do not chemically react (for example, sodium and chloride react to form sodium chloride commonly known as salt whereas sand and water do not create a chemical compound) or form solid solutions phased (for example, pearlite in steels). When they are combined, they create a material with improved properties, for instance, increased specific strength. The reason for their use over traditional materials is they improve the properties of their base materials and are applicable in many situations. Concrete is the most common artificial composite material of all and typically consists of loose stones (aggregate) held with a matrix of cement. When one looks at the structure of concrete under a microscope, he or she can easily distinguish stones and the surrounding cement matrix.

There are two distinctive groups of composite work materials: metal and polymer based. The former is known as metal-matrix composites (MMCs) whereas the latter are polymer-based composite materials mainly carbon fiber-reinforced plastics (CFRP). Both groups present challenges in their machining particularly when high quality of the machined parts and high productivity and efficiency are required.

4.1.1 MMCs

MMCs offer a high strength-to-weight ratio, high stiffness, and good damage resistance over a wide range of operating conditions, making them an attractive option in replacing conventional materials for many engineering applications. Typically, MMCs are aluminum, titanium, copper, and magnesium alloys, while the reinforcement materials are silicon carbide, aluminum oxide, boron carbide, graphite, etc., in the form of fibers, whiskers, and particles. If one pays close attention to the Al–Si phase diagram and micrographs (Astakhov and Patel 2018), then he or she can conclude that although HSAA is termed as alloys (consisting of solid, element-indistinguishable, phases) in all technical documents, standards, and research papers, they are not actually alloys but rather MMCs.

Aluminum die-casting alloys are lightweight and offer good corrosion resistance, ease of casting, good mechanical properties, and dimensional stability. They are widely used as foundry alloys for a variety of different applications. For example, engine blocks and pistons for air compressors employed in the automotive industry are cast from Al–Si-based alloys. Casting alloys are distinguished from wrought alloys that contain 95% or more aluminum and are not used for castings but are used for applications such as can stock, gutters, siding, airplane skins, and so on. In the

DOI 10.1201/9781003263296-4

automotive industry, transmission and engine components are made of high-silicon-aluminum alloys (hereafter, HSAAs), which have a high strength-to-weight ratio.

Aluminum is cast at a temperature of 650°C (1,200°F). It is alloyed with silicon (9%) and copper (3.5%) to form the Aluminum Association 380 alloy (UNS A03800). Silicon increases the melt fluidity and reduces machinability. Copper increases hardness and reduces ductility. By greatly reducing the amount of copper (less than 0.6%), the chemical resistance is improved making AA 360 (UNS A03600) well-suited for use in marine environments and in automotive transmissions (valve bodies, case, and torque converter housings). HSAAs with more than 13% Si are used in automotive transmissions (pump cover) and engines (for cylinder castings).

Machining of HSAA presents a great challenge due to their unique properties (Tomac and Tonnessen 1992, Hung et al. 1996, El-Gallab and Sklad 1998, Andrewesa et al. 2000). The real reason for that is discussed by Astakhov and Patel (2018). This reason is that HSAA includes two distinctive components: (1) aluminum as the matrix metal and (2) silicon as reinforcement, each of which has its own microstructure and interfaces between them. The former causes adhesive wear of carbide cutting tools whereas the latter causes abrasive wear.

A simplified mechanism of adhesive wear can be represented as follows. Strong adhesion takes place between surfaces free of oxides under high contact pressure. The harder the contact pressure the rougher the tool contact surface, the stronger the adhesion bonding. Due to the high plastic deformation of the chip and high contact pressures at the tool–chip and tool–workpiece interface, MWF cannot penetrate into these interfaces (Astakhov and Joksch 2012) regardless of its extreme pressure and antiwear additives in it. As a result of adhesion of the work material (converted into the chip) and the tool rake face, the so-called built-up edge (commonly referred to as BUE in the literature on metal machining) is formed. BUE is not stable in metal cutting so it changes within each cycle of chip formation (Astakhov 1998/1999). As BUE adheres to the rake face, the adhesion causes mechanical bonding (as glue with a piece of paper). When BUE is periodically removed by the moving chip (as its height becomes sufficient), it brings a small piece of the tool material with it (as the glue is removed from the paper). The process repeats itself frequently so the tool becomes worn by this process. As adhesive wear progresses, the rake face becomes rougher and the cutting edge more round so the extent and height of BUE grows as the strength of adhesion bonds increases. This may cause local chipping or even bulk breakage of the tool (Astakhov 2014). All this causes rapid degradation of the cutting ability of the carbide cutting tools.

Abrasive wear is caused by hard particles in the HSAA structure as they are mainly silicon particles. This is because aluminum has zero solid solubility in silicon at room temperature so that Al–Si phase diagram differs from the "standard" phase diagrams. There is no beta phase and so this phase is "replaced" by pure silicon particles, which are non-lamellar in form and appear in micrographs as separate flakes although studies have shown that the flakes are, in fact, interconnected three-dimensionally. Moreover, in reality, real die-casting HSAAs are supplied to automotive plants containing clusters of sludge. The presence of this sludge and silicon particles makes HSAA highly abrasive that causes abrasive tool wear. As a result of this issue, tool life is significantly shortened causing high tooling cost per unit (a machined part). Even when the most advanced nanocoatings are used, the silicon carbide fibers quickly dull the cutting edge of the tungsten carbide tooling. Therefore, when it comes to HSAA drilling operations, tungsten carbide tools are used only for roughing operations and when high accuracy of machined holes is not required.

4.1.2   CFRP

Fiber-reinforced polymer (FRP) composites including CFRP are a class of materials that offer numerous advantages over monolithic metals and other homogeneous materials. CFRPs consist of a wide range of composite materials with different fiber types, fiber orientation, fiber content, and matrix materials. Due to their greater strength-to-weight ratio, the composites are widely used in various structures and components. The aerospace industry is making a major effort to incorporate an increasing number of composite materials into various components and structures. Each Boeing 787 Dreamliner, for example, is made up of (by weight) 50% composite, 20% aluminum, 15% titanium, 10% steel, and 5% other materials. By volume, the aircraft is 80% composite. Each 787 contains approx. 35 tons of CFRP, a good portion of it stacked in conjunction with aluminum or titanium alloys (Meguid and Sun 2005). Aircraft designers and manufacturers like stack materials because they combine the high strength of metals with the low weight and corrosion resistance of composites. The variety of stack materials is increasing nearly as fast as the applications, replacing aluminum honeycomb materials, which consist of honeycomb paper sandwiched between layers of aluminum. Often, stacks have various layers of composite, or a composite/metal stack combination, or foam or other core materials and are then wrapped with composites. Some stack materials may also incorporate a thin copper mesh designed to protect against lightning strikes. Overall stack thicknesses can vary from less than 6 to 60 mm (Destefani 2002).

As the use of such materials has been steadily increasing in many industries, there is an increasing need for a cost-effective method of producing high-quality holes in such materials with dimensions that are within narrow tolerances. Delamination generally represents the main concern when drilling composites (see Chapter 2, Figure 2.32), because of the lowering of fatigue strength as well as a poor assembly tolerance. The thrust force has been widely cited by various authors as being responsible for delamination; thus, the drill geometry should be developed to reduce the axial (thrust) force, thus reducing delamination. Drill design and application engineers rarely see anything but test panels and for proprietary reasons may have only a generic idea of the materials making up the stack so that the old-fashioned trial-and-error method combined with experience is used mostly in the development of drills for such applications. Various drill geometries and tool materials have been tested for the drilling of composite materials. Polycrystalline diamond (PCD) drills found wide application in drilling various FRPs.

It was found that in drilling of FRP almost pure abrasion wear to be the case as carbon fibers are very abrasive. Moreover, these fibers are loosely supported by a relatively soft matrix so that delamination as fiber peel-up at the drill entrance and push-out at the exit is of prime concern. Heat generated in drilling operations causes the softening and re-solidification of the matrix material that has different thermal properties than the fiber is another problem that occurs in drilling of FRP. The low thermal conductivity and sharp temperature gradients in FRP lead to thermal damages and burning of the matrix.

4.1.3   DIAMOND TOOL AS THE MOST FEASIBLE OPTION

The most feasible way to solve the above-discussed problems with drilling of MMC and FRP is the application of diamond tools. The use of the PCD tool material is beneficial in drilling both MMC and FRP as almost pure abrasion wear was found to be the case in drilling such materials. This, however, is the only similarity, while there are a number of differences. The major differences are as follows:

- The chip in drilling FRP is powderlike so that its removal does not present a problem. In drilling MMC, a great amount of the chip is formed, which should be transported over the chip flutes out of the hole being drilled.
- Aggressive drill geometry is normally used for FRP to reduce the drilling torque and axial force to reduce delamination. MMCs are generally much stronger so that the drilling torque and axial force are much higher, which imposes restrictions on the drill design.
- The drilling temperature is normally much greater in drilling MMC as the formed chip is heavily deformed due to highly ductile matrix material and due to friction of this formed chip in its sliding over the tool–chip interface.
- MMCs possess the low elasticity modulus and high thermal conductivity that causes great spring back of the wall of the hole being drilled and thus friction on the drill margins.

- The need to cool down the machining zone and facilitate chip transportation calls for the use of a high-pressure internal MWF (coolant) supply in drilling MMCs, while MWF is not normally used in drilling FRPs.
- The tolerances on holes (diameter, shape, location, etc.) are normally much tighter in drilling MMCs.

The listed differences clearly indicate that a PCD drill used with great success for drilling FRP cannot be *automatically* applied for drilling MMCs although a number of such attempts are known. They are discussed later in this chapter.

4.1.4 APPLICATION PARTICULARITIES

Over the last few decades, unattended machining centers and manufacturing cells have been rapidly developed for factory use in the automotive industry. As well understood by specialists in the automotive industry, the principal difficulty preventing automated, unattended, and around-the-clock operations is the cutting tool. Short tool life, a great scatter in tool life, lack of reliable data, and the lack of effective sensors to monitor unmanned production systems are major contributors to the problem. Worn or fractured tools result in the manufacture of products outside the specifications and significant scrap of almost finished parts. Moreover, breakage of the cutting tool can cause damage to the machine tool itself. This issue is particularly problematic in high-cost ceramic-bearing spindles. These factors invariably lead to increases in manufacturing costs, loss of manufacturing capacity, and unnecessary use of energy, materials, and labor.

Among the five basic types of tool materials used in the modern industry, PCD tools have rapidly advanced in the machining of aluminum alloy die castings in the automotive industry, machining of difficult-to-machine materials in the aerospace industry and biomedical engineering (Coelho et al. 1995, Li et al. 2020). Today, practically all finishing *hole* operations and all flat surfaces of the aluminum alloy engine parts and parts of transmissions are machined using these tools. PCDs allow high-speed aluminum machining that increases productivity, efficiency, and quality of the machined part. The high efficiency of these tools comes with a high cost. The costs of PCD tools are three to five times higher than that of carbide tools for similar applications in the automotive industry, and PCD tools are highly sensitive to any inaccuracy (runout, alignments, etc.) in machine tools; thus, special machine tools are required for their effective use. However, when a PCD tool is made and run properly, its tool life is up to tenfold higher than that of the best (the best carbide grade and optimized tool design) carbide not to mention mirror-shining surface finish of the machined holes and ability to maintain tight tolerances over the whole tool lifetime. The latter cannot be achieved with cemented carbide tools even in principle. In modern setting of HSM, the difference in the performance of cemented carbide and PCD tools is even greater. As a result, PCD tools are widely used in the automotive industry.

4.2 DIAMOND: WORD ORIGIN AND EARLY HISTORY

The name *diamond* is derived from the ancient Greek word αδάμας (*adámas*), meaning *proper, unalterable, unbreakable,* and *untamed,* and from d– (a–), *un–* + δαμάω (*damáō*), meaning *I overpower* and *I tame.* Naturally occurring diamond is a relatively rare polymorph of carbon characterized by a 3D arrangement of tetrahedrally coordinated carbon atoms. On the Earth's surface, diamonds occur in several major kinds of deposits: primary and secondary (both alluvial and littoral). In primary deposits, they are enclosed in host rocks of kimberlite or lamproite that form *pipes* that are downward-tapering, cone-shaped structures of igneous (volcanic) origin. Subsequent erosion of these pipes and fluvial transport of the diamonds lead to the formation of alluvial deposits in river beds and river terraces. The final resting place of diamonds is in littoral or ocean floor deposits.

Separation of diamonds from their enclosing or associated rocks includes crushing, screening, and sieving, use of grease belts, suspension in heavy or dense media, and sorters that use the luminescence of diamonds when they are exposed to x-rays. Diamond crystals are sorted into many categories depending on size, color, clarity, and shape for valuation purposes. Both natural and synthesized diamonds have the highest hardness of all known materials, the highest thermal conductivity at room temperature, a high refractive index and optical dispersion, a low thermal expansion, and a relatively high inertness to chemical attack. This unique combination of properties permits diamond to be foremost in certain applications: as a highly prized gemstone; industrially as an important abrasive unsurpassed in certain cutting, drilling, sawing, machining, grinding, and polishing operations for many materials; and in electronic and optical applications as a heat sink and window material, respectively.

Diamonds were first discovered in ancient times in India and Borneo and later in Brazil in the early 1700s, in alluvial deposits where flowing water had sorted minerals on the basis of density and toughness. These alluvial concentrations of diamonds are derived from primary source rocks by erosion and fluvial transport, where the tumbling action in the transported sediments often concentrates on the better-quality diamond crystals by preferentially destroying flawed ones. These processes have occurred over long periods of geologic time in all rivers draining primary sources, and particularly good-quality alluvial diamonds are found in rivers in Angola, Brazil, and India and in areas of southern and western Africa. The Orange River system in South Africa has transported diamonds from inland primary sources to the Atlantic Ocean, where ocean currents have distributed the diamonds along both the littoral zone and the ocean floor off of South Africa and Namibia. Exploration of alluvial deposits can be done by panning of stream sediments or by drilling, pitting, and trenching of streambed and terrace deposits, in conjunction with a search for the heavy mineral assemblages that accompany diamond. Recovery of diamonds from the ocean floor requires the use of sophisticated underwater equipment to collect the diamonds and return them to a processing ship. Alluvial deposits account for approx. 20% by weight (but 41% by value) of the world's annual production of diamonds.

In 1772, Antoine Lavoisier, the great French chemist, pooled resources with other chemists to buy a diamond, which they placed in a closed glass jar. They focused the sun's rays on the diamond with a remarkable magnifying glass and saw the diamond burn and disappear. Lavoisier noted that the overall weight of the jar was unchanged and that when it burned, the diamond had combined with oxygen to form carbon dioxide (Krebs 2006). He concluded that diamond and charcoal were made of the same element – carbon. In 1779, Carl (Karl) Scheele showed that graphite burned to form carbon dioxide and so must be another form of carbon (Partington 1962). In 1796, the English chemist Smithson Tennant established that diamond was pure carbon and not a compound of carbon; it burned to form only carbon dioxide. Tennant also proved that when equal weights of charcoal and diamonds were burned, they produced the same amount of carbon dioxide (Barnard 2000). Henry and Lawrence found carbon to embody three separate structures: cubic, hexagonal, and amorphous. The composition of natural diamond actually consists of approx. 99% carbon 12 and 1% carbon 13 (Davis 1993).

Whether society's infatuation with diamonds rests in its inherent beauty, its list of useful industrial properties, or because of the love, excellence, and purity it has come to represent, our society undeniably values these diamonds with great worth. However, in this technologically driven age, our values come into question as scientists introduce synthetic diamonds into the market. In other words, diamonds like bronze and steel, which were once a scarce commodity, have now become ordinary, reproducible materials. While the news of such may dismay some jewelers and gem connoisseurs, contrasting methods for creating diamonds may also encourage innovation from other industries as they can begin utilizing diamond's material properties. Regardless of whether or not we positively welcome their arrival, synthetic diamonds already have, are, and will invariably shape our world and thereby require further examination.

In order to replicate a diamond, the properties of the diamond itself must be examined. Diamonds are in part so valuable because of the sheer number of useful properties they possess. Spears and Dismukes comment on their superior properties saying, "Choose virtually any characteristic property of a material – structural, electrical, or optical – and the value associated with diamond will almost always represent an extremist position among all materials considered for that property" (Spear and Dismukes 1994). Simply stated, in addition to the aesthetic appeal of diamonds, an array of practical uses also exists. Transparent and lustrous, they catch the eye with their unique appearance and magnificent beauty. This may mislead the viewer to also believe they are delicate; however, diamond, an allotrope of carbon, is also the hardest natural material on Earth.

4.3 STRUCTURE AND IMPORTANT PROPERTIES

Figure 4.1 shows a simplified phase diagram of carbon. It shows that diamond is a stable form of carbon at very high pressures and temperatures. Under ordinary conditions for temperature and pressure, near 0.1 MPa and room temperature, diamond may be considered a metastable form of carbon. Even though it is not at the minimum energy state, it does not spontaneously convert to graphite (Asmussen and Reinhard 2002).

4.3.1 STRUCTURE AND BONDING IN GRAPHITE AND DIAMOND

Bonding and structures of graphite and diamond are briefly considered in this section because it is of high importance in diamond quality control using the Raman spectroscopy as discussed later in this chapter.

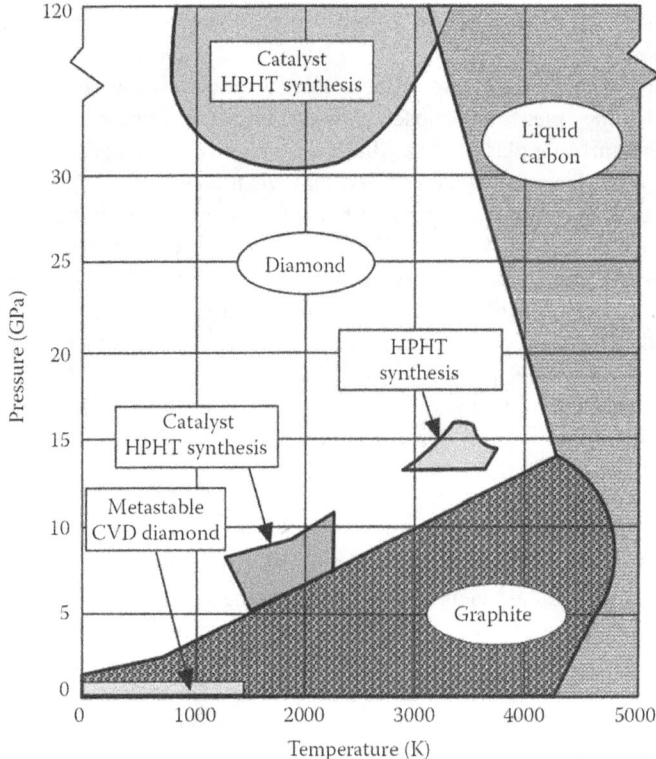

FIGURE 4.1 Carbon phase diagram.

4.3.1.1 Orbitals

The electron structures and bonds of graphite and diamond are considered in this section in a simplified way to the extent important for further consideration. In other words, only notions and definitions used in the inspection of diamond quality in PCD cutting tool manufacturing and application are considered and explained in an entirely new manner.

An atom consists of two basic parts: the nucleus and the electrons. The nucleus is the central core of an atom and is made up of protons and neutrons. Electrons are very light, negatively charged particles that spin about the positively charged nucleus over certain orbits. Early models of the atom depicted the electrons circling the nucleus in fixed orbits, much like planets revolving around the sun. When a planet moves around the sun, its definite path, called an orbit, can be plotted. A drastically simplified view of the atom depicted in many books looks similar, in which the electrons orbit around the nucleus. The truth is different; electrons, in fact, inhabit regions of space known as *orbitals*. Orbits and orbitals sound similar, but they have quite different meanings (Clark 2020, August 15).

Orbital in chemistry or physics defines a three-dimensional region where the probability of finding an electron is maximum. There are four basic types of orbitals: *s*, *p*, *d*, and *f*. Only *s* and *p* orbitals are considered relevant in the current topic. The *s* orbital is a sphere around the atomic nucleus. Within the sphere, there are shells in which an electron is more likely to be found at any given time. The *p* orbital has a dumbbell shape and is oriented in a particular direction. At any one energy level, there are three equivalent *p* orbitals that point at right angles to each other (p_x, p_y, p_z). The shape of *s* and *p* orbitals is shown in Figure 4.2. As with the *s* orbital, the *p* orbital describes a region in space around the nucleus in which an electron may be found with the highest probability. Some specialists think that the "*s*" stands for "spherical" and "*p*" stands for "polar" because these imply the shapes of the *s* and *p* orbitals.

4.3.1.2 Hybridization

The special structure of the atom is called hybridization. In chemistry, orbital hybridization is the concept of mixing atomic orbitals to form new hybrid orbitals (with different energies, shapes, etc., than the component atomic orbitals) suitable for the pairing of electrons to form chemical bonds in valence bond theory.

The normal binding behavior of molecular orbitals is the binding of the same kind of orbital, i.e., two *s*-orbitals or two similar *p* orbitals bind together in an antibonding and bonding way depending on the sign of the orbital. *For carbon atoms, however, the binding of the same kind of molecular orbitals is not the case.* They form bonds by mixing different orbitals, namely *s*- and *p* orbitals. Because it is important for current deliberation, sp^2- and sp^3-hybridization of carbon are considered as related to graphite and diamond, respectively.

In the sp^2 hybridization, one *s* orbital is mixed with two *p* orbitals to form three sp^2 hybridized orbitals. Each of these hybridized orbitals has 33% *s* character and 67% *p* character. These sp^2 hybridized orbitals are oriented with a bond angle of 120°, in a trigonal planar (triangular)

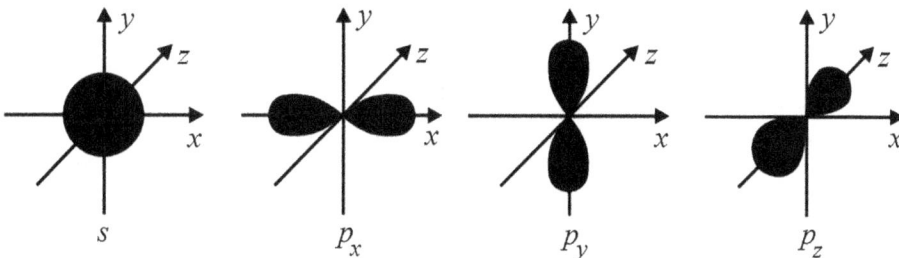

FIGURE 4.2 Shape of *s* and *p* orbitals.

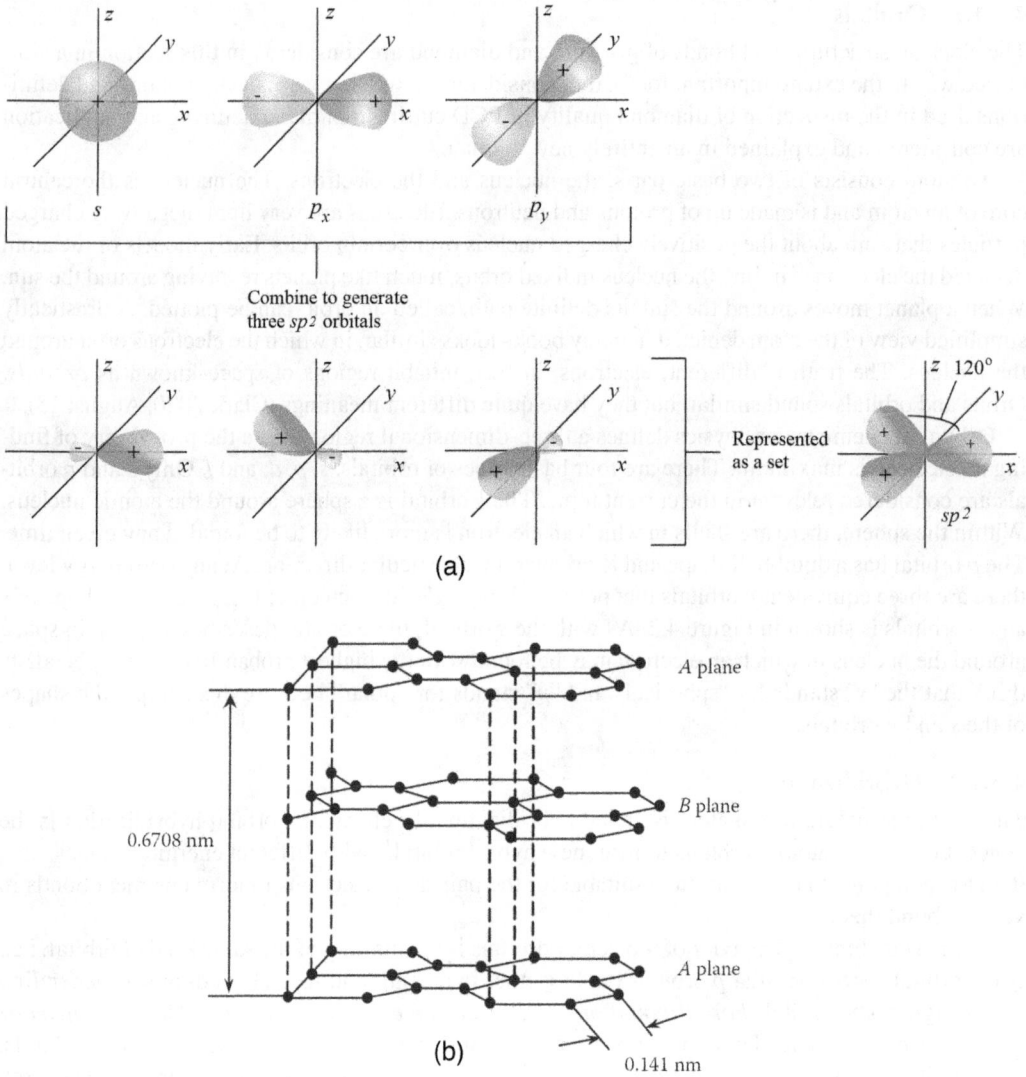

FIGURE 4.3 sp^2 hybridization (a) and schematic of graphite hexagonal crystal structures (b).

geometry. The remaining p orbital is unchanged and perpendicular to the plane of the hybridized orbitals. An example of sp^2 hybridization is shown in Figure 4.3a.

The structure of graphite is shown in Figure 4.3b. In the graphite lattice structure, the carbon atoms form continuous hexagons in staked basal planes. Within each plane, the carbon atom is strongly bonded to its three neighbors with a covalent bond having a bond strength of 524 kJ/mol. This atomic bonding is threefold coordinated as sp^2. The spacing between basal planes (0.335 nm) is larger than the spacing between atoms in plane (0.142 nm). Such a configuration results in a large anisotropy in the crystal (Pierson 1999, Purniawan 2008).

The sp^3 orbitals are formed (hybridized) from 1 s and 3 p orbitals as shown in Figure 4.4a. The energies of these four orbitals are generally averaged forming four equal orbitals. These orbitals are spread out uniformly in space forming a tetrahedral geometry. The tetrahedral shape allows the easy formation of direct sharing and overlap of electron density. A simplified model of sp^3 hybridization is shown in Figure 4.4a. As can be seen, the orbitals of diamond are bonded to the

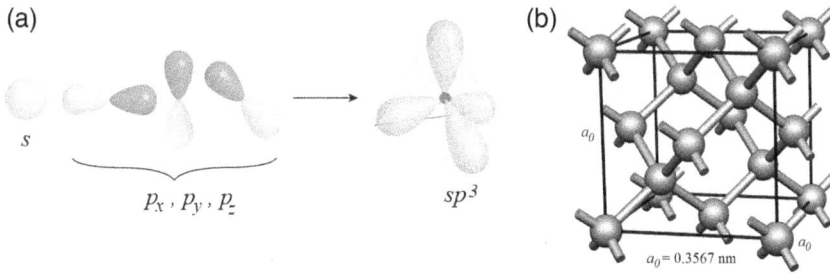

FIGURE 4.4 sp^3 hybridization (a) and schematic of diamond octahedron crystal structures (b).

orbital of four other carbon atoms with a strong covalent bond (i.e., the atoms share pairs of electrons) to form a regular tetrahedron with equal angles (109°28″). Unlike graphite, the structure of diamond is isotropic. Each diamond tetrahedron combines with four other tetrahedrons to form a strongly bonded, 3D, and entirely covalent crystalline structure, with a small bond length of 0.154 nm and a high bond energy of 711 kJ/mol or 170 kcal (Pierson 1999, Purniawan 2008). The lattice constant of the unit cell shown in Figure 4.4b of the FCC form of diamond is 0.3567 nm with a density of 3.52 g/cm^3.

4.3.2 Important Properties

Diamond is thermodynamically unstable at ambient conditions, but unless heated to 650°C in air (oxidation to CO_2) or 1,700°C in vacuum or an inert atmosphere (graphitization), it will remain as diamond indefinitely. To convert graphite directly to diamond requires very high pressures and temperatures (see Figure 4.1). For the synthesis of industrial diamond, these conditions can be lowered by the presence of metal solvents/catalysts. More recent work on the thermodynamically stable region for diamond indicates a positive slope for the melting curve instead of the negative slope deduced previously on the basis of analogy to the P-T diagrams for Si and Ge (Bundy 1989).

Diamond is chemically inert to inorganic acids but can be etched in oxidizing molten salts such as KNO_3 at 600°C. Carbon as diamond or graphite is soluble in several metals, particularly Fe, Ni, Co, and other Group VIII, the rightmost group in the Periodic Table, elements. This imposes a serious limitation on the use of diamond for machining alloys based on these materials in spite of the high hardness of diamond. At the temperatures developed at a cutting tool tip, diamond reacts with these metals. However, similar reactions facilitate the bonding of diamond to metal in diamond tool making because of the wetting of diamond by carbide-forming metals such as W, Ta, Ti, and Zr (Wilks and Wilks 1991, Field 1992).

Diamond is the hardest material known because of its combination of a 3D arrangement of tetrahedrally coordinated bonds with a bond distance of 0.1567 nm. The Knoop hardness (K) is in the range of 68.7–98.1 kN/m^2 (7,000–10,000 kgf/mm^2). Harder materials have been postulated on the basis of bond distance but so far have not been found or made in a useful form (Liu and Cohen 1989). Diamond is twice as hard as cubic BN (K is approx. 44 kN/m^2), which in turn is twice as hard as SiC (K is approx. 23.5 kN/m^2).

Although hard, diamond is also very brittle and cleaves readily on the (111) plane and also on other planes under certain conditions. The ability of diamond to fracture is useful in cutting and grinding applications because the crystals are self-sharpening. For other procedures such as sawing rocks or concrete, tougher diamonds are desirable (Wilks and Wilks 1991). The uncontrolled fracture of a gemstone is disastrous, but in the cutting and polishing industry, large diamonds are first reduced to desired sizes by controlled cleavage or laser cutting to avoid the time-consuming process of sawing (Bruton 1970).

TABLE 4.1

Mechanical Properties of Diamond and Other Hard Materials (GPa)

	Diamond	Cubic BN	Cemented WC	SiC
Bulk modulus	440	370		211
Compressive strength	8.69		4–6	
Modulus of elasticity, E	950–1,100	890	460–675	

Measurement of the mechanical properties of diamond is complicated, and references should be consulted for the various qualifications (Bruton 1970, Wilks and Wilks 1991). Table 4.1 compares the theoretical and experimental bulk modulus of diamond to that for cubic BN and for SiC and compares the compressive strength of diamond to that for cemented WC and the values for the modulus of elasticity E to those for cemented WC and cubic BN.

The thermal conductivity value of 2,000 W/(mK) at room temperature for type IIa natural diamond is about five times that of Cu, and recent data on 99.9% isotopically pure C^{12} type IIa synthesized crystals are in the range of 3,300–3,500 W/(mK) (Field 1992). This property combined with the high electrical resistance makes diamond an attractive material for heat sinks for electronic devices. More impure forms of diamond (type Ia) have lower thermal conductivities (600–1,000 W/(mK)).

The average value of the coefficient of linear thermal expansion of diamond is 1.18×10^{-6} (m/m K). The relatively low expansion combined with the low reactivity of diamonds, except for carbide formation, leads to some challenges in making strong bonds between diamond and other materials.

Diamond is an electrical insulator (approx. 1,016 Ω/cm) unless doped with boron when it becomes a p-type semiconductor with a resistivity in the range of 10–100 Ω/cm. n-Type doping has been claimed but is less certainly established. The dielectric constant of diamond is 5.58.

4.4 APPLICATIONS

The first extensive use of diamond in the industry was a diamond powder mixed with olive oil. This abrasive slurry was placed on the horizontal face of a spinning cast iron plate to polish rough diamonds (held under pressure against the face) into faceted gems. The beginnings of this art are lost to history but probably started about AD 1500. A major use of industrial diamond today is in the grinding and finishing of various cemented carbide parts and cutting tools. As the cemented carbide industry has expanded (growth has been remarkable since its introduction), the growth of the diamond wheel market has correspondingly increased. In 1935, only about half of all diamonds mined were used in industry. Now, this figure has increased to about 80%.

The selected applications of diamond are based on its high hardness, wear resistance, ability to self-sharpen as it cleaves, thermal conductivity, and chemical inertness. Loose abrasive grain is used in lapping and polishing operations, for example, on the scaife used for faceting gem crystals and polishing rock materials for monuments and buildings. Single-point tools for engraving, turning, or cutting applications are made by mounting an individual diamond in metal holders (Grodzinski 1953, Smith 1974, Wilks and Wilks 1991). Other uses for single diamonds are as phonograph needles, bearings, surgical knives, and wire dies. The differential hardness of diamond is the cause of the non-uniform wear, or loss of roundness, of a wire die. Sintered (polycrystalline) diamond is used for wire dies, and the anisotropic wear problem disappears. For many applications such as grinding, cutting, drilling, and sawing, abrasive grains are mounted in resin, metallic, or vitreous bonding materials on wheels, disks, or saw blade sectors. Metallic bonding can be accomplished by electroplating as well. The grooving of concrete runways and roads is a well-known example of the use of impregnated diamond saw blades. An interesting new development is as beads, which are strung on long wires for sawing blocks of stone from quarries. About 90% of industrial diamond is nowadays synthesized.

4.5   TYPES

There are two types of diamonds: natural and synthetic diamonds. Both can be either single-crystal diamond or polycrystalline diamond (known as PCD). Natural PCD abrasives have long been used for cutting and drilling very hard materials, and scientists continue to explore methods of replicating the properties of carbonado, one of the more industrially important varieties of PCD. Carbonado occurs as black, irregularly shaped, polycrystalline aggregates with grain sizes ranging from less than 1 to several hundred microns (Jeynes 1978, Dismukes et al. 1988). Unlike non-porous microcrystalline diamond aggregates found in kimberlites (e.g., stewartite and framesite), carbonado is recovered only from alluvial deposits in Brazil and the Central African Republic. Carbonado exhibits the extreme hardness and high thermal conductivity of single-crystal diamond, but it is much less vulnerable to catastrophic cleavage. The random crystallographic orientation that is typical of carbonado and other natural and synthetic PCD combines with a complex defect microstructure to minimize crack propagation (Lammer 1988). In addition, grinding materials composed of PCD tend to wear more uniformly than single-crystal diamonds, whose hardness varies with crystallographic orientation. The tight diamond–diamond bonding in carbonado suggests that it is a naturally *sintered* diamond (Hall 1970).

4.6   SYNTHETIC DIAMOND

4.6.1   HISTORY

The 1797 discovery that diamond was pure carbon catalyzed many attempts to make artificial diamonds. If the common carbon of commerce could be turned into a diamond, a millionfold increase in the value of the starting material would be obtained. There was, in addition, an interest in the scientific achievement. The earliest claim of success (C. Cagniard de la Tour) was made in the year 1823. The diamond synthesis problem has attracted the interest of thousands. Those pursuing the problem have ranged from charlatans through rank amateurs to the world's greatest scientists. Included among the great are Boyle, Bragg, Bridgman, Crookes, Davey, Despretz, Friedel, Liebig, Ludwig, Moisson, Parsons, Taman, and Wohler.

The literature on diamond synthesis is not very extensive. Most of the work has gone unpublished. Many of the world's large industrial organizations have considered the problem and have spent millions of dollars. So many years passed without success that those working on the problem felt embarrassed to admit that they were so engaged. Another aspect of the problem was the chicanery and fraud. Quite a number have claimed to possess a procedure for converting graphite into diamond, invited the unwary to invest their money, and then vanished.

The earliest successes were reported by James Ballantyne Hannay in 1879 (Hannay 1879) and by Ferdinand Frédéric Henri Moissan in 1893 (Royère 1999). Their method involved heating charcoal at up to 3,500°C with iron inside a carbon crucible in a furnace. Whereas Hannay used a flame-heated tube, Moissan applied his newly developed electric arc furnace, in which an electric arc was struck between carbon rods inside blocks of lime. The molten iron was then rapidly cooled by immersion in water. The contraction generated by the cooling supposedly produced the high pressure required to transform graphite into diamond. Moissan published his work in a series of articles in the 1890s (Moissan 1994).

Despite these early successes, many following scientists experienced difficulty in reproducing such results. Spending nearly 30 years of his life and over £30,000 pounds, British scientist Charles Algernon Parsons went through great lengths in order to reproduce a high-pressure high-temperature (hereafter HPHT) diamond. While claiming to have produced HPHT diamonds, overtime he reluctantly admitted that no scientist, past or present, could create synthetic diamonds and that the best they could create were spinels, or a simple class of minerals crystallized in an octahedral form (Davis 1993).

Despite past failures and inspired by the vast industrial uses of diamonds especially during wartime, General Electric (GE) resumed the synthetic diamond project in 1951. Schenectady Laboratories of GE and a high-pressure diamond group formed. The first authenticated synthesis of diamond was made on December 16, 1954 by H. Tracy Hall. It should be noted that this synthesis was unique in several aspects when compared to prior claims:

1. The diamonds were grown in a very short time; 15–120 seconds.
2. The yield was high; that is, a significant portion of the graphite starting material was converted so that the diamonds formed could be seen by the naked eye without being separated from the starting material. Also, the diamonds, though tiny (up to 350 µm across) were obtained in sufficient quantity to be felt and held in the hand and could be heard to "scratch" as they were drawn across a piece of glass. These perceptions of the diamonds by the unaided physical senses were remarkably satisfying.
3. Hundreds of crystals were grown, many of them intergrown with each other and some were twinned. This result contrasts sharply with the way diamonds are found in nature.
4. The discoverer readily duplicated his successful experiment a dozen times, and a short time later the experiment was duplicated by others.
5. The process was immediately recognized as being commercially feasible; indeed, less than 3 years after discovery, diamond grit was being sold on the commercial market.

4.6.2 SYNTHETIC DIAMOND BRIEF CLASSIFICATION

A simple classification of synthetic diamond-cutting tool materials is shown in Figure 4.5. In machining where the wear mechanism is mainly abrasion, PCD made using HPHT synthesis has already proved itself to be a superior tool material. Over the last 20 years, PCD has been accepted for volume production of hypereutectic aluminum-silicon alloy components in the automotive industry, the machining of non-ferrous alloys of copper, the machining of abrasive plastics and plastic composites such as glass fiber and printed circuit boards, as well as volume machining of wood composites, such as chipboard.

A great number of attempts to use chemical vapor deposition (CVD)-synthesized diamond in cutting tools were with limited success. Although this material has superior properties such as higher hardness/wear resistance and much higher thermal stability (discussed below), its low toughness causes problems in tool manufacturing, handling, and applications as discussed later in this chapter. Therefore, this chapter considers PCD tool material and its use in drilling tools.

4.6.3 SYNTHETIC DIAMOND TECHNOLOGY: HIGH-PRESSURE EQUIPMENT

Early in the 1930s, devices known as piezometers were developed. They include steel thick-walled cylinders equipped with plungers. Pressures up to 0.3 GPa were created using such devices. The

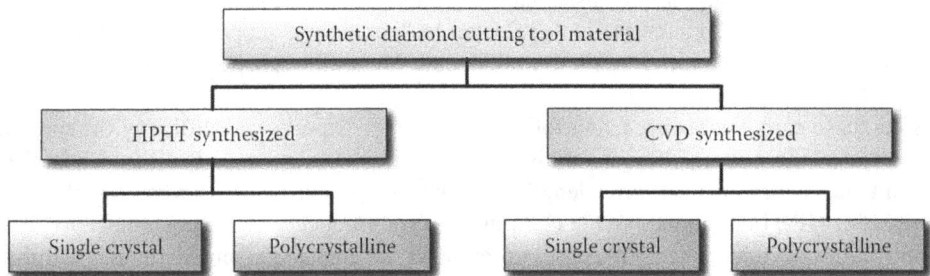

FIGURE 4.5 Simple classification of synthetic diamond-cutting tool materials.

great American scientist and Nobel Prize winner Percy Bridgman pioneered the development of high-pressure creating devices known as *Bridgman anvil* (Prikhna 2008). Bridgman placed a piezometer into the cavity of a great hydraulic cylinder and obtained a two-stage apparatus, in the internal cavity of which a pressure up to 3 GPa was attained. The hardening of many pressure-transmitting liquids at higher pressure became an obstacle to further increase the pressure in apparatuses of this type. Therefore, Bridgman had to use hard minerals like pipestone or pyrophyllite as a pressure medium. It is remarkable that the higher pressures were possible to generate in such a simple device.

Howard Tracy Hall first realized previous high-pressure creating devices failed in two main areas: the pistons and the cylinders. Without finding a solution to this problem, the device would not be able to create an environment suitable for creating the pressure needed. He solved the problem by eliminating the bottom of the bore and using two tops *back to back*. Hall also realized that the steel currently available could not withstand the pressure that would be applied. Instead, he used cemented carbide, which was necessary to create a device strong enough to withstand the pressure. Hall called his device a *belt* apparatus (US Patent 2,947608, 1960; failed August 29, 1955 – diamond synthesis) (Hall 1960). Subsequently, three types of equipment are the inventions of H. Tracy Hall. They are the belt, the tetrahedral press, and the cubic press shown in Figure 4.6 (Hall 1982). The further development of high-pressure equipment with detailed explanations and figures is presented by the author earlier (Astakhov 2014).

4.6.4 Synthetic Diamond Technology: The Synthesis

The cubic cell of a 200-ton cubic press similar to that depicted in Figure 4.6 is shown in Figure 4.7. The body of the cell is made of pyrophyllite, and the edge of the cube is 16 mm, which is 25% longer than the edge of the corresponding square anvil face. The exterior surface of the cell is painted with a suspension of red iron oxide in water and is dried in an oven at 120°C for 0.5 hour before use (Hall 1982).

The central hole in the pyrophyllite is approx. 6 mm in diameter and the counterbores on each end are 11 mm in diameter by 4.2 mm deep. The pyrophyllite thermal insulation plug is located within the steel current ring. The molybdenum disks are located at the bottom of each counterbore. The central is completely filled with alternating disks of graphite and pure nickel.

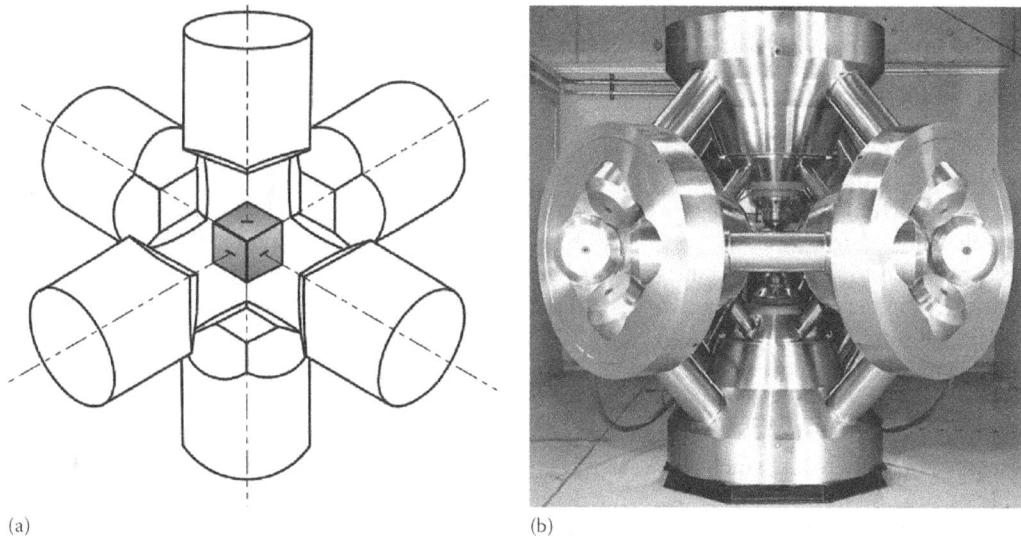

(a) (b)

FIGURE 4.6 (a) Cubic anvils and cell (schematics) and (b) its modern design.

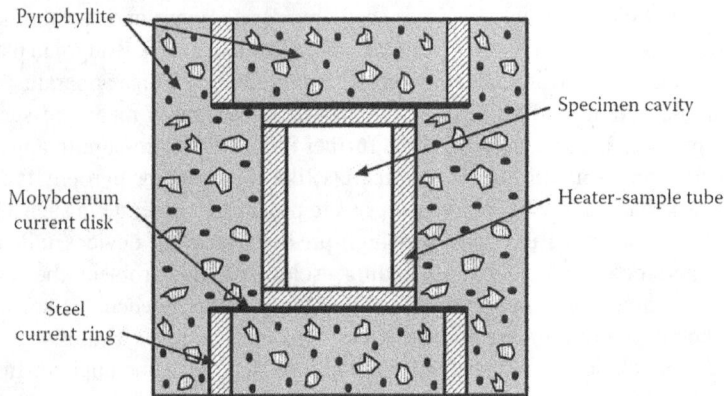

FIGURE 4.7   Cubic cell (schematic).

The first treatment consists of concentrated sulfuric acid (H_2SO_4) to which about 10% by weight of sodium or potassium nitrate is added. This mixture and the sample are heated strongly on a hot plate within a shielded fume hood for about an hour. When an appropriate temperature is reached, *clouds* of graphite particles will be streaming upward from the sample. Eventually most of the graphite is oxidized and disappears. At this point, the liquid is carefully decanted and discarded. Following this, concentrated hydrochloric acid (HCl) is added to the sample and they are heated lightly on the hot plate. HCl is then decanted and the residue is washed with water. The diamonds will be clean enough at this point to be observed under 10x power magnification. Typical diamond morphological features will be evident. For additional cleaning, the H_2SO_4/$NaNO_3$ and HCl treatments should be repeated until all the graphite and nickel are gone. There will still be some residue from the pyrophyllite. This can be removed by prolonged treatment with concentrated HF solution or by fusion with NH_4F.

In industrial applications, there are a number of quality procedures to obtain the final product – diamond powder similar to that shown in Figure 4.8. Diamond powders can be of different quality. Some grades can be used for diamond grinding wheels, while others are subjected to further HPHT processed to make PCD disks to be used in PCD cutting tools.

4.7   BLANKS FOR DRILLING TOOLS

4.7.1   PDC AND PCD

The vast majority of PCD drilling tools are actually PCD-tipped tools made from PCD disks manufactured using the HPHT process. There are two types of such blanks: polycrystalline diamond compacts (PDCs) and PCD disks. PDCs shown in Figure 4.9 are used to manufacture the cutting elements of drilling tools (called drill bits shown in Figure 4.10) for oil and gas drilling PCD disks are used to make PCD cutting tips (Figure 4.11) for drilling tools similar to that shown in Figure 4.12 in metalworking industry (MWI). Although these two types of blanks were developed practically simultaneously, PDC and thus PDC drill bits experience a much greater advancement rate due to high demands for advancement in oil and gas drilling (OGD) compared to the MWI.

PDC cutters, referred to at the time by the trade name Stratapax, were first developed by GE in 1973. They utilized GE's earlier invention, monocrystalline man-made (synthesized) diamond discussed in Section 4.5, which was loaded into a pressure cell with cemented carbide substrate and repressed to produce a compact of 13 mm diameter and 3.3 mm length that incorporated a 0.5 mm thick diamond table. Over the next 30 years, a number of significant developments were made to improve both PDC design and materials properties and the design of drill bits.

FIGURE 4.8 Morphology of diamond powder.

FIGURE 4.9 PDCs.

The first difficulty was the reliable mounting of PDC cutters in bit bodies. Brazing techniques and practice of the day frequently led to debilitating cutter loss and failed runs. Post-mount press-fit cutters that were deployed in steel-body bits were prone to fracture breakage of the post at the mounting point and to loss through erosion of the steel bit body. Improved brazing and mount pockets provided solutions.

FIGURE 4.10   Modern drill bit with PDC.

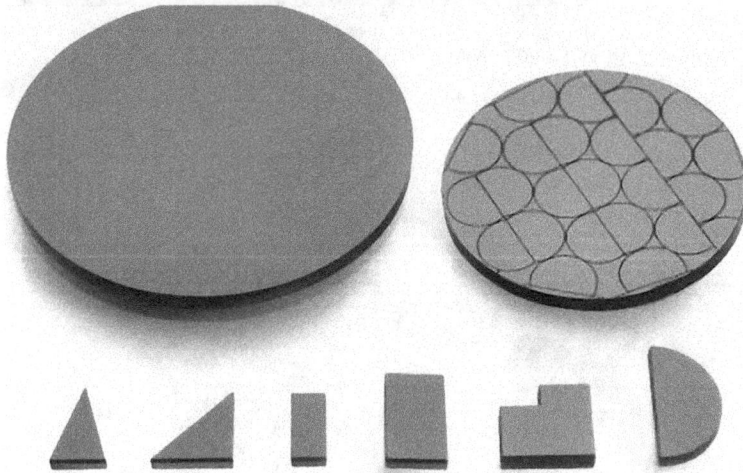

FIGURE 4.11   PCD disk blank and a variety of PCD tips made out of this blank.

FIGURE 4.12   Typical PCD-tipped multistage drill.

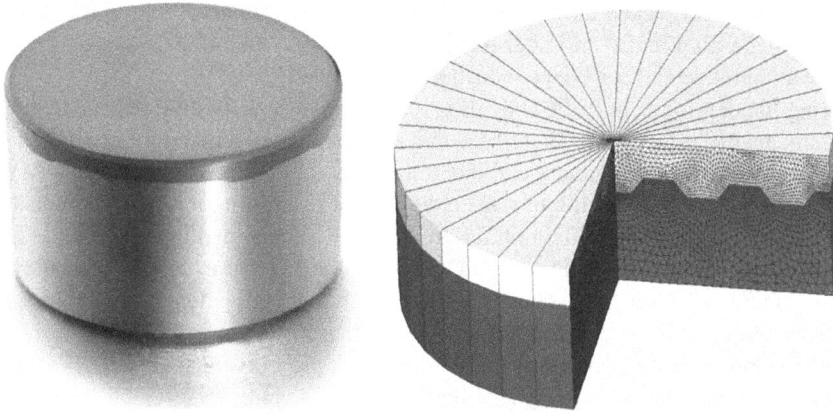

FIGURE 4.13 PDC with non-planar diamond–substrate interface.

The next major hurdle was reducing diamond table (layer) delamination from the substrate under impact loading. The initial solution to this problem was the development of a non-planar diamond–substrate interface by providing grooves and other patterns on the face of the cemented carbide substrate (e.g., as shown in Figure 4.13). At this stage of development, PDC bits had increased their market presence to about 15% of all footage drilled and were considered by many to have neared the peak of their potential development.

The major advancement, however, was made by reducing the so-called bit whirl, that is, self-regeneration off-center rotation condition as the prime source of the delamination. It was reduced by the design of force-balanced, symmetrical-bladed, spiral-edged drill bits. These improvements made by the late 1990s led to a significant increase in the use of such tools – by this time, PCD drill bits accounted for about 46% of all footage drilled in OGD.

When design and manufacturing problems of PDC drill bits were solved, the coolant parameters and its delivery were optimized, and drilling cycle programming improved, common failures due to bit balling, cutter loss, and impact damage were mitigated. The next logical step in the further improvement of PDC drill bit performance was closer consideration of PCD layer properties as related to its performance in the normal wear mode, which is abrasive wear. It was realized that PDC cutters were subject to thermal damage and accelerated wear at the cutting tip, due to a large amount of the residual cobalt catalysis remaining in the interstitial matrix of the cutters' PCD face. By reducing the cobalt content in the outermost layer of the diamond table, the cutters' abrasion resistance and thermal stability were significantly improved, allowing PDC bits to compete economically with roller-cone bits in even more applications. In 2010, PDC bits accounted for an astonishing 65% of footage drilled in the OGD applications and still not do appear to have peaked in their development. More research than ever is going into PDC bits and especially into PDC cutters.

The development of metal cutting tools with PCD tips and thus PCD disks as blanks for such tips was much slower due to a number of reasons. The prime reason for that was unavailability of drilling systems capable of running PCD drilling tools efficiently. Low available spindle speeds, low system rigidity, and excessively poor properties and maintenance of MWF were the prime reasons. Moreover, a widespread almost-religious belief is that PCD is only suitable for machining aluminum and other non-ferrous materials because of the so-called affinity of diamond, which is essentially carbon in PCD and carbon in steels so that PCD *dissolves* in steels. As such, carbon in cemented carbides (WC) and that in steels mysteriously does not cause any *affinity* problems. As discussed by the author earlier (Astakhov 2014), carbon as diamond or graphite is soluble in several metals (nothing to do with affinity), particularly Fe, Ni, and Co when the temperature is high enough. However, this temperature is much higher than that leading to structure failures of

PCDs. The development picked up its pace only when a high volume of material-matrix composites (MMS) (silicon-aluminum) and fiber-reinforced composite materials became the materials of choice in the automotive and aerospace industries, respectively.

Comparing the development of PCD drilling tools used in MWI with those made in OGD, one can see significant differences. The first one is that there is a great demand for productivity of drilling in OGD as the costs of equipment and labor are high. It is not nearly the case in MWI, where effective managers have been trying to reduce the cost of machining through the reduction of the tool cost while paying almost no attention for productivity of PCD drilling operations. The second significant one is the volume of PCD used in OGD compared to that in MWI – the ratio is 95:5.

As a result, PCD tool material manufacturers pay much more attraction through a high volume of R&D activities, testing new designs, and technologies of their materials for OGD. The level of publication on the development/testing/application of PDC is much higher compared to publication on PCDs in MWI as they consider the most important aspects of the PDC in terms of tool material, tool design, and application conditions. For example, Gu and Huang (2016) found out that controlling the content of Co in PCD is not only a vital problem existing in the PDC industry but also an effective way to improve drilling performance; Paggett et al. (2002) studied thermal residual macrostresses and their gradients in a series of PDCs using neutron diffraction. Sintering process parameters, thermal treatments, and bonding were all found to affect residual stress levels and stress gradient characteristics. Measured average in-plane stress gradients are shown to differ substantially in some cases from linear elastic predictions; Mashhadikarimi et al. (2021) obtained a triple-layer PDC via the HPHT sintering method. A thin layer of WC with 20 wt% Nb/Ni was used as an interface between the top layer of PDC with pure Nb binder and WC 10 wt% Co substrate. The overall results showed that this new kind of PDC can be produced successfully using pure niobium as a binder for the PDC layer and no sign of graphitization was detected. It was also found that using an interface having the resemblance to both substrate and sintered diamond body caused good adhesion between layers that can potentially enhance the performance and durability of PDC; Kanyanta et al. (2014) studied the fatigue behavior and failure of PDC cutting tools under cyclic impact loading. It was found that PDC cutters with a coarser-grained microstructure exhibited up to 70% better impact fracture resistance than their fine-grained counterparts. Their fatigue endurance limit was also about 10%–15% higher; Yahiaoui et al. (2016) studied the wear behavior of conventional and graded PDCs. Different cutters were considered regarding their diamond grain size and the HPHT conditions of the manufacturing process. On the base of these cutters, a cobalt graduation process was performed on the WC-Co substrates by reactive imbibition. A quality factor developed in previous studies was calculated to evaluate cutters' wear performances. The results showed that a controlled HPHT process can act on the wear resistance certainly by improving the diamond grain boundary cohesion. Unexpectedly, the diamond granulometry appeared to be a secondary factor influencing the wear resistance. The reactive imbibition clearly increased the wear resistance, even for cutters with coarse diamond grains (i.e., potentially impact resistant); Boland et al. (2010) developed an inexpensive abrasive wear test to assess the quality of a range of diamond composite materials. The abrasive resistance of commercially available diamond composites is observed to span over four orders of magnitude. Microstructural and phase distribution studies of these composites revealed poor quality control during the manufacture of some diamond composites.

The third reason is in high cost including time lost and tool cost of changing a tool in OGD. This is not nearly the case in MWI. As a result, manufacturing management in OGD took full control on the performances of PDC tools driving further research and development of their tools. Unfortunately, this is not nearly the case in MWI.

4.7.2 Available PCD Blanks/Disk

Figure 4.14a shows the geometry parameters of PCD disks. Various PCD disk manufacturers establish different standards on the disk diameter, D_{db} (Figure 4.14b), PCD layer thickness, t_{dl}, and disk thickness, H_{db}. Obviously, various PCD grain sizes and grades are available, for example:

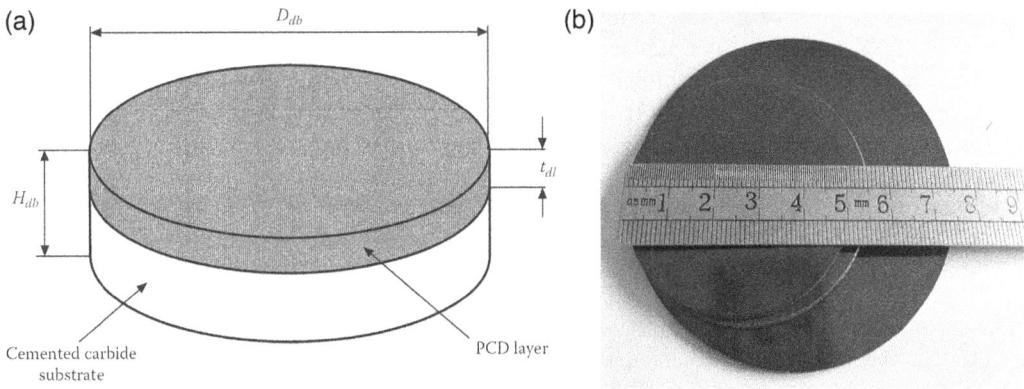

FIGURE 4.14 PCD disk: (a) geometry parameters and (b) range of diameters.

1. Element 6 Co (hereafter E6) makes all disks of 70 mm dia. six grades starting with the most popular in industry-grade CMX850 are available. PCD layer structure/morphology is shown for each grade so that an end user can verify the actual grade he or she purchased. For each grade, the available PCD layer thicknesses, t_{dl}, are given with tolerances. For example, for CMX850 grade, the PCD layer thicknesses can be 0.3 mm (the actual range is 0.20–0.45 mm), 0.5 mm (the actual range is 0.35–0.65 mm), and 1.0 (the actual range is 0.83–1.117 mm). The available disk thicknesses, H_{db}, for each grade and PCD layer thickness are also given with tolerance ±0.05 mm. It is pointed out that a tighter overall height tolerance ±0.025 mm is available for selected CMX and CTB disks. Moreover, extensive information on the application ranges of the available PCD grades including the type of metal cutting operation, cutting speed, feed, and depth of cut is provided.
2. ELJIN Co makes PCD disks of 60.0 and 13.0 dia., although it is pointed out in some company literature that 75 mm diameter disks are available in CM, CMW, and CXL grades. Disks for all grades are made with the same PCD layer thickness, t_{dl}=0.5 mm. The available disk thicknesses, H_{db}, are 1.6, 2.0, 2.4, 3.2, and 4.8 mm. No tolerances on the listed parameters are given. Six standard and two premium grades of PCD are listed and the guide for the grade selection is available. Operation cutting conditions for each grade are provided for turning and milling.
3. SANDVIK HYPERION Co makes all disks with 59.1 mm diameter with 58.0 mm usable diameter. The company offers six PCD grades. Similarly to E6 Co, PCD layer structure/morphology is shown for each grade so that an end user can verify the actual grade he or she purchased. The available PCD layer thickness, t_{dl}, and total disk thicknesses, H_{db}, are provided. For example, for the most popular (yes, least expensive, why pay more for premium grades as a PCD grade is not a part of PCD drilling tool drawings) grade 1200P: for the nominal t_{dl}=0.3 mm (the actual range is 0.2–0.45 mm) and H_{db}=1.0, 1.6, 2.0, 2.4, and 3.2 mm with tolerance ±0.05 mm; for t_{dl}=0.55 mm (the actual range is 0.4–0.8 mm) and H_{db}=0.8, 1.2, 1.6, and 3.2 mm with tolerance ±0.05 mm. The machining parameters guidelines are given for various work materials.
4. Sumitomo Co makes all disks of 55 mm dia. with the layer thickness of 1, 1.5, 2.5, and 3.5 mm. It offers five grades, F, S, M, C, and E having different diamond content and grain size. Very general application notes for these grains are provided.

It is important to know that not all the disk surfaces should be used as areas close to the edge may not have the same properties/PCD structure. However, only SANDVIK HYPERION Co informs end user that their disks with 59.1 mm diameter have only 58.0 mm usable diameter.

PCD disks supplied to cutting tool manufacturers have the PCD side mirror polished although just ground PCD disks are also available (Astakhov 2014). Cutting tool manufacturers should avoid grinding/electrical discharge machining (EDM) this polish side used as the rake face of the tool. The reason for that is explained later in this chapter.

The quality of PCD disks including consistency and residual stresses after sintering and other aspects are not specified by PCD blank manufacturers. The cost of disks of a similar diameter can differ more than tenfold depending on a particular PCD disk with no explanations provided.

4.7.3 Manufacturing of PCD Disks

PCD disks are produced by high-pressure sintering techniques using a solvent-catalyst metal such as cobalt (Co) at pressure 6 GPa and temperature from 1,673 to 1,873 K. Mixed powder methods and infiltration methods are common sintering techniques. General Electric was the first company that developed one of the previous infiltration methods in 1972 and then industrialized PCD blanks. In GE's method, diamond powder is sintered and bonded to a WC-Co substrate by infiltration of Co from the substrate using HPHT presses similar to the above-described process used in manufacturing PCD powder.

Figure 4.15 shows the assembly of the press for sintering, and Figure 4.16 shows details of a diamond powder disk and WC-Co substrate wrapped up in a tantalum (Ta) shell. Such a wrapping is then placed in the assembly composed of a solid pressure salt (NaCl) surrounded by the crucible and graphite heater. Formation stages of a common PCD disk during HPHT sintering are shown in Figure 4.17. First, pressure is raised to its nominal level with little or no heating. During this stage, all the crystals are being pushed against each other with increasing force. Many diamond particles are sliding relative to each other, and many are cracking into two or more fragments with the overall effect of increasing powder density (Uehara and Yamaya 1990). A coarser powder presents a higher degree of crushing than a fine one as the former includes a much smaller average number of contact points and thus much higher contact stresses (compared to fine powders) that cause the described crushing.

After the crushed and compacted powder is under full pressure, the temperature is raised at a heating rate of 60°C/min to its nominal value (approx. 1,600°C). The pressure–temperature condition is always kept in the thermodynamically stable region of diamond (see Figure 4.1) above the eutectic temperature of carbon and Co. As the diamond powder is packed against a WC-Co substrate, cobalt is the source of the catalyst metal that promotes the sintering process. When the cobalt reaches its melting temperature of 1,435°C at 5.8 GPa, it is instantaneously squeezed into the

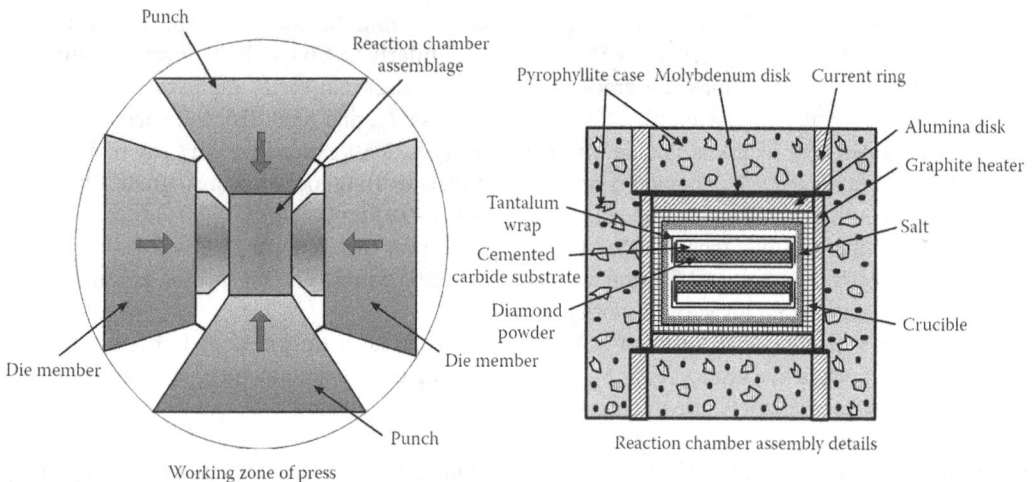

FIGURE 4.15 Apparatus used in PCD disks sintering.

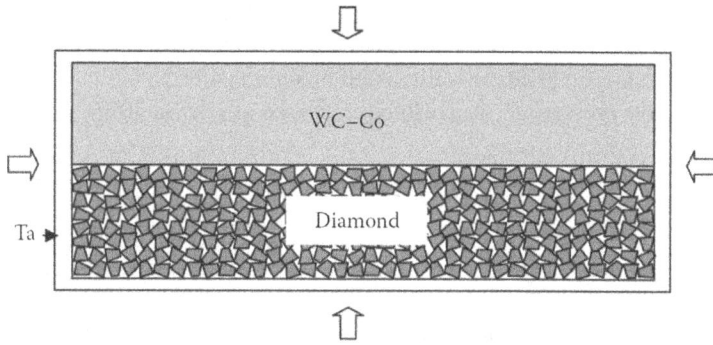

FIGURE 4.16 Diamond powder and WC-Co substrate disk wrapped up in tantalum (Ta) shell (the capsule).

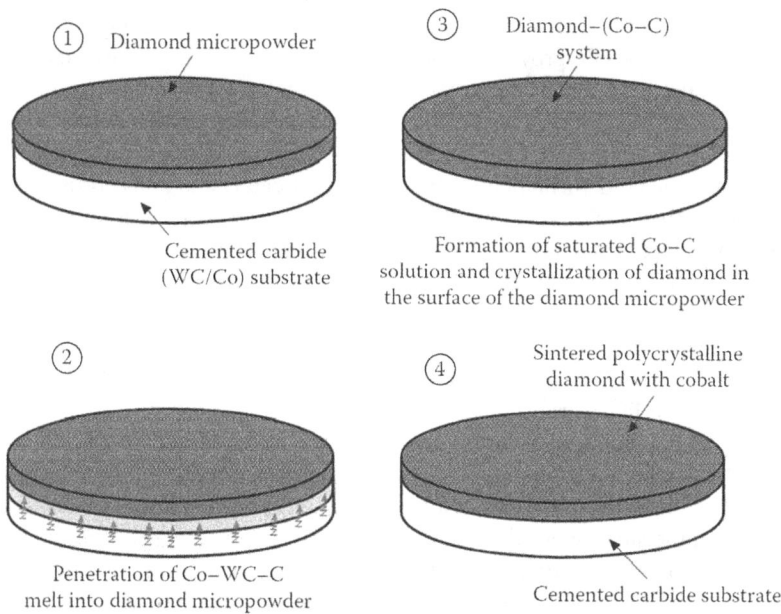

FIGURE 4.17 Formation stages of a common PCD disk during HPHT sintering: (1) start of the process with applying high pressure, (2) melting and infiltration stage, (3) phase formations, and (4) finish of the process.

open porosity left in the layer of compacted diamond powder. At this point, the sintering process takes place through a mechanism of carbon dissolution and precipitation. Technically, this process is defined as a pressure-assisted liquid-phase sintering. The driving force for the densification under an extreme pressure is determined by the pressure itself and also by the contact area relative to the cross-sectional area of the particles. The reaction speed is proportional to the temperature and to the average effective pressure P_{ae}, which is the actual contact pressure between particles, as expressed in the equation (Bellin 2010a):

$$\log\left(d\rho_{ap}/dt\right) \approx -D_l P_{ae}/(RT) \tag{4.1}$$

where ρ_{ap} is the powder apparent density, D_l is the carbon diffusivity in the molten metal catalyst, R is the ideal gas constant, and T is the temperature.

It follows from Eq. (4.1) that the sintering process is faster if both the contact pressure and temperature are increased. On the other hand, when temperature and/or process time are insufficient, cobalt infiltration would not be completed resulting in weaker bonds of diamond grains. For this condition, the author has introduced the term "undercooked" PCD used in describing the failures of the veined and sandwich-type PCD drills discussed in Section 4.11.2.

The average effective pressure P_{ae} can also be expressed as (Bellin 2010)

$$P_{ae} \approx \frac{4a_{ps}^2}{Z_{sp}r_{sp}} P_{ex-s} \qquad (4.2)$$

where a_{ps} is the average particle size, r_{sp} is the radius of the contact area between two spherical particles, Z_{sp} is the number of surrounding particles, and P_{ex-s} is the external pressure applied to the system.

It follows from Eq. (4.2) that smaller grain size and better packing result in lower contact pressure. Therefore, in sintering PCD grades with small average grain size (AGS), higher pressure and temperature are required to accomplish the sintering process.

The result of the process is a disk consisting of the WC-Co substrate and the diamond layer of a certain thickness strongly bonded with this substrate. In this disk, PCD composite is a fully dense mass of randomly oriented, intergrown micron-size diamond particles that are sintered together in the presence of a metallic catalyst phase, usually cobalt. Small pockets of the catalyst phase, which promotes the necessary intergrowth between the diamond particles, are left behind within the composite material. Figure 4.18 shows SEM micrographs of the diamond grains before and after sintering.

It is believed that plastic deformation in the presence of catalyst is induced in the particles and diamond–diamond strong bonding occurs during sintering. This is the strongest known bonding causing PCD high hardness and strength. Note that the catalysis, most commonly Co, just accelerates the formation of bonds. In this sense, it is highly improper to call Co in PCD as the binder as routinely occurs in the professional and sometimes scientific literature as well as in many promotional/sales materials, presentations, and web sources in the tool industry. As discussed later, after diamond-to-diamond crystal bonds form a continuous structure similar to that shown in Figure 4.4b, Co can be removed from PCD structure to improve its thermal stability.

The formation of diamond-to-diamond crystal bonds is explained by the Ostwald ripening mechanism. That is, some of the diamond particles dissolve into molten Co–C(–Zr) in the sintering stage (Co penetration stage) at high temperatures and high pressures. The dissolved diamond precipitates again on the surface of the remaining diamond particles by the driving force caused by the difference in the solubility of diamond due to different curvatures of the particles and results in direct diamond–diamond bonding. It is considered that the previous solution and reprecipitation of diamond and Co penetration into the diamond powder are accelerated by the fine diamond particles generated in the pressure-rise stage.

(a)   (b)

FIGURE 4.18   SEM micrographs of the diamond grains before (a) and after (b) sintering.

Another explanation of the diamond sintering process is based on the formation of graphite during the heating stage. According to this theory, locally low-pressure regions are inevitably generated in the powder compact due to defects such as voids. Transformation of diamond into graphite occurs during the heating stage in the low-pressure regions. The graphite dissolves into molten Co more easily than diamond, and the direct bonding of diamond is formed by the precipitation of graphite as diamond. To confirm this process, some experiments were carried out at 6 GPa pressure and up to 1,600°C using 10–15 and 2–4 µm grade powders. Neither Co nor graphite is detected by XRD or energy-dispersive x-ray spectroscopy (known as EDS) in the diamond layers of either 10–15 or 2–4 µm grades sintered at 1,500°C for 10 minutes. Therefore, it was concluded that transformation of diamond into graphite does not occur in the HPHT sintering process (Uehara and Yamaya 1990). However, this result is not conclusive as a very short sintering time (10 minutes instead of the usual 1 hour) and an excess of Co (an additional Co plate placed on the top of the compact) were used in the test.

Therefore, it can be concluded that the phenomena of formation diamond–diamond bonds are not clearly demonstrated yet. More detailed and sophisticated studies are needed to understand and then to optimize the process, which in its present stage is fully empirical/phenomenological. No tests/inspection methods are offered to optimize the strength of diamond–diamond bonds as well as their possible degradation during manufacturing of PCD disk and/or in tool service. In the author's experience, this is one of the major issues in evaluation of the quality of PCD blanks made by various manufacturers; in evaluation of the quality of PCD drilling tools after manufacturing; in finding the real root cause of failure of PCD drilling tools in operation. Needless to say that PCD blanks' and PCD tools' manufacturers are not keen to develop such tests.

The Ta foil used as the wrapper in the process is changed into a TaC layer containing Co after sintering.

4.7.4 POWDER MIX

The powder used in the sintering consists of the diamond powder and some other materials added for specific purposes. The presence of the latter is rarely mentioned in the literature as these additions in both the composition and the amount are closely guarded process secrets specific to a particular PCD blank manufacturer.

The quality of the feedstock diamond powder (powder mix) in this compact is of utmost importance in determining the final properties and consistency of the sintered product. Usually, the micronized diamond powder used in PCD manufacturing is a by-product of the industrial diamond powder synthesis process, where crystals of at least 100 µm size are produced mainly for stone cutting market (e.g., diamond saw blades, diamond wheels, and loose diamond powder). There are a great variety of industrial diamond powders available on the market with an extremely wide quality range. Depending on the press cycle parameters (mainly pressure and temperature), crystals can be grown with different shapes (e.g., cubic rather than octahedral). Furthermore, if the crystals are growing too fast within the molten catalyst bath, it is possible to find metal inclusions buried deep inside the crystal at the end of the cycle.

Figure 4.19 shows SEM micrographs of starting diamond powders (Uehara and Yamaya 1990). As can be seen, a range of sizes rather than a particular size (grade) is normally assigned to the powder. Figure 4.20 shows particle size distribution in 2 and 4 µm graded commercial diamond powder of good quality (Shin et al. 2004). The better the quality of the diamond powder, the narrower the range of particles distributing in terms of their shapes and sizes.

Earlier researchers (e.g., Uehara and Yamaya 1990) reported that the grain size of diamond becomes smaller after sintering. They attribute this phenomenon to crushing of the starting powders by friction among the particles during the pressure-rise stage. To confirm this, diamond particles were observed after pressing at 5.5 GPa. As shown in Figure 4.21, all powder sizes used in the test were crushed down. The crushing was particularly remarkable in larger grades and a larger number of crushed fine particles smaller than 1 µm were observed in the crushed powders. This idea is fully

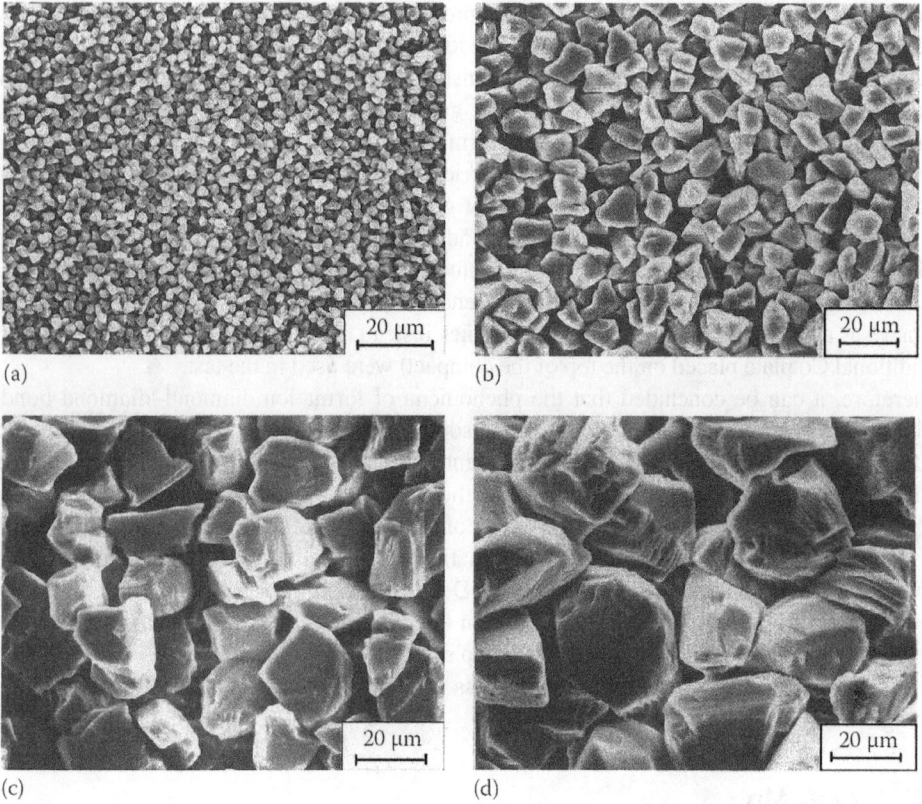

FIGURE 4.19 SEM micrographs of starting diamond powders. Grades, microns: (a) 3–6, (b) 10–15, (c) 22–36, and (d) 40–60.

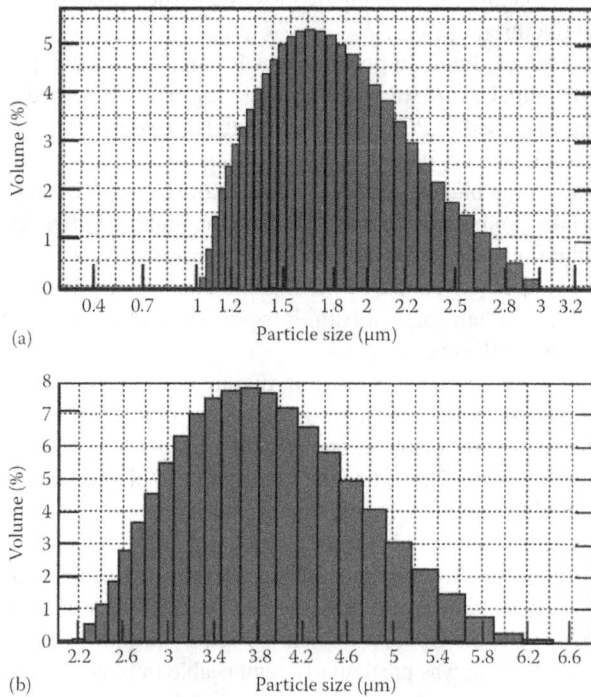

FIGURE 4.20 Particle size distribution of (a) 2 μm- and (b) 4 μm-graded diamond powder.

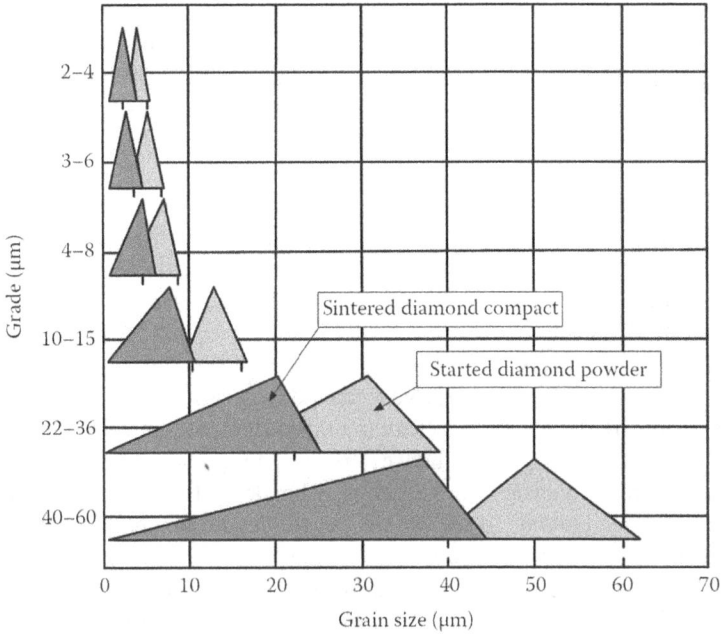

FIGURE 4.21 Schematic diagram showing the relationship between the grain size of diamond before and after sintering.

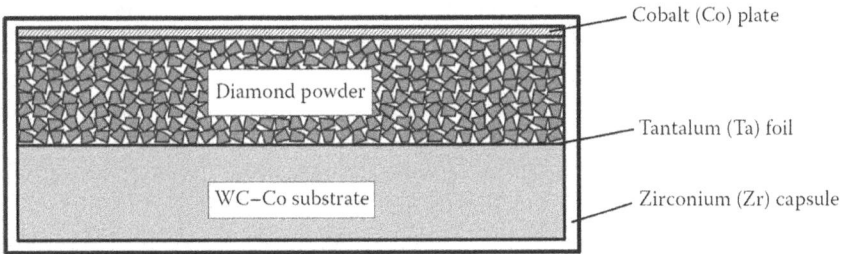

FIGURE 4.22 Sample capsule used by Uehara and Yamaha. (Uehara and Yamaya 1990.)

accepted in the consideration of PDC for gas and oil drilling bits (Bellin 2010). Note that Uehara and Yamaha (1990) used the sample capsule with an additional cobalt disk placed on the top of diamond powder and zirconium capsule while the tantalum foil was placed between the powder and the substrate as shown in the author's representation in Figure 4.22.

Later researchers, however, observed the opposite effect, that is, diamond grain growth, referred to as the abnormal grain growth (AGG) in PCD during HPHT sintering (Shin et al. 2004). Some grains preferentially grow to a size of several hundreds of micrometers that later on makes it impossible for the PCD disk to be cut by the wire EDM. It has been generally accepted that AGG can occur in the presence of liquid phases (Kang 2005). The coarsening, or growth, of solid grains during liquid-phase sintering is usually explained in terms of the Ostwald ripening (Lifshitz, Slyozov, and Wagner theory). Faceted particles are known to undergo AGG. Park et al. (1996) suggested a model in which, for grains dispersed in a liquid matrix being faceted with flat surface planes, the 2D nucleation would take place, resulting in AGG of certain crystals that are bigger than a specific minimum radius.

AGG in the fine-grained PCD can be controlled by the addition of fine WC powder. Hong et al. (1988) reported behavior of cobalt infiltration and AGG during the sintering of diamond on a cobalt substrate. Akaishi et al. (2001) studied AGG in the synthesis of fine-grained PCD compact and its microstructure. They found that the addition of a small amount of cubic boron nitride powder suppressed AGG. Yu et al. discussed methods to suppress AGG of diamond in their sintering of 0.5–1.5 μm diamond powder using Ni-Zr alloy as a sintering aid under HPHT (Shin et al. 2004). The best-reported results are achieved when the Co powder (approx. wt. 5%) of 1.5 μm in size and 0.8 μm WC powder as much as 0.35 wt% as a grain growth controlling agent are added while sintering took place using WC-10 wt% Co substrates (Shin et al. 2004).

Most available PCD blanks for drilling tools produced today have diamond grain size after HPHT sintering (*as-sintered*) of 1–30 μm. Finer, uniform, as-sintered diamond grain sizes, for example, of about 0.1–1 μm (referred to as submicron grades) have proven challenging to produce commercially using a common HPHT process where the compact of the diamond powder is placed on the top of a carbide substrate. This is not only because of AGG as described previously but primarily due to the low packing density of submicron diamond particles that cause problems during loading of shielding enclosures and HPHT processing. Very fine pores between the submicron diamond grains in the initial diamond particle mass are difficult to uniformly penetrate with catalyst metal, leading to incomplete formation of strong bonds between diamond particles. As a result, the high surface areas of submicron diamond powders are not properly bonded. This is the actual cause for AGG.

One of many possible solutions to the problem is to blend diamond particles and particles of catalyst metal together. As such, the size of catalyst particles should be less than that of diamond particles (Russell et al. 2016). Then, the resultant blend is sintered using a pressure and a temperature for a time sufficient to affect intercrystalline bonding between adjacent diamond particles. The catalyst metal, normally cobalt, may be about 0.5–15 wt% of diamond powder blend.

The foregoing consideration reveals that the diamond powder is a mix that may contain a lot of different technological additives to promote sintering (e.g., Co powder) and to prevent AGG (e.g., WC and/or Ni-Zr powder). The amount of these additives as well as their geometrical and physical characteristics is not disclosed in the PCD blank characterization. As a result, PCD cutting elements made of the same PCD grain size by different PCD blank manufacturers may exhibit considerably different machinability (e.g., grindability, sensitivity to heat, EDMability, etc.) in the cutting tool manufacturing and cutting performance in machining (e.g., tool wear, sharpness of the cutting edge, and adhesion to the work material).

4.7.5 Grain Size

A single-crystal diamond is extremely hard and has very high abrasion resistance and thermal stability. At the same time, it is a highly anisotropic material, that is, the properties of a single-crystal diamond are not the same depending on the crystallographic plane in which they are measured. This allows the natural gemstones to be cut along specific *cleavage* planes where the energy required to split the crystal is at its minimum. In a diamond-sintered compound, all the weak crystallographic planes are randomly oriented so, at a microscopic scale, this compound behaves isotopically (the properties do not depend on the load direction) with improved impact strength.

In selecting a particular grain size for a given application, it can be assumed that the smaller the size of the diamond crystals sintered together, the higher the abrasion resistance of a PCD cutting element. Moreover, such a cutting element can be easily EDMed and then finish ground (polished) to obtain a cutting edge of high quality with minimum serrations. When high requirement to surface roughness is the case, fine-grained PCDs (less than 1 μm) should be used. For normal PCD-tipped reamers, 5–10 μm grain size is a common choice. Fine-grained PCDs are normally more expensive because:

- Initial diamond powder should be of good quality with a narrow distribution of particle sizes (Figure 4.20);
- Special recipes for powder compact and tight control of the sintering process parameters are required to prevent AGG;
- Considerable higher pressure and temperature (compared to coarse-grained PCDs) are needed for their sintering.

However, these good properties come at the expense of lower impact strength compared to coarse-grained PCD grades.

Coarse grades are normally used for interrupted and/or high-impact cutting conditions, for example, in milling where PCDs with 30 μm or even higher grain sizes are recommended by PCD manufacturers. Tool manufacturing, particularly EDMing and grinding coarse-grained PCDs, presents problems and generally requires much more time to accomplish which also adds significantly to the tool manufacturing cost.

Figure 4.23 shows tool life and tool edge roughness values for fine (2 μm)-, medium (10 μm)-, and coarse (25 μm)-grained PCD tools with edges prepared by spark erosion and used to edge mill a ceramic-impregnated layer (Cook and Bossom 2000). When machining under harsh conditions, the differences in PCD wear resistance become more evident, therefore coarse-grained PCD tends to be used in such conditions. When machining under moderately harsh conditions, the PCD wear resistance is of less importance, and factors such as edge quality/surface finish must be considered; therefore, medium- and fine-grained PCD grades tend to be used.

The discussed relationship is not completely linear (Cook and Bossom 2000) even under harsh machining conditions; the tool life of PCD can deteriorate if grain size is increased significantly beyond approx. 25 μm. Figure 4.24 shows the performance of various PCD grain size products in milling a ceramic-impregnated surface of a flooring board (Cook and Bossom 2000). As can be

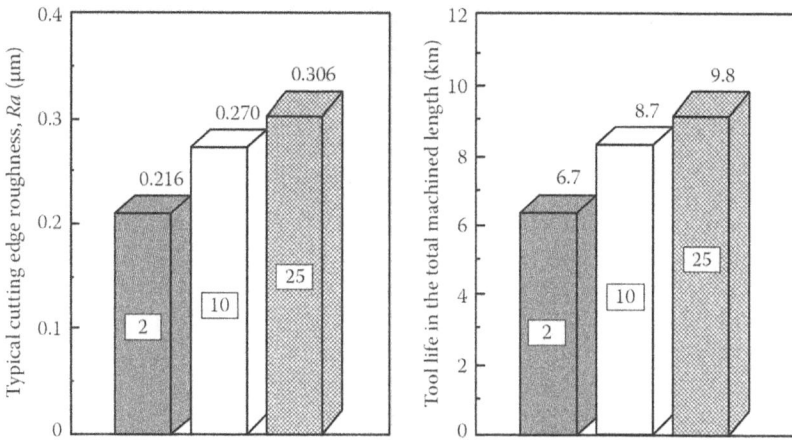

FIGURE 4.23 Relationship between PCD grain size, edge roughness, and tool life in end milling.

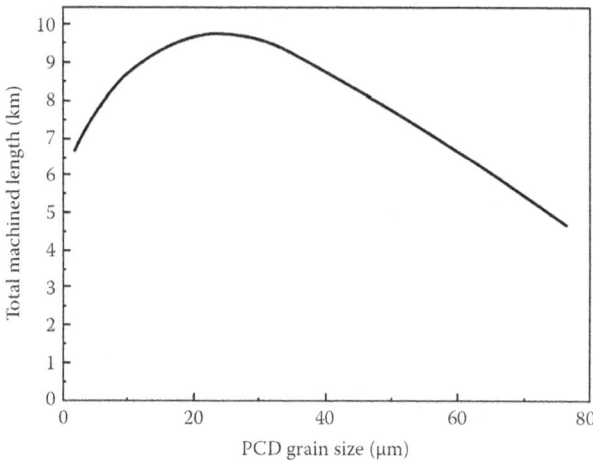

FIGURE 4.24 The performance of various PCD grades milling ceramic-impregnated high-pressure laminates flooring board.

seen, the tool life achieved with an ultra-coarse-grained PCD (75 µm) is lower than that achieved with coarse- and medium-grained products. While an ultra-coarse PCD has the theoretical abrasion resistance required for increased performance, the coarseness of the particles results in a substantially rougher cutting edge, which has a significant negative influence in contributing to overall tool performance. Therefore, by maintaining grain size at 25 µm and modifying the conditions of diamond synthesis, it has proven possible to produce a product with increased abrasion resistance but without reducing the edge quality characteristics.

Although the above-presented test data on the influence of the grain size on tool life seems to be in the direct contradiction with the assumption made about higher wear resistance of fine grain, it is not so in reality. This is because the test data presented above and often used in many trade publication/sales presentations are application-specific due to the following:

- The obtained in end milling, i.e., in highly interrupted cut so that impact resistance of PCD was of high importance. Greater tool life for coarse grade shown in Figure 4.23 can be explained by the higher impact resistance of these grades.
- The tool was finished with electrical discharge grinding (EDG), the method not preferable for fine grades (discussed later in this chapter).
- The work material is ceramics, i.e., a very brittle material having no solid abrasive faces and significant springback. As such, there is no chip sliding on the rake and significant friction on the flank faces so that the PCD abrasion resistance was not even a factor in the tool life consideration.
- Quality of the machining was not considered as one cannot make a high-quality surface with an edge roughness of 0.306 mm, i.e., with an excessively dull edge.

The author's experience shows that only fine PCD grade finished with grinding to obtain sharp cutting edge should be used in PCD drilling tools when quality, productivity, and efficiency of machining operations are considered. This is particularly true when the work material contains abrasive solid phase as, for example, HSAAs are widely used in the automotive industry. However, it often happens that drilling tools should cut through crossed holes, veins, or experience some hard inclusions in the work material (Astakhov and Patel 2018). In this case, a suitable combination of edge quality, abrasive wear resistance, and impact strength is highly requested.

When designing a PCD grade for a specific application, it is possible to mix together diamond powders with considerably different average particle sizes and dimensional statistical distributions. It is believed that such mixing results in the achieving of a good degree of powder packing as it minimizes the empty spaces between crystals and favors a good sintering process during the HPHT cycle, delivering a PCD with superior toughness and abrasion resistance. Figure 4.25 shows a grain

FIGURE 4.25   Secondary electron micrograph of PCD containing the grains of two distinctive sizes.

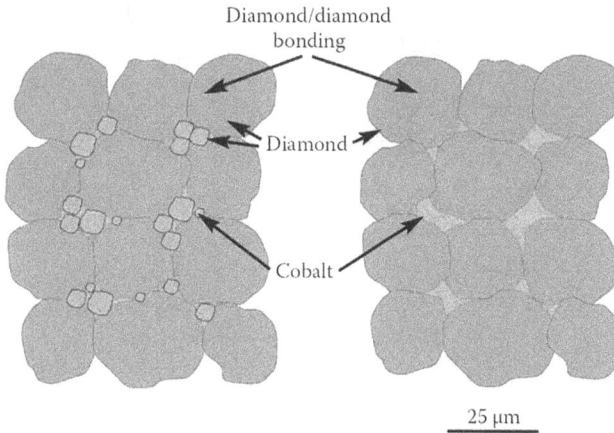

FIGURE 4.26 Schematics of the packing density of CTM 302 and CTB025 PCD grades (Element 6 Co.).

size distribution of the diamond phase consisting of a coarse fraction with an AGS of 35 μm with the intergranular space filled with a fine grain fraction with an average size of 6 μm (Boland et al. 2010).

One of feasible improvement in PCD property can be achieved with the so-called multimodal PCD grades. E6 Co., one of the major suppliers of PCDs for the metal cutting industry, markets CTM 302 PCD grade. CTM contains a proprietary mix of micron diamond particle sizes between 30 and 2 μm. Figure 4.26 shows schematics of the packing density of CTM 302 and CTB025 PCD grades. A combination of this mix and carefully controlled sintering conditions increases the diamond packing density, which increases wear resistance. The improved packing density in turn results in a higher degree of contiguity between diamond grains, thereby enhancing the chipping resistance of the PCD cutting elements. An added advantage of the increased packing density is the quality of the ground edge, which is superior to a normal coarse-grained PCD. Filling the areas between coarse diamond grains with finer diamond yields a continuous as opposed to a rougher more irregular cutting edge. E6 claims improvement in the wear resistance of CTM 302 grade compared to the usual CTB025 grade although the author's experience shows that this is not always the case. The problem is not in the grade itself but in the experience of PCD tool manufacturers to finish up this grade as it requires the proper processing to achieve the required quality of the cutting edge and surface roughness of the ground flank face. In other words, they should learn how to finish up this multimodal grade in tool manufacturing.

The selection of the proper grain size is not a simple or straightforward task because the grain size affects the performance of a given cutting tool. This is because of the following:

- PCD blanks even of the same grain by different manufacturers are not the same in terms of composition (and thus properties/performance) of PCD layer. Consistency of PCD blank quality is of prime concern in many PCD applications for drilling tools.
- The quality of the cutting edge depends not only on the grain size but also on the manufacturing operations (and their regimes) used to make this edge, namely, EDMing, EDG, grinding, polishing, and lapping.
- The performance of PCD of a particular grain size depends on the work material, that is, its properties, chemical composition, inclusions, and porosity.
- The performance of PCD depends on the machining regime chosen for a drilling/reaming application and on the properties of the drilling system. Particularly, the cutting speed and the system rigidity are of prime concern.
- In HSM, the coolant properties and its application technique make significant contributions to PCD performance.

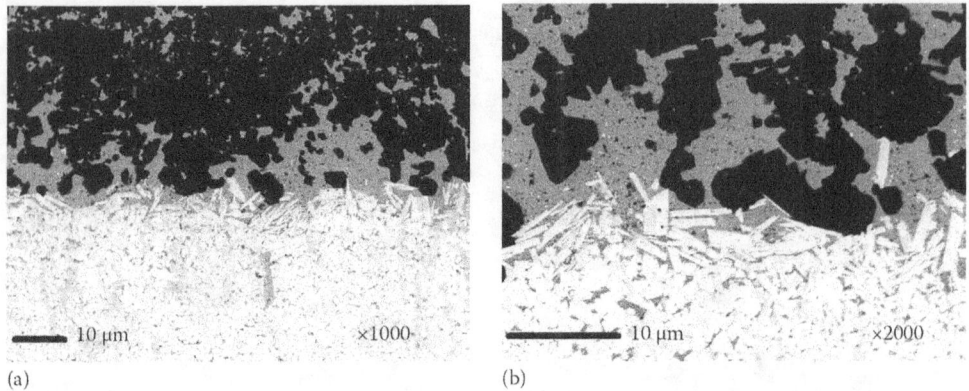

FIGURE 4.27 Micrographs of the PCD-carbide substrate interface at different magnifications: (a) ×1,000 and (b) ×2,000.

4.7.6 INTERFACES

There are three interfaces in a typical PCD cutting element: (1) the top of the PCD layer that eventually becomes the rake faces of the tool, (2) the bottom of the carbide substrate that is used to attach this cutting element to the tool body, and (3) the interface between the sintered diamond layer and the cemented carbide substrate. In the literature on PCD, the prime attention is paid to the third interface while the other two are left aside although they play an important role in PCD material performance in cutting tools. Note also that these interfaces are not considered neither in the literature on PCD tool design nor in PCD blanks selection and their quality considerations as well in the failure analyses of PCD cutting tools.

The prime attention to the interface between the sintered diamond layer and the cemented carbide substrate is paid because this interface not only provides the necessary strength so that the cutting element can manage the static and dynamic shear loads that otherwise would cause the diamond layer to delaminate, but it also has to handle the residual stresses that arise within both the substrate and the PCD layer as a consequence of the HPHT sintering process. Moreover, as clearly seen in Figure 4.27, this interface is rich in cobalt, which creates some problems in EDMing PCD blanks as discussed later in this chapter.

As discussed previously, PCD blanks are produced by the HPHT process. Temperatures from 1,350°C to 1,800°C and pressures from 5 to 9 GPa with an isostatic holding are used in this process that ensures the formation of a strong PCD layer and a reliable bonding of it with a carbide substrate. At the same time, the high thermobaric parameters cause considerable residual stresses in a finished product because of differences in bulk modulus and thermal expansion coefficients between PCD and carbide. Residual stresses are generated during the cooling stage when the PCD layer is already fully sintered. The thermal expansion coefficients mismatch between PCD layer and substrate causes the carbide to shrink more than the top PCD layer, forcing it to bend outward. The tensile state of stress within the diamond layer becomes worse as its thickness increases, leading in some instances to a spontaneous delamination of the top layer (Bellin 2010).

It is important for the sintering process optimization and PCD blanks processing to determine the level of residual stresses in the PCD layer and in the carbide substrate. Using the micro-Raman spectroscopy, Vohra et al. (1996) measured the compression residual stresses up to 1.3 GPa in the diamond layer and of tensile stresses from 0.87 to 0.95 GPa in the carbide substrate. Krawitz et al. (1999) measured the compression residual stress of 0.47 GPa in a diamond layer by neutron diffraction. According to Paggett et al. (2002), the mean stress in a diamond layer plane ranges from 0.25 to 0.58 GPa.

It is evident that the previous data should be considered as the indicative ones in view of the difference in PCD blanks geometry and production technology as well as in the place, method, and accuracy of the stress measurements. However, the authors of the referred papers agree that residual thermal stresses bring about an essential decrease in the PCD blank mechanical strength. According to Chen and Xu (2010), up to 70% of PCD blank premature failures, including brittle fracture of the diamond layer and substrate, stratification, spallings, and cracking are caused just by residual stresses. Therefore, the analysis of residual stresses is a necessary stage in designing PCD blanks, and the development of methods to decrease these stresses is a line of investigation, which is promising for the improvement of the cutting insert quality. These methods include techniques (temperature gradient, annealing, stabilization, etc.) used both in the sintering process and after its completion. It explains the above-discussed difference in the cost of PCD disks from various manufacturers. As even advanced PCD tool manufacturers have no means to control the real quality of PCD blanks, the only proper way is not to buy the cheapest PCD grades but rather work closely with leading PCD blanks' suppliers to obtain high-quality products and to gain understanding through sharing technical information between PCD manufacturers and PCD tool manufactures to improve PCD overall performance in tool applications for various work materials.

4.7.7 Quality Control

4.7.7.1 Common Practice

Normally, PCD disks are supplied with no information on their quality although the manufacturing lot number is available for each PCD disk so that a user can hypothetically request the PCD disk quality data should a problem occur in manufacturing. An exception is E6. According to catalog of E6, it "supplies a unique ultrasonic scan depicting the PCD layer profile. The PCD scan indicates a 'North Point', which matches a 'North Point' laser marked on the disk, allowing users to optimize the cutting areas." Two examples are shown in Figure 4.28. It can be seen that only variation of PCD layer thickness is indicated on the scans. Although, this information can be theoretically useful if the pockets for PCD inserts in carbide bodies can be adjusted correspondingly, which is rather an ambitious goal. It is not very practical as PCD tips for drilling tools, where the same pockets are used on tool re-tipping.

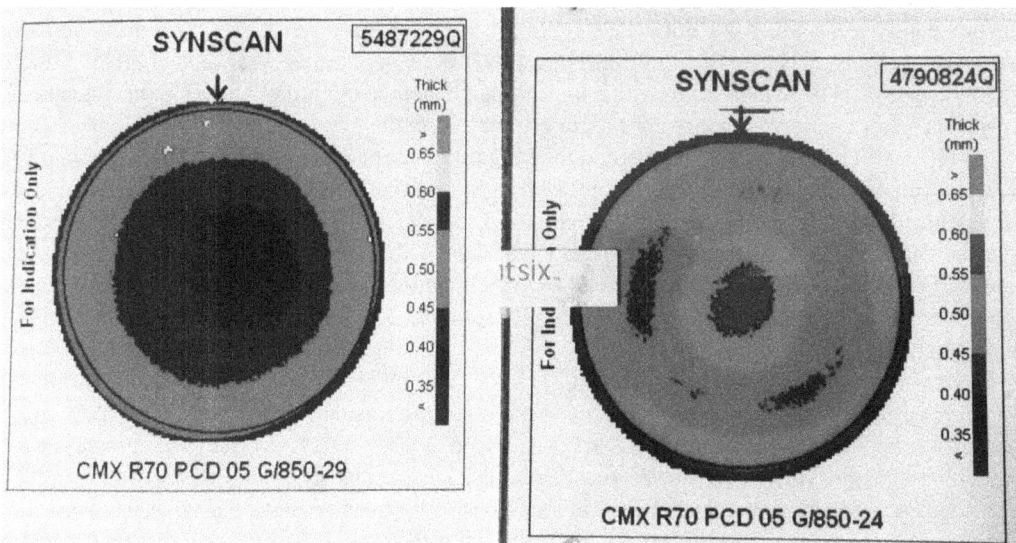

FIGURE 4.28 C-scans from E6 Co.

4.7.7.2 Laboratory Testing

Laboratory testing of PCD requires a well-equipped laboratory and highly qualified/trained person-nel preferably with advanced degrees in composite tool materials. This is because the preparation of specimens for testing involves a series of rather complicated procedures and interpretation of the test results is subjective mainly based on advanced knowledge and vast experience in the field. For example, when SEM is used to image the surface topography (secondary electron imaging) and composition of a PCD sample (backscatter electron imaging), backscatter imaging mode is utilized. In this imaging mode, diamond appears black, and tungsten white with the binder phases being a mid-gray in color. It requires a lot of knowledge for proper interpretation of the obtained results, which are not numerical, i.e., cannot be quantified.

The common laboratory testing involves the image analysis technique. Using this technique, the following can be determined:

- Diamond grain size (μm) (mean, standard deviation, distribution, and maximum size using the standard Saltikov method) (Gublin 2008, Jeppsson et al. 2011)
- Cobalt pool size (pm) using the standard Saltikov method (Gublin 2008, Jeppsson et al. 2011)
- Diamond content area (in % of area)
- Diamond MFP (μm)
- Cobalt MFP (μm)
- Diamond contiguity (in % of area)
- WC-phase area (in % of area)
- WC pool size (μm)
- Cobalt next neighbor distance (μm).

Being important to characterize PCD structure, these characteristics are not yet correlated with the performance of PCDs as the tool material. Nevertheless, they are supplied by PCD manufactur-ers to PCD tool manufacturers/users in the course of a failure analysis, which does not provide a significant help in finding the root cause of the failure as the latter cannot make any use of these characteristics.

Unfortunately, there are no standard manufacturing testing methods available for PCD tool materials used in the cutting tools. Perhaps, the most convenient, cost-effective, and reliable test for the quality of PCD/CVD components is a wear test. The pin-on-disk or pin-on-drum abrasive wear tests are unsuitable for such ultrahard materials because the tool wear is negligible. A more aggressive test has been devised in which the cutting element is monitored during a turning/cutting operation on an industry-standard corundum grinding wheel.

A very useful laboratory testing method for PCD disk is ultrasonic scanning (Bellin 2010). It is a non-contact screening method offering unique insight into the PCD cutter or blank. Its advantages include the detection of void formations, delaminations, cracks, fractures, as well as other hidden internal defects within inherently susceptible materials. The C-scan mode provides a planar view image at any specific depth. Much like the MRI machine used to examine patients at hospitals, the separation of the PCD layer can also be tested; illustrations of delamination and void formation can be generated.

Figure 4.29 shows C-scan images of PCD disks (Worldwide Solids, a division of Worldwide Superabrasives LLC) made using an acoustic emission microscope Sonoscan D9600. As can be seen, voids and delamination can be clearly visualized. Moreover, PCD layer thickness variation, bond interface uniformity (PCD layer to carbide), disk surface flatness, pooling, and pitting can be inspected. A problem is that such a scan is time-consuming so it cannot be used as an inspection method in mass production of PCD disks. However, when the process is stable, a periodic inspection of PCD disks made assures the required quality in full analogy with selected parts inspection in the automotive industry. Unfortunately, this is not nearly the case in the tool industry.

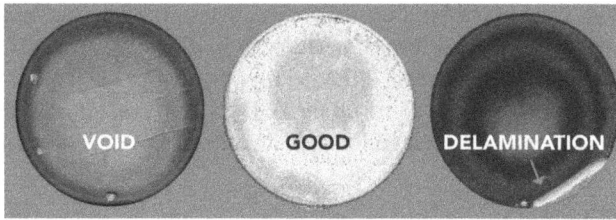

FIGURE 4.29 C-scan images of PCD disks by Worldwide Solids, a division of Worldwide Superabrasives LLC.

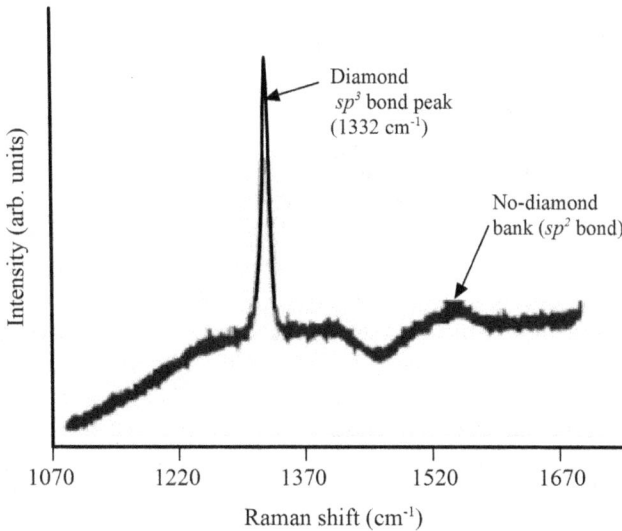

FIGURE 4.30 Raman spectra obtained using 532 nm wavelength for a microcrystalline diamond.

Although the use of the ultrasonic scanning technique provides some information on PCD disk quality useful for its application, it does not reveal the properties/quality of the internal structure of PCD, e.g., the strength of PCD bonds, microdefects, etc., i.e., to reveal what is the root cause of the defects found so that the disk manufacturing process can be optimized to minimize the occurrence of the listed defects. Moreover, the ultrasonic technique does not allow to compare the quality of PDC/PCD cutters from various manufacturers. Therefore, a much more sophisticated technique, for example, X-ray diffraction of Raman spectroscopy, is used (Jaworska et al. 2014, Abbas and Musa 2019).

Raman spectroscopy is a popular technique for the analysis of diamond graphitization (Jaworska et al. 2014, Zhao et al. 2015, Abbas and Musa 2019, Auciello and Aslam 2021). Raman spectroscopy was named in the honor of its inventor, C.V. Raman, who, together with K.S. Krishnan, published the first paper on this technique (Raman and Krishnan 1928). Its basic principles and theory are well-developed (Larkin 2011, Xiao 2014). A number of Raman spectroscopes are commercially available at a relatively low cost (starting from $5,000) so that even a small PCD tool manufacturer can afford it.

Raman spectroscopy is a robust identification method for different allotropes of carbon. Focusing a laser beam of 2 μm on the surface of the area of interest of the PDC/PCD insert, a Raman spectrometer is used to observe the scattered light by the specimen. As sp^3 bonds attributed to diamond and sp^2 bonds attributed to graphite (see Section 4.3.1), we have different peaks in a plot of the Raman-Shift wavelength vs. the intensity. Genuine diamond shows only a single Raman line at 1,332 cm^{-1} (known as the sp^3 or D peak) at room temperature; natural single-crystal graphite exhibits a single Raman, the so-called sp^2 or G peak at 1,580 cm^{-1} that is associated with the carbon–carbon stretching mode. Figure 4.30 shows an example of Raman spectra.

The quality of diamond is assessed by the sp^3/sp^2 ratio determined by the height ratio of the corresponding intensity peaks. The phase transformation from diamond carbon to amorphous carbon takes place during diamond processing/use if the sp^3/sp^2 ratio decreases. Therefore, taking a Raman spectrogram helps to understand the graphitization PCD in its manufacturing using various operations/regimes. A great example is presented for such an assessment by Rahim et al. (2015b).

4.8 THERMAL STABILITY

4.8.1 EFFECT OF TEMPERATURE ON PCD STRUCTURE

Thermal stability of PDC/PCD is its Achilles heel, i.e., is a critical weakness in spite of the overall high strength and hardness of this material, which can lead to downfall at any stage of its processing and cutting tool application. The vital importance of taking care of thermal stability of PCD cannot be overstated. Therefore, a real understanding of the issue, which is not nearly the case in the PCD tool industry and, moreover, poorly covered in the literature on the subject, should lead to proper manufacturing and implementation of PCD tools.

The previously described PCD tool material is known to be thermally unstable. In other words, once a PCD layer is formed, it starts its reverse transformation into carbon. However, the reverse reaction is extremely slow at room temperature. This is not the case, however, at high-temperature involved manufacturing of PCD tools and in application, particularly in HSM, and when the work material is highly abrasive, that is, creates a lot of heat due to friction at the tribological interfaces or when the work material is of low thermal conductivity as, for example, fiber-reinforced plastic materials. The author's experience shows that the stage of manufacturing of PCD tools is of prime concern as at this stage the overheat of PCD routinely occurs. It is to emphasize that such an overheat causes up to 80% of PCD tool premature failures whereas only 20% is caused by overheat in tool operation.

Commonly PCD suffers from thermal instability when exposed to high temperatures in excess of 700°C (Sneddon and Hall 1987, Mehan and Hibbs 1989, Lin et al. 1992, Li et al. 2016, Jivanji et al. 2023), the critical temperature at which degradation begins is 400°C as discussed below. It is recognized that this instability leads to a reduction in wear performance and premature failure of PCD tools (Westraadt et al. 2015). It is also noted that during tool manufacturing and tool performance, PCD experiences high temperatures twice. The first time is on brazing of PCD and finishing, where high-temperature brazing is used to fix the material to the tool body and to finally shape PCD inserts. The second time is during drilling or machining operations where frictional forces at the contact surface are converted into heat energy during operation (Bellin 2010, Jianxin et al. 2011, Westraadt et al. 2015).

In the author's opinion, there are two major drawbacks in the consideration of PDC thermal stability. The first one is that the influence of thermal degradation on brazing was not considered in the corresponding literature. Instead, tests on thermal stability are performed as cutting tests (Westraadt et al. 2015). Moreover, every attempts to develop the cutting theory based upon the metal cutting models is known (Che et al. 2012a, b) although rock drilling has a little correlation with metal cutting (no plastic deformation of the layer being removed as the major source of heat in metal cutting Astakhov and Xiao 2016). The results obtained are qualitative with little practical significance. The second drawback is that the influence of PDC finishing operation as, for example, EDMing and EDG are not considered as contributors to thermal degradation.

Based on the analysis of the publication on thermal stability of PCD, Astakhov (2014) and Westraadt et al. (2015) pointed out that two different hypotheses have emerged as a means of explaining the influence of the catalysis on the thermal stability of PCD:

1. Thermal degradation due to differential thermal expansion characteristics between the interstitial catalyst material and the intercrystalline bonded diamond. The metallic sintering catalysis (e.g., cobalt) has a greater coefficient of thermal expansion (α) than diamond. At elevated temperatures, cobalt ($\alpha = 14 \times 10^{-6}$/°C) expands more than the rigid diamond skeleton ($\alpha = 1 \times 10^{-6}$/°C) causing stresses that are thought to contribute to cracking and

microchipping during machining operations (Mehan and Hibbs 1989, Lin et al. 1992, Bellin 2010). Such differential thermal expansion is known to occur at temperatures of about 400°C, causing ruptures to occur in the diamond–diamond bonds and resulting in the formation of cracks and chips in the PCD structure. Chipping and microcracking of cutting edges occur as a result as discussed by Miess and Rai (1996), Vandenbulcke and De Barros (2001), and Chen et al. (2007). Note that PCD material manufacturing specialists do not agree with this hypothesis.

2. Thermal degradation due to the presence of cobalt in the interstitial regions and the adherence of cobalt to the diamond crystals. Specifically, cobalt is known to cause an undesired catalyzed phase transformation in diamond converting it to carbon monoxide, carbon dioxide, or graphite, or causing degradation of the strong diamond sp^3 bond into weak sp^2 bond (see Section 4.3.1) with increasing temperature as discussed by Fedoseev et al. (1986), Coelho et al. (1995), Cook and Bossom (2000), and Shimada et al. (2004). The conversion process involves a density change from 3.6 to 2.0 g·cm^{-3} as the diamond is converted to graphite. This volume increase causes stress on the weakened diamond network, which leads to cracking of PCD (Westraadt et al. 2015).

Needless to say that PCD materials science specialists support the second hypothesis. They just cannot accept the first hypothesis arguing by putting forward a counter argument to the thermal expansion hypothesis that the cobalt metal might plastically yield before it could apply enough stress to cause cracking of the diamond structure. This argument is reputedly supported by the results of pin-on-disk experiments performed at high temperatures that showed that cracking of the PCD microstructure only occurred at temperatures above 700°C, where graphitization of the material could be a contributing factor (Jianxin et al. 2011). Specialists in rock and metal cutting are proponents of the first hypothesis pointing out that the pin-on-disk test technique is not relevant in cutting and that thermal degradation of PDC starts at much lower than 700°C temperatures (Bellin 2010, Astakhov 2014). The experimental techniques used by PCD materials science specialists are not capable of detecting weakening of diamond–diamond bonds. It can only detect PCD thermal degradation of grain boundaries when it becomes really noticeable, i.e., when the temperature is more than 700°C.

As pointed out by Westraadt et al. (2015), other factors that may affect PCD thermal stability are considered to a lesser degree. Among them, oxidation of the diamond, oxidation of cobalt, the macro-residual stress state between the diamond layer and the tungsten carbide backing, and chemical reactions between the rock/cooling fluids used during operations.

4.8.2 Improving Thermal Stability of PCDs

Although applications of PCD tools are rapidly developing, standard PCD grades have a substantial disadvantage with their low thermal stability. In the meantime, in OGD, drill bits equipped with thermally stable polycrystalline diamond (TSPD) elements demonstrate high performance. On the basis of analysis of different publications (Akaishi et al. 1990, 1991, 2001, Wilks and Wilks 1991, Field 1992, Tomlinson and Clark 1992, German et al. 2005, Kang 2005, Boland et al. 2010, Chen and Xu 2010, Liu et al. 2017) in which there are descriptions of properties of the PCD inserts and TSPD elements, a comparison of the basic physical and mechanical properties of these superhard materials can be made. The data show (Table 4.2) that the most important operating mechanical properties of the diamond-containing layer of PCD inserts and TSPD elements are related to each other, such as hardness and wear resistance. Characteristic values of strength of PCD inserts are better than TSPD elements. However, the latter keep their own mechanical properties up to 1,200°C while mechanical properties of PCD blanks quickly decrease after heating at temperatures above 700°C. A need is felt to develop a technology to manufacture a two-layer composite of diamond-containing/cemented tungsten carbide substrate, which has a combination of high wear resistance and high thermostability.

TABLE 4.2

Basic Properties of Standard PCD Inserts, Thermally Stable PCD Elements, Synthetic Monocrystalline Diamond and Cemented Tungsten Carbide

Property	Standard PCD Insert	Thermal Stable PCD Element	Synthetic Monocrystalline Diamond	Cemented Tungsten Carbide (WC/15%Co)
Density (g/cm³)	3.90–4.10	3.42–3.46	3.51–3.52	13.80–14.10
Knoop microhardness (GPa)	50–60	50–55	70–105	15–18
Compressive strength (GPa)	7.5–8.5	4.0–5.5	7.0–9.0	12.0–14.0
Fracture toughness (MPa m$^{1/2}$)	9.0–10.0	7.0–8.0	3.0–4.0	11.0–12.0
Young's modulus (GPa)	810–850	920–960	1,000–1,150	70–100
Thermal conductivity (W m^{-1} K^{-1})	550–750	120–250	700–1,500	70–100
Wear resistance coefficient[a] (mm)	0.15–0.20	0.20–0.25	–	>2.0
Thermal stability (°C)	700	1,200	1,200	1,400

[a] Value of linear wear of inserts after sandstone block testing.

Up to date, attempts to improve thermal stability of PCD can be generally classified as follows:

1. Removing the solvent-catalyst phase from the PCD material, either in the bulk of the PCD layer or in a volume adjacent to the working surface of the PCD tool (where the working surface typically sees the highest temperatures in the application because of friction events). This is commonly known as *cobalt leaching*.
2. Two-step sintering.
3. Modify the catalysis by adding materials to cobalt powder. For example, Jivanji et al. (2023) studied ZrB_2 additions on the microstructure, thermal stability, and thermo-mechanical wear resistance of PCD showing that the onset of bulk graphitization occurred above 1,000°C. However, lowered abrasion wear resistance of PCD was observed due to the reduced diamond contiguity (diamond–diamond intergrowth).
4. Use the catalysis made of ceramic matrix materials such as SiC instead of cobalt. According to Westraadt et al. (2015), various ceramic binder systems have been investigated for PCD; these include SiC, TiC, and MAX phases. The major drawback of this approach is the lack of an intergrown diamond network. As a result, ceramics matrix PDC composites have significantly reduced mechanical properties (toughness, hardness, wear resistance) when compared to conventional PCD.
5. Sintering the diamond in the absence of the cobalt binder (Irifune and Sumiya 2014). It became possible with the availability of nanodiamond particles (Shenderova 2014). Ultrahard nano-polycrystalline diamond (NPD) material is formed by directly converting graphite/carbonaceous material at very high pressures (15 GPa) and temperatures (2,300°C) into diamond. The results of the cutting tests carried out by Harano et al. (2012) showed that the allowable cutting speed was approximately 10–20 times higher than that with PCD tools in turning of Al–Si alloy and 3–5 times higher than that with PCD tools on turning of cemented carbide. Moreover, detailed observations of the cutting edge after the cutting tests indicated that the excellent cutting performance of the tool was the result of its isotropic feature and fine texture without any binder.

The essence of cobalt leaching is as follows (US Patent Nos. 4,224,380 and 4,288,248). After finishing PCD conventionally, the metallic phase is removed from the compact by acid treatment, liquid zinc extraction, electrolytic depleting, or similar processes, leaving a compact of substantially 100%

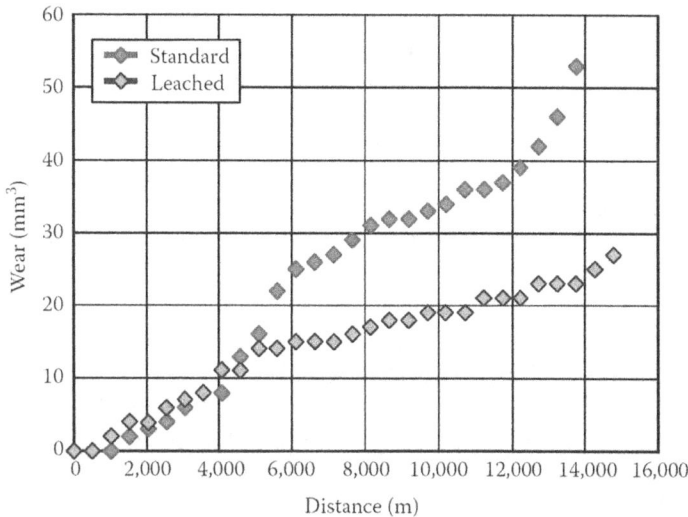

FIGURE 4.31 Wear curves for leached and non-leached PDC cutters subjected to a severe wear test.

abrasive particles that remove the cobalt phase through an acid etching process called leaching. For drilling applications, cobalt is removed up to 200 μm deep into the PCD layer. As the cobalt phase remains inside the PCD layer, the loss of overall strength is not as significant as with TSPD manufactured using other known methods.

Leaching a thin layer at the working surface dramatically reduces diamond degradation and improves the tool's thermal resistance. First, with no cobalt, the diamond–diamond bonds remain strong as little graphitization and cracking due to mismatch in thermal expansion properties occurs at high cutting temperatures. Second, the heat conduction of the contact diamond surface increases; thus, the heat is transmitted faster from the tool–chip and tool–workpiece that lowers the maximum cutting temperature.

Figure 4.31 shows wear curves for leached and non-leached PDC cutters subjected to a severe wear test (Bellin 2010). As can be seen, after a sliding distance of about 5,000 m, the volumetric wear rate of the unleached cutter increased dramatically. The leached PDC cutter, in contrast, maintained a relatively constant wear rate for about 15,000 m.

It has been found (US Patent Nos. 4,224,380 and 4,288,248) that the leached PDC can withstand exposure to temperatures up to 1,200°C–1,300°C without substantial thermal degradation. SEM analysis revealed that the non-leached samples exhibited many different characteristics when compared to the leached samples. The metallic phase begins to extrude from the surface between 700°C and 800°C as viewed under 2,000× magnification. As the temperature was increased to 900°C, the samples cracked radially from the rounded cutting edge to the center of the sample. The leached samples did not exhibit this behavior but were relatively unchanged until 1,300°C. The diamond layers were clean at 1,200°C, but at 1,300°C, the edge looked rounded and fuzzy at 20× magnification. The images taken at 1,000× magnification showed an etched surface with many exposed crystals due to thermal degradation of the surface.

Sumitomo Co in its online catalog presented WD800 (Heat-Resistant type) PDC so it is obviously made for OGD. According to the catalog, the WD800 Series are self-supported thermally stable blanks. These blanks are especially suited for applications that require high temperatures in mounting and high heat resistance in drawing. It is stated in WD800 Series, "binder (it should be a catalysis, *aut*.) metal removed from WD700 series, is high heat resistant material." A rather blurry micrograph is shown with no additional explanations. This grade is not offered for PCD disks for MWI.

There are several important issues associated with this approach (cobalt leaching) to achieve improved thermal stability. The prime concern is that a continuous network of empty pores resulted from the leaching, possessing a substantially increased surface area, which can result in increased vulnerability to oxidation (particularly at higher temperatures). This can then result in reduced strength of the PCD cutter at high temperatures. The second concern is the time needed for leaching, for example, according to US Patent No. 4,288,248 between 8 and 12 days, and strong chemicals used, such as hot concentrated acid solutions.

Two-step sintering results in the formation of a bi-layered sintered PCD disk, which includes a thermally stable top layer. According to US Patent No. 4,944,772, a leached PCD compact and cemented carbide support is separately formed. An interlayer of unsintered diamond crystals (having the largest dimension of 30–500 μm) is placed between the carbide and thermally stable PCD (TSPCD) layer. A source of catalyst/sintering aid material is also provided in association with this layer of interposed crystals. This assembly is then subjected to the HPHT process, sintering the interlayer and bonding the whole into a bi-layered supported compact. In this application, appreciable re-infiltration of the TSPCD layer is not seen as advantageous, but the requirement for some small degree of re-infiltration is recognized in order to achieve good bonding.

4.8.3 PCD Overheating – Results

The measures to improve the thermal stability of PDC/PCD discussed in the previous section are used exclusively in OGD whereas practically nothing has been attempted to improve the thermal stability of PCDs used in MWI. The chief problem for the latter is that this issue with PCD is totally ignored in the design, manufacturing, and application of PCD cutting tools in MWI. In the author's opinion, no further progress with PCD tool in MWI in terms of their productivity and efficiency can be achieved unless the problem with thermal stability is fully recognized and practical measures to address this problem at different levels are developed and implemented. As such, no need to invent something new – rather the experience already gained with PDC tools in OGD should be used. Unfortunately, there is no known to the author attempts were made in this direction

PCD tools in MWI fail due to the loss of thermal stability of PCD cutting inserts/tips and their exposure to high temperatures caused by overheating of these tools. Analyzing a great number of PCD tool premature failures, the author developed the following classification of PCD tool overheating levels:

Light overheating of PCD. This kind of overheating causes the development of a network of microcracks in the PCD structure that make this structure weaker. When the stresses of cutting are applied, the chipping of the PCD structure takes place. It could range from chipping of relatively small portions of the PCD structure as shown in Figure 4.32 to severe chipping of the whole PCD layer up to the carbide substrate as shown in Figure 4.33.

FIGURE 4.32   Chipping of PCD due to *light* overheating.

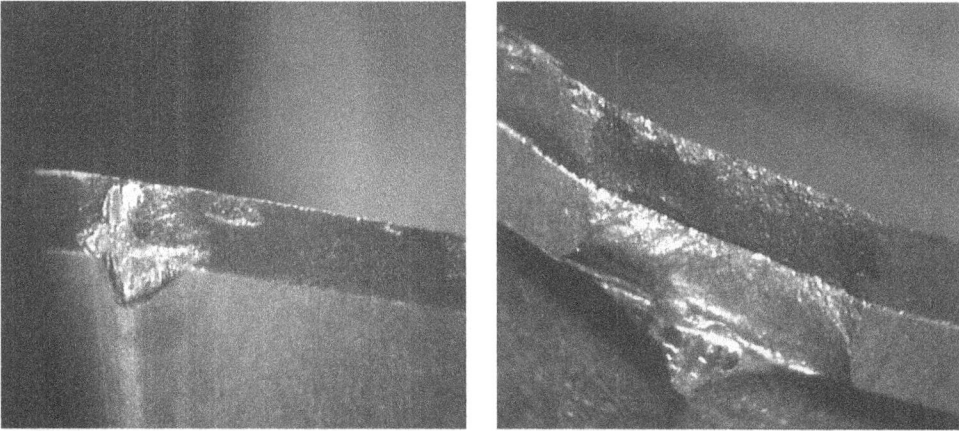

FIGURE 4.33 Example of micro-flaking.

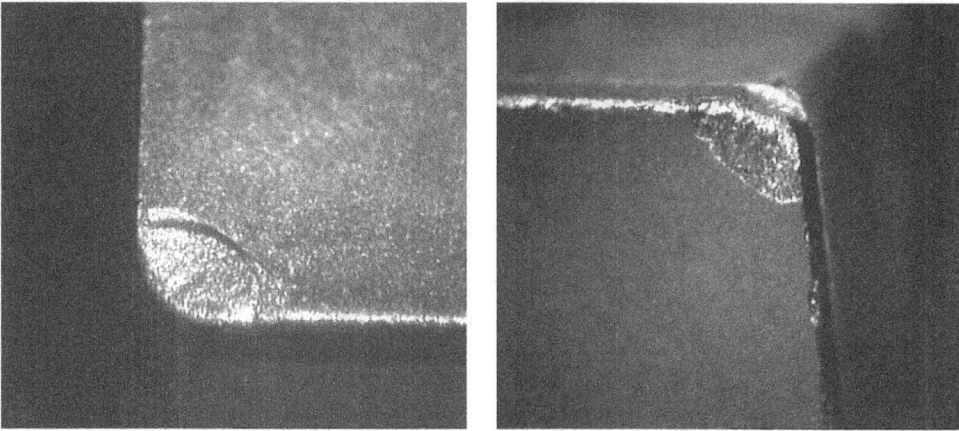

FIGURE 4.34 Example of medium flaking.

Medium overheating of PCD. Although there are a number of possible outcomes of this kind of overheating, the common denominator is the PCD layer flaking of different sizes that range from micro-flaking (Figure 4.34) and medium flaking (Figure 4.35) to severe layered flaking (Figure 4.36).

Severe overheating of PCD. Severe overheating of PCD leads to plastification of the PCD layer, which can be local as shown in Figure 4.37 or even global as shown in Figure 4.38. As such, the PCD layer behaves as a highly plastic material with the corresponding wear pattern developed after just a few working cycles.

4.8.4 PCD Overheating – Causes

Overheating occurs mainly in tool manufacturing whereas overheating in tool use occurs only when a tool is not designed and/or made properly. The operations used in the manufacturing of PCD tool are of prime concern. PCD tools usually are fabricated in three steps: (1) PCD blanks cutting into small inserts as shown in Figure 4.11, (2) PCD inserts brazing into the tool body, and (3) roughening/trimming and then finishing of PCD inserts into the required dimension, geometry, and surface integrity including surface roughness.

FIGURE 4.35   Examples of severe localized layered flaking.

The critical manufacturing operations in terms of thermal damage to PCD are brazing, electrical discharge machining including both types known as wire electrical discharge machining (commonly referred to as EDM) and electrical discharge grinding (commonly referred to as EDG), and abrasive grinding, commonly referred to as grinding. Out of these operations, brazing is probably the greatest culprit followed by EDM/EDG. These operations induce significant stress into PCD inserts and, what is most important, cause damage to the PCD surface as transformation of strong diamond sp^3 bond into weak graphite sp^2 bond takes place. This is known as graphitization of PCD surface layer that leads to lower tool life and/or to fracture of PCD inserts.

Unfortunately, brazing, EDM/EDG, and grinding of PCDs did not receive much coverage in the scientific literature. Trade literature and educational material of PCD and PCD tool manufacturing companies cover these operations to the extent of selling a particular product/process, i.e., revealing only a particular aspect of the whole story, and thus leaving others uncovered. Although this book cannot completely cover particularities of brazing and EDM/EDG/Grinding operations used in the manufacturing of PCD drilling tools, it covers the major issues/points of attention related to these operations.

4.9   BRAZING

In the author's opinion, the problem with proper brazing of PCD drilling tools, particularly of complex design having a number of PCD tips/elements, can only be completely solved when thermal stable grades of PCD are available for such tools.

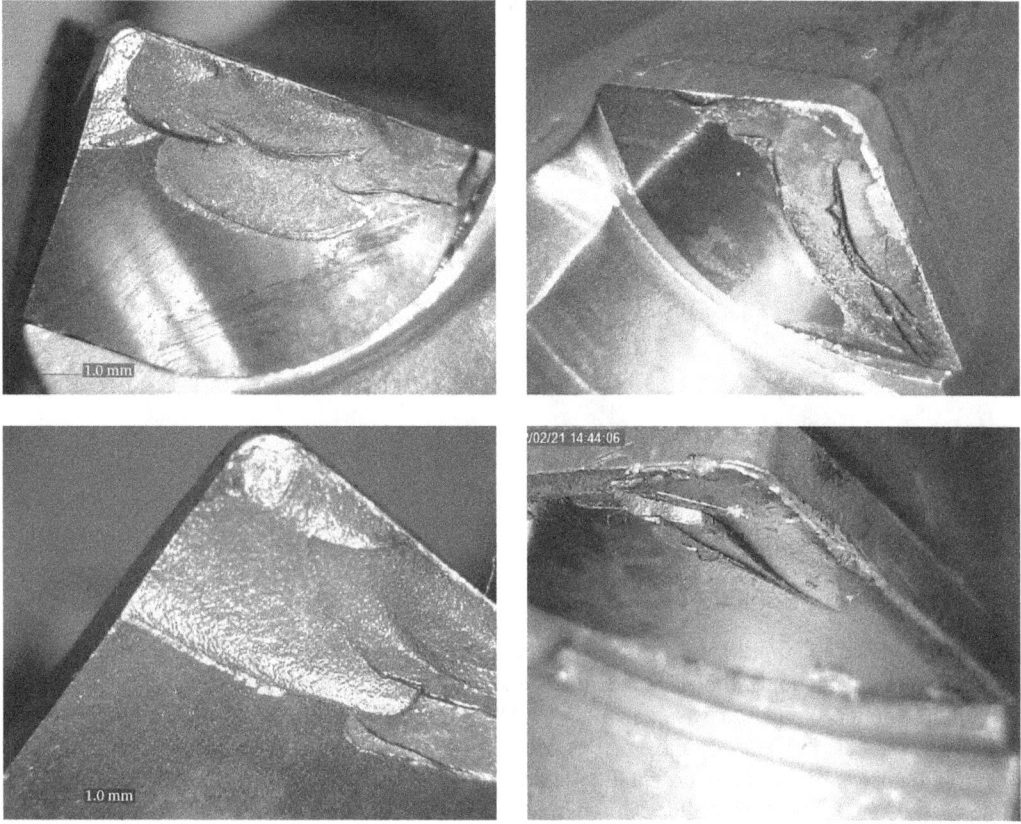

FIGURE 4.36 Examples of severe global layered flaking.

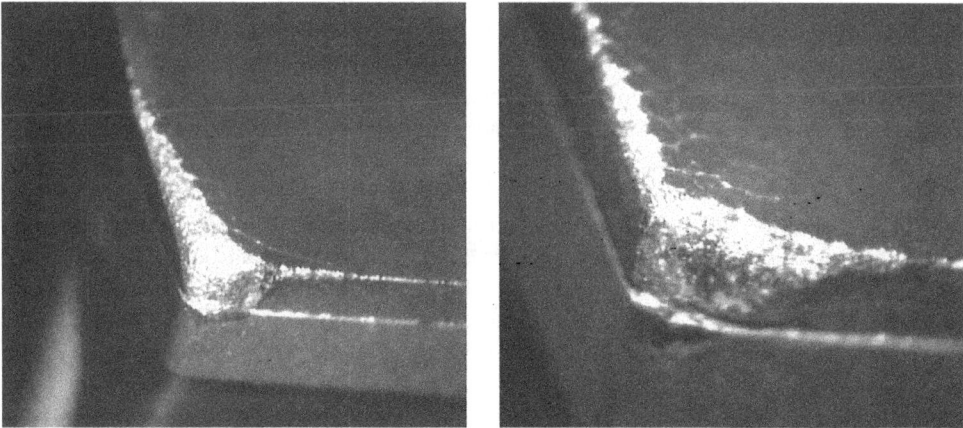

FIGURE 4.37 *Plastification* of PCD layer I.

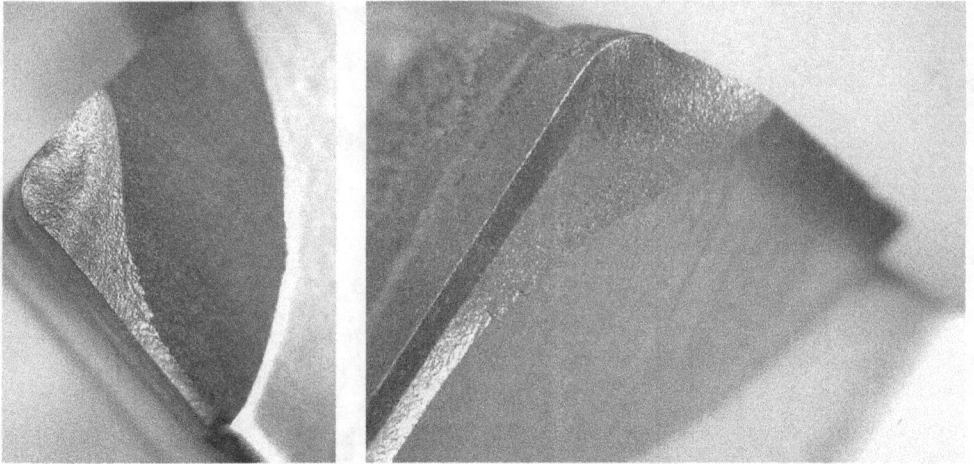

FIGURE 4.38   *Plastification* of PCD layer II.

4.9.1   EXAMPLES OF POOR BRAZING

Figure 4.39 shows examples of poor brazing quality. Spilled brazing filler material (BFM) shows that an excessive amount of the BFM was used, and thus, excessive heat was applied on brazing. As a result, heat-induced damage was done to the PCD tool material so that the tools failed prematurely. Even when the heat introduced on brazing is not excessive but the amount of the brazing filler is, as shown in Figure 4.39, the PCD tool quality deteriorates as the PCD insert rests on the relatively soft BFM rather than on the rigid tool body often made of sintered carbide for structural rigidity. In other words, such brazing undermines the whole purpose of the expensive carbide body. The rule of thumb for brazing is as follows: the proper brazing should not be distinguished with the naked eye as no separate layer of BFM should be visually formed. Rather, BFM should only fill out the valleys on the surfaces of the PCD insert (the carbide substrate side) and those in the PCD pocket in the tool body.

On the other hand, when the heat applied on brazing is insufficient and when a low-quality (e.g., oxidized as left open for days) brazing flux or its insufficient amount is used and when the PCD insert and the pocket in the tool body are not freshly cleaned, plus the improper grade of BFM (a cheap, high melting temperature alloy) is used, then the so-called *dry brazing* is a result of the listed brazing conditions. Figure 4.40 shows a typical appearance of the pocket in the tool body and PCD insert that failed due to dry brazing.

In the author's opinion, the prime cause of poor brazing is the use of induction or even torch brazing of PCD inserts with no temperature control. Experience shows that PCD inserts of any type cannot be brazed successfully without special processes, for example, vacuum (inert gas) brazing furnaces. Proper cleaning and fluxing procedures should be established and religiously maintained. Unfortunately, only a handful of PCD tool manufacturers conditionally justify these requirements. Proper PCD insert brazing for HP drilling tools is yet to be seen.

4.9.2   LITERATURE SUPPORT

Brazing as a manufacturing operation is used for centuries so that a number of books and papers are written on the subject. The existing literature including books on brazing (Roberts 2013, 2016, Sekulić 2013) considers brazing technology of PDC stating that temperatures up to 1,000°C is normal for brazing. The corresponding BFMs and brazing practices are recommended. Although PCD is mentioned, the essence of PCD thermal stability is not considered properly.

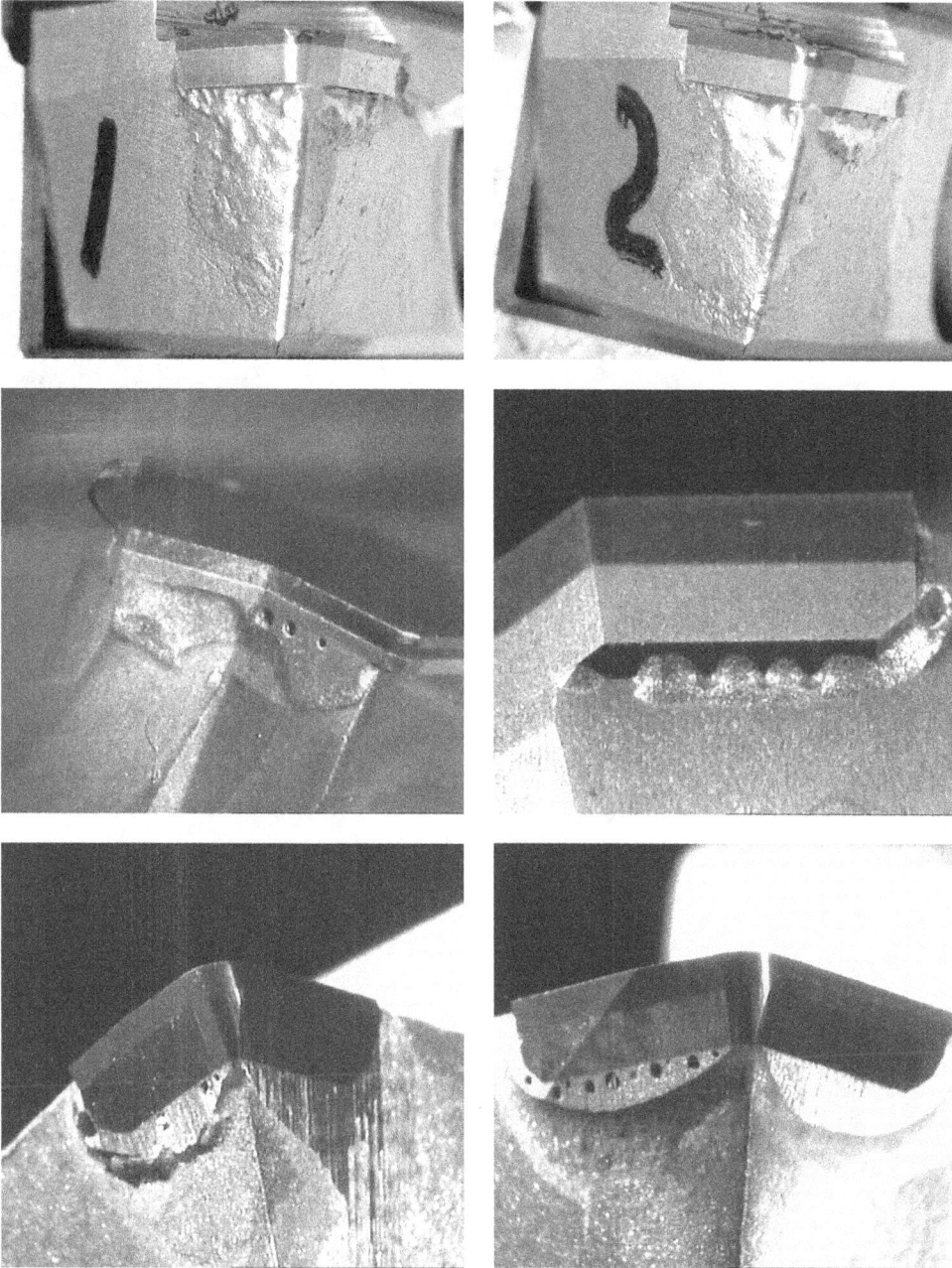

FIGURE 4.39 Examples of poor brazing quality where there is an excessive amount of the BFM, and thus, too much heat was applied.

4.9.3 FACTORS TO BE CONSIDERED

The brazing of PCD has some specifics so that this operation, first of all, should follow the known rules and consider the known factors particular to brazing (Roberts 2013). There are seven fundamental factors to be considered:

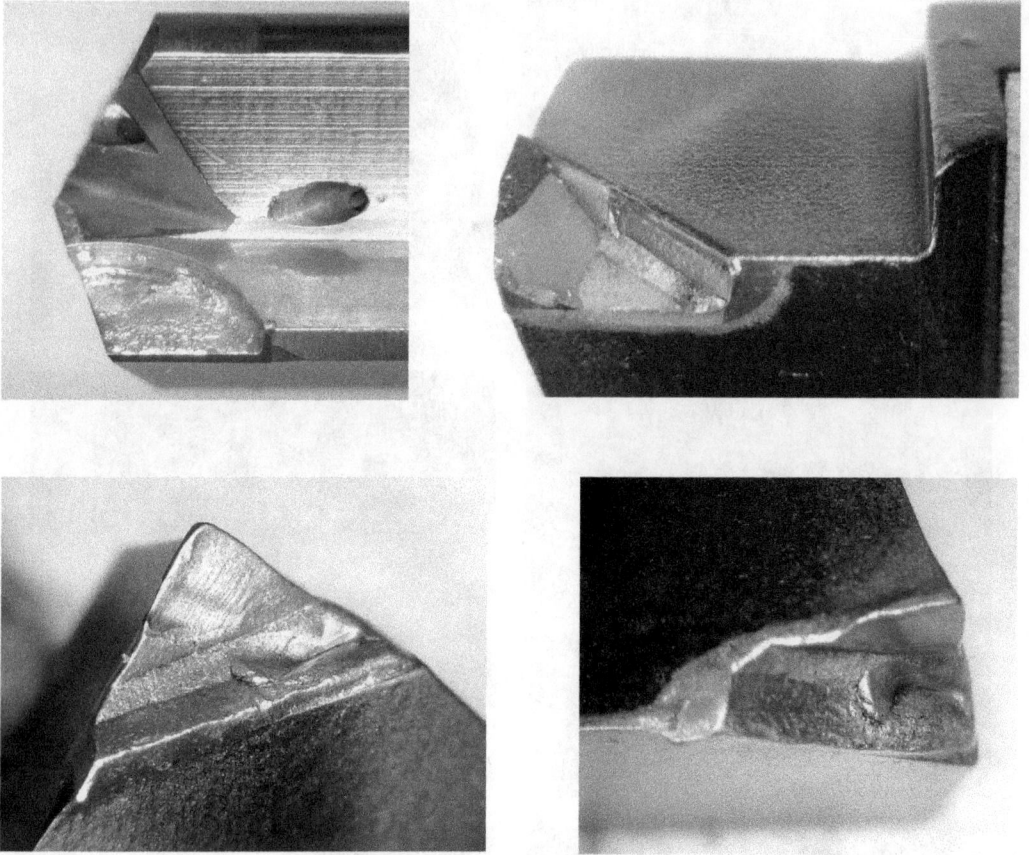

FIGURE 4.40 Typical appearance of the pocket in the tool body and PCD insert that failed due to dry brazing.

1. Design of the pockets for PCD inserts.
2. Heating process/brazing technology.
3. BFM – selection of the type, grade, and application technique.
4. BFM placement.
5. Joint and components clearness.
6. If used, flux selection and application.
7. Temperature control.

Design of the pockets for PCD inserts presents some challenges to PCD cutting tool designer as not much information is available on the matter. Although Roberts (2016) gives some recommendations on subject, most of his recommendations (shown in Figure 9.6 of his book Roberts 2016) are out-dated. Some of the particular and potentially important are overhang of PCD inserts, conformance of the shape of the PCD insert to that of the carbide/steel body, and unavoidable gap between the PCD layer and carbide body.

Overhang of PCD inserts is a common issue in the design of PCD tools. Figure 4.41a shows examples of PCD drilling tool with a significant overhang of PCD insert. Figure 4.41b shows a simple force model where the overhand distance o_h is shown as presented in the literature (Roberts 2016). It is recommended that this overhang should be less than 100 μm. In reality, the overhang distance o_{h1}, the distance between the point of application of the cutting force F_c acting on the

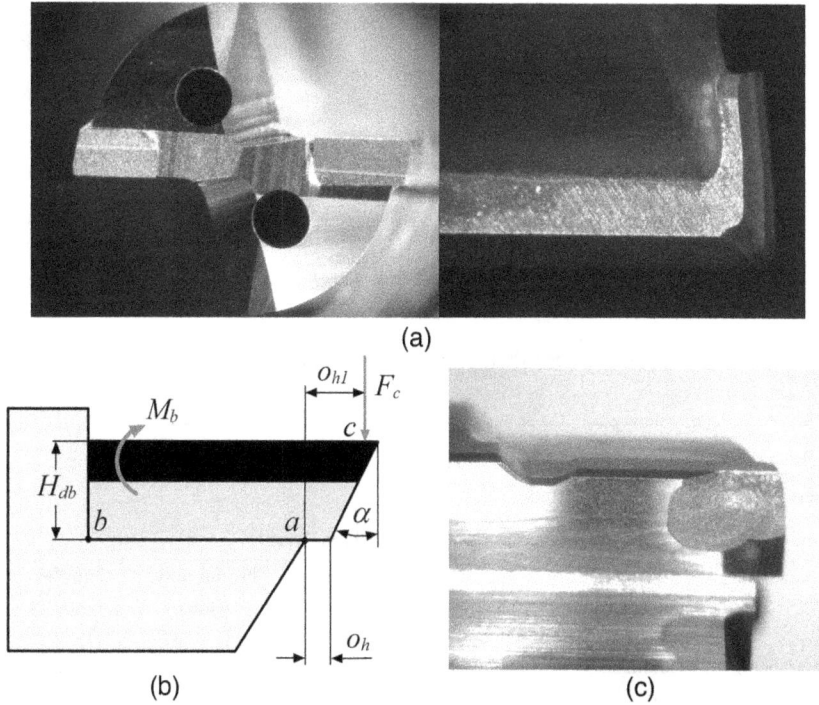

FIGURE 4.41 Overhang of PCD inserts: (a) tool with significant overhang, (b) force model, and (c) common consequence of overhang.

rake face and support point a, should be considered. This is because this force causes the bending moment $M_b = F_c \cdot o_{hl}$, which, in turn, causes bending/tensile stresses over the braze joint represented by length ab in Figure 4.41b. As both, the braze joint and PCD layer have weak resistance to such stresses, chipping of PCD layer, as shown in Figure 4.41c, or fail of the braze joint, as shown in Figure 4.40, often occurs. Therefore, the proper design of the pocket for a PCD insert should have a zero overhang distance o_h and, when the clearance angle α is great as in modern high-penetration rate PCD drilling tools, provide design measures to alter the binding/tensile stresses over the braze joint into the compression stresses.

Steps on the rake face between PCD inserts and the carbide body are common in the design and manufacturing of PCD tools. Apparently, there should be no steps/conformances with the use of a CNC EDMing machine to make the pockets in carbide bodies and to pre-cut of PCD inserts from a PCD blank/disk. In reality, many PCD tool manufacturers use PCD blank from different suppliers for the same tool. As discussed in Section 4.7.2, PCD layer thickness, t_{dl}, and disk thickness, H_{db}, can be different even with one PCD plank supplier as well as with various PCD plank suppliers. Moreover, there tolerances on these thicknesses can also significantly add to the discussed problem. Figure 4.42 shows steps in the flute on straight-flute multistage twist drills. Such steps disturb normal chip transportation over the flute causing chip clogging in high-penetration-rate drilling.

The problem with an unavoidable gap between the PCD layer and carbide body is probably the most severe problem in application of high-penetration rate of PCD-tipped tools. It is discussed in Section 4.11.1.

Heating process/brazing technology includes heat sources for brazing. These sources include torch brazing, induction brazing, continuous furnace, retort or batch furnace, and vacuum furnace discussed in great detail in the literature. There are various considerations to take into account when selecting the brazing process for a particular application as presented by Schwarts (2003). Some of these considerations are listed in Table 4.3. Although all of these heating methods are used in the

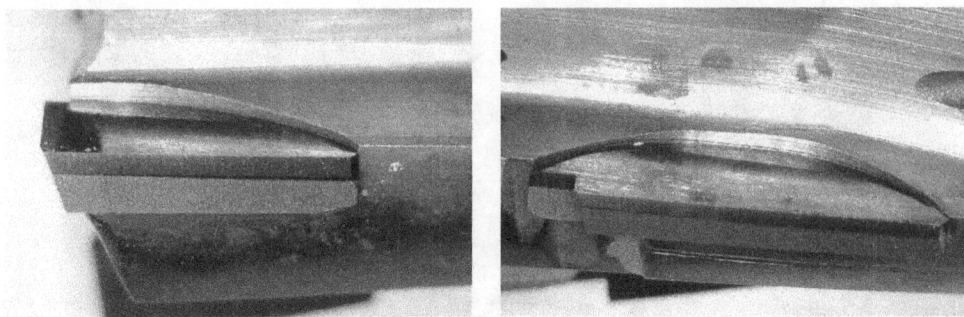

FIGURE 4.42 Steps in the flute between PCD inserts and carbide body due to their poor conformance.

TABLE 4.3
Characteristics of Heating Methods

			Characteristics			
Method	Capital Cost	Usage Cost	Flux Required	Production Rate	Complex Geometry	Feasible Sizes
Torch	Large	Medium/High	Yes	Medium	Low	Large
Furnace (atmosphere)	Medium/High	Medium/High	Yes/no	High/low	High	Large
Furnace (vacuum)	High	Low	No	High/low	High	Large
Induction	Medium/High	Medium/High	Yes/no	Low	Medium	Medium
Infrared	Medium	Low	Yes/no	Low	High	Medium

manufacturing of PCD tools, the most widely used methods in the brazing of PCDs are induction/induction furnaces and vacuum furnaces.

In a basic induction heating setup, a solid-state RF power supply sends an AC through a copper coil, and the part to be heated is placed inside the coil. The coil serves as the transformer primary and the part to be heated becomes a short circuit secondary. When a metal part is placed within the induction coil and enters the magnetic field, circulating eddy currents are induced within the part. These eddy currents flow against the electrical resistivity of the metal, generating precise and localized heat without any direct contact between the part and the coil. Temperature uniformity is achieved through coil design, which is of high importance in PCD tool brazing. The most effective uniformity can be achieved in round parts. Due to the nature of electrical current path flow, sharp edges could preferentially heat if the proper coil design is not used.

For special PCD drilling tools, a special fixture together with the application-specific induction coil design/geometry should be made for high-quality tools. Figure 4.43 shows the Elmec Co temperature-controlled brazing process, preventing thermal damage of PCD inserts and ensuring the adherence of the PCD tip to the substrate.

For PCD brazing operations, an induction-based system should be fitted with an optical pyrometer to facilitate closed-loop temperature control. Although it is advertised that through proper selection of the induction heating power supply, temperature controller, and optical pyrometer, temperature control of ±1°C is routinely achievable; this temperature is measured only at one point of the PCD insert rake face so that the actual temperature of BFM is unknown and thus should be found out using the trial-and-error method. As such, the finding is good only for a given thickness, shape, and make (as PCD blanks made by different manufacturers may have substantially different amounts of cobalt) of the PCD insert.

FIGURE 4.43 Elmec Co brazing of high-quality PCD drilling tool (Courtesy Elmec Co.).

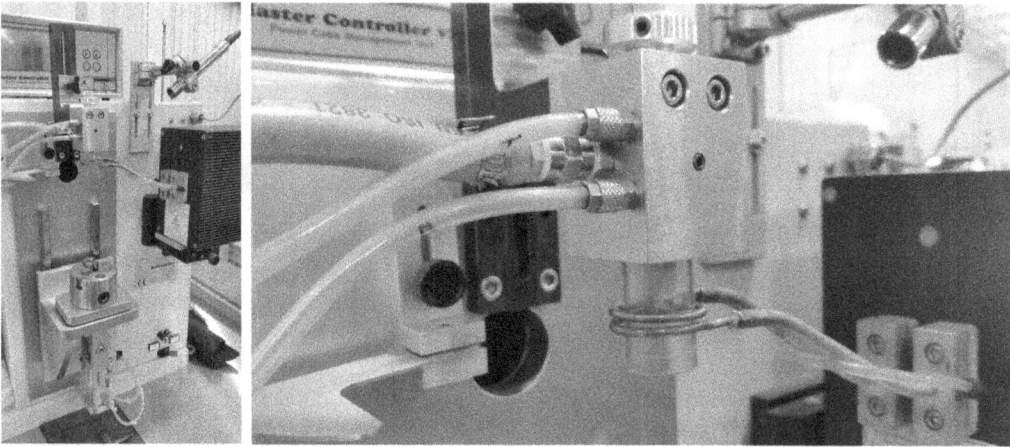

FIGURE 4.44 Induction coil located outside of a quartz tube.

Induction coils can be placed into both high vacuum and inert atmosphere systems. Alternatively, the induction coil can be located outside of a quartz tube as shown in Figure 4.44. In this setup, the quartz tube houses the part and is atmospherically controlled. The electromagnetic energy from the induction coil penetrates through the quartz tube and couples to the part. There are no losses through the quartz tube, because the quartz tube is not heated. Such a setup allows to braze not only PCD inserts with carbide substrates but also PCD inserts with no substrates for cross and full-head PCD tool as discussed in Sections 4.11.2 and 4.11.3.

Vacuum brazing is brazing in a high vacuum environment that provides the best process control and produces the cleanest parts, free of any oxidation or scaling. In a vacuum system, parts are heated in a fully enclosed, stainless steel chamber, which can be pumped down to 10^{-6} Torr. Special fixturing can be designed for automatic part loading and unloading, and quartz viewports can provide access for infrared temperature sensing of each individual part.

According to FUNIC Co (Center 2018), the vacuum brazing process is a two-stage process, namely, preparation work before entering the furnace and vacuum brazing process in the furnace. The preparation stage includes the selection of an appropriate vacuum BFM paste and then its application to the PCD insert pocket(s). Then, the tool is placed in a pre-brazing furnace so that the

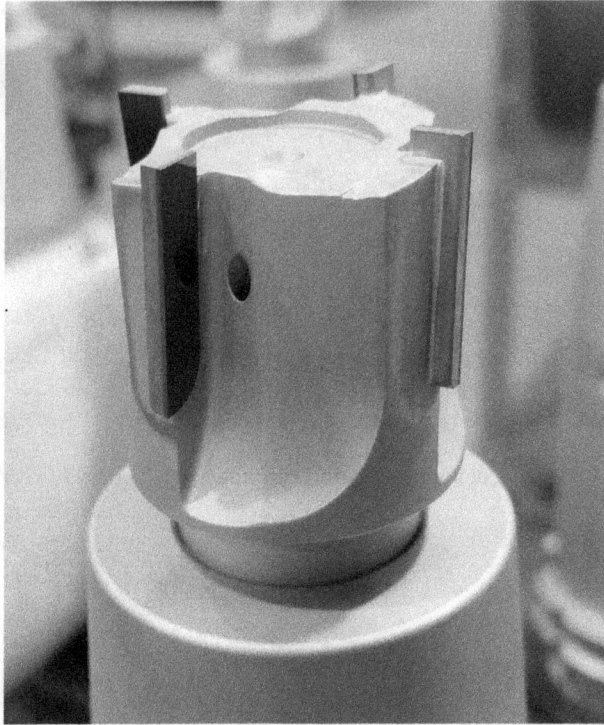

FIGURE 4.45 Appearance of a PCD-tipped tool after brazing.

insert in a fixed position can be cured by the brazing paste. The second stage includes placing the processed PCD cutting tool products into a vacuum furnace, sealing it and vacuum continuously, and then heating, melting brazing paste, completing brazing action, and cooling before removing from the furnace. When carried out according to the proper procedure, the BFM is evenly distributed to all parts of the brazing seam under the action of capillary pipette, so as to achieve the ideal brazing effect.

A number of low-melting point paste silver-copper-zinc BFM suitable for vacuum brazing are available today. For example, Meta-Braze™ 450 Braze Paste that has a melting range of 660°C–705°C and Meta-Braze™ 049 Braze Paste that has a melting range of 670°C–690°C developed for PCD brazing are offered by Meta Braze Co; Silver-flo™ 452 that has a melting range of 640°C–680°C developed for PCD brazing is offered by Johnson Matthey Metal Joining Co, etc.

4.10 ROUGHING/TRIMMING AND FINISHING

After brazing, the tool looks as shown in Figure 4.45 so it has to be roughed/trimmed and then finished to the final tool geometry and surface quality of its PCD inserts. To do these operations, a number of processes are used today in industry. The essence and feasibility of these processes are discussed in this section.

4.10.1 EDM

4.10.1.1 Principle and Cutting Mechanism

The basic EDM process is really quite simple. EDM is a controlled metal-removal process that is used to remove metal by means of electric spark erosion. In this process, an electric spark is used as the cutting tool to cut (erode) the workpiece to produce the finished part to the desired shape.

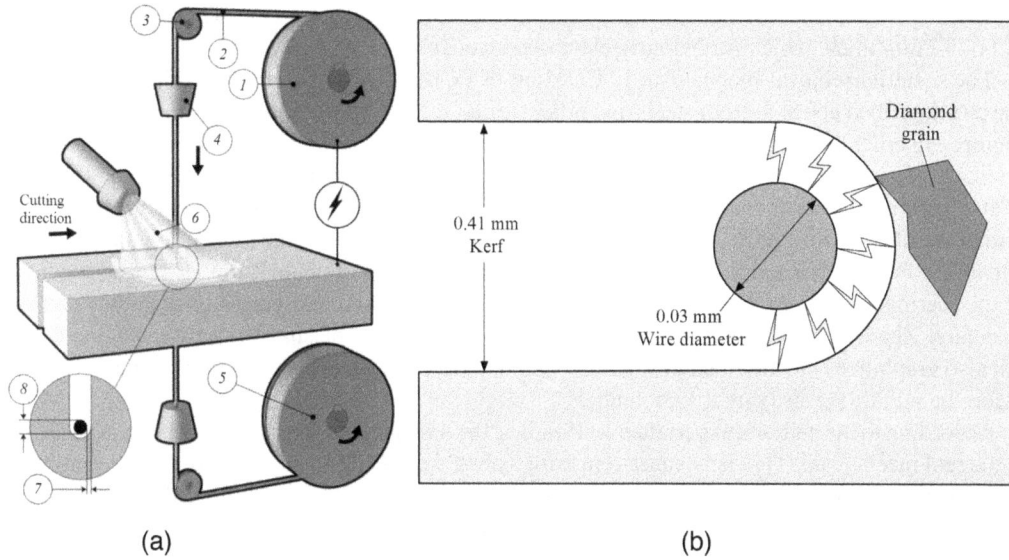

FIGURE 4.46 Schematic of EDM process arrangement (a) and mechanism of EDM cutting (b).

The metal-removal process is performed by applying a pulsating (ON/OFF) electrical charge of high-frequency current through the electrode to the workpiece. This removes (erodes) very tiny pieces of metal from the workpiece at a controlled rate (Bisaria and Shandilya 2015).

An electrical spark is created between an electrode and a workpiece. The spark is visible evidence of the flow of electricity. This electric spark produces intense heat with temperatures reaching 8,000°C–12,000°C, melting almost anything. The spark is very carefully controlled and localized so that it only affects the surface of the material. The spark always takes place in the dielectric of deionized coolant/water, which acts as a coolant and flushes away the eroded metal particles.

A schematic of EDM process arrangement is shown in Figure 4.46a. The wire supply spool *1* is CNC controlled and thus supplied the wire *2* through wire tension wheel *3* and guide *4* sapphire or diamond to prevent abrasion. The wire tension wheel and guide are to keep the wire straight with a high value of wire tension. Normal wire velocity varies from 0.1 to 10 m/min. The wire is finally collected by wire take-up spool *5*. A power supply delivers a high-frequency pulse of electricity to the wire as cathode and the workpiece as anode. When the wire is placed close to the workpiece, spark discharge occurs, which causes material removal/evaporation from the workpiece due to high temperatures. The dielectric coolant *6* supplied with a generous flow rate is used to create working conditions between the wire and workpiece (a dielectric gap). The spark gap, referred to as the kerf *7*, causes the machined contour to be slightly larger than the diameter of wire *8*. A commonly used wire 0.03 mm usually creates 0.041 mm kerf as shown in Figure 4.46b. Thinner wires create smaller kerfs.

The wire should never contact the workpiece. If it happens, however, there would be a short circuit and no cutting (Sommer and Sommer 2005). The wire electrode cuts by means of spark erosion due to high temperatures as the condition for the so-called microplasmas is formed by the process arrangement. This is the mechanism of common EDM cutting.

4.10.1.2 EDMing of PCD: Mechanism and Results

High hardness of PCD makes the manufacturing of PCD tools with conventional abrasive grinding very difficult. The material removal rate of abrasive grinding of PCD tools is only 0.002 mm³/s (Li et al. 2021). EDM is another approach to machine PCD tools by eliminating the cutting force, increasing the production flexibility, and reducing the production cost. The presence of 2%–8% metallic cobalt catalyst phase in PCD makes it conductive and ensures the feasibility of using EDM for making PCD tools. The PCD layer is joined to a cemented tungsten carbide substrate layer that is 1.6–3.2 mm thick, which is obviously electroconductive.

The material removal mechanism in EDMing of PCD, however, is considerably different than the one described above for tool steels and other similar matter. Understanding of this mechanism is required in order to fully comprehend the thermal damage development in EDMing. The scarcity of relevant information on the removal mechanism creates a challenge for further exploration. Rahim et al. (2015b) made some comparisons. It is discussed that Olsen et al. (2004) reported that the PCD grains are detached from the surface due to the preferential erosion of the highly conductive cobalt network. On the other hand, Kozak et al. (1994) believed that the cobalt binder is fractured due to the thermal stress generated by the process, which causes diamond grains to separate from the structure. Zhang et al. (2013) thought that the high-temperature plasma not only melts the cobalt but also graphitizes the diamond grains. The graphite consequently dissolves into the molten cobalt before being flushed away by the coolant.

According to the author's experience and studies, the mechanism suggested by Zhang et al. (2013) is the real mechanism. This is because removing cobalt from PCD structure, e.g., by its leaching as described in Section 4.7.2 does not compromise the structure and thus strength of PCD. Therefore, when the wire faces a PCD crystal as shown in Figure 4.46b, the only way to advance further is to "burn" this crystal having strong diamond sp^3 bonds and strong diamond crystal-to-crystal bonds into amorphous graphite having weak sp^2 bonds and very weak bonds with surrounding crystals as discussed in Section 4.3.1. After this, this amorphous graphite can be easily removed by the applied coolant so the wire can advance further. The greater the size of PCD crystal known as the grain size, the more energy and time need for transformation of PCD crystals to graphite, the more difficult for the wire to advance. It explains the known problems with EDMing coarse-grained PCDs compared to fine grains. The conclusive supporting evidences for this mechanism derived from SEM image analysis, Raman spectroscopy, and chemical analysis were presented by Rahim et al. (2015b) and Kui et al. (2017). These evidences show the existence of a significant, namely 40–100 μm (depending on EDMing regime and conditions), graphitized layer on EDM machined surface.

Another problem with EDMing of PCDs was pointed out by Pisarciuc (2013) and Astakhov (2014) is that PCDs for cutting tools consist of two layers, namely, the PCD layer and carbide layer sintered together as shown in Figure 4.17. These layers have different electrical and thermal properties that make EDMing of PCDs much more difficult compared to conventional homogeneous materials in terms of both the speed of processing and process stability (Spur et al. 1988).

As a result, it is not easy to establish an acceptable position of the wire and, thus, establish an optimal working regime to be suitable for both carbide and PCD layers. If cobalt is not evenly dispersed (as is the case in low-quality PCD products), an agglomeration of diamond grains can cause the wire to break or deviate as the wire tries to erode a material that it cannot.

As mentioned above, coarse-grained PCD grades are inherently more difficult to cut because of their relatively large grain size. EDMing of such grade causes wire jumping leaving visible "feed" lines on the EDMed surface as can be seen in Figure 4.47a. Such feed lines always make the cutting edge serrated causing its excessive radius as can be seen in Figure 4.47b. In operation, this radius rapidly increases as shown in Figure 4.47c due to graphitization of EDMed surface. When overaggressive EDMing is used, the whole face can collapse in tool operation as shown in Figure 4.47d.

No matter what the electrical parameters were adopted, a groove or notch was found at the separation limit of two layers as shown in Figure 4.47a–d. The explanations for its occurrence are based on the fact that cobalt concentration is higher in the transition zone between the carbide substrate and PCD layer as clearly seen in Figure 4.27. The low resistance of the cobalt to thermal erosion, compared with the other components, gives rise to increased material removal in this area.

Unfortunately, many PCD tool manufacturers and even some leading wire EDM machine suppliers try to deny this obvious fact observed by the author on a daily basis. They maintain, with no supporting evidence, that the difference in performance is small (5%–15%) so that the higher cost and longer lead time required to grind PCD cannot be justified. They maintain that EDM cuts PCD

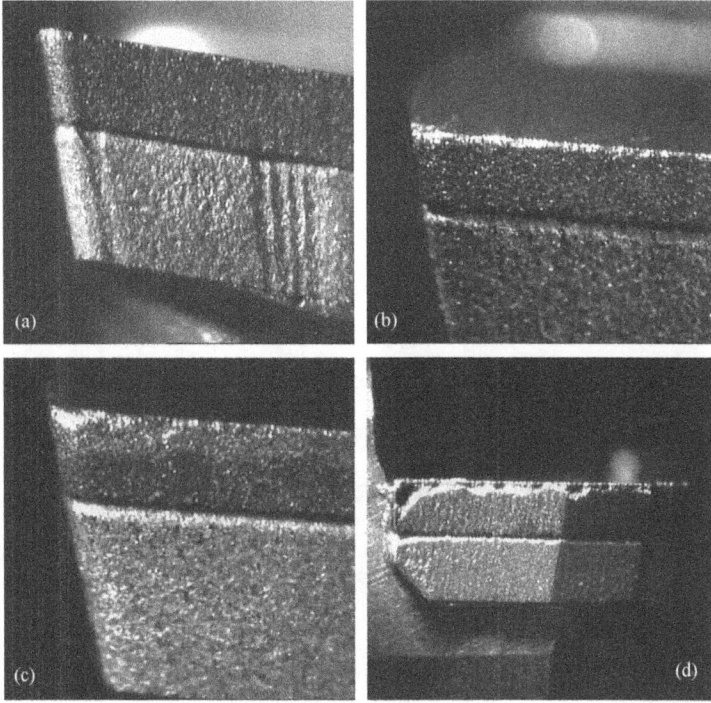

FIGURE 4.47 Typical flaws in PCD EDMing and its consequences: (a) wire jumping marks, (b) excessive cutting edge radius, (c) failure of the cutting edge due to graphitization, and (d) failure of the flank face due to graphitization.

efficiently and to an edge quality that is close to what grinding can achieve because of advances in the EDM generator, the circuitry that controls the impulses of electrical energy to create an effective spark. Developments in solid-state technology provide a more uniform and reliable *square wave form* to the individual impulses. According to the author's experience, it is not true. The latest EDM machines include a polishing capability (the embedded additional polishing diamond wheel) that provides polishing after EDMing to improve the quality of the cutting edge, that is, reduce its micro-serrations.

4.10.2 Electric Discharge Grinding

As pointed out by Thoe et al. (1996), EDG has nothing whatsoever to do with conventional abrasive grinding: the name is a misnomer. Unlike electrolytic grinding where workpiece material is removed by both a diamond abrasive and an electrochemical action, EDG relies purely on the erosive effect of a spark discharge. The term EDG stems from the similarity in overall appearance and arrangement of the wheel electrode and tool in EDG point grinding setup as shown in Figure 4.48. It is to say that modern EDG machines for PCD tool manufacturing mimic the abrasive grinding machine arrangement, i.e., the "grinding wheel" and PCD tool relative location. There are a number of EDG drilling tools and point grinding machines that are commercially available, for example, ANCA RX& Diamond, Walter Helitroninc Raptor Diamond, and Vollmer VHybrid 260.

A few important features of any EDG machine are to be considered. For example, automatic monitoring and control of the gap between the tool and the workpiece; maintaining the optimum erosion speed and spark control. For example, ANCA developed Adaptive Spark Control to prevent premature tool wear during heavier roughing operations while still optimizing the erosion process. The real heart of any EDG machine, however, is the erosion generator, which defines the whole EDG process.

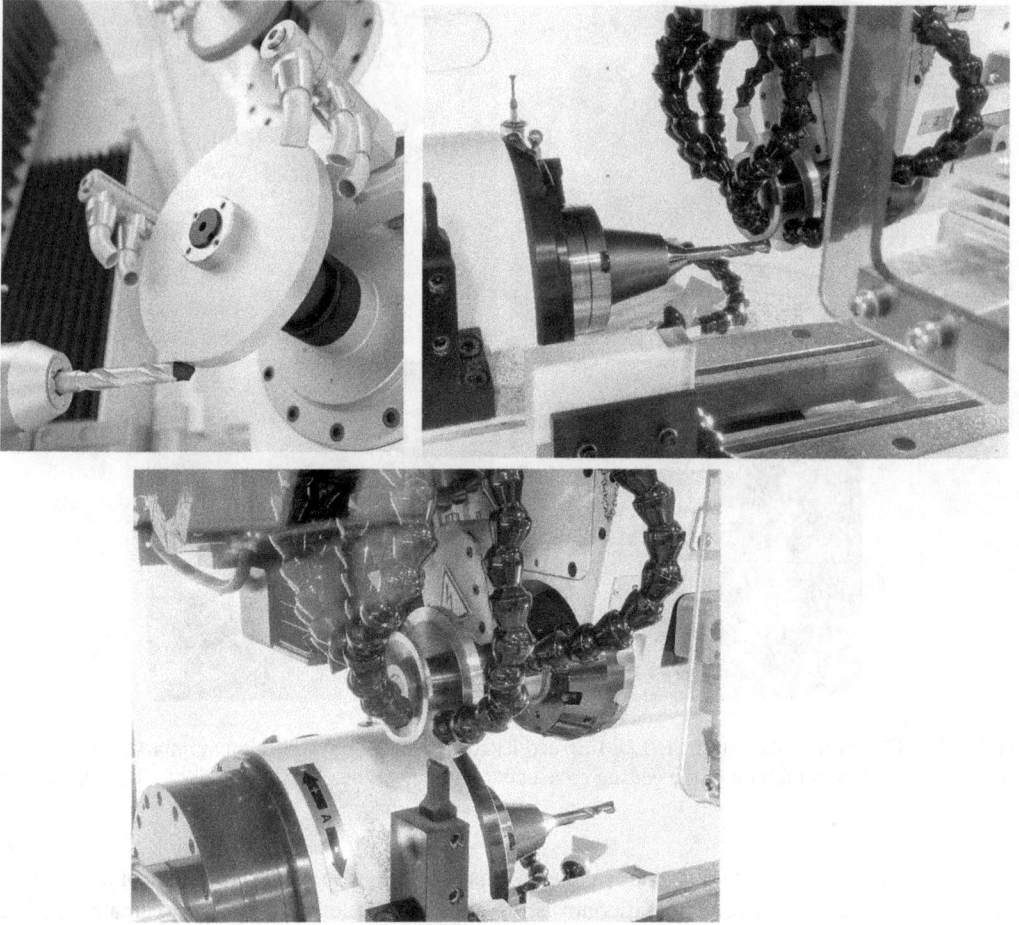

FIGURE 4.48   Appearance and arrangement of the wheel electrode and tool.

For example, in the ANCA EDG machine, the actual EDG process is controlled by the ANCA Motion SparX erosion generator so that the material is removed with high energy density, offering the user better feed rates as well as surface quality and significantly reduced cycle times. Ultra-low energy pulses, on the other hand, help to achieve exceptional ultra-fine machining thanks to SparX.

In reality, EDG is one of the variants EDM processed based on removing unwanted material from a part by means of a series of recurring electrical discharges (created by electric pulse generators in microseconds) between a rotary tool called disk commonly made of tungsten-copper alloy and the work material in the presence of a dielectric fluid/coolant. The sparks are generated by the pulse generator and happen in small sparking gaps of between 15 and 20 μm (Rahim 2015). During erosion, the dielectric coolant is flushed into the gap to provide better disruptive strength. Combined with the rotating electrode process, higher debris removal efficiency compared to EDM is achieved. In addition, the supplied coolant quickly lowers the temperature at the erosion surface during spark intervals (Rahim 2015). This shows that the material removal mechanism in EDG is the same as in EDM with all the above-discussed consequences (see Section 4.10.1.2).

Probably, the best comparison of the results of EDG and abrasive grinding of PCD tools is made by Rahim et al. (2015a). They concluded that both grinding and erosion processes are capable of producing similar visible surface quality (*Ra*) and sharpness value (the radius of the cutting edge). Although the visible quality (surface roughness and sharpness) is closely similar for both processes, it does not guarantee that similar surface qualities (residual stress and graphitization level) have been achieved. Overall, the ground PCD produces lower graphite than the erosion process due to lower processing temperatures.

Li et al. (2021) showed that when both EDG and abrasive grinding processes are optimized, the abrasive grinding is capable of producing lower surface roughness and sharper cutting edge compared to EDG for various commercial PCD grades. Moreover, a closer examination of the surface morphology after abrasive grinding and EDG showed the following:

1. Cobalt can be found on the ground surface of the unimodal CTB002 PCD grade, while large-sized diamond grains were found on the ground surface of the multi-model CTM302 grade (see Figures 4.25 and 4.26). No clear cracks or scratches were found on the flank faces of tools machined by abrasive grinding.
2. A "reconsolidated layer" with craters and voids was found on the surface of the tool surface machined by EDG. This layer contributed to the larger surface roughness of the eroded tools. The thickness of the two reconsolidated layers is around 4 μm regardless of the type of PCD materials.
3. In testing ground and EDG turning tools in turning of titanium alloy Ti6Al4V, it was found that the BUE of the eroded tools was more severe compared with that of ground tools. Moreover, BUE for multimodal CTM302 grade was thicker due to the formation of the huge crater caused by tip breakage.

The author's experience and testing of EDG-finished PCD tools in machining of various grades of aluminum alloys shows that the "dimpled" surface is covered by a thin film of aluminum causing a significant increase in friction forces over the tool–chip and tool–workpiece interfaces (Astakhov 2014). As such, the cutting force/drilling torque increases up to 30% (as registered by machine controllers) compared to the tool of exactly the same design/geometry made of carbide. This sets additional constraints on the tool design and selection of its optimal geometrical parameters. This issue is discussed further in this chapter.

In industry, acceptable PCD tools have been judged only by visible surface quality such as surface roughness and edge sharpness. As such, residual stress and graphitization levels inside the PCD tool are not evaluated although the finish process parameters as the whole, and EDG parameters in particular (Rahim 2015), combined with a particular PCD grade, significantly affect the quality of PCD tools. Thus, this highlights the importance of the evaluation of these invisible quality indices in assessing the performance of PCD tools.

4.10.3 Laser Cutting of PCD

The proper term of laser cutting/laser processing of PCD is Pulsed Laser Ablation (PLA) (Pacella et al. 2014). The technical meaning of the verb "ablate" is to gradually remove material from or erode (a surface or object) by melting, evaporation, etc. Laser ablation or photoablation is the process of removing material from a solid (or occasionally liquid) surface by "vaporizing" material from surfaces at the micron level. As no electrical conductivity of the material is required, PLA can be used to cut natural, CVD, and PCD diamonds.

It should be clear that material vaporization of diamond (natural, CVD) and diamond and cobalt in PCD requires extremely high local temperatures so graphitization of PCD should be expected. Even in PLA of monocrystalline CVD diamond grown in microwave plasma, i.e., with no cobalt that promoted graphitization, Raman spectroscopy shows transformation of strong diamond sp^3 bonds into amorous graphite sp^2 bonds over the surface area up to 10 μm (Kononenko et al. 2013).

There are three types of PLA machines that differ by the laser's pulse duration of the laser used: nanosecond, picosecond, and femtosecond lasers. The quality and productivity of these machines differ dramatically.

The dominating type of lasers used today in PCD manufacturing is nanosecond lasers because it was developed first and, what is most important, it has much higher productivity compared to the other two types of lasers in PCD machining. The latter is the most attractive feature as viewed by many cutting tool manufacturers. According to the author's experience and testing, the machine with such a type of laser can be used only for roughing followed by post-process treatment of the machined surface in terms of achieving the required surface roughness and removing graphitized layer. Unfortunately, this is not always the case in practice where even high-precision drilling tools are supplied with the surface quality as shown in Figure 4.49a. Many users never look at the tool under the microscope so they do not see it. Besides the rough surface and compromised quality of the cutting edge, a great graphitized layer is always there as shown in Figure 4.49b as reported by Wu et al. (2014). As can be seen, the intensity of the graphite sp^2 bonds is much higher than that

(a)

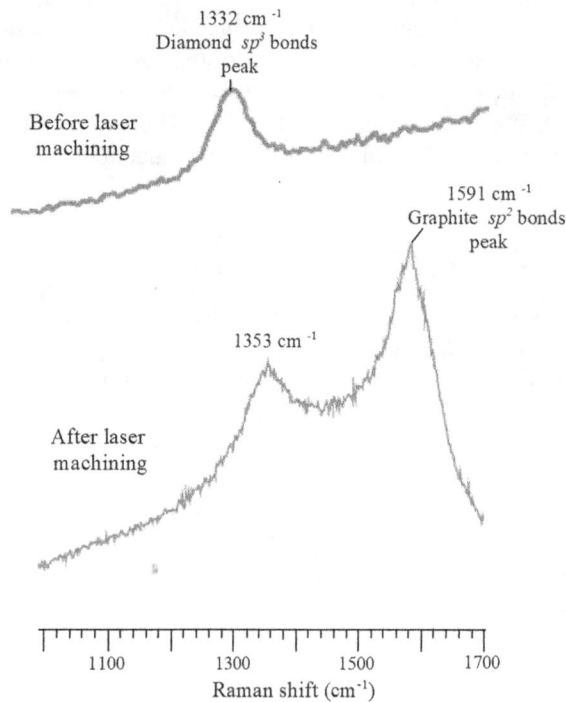

(b)

FIGURE 4.49 Flank face after nanosecond laser cutting: (a) typical appearance and (b) Raman spectra.

of diamond sp^3 bonds. Similar spectra were reported by Zang et al. (2007) and by Eberlea and Wegenera (2014). Rather rough PCD machinated surface and even kerf are reported by Chang et al. (2021). Significantly lower tool life, lower machining productivity, and quality of machined surfaces/features compared to other common types of finishing are direct results.

Much better appearance of machined surface and cutting edge quality improvements can be achieved using nanosecond laser with laser polishing techniques. Wu and Wang (2010) presented an approach of laser surface processing in terms of a review on laser polishing techniques using ultra-violet wavelengths in the nanosecond pulse-width regime. The polishing mechanism achieved by localized heat caused high-temperature oxidation and removed asperities from the diamond surface (Pacella 2020). Harrison et al. (2005) pointed out that the best results in polishing were obtained with different laser settings and gas pressure for the polishing phase using the following three-stage procedure: (1) basic cutting pass, (2) polishing pass, and (3) final cutting pass. The residual graphitization and/or tool life improvement after such a polishing were not studied.

Ultrafast lasers seem to offer a better option compared to nanosecond lasers. They are divided into two main categories:

- A *picosecond* (PS) laser emits optical pulses with a pulse duration of around 10 PS – just over one trillionth (10^{-12}) of a second, or one millionth of a microsecond.
- A *femtosecond* (FS) laser emits pulses that are one around 400 FS, less than one trillionth of a second in duration.

It is believed that picosecond and femtosecond lasers offer better-quality material processing possibilities because the laser's pulse duration is less than the target material's conduction time when the latter is the case. It is thought that cold machining of PCD blanks/cutting inserts is possible because the material being removed by sublimation. However, these systems are expensive – the typical price of an ultra-short pulse system is in the range of $400,000 and above, so the laser and system choice must be carefully considered.

There is a significant limitation on the use of these lasers, i.e., these lasers are most suitable for processing material thicknesses of less than 0.25 mm. Obviously, thicker materials can be processed, but the cycle time becomes very long. Therefore, users should clearly understand their return on investment so they can justify the investment in terms of cost reduction, unique processing capabilities, and so on. The final decision on which one to select is made after an iterative process, which usually includes a number of short runs followed by quality/productivity assessment of the results at the laser machine manufacturer's place.

According to the author's experience, tool manufacturing picosecond PCD laser machines can be a good choice, for example, EWAG Laser Line machines having pulse duration <15 ps, wavelength 1,064 nm, and 50 W power laser. Comparison of the ground and picosecond laser finished PCD insert in machining of CFRP shows that the process time and tool life of these inserts are almost the same for fine grain (2–4 μm) PCD but coarse-grained PCD (25 μm) still require finish grinding to achieve the required quality of the cutting edge (Dolda et al. 2013). For more challenging applications, Fraunhofer IPT has developed a novel finishing method for PCD cutting tools, which combines laser processing and grinding (Brecher et al. 2014). According to this method, short pulse laser ablation is used to completely remove a PCD surface layer of about 180 μm. To achieve the required tool quality/integrity, the remaining 20 μm are removed by subsequent grinding. The obtained results demonstrate that the combination allows reducing the finishing time by more than 50%. Therefore, laser technology has a high potential to make the finishing process more efficient and significantly lower manufacturing costs.

Femtosecond lasers offer precise and near-damage-free processing (Li et al. 2012, Ali et al. 2021) but, at this stage of development, their capability makes them unsuitable for material removal rate (i.e., productivity/process economy) feasible in diamond drilling tool manufacturing.

4.10.4 Abrasive Grinding

Abrasive grinding is the far best method of PCD finishing compared to any other methods provided that it is accomplished under optimal conditions (Liu and Tso 2003). As discussed by Bergs et al. (2020), the machining of the diamonds based on abrasive material removal is, when optimized, the best finishing process. Cracks in the diamond crystal could only be identified sporadically after grinding. There were no signs of plastic deformation in the PCD structure. A transmission electron microscope analysis of the ground surface shows that there is no thermally induced conversion of diamond into graphite or combustion of diamond, i.e., graphitization.

The problem with abrasive grinding is the process economy. Low G ratios (volumetric material removal divided by the volumetric diamond wheel loss), high cutting force, and high wheel cost pose the main challenges to conventional grinding production of PCD parts (Cao and Zhang 2004). Experiments show that the G-ratio of conventional grinding of PCD tools is between 0.015 and 0.025 and MRR is between 0.226 and 0.886 mm³/min, depending on different grind sizes and structures (Liu and Tso 2003).

Therefore, abrasive grinding should be used only for the final finishing of PCD tools, i.e., after roughing by EDM or EDG. Obviously, the depth of cut in abrasive grinding should be sufficient to remove the damaged layer left by the presiding operations. A special grinding wheel suitable for PCD grinding should be used. As discussed by Liu and Tso (2003), vitrified bond diamond wheels are the most suitable for grinding PCD inserts. The resin bonding is so soft that it is easily deformed while grinding and therefore cannot grind the PCD effectively. The metal bond wheels, on the other hand, have a too-strong holding capacity of diamond grits to sharpen grits in time and display a poor grinding efficiency. The diamond wheel with the grit mesh number 1,000 is the optimal choice to get a balance between grinding efficiency and quality.

Obviously, additional finishing operations such as grinding require additional manufacturing time, often special additional machines and expensive grinding wheels. As a result, the tool cost unavoidably increases which presents a major problem with the mindset of the purchasing department, particularly in the automotive industry. All arguments that the cost-per-part significantly decreases, tool life noticeably increases (2–3 times), quality of drilling operations improved etc. do not produce, unfortunately, much impression. The author always wonders how PCD drilling tools made their way under such conditions as they are significantly more expensive than carbide and HSS drilling tools.

4.11 PCD DRILLING TOOLS DESIGN

4.11.1 PCD-Tipped Drilling Tools

4.11.1.1 Common Design

Since PCD blanks became commercially available, multiple drilling tools for machining non-ferrous materials were designed and implemented. Figure 4.50 shows a two-stage PCD-tipped reamer with an internal high-pressure MWF supply through properly designed nozzles. Because the cost of earlier PCD blanks was high, initial applications of PCD were mainly for expensive finishing tools, for example, reamers where both surface finish and diametric accuracy were of prime concern. Those years, specialist learned how to braze PCD tips and how to trim and finish grind PCD-tipped tools through the painful trial-and-error method.

Eventually, with the further development of the technology and lowering cost of PCD blanks, PCD-tipped drills found their way to the shop floor. A typical PCD-tipped straight-flute drill with internal high-pressure MWF supply is shown in Figure 4.51. It has a cemented carbide body and the two PCD tips brazed in the pockets made (normally EDMed) on the periphery of its terminal end

FIGURE 4.50 Two-stage PCD-tipped reamer.

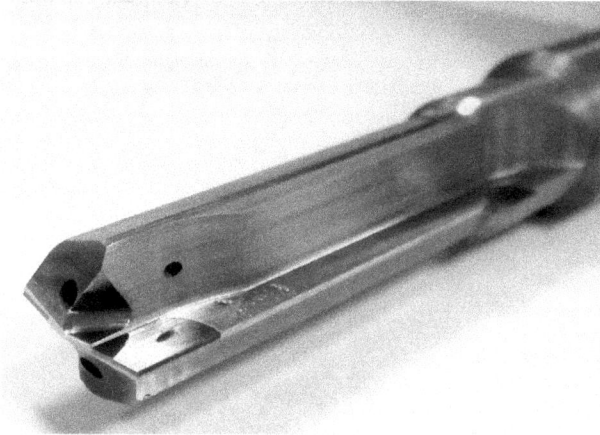

FIGURE 4.51 PCD-tipped straight-flute drill.

and two PCD tips brazed in the pockets of the second stage. As can be seen, such a drill has two combined major cutting edges (a.k.a. the lips), each consisting of the PCD and carbide portions. The idea behind this design was crystal clear – as maximum cutting speed and thus tool wear (under proper drill design and normal drilling conditions) occur at the drill periphery regions, the placement of a PCD tip in this region of the major cutting edge should solve the problem, that is, it should help to increase tool life while enjoying the high quality of PCD machining. Soon, however, it was found that this is not always the case.

When PCD-tipped drills were used at relatively low (for the PCD tool material) cutting speeds and moderated penetration rates, their performance was satisfactory. The advantages of switching to the PCD tool material were marginal as the gain in tool life was overshadowed by the increased cost of the tool. Moreover, it was soon found that such drills are more sensitive to the conditions of the drilling system, particularly to the accuracy of the tool holders and machine spindles, imperfections in castings (e.g., inclusions) (Astakhov and Patel 2018), and the accuracy of the tool presetting. The next generations of the drilling systems were designed and made to support the performance of the PCD tool material in the hope to gain the maximum advantages of PCD-tipped. Rigid and powerful high-speed spindles, internal high-pressure through-tool MWF supply, advanced machine controllers, high-accuracy tool holders, laser tool presetting machines, etc., were introduced to support a high-penetration rate and thus high efficiency of these drills. Soon, however, it was found that PCD-tipped drills of traditional designs and manufacturing accuracy have two inherent problems.

The first problem is that the tool life of the carbide portion of the drill often defines tool life. An example is shown in Figure 4.52 where a PCD-tipped drill for machining a high-silicon automotive aluminum alloy is shown at the end of its tool life. As can be seen, the carbide portion of the drill is excessively worn, which caused radial tool vibrations that, in turn, caused the chipping of PCD tips. Moreover, on tool re-sharpening, a great amount of the carbide portion should be ground off to restore the normal working condition of this portion. This is not feasible for the PCD inserts as they are much harder and can easily be damaged by the heat generated in grinding. The common way to deal with the problem is to EDM or EDG a worn tool to restore their working condition. As an example, Figure 4.53 shows the drill reconditioned by EDG with no subsequent grinding/polishing of at least the carbide portion. Although it may appear the simplest and thus the fastest/economical method of drill reconditioning, the performance of such a drill is inferior to that finished with the grinding wheel as carbide flank faces finished with EDG do not perform well.

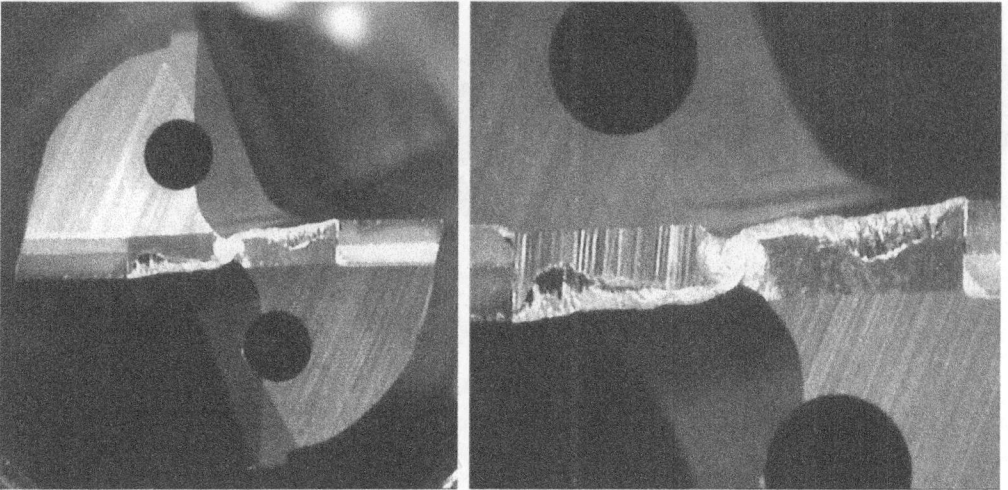

FIGURE 4.52   Showing that the carbide portion often defines the tool life of PCD-tipped drills.

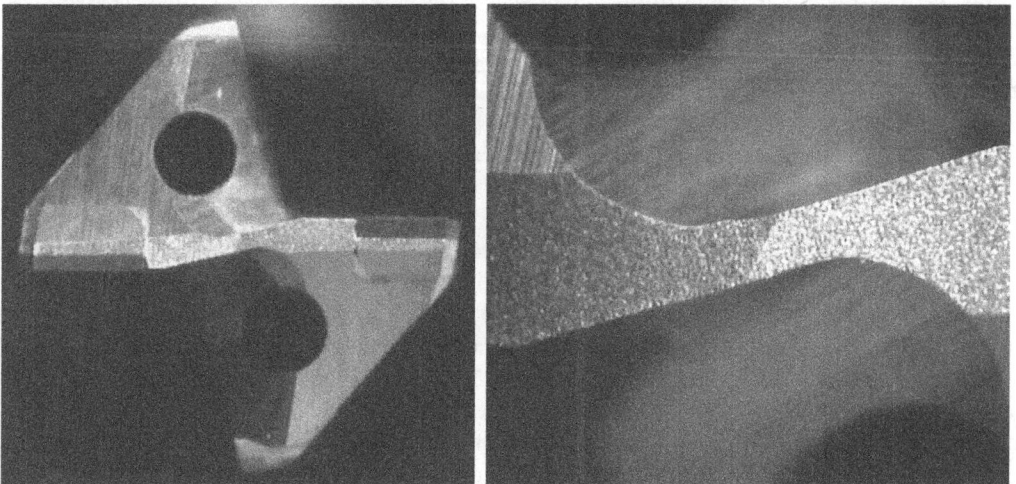

FIGURE 4.53   Showing a PCD-tipped drill reconditioned using EDG.

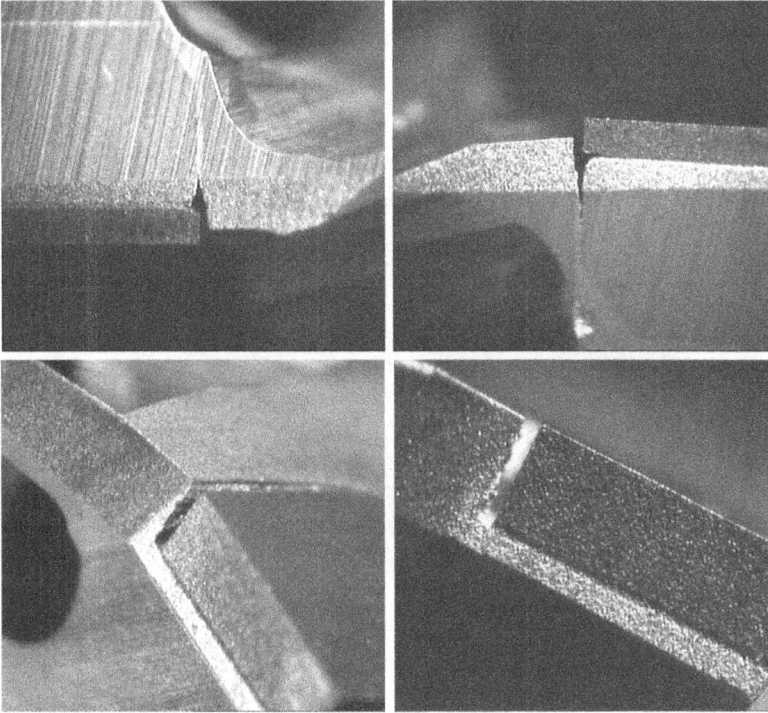

FIGURE 4.54 Excessive gaps between the PCD-brazed inserts and the carbide body in a reconditioned PCD-tipped drill.

The second problem with PCD-tipped drills is probably the oldest, yet the most severe, problem that requires solution in HSM high-efficiency drilling. The essence of the problem is an unavoidable gap between the PCD inserts and the carbide body as shown in Figure 4.54. This gap occurs because of the following:

- When a PCD insert is brazed into a tool body using a common brazing technology (e.g., the induction brazing), the BFM does not adhere to the PCD layer, that is, the only carbide substrate is actually brazed to the drill body so that a gap between the PCD layer and the pocket is unavoidable.
- When a tool is re-tipped, it is difficult to remove the remaining BFM from the pocket of the carbide body so that when new PCD inserts are brazed, their position may not be as intended. One may wonder, however, why not to apply more heat and thus to melt the remaining of BFM. The problem is that the amount of heat in PCD brazing is always limited to avoid overheating of the PCD tool material.

When a tool with excessive gaps between the PCD tips and the drill body (similar to that shown in Figure 4.54) is used in high-speed drilling of a highly abrasive work material, the work material goes into the discussed gaps, "grinds" down the remaining BFM, and then creates a splitting wedge that may cause chipping of the PCD layer as illustrated in Figure 4.55.

Another inherent problem with PCD-tipped drilling tools is the variation of the thickness of PCD inserts from one manufacturing batch of PCD blanks to the next. Because not much can be done to increase this thickness H_{db} (see Figure 4.14) when it is smaller than the depth of the pocket, and it is not feasible to grind each insert when this thickness is greater than the depth of the pocket

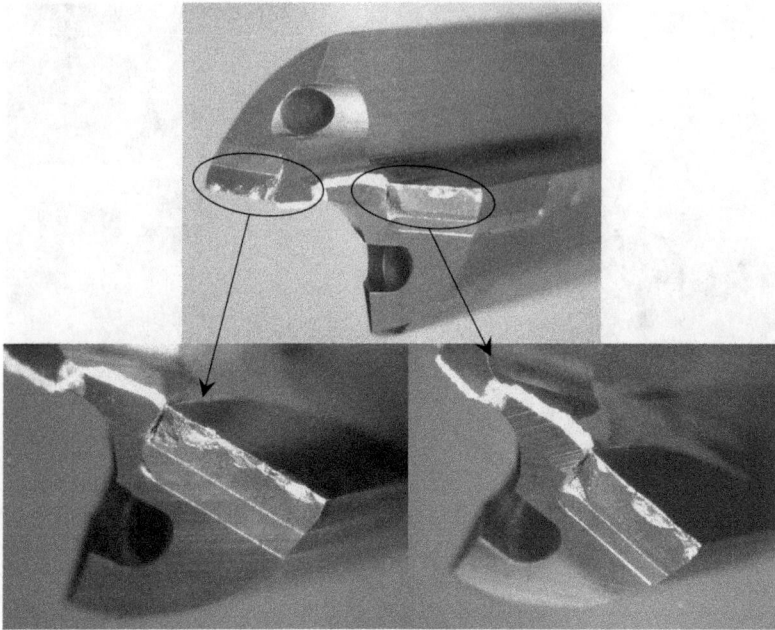

FIGURE 4.55   Chipped PCD portions of PCD inserts as a result of work material flow into the gaps between the PCD inserts and the carbide body.

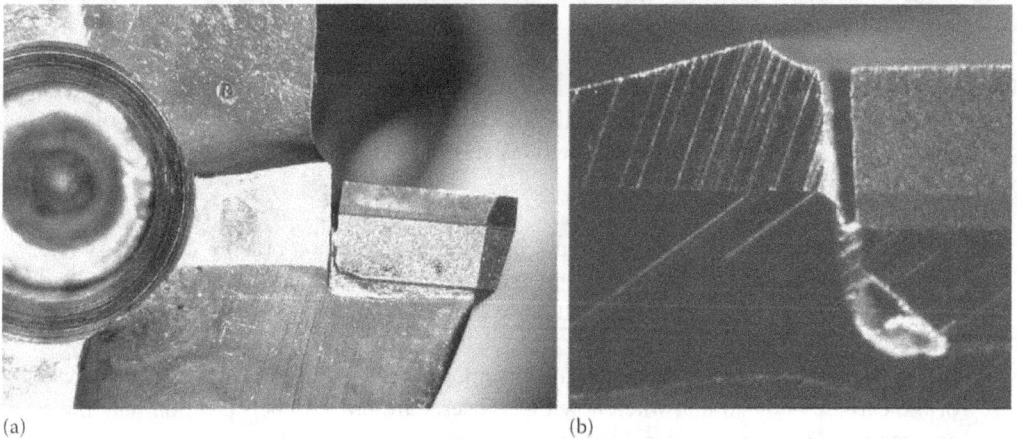

(a)   (b)

FIGURE 4.56   Step on the rake face combined with the gap between the PCD insert and the carbide body: (a) in a reamer and (b) in a drill.

made in the tool body, steps on the rake face are often found on the PCD-tipped tool. Such steps can be seen in Figures 4.54 and 4.55. Figure 4.56 shows a step on the rake face combined with the gap between the PCD insert and the carbide body. Figure 4.42 shows steps in the flute on a helical multistage drill. Such steps disturb normal chip transportation over the flute causing chip clogging in high-penetration-rate drilling.

According to the author's experience, the root cause of the listed problem with PCD-tipped drilling tools is improper design and tolerances assigned by tool drawings, inferior manufacturing quality, and lack of proper inspection of these tools. It is to say that the listed items are the same as for standard carbide drills with no accounting on the specific properties of PCD and particularities of

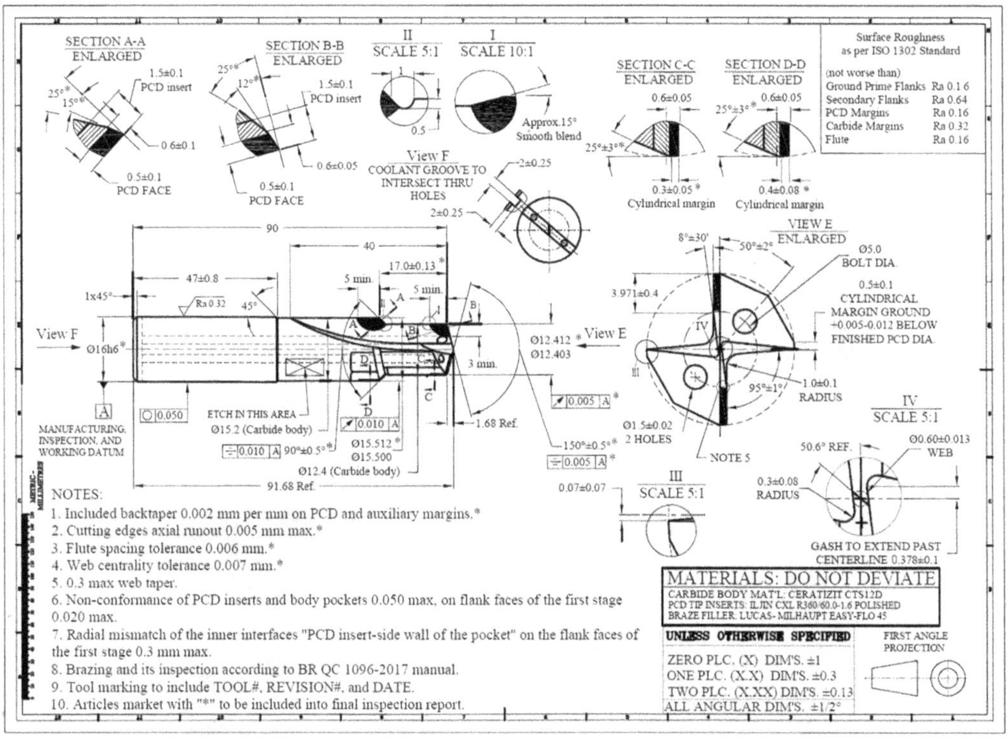

FIGURE 4.57 Example of the proper drawing of a PCD-tipped drill.

the finishing process used for PCD tips. Application experience shows that PCD-tipped drills can be used successfully if the listed issues are adequately addressed. As the quality of such a drill begins with a proper drawing, Figure 4.57 shows an example of proper drawings of a PCD-tipped drill with many intricate details, which normally do not appear on carbide drill drawings.

4.11.1.2 Advanced Design

The problem with the gap between the PCD inserts and carbide body can be solved in a number of ways to be developed. The objective of any new design of PCD-tipped drill is to exclude the gap from cutting, i.e., to eliminate the problem discussed in Section 4.11.1.1. One promising solution is the four-flute combined drill (FFCD), the concept of which was developed by Star Cutter Co. and tested, optimized, and implemented by PSMi Corp.

According to the FFCD design concept shown in Figure 4.58a, the carbide portion of the drill is revolved by an angle with respect to its PCD portion. As can be seen, each portion of the drill has its own chip flute, namely, fluted F1C and F2C are for the carbide cutting edges, whereas flutes F1P and F2P are for the PCD cutting edges/inserts. As such, the outside diameter (OD) of the carbide portion overlaps the gaps between the PCD inserts and carbide body. As can be seen in Figure 4.58a, Gap 1 and Gap 2 are not involved in cutting so that no aluminum can flow into these gaps. Figure 4.58b shows the appearance of the tool made for real applications.

Besides the elimination of the "gap" problem, another significant advantage of the described design is that sharpening and subsequent re-sharpening of the carbide and PCD portions can be made using a different operation most suitable for the corresponding tool materials as, for example, grinding for the carbide portion and EDGing for the PCD portion. Moreover, these two portions can be shaped with different flank and rake face geometries and different point angles which are impossible with the current design.

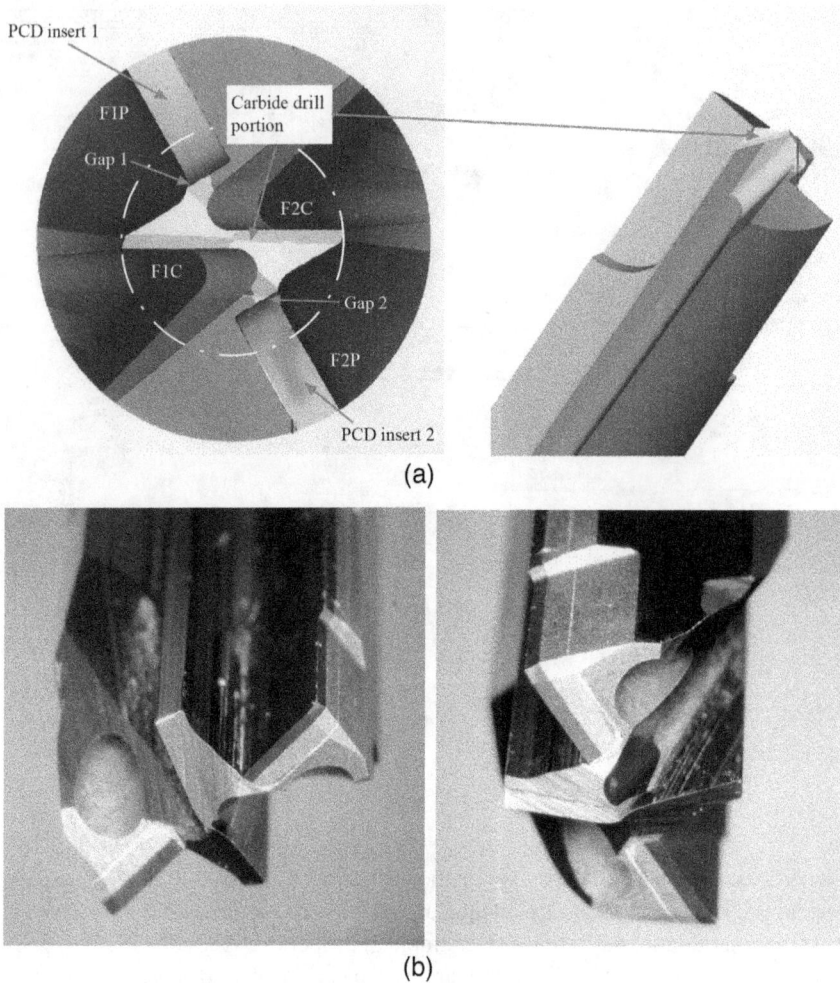

FIGURE 4.58   FFCD: (a) concept/model and (b) tool appearance.

Our testing and the implementation practice of FFCDs revealed that their design should be optimized for a particular application to achieve the best results in terms of drilled holes roughness, position, and shape. Figure 4.59a shows the most important parameters of optimization. In this figure, points *1* and *2* are the interception points at which the cutting by the carbide portion of the drill is taken over by the PCD inserts. The location of these points defined by diameter d_c is determined by the carbide portion point angle Φ_c and axial distance l_c. Another important issue is the angle of the carbide portion with respect to PCD insert, which defines the corresponding flutes cross-sectional areas and locations. Figure 4.59b shows that this angle is not to equal 90° as per the initial design shown in Figure 4.58.

4.11.2   Full-Face (Cross) PCD Drills

The discussed disadvantages of PCD-tipped drills have been known for a long time, practically since the beginning of their practical use. Therefore, a number of attempts (with some great successes) were made in the development of the so-called full-face (cross-PCD) drills. These can be broadly divided into two major design (technological) approaches: (1) PCD is sintered in a part of the drill body and (2) a fully sintered PCD segment(s) is brazed into the drill body.

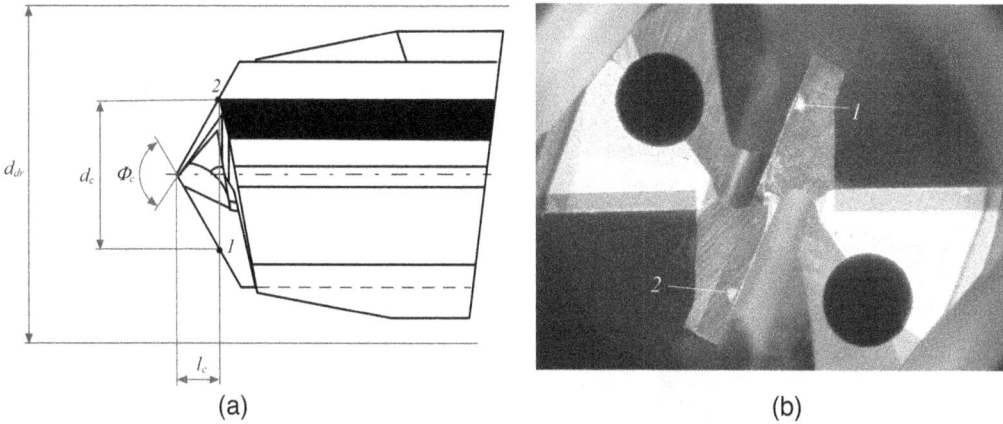

FIGURE 4.59 FFCD: (a) optimization parameters and (b) interception points *1* and *2* on the flank face.

4.11.2.1 First Approach: PCD Is Sintered in a Part of the Drill Body

A number of different technologies were developed to realize this approach. They are described by the author earlier (Astakhov 2014). Today, most of them have only historical significance. This section briefly discusses only two of them because they have some limited use in industry as other much advanced PCD drill designs are taking over. Another reason to keep some descriptions of these old design/technologies is based on the known saying: Lessons from the past may not always ward off doom, but they can provide insights into the present and even the future.

4.11.2.1.1 Veined Drill

The essence of multiple design variations of the so-called veined drill idea pioneered by Precorp Co. (Spanish Fork, UT) is in sintering a PCD layer inside a carbide blank to form the nib. The simplest nib design is shown in Figure 4.60a initially developed for printed circuit board PCD drills in the range of 0.15–3.2 mm diameter according to US Patent 4,713,286 (1987).

Precorp Co. used a seven-step process to produce its veined drills. These are shown in Figure 4.60b. The process begins with a solid-carbide blank, which is then slotted. Diamond powder is inserted into the slot of the blank. The blank is then placed in the press and subjected to the HPHT sintering process. The finished PCD-embedded blank, referred to as a nib, is then brazed to a carbide blank a good distance from the cutting edge, depending on the drill diameter. This creates the specified length of the tool to be completed. The particularities of the drill geometry are then ground to produce the finished PCD-veined drill.

The spacing of the braze interface keeps it away from the heat generated in the cutting zone, thereby preventing the PCD portion from overheat/graphitization. This feature alone can be reason enough to explore the application of a PCD-veined drill against a more common PCD-tipped variety.

PCD-veined drills were originally designed for the electronics and aerospace industries, but further application testing has brought this technology to the attention of other industries to drill green carbide, aluminum MMCs, reinforced ceramic composites, and various grades of aluminum. Unfortunately, some limiting success was achieved for these materials as it was found that tool life is not as high as was expected. The reason for this is discussed in Section 4.7.3 ("undercooked" PCDs) and in the next section.

4.11.2.2 Cross-PCD Drills Designed/Made with Sandwich Sintered PCD Wafer

The sandwich-type PCD drill technology can be considered the development of the previously discussed intersintered drill technology. This approach was pioneered by the development of the sandwich PCD drill design in the early 1980s (US Patent No. 4,627,503, 1986). The drill includes

(a)

| Carbide blank (the nib) | Slotting the nib (applying the vein) | Filling with diamond grade mix (powder) | HPHT sintering | Cleaning the finished nib | Brazing to the tool shank | Finish grinding the required tool configuration |

(b)

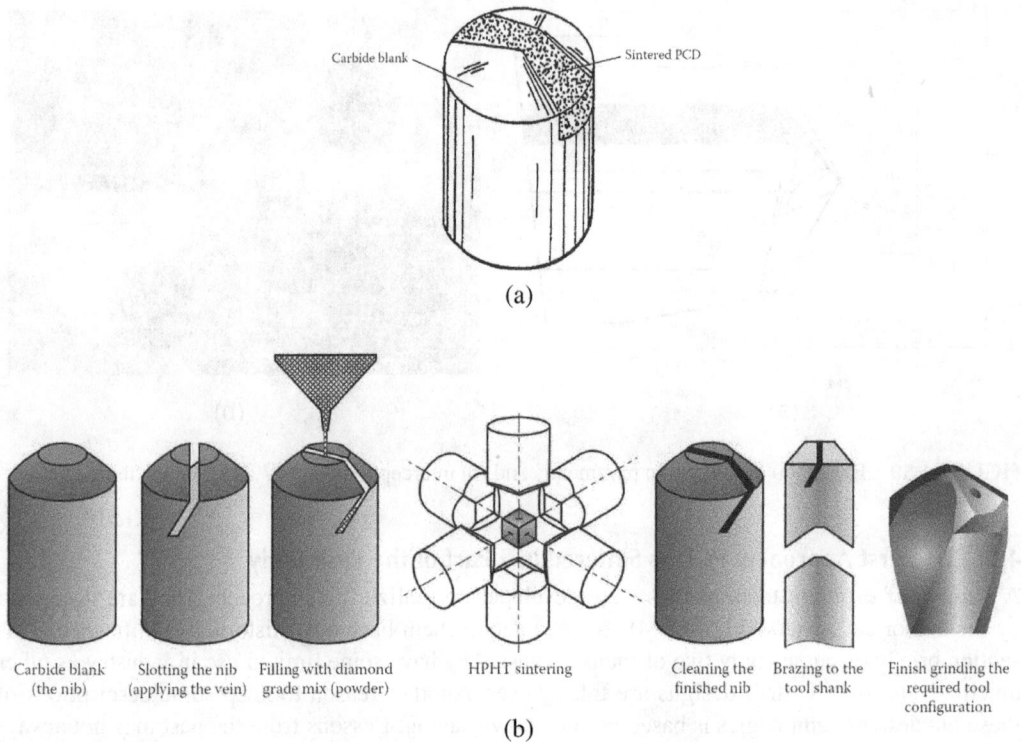

FIGURE 4.60 This figure shows: (a) Design of the nib according to US Patent 4,713,286 (1987) (a) and (b) basic steps in manufacturing a modern veined drill used by Precorp Co.

a sintered PCD wafer having a PCD layer sintered together with two carbide substrates as shown in Figure 4.61a. Such blanks are readily available in the market nowadays. As the wafer has two cemented carbide substrates, it can be assembled with the drill body having an axial slot and then brazed to this body. After the brazing, the drill is finish trimmed and ground.

The major drawback of this approach is a need to grind a substantial amount of carbide and then PCD to achieve the desired drill geometry. Such grinding was a huge problem at the time when this design was developed. The second problem is that a significant amount of cobalt should be placed in the mix (grade powder) to assure sintering as the two carbide substrates act as the shields limiting the pressure and temperature needed to fully sinter a *standard* (low cobalt) grade powder. The resultant product (PCD) is not as strong as that *normally* sintered. Thirdly, relatively thin carbide substrates did not provide sufficient protection to the PCD layer on brazing, so it often happens that the PCD layer is overheated with the above-discussed consequences. Note that the modern technologies (sintering, EDMing, grinding) reduce the severity of these problems.

Figure 4.61b shows typical blanks for manufacturing such drills that are now commercially available from various PCD manufacturers. The quality (and often cost) of these blanks may differ significantly, which affects drill performance in HP applications. A detailed experimental comparison of the quality of the blanks made by two well-known companies was discussed earlier (Astakhov 2014).

The analyses of performance of drills made using pure veined technology and sandwich-type drills show that some important issues cannot be resolved in principle in manufacturing of these drill types:

1. A significant amount of carbide and PCD is ground off to bring the veined blank to the final shape of the drill. This grinding damages PCD tool material developing a network of microcracks.

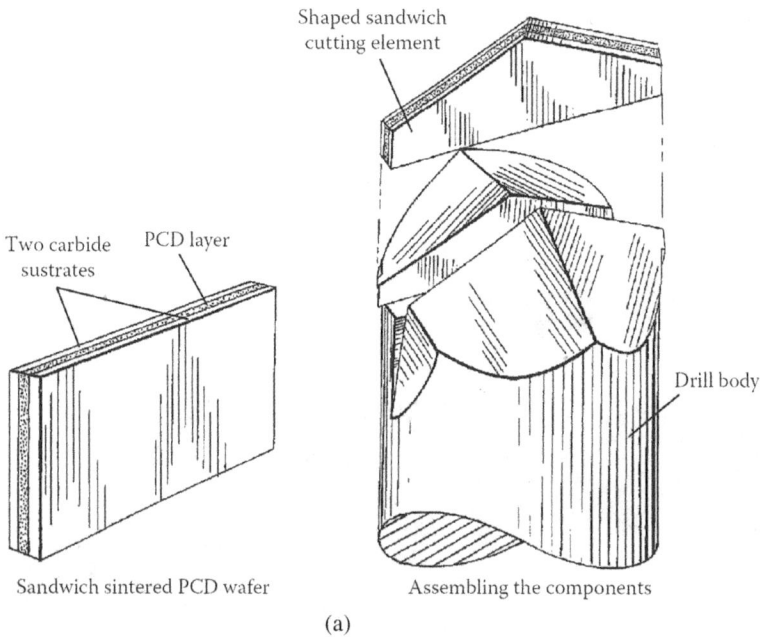

Two carbide sustrates   PCD layer

Shaped sandwich cutting element

Drill body

Sandwich sintered PCD wafer   Assembling the components

(a)

(b)

FIGURE 4.61 Sandwich-type PCD drill: (a) sandwich sintered PCD wafer and assembling the components of the drill and (b) modern PCD blanks/wafers.

2. The full sintering of PCD requires high temperature and isostatic pressure. However, when the sintering is carried out with a carbide nib (veined drills) or between two carbide plates (sandwich drills), these pressures and temperatures are much lower than those in normal PCD sintering. This forces PCD blank manufacturers to use an extensive amount of cobalt (up to five times more than in normally sintered PCDs). Pools (a.k.a. lakes) of cobalt can often be seen on PCD surfaces after grinding.

1. BUE still persists on this drill type although in a different form. When such drills are used, the ground rake face is covered by a thin layer of aluminum that eventually develops BUE with all previously listed consequences.

2. Because different parts of PCD do not have the same exposure to sintering temperature and pressure, the properties of the PCD vein layer vary along its length deteriorating to the drill's center.

FIGURE 4.62   Typical pattern of breakage of sandwich-type PCD drill in machining of HSAAs.

3. The relatively low strength of PCD sintered in this way combined with high forces causes drill breakage in the center as shown in Figure 4.62. All the attempts to improve the manufacturing quality of these drills by better grinding of the rake face, improving the drill geometry, etc., were of little help.

4.11.2.2   Design and Manufacturing of Cross-PCD Drills

Figure 4.63 shows the design concept aimed to reduce the grinding stock on PCD (US Patent No. 4,527,643, 1995) and to fully expose the PCD part allowing its proper sintering condition. It can be referred to as the first cross-PCD drill. As can be seen, a PCD layer is sintered on the substrate top forming two major cutting edges, one chisel, and two side margins. This design concept, however, was not used widely in practice for a variety of primary technological reasons. However, this design concept was used in the development of more feasible one-time designs described earlier (Chapter 5 in Astakhov (2014)).

Despite multiple successful implementations, frequent failures of such drills were reported. These failures occurred due to far from perfect EDMing, brazing, and grinding of PCD as many modern technologies to perform these operations were not yet fully developed at the time. EDM and grinding machine tool builders, brazing professionals, and tool application specialists learned on the go. The major obstacle of such learning was a rather limited market for PCD drills in the automotive and aerospace industries. This was, unfortunately, combined with a relatively high cost of PCD drills. A limited number of tool manufacturing companies were capable of producing such drills efficiently maintaining their quality and consistency.

The De Beers Technical Service Centre pioneered the development of PCD drills with freestanding PCD and worked out a method for their manufacturing (Sani 1994). This method of manufacture became possible due to the development of a brazing procedure of freestanding PCD layers (i.e.,

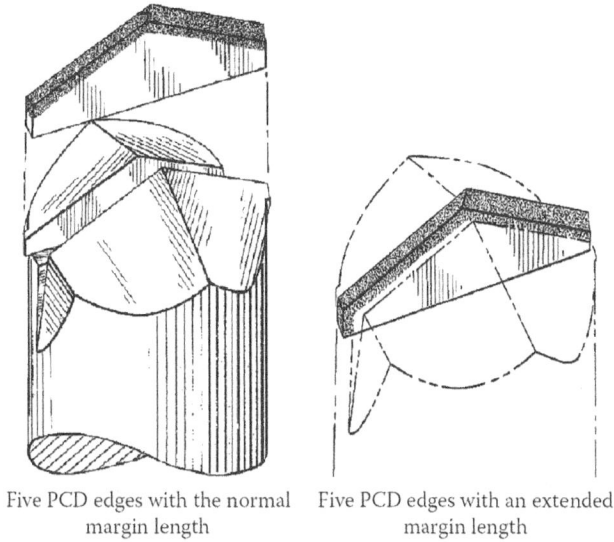

Five PCD edges with the normal   Five PCD edges with an extended
        margin length   margin length

FIGURE 4.63   The first cross-PCD drill.

without an integral carbide backing) onto carbide and steel. In the course of this work, freestanding PCD was successfully brazed, in the presence of argon, using a titanium-activated braze alloy under inert atmosphere/inert gas. To manufacture PCD-tipped twist drills, standard HSS and carbide drills are modified by machining a pocket at the tip of the drill and then brazing an unbacked (free-standing) PCD segment into the pocket. The pocket can be produced by either grinding or EDMing. Unlike most previous designs, the pocket is not cut square to the axis of the drill but parallel to the cutting edges. This produces a pocket in the form of an inverted *V* as shown in Figure 4.64a.

The reason for machining the pocket in this form is twofold:

1. To braze the PCD tip, the *V*-shaped segment is simply placed into the pocket that, by virtue of its matching geometry, causes the PCD tip to be automatically centered. Brazing is carried out by pressing the segment against the side of the pocket, while maintaining gentle pressure on the apex. Using this arrangement, one does not need a special setting-up operation. In fact, once the segment and the pocket are cut to the correct geometries, a perfect braze can be guaranteed each time (Figure 4.64b).
2. Apart from locating the segment, the *V*-shaped pocket has a dramatic effect on the amount of material that needs to be removed from the flute area. By eliminating the PCD from certain areas, standard grinding instead of sophisticated flute grinding can be used to produce the necessary angles in the flute area.

This drill design and its manufacturing steps were much ahead of the time so they have not attracted much attention from PCD drill users and manufacturers because unbacked PCDs (with no substrate) were not yet common and the brazing procedure was cumbersome for most of the tool manufacturers. A great potential of this drill is not recognized until now. Therefore, it is instructive to list the major advantages of this design concept:

1. Any grade of PCD, including leached and thermally stable as well as CVD grades, can be used.
2. The force balance is preferable compared to other drill designs.
3. Practically all known drill point geometry/designs can be then applied to make application-specific drills.
4. Minimum trimming/grinding of the brazed PCD insert is required.

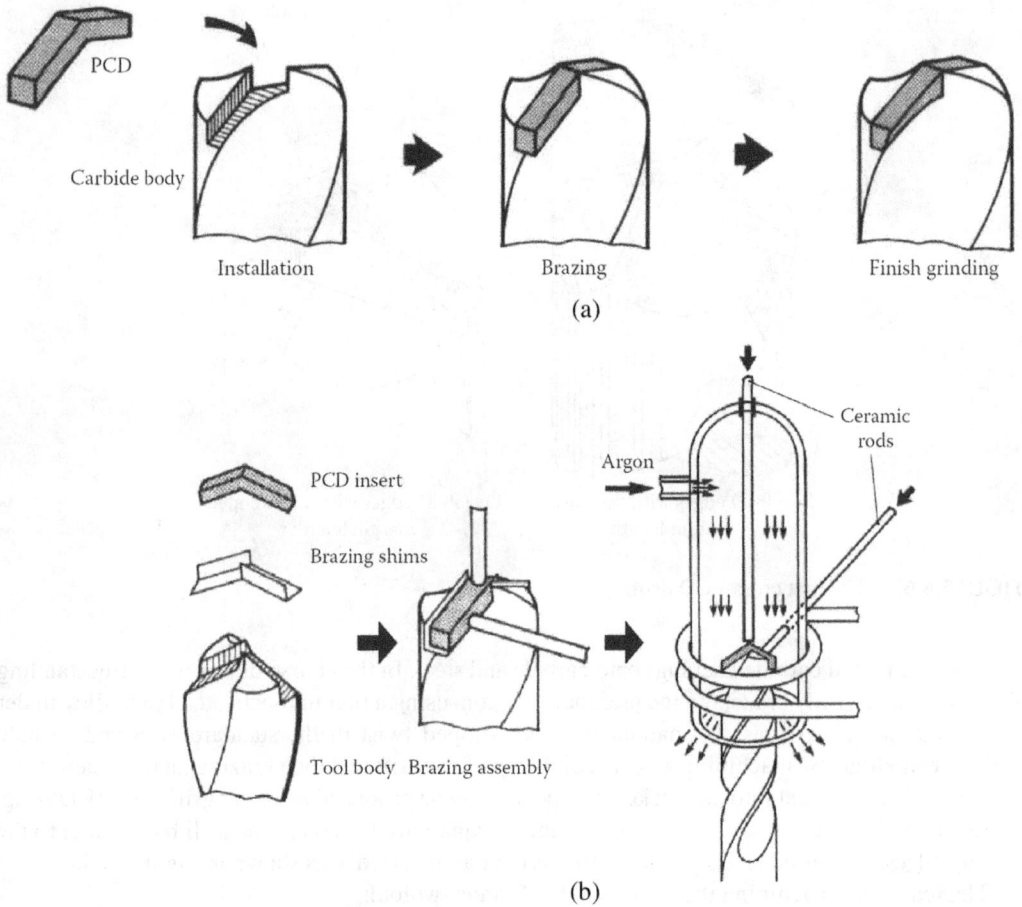

FIGURE 4.64 PCD-tipped drill manufacture using freestanding PCD: (a) principle and (b) steps involved in the brazing procedure.

With further development of PCD technology, a great availability of various freestanding PCD shapes and sizes has become commonplace. They are available from many companies over the world. The capability of multi-axis EDM, EDG, and point grinding machines also increases significantly. All this resulted in the development of the modern versions of cross-PCD drill designs. The general principle of such a modern cross-PCD drill design is shown in Figure 4.65a. As can be seen, a PCD insert of an appreciated size is placed in the slot made in the carbide body and then brazed using the vacuum-furnace technology. Figure 4.65b shows a modern cross-PCD drill made by Fullerton Co and Figure 4.65c shows a modern cross-PCD drill made by Lach Diamond Inc.

The geometry of the drill point, particularly the geometry of the chisel edge, plays a crucial role in the performance of cross-PCD drills as the fracture toughness and thus bending strength of PCD tool material are low compared to carbide and HSS tool materials. It is to say that such a geometry should be tailored for a given application accounting for the properties of the work material and machining regime. When the work material has high strength, the chisel edge region should be protected from chipping. In this case, the so-called pyramidal point grind shown in Figure 4.66a is recommended for work material of high strength. As can be seen, the rake face of the chisel edge is formed by the tertiary flank face. Figure 4.67 shows essential features of this face, i.e., its tertiary flank face (20°/25°) and its orientation angle (45° ± 5°). As discussed by the author earlier (see Section 4.5.8.4 in Astakhov (2014)), these parameters are application-specific. For the considered case, they are optimized for HSAA (ANSI/AA B390.0). With such parameters, the drills made

(a)

(b)

(c)

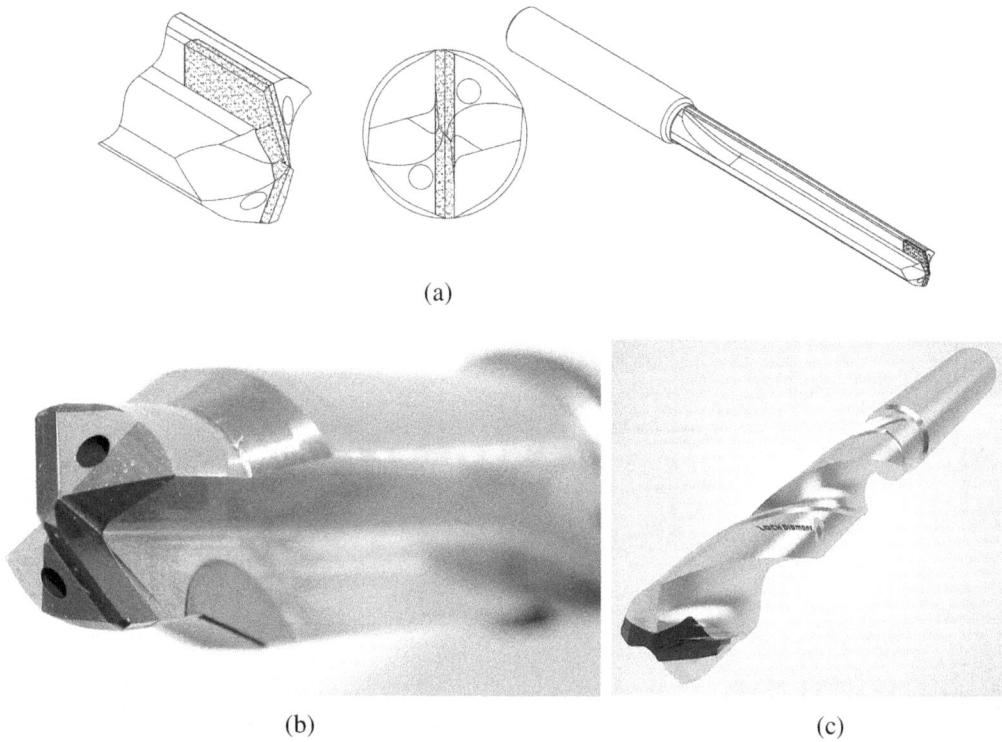

FIGURE 4.65 Modern cross-PCD drills: (a) general principle, (b) drill by Fullerton Co., and (c) drill by Lach Diamond Inc.

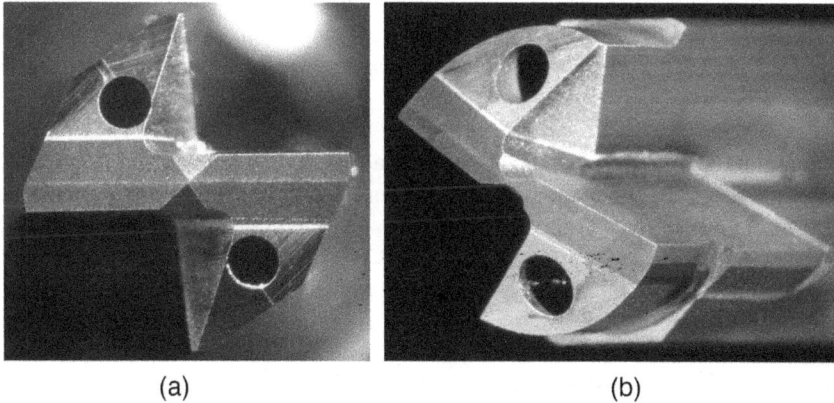

(a)

(b)

FIGURE 4.66 Point geometry of cross-PCD drills: (a) the pyramidal point grind for work material of high strength and (b) for ductile work materials.

according to the geometry shown in Figure 4.67 are reliable, high-productivity, and cost-efficient tools as conclusively proved by their application experience in the setting of a number of modern automotive plants.

When the work material is ductile, greater cutting feeds are used to achieve greater productivity of drilling. As such, however, a greater deformation of the chip takes place, i.e., a greater chip compression ratio (see the Appendix). As a result, the chip, generated by the chisel edge, has problems passing the gap between the rake face of the chisel edge and the bottom of the hole being drilled as discussed

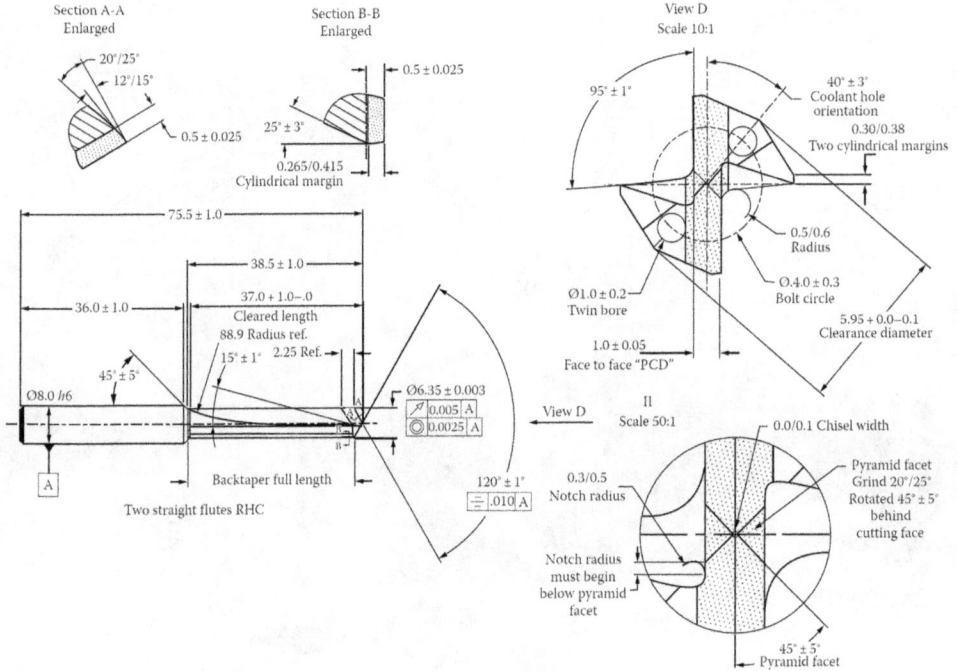

FIGURE 4.67   Essential features of the drawing of the pyramidal point grind drill.

in Section 2.1.5.11.6 (Chapter 2). To solve this problem, the split point grind shown in Figure 4.66b is recommended with all its advantages discussed in Section 2.1.5.11.10 (Chapter 2). Essential features of the drawing of such a drill are shown in Figure 4.68. Application experience of this point grind reveals that the most important feature is the distance between two parts of the chisel edge. This distance is shown in VIEW I to be 0.100±0.025 mm as optimized for ANSI/AA A380.0.

The author's experience in the design and implementation of this kind of drill allows to state the following:

1. The manufacturing of this drill design is mainly limited due to the definite lack of understanding of their advantages from the users and tool manufacturers. Moreover, many tool manufacturers are reluctant to invest in vacuum furnaces required to manufacture such tools.
2. The performance of such drills is very sensitive to their proper application-specific design and accuracy of tool manufacturing, and even small details of application conditions. In other words, such drills are not forgiving compared to other kinds of PCD drill designs. Lack of understanding of the particularities of the application conditions for such drill is the major hurdle today in wide applications of modern cross-PCD drills.
3. Performance including tool life and accuracy of drilled hole achieved with this drill design strongly depends on the grain size of PCD insert/wafer. The most common mistake of many PCD drill manufacturers is to use coarse-grained wafers (normally up to 25 μm) to achieve better EDMing, EDGing, and grindability to speed up the finishing processes. As discussed earlier in this chapter, it is impossible to achieve a sharp cutting edge and surface roughness of the tool–chip and tool–workpiece contact surfaces (interfaces) that may ruin tool performance. Application experience shows that the grain size of the PCD wafer should not be greater than 5 μm.

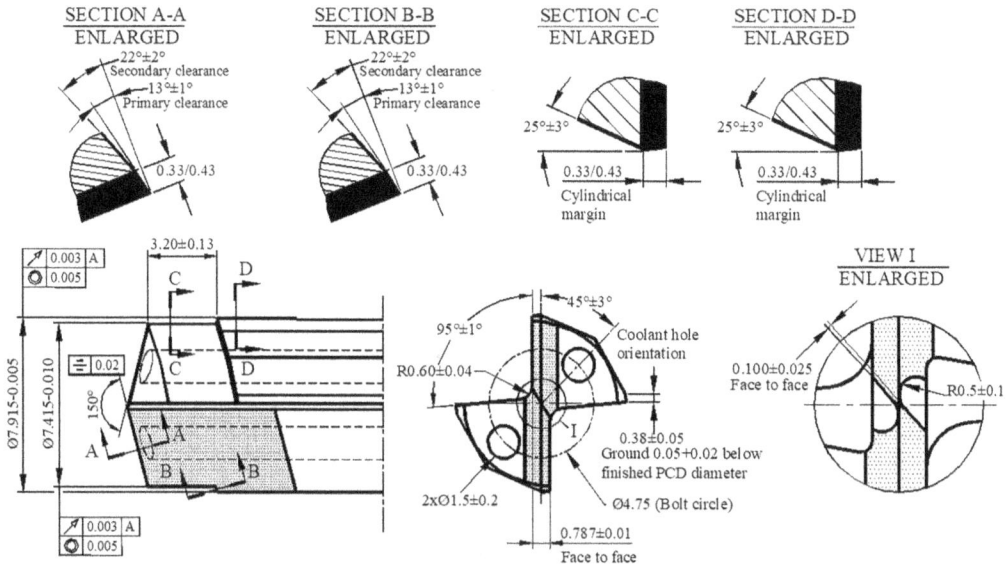

SECTION A-A ENLARGED

SECTION B-B ENLARGED

SECTION C-C ENLARGED

SECTION D-D ENLARGED

VIEW I ENLARGED

LIP HEIGHT VARIATION: 0.005 max
FLUTE SPACING: 0.08 max
WEB SYMMETRY: 0.08 max
CHISEL EDGE CENTRALITY: 0.01 max
Backtaper: 3 microns per mm.

FIGURE 4.68 Essential features of the drawing of the split point grind drill.

4. When properly designed and made, this kind of drill outperforms any other types of PCD-tipped drills allowing much higher drilling productivity (the penetration rate) and quality of machined holes. These advantages significantly overlap the greater cost of cross-PCD drills.

4.11.3 FULL-HEAD PCD DRILLS

The attempt to use a cylindrically-shaped PCD block with the carbide substrate on the drill face started in the 1980s. One of the earlier drill designs is shown in Figure 4.69 (US Patent No. 4.697,971, 1987). In this simplest design, a PCD nib having a rather thick layer of PCD and carbide substrate is brazed on the end of a correspondingly shaped drill body. After cleaning, the brazed blanks are fluted and then pointed to obtain the final desired drill geometry. Note that the patent describes the brazing procedure and fixtures in great detail as a chance to overheat the diamond layer with the brazing technology of the 1980s was great.

Enjoying the full-PCD face as a lucrative advantage, this drill design, however, requires a substantial removal of the PCD material that is both wasteful and ineffective as grinding of a great amount of PCD is expensive and time-consuming. Moreover, the heat generated on grinding can be equally (to brazing) harmful for a PCD layer that may reduce the quality of the drill.

Nowadays, PCD round tool blanks are widely available from various PCD manufacturers. Normally, their diameter is limited by 8 mm whereas the length of the PCD and carbide portions varies from one manufacturer to the next with various grain sizes. For example, Hyperion Co offers such blanks having grain sizes: 1.7, 5, and bimodal 25/4 μm; diameter range 1.5–8 mm, PCD layer length 2.5 and 4 mm, and overall length 13 mm. Figure 4.70 shows some examples of the modern full-PCD head drill.

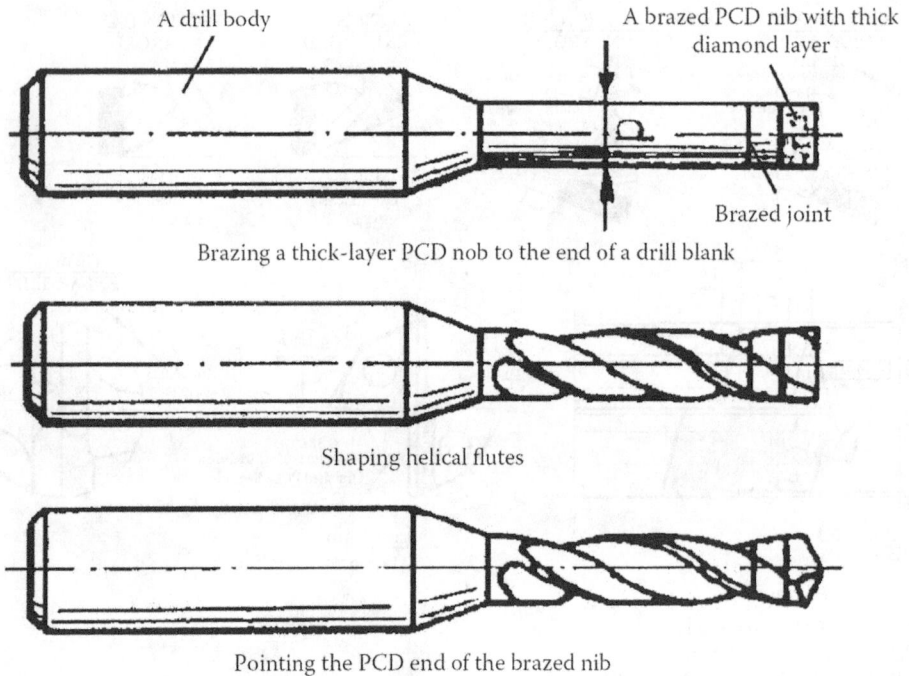

FIGURE 4.69 Earlier drill design with the full-PCD head.

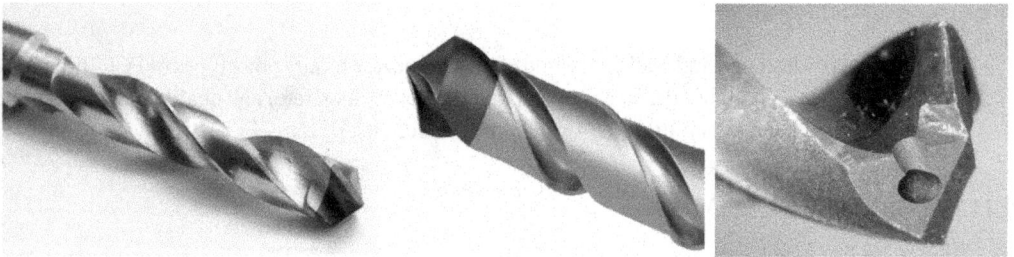

FIGURE 4.70 Examples of the modern full-PCD head drill.

REFERENCES

Abbas, R. K., Musa, R. M. (2019). "Using Raman shift and FT-IR spectra as quality indices of oil bit PDC cutters." *Petroleum* **5**: 329–334.

Akaishi, M., Kanda, H., Yamaoka, S. (1990). "Synthesis of diamond from graphite-carbonate systems under very high temperature and pressure." *Journal of Crystal Growth* **104**: 578–581.

Akaishi, M., Ohsawa, T., Yamaoka, S. (1991). "Synthesis of fine-grained polycrystalline diamond compact and its microstructure." *Journal of the American Ceramic Society* **74**: 5–10.

Ali, B., Litvinyuk, I. V., Rybachuk, M. (2021). "Femtosecond laser micromachining of diamond: Current research status, applications and challenges." *Carbon* **179**: 209–226.

Andrewesa, C. J. E., Feng, H.-Y., Laub, W. M. (2000). "Machining of an aluminum/SiC composite using diamond inserts." *Journal of Materials Processing Technology* **102**(1–3): 25–29.

Asmussen, J., Reinhard, D. K. (2002). *Diamond Films Handbook*. New York, Marcel Dekker.

Astakhov, V. P. (1998/1999). *Metal Cutting Mechanics*. Boca Raton, FL, CRC Press.

Astakhov, V. P. (2014). *Drills: Science and Technology of Advanced Operations*. Boca Raton, FL, CRC Press.

Astakhov, V. P., Joksch, S., Ed. (2012). *Metal Working Fluids for Cutting and Grinding: Fundametals and Recent Advances*. London, Woodhead.

Astakhov, V. P., Patel, S. (2018). Efficient drilling of high-silicon aluminum alloys: Fundamentals and recent advances. *Drilling Technology: Fundamentals and Recent Advances*, J. P. Davim. Berlin, Germany, De Gruyter Oldenbourg.

Astakhov, V. P., Xiao, X. (2016). The principle of minimum strain energy to fracture of the work material and its application in modern cutting technologies. *Metal Cutting Technology*. P. J. Davim. Berlin, De Gruyter 1–35.

Auciello, O., Aslam, D. M. (2021). "Review on advances in microcrystalline, nanocrystalline and ultrananocrystalline diamond films-based micro/nano-electromechanical systems technologies." *Journal of Materials Science* **56**: 7171–7230.

Bellin, F., Dourfaye, A., King, W., Thigpen, M. (2010a). "The current state of PCD bit technology. Part 1: Development and application of polycrystalline diamond compact bits have overcome complex challenges from the difficulty of reliability mounting PDC cutters in bit bodies to accelerate thermal wear." *World Oil* **231**: 41–46. Cited now

Bellin, F., Dourfaye, A., King, W., Thigpen, M. (2010b). "The current state of PCD bit technology. Part 2: Leaching a thin layer at the working surface of a PCD cutter to remove the cobalt dramatically reduces diamond degradation due to friction heat." *World Oil* **231**(September): 41–46.

Bellin, F., Dourfaye, A., King, W., Thigpen, M. (2010c). "The current state of PCD bit technology. Part 3: Improvements in material properties and testing methods are being pursued to make PDC the cutter of choice for an increasing variety of applications." *World Oil* (November): 67–71. https://www.worldoil.com/magazine/2010/november-2010/.

Bergs, T., Muller, U., Vits, F., Barrg, S. (2020). "Grinding wheel wera and material removal mechnisms during grinding of polycrystalline diamond." *Procedia CIRP* **93**: 1520–1525.

Bisaria, H., Shandilya, P. (2015). Machining of metal matrix composites by EDM and its variants: A Review. *DAAAM International Scientific Book 2015*. B. Katalinic. Vienna, Austria, DAAAM International: 267–282.

Boland, J. N., Li, X. S., Rassool, R. P., Hay, D. (2010). "Characterisation of diamond composites for tooling." *Journal of the Australian Ceramic Society* **46**(1): 1–10.

Brecher, C., Emontsa, M., Hermania, J-P., Storms, T. (2014). "Laser roughing of PCD." *Physics Procedia* **56**: 1107–1114.

Bruton, E. (1970). *Diamonds*. Philadelphia, PA, Chilton Book.

Bundy, F. P. (1989). "Pressure-temperature phase diagram of elemental carbons." *Physica A* **156**: 169–178.

Cao, F., Zhang, Q. (2004). "Neural network modelling and parameters optimization of increased explosive electrical discharge grinding (IEEDG) process for large area polycrystalline diamond." *Journal of Materials Processing Technology* **149**: 106–111.

Center, F. R. D. (2018). Research on PCD vacuum brazing technology. Press release by FUNIK R&D Center https://en.funik.com/faq/Research-on-PCD-vacuum-brazing-technology.html.

Chang, F.-Y., Hsu, C-F., Lu, W-H. (2021). "Nanosecond-fiber laser cutting and finishing process for manufacturing polycrystalline diamond-cutting tool blanks." *Applied Science* **11**: 5871.

Che, D., Guo, P., Ehmann, K. (2012a). "Issues in polycrystalline diamond compact cutter-rock interaction from a metal machining point of view-Part II: Bit performance and rock cutting mechanics." *Journal of Manufacturing Science and Engineering* **134**: 064002.

Che, D., Guo, P., Ehmann, K., (2012b). "Issues in polycrystalline diamond compact cutter-rock interaction from a metal machining point of view-Part I: Temperature, stresses, and forces." *Journal of Manufacturing Science and Engineering* **134**: 064001.

Chen, F., Xu, G. (2010). "Thermal residual stress of polycrystalline diamond compacts." *Transactions of Nonferrous Metals Society of China* **20**: 227–232.

Chen, Y., Zhang, L. C., Arsecularatne, J. A. (2007). "Polishing of polycrystalline diamond by the technique of dynamic friction." *International Journal of Machine Tool and Manufacturing* **47**: 1625–1624.

Clark, J. (2020, August 15). "Atomic Orbitals." from https://chem.libretexts.org/@go/page/3611.

Coelho, R. T., Yamade, S., Aspinwall, D. K., Wise, W. L. H. (1995). "The application of polycrystalline diamond (PCD) tool materials when drilling and reaming aluminium based alloys including MMC." *International Journal of Machine Tools and Manufacturing* **35**(5): 761–774.

Cook, M. W., Bossom, P. K. (2000). "Trends and recent developments in the material manufacture and cutting tool application of polycrystalline diamond and polycrystalline cubic boron nitride." *International Journal of Refractory Metals & Hard Materials* **18**: 147–152.

Davis, R. F. (1993). *Diamond Films and Coating*. Park Ridge, NJ, Noyes Publications.

Destefani, J. (2002). "Cutting tools 101: Geometries." *Manufacturing Engineering* **129**(10): 25–42.

Dismukes, J. P., Gaines, P. R., Witzke, H., Leta, D. P., Kear, B. H., Behal, S. K. (1988). "Demineralization and microstructure of carbonado." *Material Science and Engineering A* **105/106**: 555–563.

Dolda, C., Henerichsb, M., Gilgenb, P., Wegenera, K. (2013). "Laser processing of coarse grain polycrystalline diamond (PCD) cutting tool inserts using picosecond laser pulses." *Physics Procedia* **41**: 610–616.

Eberlea, G., Wegenera, K. (2014). "Ablation study of WC and PCD composites using 10 picosecond and 1 nanosecond pulse durations at green and infrared wavelengths." *Physics Procedia* **56**: 951–962.

El-Gallab, M., Sklad, M. (1998). "Machining of Al/SiC particulate metal matrix composites. Part I. Tool performance." *Journal of Materials Processing Technology* **83**(2): 151–158.

Fedoseev, D. V., Vnukov, S. P., Bukhovets, V. L., Anikin, B. A. (1986). "Surface graphitization of diamond at high temperatures." *Surface and Coatings Technology* **28**: 207–214.

Field, J. E. (1992). *The Properties of Natural and Synthetic Diamond*. London, Academic Press.

German, R. M., Smid, I., Campbell, L. G., Keane, J., Toth, R. (2005). "Liquid phase sintering of tough coated hard particles." *International Journal of Refractory Metals & Hard Materials* **23**(4–6): 267–272.

Grodzinski, P. (1953). *Diamond Technology*. London, N.A.G. Press.

Gu, J., Huang, K. (2016). "Role of cobalt of polycrystalline diamond compact (PDC) in drilling process." *Diamond & Related Materials* **66**: 98–101.

Gublin, Y. (2008). "On estimation and hypothesis testing of the grain size distribution by Saltykov method." *Image Analysis and Stereology* **27**: 163–174.

Hall, H. T. (1960). "Ultra-high-pressure, high-temperature apparatus: The 'belt'." *Review of Scientific Instruments* **31**: 125–131.

Hall, H. T. (1970). "Sintered diamond: A synthetic carbonado." *Science* **169**: 868–869.

Hall, H. T. (1982). Diamonds, synthetic. *Encyclopedia of Chemical Processing and Design*, Volume 15. A. W. C. J. J. McKetta. New York, Marcel Dekker: 410–435.

Hannay, J. B. (1879). "On the artificial formation of the diamond." *Proceedings of the Royal Society of London* **30**(200–205): 450–461.

Harano, K., Satoh, T., Sumiya, H. (2012). "Cutting performance of nano-polycrystalline diamond." *Diamond & Related Materials* **24**: 78–82.

Harrison, P., Henry, M., Wendland, J. (2005). "Enhanced cutting of polycrystalline diamond with a q-switched diode pumped solid state laser." *International Congress on Applications of Lasers & Electro-Optics* **2005**(1): 202.

Hong, S.M., Akaishi, M., Kanda, H., Osawa, T., and Yamaoka, O. (1988). "Behaviour of cobalt infiltration and abnormal grain growth during sintering of diamond on cobalt substrate." *Journal of Materials Science Letters* **23**: 3821–3826.

Hung, N. P., Boey, F. Y. C., Khor, K. A., Phua, Y. S., Lee, H. F. (1996). "Machinability of aluminum alloys reinforced with silicon carbide particulates." *Journal of Materials Processing Technology* **56**(1–4): 966–977.

Irifune, T., Sumiya, H. (2014). Nanopolycrystalline diamond without binder and its application to various high-pressure apparatus. *Comprehensive Hard Materials*. V. K. Sarin & C. E. Nebel, Vol. 3, Amsterdam, Elsevier Science: 173–191.

Jaworska, L., Szutkowska, M., Klimczyk, P., Sitarz, M., Bućko, M., Rutkowski, P., Figiel, P., Łojewska, J. (2014). "Oxidation, graphitization and thermal resistance of PCD materials with the various bonding phases of up to 800°C." *International Journal of Refractory Metals & Hard Materials* **45**: 109–116.

Jeppsson, J., Mannesson, K., Borgenstam, A., Agren, J. (2011). "Inverse Saltykov analysis for particle-size distributions and their time evolution." *Acta Materialia* **59**(3): 874–882.

Jeynes, C. (1978). "Natural polycrystalline diamond." *Industrial Diamond Review* **39** (January): 15–28.

Jianxin, D., Hui, Z., Ze, W., Aihua, L. (2011). "Friction and wear behaviour of polycrystalline diamond at temperatures up to 700°C." *International Journal of Refractory Metals and Hard Materials* **29**: 631–638.

Jivanji, M., Forbes, R. P., Sithebe, H., Westradt, J. E. (2023). "Effect of ZrB2 additions on the thermal stability of polycrystalline diamond." *International Journal of Refractory Metals and Hard Materials* **113**: 106202.

Kang, S. J. L. (2005). *Sintering: Densification, Grain Growth and Microstructure*. Amsterdam, the Netherlands, Elsevier.

Kanyanta, V., Dormer, A., Murphy, N., Invankovic, A. (2014). "Impact fatigue fracture of polycrystalline diamond compact (PDC) cutters and the effect of microstructure." *International Journal of Refractory Metals and Hard Materials* **46**: 145–151.

Kononenko, T. V., Khomich, A. A., Konov, V. I. (2013). "Peculiarities of laser-induced material transformation inside diamond bulk." *Diamond & Related Materials* **37**: 50–54.

Kozak, J., Rajurkar, K. P., Wang, S. Z. (1994). "Material removal in WEDM of PCD blanks." *Journal of Manufacturing Science and Engineering* **116**: 363–369.

Krawitz, A. D., Winholtz, R. A., Drake, E. F., Griffin, N. D. (1999). "Residual stresses in polycrystalline diamond compacts." *International Journal of Refractory Metals & Hard Materials* **17**: 117–122.

Krebs, R. E. (2006). *The History and Use of Our Earth's Chemical Elements: A Reference Guide*. London, Greenwood Press.

Kui, L., Hu, W., Tong, N., Senthil, K. (2017). EDM wire cutting of polycrystalline diamond (PCD) for customized cutting tool fabrication. *Proceedings of the 20th International Symposium on Advances in Abrasive Technology*, Okinawa, Japan.

Lammer, A. (1988). "Mechanical properties of polycrystalline diamond." *Materials Science and Technology* **4**: 949–955.

Larkin, P. (2011). *Infrared and Raman Spectroscopy: Principles and Spectral Interpretation*. Waltham, MA, Elsevier.

Li, G., Rahim, M. Z., Pan, W., Wen, C., Ding, S. (2020). "The manufacturing and the application of polycrystalline diamond tools - A comprehensive review." *Journal of Manufacturing Processes* **56**: 400–416.

Li, G., Wu, G., Pan, W., Rahman Rashid, R.A, Palanisamy, S., Ding, S. (2021). "The performance of polycrystalline diamond (PCD) tools machined by abrasive grinding and electrical discharge grinding (EDG) in high-speed turning." *Journal of Manufacturing and Materials Processing* **5**: 34.

Li, J., Yue, W., Wang, C. (2016). "Microstructures and thermal damage mechanisms of sintered polycrystalline diamond compact annealing under ambient air and vacuum conditions." *International Journal of Refractory Metals and Hard Materials* **54**: 138–147.

Li, Z. Q., Wu, Q., Wang, J. (2012). "Ultrashort pulsed laser micromachining of polycrystalline diamond." *Advanced Materials Research* **497**: 220–224.

Lin, T., Hood, M., Cooper, G. A., Li, X. (1992). "Wear and failure mechanisms of polycrystalline diamond compact bits." *Wear* **156**(1): 133–150.

Liu, A. Y., Cohen, M. L. (1989). "Prediction of new low compressibility solids." *Science* **245**: 841–842.

Liu, S., Han, L., Zou, Y., Zho, P., Liu, B. (2017). "Polycrystalline diamond compact with enhanced thermal stability." *Journal of Materials Science & Technology* **33**: 1386–1391.

Liu, Y.-K., Tso, P.-L. (2003). "The optimal diamond wheels for grinding diamond tools." *International Journal of Advanced Manufacturing Technology* **22**: 396–400.

Mashhadikarimi, M., Medeiros, R. B. D., Barreto, L. P. P., Gurgel, D. P., Gomes, U. U., Filgueira, M. (2021). "Development of a novel triple-layer polycrystalline diamond compact." *Diamond & Related Materials* **111**: 108182.

Meguid, S. A., Sun, Y. (2005). "Intelligent condition monitoring of aerospace composites: Part I - Nano reinforced surfaces & interfaces." *International Journal of Mechanics and Materials in Design* **2**(3–4): 37–52.

Mehan, R. L., Hibbs, L. E. (1989). "Thermal degradation of sintered diamond compacts." *Journal of Materials Science* **24**: 942–950.

Miess, D., Rai, G. (1996). "Fracture toughness and thermal resistance of polycrystalline diamond compacts." *Materials Science and Engineering A* **209**: 270–276.

Moissan, H. (1994). "Nouvelles expériences sur la reproduction du diamant." *Comptes Rendus de l'Académie des Sciences (Paris)* **118**: 320–341.

Olsen, R. H., Dewes, R. C., Aspinwall, D. K. (2004). "Machining of electrically conductive CVD diamond tool blanks using EDM." *Journal of Materials Processing Technology* **149**(1–3): 627–632.

Pacella, M. (2020). "Laser finishing of polycrystalline diamond as strengthening mechanism." *Procedia CIRP* **87**: 240–244.

Pacella, M., Axinte, D. A., Butler-Smith, P. W., Daine, M. (2014). "On the topographical/chemical analysis of polycrystalline diamond pulsed laser ablated surfaces." *Procedia CIRP* **13**: 387–392.

Paggett, J. W., Drake, E. F., Krawitz, A. D., Winholtz, R. A., Griffin, N. D. (2002). "Residual stress and stress gradients in polycrystalline diamond compacts." *International Journal of Refractory Metals & Hard Materials* **20**: 187–194.

Park, Y. J., Hwang, N. M., Yoon, D. Y. (1996). "Abnormal growth of faceted (WC) grains in a (Co) liquid matrix." *International Journal of Refractory Metals & Hard Materials* **20**: 187–194.

Partington, J. R. (1962). *A History of Chemistry*. London, Macmillan & Co.

Pierson, H. O. (1999). *Handbook of Chemical Vapor Deposition: Principles, Technology and Applications*. New York, Noyes Publications.

Pisarciuc, C. (2013). "Study of process parameters at electrical discharge machining of polycrystalline diamond." *Nonconventional Technologies Review* **XVII**: 54–58.

Prikhna, A. I. (2008). "High-pressure apparatuses in production of synthetic diamonds (Review)." *Journal of Superhard Materials* **30**(1): 1–15.

Purniawan, A. (2008). Deposition and characterization of polycrystalline diamond coated on silicon nitride and tungsten carbide using microwave plasma assisted chemical vapour deposition technique Master of Engineering Universiti Teknologi Malaysia.

Rahim, M. Z. (2015). Research on Electrical Discharge Machining of Polycrystalline Diamond. PhD, RMIT University.

Rahim, M. Z., Ding, S., Li, G., Mo, J. P. T. (2015a). "Electrical discharge grinding versus abrasive grinding in polycrystalline diamond machining." *Jurnal Teknologi* **74**(10): 79–87.

Rahim, M. Z., Ding, S., Mo, J. (2015b). "Electrical discharge grinding of polycrystalline diamond-effect of machining parameters and finishing in-feed." *Journal of Manufacturing Science and Engineering* **137**: 021017.

Raman, C. V., Krishnan, K. S. (1928). "A new type of secondary radiation." *Nature* **121**: 501–502.

Roberts, P. M. (2013). *Industrial Brazing Practice*. Boca Raton, FL, CRC Press.

Roberts, P. M. (2016). *Introduction to Brazing Technology*. Boca Raton, FL, CRC Press.

Royère, C. (1999). "The electric furnace of Henri Moissan at one hundred years: Connection with the electric furnace, the solar furnace, the plasma furnace?" *Annales pharmaceutiques françaises* **57**(2): 116–130.

Russell, W. C., Sowers, S., Webb, S., Raghavan, R. (2016). Polycrystalline diamond material with extremely fine microstructures. United States Patent, Diamond Innovations, Inc. 9,403,137 B2.

Sani, M. N. (1994). "Further developments in PCD-tipped drills." *Industrial Diamond Review* **194**: 6–7.

Schwarts, M. M. (2003). *Brazing of Cutting Materials*. Metals Park, Ohio, ASM International.

Sekulić, D. P., Ed. (2013). *Advances in Brazing: Science, Technology and Applications*. Cambridge, Woodhead Publishing Limited.

Shenderova, O. A. (2014). Production of nanodiamond particles. *Comprehensive Hard Materials*. D. M. C. Nebel, Miguel, L., Sarin, V. Amsterdam, Elsevier Science. **3:**143–171.

Shimada, S., Tanaka, H., Higuchi, M., Yamaguchi, T., Honda, S., Obata, K. (2004). "Thermo-chemical wear mechanism of diamond tool in machining of ferrous metals." *CIRP Annals-Manufacturing Technology* **53**: 57–60.

Shin, T., Oh, J., Oh, K., Lee, D. (2004). "The mechanism of abnormal grain growth in polycrystalline diamond during high pressure-high temperature sintering." *Diamond and Related Materials* **13**: 488–494.

Smith, N. R. (1974). *User's Guide to Industrial Diamonds*. London, Hutchinson Benham.

Sneddon, M. V., Hall, D. R. (1987). Recent advances in polycrystalline diamond (PCD) technology open new frontiers in drilling. SPE Annual Technical Conference and Exhibition. Dallas, Texas, SPE.

Sommer, C., Sommer, S. (2005). *Complete EDM Handbook*. Houston, TX, Advance Publishing.

Spear, K., Dismukes, J. (1994). *Synthetic Diamond: Emerging CVD Science and Technology*. Pennington, NJ, John Wiley & Sons.

Spur, G., Puttrus, M, Wunsch, U. W. (1988). "Wire EDM of PCD." *Wire EDM of PCD* **48**: 264–266.

Thoe, T. B., Aspinwall, D. K., Wise, M. L. H., Oxle, I. A. (1996). "Polycrystalline diamond edge quality and and surface integrity following electrical discharge grinding." *Jounal Materials Processesin Technology* **56**: 773–785.

Tomac, N., Tonnessen, K. (1992). "Machinability of particulate aluminium matrix composites." *Annals of the CIRP* **41**(1): 55–58.

Tomlinson, O. N., Clark, I. E. (1992). "Syndax3 pins-New concepts in PCD drilling." *Industrial Diamond Review* **52**(3): 109–111.

Uehara, K., Yamaya, S. (1990). High pressure sintering of diamond by cobalt infiltration. *Science and Technology of New Diamond*. O. F. S. Saito, M. Yoshikawa, Tokyo, KTK Scientific Publishers: 203–209.

Vandenbulcke, L., De Barros, M. I. (2001). "Deposition, structure, mechanical properties and tribological behavior of polycrystalline to smooth fine-grained diamond coatings." *Surface and Coatings Technology* **146–147**: 417–424.

Vohra, Y. K., Catledge, S. A., Ladi, R., Rai, G. (1996). "Micro-Raman stress investigations and x-ray diffraction analysis of polycrystalline diamond (PCD) tools." *International Journal of Refractory Metals & Hard Materials* **26**: 232–241.

Westraadt, J. E., Sigalas, I., Neethling, J. H. (2015). "Characterisation of thermally degraded polycrystalline diamond." *International Journal of Refractory Metals and Hard Materials* **48**: 286–292.

Wilks, J., Wilks, E. (1991). *Properties and Applications of Diamond*. Oxford, Butterworth-Heinemann.

Wu, Q., Wang, J. (2010). "Development in laser polishing of polycrystalline diamond tools." *Advanced Materials Research* **135**: 1–6.

Wu, Q., Wang, J., Huang, C. (2014). "Analysis of the machining performance and surface integrity in laser milling of polycrystalline diamonds." *Proceedings of the Institution of Mechanical Engineers, Part B: Journal of Engineering Manufacture* **228**: 903–917.

Xiao, H. (2014). In situ tribo-electrochemical characterization of diamond-containing materials. PhD, Texas A&M University.

Yahiaoui, M., Paris, J.-Y., Delbé, K., Denape, J., Gerbaud, L., Colin, C., Ther, O., Dourfaye, A. (2016). "Quality and wear behavior of graded polycrystalline diamond compact cutters." *International Journal of Refractory Metals and Hard Materials* **56**: 87–95.

Zhang, G. F., Zhang, B., Deng, Z. H., Chen, J. F. (2007). "An experimental study on laser cutting mechanisms of polycrystalline diamond compacts." *Annals of the CIRP* **56**(1): 201–204.

Zhang, Z., Peng, H., Yan, J. (2013). "Micro-cutting characteristics of EDM fabricated high-precision polycrystalline diamond tools." *International Journal of Machine Tools and Manufacture* **65**: 99–106.

Zhao, Y., Yue, W., Lin, F., Wang, C., Wu, Z. (2015). "Friction and wear behaviors of polycrystalline diamond under vacuum conditions." *International Journal of Refractory Metals and Hard Materials* **50**: 43–52.

Appendix
Basics of the Tool Geometry

And one more thing.

> **Steven Paul Jobs (1955–2011), an American business magnate, industrial designer, media proprietor, and investor**. *He used the phrase since 1999 before unveiling products such as the iMac and iPod at computer shows.*

A.1 BASIC TERMS AND DEFINITIONS

The geometry and nomenclature of cutting tools, even single-point cutting tools, are surprisingly complicated subjects (Astakhov 1998/1999, 2010). To simplify the common consideration, this section aims to give the proper unambiguous definitions of the components involved, present a step-by-step representation of the tool geometry articles, and then explain their significance in metal cutting operations. It also aims to clean up old notions and lay-language terminology that unfortunately routinely appear in the trade, engineering, and even scientific publications presenting a great deal of confusion as they are not suitable in modern manufacturing.

The simplest cutting operation is one in which a straight-edged tool moves with a constant velocity in the direction perpendicular to the cutting edge of the tool. This is known as the 2D or orthogonal cutting illustrated in Figure A.1. This model includes the following components of the cutting system (Astakhov and Shvets 1998): the workpiece, the cutting tool, and the chip. Each of these components has some surfaces and characteristics important in tool geometry considerations and in the cutting process as the tool geometry should always be considered in close relations to this process.

As pointed out by Taylor (1907), while a metal cutting tool looks like a wedge, its function is far different from that of a wedge. The flank face is never (at least theoretically) allowed to touch the workpiece. As pointed out by Astakhov (2010), this distinguishes a metal cutting tool (regardless of what the work material is actually cut) from other cutting tools such as scissors, knife, punch, hatchet, etc. *Therefore, the most distinguishing feature of any cutting tool is the clearance angle α, which clears the motion of the cutting edge.*

Although it is true that the cutting operation can be easily understood in terms of orthogonal cutting parameters shown in Figure A.1, so the basic cutting terminology and the theory of metal cutting were developed for orthogonal cutting, it is not always easy to correlate the presented terms with real machining operations such as with turning shown in Figure A.2. Figure A.3 helps to correlate the terminology used in orthogonal cutting and turning.

A.1.1 WORKPIECE SURFACES

In orthogonal cutting (Figure A.1), the two basic surfaces of the workpiece are considered:

- Work surface is the surface of the workpiece to be removed by machining.
- Machined surface is the surface produced after the cutting tool pass.

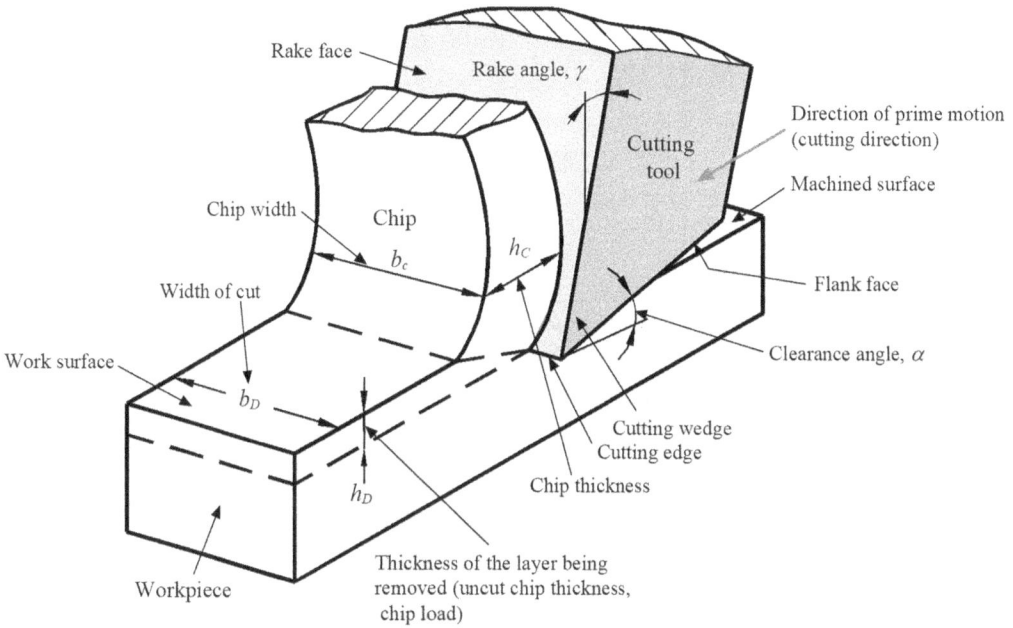

FIGURE A.1   Visualization of the basic terms in orthogonal cutting.

FIGURE A.2   Arrangements of the components in turning.

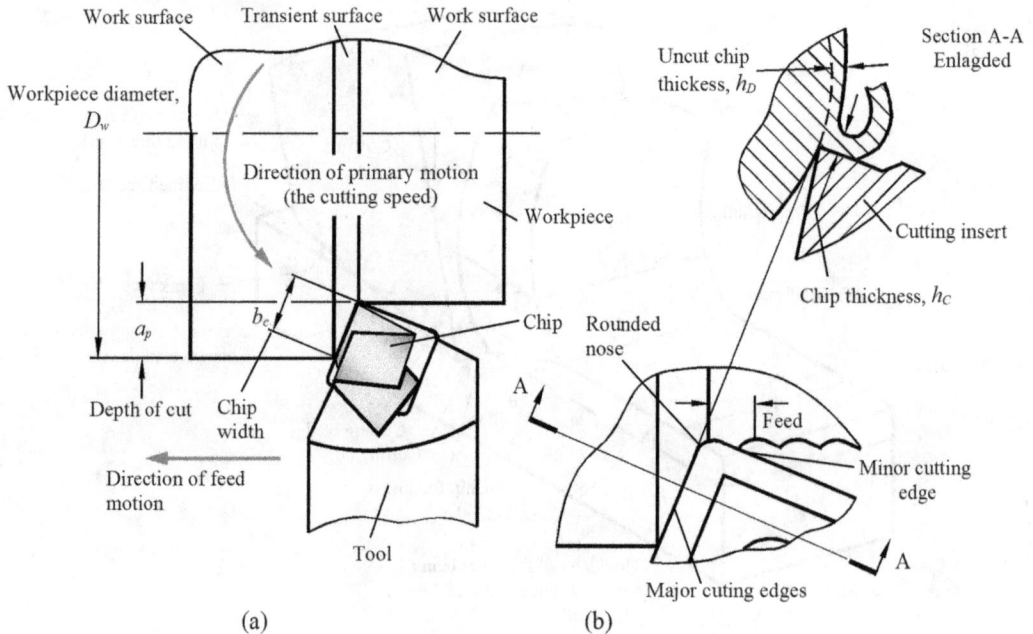

FIGURE A.3   Visualization of basic terms in turning: (a) general view and (b) enlarged cutting portion.

In many practical machining operations, an additional surface, namely, the transient surface is considered. The transient surface is the surface being cut by the major cutting edge (Figure A.3a). Note that the transient surface is always located between the work surface and the machined surface. Besides simple shaping, planning, and broaching, the cutting edge does not form the machined surface in many real cutting operations. As can be clearly seen in Figure A.3b, the machined surface is formed by the tool corner and minor cutting edge. Unfortunately, not much attention is paid to these two important articles of the tool geometry although their parameters directly affect the integrity of the machined surface including the surface roughness and machining residual stresses. Misunderstanding of the previously discussed matter causes a great mismatch in the results of known modeling of the cutting process and reality as the tribological conditions on the tool flank face (over the tool–workpiece tribological interface) including forces, stresses, and temperatures are considered to be important in the formation of the integrity of the machined surface. As clearly seen in Figure A.3, it is not so as these tribological conditions actually affect the formation of the transient surface quality which is not of importance.

A.1.2   TOOL SURFACES AND ELEMENTS

The design components of the cutting tool are defined as follows:

- Rake face is the surface over which the chip, formed in the cutting process, slides.
- Flank face is the surface(s) over which the machined surface passes.
- Cutting edge is a theoretical line of intersection of the rake and the flank surfaces.
- Cutting wedge is the tool body enclosed between the rake and the flank faces.
- Shank is the part of the tool by which it is held.

Understanding the previously introduced terms is important for further considerations. As introduced, the cutting edge is a line, in general, a 3D line that does not have any "meat" attached. Therefore, it cannot have strength, temperature, wear, chipping, and other physical characteristics of a 3D body although these characteristics are routinely prescribed to the cutting edge in the professional and even in scientific literature. In reality, the cutting wedge and its corresponding surfaces (the rake and flank) should be considered in the modeling of the cutting process, cutting tool design, and optimization of machining operations. Note that the notion of the cutting wedge was first introduced by Taylor (1907)

A.1.3 Types of Cutting

Orthogonal cutting is the type of cutting where the straight cutting edge of the wedge-shaped cutting tool is at the right angle to the direction of cutting as shown in Figure A.4a. The additional distinctive features of orthogonal cutting are as follows:

- The cutting edge is wider than the width of cut.
- No side spread of the layer being removed occurs on its transformation into the chip.
- Plane strain condition is the case, that is, a single slice (by a plane perpendicular to the cutting edge) of the model shown in Figure A.4a can be considered in the analysis of the chip formation model.
- The cutting edge does not pass the previously machined surface by this cutting edge, so there is no influence of the previous cutting passes on the current pass. This is not the case in many real machining operations, for example, drilling operations, because the temperatures and machining residual stresses built on the previous pass might significantly affect the cutting conditions on the current pass. Moreover, this influence depends on many cutting parameters such as the rotational speed of the workpiece (which defines the time

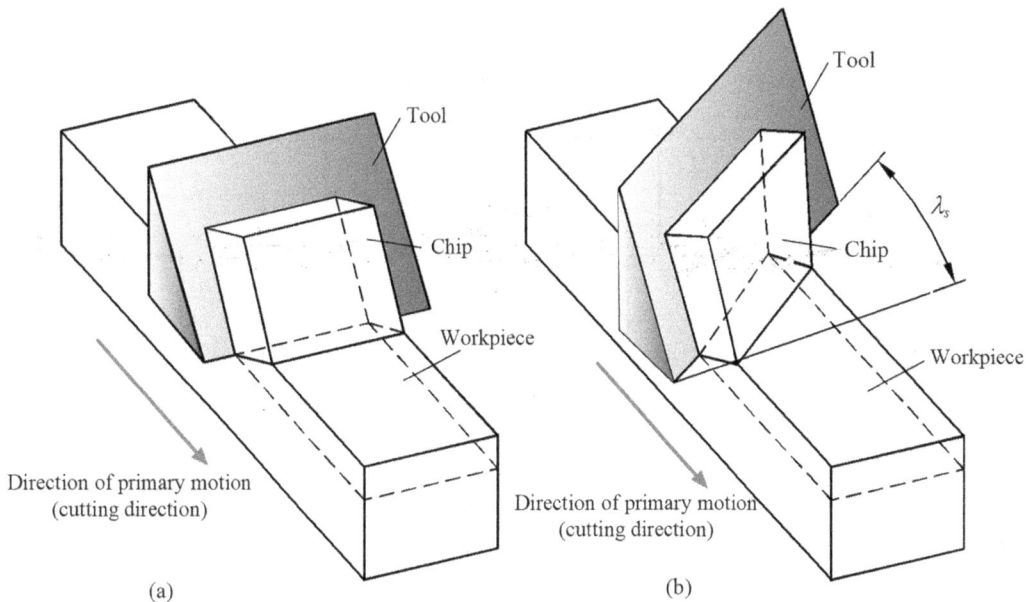

FIGURE A.4 Orthogonal (a) and oblique (b) cutting.

difference between two successive positions of the cutting edge and the intensity of the residual heat) and axial feed (which defines the machining residual stresses left from the previous pass of the cutting tool).

Oblique cutting is the type of cutting where the straight cutting edge of the wedge-shaped cutting tool is not at the right angle to the direction of cutting as shown in Figure A.4b. Figure A.4 illustrates the difference between orthogonal and oblique cutting. In orthogonal cutting, the cutting edge is perpendicular to the direction of the primary motion, while in oblique cutting, it is not. The angle that the straight cutting edge makes with the direction of the cutting speed is known as the cutting edge inclination angle λ_s. The plastic deformation of the layer being removed in oblique cutting is more complicated than that in orthogonal cutting (Shaw 2004).

Free cutting is the type of orthogonal or oblique cutting when only one cutting edge is engaged in cutting. Although this definition is widely used in the literature on metal cutting (Zorev 1966, Armarego and Brown 1969, Shaw 1984, Astakhov 1998/1999), it does not provide the proper explanation to the idea of free cutting. For example, if a cutting edge is not straight, it does not perform free cutting. In contrast, a number of cutting edges can be simultaneously engaged in cutting in surface broaching, but each edge is engaged in free cutting. In the definition, free means that the elementary chip flow vectors from each point of the cutting edge are parallel to each other and do not intersect any other chip flow vectors. An example of free cutting is shown in Figure A.5a. If more than one adjacent cutting edges are involved in cutting (Figure A.5b shows an example of two cutting edges) or when the cutting edge is not straight (Figure A.5c), the chip flows formed at different cutting edges or at different points of the same cutting edge cross each other, creating chip flow interface and thus causing greater chip deformation and cutting force than in free cutting.

A.1.4 IMPORTANT NOMENCLATURE OF A SINGLE-POINT TOOL

This section briefly presents some visualization and definitions of some important articles on a single-point tool relevant to considerations of its geometry. Figure A.6a shows an example of a modern

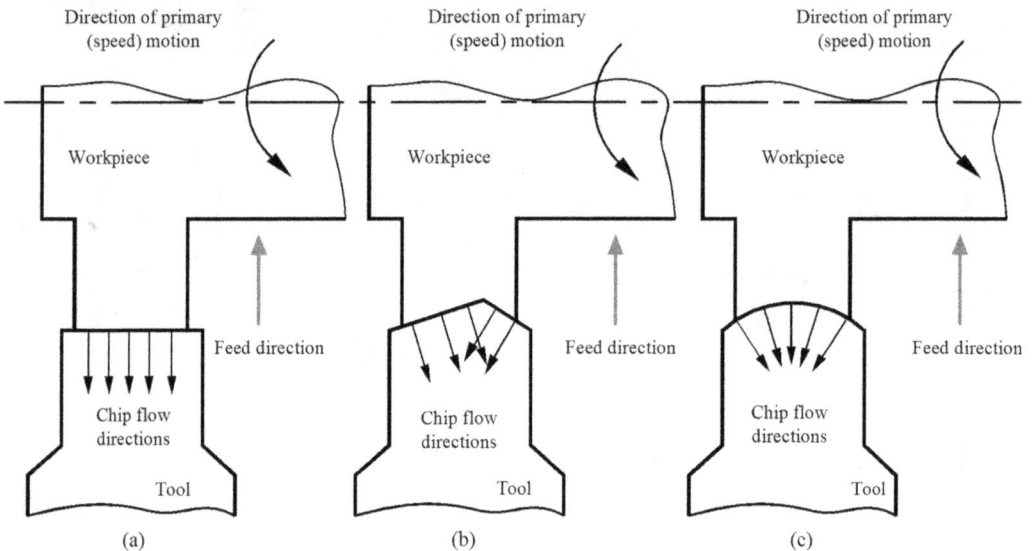

FIGURE A.5 Model showing (a) free and (b) non-free cutting with two cutting edges and (c) non-free cutting with a rounded cutting edge.

(a)

(b)

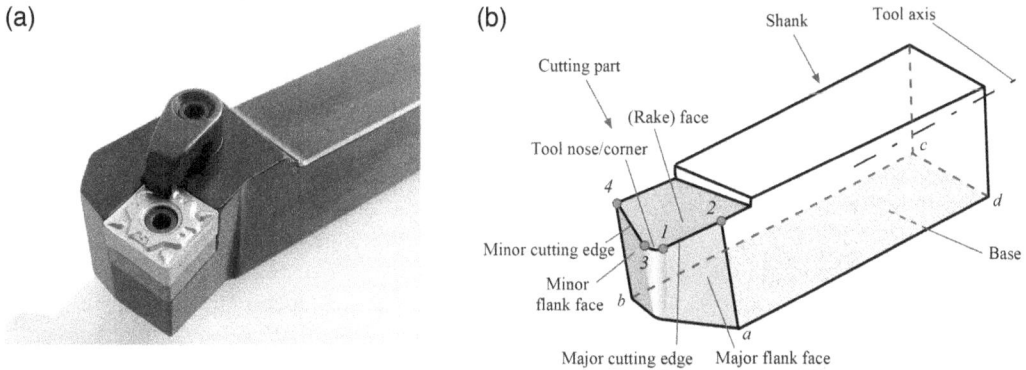

FIGURE A.6   Modern single-point cutting tool (a) and its simplification in considerations of tool geometry articles (b).

single-point tool, and Figure A.6b shows its simplification used in the tool geometry analysis as well as visualization of relevant articles. These articles are defined as follows:

- *Shank*: the part of the tool by which is it held.
- *Cutting part*: the functional part of parts of the tool each composed of chip-producing elements; cutting edges, rake and flank faces are therefore elements of the cutting parts.
- *Base*: a flat surface of the tool shank, parallel, or perpendicular to the tool reference plane (defined further), useful for locating or orientating the tool in its manufacturing, sharpening, and measurement. In Figure A.6, the base is a plane *abcd*.
- *Tool axis*: standard ISO 3002/1 does not define the axis for the single-point tool shown in Figure A.6b whereas standard ANSI B94.50 does. It is defined as an imaginary straight line with defined geometrical relationships to the locating surfaces used for manufacturing and sharpening of the tool and for holding the tool in use. Generally, the tool axis is the centerline of the shank as shown in Figure A.6b. It is usually parallel or perpendicular to the locating surface, although it could be the center line of a conical surface as in the case of a taper shank. When it is not obvious, the tool axis must be defined by the designer.
- *Face*: the surface or surfaces over which the chip flows/slides. When the face is composed of a number of surfaces inclined to one another, these are designated first face, second face, etc. starting from the cutting edge. In the author's opinion, the proper term should be "the rake face" to distinguish it from other tool faces. Moreover, the identification of the articles of the cutting part should begin with the identification of the rake face bearing in mind the origin of the term discussed in Astakhov (2010).
- *Flank*: the tool surface or surfaces over which the surface produced on the workpiece passes. When the flank is composed of a number of surfaces inclined to one another, these are designated first flank, second flank, etc. starting from the cutting edge. These surfaces may be called lands and unless otherwise specified, it is assumed that these are associated with the major cutting edge. In the author's opinion, it is much better to call this feature as the flank face containing the major cutting edge as the primary flank, then the flank face to follow is the secondary flank, and then the next flank face is tertiary flank.

    Where it is necessary to distinguish the flanks associated with the major and minor cutting edges, that part of the flank which intersects the face to form the major cutting edge is called the major flank and that part of the flank which intersects the face to form the minor cutting edge is called the minor flank, for example, major first flank, minor first flank, etc.
- *Cutting edge*: that edge of the face which is intended to perform cutting.

- *Tool major cutting edge*: that entire part of the cutting edge which commences at the point where at least a portion is intended to produce the transient surface on the workpiece. In the case of tools having a sharp corner at which the value of κ_r (explained further) may be considered to pass through zero, the major cutting edge commences at that corner. In the case of tools for which the value κ_r does not decrease to zero at any point of the cutting edge, the entire cutting edge is the major cutting edge as, for example, in the case of a slab milling cutter.
- *Tool minor cutting edge*: is the remainder of the cutting edge, if any, and where present commences at the point of the cutting edge where κ_r is zero but extends from this point in the direction away from the major cutting edge. It is not intended to produce any of the transient surface on the workpiece. Some tools may have more than one tool minor cutting edge as, for example, in the case of a cutoff tool.
- *Working major cutting edge* (Figure A.7): that entire part of the cutting edge which commences at the point where at least a portion is intended to produce the transient surface on the workpiece. In the case of tools having a sharp corner at which the value of κ_{re} (explained further) may be considered to pass through zero, the major cutting edge commences at that corner. In the case of tools for which the value κ_{re} does not decrease to zero at any point of the cutting edge, the entire cutting edge is the major cutting edge as, for example, in the case of a slab milling cutter.
- *Working minor cutting edge* (Figure A.7): is the remainder of the cutting edge, if any, and where present commences at the point of the cutting edge where κ_{re} is zero but extends from this point in the direction away from the major cutting edge. It is not intended to produce any of the transient surface on the workpiece. Some tools may have more than one tool minor cutting edge as, for example, in the case of a cut-off tool.

  Note: A distinction must be made between the tool major cutting edge and the working major cutting edge because at points at which κ_r and κ_{re} can be considered to be zero are not, in general coincident.

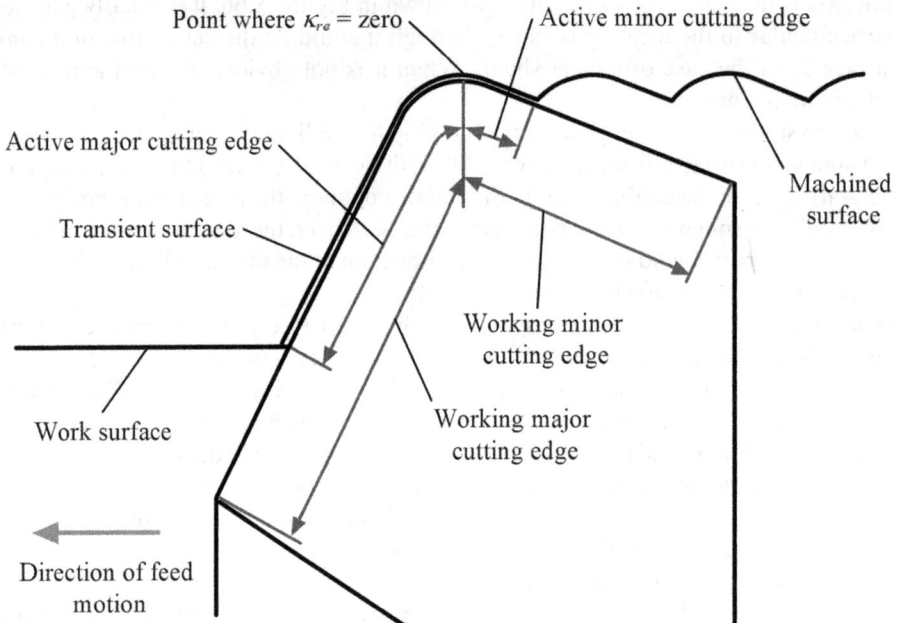

FIGURE A.7   Illustration of various terms related to cutting edges.

- *Active cutting edge*: that portion of the working cutting edge which is actually engaged in cutting at a particular instant generating both the transient and machine surface on the workpiece.
- *Active major cutting edge* (Figure A.7): the portion of the active cutting edge measured along the cutting edge from the point of intersection of the cutting edge and work surface to the point on the working cutting edge at which the working cutting edge κ_{re} angle may be considered to be zero.
- *Active minor cutting edge* (Figure A.7): the portion of the active cutting edge measured along the cutting edge from the point at which the working cutting edge κ_{re} angle may be considered to be zero to the point of intersection of the working minor cutting edge and the machined surface
- *Corner*: the relatively small portion of the cutting edge at the junction of the major and minor cutting edges; it may be curved, straight, or actual intersection of these cutting edges:
  Rounded corner is a corner having a curved cutting edge.
  Chamfered corner is a corner having a straight cutting edge.
- *Selected point on the cutting edge*: a point selected on any part of the cutting edge to define, for example, the tool or working angles at that point. The selected point may be on the major cutting edge or on the minor cutting edge. When the selected point is so chosen as to be located on the major cutting edge, the planes and angles associated with this point are so designated.
- *Tool profile* is the curve formed by the orthogonal projection of the tool cutting edge on any desired plane. Normally, this profile is defined in the tool reference plane. If it is to be defined in another plane, this shall be clearly specified.

A.2 SYSTEMS OF CONSIDERATION OF THE CUTTING TOOL GEOMETRY

A.2.1 STANDARD AND PROPOSED SYSTEMS

Two basic standards provide definitions of the cutting tool geometry: ISO 3002/1 (1982)/ reviewed and confirmed in 2018 "Basic quantities in cutting and grinding. Part 1: Geometry of the active part of cutting tools – general terms, reference systems, tool and working angles, chip breakers" and ANSI B94.50 (1975) "Basic nomenclature and definitions for single-point cutting tools. 1975 (R2003)." A simple comparison of these standards shows that the ISO standard is more advanced as it contains many more clear and functional definitions. Moreover, the basic notions of the ISO standard are well explained and shown with multiple examples as applied to various cutting tools while the ANSI standard concentrates only on single-point cutting tools. Both standards, however, failed to answer a simple yet most important question: "Why one should know the tool geometry?" The educated and thus complete answer to this question is not simple and straightforward as it involves some understanding of the metal cutting theory. The simplest answer, however, can be thought of as follows: "To be able to reproduce THE SAME tool geometry from one tool re-sharpening to another, from one cutting insert to the next, etc." so in the author's opinion it should be stated as both standards (Astakhov 2010).

The ISO Standard is widely used worldwide while the ANSI standard is used in parallel with the ISO Standard in North America. Similarity of some terms and definitions and differences of others creates a number of confusions in publications of various types starting from textbooks and research papers and finishing with flyers of various tool companies for new tools (Astakhov 2010). The further considerations are mainly based on the notions and definitions of the ISO standards (with some corrections of obvious flaws) although some important information from ANSI B94.50 is also given.

Both geometry standards discuss two systems of consideration of the cutting tool geometry, namely, the tool-in-hand and tool-in-use systems (hereafter, T-hand-S and T-use-S, respectively).

The former relates to the so-called static geometry, while the latter is based on the consideration of tool motions with respect to the workpiece. T-hand-S is needed for defining the geometry of the tool for its design, manufacturing, and inspection/measurement. T-use-S is needed for specifying the geometry of the cutting tool when it is performing a cutting operation.

A note is given in Standard ISO 3002/1 that reads: a third reference system of planes called the machine reference system is required to define the orientation of a cutting tool with respect to the machining tool. This reference system is defined in ISO 3002/2. ANSI B94.50 points out the existence of such a system which is defined by ISO. According to ISO 3002/2, this system aims to relate the articles of the tool geometry in T-hand-S and T-use-S with a machine reference system of axis defined in ISO 841. This system can be referred to as the tool-in-machine (T-mach-S) that is needed for defining the geometry of the tool accounting for its actual tool position in the machine tool and actual directions of its motions in such a setting.

In the author's opinion, T-hand-S and T-use-S are insufficient for the proper consideration of cutting tool geometry as discussed later. Another system, namely, the tool-in-holder system (T-hold-S), which was previously incorrectly referred to as the tool-in-machine system (Astakhov 1998/1999, 2010), should also be considered. The introduction of an additional system of consideration may be thought by some as a kind of overcomplicating of the cutting tool geometry and its practical applications so it is suitable only for ivory academicians as it has little practical value at the shop floor level. In the author's opinion, the opposite is actually the case. Namely, misunderstanding and, as a result, misusing the tool geometry articles in the previously mentioned systems leads to improper selection of the tool geometry parameters and prevents meaningful optimization of practical machining operations. This is especially true for drilling tools.

A.2.2 WHAT DOES IT MEAN "THE TOOL GEOMETRY"?

To make it clear for further considerations, one needs to understand the difference between the geometry of the tool and the tool geometry although grammatically and semantically these two appear to be the same. It is not so in reality. The geometry of the tool includes many geometrical articles, for example in drilling tool, it includes the flute length of the drilling tool, step length, shank diameter, and so on. The tool geometry of the tool cutting edges, i.e., the articles affecting the cutting conditions of this tool. Obviously, the geometry of the cutting tool includes the cutting tool geometry as its part.

The prime article in the considerations of the cutting tool geometry is *the cutting edge*. As defined above (Section A.1.2), a cutting edge is actually a line of intersection of the rake and flank faces so that its shape is entirely defined by the shape of these surfaces. *In metal cutting, the geometry of the cutting edge is attributed with a set of certain angles, namely the clearance, rake, tool cutting edge, and cutting edge inclination angles.* Therefore, these angles in terms of their definitions, meaning, and significance to tool performance should be considered, and thus unambiguously understood by anyone involved in the cutting tool design, development, research, and application.

The geometry of the cutting edge can be thought of at two levels: macrogeometry and microgeometry both considered in Chapter 2. Note also that if a tool has several cutting edges, each cutting edge should be considered separately.

A.3 REFERENCE SYSTEM

A.3.1 GENERAL

According to both the ISO and ANSI standards, reference systems are introduced as systems of planes for defining and specifying the angles of a cutting tool. The planes used in T-had-S are termed tool-in-hand planes; their titles (with two exceptions defined by ISO 3002/1) each include the word "tool." The planes in T-use-S are termed tool-in-use planes; their title (with one exception defined by ISO 3002/1), all include the word "working."

Because the articles of the tool geometry may vary from point to point along the cutting edge, each plane, in general, is therefore defined with respect to a selected point on the cutting edge. Obviously, a point can be selected either on the major cutting edge or on the minor cutting edge.

Each plane is provided with a symbol consisting of P with suffix indicating the plane identity (for example P_s, the tool cutting edge plane). When reference plains are defined for a point of the minor cutting edge, the appropriate symbol bears a prime (for example P_s', the tool minor cutting edge plane).

When the cutting edge, rake or flank face, is curved, the tangent or tangent plane through the selected point should be used in the reference system of planes.

The author wants to prevent attention of potential readers/tool specialists that each plane in the reference system for the analysis of tool geometry in T-hand-S, T-hold-S, T-use-S, and T-mach-S are the planes of measurement so they should be regarded as such, i.e., should obey the definition of the plane of measurement (the definition and properties of a plane of measurement are given in Chapter 1 in Astakhov (2024)).

A.3.2   REFERENCE PLANE

The selection of the reference plane in any system of reference planes is the first and key step as all other planes directly (except for only the cutting-edge normal plane) are defined with respect to this plane. Unfortunately, the existing tool geometry standards do not provide the proper self-obvious definition of this plane, not even close. Both standards ISO 3002/1 and ANSI B94.50 provide identical definitions of the reference plane as

> a plane through the selected point of the cutting edge, so chosen as to be either parallel or perpendicular to a plane or axis of the tool convenient for locating or orienting the tool for its manufacture, sharpening, and measurement. The plane must be chosen and defined for each individual type of the cutting tool so that it meets the conditions prescribed above and in generally oriented perpendicular to the assumed direction of primary motion. For ordinary lathe, planner of shaper tools it is a plane parallel to the base of the tool, for a vertical shank or tangential tool or for a horizontal planner tool it is a plane perpendicular to the tool axis. For side and face milling cutter, for drills or screwing tap it is a plane containing the tool axis.

In the author's opinion, this is the misleading definition as it does not provide a clear instruction to the tool designer of how to assign the reference plane properly. Reading this definition, one may wonder what it that *either parallel or perpendicular to a plane or axis of the tool convenient for locating or orienting the tool for its manufacture, sharpening, and measurement* – convenient to whom? *Parallel or perpendicular* are two entirely different planes. *For side and face milling cutter, for drills or screwing tap it is a plane containing the tool axis* – there are an infinite number of planes containing this axis – which one plane to choose as the reference plane for further tool geometry considerations?

In the author's opinion, in any system of considerations, the reference plane should be defined as *the plane perpendicular to the velocity vector* **v** *at the selected point on the cutting edge.* For example, the definition of the reference plane for point *i* of the major cutting edge *2–3* is shown in Figure A.8. In T-hand-S, this vector represents the assumed direction of the primary motion; in T-hold-S, this vector represents the actual direction of the primary motion; in T-use-S, this vector represents the resultant cutting direction, which, as explained later, include the primary and feed motions; in T-mach-S, this vector represents the resultant cutting direction accounting for the actual setting of the considered cutting edge in the machine.

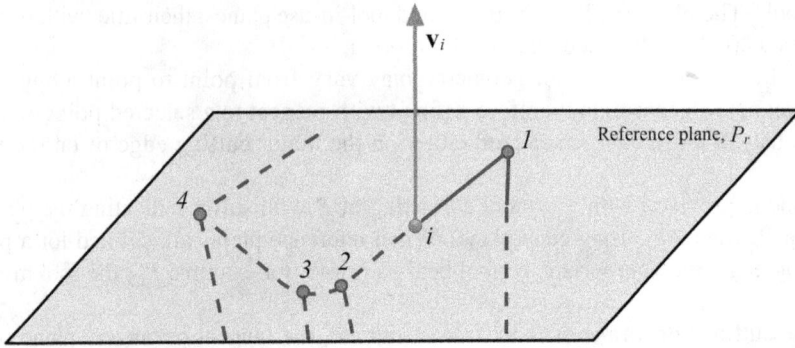

FIGURE A.8 Notion of the reference plane.

A.4 TOOL-IN-HAND SYSTEM (T-HAND-S)

As discussed above, T-hand-S is needed for defining the geometry of the tool for its design, manufacturing, and inspection/measurement. The major and controversial word in the defining reference planes in this system is "assumed," which makes all considerations of the cutting tool geometry a kind of uncertain as the reference systems in other listed systems are defined with respect to the planes in T-hand-S. Therefore, a very detailed, step-by-step consideration with plenty of simple illustrations of the reference planes and tool geometry articles are presented in this section in hope to eliminate or, at least, reduce the "assumed" portion of the consideration of the cutting tool geometry in T-hand-S.

A.4.1 REFERENCE PLANE, P_r

The reference plane P_r in T-hand-S is the plane perpendicular to the direction of the assumed direction/cutting speed in this system. That is why is referred to as the assumed reference plane. Although it was never recognized and explained how to set this direction, the proper explanation, which generalizes the basic notions of the above-discussed tool geometry standard, literature on the subject, and practice of tool design, manufacturing, and inspection/measurement, is as follows. Figure A.9 shows the first stage in the design of a simple tool. In developing this design, the tool designer assumes the workpiece this tool is going to cut and the location of this tool relative to the axis of rotation of this workpiece. Once this basic assumption is made, then the directions of primary and feed motions in T-hand-S can be set as shown in Figure A.9. As can be seen in this figure, all points of the cutting edge *1–2* including the selected point *i* have the same direction of the primary motion, i.e., the cutting speed as this cutting edge is set on the centerline. This is shown in Figure A.9 as vector v_i for the sleeted point *i*. As explained further, this setting is *the only way to define the angles of the cutting edge, the rake, clearance, tool cutting edge, and inclination angle independently* in T-had-S and then find simple geometrical relationships between these angles.

Once the direction of the primary motion and thus the reference plane are defined, this reference plane can be drawn through the selected point on the cutting edge. Figure A.10a shows that a point on the cutting edge is selected, and Figure A.10b shows that the reference plane P_r is drawn through this point. It should be pointed out the following:

- In T-hand-S, plane P_r can be moved along the direction of the prime motion (i.e., in the direction defined by the vector **v**) in any location convenient for the tool geometry considerations/analysis as only projections of the components of the tool are used for the considerations in this plane.
- Planes P_r for all points of the cutting edge are parallel to each other as the direction of the primary motion is the same for all points of this edge.

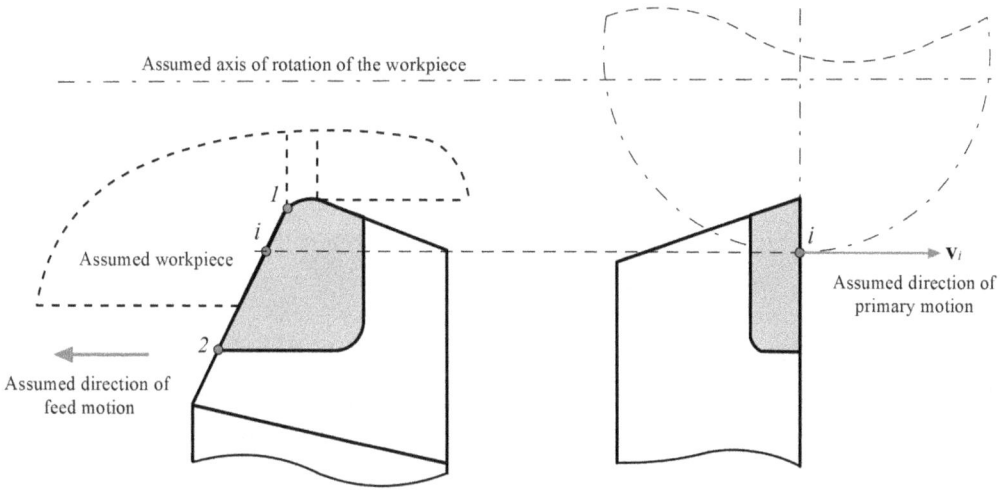

FIGURE A.9 Setting the direction of primary and feed motions in T-hand-S.

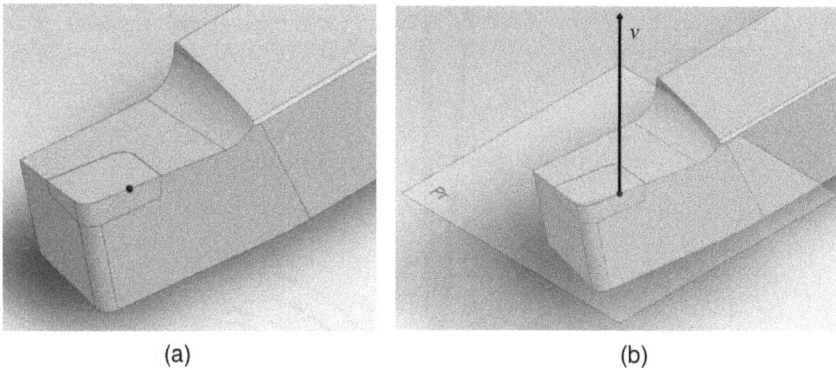

(a)   (b)

FIGURE A.10 Selected point on the cutting edge (a) and the reference plane P_r through this point (b) (Courtesy Professor Jose J. Outeiro).

A.4.2 WORKING PLANE, P_f

Working plane P_f is the plane through the selected point of the cutting edge defined as to be perpendicular to the reference plane P_r and containing the assumed direction of feed motion (see Figure A.9) designated as the feed rate or speed of feed, v_f (see Chapter 1).

Once the direction of the feed motion is defined, the working plane P_f can be drawn through the selected point of the cutting edge. Figure A.11a shows the direction of feed applied to the selected point of the cutting edge, and Figure A.11b shows that the working plane P_f is drawn through this point perpendicular to the reference plane P_r. It should be pointed out the following:

- In T-hand-S, plane P_f can be moved parallel to the direction parallel to the feed motion in any location convenient for the tool geometry considerations/analysis as only projections of the components of the tool are used for the considerations in this plane.
- Planes are P_f for all points of the cutting edge that are parallel to each other as the direction of the feed motion is the same for all points of this edge.

Figure A.12 shows the assumed directions of the tool motions in T-hand-S for the selected point of the cutting edge and the introduced reference and working planes for further clarification.

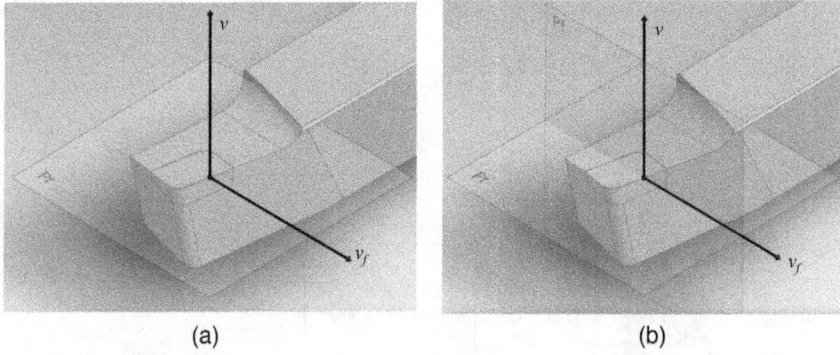

FIGURE A.11 The assumed direction of feed motion is applied to the selected point of the cutting edge (a) and the working plane P_f through this point (b) (Courtesy Professor Jose J. Outeiro).

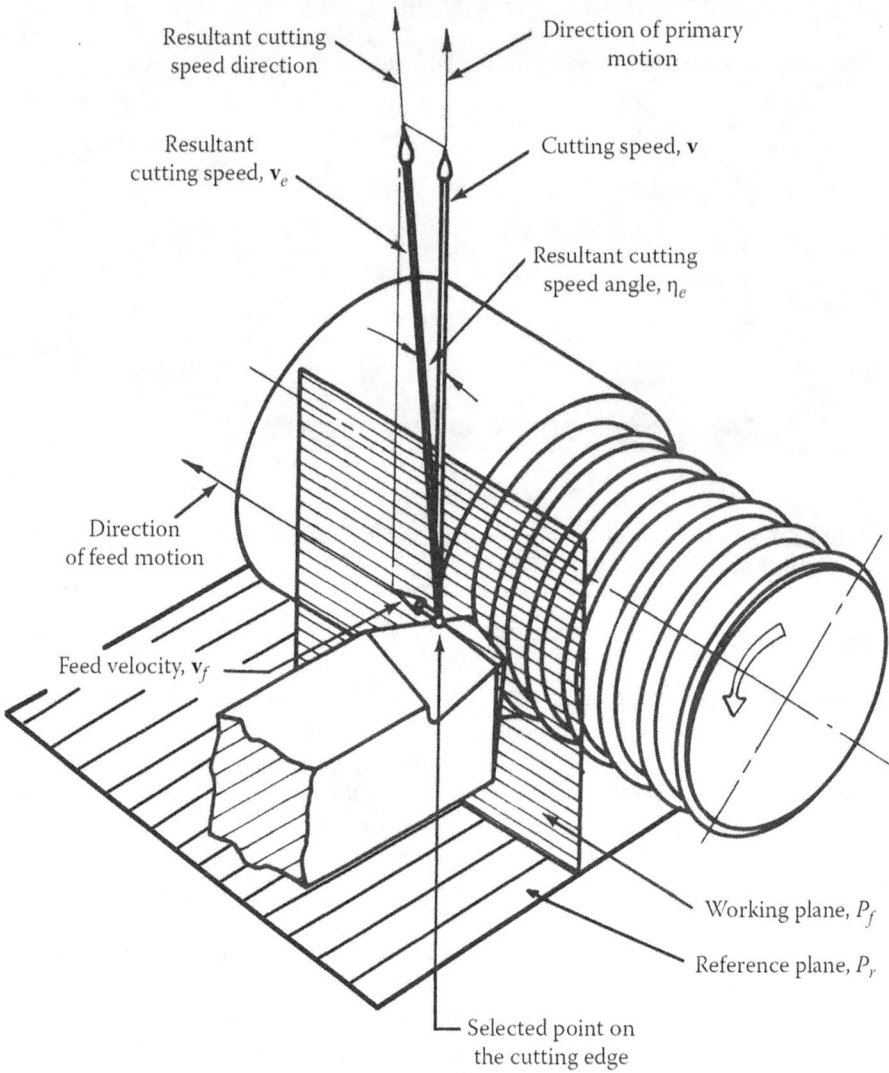

FIGURE A.12 Assumed directions of the tool motions in T-hand-S for a selected point of the cutting edge and the introduced reference and working planes.

A.4.3 BACK PLANE, P_p

Back plane P_p is defined as a plane through the selected point on the cutting edge and perpendicular both to the tool reference plane P_r and to the tool working plane P_f. Figure A.13 shows the graphical representation of the tool back plane P_p. As can be seen, this plane is defined as a plane through the selected point on the cutting edge and perpendicular both to the tool reference plane P_r and to the tool working plane P_f. Because the back plane is defined through the reference and working planes that are parallel for points of the cutting edge, the back planes for all points of the cutting edge are also parallel to each other.

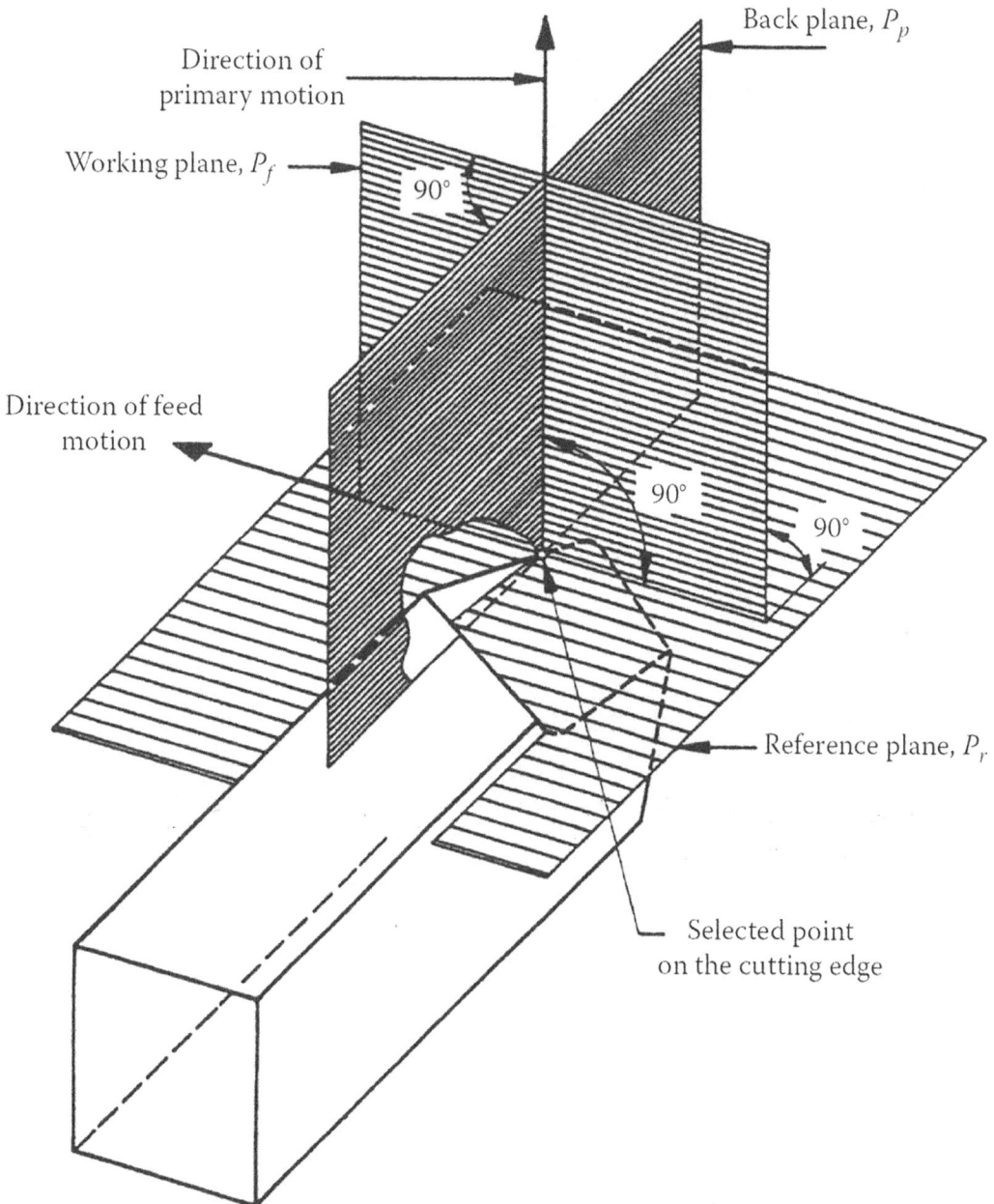

FIGURE A.13   Graphical representation of the tool back plane P_p.

A.4.4 TOOL CUTTING EDGE PLANE, P_s

Tool cutting edge plane P_s is defined as a plane that contains the cutting edge and is perpendicular to P_r as shown in Figure A.14. If the major cutting edge is straight, all its points lie in this plane so that the tool cutting edge plane is the same for any point of the cutting edge. If the major cutting edge is not straight, then the tool cutting-edge plane should be determined for each point on the curved cutting edge, thus being the plane that is tangent to the cutting edge at the point of consideration and perpendicular to the reference plane.

A.4.5 TOOL ORTHOGONAL, P_o AND TOOL NORMAL, P_n PLANES

Assumed tool orthogonal plane P_o is defined as a plane through the selected point on the cutting edge and perpendicular both to the tool reference plane P_r and to the tool cutting edge plane P_s. The relation of this plane to the tool cutting edge plane implies that if the major cutting edge is straight, the orthogonal planes for various points of this edge are parallel to each other. If the major cutting edge is not straight, then they are not.

Assumed tool normal plane P_n is defined by ISO 3002/1 as the plane normal to the cutting edge at the selected point on the cutting edge. In the author's opinion, a more precise definition accounting for the way the tool geometry articles are measured in this plane is as follows: it is a plane through the selected point on the cutting edge and perpendicular to the tool cutting edge plane through this point. Note that by ISO 3002/1 definition, this is the only plane in the reference system that is not defined by the above-introduced reference plane. Obviously, when the cutting edge is straight, the tool's normal planes are parallel to each other.

Figure A.15 shows the graphical representation of the tool cutting edge, orthogonal and normal planes.

A.4.6 REFERENCE SYSTEM FOR THE MINOR CUTTING EDGE

Similarly to points of the major cutting edge, a system of reference planes can be defined for points of the minor cutting edges. Figure A.16 shows the graphical representation of the tool reference, cutting edge, and normal planes for the selected point of the minor cutting edge.

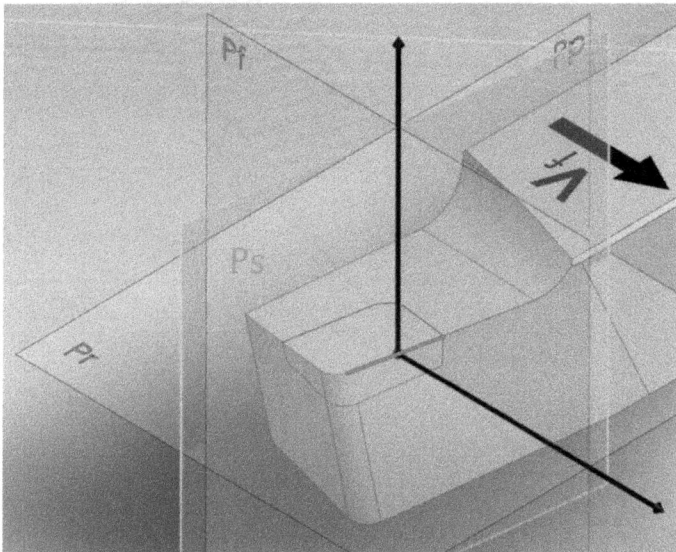

FIGURE A.14   Graphical representation of the tool cutting edge plane P_s (Courtesy Professor Jose J. Outeiro).

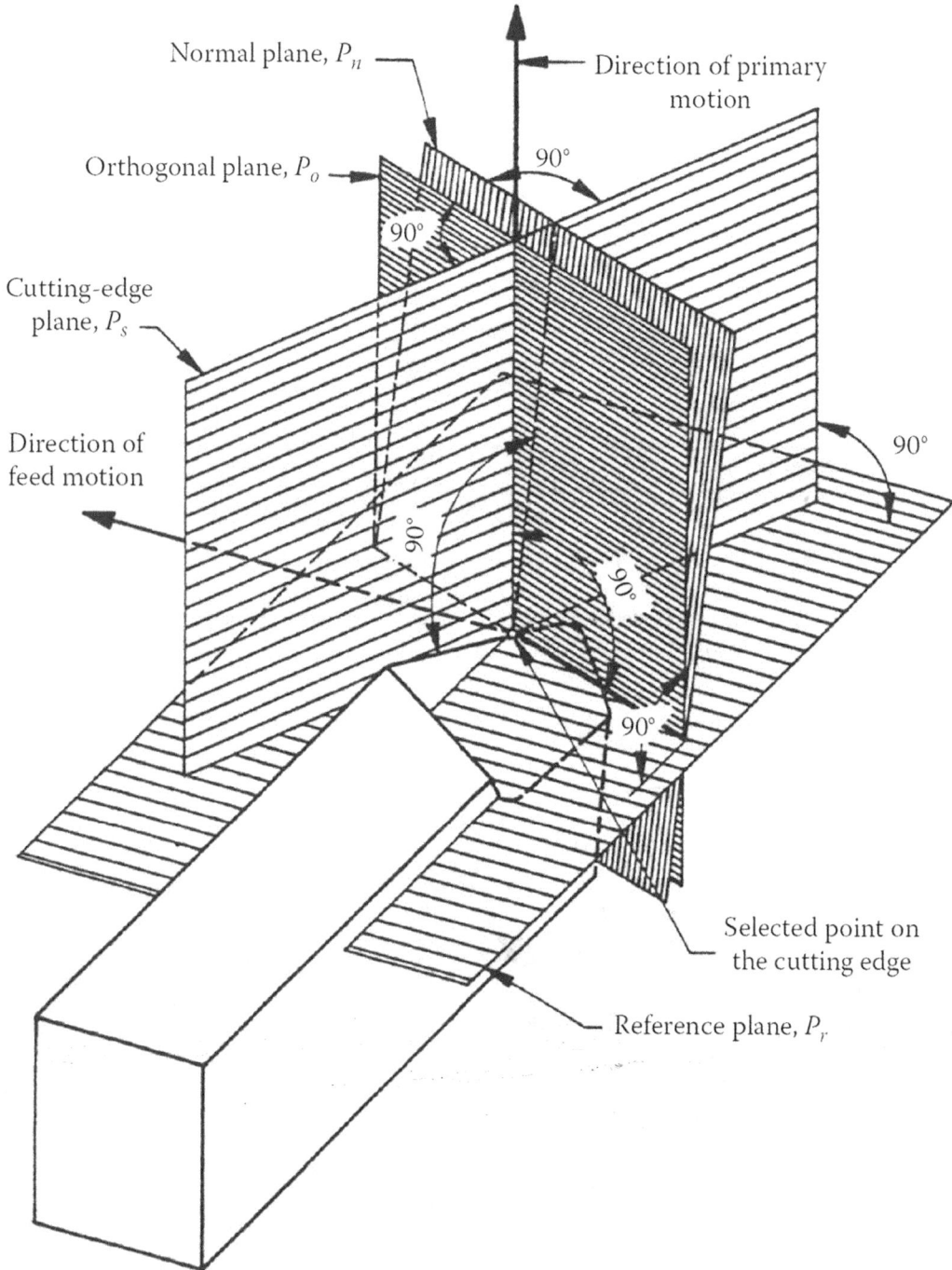

FIGURE A.15   Graphical representation of the tool cutting edge, orthogonal, and normal planes.

A.4.7   Angles in the Reference Plane P_r

As soon as the reference planes are defined, the relevant angles of the cutting tool geometry in this plane can be considered. As such, a particular plane of measurement in the above-introduced reference system in T-hand-S is used and angular articles are defined as angles between the corresponding lines of intersection of these planes known as traces. Figure A.17 shows the graphical

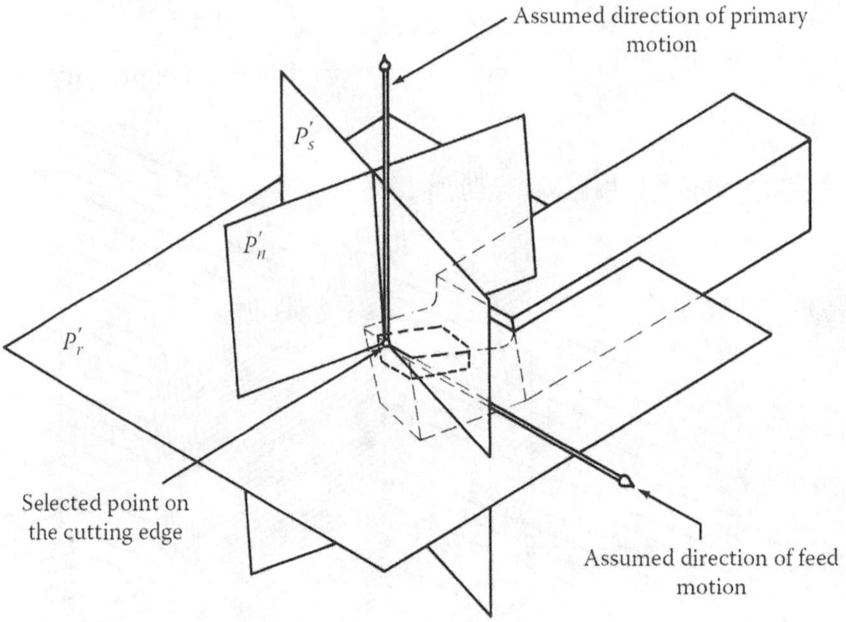

FIGURE A.16 Graphical representation of the tool reference, cutting edge, and normal planes for the selected point of the minor cutting edge.

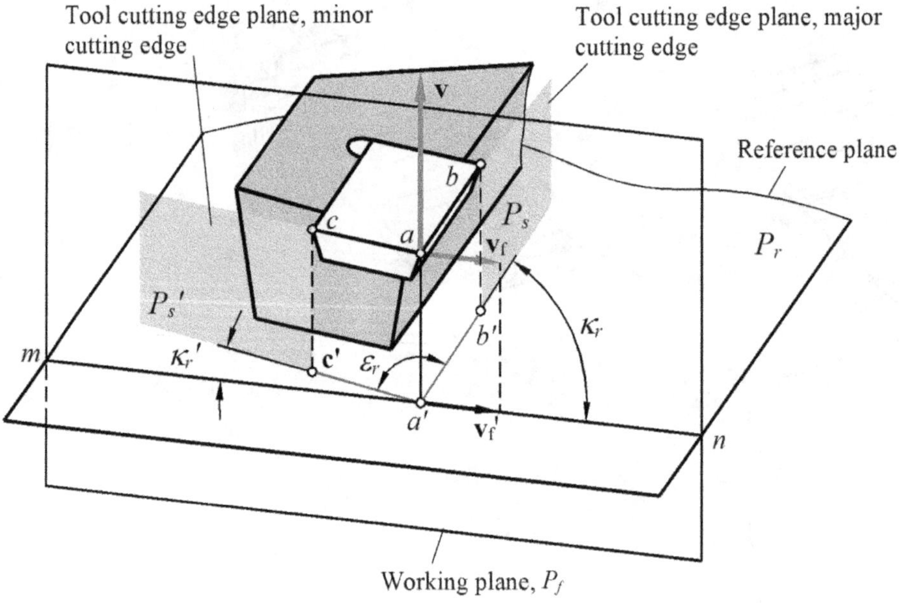

FIGURE A.17 Graphical representation of the formation of angles in the reference plane.

representation of the formation of angles in the reference plane. In this figure, planes P_s and P_s' project the major cutting edge ab into $a'b'$ and minor cutting edge ac in $a'c'$ into the reference plane P_r. The latter is moved parallel to itself in a location convenient for the current presentation. Line mn is the line of intersection of the working P_f and reference P_f planes called the trace of P_f in P_f. Obviously, the vector of the speed of feed (the feed rate) $\mathbf{v}_f$ is projected on this trace as $\mathbf{v}_f'$. Using such a construction, many books and other publications on metal cutting/cutting tools define the tool cutting edge angle κ_r as the angle between the projection of the major cutting edge into the reference plane and the feed direction; the tool included angle ε_r as the angle between the projections of the major and minor cutting edges into the reference plane and so on. Although such a representation is correct for a particular case, it makes it bulky/blurry for the consideration of other tool angles in the above-introduced planes.

The following angles and parameters are defined in the reference plane as follows:

- *Tool cutting edge angle* κ_r as the acute angle between the traces (lines of intersection) of the tool working plane P_f and the tool cutting-edge plane P_s in the reference plane P_r as shown in Figure A.18. Angle κ_r is always positive, and it is measured in the counterclockwise direction from the trace of the working plane P_f in the manner shown in Figure A.18. In this figure, this angle is shown for a selected point m on the major cutting edge. Because the tool cutting edge plane for each point of this cutting edge is the same, this angle is the same for the whole cutting edge.
- *Tool minor cutting edge angle* κ_{r1} is the acute angle between the traces (lines of intersection) of the tool working plane P_f and the tool minor cutting-edge plane P_s' in the reference plane P_r as shown in Figure A.18. Angle κ_{r1} is always positive, and it is measured in the clockwise direction from the trace of the working plane P_f in the manner shown in Figure A.18. In this figure, this angle is shown for a selected point n on the major cutting edge. Because the tool minor cutting edge plane for each point of this cutting edge is the same, this angle is the same for the whole cutting edge.
- *Tool approach angle* ψ_r (the lead angle is termed in as ANSI B94.50) is the acute angle that the trace of the tool cutting-edge plane P_s in P_r makes with the trace of the tool back plane P_p as shown in Figure A.18.
- *Tool included angle* ε_r is the angle between the traces of P_s and P_s' in P_r.
- Tool corner/nose radius r_ε is the radius between the projections of the major and minor cutting edges into the reference plane.

It directly follows from Figure A.18 that

$$\kappa_r + \varepsilon_r + \kappa_{r1} = 180°$$ (A.1)

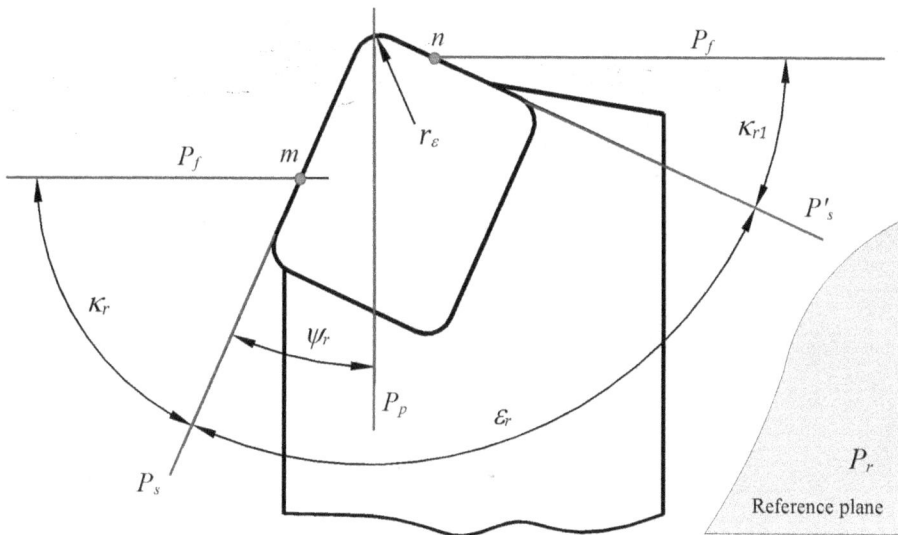

FIGURE A.18 Tool geometry articles defined/measured in the reference plane.

When the cutting edge is not straight, the tool cutting edge angles vary from one point of this edge to the next because the tool cutting edge plane is tangent to any considered point on this edge. For the common round cutting insert shown in Figure A.19, κ_r at point a of the major cutting edge ab is zero. This angle at point b is defined as the angle between the tangent to the projection of the round major cutting edge into the reference plane at point b and the working plane (the line of its intersection with the reference plane). The tool cutting edge angle κ_{r1} of the minor cutting edge ac is also zero at point a and at point c, the last contact point of the major cutting edge and the machined surface is defined as the angle between the tangent to the projection of the round minor cutting edge into the reference plane at point c and the working plane.

A.4.8 Angles in the Working and Back Planes

Figure A.20 shows the angles defined/measured in the working P_f and back P_p planes. In this figure, the tool is shown in the reference plane P_r. Section A-A in the drawing represents the section through the selected point m of the major cutting edge by the working plane, P_f and section B-B – by the back plane, P_p. The angles defined in the working plane are as follows:

- Both the ISO and ANSI standards define the tool side rake as the angle between the rake face and the tool reference plane face measured in the assumed working plane P_f. In the author's opinion, this definition is not strict and general. The proper definition and the name of this angle are as follows: the rake angle in the working plane γ_f is the angle between trace of P_r and trace of the rake face (the line of intersection of the rake face and the working plane) in the working plane P_f, which is the plane of measurement. Further definitions of the tool angles will follow this way.
- The clearance angle in the working plane α_f is the angle between trace of P_s and trace of the flank face (the line of intersection of the flank face and the working plane) in the working plane P_f.
- The wedge angle in the working plane β_f is the angle between traces of the rake and flank faces in the working plane P_f.
- The rake angle in the back plane, γ_p is the angle between trace of P_r and trace of the rake face (the line of intersection of the rake face and the working plane) in the back plane P_p.

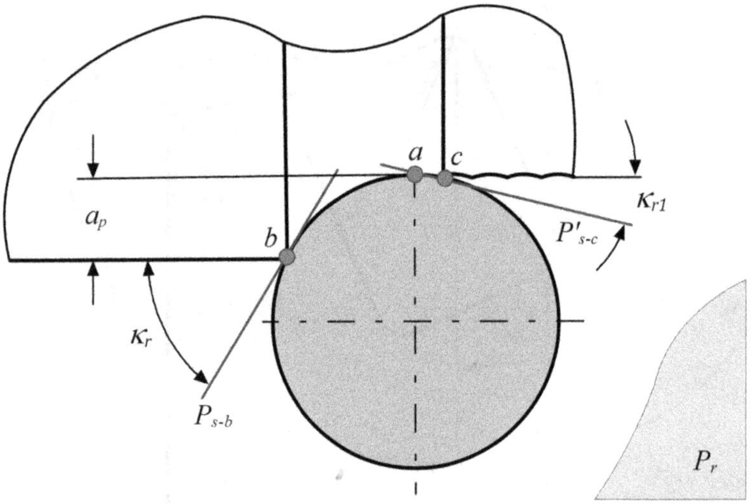

FIGURE A.19   Tool cutting edge angles of the major and minor cutting edges for a round cutting insert.

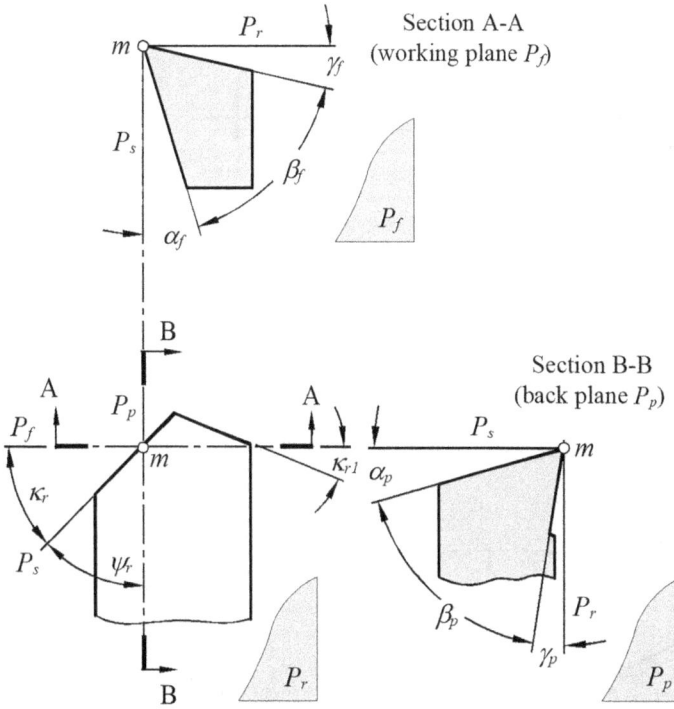

FIGURE A.20 Angles in the working P_f and back P_p planes.

- The clearance angle in the back plane α_p is the angle between trace of P_s and trace of the flank face (the line of intersection of the rake flank and the back plane) in the back plane P_p.
- The wedge angle in the back plane β_p is the angle between traces of the rake and flank faces in the back plane P_p.

No signs are assigned to angles α_p and β_p as they are always positive in any plane in any reference system used for defining tool geometry whereas the rake angles γ in various reference planes can be positive, neutral, or negative. Figure A.21 explains the meaning of the sign of the rake angle in any plane of measurement, namely working, back, orthogonal, and normal. By definition, the rake angle is positive when the trace of the rake face in the considered plane of measurement is below the trace of the reference plane P_r in this plane. When the discussed traces are coincident, the rake angle is neutral or simply zero. The rake angle is negative when the trance of the rake face is above the trace of the reference plane in the corresponding plane of measurement.

When the traces of the rake and/or flank faces are curved in any standard (working, back, orthogonal, normal) or non-standard plane of measurement, then the assigning/measurement of the corresponding rake and clearance angle should follow the procedure described in Chapter 1 in Astakhov (2024). Figure A.22a shows the case where the trace of the rake face is a curve in a certain plane of measurement P_i drawn through a point m of the cutting edge. As such, the rake angle γ_i of the cutting edge at this point is defined as the angle between the tangent to the trace of the curved rake face drawn at point m and the trance of the reference plane P_r. When the flank face is a curved surface, the clearance angle of the cutting edge at the selected point m is defined as the angle between the tangent to the trace of the curved flank face drawn at point m and the trance of the tool cutting edge plane P_s as shown in Figure A.22b.

The optical measurement of the rake and clearance angles when the rake and flank faces are curved presents a problem, particularly with measurements of the clearance angle as discussed in Section 2.1.3 in Chapter 2.

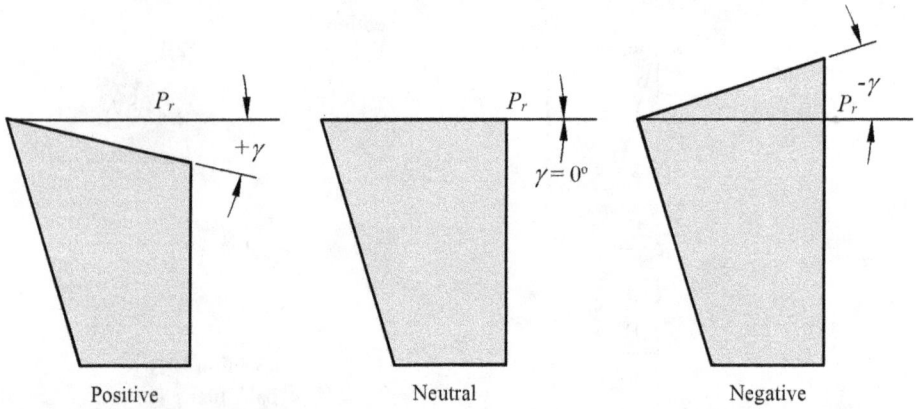

FIGURE A.21   Sign of the rake angle.

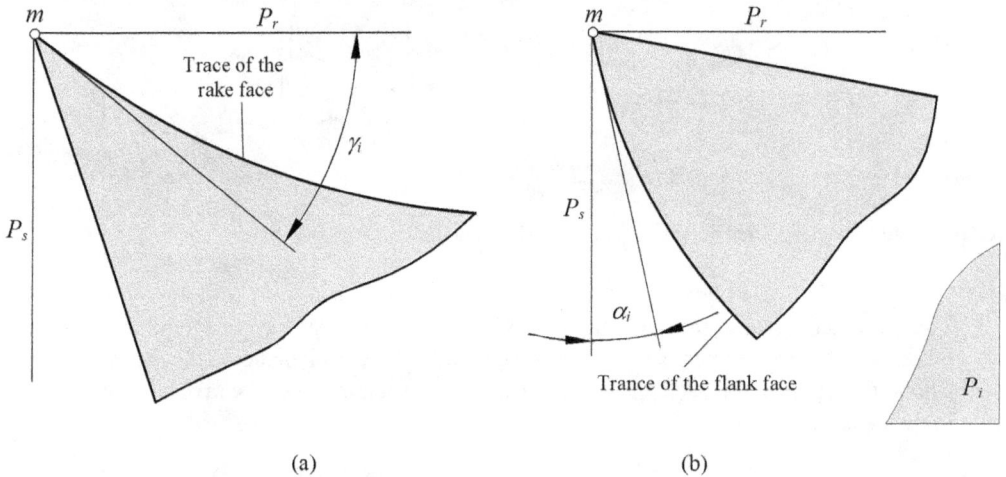

(a)   (b)

FIGURE A.22   Definitions of (a) the rake angle γ_i when the rake face is curved and (b) the clearance angle α_i when the flank face is curved in any standard or non-standard plane of measurement P_i.

A.4.9   ANGLES IN THE ORTHOGONAL PLANE

Figure A.23 shows the angles in the orthogonal plane P_o. In this figure, the tool major and minor cutting edges as shown by their projections into the reference plane P_r. Section C-C in the figure represents the section through the selected point m of the projection of the major cutting edge by the orthogonal plane, P_o. The angles defined in the orthogonal plane are as follows:

- The rake angle in the orthogonal plane γ_o is the angle between trace of P_r and trace of the rake face (the line of intersection of the rake face and the orthogonal plane) in the orthogonal plane P_o, which is the plane of measurement.
- The clearance angle in the orthogonal plane α_o is the angle between trace of P_s and trace of the flank face (the line of intersection of the flank face and the orthogonal plane) in the orthogonal plane P_o.
- The wedge angle in the orthogonal plane β_o is the angle between traces of the rake and flank faces in the orthogonal plane P_o.

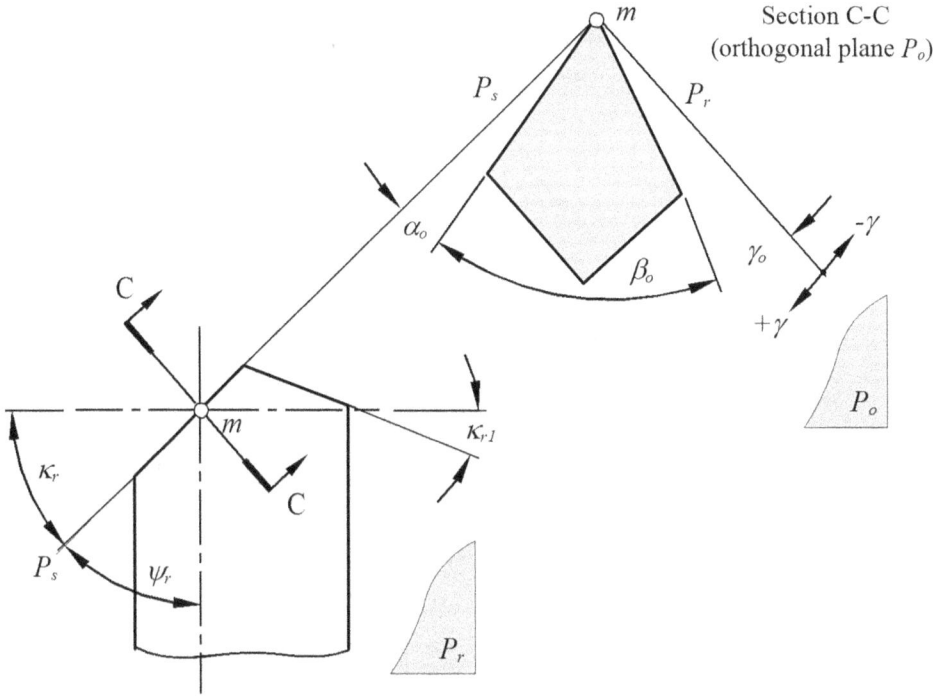

FIGURE A.23   Angles in the orthogonal plane, P_o.

A.4.10   ANGLES IN THE CUTTING EDGE AND NORMAL PLANES

Figure A.24 shows the angles in the tool cutting edge P_s and normal P_n planes. View E in this figure is the tool cutting edge plane P_s where the true inclination of the major cutting edge ab can be observed as ab belongs to this plane by its definition, which is the plane of measurement for the inclination of the cutting edge ab. This inclination is characterized by the tool inclination angle λ_s. It is defined as the angle between the projection of the major cutting edge into the tool cutting edge plane and the trace of the reference plane P_r in this plane. In full analogy with the rake angle (see Figure A.21), this angle is positive when the cutting edge is located below the trace of the reference plane P_r; negative if it is located above this trance, and zero when the cutting edge ab coincides with the trace of P_r in P_s.

Section D-D in Figure A.24 represents the normal plane P_n as it is a plane through the selected point m on the cutting edge and perpendicular to the tool cutting edge plane through this point. In this plane of measurement, the following angles are defined:

- The rake angle in the normal plane, γ_n is the angle between trace of P_r and trace of the rake face (the line of intersection of the rake face and the normal plane) in the normal plane P_n, which is the plane of measurement.
- The clearance angle in the normal plane α_n is the angle between trace of P_s and trace of the flank face (the line of intersection of the flank face and the normal plane) in the normal plane P_n.
- The wedge angle in the normal plane β_n is the angle between traces of the rake and flank faces in the normal plane P_n.

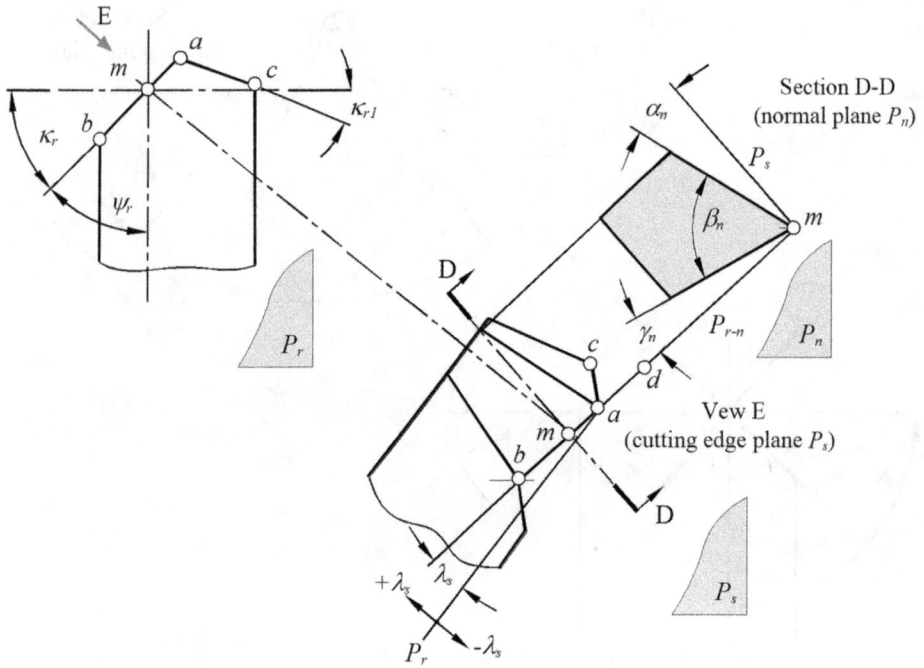

FIGURE A.24 Angles in the tool cutting edge P_s and normal P_n planes.

A.4.11 MINOR CUTTING EDGE

Although both standards ISO 3002/1 and ANSI B94.50 suggest that the same, as for the major cutting edge, system of reference planes can be considered for the minor cutting edge as shown in Figure A.16, it is not so in practice because

- The location of the rake face is set by the angles defined for the major cutting edge so that, besides very special cases not occurred in drilling tools, the only clearance angle of this edge can be applied independently.
- The active length of the minor cutting edge is small compared to its length ac shown in Figure A.25. As such, the inclination angle of this edge does not play any significant role so the only normal orthogonal angle of this angle is considered.

Figure A.25 shows the graphical interpretation of the orthogonal clearance angle of the minor cutting edge. As shown, it is defined as the angle between the trace of the tool minor cutting-edge plane P_s' and the trace of the flank plane in the orthogonal plane P_o' drawn through the considered point n on the minor cutting edge perpendicular to the reference plane P_r.

A.4.12 GEOMETRICAL RELATIONSHIP AMONG ANGLES IN T-HAND-S

Simple relationships exist among the considered angles in T-hand-S (Astakhov 2010):

$$\tan \gamma_n = \cos \lambda_s \tan \gamma_o \tag{A.2}$$

$$\tan \alpha_p = \frac{\tan \alpha_o}{\cos \kappa_r} \tag{A.3}$$

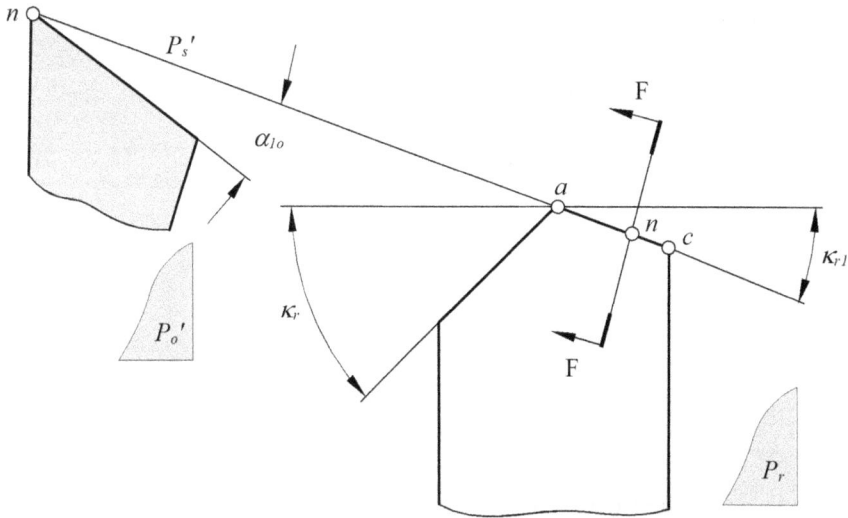

FIGURE A.25 Orthogonal clearance angle α_{lo} at point n of the minor cutting edge.

$$\tan \alpha_f = \frac{\tan \alpha_o}{\sin \kappa_r} \tag{A.4}$$

$$\cot \alpha_n = \cos \lambda_s \cot \alpha_o \tag{A.5}$$

$$\tan \lambda_s = \sin \kappa_r \tan \gamma_p - \cos \kappa_r \tan \gamma_f \tag{A.6}$$

$$\tan \gamma_f = \tan \gamma_o \sin \kappa_r + \tan \lambda_s \cos \kappa_r \tag{A.7}$$

$$\tan \gamma_p = \tan \gamma_o \cos \kappa_r - \tan \lambda_s \sin \kappa_r \tag{A.8}$$

$$\tan \gamma_o = \cos \kappa_r \tan \gamma_p + \sin \kappa_r \tan \gamma_f \tag{A.9}$$

$$\cot \alpha_o = \cos \kappa_r \cot \alpha_p + \sin \kappa_r \cot \alpha_f \tag{A.10}$$

It must be stated, however, that Eqs. (A.6)–(A.10) apply only when the cutting edge angle κ_r is less than or equal to 90°.

Example A.1

Problem

The optimal cutting performance of a single-point tool for turning was found when this tool has the following geometry: normal clearance angle $\alpha_n = 12°$, normal rake angle $\gamma_n = 8°$, cutting edge inclination angle $\lambda_s = 10°$, and tool cutting edge angle $\kappa_r = 60°$. Find the corresponding angles in T-hand-S in the orthogonal, back, and working planes that were used in tool design and manufacturing.

Solution

The clearance angle in the orthogonal plane is calculated using Eq. (A.5) as

$$\alpha_o = \arctan \left(\cos \lambda_s \tan \alpha_n \right) = \arctan \left(\tan 10° \tan 12° \right) = 11.82°$$

The rake angle in the orthogonal plane is calculated using Eq. (A.2) as

$$\gamma_O = \arctan\left(\frac{\tan\gamma_n}{\cos\lambda_s}\right) = \arctan\left(\frac{\tan 8°}{\cos 10°}\right) = 8.12°$$

The clearance angle in the assumed working plane is calculated using Eq. (A.4) as

$$\alpha_f = \arctan\left(\frac{\tan\alpha_o}{\sin\kappa_r}\right) = \arctan\left(\frac{\tan 11.82°}{\sin 60°}\right) = 13.59°$$

The clearance angle in the back plane is calculated using Eq. (A.3) as

$$\alpha_p = \arctan\left(\frac{\tan\alpha_o}{\cos\kappa_r}\right) = \arctan\left(\frac{\tan 11.82°}{\cos 60°}\right) = 22.71°$$

The rake angle in the assumed working plane is calculated using Eq. (A.7) as

$$\gamma_f = \arctan\left(\tan\gamma_o \sin\kappa_r + \tan\lambda_s \cos\kappa_r\right) = \arctan\left(\tan 8.12°\sin 60° + \tan 10°\cos 60°\right) = 11.96°$$

The rake angle in the back plane is calculated using Eq. (A.8) as

$$\gamma_p = \arctan\left(\tan\gamma_o \cos\kappa_r - \tan\lambda_s \sin\kappa_r\right) = \arctan\left(\tan 8.12°\cos 60° - \tan 10°\sin 60°\right) = -4.65°$$

A.4.13   DRILLING TOOL GEOMETRY IN T-HAND-S

A.4.13.1   T-hand-S System of Planes for Drilling Tools

As for single-point tools, the basis of the definition of the T-hand-S system of planes starts with the definition of the reference plane as all other planes are defined in the well-established relations to this plane. It is to say that once the reference plane is defined, the definitions of other planes come "automatically" as discussed in Section A.4. As discussed in Sections A.3.2 and A.4.1, the reference plane is defined as perpendicular to the assumed direction of the cutting speed. Figure A.26 shows the definition of the reference plane P_r for the selected point i on the major cutting edge 1–2 of a drill. Figure A.26 shows other planes, namely the working plane, P_f, the back plane, P_p, and the

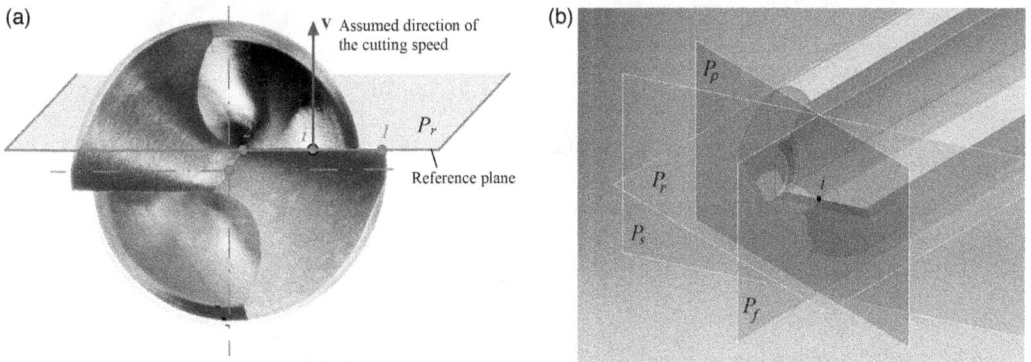

FIGURE A.26   T-hand-S system of planes for drilling tools: (a) visualization of the reference plane, (b) other planes defined in the relation to the reference plane.

tool cutting edge plane, P_s defined exactly in the same way as discussed in Section A.4. Note that all geometrical relationships among the angles in these planes discussed in Section A.4.12 are the same for drilling tools. They are extensively used in the consideration of the drilling tools geometry in Chapter 2.

A.4.13.1   Point and Half-Point Angles

The point angle Φ_p is defined in the reference plane as the angle between projection of the major cutting edges in this plane as shown in Figure A.27. The half-point angle $\varphi = \Phi p/2$ is the angle of the projection of the considered cutting edge into the reference plane P_r and the drill longitudinal axis or the feed direction. Although in many other tools, the tool cutting edge angle κ_r is considered to determine the edge orientation in the zx plane. The half-point φ and point Φ_p angles are found to be more convenient for drilling tool analyses than the standard tool cutting edge angle κ_r discussed in Section A.4.7. Obviously, $\varphi = \kappa_r$.

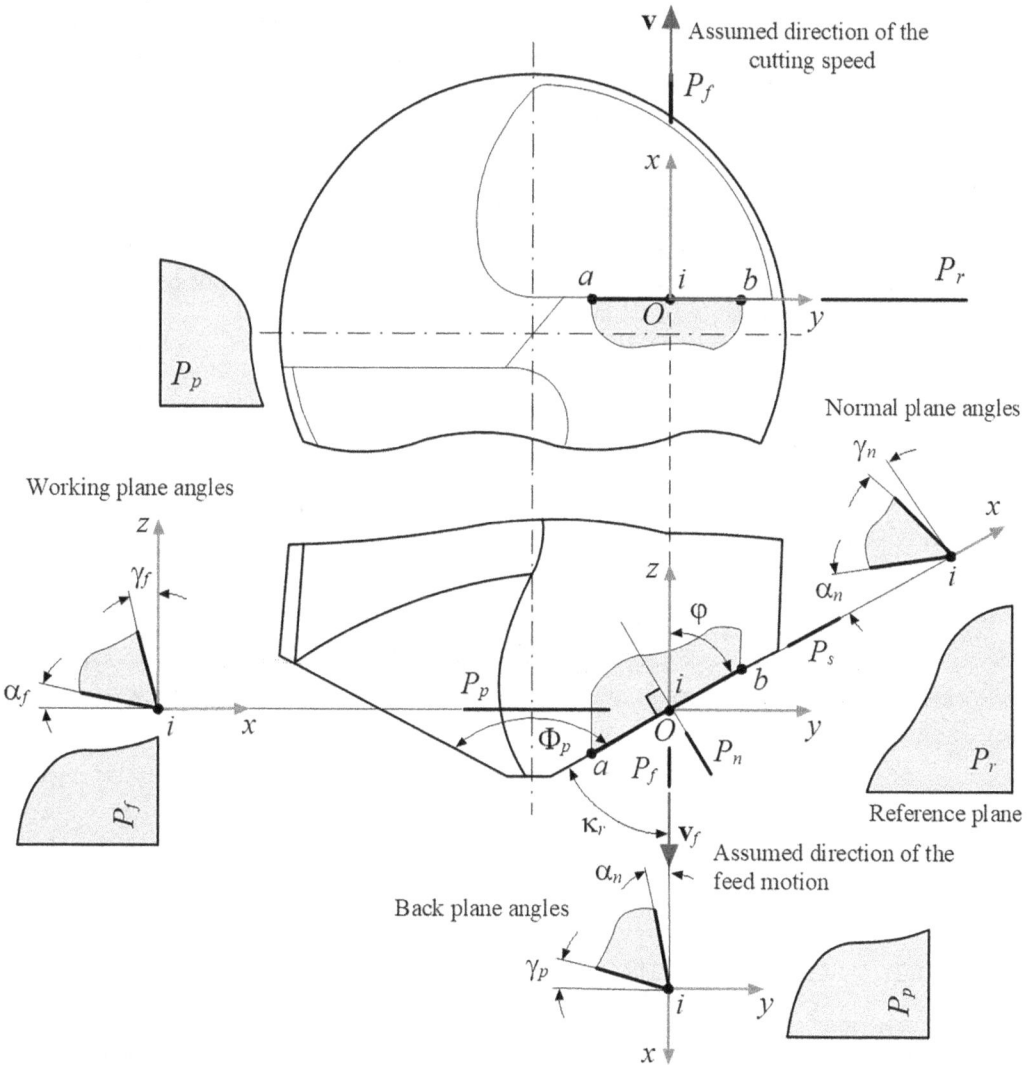

FIGURE A.27   Point angle and T-hand-S rake and clearance angles in the above-introduced planes.

A.4.13.2 Rake and Clearance Angles

This section considers the rake and clearance angles in T-hand-S of the major cutting edge. Figure A.27 shows a cutting tooth of a drill having cutting edge *ab*. A point of consideration *i* is selected on this edge for further analysis. The T-hand-S right-hand *xyz* Cartesian coordinate system is set as follows: its origin *O* is in point *i* selected on cutting edge *ab*, the *z*-axis is along the same direction as that of the assumed feed motion (vector $\mathbf{v}_f$ in Figure A.27), the *x*-axis is along the direction of the assumed direction of the cutting speed (the primary motion) (vector $\mathbf{v}$ in Figure A.27), and the *y*-axis is perpendicular to the *x* and *z* axes as shown in Figure A.27.

The tool normal plane P_n is defined as a plane through the selected point on the cutting edge and normal to the cutting edge at the selected point on the cutting edge as defined in Section 4.5.

The following three angles are of prime concern in T-hand-S:

1. The normal clearance angle α_n is the angle between the tool cutting edge plane P_s and the tool flank plane as shown in Figure A.27 measured in the normal plane P_n. It is clear that if the flank surface is not a plane, then the plane tangent to the curved flank surface at the considered point on the cutting edge is used instead of the flank plane.
2. The normal rake angle γ_n is the angle between the rake face and the reference plane P_r measured in the normal plane P_n as shown in Figure A.27. It is clear that if the rake face surface is not a plane, then the plane tangent to the curved rake face surface at the considered point on the cutting edge is used instead of the rake plane.

The great importance of the introduced normal angles is rather simple – these angles are indicated on the tool drawing and thus used in the manufacturing of the tool points, tool re-sharpening, and tool inspection. They, however, should not be used even to the first approximation to judge tool performance through application of the mechanics/physics of metal cutting.

The inspection of the normal rake and flank angles using modern tool geometry inspection machines is not always easy as many of them have some difficulties in measuring the normal rake and clearance angles directly. Instead, they are capable of measuring the so-called side (often referred to as axial) and back (sometimes referred to as radial) angles. These angles shown in Figure A.27 are in the working and back planes, respectively. The clearance α_f and rake γ_f angles in the working plane as well as clearance α_p and rake γ_p angles in the back plane are determined exactly the same as discussed in Section 4.8.

To carry out a proper inspection, that is, to compare the results of inspection with those shown on the tool drawing, one needs to know the following simple relationships between the measured and the tool angles shown in the tool drawing. For example, if the rake γ_f and clearance α_f in the T-hand-S working plane are measured (the common case for many tool inspection machines), then the corresponding normal angles indicated on the drawing can be calculated as (Astakhov 2010)

$$\gamma_n = \arctan\left(\frac{\tan\gamma_f}{\sin\varphi}\right) \qquad (A.11)$$

$$\alpha_n = \arctan\left(\tan\alpha_f \sin\varphi\right) \qquad (A.12)$$

A.5 TOOL-IN-HOLDER SYSTEM (T-HOLD-S)

The two most important systems of the consideration of the drilling tool geometry are the tool-in-hand system (T-hand-S) and the tool-in-holder system (T-hold-S). The drilling tool geometry parameters in T-hand-S are used in the tool design, and thus assigned in the tool drawing, used in tool manufacturing and its inspection as explained in Chapter 2. The geometry parameters

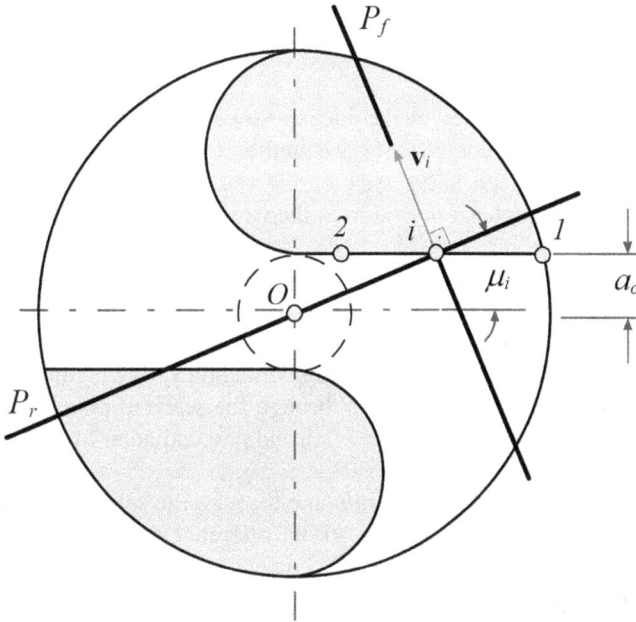

FIGURE A.28 T-hold-S reference P_r and working P_f planes for the considered point I of the major cutting edge 1–2.

of a drilling tool in T-hold-S define the working conditions of this tool (discussed in Chapter 2), i.e., its performance including tool life, allowable penetration rate/productivity, and quality of the machined holes. Obviously, the geometry parameters in these systems are uniquely related through particularities of the drilling tool design.

Compared to T-hand-S where the assumed direction of the cutting speed is considered as discussed in Section A.4.1 for single-point tools and in Section A.4.13 for drilling tools, the actual direction of the cutting speed is considered in T-hold-S. As discussed in Chapter 2, the major cutting edge 1–2 in an actual drill is located above the centerline by distance a_o as shown in Figure A.28. Considering a point i on this cutting edge, one can see that its radius is Oi and the actual cutting speed $\mathbf{v}_i$ is perpendicular to this radius. Therefore, the reference plane P_r is drawn through this point i perpendicular to the actual direction of the cutting speed $\mathbf{v}_i$. From the above definition, it follows that the reference plane in T-held-S is just "tilted" by the cutting speed location angle μ_i (Figure A.28) to the reference plane in T-hand-S. The other planes are also "tilted" by the same angle as all of them are defined with respect to the reference plane in this system.

As angles μ_i vary along the cutting edge 1–2, T-hold-S should be defined for each point of this edge so that the geometry articles of the drill vary in T-hold-S from one point to the next whereas in T-hand-S they are the same. The correlations between the articles of tool geometry for round tools in T-hand-S and T-hold-S are provided in Chapter 2.

A.6 TOOL-IN-MACHINE SYSTEM (T-MACH-S)

T-mach-S accounts for the actual cutting tool setting in the machine as this setting can be significantly different from that assumed by the tool designer. As a result, the tool geometry established in the T-hand-S is altered, i.e., one or more important tool angles are changed. This system considered by the author earlier in great detail (Astakhov 2010) is not normally relevant to drilling tools.

A.7   TOOL-IN-USE SYSTEM (T-USE-S)

A.7.1   T-use-S Reference Planes

The T-use-S considers the geometry of the cutting tool accounting for machining kinematics. As the parameters of the tool geometry are affected by the actual resultant motion of the cutting tool relative to the workpiece, a new system, referred to as the tool-in-use system (T-use-S), and the corresponding tool angles, referred to as the working angles, are established in this new system.

Figures A.29 and A.30 show the graphical representation of the following planes:

- *Working reference plane P_{re}* is defined as a plane through the selected point on the cutting edge and perpendicular to the resultant cutting direction $\mathbf{v}_e$ (see Figure A.12).
- *Working plane P_{fe}* is defined as a plane through the selected point on the cutting edge and containing the resultant cutting direction and perpendicular to the working reference plane P_{re}.
- *Working back plane P_{pe}* is defined as a plane through the selected point on the cutting edge and perpendicular both to the working reference plane P_{re} and to the working plane P_{fe}.
- *Working cutting-edge plane P_{se}* is defined as a plane tangent to the cutting edge at the selected point and perpendicular to the working reference plane P_{re}. This plane thus is parallel to the resultant cutting direction.
- *Working orthogonal plane P_{oe}* is defined as a plane through the selected point on the cutting edge and perpendicular both to the working reference plane P_{re} and to the working cutting edge plane P_{se}.
- *Cutting edge normal plane P_{ne}.* According to standards ISO 3002/1 and ANSI B94.50, this plane is identical to the cutting-edge normal plane defined in the T-hand-S, that is, $P_{ne} \equiv P_n$.

It directly follows from the above definitions that the reference plane in T-use-S is just "tilted" by the resultant cutting speed angle η_e (Figure A.12) to the reference plane in T-hand-S. The other planes (besides the cutting edge normal plane which is the same in T-use-S and T-hand-S, that is, $P_{ne} \equiv P_n$) are also "tilted" by the same angle as all of them are defined with respect to the reference plane in this system. It follows from Figure A.12 that

$$\tan \eta_e = \frac{v_f}{v} \tag{A.13}$$

Because for turning, the feed rate is $v_f = nf$ and the cutting speed is $v = \pi D_w n$ (see Chapter 1), Eq. (A.13) becomes

$$\tan \eta_e = \frac{v_f}{v} = \frac{f}{\pi D_w} \tag{A.14}$$

A.7.2   T-use-S Angles

A set of angles is defined in T-use-S. These angles have the prefix "e" in their title. Because working angles may vary from point to point along the cutting edge, the definitions of the angles refer always to the angles at the selected point on the cutting edge. When the cutting edge, face, or flank

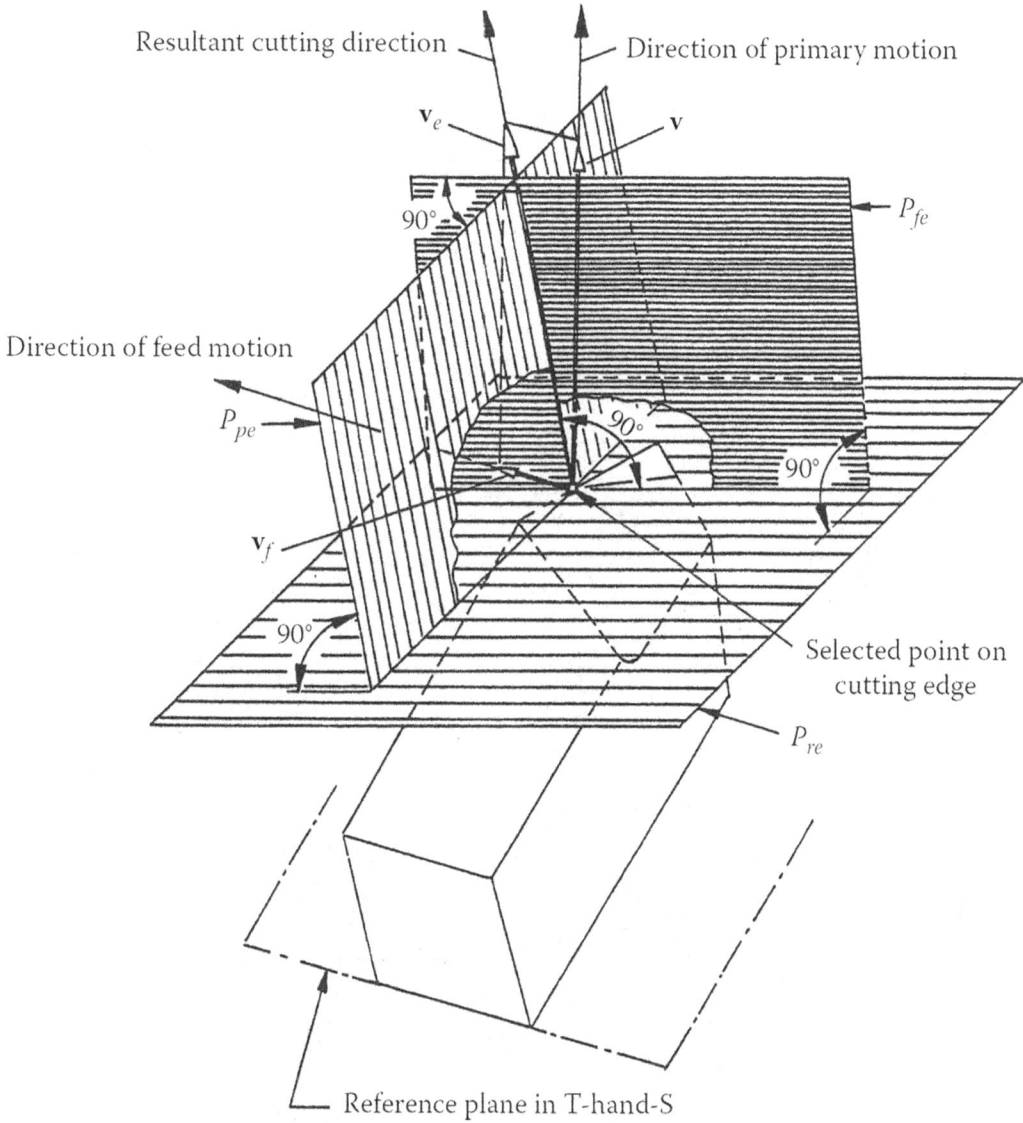

Resultant cutting direction

Direction of primary motion

$\mathbf{v}_e$

$\mathbf{v}$

$90°$

P_{fe}

Direction of feed motion

P_{pe}

$90°$

$90°$

$\mathbf{v}_f$

$90°$

Selected point on cutting edge

P_{re}

$90°$

Reference plane in T-hand-S

FIGURE A.29 Representation of the reference, working, and back planes in T-use-S.

is curved, the tangents or tangent planes through the selected point are used in the reference systems of planes to define the angles.

Each angle is specified, where appropriate, with reference to a particular cutting edge on the tool, depending upon the location of the selected point on the cutting edge. The title of the angle may include an indication of whether the selected point is located on the major or minor cutting edge (e.g., at a selected point on the major cutting edge, there is the tool normal rake, and at a selected point on the minor cutting edge, the corresponding angle is termed the tool minor cutting edge normal rake).

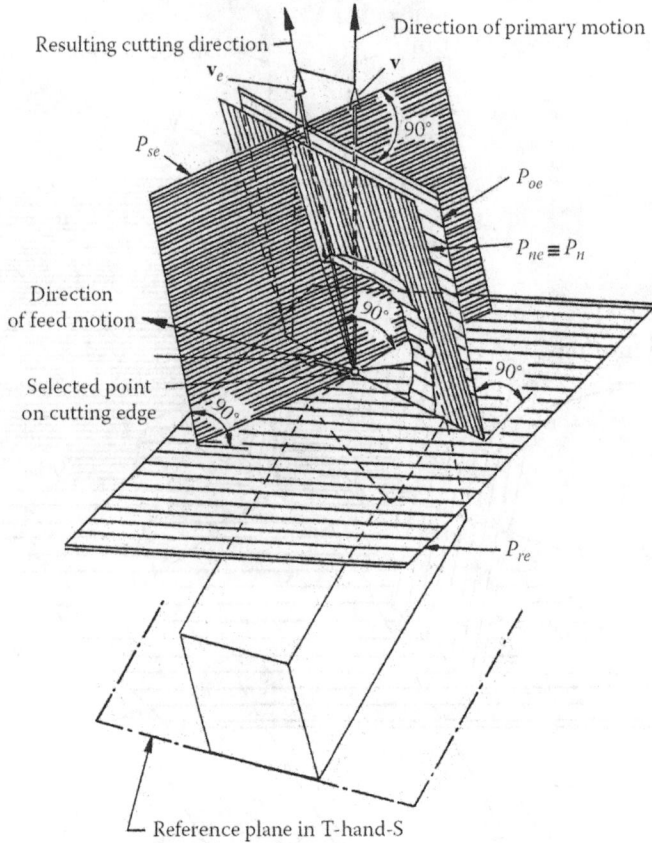

FIGURE A.30 Representation of the cutting edge, orthogonal, and normal planes in T-use-S.

The angles of the cutting tool in T-use-S are defined in these planes in the same manner as in T-hand-S with respect to the T-use-S reference planes. To comprehend the concept of working planes, consider a simple model shown in Figure A.31. As can be seen, the clearance angle in the working plane α_{fe} in T-use-S is smaller than that in T-hand-S by angle η_e.

Figure A.32 shows a stricter model. It visualizes the sense of working rake γ_{fe} and clearance α_{fe} angles in T-use-S considered in the working plane P_{fe}. The three planes namely, the tool reference plane P_r, tool cutting-edge plane P_s, and working plane P_f in T-hand-S, pass through selected point m on the cutting edge. As such by definition, tool rake γ_f and clearance α_f angles in T-hand-S are as shown in Figure A.32. In T-use-S, the reference plane P_{re} is perpendicular to the resultant cutting direction, that is, rotated counterclockwise by angle η_e with respect to P_r.

It follows from Figure A.32 that the working rake γ_{fe} and clearance α_{fe} angles are calculated as

$$\alpha_{fe} = \alpha_f - \eta_e \tag{A.15}$$

$$\gamma_{fe} = \gamma_f + \eta_e \tag{A.16}$$

As can be seen, angle γ_{fe} becomes larger and α_{fe} becomes smaller in T-use-S than those in T-hand-S.

In many practical cases of turning and boring, angle η_e is small. For example, in turning with the following parameters, $D_w=40\,\text{mm}$, $f=0.3$ mm/rev, angle $\eta_e \approx 0.14°$, which is much smaller than common tolerances on tool angles. Therefore, T-use-S angles are considered only in some special cases of machining with high feeds, for example, in thread cutting (Astakhov 2010). However, the significance of the T-use-S clearance angle is considered in Chapter 2 because the resultant cutting

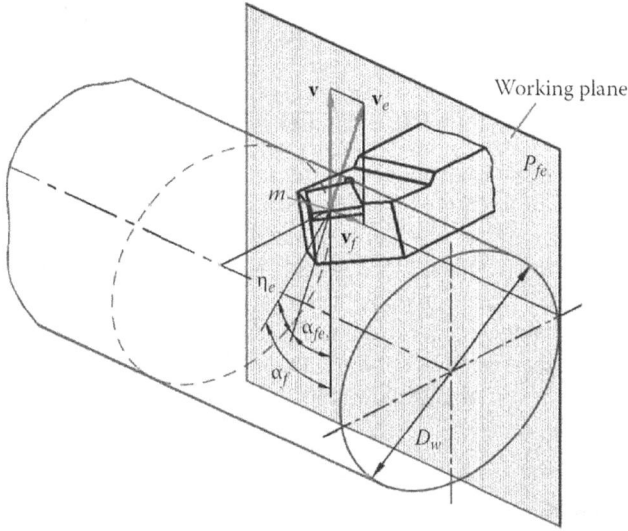

FIGURE A.31   Showing the sense of the clearance angle α_{fe} in T-use-S.

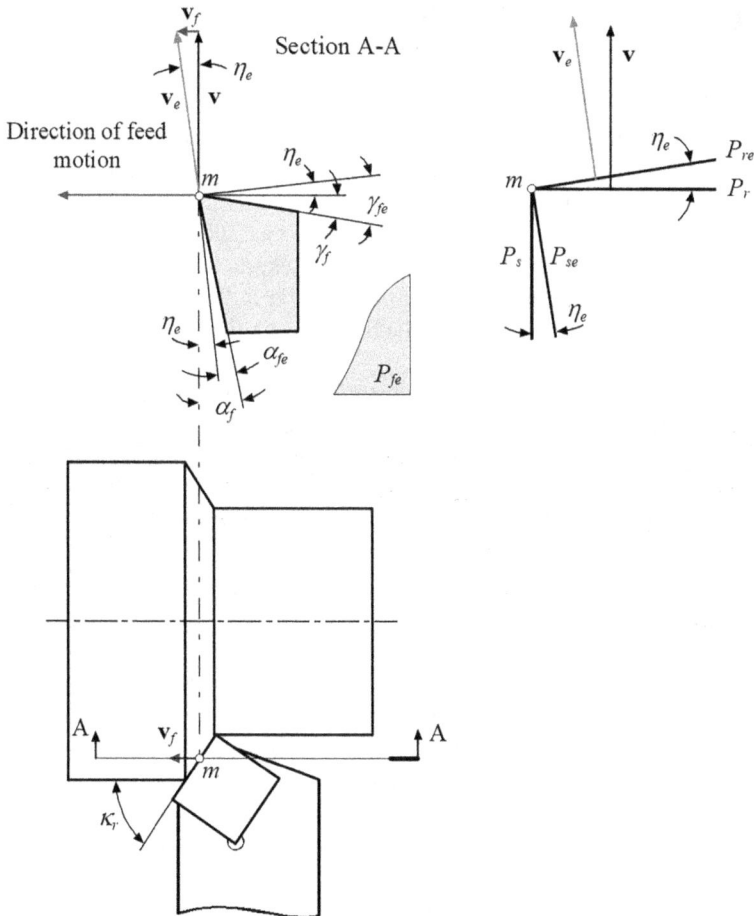

FIGURE A.32   Sense of working rake γ_{fe} and clearance α_{fe} (in the working plane) angles in T-use-S.

speed angle η_e in the regions near the drill rotational axis can be great as the cutting speed is small in these regions whereas the cutting feed is the same.

A.8   IMPORTANCE OF THE INTRODUCED ANGLES

The parameters of the tool geometry are widely discussed and used in the scientific, professional, and trade literature including catalogs of various tool manufacturers. Much less publications are available on the significance of the introduced parameters. Therefore, a question: Why does one need to know the parameters of the tool geometry?, should be answered as the proper answer to this seemingly simple question has significant theoretical and very practical implications. It should be clear that the introduced angles that constitute the cutting tool geometry have functional importance. Each angle is of concern when a certain function of the cutting tool is considered, that is, the cutting force, tool life, direction of chip flow, and theoretical surface roughness of the machined surface.

A.8.1   CLEARANCE ANGLE

One angle that stands out among other angles is *the clearance angle* because, as discussed above, this angle is the major distinguishing feature of the cutting tool. This section reveals the real significance of the clearance angle in metal cutting because many even seasoned researchers and professionals in the field have certain difficulties to fully understand this statement. Let us start explanation with defining the term "metal cutting."

A.8.1.1   Explanation of Cutting

First, by definition, fracture is PHYSICAL separation of a body into two or more parts. It takes place in many mechanical tests of various work materials (e.g., the tension test) so that the stress at which a given material breaks is called the strength of this material. In the tension test, it is termed as the ultimate tensile strength. Because many engineering real-world materials show some deformation before fracture, this deformation is characterized by the strain at fracture. For example, in the tension test, it is known as elongation at fracture. Both the ultimate tensile strength and elongation at fracture are called the basic mechanical properties of the work material (Astakhov 2018).

Second, because any *cutting* is *physical separation* of a material into two or more parts, cutting always includes fracture. However, cutting is physical separation of a material into two or more parts carried out by *a cutting tool (blade) in a controllable manner.* As discussed by the author earlier (Astakhov 2010), a number of cutting operations fall into this definition, namely splitting, cutting by a knife, shearing, punching, and metal cutting. It is of vital importance to consider some similarities and the principal differences between these operations and metal cutting in order to distinguish the process of metal cutting among other closely related manufacturing processes/operations.

The common cutting operations (besides metal cutting) are shown in Figure A.33. The easiest way to explain splitting as a separation process is to use wood splitting as an example in Figure A.33a. The first thing to understand about wood splitting is that it is different from wood cutting. Wood is a natural polymer – parallel strands of cellulose fibers held together by a lignin binder. These long chains of fibers make the wood exceptionally strong – they resist stress and spread the load over the length of the board. Furthermore, cellulose is tougher than lignin. It is easier to split a board along the grain (separating the lignin) than to break it across the grain (separating the cellulose fibers). This makes wood an anisotropic material, that is, its properties depend on the direction in which they are measured and its strength depends on the direction of load application. When one splits wood, he or she is separating the cellulose fibers from each other. This is why wood can only be split along the grain.

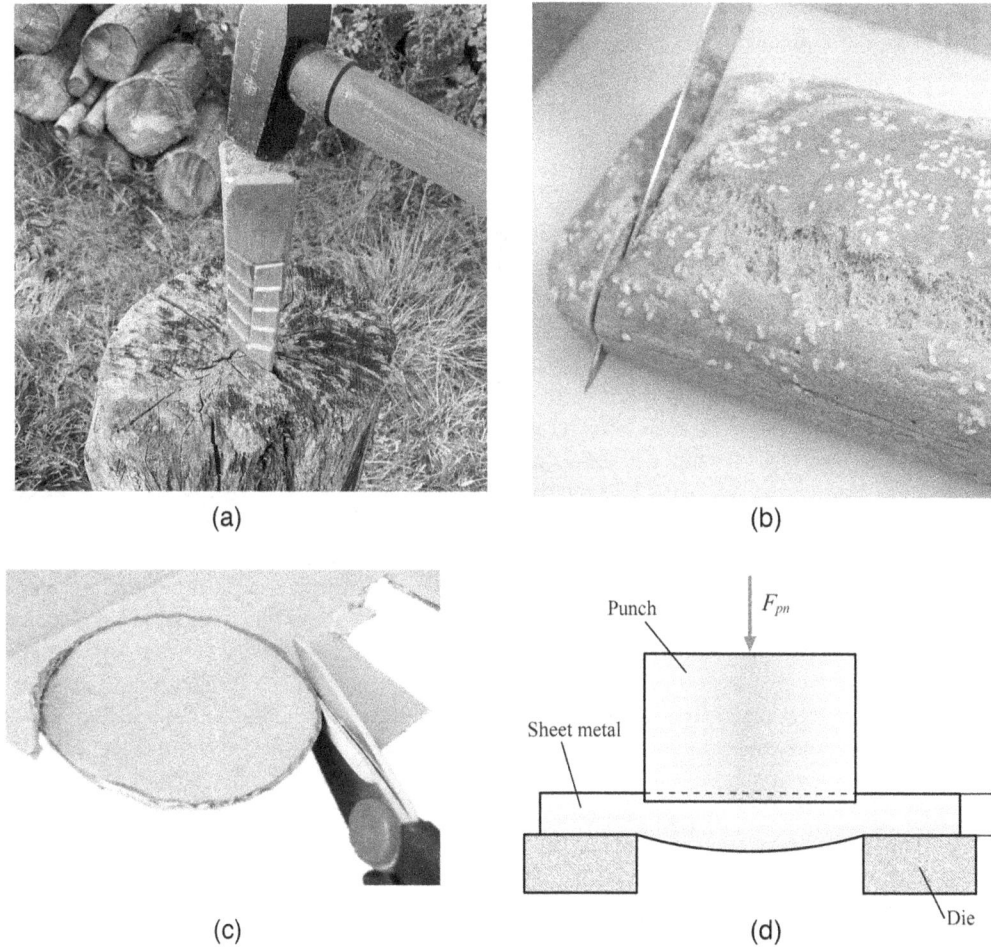

FIGURE A.33 Common cutting operations: (a) wood splitting by a wedge, (b) cutting by a knife, (c) cutting by scissors (shears), and (d) punching.

Commonly, the tool used for wood splitting is a maul or its lighter version – hatchet. For very tough and/or large pieces of wood, splitting wedges are used as shown in Figure A.33a. A wedge is used like this. First, the wedge is *set* in the wood by gently tapping it in with the maul held in one hand or a small sledge, known as a *single jack*, can be used for this task. Basically, it is desired for the wedge to dig in enough that it stands up on its own. It is sometimes helpful to hit the wood with the maul in order to make a divot in which one then starts the wedge. Once the wedge is set, one begins to hit it with the back side of a dedicated sledge hammer and drive it into the wood. The important particularities of wood splitting including the force involved, wedge geometry, and contact conditions on the wedge-wood interfaces are discussed by the author earlier (Astakhov 2014).

Slicing bread with a knife is another use of splitting (Figure A.33b). When a knife is pressed or chopped straight through an object, that object is essentially being pinched in half. One might ask why a sharp knife is more effective at this than a dull knife. This is a matter of physics – a transfer of kinetic energy, to be specific. When a knife is sharp, its edge is narrower than the edge of a dull knife. This means the point of contact between the knife and the object to be cut is much smaller. The weight of the knife combined with the force one uses in pressing down is exerted over a much smaller point of contact, magnifying that force dramatically so that the sharp knife cuts through the object more cleanly and with less force than a dull one.

Two other cutting operations shown in Figure A.33c and d are the so-called shearing operations. Shearing is the deformation and then separation of a material substance in which parallel surfaces are made to slide past one another. In shearing, one layer of a material is made to move on the adjacent layer in a linear direction due to the action of two parallel forces located at distance known as the clearance distance. A typical example of shearing is cutting with a pair of scissors (Figure A.33c). Scissors are cutting instruments consisting of a pair of metal blades connected in such a way that the blades meet and cut materials placed between them when the handles are brought together. In industry, much greater scissors, known as the industrial shears, are used for cutting a great variety of sheet metals. Shearing operations use the same principle of material separation as shown in Figure A.33d. Many sheet-metal parts are made from a blank of suitable dimension that is first removed from a large sheet or coil using a variety of manufacturing processes called shearing operations as they all are based on the shearing process. In these operations, the sheet is cut by subjecting it to shear stress typically between a punch and a die. Shearing usually starts with the formation of the shear planes and then cracks on both the top and bottom edges of the workpiece. These cracks eventually meet each other and separation occurs. The rough fracture surfaces form due to these cracks. The particularities of the punch wear and quality of blanks depend on the clearance between the punch and die. The force, energy, particularities of fracture conditions, and quality of punched holes are discussed by the author earlier (Astakhov 2010).

What unifies the listed cutting processes is that they all aim to physically separate a part of the workpiece in a controllable manner by applying an external force (forces) using a cutting blade (blades). Figure A.34 shows that the contact between the work material and the tool takes place on both working faces, that is, on the conditional rake and flank faces, which causes excessive plastic deformation of the work material and great friction forces/losses over the contact areas. This fact combined with the great plastic deformation of the machined surface and unfavorable state of stress in the deformation zone causes subpart quality in terms of surface roughness and dimensional accuracy of the machined part.

A.8.1.2 Explanation of Metal Cutting

The critical difference between metal cutting and the above-discussed cutting operations is that the cutting tool in metal cutting has only one contact face, namely the tool–chip contact face known in metal cutting as the rake face over which chip slides. As can be clearly seen in Figure A.1, the second face of the cutting tool is cleared by the clearance angle α so the machined surface is formed (at least, theoretically) by the cutting edge.

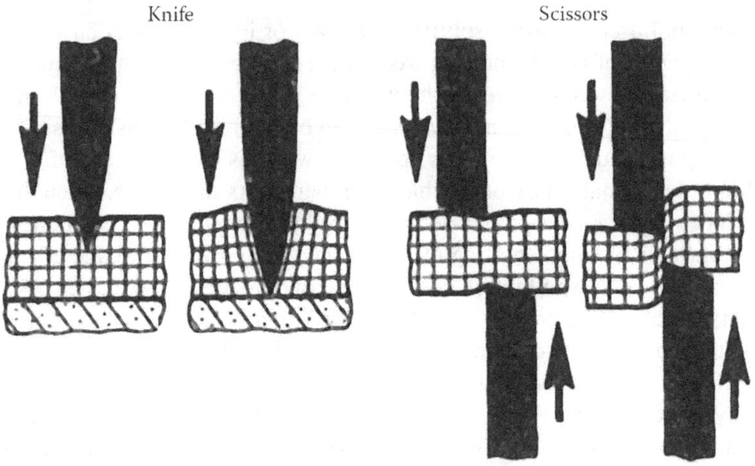

Knife   Scissors

FIGURE A.34 Showing that in both splitting and shearing operations the contact between the work material and the tool takes place on both working faces.

The positive clearance angle is the major distinguishing feature of the cutting tool, that is, it is the only feature that distinguishes the cutting tool from those tools used in closely related manufacturing processes as in splitting and shearing (Astakhov 2010). In other words, other angles of the tool geometry can assume virtually any signs and values that only affect optimality of the cutting process, while the clearance angle should always be positive – it is the only condition of the existence of the cutting process.

The previous statement is explained using the model shown in Figure A.35. This figure presents the simplest example of orthogonal cutting where the primary motion is straight having velocity **v**. The designations and meaning of the width of cut, $b_D = a_p$ and uncut chip thickness, h_D are the same as in the model shown in Figure A.1 As shown in Figure A.35a, a square bar

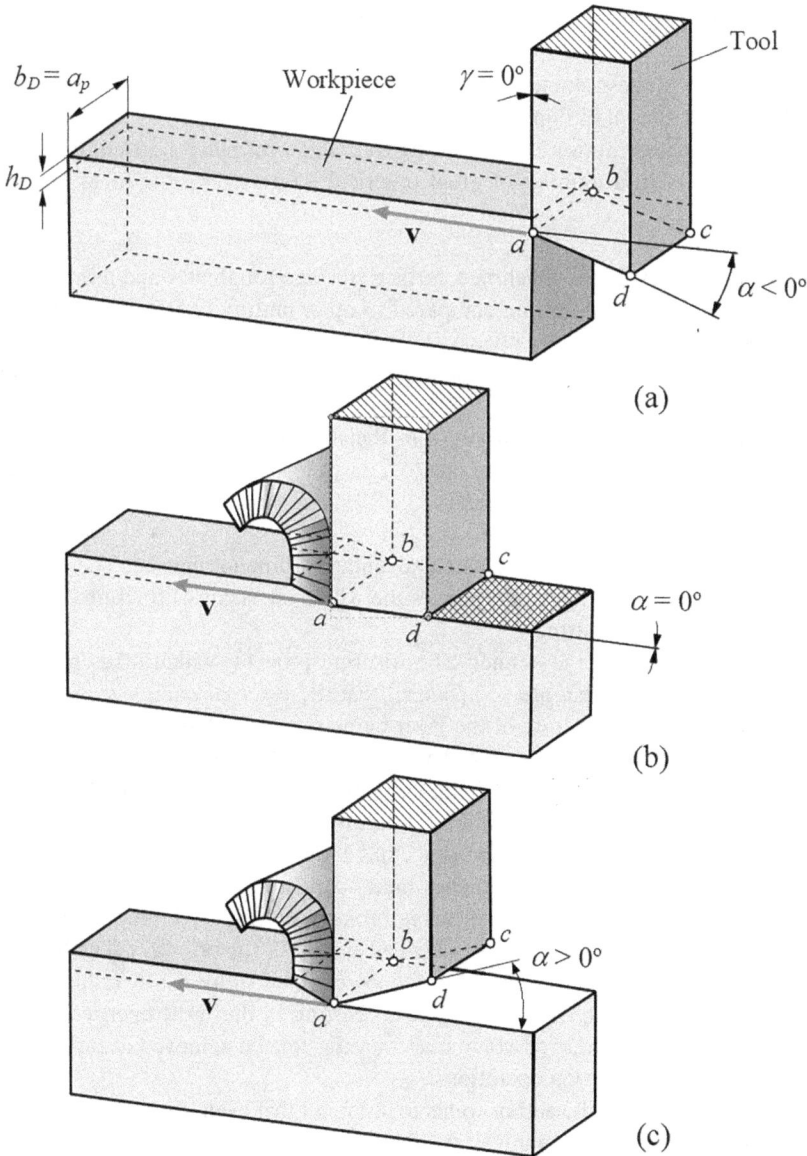

FIGURE A.35 Orthogonal cutting: (a) attempt to cut with the tool having a negative clearance angle, (b) attempt to cut with a zero clearance angle, and (c) proper cutting with a positive clearance angle.

stock is used as a tool. The side face of this bar stock is used as the rake face having cutting edge *ab*, and thus, the rake angle γ is zero for the sake of simplicity of the current considerations. The square face of this bar stock *abcd* is used as the flank face. According to the designation used in this appendix, the clearance angle of this flank face is negative, i.e., $\alpha < 0$. It is obvious that any attempt to cut with such a tool leads to breakage either the tool or the workpiece, or the whole cutting setup as even a small advance of the tool shown creates a great force, primarily in the direction of the cutting speed **v**. In other words, cutting with a negative clearance angle is virtually impossible.

When one attempts to cut using the tool with a zero clearance angle as shown in Figure A.35b, the chip begins to form. However, the normal cutting process will not last. As soon as flank face *abcd* comes into contact with the machined surface, severe rubbing takes place due to spring back of the work material (see Section 2.1.3.8.1), which makes the cutting process impossible due to severe friction force on the flank face, frictional vibrations, and high temperature as a result of heat generation on rubbing.

To make the cutting process stable, a positive clearance angle $\alpha_n > 0$ should be applied to the cutting tool in the manner shown in Figure A.35c.

The relative setting and motion of the cutting tool and workpiece particular to metal cutting (see Figure A.1) results in a number of great practical advantages. The most important are as follows:

1. Much greater quality of the machined surface (surface roughness and dimensional accuracy) is achieved in metal cutting compared to other cutting operations as the machined surface is formed by the cutting edge.
2. Much less plastic deformation/distortion of the machined surface occurs in metal cutting.
3. Much smaller machining residual stress in the machined surface (both superficial and in-depth) is found after metal cutting due to a much lower force and temperature involved in the formation of the machined surface.

One may wonder about the origin of the listed advantages of metal cutting. The reason for that is the tool–workpiece relative arrangement changes the whole physicals of fracture. It is explained by the proper definition of metal cutting:

The process of metal cutting is defined as a forming process, which takes place in the components of the cutting system that are so arranged that the external energy applied to the cutting system causes the purposeful fracture of the layer being removed. This fracture occurs due to the combined stress including the continuously changing bending stress causing a cyclic nature of this process. The dynamic interactions of these components take place in the cutting process, causing a cyclic nature of this process (Astakhov 2006, 2010, 2014).

It follows from this definition that the presence of bending moment in the deformation zone and relatively small size of this zone distinguishes metal cutting among other closely related manufacturing processes and operations (e.g., shearing process, cutting by knife, and splitting) as first revealed by F. Taylor in his classical work published in 1907 (Taylor 1907). The bending moment forms the combined state of stress in the deformation zone that significantly reduces the resistance of the work material to cutting. As a result, metal cutting is the most energy-efficient material removal process (energy per removed volume accounting for the achieved accuracy) compared to other closely related manufacturing operations.

To finish this consideration, the author wants to point out that even if wood, plastic, stone, and so on is cut using the tool–workpiece relative arrangement shown in Figure A.1, such a process/operation should be called as *metal cutting*. If the sheet metal is cut by scissors or punched in a shearing operation, such a process/operation should be called *cutting*. It can be cutting of metal, plastic, and composite material whatever the work material is.

A.8.2 RAKE ANGLE

As discussed in Section A.4, the rake angle comes in three varieties: positive, zero (sometimes referred to as neutral), and negative as shown in Figure A.36a–c, respectively. There is a great body of experimental and numerical modeling results dealing with the influence of the value and sign of the rake angle on the machining process. In the author's opinion, the role and importance of the rake angle in metal cutting is not well understood because the available data are contradictive and often misleading. Moreover, the available studies did not concern with the system consideration of the influence of the rake angle on the various outcomes of the cutting process. Rather, one outcome parameter is normally considered (e.g., the cutting force), while others (e.g., tool life) are ignored. Using these data, a practical tool/process designer cannot make an intelligent selection of the proper rake angle for a given application.

As mentioned earlier, Shaw (1988) argued that the specific cutting energy (and thus the cutting force) decreases about 1% per degree increase in the rake angle when this angle is changed from −20° to +20°. Dahlman et al. (2004) showed that by controlling the rake angle, it is possible to generate tailor-made machining residual stresses in the product. Günay et al. (2005) in their experimental study found that a change in the rake angle from 0° to +2.5° resulted in a 2% reduction of the cutting force while a change from −2.5° to 0° resulted in a 3.4% reduction. Tetsuj et al. (1999) in their test on rock cutting found that the cutting force of the bit with a 20° rake angle decreased about 30%–80% (depending upon other machining parameters), compared to that of the bit with a −20° rake angle. Moreover, an increase in cutting force with the cutting depth becomes lower with an increase in the rake angle. Gunay et al. (2006) carried out a detailed experimental study of the influence of the rake angle in machining of AISI 1040 steel. They found a very small influence that diminishes at high cutting speed. Saglam et al. (2007) carried out an extensive research program on machining of AISI 1040 steel bars hardened to HRC 40 in order to reveal the effect of tool geometry. A system consideration of the major geometry parameter was attempted as the inter-influence of these parameters was considered. It was found that an increase in the rake angle noticeably reduces the cutting force while the cutting temperature increases. It was also found that the influence of the rake angle depends on the tool cutting edge angle. More dramatic influences of the rake angle on the cutting force and temperature were found for high cutting speeds.

A complete analysis of the influence of the rake angle on the cutting process and recommendation for the selection of the so-called effective rake angle was presented by the authors earlier (Astakhov 2010). It is shown that:

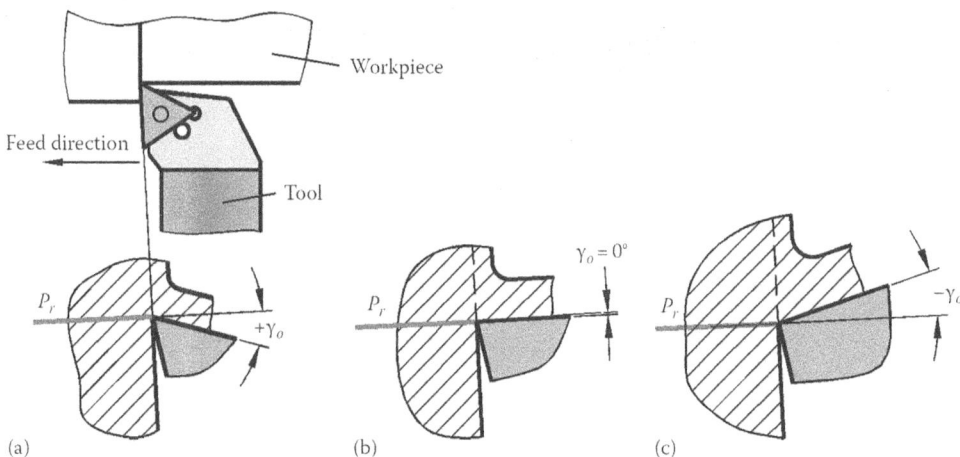

FIGURE A.36 Sense of the positive (a), neutral (b), and negative (c) rake angles in turning.

- Rake angle has a significant effect on the amount of work of plastic deformation in metal cutting.
- The effect of the rake angle is more profound at low cutting speeds although it is still significant at moderate and high cutting speeds.

Three important effects of the rake angle relevant to the consideration in this book should be considered as relevant: (1) Influence of the rake angle on the plastic deformation of the layer being removed as this deformation defines to a large extent the cutting force, temperatures in the machining zone, and tool life (Astakhov and Xiao 2008), and (2) influence of the rake angle on the tool–chip contact length (Astakhov 2006) and influence of the rake angle on force direction. These are discussed in Chapter 2.

A.8.2.1 Measure of Plastic Deformation in Metal Cutting and Tool–Chip Contact Length

When material is cut, the tool moving in the cutting direction removes a certain layer from its surface as shown in Figure A.37a. For further consideration, let's mark a rectangular *1234* (in the 2D representation shown in Figure A.37a) in the layer being removed. As shown, its length is l_D and it is known as the length of cut whereas its thickness is h_D and it is known as the uncut chip thickness h_D. This rectangular undergoes plastic deformation in the deformation zone ahead of the tool rake face and thus transforms into rectangular *1'2'3'4'* having length l_C known as the chip length and thickness h_C known as the chip thickness. As can be seen, the length of cut l_D compresses into the chip length l_C. This phenomenon was first noticed by Time in his earlier metal cutting experiments (Time 1870). He termed the ratio of the length of cut and the chip length as the chip compression ratio (CCR), i.e.,

$$\zeta = \frac{l_D}{l_C} = \frac{h_C}{h_D} \qquad (A.17)$$

As the chip width in the direction perpendicular to cutting direction does not change in cutting changes and the volume of work material does not change either, CCR is also the ratio of h_D and h_C. CCR is known as one of the most useful characteristics of the metal cutting process defining the cutting force, process efficiency, temperature involved, and so on (Astakhov and Shvets 2004, Astakhov 2006, Astakhov and Xiao 2008, 2016). CCR can also be used for

1. Determination of the chip thickness using the known uncut chip thickness according to Eq. (A.17).
2. Determination of the chip contact length, i.e., the length of the tool–chip contact, lc.

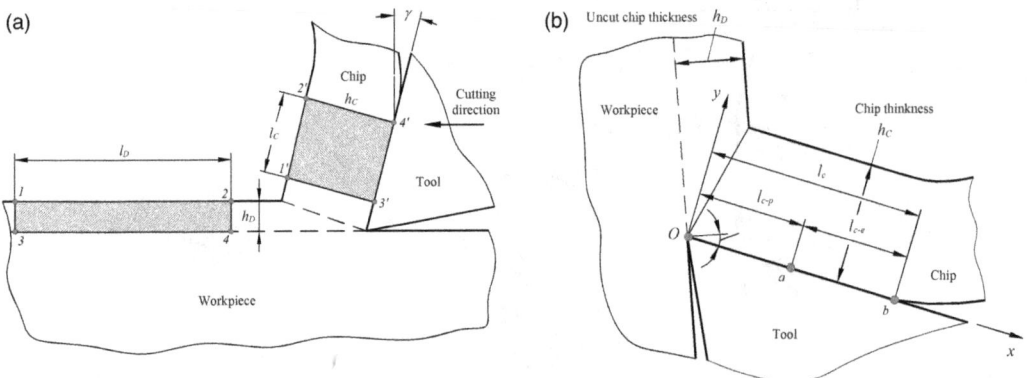

FIGURE A.37 Models: (a) chip deformation in metal cutting and (b) tool–chip contact length.

A model that depicts the concept of the tool–chip contact length l_c is shown in Figure A.37b. According to this model, the chip slides on the rake face over a certain distance in the direction of the x-axis drawn along the rake face. After reaching point b, the chip curls away due to specific stress distribution in its body (Astakhov 1998/1999). The distance from the cutting edge (point O) to point b is known as tool–chip contact length l_c. In metal cutting, tool–chip contact length l_c consists of two parts, namely the plastic part having the length $l_{c\text{-}p}$ over which the shear contact stress is equal to the material yield strength under a given state of stress, and the elastic part $l_{c\text{-}e}$ over which the shear contact stresses are in the elastic region for a given state of stress.

The tool–chip contact length l_c can be determined via CCR as follows (Astakhov 2006):

$$l_c = h_D \zeta^{k_t} \tag{A.18}$$

where $k_t = 1.5$ when $\zeta < 4$, and $k_t = 1.3$ when $\zeta \geq 4$.

As shown in Figure 2.186, CCR and thus plastic deformation in metal cutting reduce with the rake angle and the cutting speed. Reduction of CCR reduces the work/power of plastic deformation, which is the major contributor to the energy spent in metal cutting (Astakhov and Xiao 2016). The reason for the discussed influence of the rake angle is that this angle significantly affects the state of stress in the deformation zone in metal cutting (Abushawashi et al. 2017), which is actually another distinguished feature of this process.

The power spent on the plastic deformation of the layer being removed, P_{pd}, can be calculated from the CCR and parameters of the deformation curve of the work material as follows (Astakhov and Xiao 2008):

$$P_{pd} = \frac{K (1.15\zeta)^{n+1}}{n+1} v A_{ch} \tag{A.19}$$

where K is the strength coefficient (N/m^2) and n is the hardening exponent of the work material, and A_D is the uncut chip cross-sectional area (m^2), i.e., $A_D = a_p f$.

As deformation of the layer being removal in its transformation into the chip reduces, the cutting force and temperature are also reduced which increases the tool life and quality of machined surface. However, an increase in the rake also reduces the cutting tool wedge angle β (see Section A.4.8., Figure A.20), which affects heat removal from the region adjacent to the cutting edge. Moreover, reduction of CCR also reduces the tool–chip contact length (as it follows from Eq. A.18), and thus increases the tool–chip normal stresses (Astakhov 2006). The total effect (the reduction of the normal stresses due to reduction of plastic deformation of the work material and the increase of these stresses due to reduction of the tool–chip contact length) depends on many particularities of a given machining operation. Therefore, there is the optimal value of the rake angle for given cutting conditions.

A.8.2.2 Influence of the Rake Angle on Force Direction

As discussed in Chapter 2, Section 2.1.3.6, fast-spiral drills have a tendency to *dig* or *grab* or run ahead of the feed particularly when breaking through the hole. This is a result of a special case of machining with a high rake angle which should be considered here as it is relevant to drilling tools. It was noticed that when a twist drill is used for enlarging previously drilled holes in relatively soft work materials, such as brass, copper, and Babbitt, the drill jumped ahead of the feed into the hole, which caused vibration.

To understand why this happens, let's consider a simplified force model for machining with a tool having a high rake angle as shown in Figure A.38a. Two primary force components and one force occur in cutting. The first one is the force component normal to the tool rake face, known as the normal force F_n. The second component that acts along the tool fake face is known as the friction force F_f. It occurs due to sliding of the chip over the rake face. The third force factor is the normal force Q_n

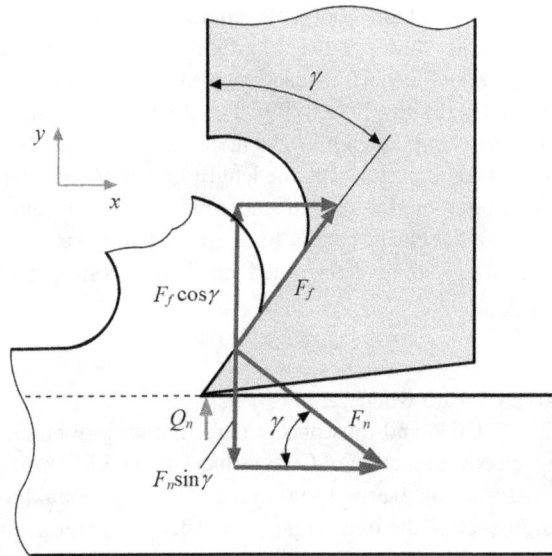

(a)

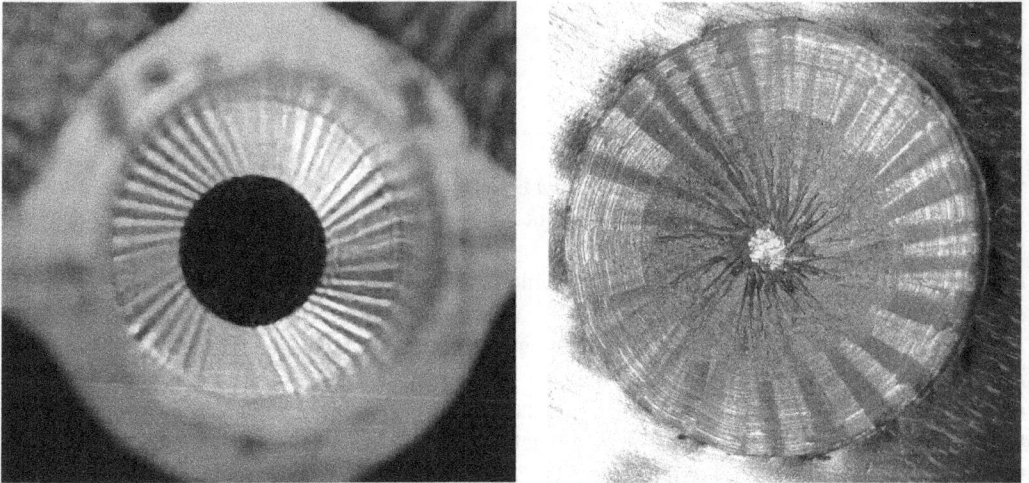

(b)

FIGURE A.38 Cutting with high rake angle: (a) Simplified force model, (b) Chatter marks on the bottom of the hole being drilled with a twist drill.

acting on the tool flank due to springback of the work material. The problem is the resultant force in the radial direction, i.e., in the y-direction. As follows from Figure A.38a, this force is calculated as

$$F_y = F_f \cos \gamma + Q_n - F_n \sin \gamma \qquad (A.20)$$

Normally, when the tool rake angle γ is small, friction force F_f is significant due to high normal stresses (force) over the tool–chip interface (Astakhov 2006), force Q_n is significant due to not sharp cutting edge (see Chapter 2 Section 2.1.4), small clearance angle, and high roughness of the flank face, this radial force is positive, it acts in the positive direction of the y-axis pushing the tool from the workpiece. However, when the tool rake angle γ is great, friction force F_f is relatively small because the rake face is well lubricated, force Q_n is small due to the sharp cutting edge (see Chapter 2 Section 2.1.4), sufficient clearance angle, and low roughness of the flank face, this radial force may become negative as the first two components of Eq. (A.20) may be smaller than the negative third one. When this is the case, the radial force acts in the negative direction of the y-direction pushing into the tool the workpiece. As a result, chatter occurs. It often happened in drilling with twist drills as discussed in Chapter 2, Section 2.1.3.6. Figure A.38b shows examples of chatter marks on the bottom of the hole being drilled with a twist drill.

A.8.3   TOOL CUTTING EDGE ANGLE

The tool cutting edge angle is often considered the most important angle of the tool geometry due to its multifaceted influence on practically all aspects of the metal cutting process and greatly affects the outcomes of a machining operation. This is because it defines the magnitudes of the radial and tangential components of the cutting force and, for a given feed and cutting depth, it defines the uncut chip thickness, width of cut, chip flow direction, and tool life. The physical background of this phenomenon can be explained as follows: when κ_r decreases, the chip width increases correspondingly because the active part of the cutting edge increases. The latter results in improved heat removal from the tool and hence tool life increases. For example, if tool life of a high-speed steel face milling tool having $\kappa_r=60°$ is taken to be 100%, then when $\kappa_r=30°$, its tool life is 190%, and when $\kappa_r=10°$, its tool life is 650%. Even more profound effect of κ_r is observed in the machining with single-point cutting tools. For example, in rough turning of carbon steels, the change of κ_r from 45° to 30° leads sometimes to a fivefold increase in tool life.

The reduction of κ_r, however, has its drawbacks. One of them is the corresponding increase in the radial component of the cutting force. As discussed in Chapter 2 (Figure 2.23), the cutting force is a 3D vector that is normally resolved into three orthogonal components, namely, the power (tangential, toque) component F_p, axial component F_a, and radial component F_r. In the metal cutting literature, these forces often designated as F_z, F_x, and F_y, respectively, referring to the machine tool coordinate system. For simplicity, these are often called tangential, axial, and radial forces, respectively. Figure A.39a depicts the force component normal to the cutting edge (often called as the horizontal force) F_n in the reference plane. As can be seen, the radial F_y and axial F_a forces are actually components of F_n. As follows from Figure A.39a, the radial and axial forces are related as

$$\frac{F_a}{F_r} = \tan \kappa_r \tag{A.21}$$

As follows from Eq. (A.21), lowering κ_r from 45° to 20° results in more than a twofold increase in the radial force that increases the bending force acting on the workpiece and thus may reduce the accuracy of machining because the rigidity of the workpiece varies along its length. When the workpiece is machined between centers, the radial force causes the so-called barreling, and when the workpiece is clamped only in the chuck, tapering may occur. Besides, because lowering κ_r leads to an increased radial force, this force often causes vibrations so that the advantages of a small cutting edge angle may become not too profound.

The tool cutting edge angle has a direct influence on the uncut (undeformed) chip cross-sectional shape, uncut chip cross-sectional area, and thus on the uncut chip thickness, which is by far the most important parameter of a machining operation because it determines (for a given work material) the cutting force, plastic deformation of the work material in its transformation into the chip, allowable feed, tool–chip contact length, etc. Therefore, the correlations between κ_r and the uncut chip thickness (known in the practice of metal cutting as the chip load) should be established.

In orthogonal cutting, the concept of the uncut chip thickness is self-obvious as it is equal to the thickness of the layer that has been removed (Figures A.1 and A.37a). Figure A.39b helps to

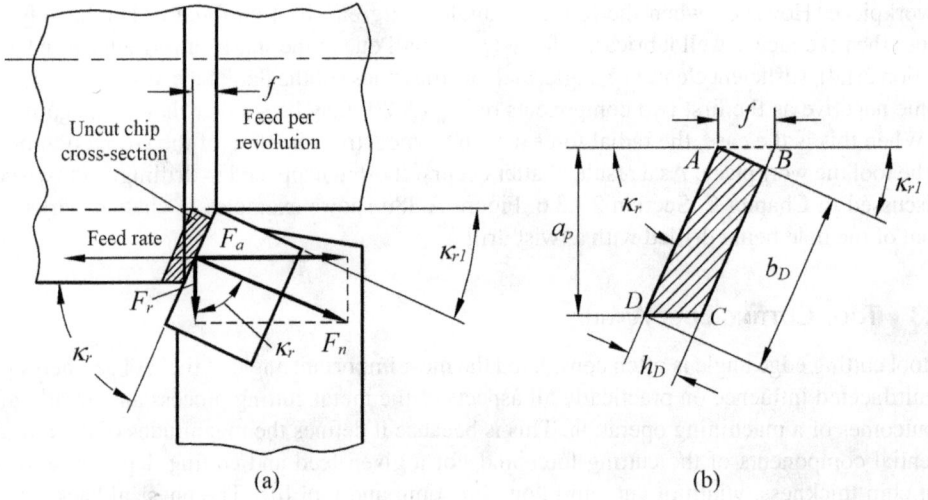

FIGURE A.39 Forces in the reference plane (a) and parameters of the uncut chip cross-sectional area (b).

comprehend the concept of the uncut chip thickness in the simplest case of turning. As seen, the uncut chip cross-sectional area is represented by polygon *ABCD*. Side *AD* is formed by the major cutting edge, side *AB* is formed by the minor cutting edge, and side *DC* is a part of the workpiece surface to be machined. The uncut chip thickness h_D is the thickness of the layer to be removed per one revolution of the workpiece as measured perpendicular to the cutting edge. As it follows from Figure A.39b,

$$h_D = f \sin \kappa_r \tag{A.22}$$

It follows from this equation that under a given feed *f*, the uncut chip thickness can be varied in a wide range by changing angle κ_r. When this angle becomes zero, the uncut chip thickness is also zero (no cutting), and when $\kappa_r = 90°$, the uncut chip thickness reaches its maximum, becoming $h_D = f$ as in orthogonal cutting. Therefore, $h_D \in (0, f)$.

The following equation for the uncut chip area correlates the feed *f* and the depth of cut a_p, with the uncut chip thickness h_D and non-orthogonal chip width b_D:

$$a_p f = h_D b_D = A_D \tag{A.23}$$

where A_D is the uncut chip cross-sectional area.

A.8.4 MINOR CUTTING EDGE

The role of the minor cutting edge and its influence on the cutting force and power consumption are seldom considered in the literature on metal cutting. At best, the influence of the tool minor cutting edge angle κ_{r1} is mentioned in the consideration of the theoretical roughness of the machined surface or the geometric component of roughness. It is obvious that out of many parameters of tool geometry, the influence of angles κ_r and κ_{r1} on the surface roughness of the machined parts is most profound as shown in Figure A.40. To explain this influence, consider the simplest case when the nose radius r_ε (see Section A.4.7, Figure A.18) is zero as shown in Figure A.40c. For this case, this roughness can easily be computed in terms of maximum peak-to-valley distance R_t as (Astakhov 2011)

$$R_t = \frac{f \sin \kappa_r \sin \kappa_{r1}}{\sin (\kappa_r + \kappa_{r1})} \tag{A.24}$$

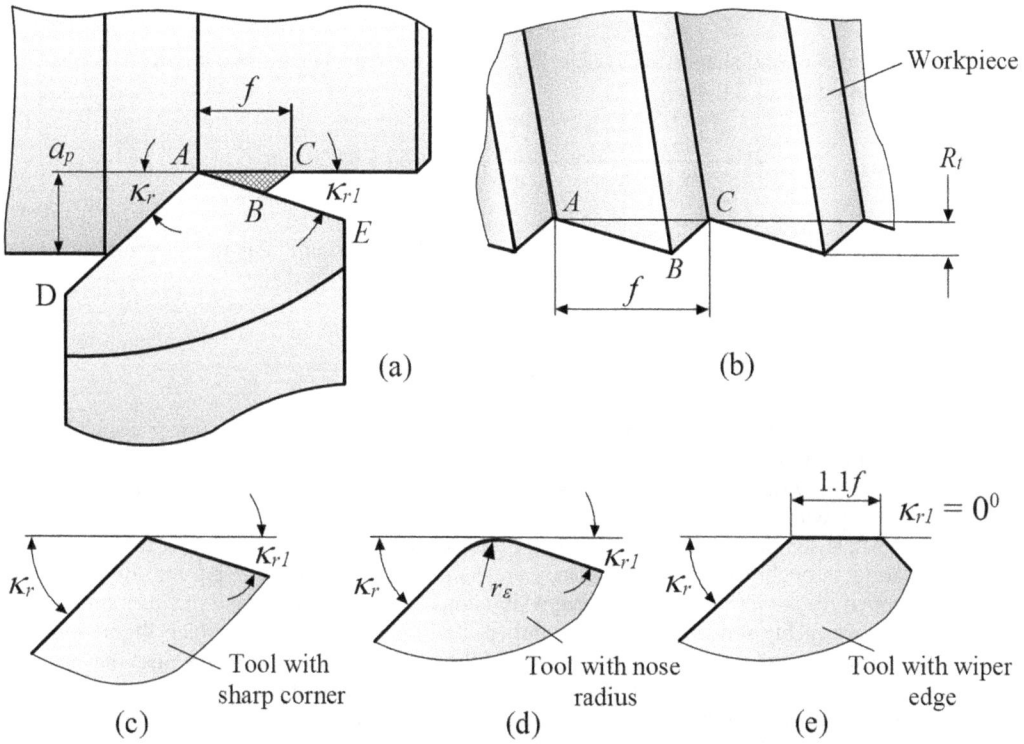

FIGURE A.40 Role of the minor cutting edge in the formation of the surface roughness: (a) roughness left on the machined surface, (b) enlarged profile of the machined surface, (c) tool with sharp corner, (d) tool with nose radius, and (e) viper geometry.

For the tool with nose radius (Figure A.40d), roughness is calculated as

$$R_t \simeq (1-\cos \kappa_{r1}) r_\varepsilon + f \sin \kappa_r \sin \kappa_{r1} \tag{A.25}$$

When the depth of cut is small, the surface profile is formed only by the round part of the cutting edge. In this case,

$$R_t = r_\varepsilon - 0.5\sqrt{4r_\varepsilon^2 - f^2} \approx \frac{f^2}{8r_\varepsilon} \tag{A.26}$$

Note that this equation can often be found in catalogs of tool manufacturers with no explanations provided.

Example A.2

Problem

Determine the maximum peak-to-valley distance (theoretical surface roughness) in turning using a single-point tool with a small nose radius (Figure A.40d), the cutting feed $f=0.25$ mm/rev and depth of cut $a_p=3$ mm. A standard diamond-shaped insert CEJN 2525M is mounted in a tool holder with $\kappa_r = 95°$ and $\kappa_{r1} = 7.5°$.

How would this distance change if a standard RCMX 1204MO round insert having a diameter of 16 mm is used instead?

Solution

With a standard diamond-shaped insert CEJN 2525M, one can calculate the maximum peak-to-valley distance using Eq. (A.24) as

$$R_t = \frac{f \sin \kappa_r \sin \kappa_{r1}}{\sin(\kappa_r + \kappa_{r1})} = \frac{0.15 \sin 95° \sin 7.5°}{\sin(95° + 7.5°)} = 0.020 \text{ mm}$$

When a standard RCMX 1204MO round insert is used, the maximum peak-to-valley distance is calculated using Eq. (A.26)

$$R_t \approx \frac{f^2}{8r_n} = \frac{0.15^2}{8 \cdot 16} = 0.0018 \text{ mm}$$

A comparison shows that the maximum peak-to-valley distance (theoretical surface roughness) decreases significantly when the latter insert is used.

To improve the surface roughness in finishing operation and when high cutting feeds are used, a tool geometry with a cleaning cutting edge which has $\kappa_{r1} = 0°$ (Figure A.40e) was introduced in the mid-1950s. It is known as the viper geometry. Although its use resulted in significant improvements in surface roughness, two drawbacks immediately became evident. First, the use of such geometry often resulted in the onset of severe vibrations (chatter), and, second, it was found that the results were highly sensitive to the location of the minor cutting edge which theoretically should be exactly parallel to the axis of rotation of the workpiece. Old, not very rigid machines and not sufficiently accurate tool posts and tools limited the wide use of the cleaning cutting edge.

Times have changed. Rigid, high-power machines and indexable close-tolerance cutting inserts have been introduced. As a result, the cutting feed can be increased to meet today's requirement for high-penetration rate/productivity, high-efficiency machining operations. However, as discussed above, in finish turning, the cutting feed has always been limited. It is a simple matter of geometry because, as shown in Figure A.40b, the cutting tool actually cuts a "thread" on the machining surface. Such a "thread" can be tolerated in a roughing turning operation while in finish turning, machinists must reduce the cutting feed to avoid leaving a rough pattern on the machined surface. Sometimes, even a slow feed and a light depth of cut cannot produce a surface roughness sufficient to meet part specifications. In these cases, the shop must rely on a finish-grinding step.

Theoretically, a machinist can use a tool with a larger nose radius to increase the feed and still obtain a smooth surface according to Eq. (A.26). The larger radius presents a broader cutting edge to the workpiece, so the tool can move further per workpiece revolution and still overlap the cut made in the previous revolution. However, increasing the nose radius to achieve a larger feed has its limits, too. First is the inability to machine shoulders at a right angle that are often required on machined parts. Second, tools with larger nose radii are more likely to onset vibrating in the cut unless very rigid and powerful machines are used.

To overcome the problem, tool manufacturers re-visited the tool geometry with a viper insert, as shown in Figure A.40e. Modern machines are normally rigid, and tool position in the holder/machine is accurate so that all the above-described limitations are overcome. As a result, nowadays, practically all leading indexable insert and turning cutting tool manufacturers offer a wide variety of inserts/tools with viper inserts. Moreover, such inserts are widely used in face milling where one or more inserts in the set are made with the viper geometry to improve surface roughness, flatness, and/or milling productivity.

A logical question to be answered is: How far is the geometrical (theoretical) roughness of the machined surface from that obtained in the real cutting? As discussed by Astakhov (2006, 2011), if the cutting process takes place at the optimal or close to optimal cutting temperature (Astakhov and Outeiro 2019), the built-up edge does not form at all, so it does not have any effect on the surface roughness, this surface roughness is practically equal to the so-called theoretical surface roughness determined by tool geometry and the cutting feed.

In the author's opinion, the term *minor* probably misleads many professionals and researchers in the field causing a common perception that this edge is not that important. It is not so. First, as discussed above, this edge actually forms the machined surface including its roughness,

roughness profile/topography, and cold working/superficial and in-depth machining residual stresses. Therefore, its working conditions should be revealed. Zorev provided a detailed analysis of the chip formation by the minor cutting edge (Zorev 1966). He studied the velocity hodograph, associated plastic deformation, and flows in this region. Using the results of this study, one can visualize the chip cross-sectional area cut by the minor cutting edge with the aid of Figure A.41. Figure A.41a shows a hypothetical single-point cutting tool having $\kappa_{r1}=90°$; that is, practically, this tool does not have a minor cutting edge. Figure A.41b shows the cross-sectional area ABC of a *tooth* of the surface profile left after this surface was machined by this tool. Real cutting tools have the minor cutting edge with $\kappa_{r1} \ll 90°$ so that the surface profile left by the cutting tool is ACD while area ADB is cut by the minor cutting edge as shown in Figure A.41c. Therefore, the major parameters of the macrogeometry and microgeometry of this edge, first of all the clearance and rake angles and the conditions of the cutting edge should be considered at the same level as for the major cutting edge in the design of cutting tools.

A.8.5 INCLINATION ANGLE

Although the sense and sign of the inclination angle λ_s is clearly shown in Figure A.24 (View E) and it is defined earlier as the angle between the cutting edge and the reference plane as viewed in the tool cutting edge plane, experience shows that there are certain difficulties and confusions in understanding this angle and its influence on the cutting process. Figure A.42 aims to clarify the issue. The inclination angle λ_s is measured in the cutting-edge plane P_s, which is perpendicular to the reference plane P_r and contains the cutting edge 1–2. Numbers 1 and 2 designate the ends of the cutting edge. As such, if point 1 (the tool corner) locates below point 2, then the inclination angle λ_s is positive; if points 1 and 2 are at the same level, then $\lambda_s=0$; and when point 1 locates above point 2, the inclination angle λ_s is negative.

The sign of the inclination angle defines the chip flow direction as shown in Figure A.42. When λ_s is positive, the chip flows to the right to the perpendicular to the cutting edge 1–2 and the angle between this perpendicular and the direction of the chip flow is $+\lambda_s$. It can potentially damage the machined surface. When λ_s is negative, the chip flows to the left to the perpendicular to the cutting edge 1–2 and the angle between this perpendicular and the direction of the chip flow is $-\lambda_s$. When $\lambda_s=0$, the chip flow direction along the perpendicular to the cutting edge 1–2 as shown in Figure A.42.

In standard tuning tools with indexable inserts, the inclination angle λ_s cannot be readily distinguished because it is rather small for standard single-point tools. This angle forms in the T-hold-S

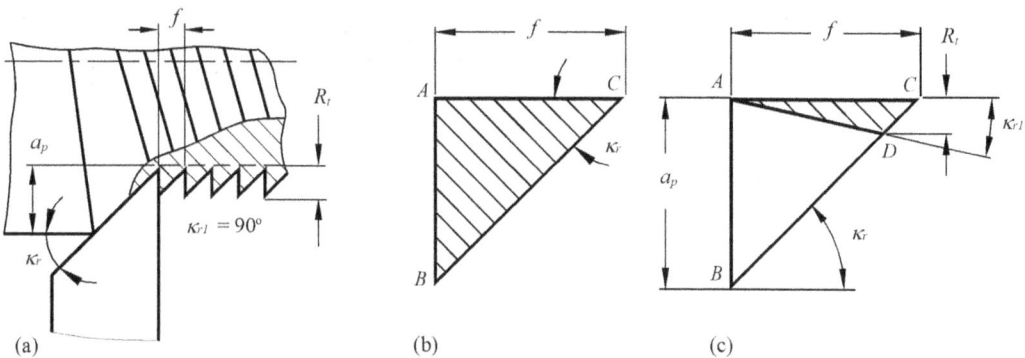

FIGURE A.41 The cross-sectional area of the chip cut by the minor cutting edge: (a) hypothetic tool having a 90° tool cutting edge angle of the minor cutting edge, (b) the cross section of the surface roughness left after cutting when the tool minor cutting edge angle is 90°, and (c) geometrical model to calculate the cross-sectional area of the chip cut by the minor cutting edge when the tool minor cutting edge angle is much less than 90°.

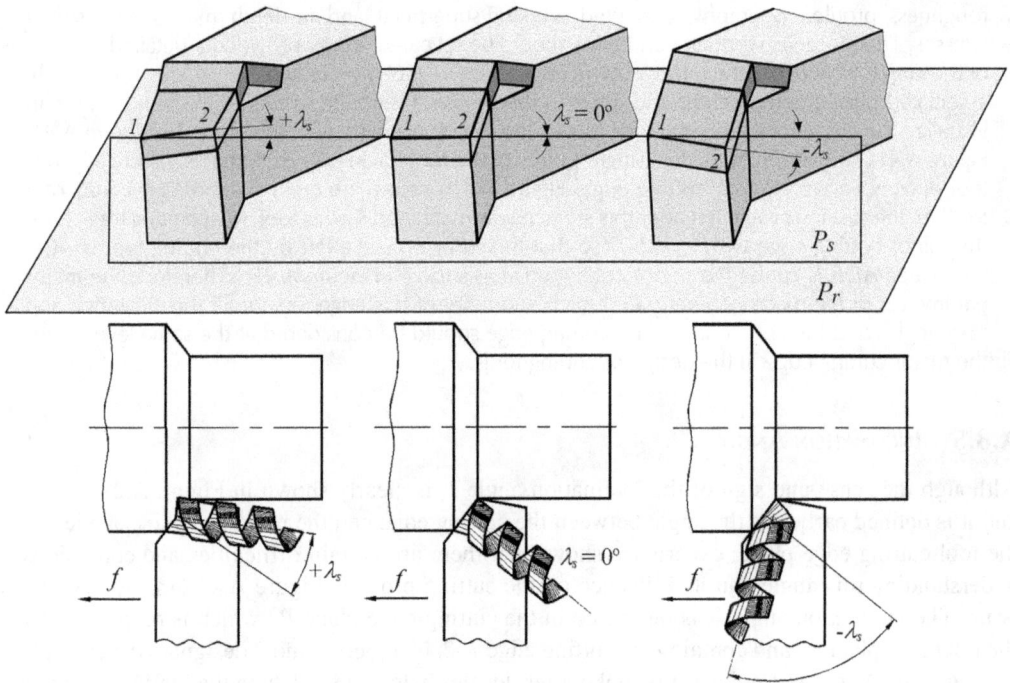

FIGURE A.42 Inclination angle: (a) sign and (b) influence on the chip flow direction.

when the insert is placed in a tool holder and the tool holder is mounted in the machine. As such, the inclination of the base face in the working plane γ_f and in the back plane γ_p is indicated in catalogs. For a common tool holder with $\kappa_r=60°$, $\gamma_f=-6°$ and $\gamma_p=-6°$. Using Eq. (A.6), one can calculate the most common inclination angle used in turning as

$$\lambda_s = \arctan\left(-\sin\kappa_r \tan\gamma_p - \cos\kappa_r \tan\gamma_f\right) =$$

$$\arctan\left(\sin 60° \tan(-6°) - \cos 60° \tan((-6°))\right) \approx 2.2° \tag{A.27}$$

REFERENCES

Abushawashi, Y., Xiao, X., Astakhov, V. (2017). "Practical applications of the "energy-triaxiality" state relationship in metal cutting." *Machining Science and Technology* **21**: 1–18.

Armarego, E. J., Brown, R. H. (1969). *The Machining of Metals*. Hoboken, NJ, Prentice-Hall.

Astakhov, V. P. (1998/1999). *Metal Cutting Mechanics*. Boca Raton, FL, CRC Press.

Astakhov, V. P. (2006). *Tribology of Metal Cutting*. London, Elsevier.

Astakhov, V. P. (2010). *Geometry of Single-Point Turning Tools and Drills: Fundamentals and Practical Applications*. London, Springer.

Astakhov, V. P. (2011). Turning. *Modern Machining Technology: A Practical Guide*. P. Davim. Oxford, Woodhead-Chandos: 1–78.

Astakhov, V. P. (2014). *Drills: Science and Technology of Advanced Operations*. Boca Raton, FL, CRC Press.

Astakhov, V. P. (2018). Mechanical properties of engineering materials: relevance in design and manufacturing. *Introduction to Mechanical Engineering*. P. Davim. Cham, Switzerland, Springer International Publishing AG: 3–41.

Astakhov, V. P. (2024). *High-Productivity Drilling Tools: Materials, Metrologyn and Failure Analysis*. Boca Raton, FL, CRC Press.

Astakhov, V. P., Outeiro, J. (2019). Importance of temperature in metal cutting and its proper measurement/modeling. *Measurement in Machining and Tribology*. P. J. Davim. London, Springer: 1–47.

Astakhov, V. P., Shvets, S. V. (1998). "A system concept in metal cutting." *Journal of Materials Processing Technology* **79**(1–3): 189–199.

Astakhov, V. P., Shvets, S. (2004). "The assessment of plastic deformation in metal cutting." *Journal of Materials Processing Technology* **146**: 193–202.

Astakhov, V. P., Xiao, X. (2008). "A methodology for practical cutting force evaluation based on the energy spent in the cutting system." *Machining Science and Technology* **12**: 325–347.

Astakhov, V. P., Xiao, X. (2016). The principle of minimum strain energy to fracture of the work material and its application in modern cutting technologies. *Metal Cutting Technology*. P. J. Davim. Berlin, De Gruyter: 1–35.

Dahlman, P., Gunnberg, F, Jacobson, M (2004). "The influence of rake angle, cutting feed and cutting depth on residual stresses in hard turning." *Journal of Materials Processing Technology* **47**(2): 181–184.

Günay, M., Korkut, I., Aslan, E., Seker, U. (2005). "Experimental investigation of the effect of cutting tool rake angle on main cutting force." *Journal of Materials Processing Technology* **166**: 44–49.

Gunay, M., Seker, U., Sur, G. (2006). "Design and construction of a dynamometer to evaluate the influence of cutting tool rake angle on cutting forces." *Materials and Design* **27**: 1097–1101.

Saglam, H., Yaldiz, S., Unsacar, F. (2007). "The effect of tool geometry and cutting speed on main cutting force and tool tip temperature." *Materials and Design* **29**: 101–111.

Shaw, M. C. (1984). *Metal Cutting Principles*. Oxford, Oxford Science Publications.

Shaw, M. C. (1988). Metal removal. *CRC Handbook of Lubrication: Theory and Practice of Tribology, Volume II: Theory and Design*. E. R. Booser. Boca Raton, FL, CRC Press: 335–356.

Shaw, M. C. (2004). *Metal Cutting Principles*. 2nd Edition. Oxford, Oxford University Press.

Taylor, F. W. (1907). "On the art of cutting metals." *Transactions of ASME* **28**: 70–350.

Tetsuji, O., Hirokazu, K., Shigeo, M. (1999). "Influence of clearance and rake angles on the cutting force of rock." *Journal of NIRE (National Institute for Resources and Environment)* **2**(3): 73–80.

Time, I. (1870). *Resistance of Metals and Wood to Cutting (in Russian)*. St. Petersbourg, Russia, Dermacow Press House.

Zorev, N. N. (1966). *Metal Cutting Mechanics*. Oxford, Pergamon Press.

Index

For Product Safety Concerns and Information please contact our EU
representative GPSR@taylorandfrancis.com
Taylor & Francis Verlag GmbH, Kaufingerstraße 24, 80331 München, Germany

www.ingramcontent.com/pod-product-compliance
Lightning Source LLC
Chambersburg PA
CBHW080121220326
41598CB00032B/4912

9 781032 203546